Anatomy for Surgeons: Volume 1 **THE HEAD AND NECK**

Anatomy for Surgeons: Volume 1

THE HEAD AND NECK

Third Edition

W. Henry Hollinshead, Ph.D.
Professor Emeritus of Anatomy,
Mayo Graduate School of Medicine
(University of Minnesota),
Rochester, Minnesota;
Visiting Professor Emeritus of Anatomy,
University of North Carolina
School of Medicine,
Chapel Hill, North Carolina

With 489 Illustrations

HARPER & ROW, PUBLISHERS
PHILADELPHIA
Cambridge
New York
Hagerstown
San Francisco
London
Mexico City
São Paulo
Sydney
1817

Acquisitions Editor: John de Carville
Sponsoring Editor: Richard Winters
Manuscript Editor: Rachel Bedard
Indexer: Marci Serepy
Art Director: Maria Karkucinski
Designer: Arlene Putterman
Production Assistant: Barney Fernandes
Compositor: American-Stratford Graphic Services, Inc.
Printer/Binder: The Murray Printing Company

The author and publisher have exerted every effort to ensure that drug selection and dosage set forth in this text are in accord with current recommendations and practice at the time of publication. However, in view of ongoing research, changes in government regulations, and the constant flow of information relating to drug therapy and drug reactions, the reader is urged to check the package insert for each drug for any change in indications and dosage and for added warnings and precautions. This is particularly important when the recommended agent is a new or infrequently employed drug.

Printed in the United States of America

ANATOMY FOR SURGEONS: VOLUME 1, The Head and Neck, third edition.

5 6 4

Library of Congress Cataloging in Publication Data

Hollinshead, W. Henry (William Henry), date.
Anatomy for surgeons.
Includes bibliographical references and index.
Contents: —v. 1. The head and neck. —v. 3.
The back and limbs.
1. Anatomy, Surgical and topographical.
I. Title. [DNLM: 1. Anatomy. 2. Surgery. WO 101 H741a]
QM531.H692 1982 611′.9 81-6909
ISBN 0-06-141266-X (v. 3) AACR2
0-06-141264-3 (v. 1)

Contents

Preface

This third edition of Volume 1 of *Anatomy for Surgeons* attempts to present, as did the first and second editions, anatomic facts and concepts concerning the head and neck that are of particular interest to the surgeon. It is not intended to be a complete descriptive anatomy of these parts but is designed to serve both as a ready reference in which the surgeon can find descriptions of the basic anatomy and as a useful review of numerous, sometimes minute, anatomic details gathered from many sources. No attempt has been made to describe the indications for, or detailed technique of, specific operations, for these are matters that belong to surgery and not to anatomy. Particular care has been taken throughout this volume, however, to relate anatomic and physiologic details and concepts to underlying surgical procedures.

Most of the anatomy described in the second edition is still both valid and pertinent, for basic anatomy does not change. However, newer techniques, particularly the widespread use of the surgical microscope, have both enlarged our knowledge of anatomy and emphasized the practicality of such knowledge. Additions, improvements, and refinements in our understanding of anatomic details and in diagnostic and operative procedures on the head and neck have made revision of this book necessary. The basic descriptions have been carefully scrutinized. Some have been rearranged or partly rewritten in the interest of greater clarity and accuracy, and numerous minor revisions have been made to incorporate more recent findings and clinical applications.

Almost every chapter contains observations and interpretations that were not available at the time the second edition was written. To cite only a few, there are new studies of the accessory rootlets of the trigeminal nerve, relationships in the cavernous sinus, the course of the sympathetic fibers to the superior tarsal muscle, functions of the muscles of the tympanic cavity and of the muscles of mastication, relations of the nasal cartilages, effects of neurectomy on intractable parotitis, routes by which parasympathetic fibers reach the submandibular and sublingual glands, movements of the arytenoid cartilages, and the origin of the calcitonin-producing cells of the thyroid gland.

The Nomina Anatomica (NA), which was adopted in 1955, was used in the preceding edition of this book and is used here. Because many clinicians may be more familiar with the Basel Nomina Anatomica (BNA), however, I have also attempted to include the more commonly used BNA synonyms at least once if they differ appreciably from those of the NA.

I remain deeply indebted to those former colleagues at the Mayo Clinic who were so helpful in the two previous editions. They are listed in each of those editions, and I regret that I could not avail myself of their counsel in the current one.

I am also indebted to my publishers, Harper & Row.

W. Henry Hollinshead
Chapel Hill, North Carolina

Chapter 1
THE CRANIUM

Scalp

The three layers of the scalp (Fig. 1-1)—the skin, the subcutaneous tissue, and the aponeurotic layer (galea aponeurotica, epicranial aponeurosis) with its associated epicranial muscles—are intimately fused and move as a unit with the contraction of the muscles. The dense subcutaneous tissue of the scalp contains the larger blood vessels and nerves, and its strong retinacula not only unite the skin and the galea, but afford support to the blood vessels. These layers are separated from the external periosteum or pericranium by a loose fourth layer that allows easy movement of the scalp and also ready spread of fluid or infections beneath it. Because of the latter fact, the subaponeurotic layer is sometimes referred to as the danger space of the scalp; fluids contained therein find exit with difficulty, typically into the periorbital connective tissue. The periosteum is fairly tightly attached to the bone, especially at the sutures; it is through the loose subaponeurotic tissue that separation occurs most easily in tears or surgical reflections of the scalp.

NERVES AND VESSELS

The nerves and blood vessels of the scalp run up into it from below, the larger ones on the forehead and in the temporal and occipital regions; although the blood vessels anastomose so freely with each other that there is relatively little danger of reducing markedly the blood supply to an area of scalp unless a skin flap with an extremely narrow inferior pedicle is produced, the nerve supply must be taken into consideration when incisions for skin flaps are planned, or else an area of denervated or partially denervated scalp, with resulting numbness or paresthesia, will be produced.

Since the arteries of the scalp anastomose freely with each other and with those of the opposite side, they form a part of the potential collateral pathway available after ligation of the external or common carotid artery on one side. It is because of these abundant anastomoses, also, that wounds of the scalp involving the subcutaneous tissue typically show arterial bleeding from both cut surfaces. The profuseness of the bleeding is further contributed to by the fact that the blood vessels lie largely in the dense subcutaneous tissue, by which they are supported and to which they are attached, and tend therefore to be held open by this tissue when they are cut. The arteries of the scalp also send small twigs to the underlying bones of the skull.

The nerves to the scalp are branches of cranial and spinal nerves (Fig. 1-2), the spinal nerves being derived both from the cervical plexus and from the posterior primary rami of

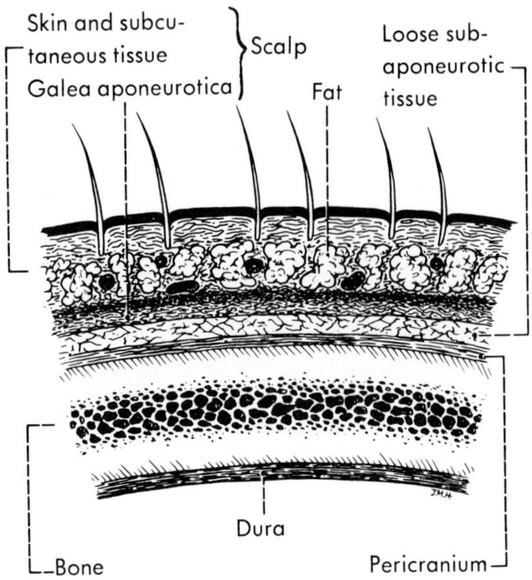

Fig. 1-1. The scalp.

the upper cervical nerves. Generally speaking, these nerves accompany the blood vessels of the scalp, which are derived largely from branches of the external carotid, but anteriorly are from the ophthalmic branch of the internal carotid.

Since the nerves of the scalp approach it from all directions and have, of course, a considerable overlap, it is usually impossible to produce adequate anesthesia of any large area of the scalp by a purely local injection; rather, the area in which anesthesia is desired must be ringed by a whole series of injections. Further, the nerves, like the vessels, run mainly in the subcutaneous tissue, and the solution must therefore be placed in this layer rather than in the subaponeurotic layer where it could spread much more easily but would be separated from the nerves by the tough galea aponeurotica.

Fig. 1-2. General areas of distribution of the main nerves to the scalp, and the positions of the principal arteries. The unlabeled arteries, from left to right, are the occipital, posterior auricular, and superficial temporal branches of the external carotid, and the supraorbital and supratrochlear branches of the ophthalmic.

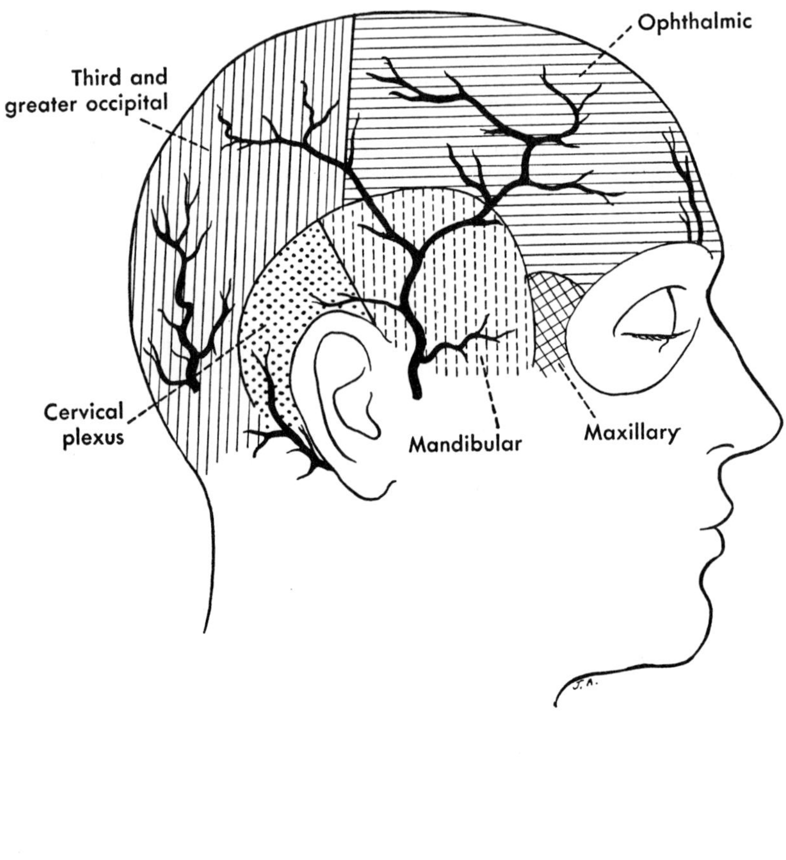

Supraorbital Nerve and Artery

The supraorbital nerve (from the ophthalmic branch of the trigeminal) and the supraorbital artery (from the ophthalmic artery) emerge through the supraorbital notch or foramen to turn sharply around the upper margin of the orbit and be distributed to the scalp. At first they lie deep to the frontalis muscle but subsequently pierce this to run in the scalp's dense subcutaneous tissue. The small supraorbital artery forms anastomoses with the other vessels of the scalp, while the supraorbital nerve supplies a very large area on the scalp, extending backward over the vault of the skull to approximately the interauricular line or slightly beyond it. Medial to the supraorbital nerve and artery are the smaller supratrochlear nerve and the supratrochlear (frontal) artery, also branches of the ophthalmic nerve and artery; the nerve supplies a limited medial region of the forehead.

Because of the extensive distribution of the supraorbital nerve, frontal skin flaps or curved incisions over the front of the vault of the skull should be large to avoid an area of denervated tissue.

Auriculotemporal Nerve

Laterally, in front of the ear, the auriculotemporal nerve (a branch of the mandibular division of the fifth nerve) extends upward with the superficial temporal vessels to supply the chief portion of the temporal region of the side of the head as high as the superior temporal line on the skull—that is, especially over the area of the origin of the temporalis muscle. More anteriorly, supplying skin behind the orbital process of the zygomatic bone, but not extending so far upward, is the zygomaticotemporal branch of the maxillary division of the fifth cranial nerve. To avoid interference with this nerve supply and with the superficial temporal artery whose main branches, parietal and frontal, spread out in this same area, temporal skin flaps should be U- or horseshoe-shaped, with the base widely attached below.

Great Auricular Nerve

Although a small and probably variable area of skin behind the ear and over the mastoid process is innervated by the auricular branches of the seventh, ninth, and tenth cranial nerves (Chap. 3), the larger nerve distributed here, and extending upward to the scalp, is the great auricular from the cervical plexus (C2 and C3). The artery in this location is the posterior auricular branch of the external carotid, which roughly parallels the nerve. More posteriorly over the mastoid process is the mastoid branch of the occipital artery, which aids in supplying the scalp and also sends a branch into the bone to supply mastoid cells.

Occipital Nerves

Behind the area of distribution of the great auricular nerve is the *lesser occipital nerve,* also ascending from the cervical plexus (primarily second cervical). Small skin flaps over the mastoid region may be turned down by a U-shaped incision passing upward from the ear to the temporal line and then curving backward and downward along this line; larger reflections of scalp here will usually involve the distributions of the auriculotemporal or occipital nerves.

The *greater occipital nerve* is derived largely from the dorsal (posterior) primary branch of the second cervical nerve; it typically appears in the suboccipital region at about the lateral border of the trapezius muscle and becomes subcutaneous by piercing the fascia between the attachments of the trapezius and the sternocleidomastoid muscles to the skull; here it is joined by the occipital artery, and is distributed over the posterior part of the cranial vault to the vicinity of the interauricular line. Its area of distribution here overlaps that from the supraorbital branch of the ophthalmic, and laterally, at the level of the temporal line, overlaps the areas of distribution of the great auricular and lesser occipital nerves. Medial to the greater occipital nerve the *third occipital* (representing the dorsal ramus of C3) extends also upward to the scalp, but for a relatively short distance.

Straight midline incisions, and curved ones along the temporal lines, lie in general between nerves; incisions connecting the two necessarily denervate some scalp, but if the skin flap is turned downward with its base left attached in the occipital region its innervation can be preserved.

VEINS

The veins of the scalp parallel the arteries. They receive many of the emissary veins of the cranium, and through these communicate with the cranial venous sinuses. The usual *emissary veins* are the parietal that penetrate the parietal bones posterior to their midpoints a little on each side of the midline, connecting the occipital veins and the superior sagittal sinus; one or more mastoid veins that penetrate the posterior aspect of the base of each mastoid process and connect the sigmoid sinuses to the occipital or posterior auricular veins; and the condylar (condyloid) veins, not as constant, that traverse the condylar (condyoid) canals and connect the confluence of the sinuses to the suboccipital venous plexus. Occasionally there is an unpaired occipital emissary vein penetrating the external occipital protuberance and connecting the confluence of the sinuses to an occipital vein.

Arteriovenous malformations, more common in the head than elsewhere, occur more frequently in the scalp; they may be continuous through the skull with similar endocranial formations, or formed by dural arteries that penetrate the skull—in either case, of course, being supplied by the external carotid system.

LYMPHATIC VESSELS

The lymphatic vessels from the frontal region of the scalp drain downward and backward into the superficial parotid (preauricular) lymph nodes; those from the parietal and temporal regions pass downward both in front of and behind the ear to end in the superficial parotid, superficial cervical, and retroauricular (mastoid) nodes, or bypass these and end directly in upper nodes of the deep cervical group; those from the occipital region drain for the most part into both the occipital and upper deep cervical nodes, but one large vessel from this region is said to follow the posterior border of the sternocleidomastoid muscle to reach lower deep cervical nodes.

SENSITIVITY OF THE SCALP

Ray and Wolff, in their study of the pain-sensitive structures in the head, found that although the skin of the scalp is sensitive to all the usual forms of stimulation to which skin is elsewhere, the *galea aponeurotica* is sensitive only to pain; where the extracranial blood vessels are in close contact with the galea the pain sensitivity was found to be greater than at other points. Like the galea, the fascia over the temporal and occipital muscles, and these muscles themselves, were reported as sensitive only to pain, this pain being appreciated near the point of stimulus.

The *periosteum* on the outside of the skull varied in sensitivity, that over the vertex being entirely insensitive; there was a general increase in sensitivity just over the eyebrows, low in the temporal regions, and low in the occipital region—that is, in the areas in which the nerves were approaching the scalp. Even here, however, the periosteum apparently is not particularly sensitive, since stripping of periosteum around the base of the skull produced only moderate pain in the neighborhood of the point of stimulation.

The *arteries* of the scalp were all found to be sensitive to pain, while the veins were either less or not at all sensitive.

The *bony skull* itself, as also its veins, was found to be insensitive to all types of stimulation.

Calvaria

The bony vault of the cranium, the calvaria, consists, from before backward, of the unpaired frontal bone, the paired parietals, and the unpaired occipital bone (Fig. 1-3). Laterally, the greater wings of the sphenoid bone,

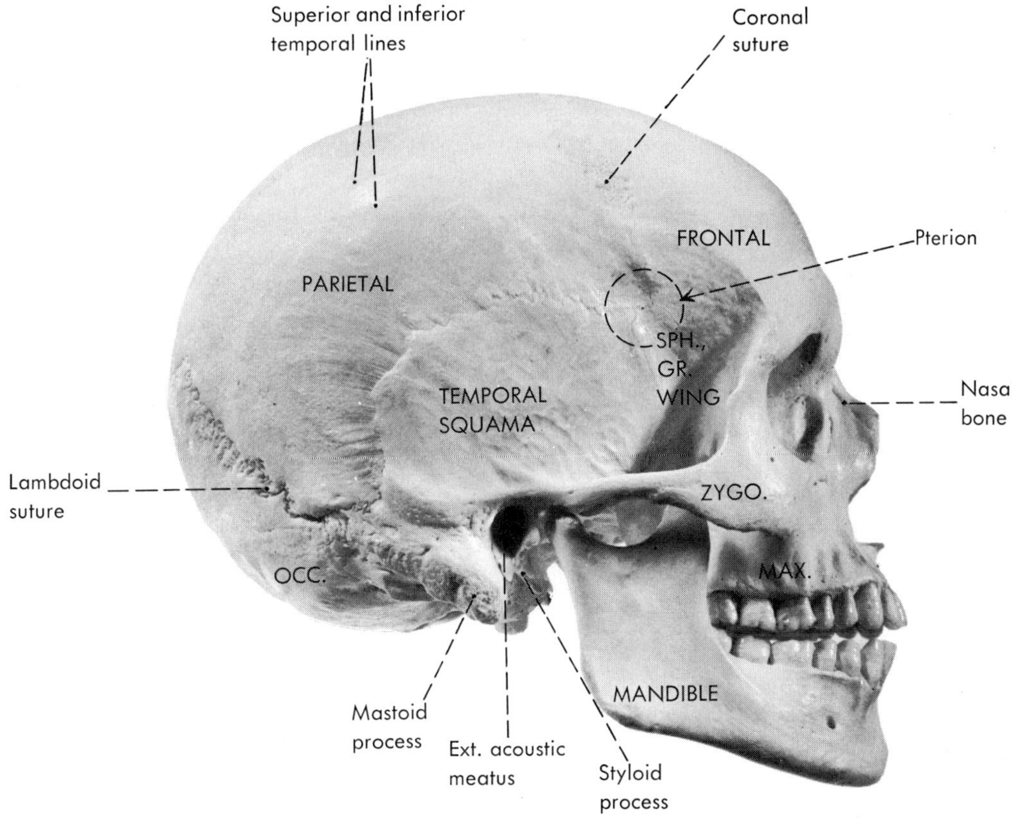

Fig. 1-3. Lateral view of the skull. *MAX., OCC., SPH. GR. WING,* and *ZYGO.* are the maxillary and occipital bones, the greater wing of the sphenoid, and the zygomatic bone.

and posterior to them the temporal bones, complete the sides of the brain case.

The calvaria is, of course, covered with periosteum on both its outer and inner surfaces; the inner periosteum is fused to the dura, forming its outer layer. The periosteum of the skull is, however, markedly deficient in osteogenic power in the adult, as compared with the periosteum of long bones; relatively little regeneration of the bones of the skull may be expected following a craniotomy unless the bone flap is replaced. Areas of bone removed and replaced in the temporal region usually heal better than bone elsewhere, presumably because of the greater blood supply furnished them by the deep temporal arteries.

The bones of the vault of the cranium show three distinct layers: hard external and internal laminae (outer and inner tables) and a cancellous middle layer or diploë. The inner table is distinctly thinner than the outer and may be fractured by blows that leave the outer table intact, thus rendering it more difficult to diagnose fractures of the skull.

Surgical approaches to the cerebral hemispheres, both diagnostic and operative, are usually made through the calvaria, as are also approaches to the other contents of the anterior and middle cranial fossae. A preliminary step is trephining the calvaria at a suitable and relatively safe spot; through the trephine opening a needle or other small instrument can be inserted or, if a larger opening is desired, several trephine openings can be made and connected by sawing through the intervening bone.

In planning trephine operations on the calvaria, factors to be considered are not only the relations of the calvaria to the various subdivisions of the cerebral hemispheres, but also the positions of the cranial venous sinuses— the superior sagittal and the transverse sinuses—related to the calvaria, and the position of the frontal sinus and the mastoid cells.

Thickness of Skull

The bony vault of the skull is relatively thick, averaging perhaps about 5 mm, but average measurements of this are of little use to the surgeon. The thickness of the skull varies considerably with the individual and in addition shows regional variations. It thus becomes much thicker in the region of the external occipital protuberance and particularly thin in the temporal region, where it is, however, in part protected by the overlying temporal muscle. The thinness of the temporal region is taken advantage of in "turning" frontal and temporal bone flaps, by sawing through the thicker bone and leaving the base of the flap in the temporal region, where it can be relatively easily fractured.

DIPLOIC VESSELS

The diploë is supplied by small but numerous diploic branches from arteries both on the external and internal surfaces of the skull—that is, from arteries of the scalp and from those of the dura mater. The diploic veins anastomose freely with each other, but usually form five chief veins on either side: the frontal, anterior temporal (typically two), posterior temporal, and occipital diploic veins.

The *frontal* diploic (Fig. 1-4*A*) veins communicate through the inner table with the superior sagittal sinus, but empty primarily downward into the supraorbital tributary to the ophthalmic vein, emerging from the skull through a small foramen located in the roof of the supraorbital notch. The two *anterior temporal* veins (Fig. 1-4*B*) usually pass downward in front of and behind the coronal suture, respectively, to communicate both with the veins of the temporal muscle and with the sphenoparietal sinus. The *posterior temporal* diploic vein, in the parietal bone, drains downward to the mastoid region where it perforates the inner lamina to join the transverse sinus. The *occipital* diploic vein (Fig. 1-4*C*) empties either externally into the occipital vein, or internally into the transverse sinus. Thus, the diploic veins may drain either into veins of the scalp or into the dural venous sinuses, or may form communications between the two. Through them an osteomyelitis originating, for instance, from the frontal sinus may involve not only the frontal bone but also the scalp and the dura.

Jefferson and Stewart, in their study of the diploic veins, found that generalized enlargement of these veins, as sometimes seen in roentgenograms, is no indication of increased intracranial pressure, and it is sometimes stated that such enlargement is a usual concomitant of increasing age; they found enlargement of the grooves for the middle meningeal vessels to be much more often associated with increased intracranial pressure.

Cranial Contents in General

Except for the relations to the cranial vault, the relationships of many of the components of the cranial contents—for instance, those of nerves and vessels to each other—can be more usefully discussed in connection with the individual cranial fossae. The following account is therefore a general one; more specific details will in many instances be found in connection with the descriptions of the cranial fossae.

DURA MATER

Where it is in contact with the bones of the skull the cranial dura mater, the general relations of which are shown in Figure 1-5, serves also as the internal periosteum of the skull. Its outer part is therefore continuous with the outer periosteum (pericranium) through the various foramina of the skull; the remainder is the equivalent of the spinal dura, with which it is continuous at the foramen magnum.

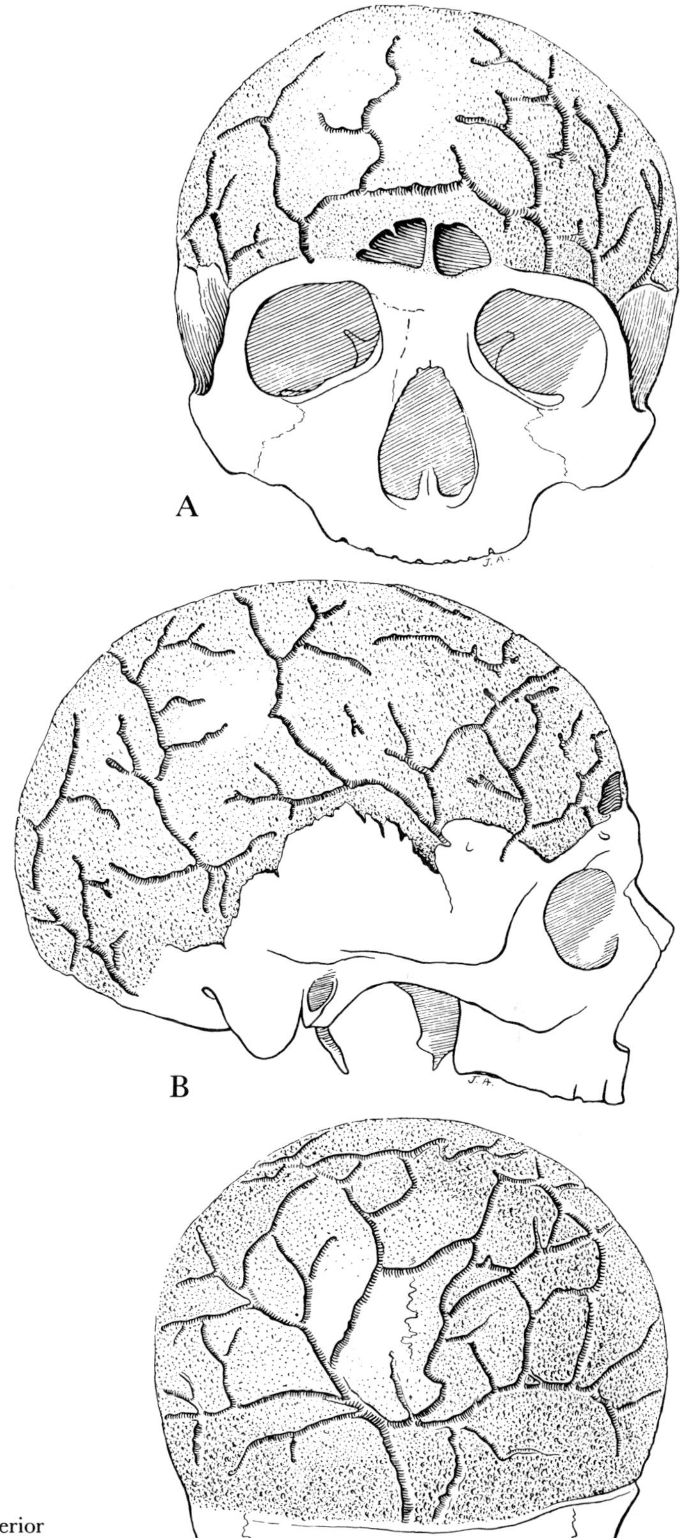

Fig. 1-4. *A.* The diploic veins from the anterior aspect. *B.* The diploic veins from the lateral aspect. *C.* The diploic veins from the posterior aspect. (After Jefferson G, Stewart D: Br J Surg 16:70, 1928)

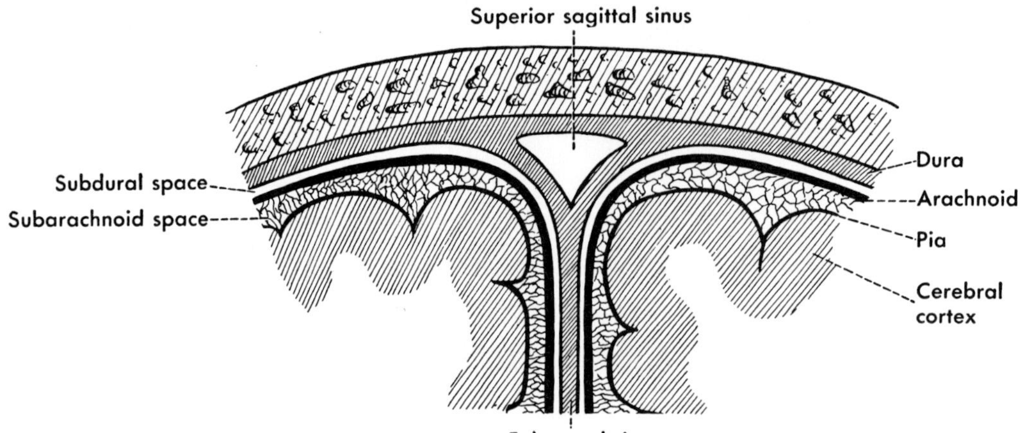

Fig. 1-5. General relation of the three meninges to each other and to the cerebral hemispheres.

Cranial Epidural Space

Although the spinal epidural (extradural) space is a large one, occupied by fat and venous plexuses, it ends above at the foramen magnum where the dura and the cranial periosteum fuse; the cranial epidural space is neither continuous with the spinal epidural space nor the homologue of it (Fig. 1-6). It is, rather, a potential space outside the inner periosteum, made possible by the loose attachment of this to the skull. However, just as the major veins about the spinal cord (the internal vertebral venous plexuses) lie between the periosteum and the dura within the vertebral canal, so do the larger venous channels draining the brain lie within the cranial dura (not, therefore, as about the cord, in the surgical epidural space). These channels, the dural venous sinuses, thus receive support form the dura in which they lie; because of this support, they have no walls of their own other than endothelium.

In addition to the venous sinuses, the dura also contains meningeal arteries and nerves, and some of the cranial nerves (notably those to the musculature of the eyeball) run for some distance in the dura.

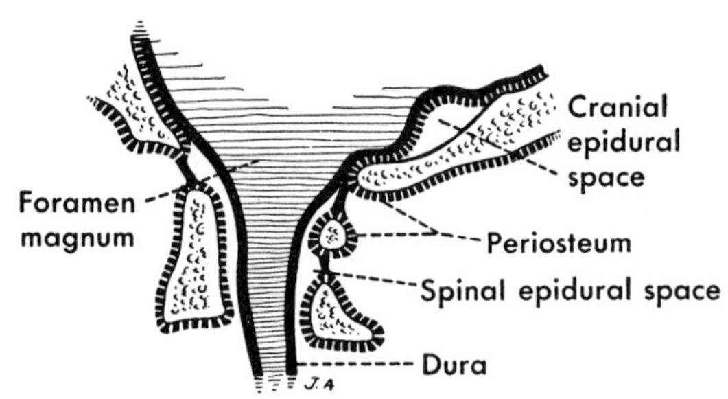

Fig. 1-6. Difference between the spinal and the cranial epidural spaces. The cranial dura has been elevated to form an epidural space.

Strength of Attachment

The strength of attachment of the dura to the skull varies considerably; the strongest attachment to the vault of the cranium is in the midline, above the superior sagittal sinus; there is also some attachment to the sutures and some along branches of the middle meningeal artery, but otherwise the attachment to the vault is weak, making removal of bone flaps relatively easy and allowing the development of massive epidural hemorrhages.

In contrast, the attachments of the dura into the base of the skull are relatively strong; in the anterior fossa the dura is attached especially strongly to the crista galli and the cribriform plate, and about the optic canal (Fig. 1-7). In the middle fossa the main attachments are about the edges of the numerous foramina through which nerves and vessels enter and leave this part of the cranium, especially the superior orbital fissure, the foramen rotundum, the foramen ovale, and

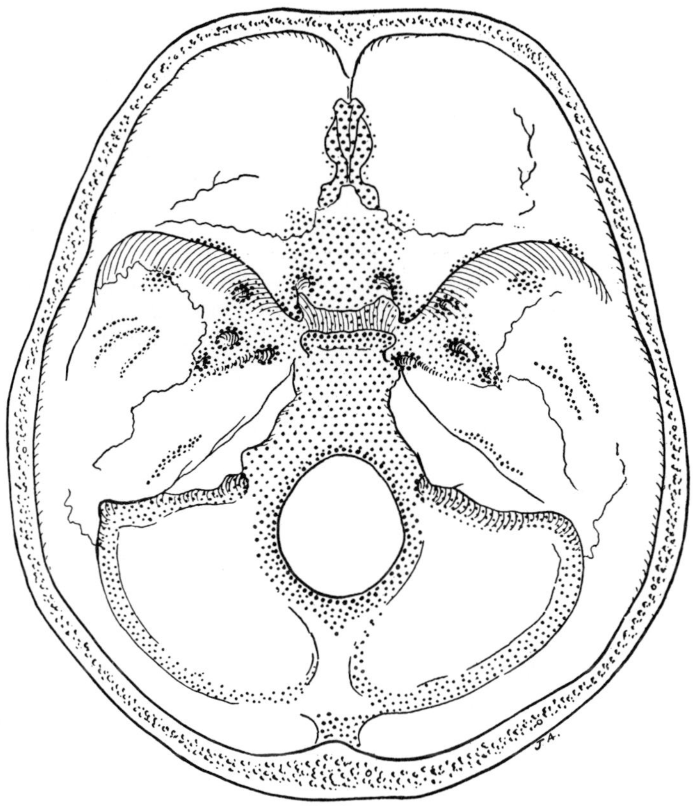

Fig. 1-7. Areas of stronger attachment of the dura to the base of the skull. (After Walker AE: Anat Rec 55:291, 1933)

the foramen lacerum. In the posterior cranial fossa the dura is attached especially to the basal portion of the sphenoid bone, to the margins of the foramen magnum and the jugular foramen, and to the internal acoustic meatus; it is less firmly attached about the venous sinuses.

Subdural Space

The dura's inner surface is lined by fibroblastic tissue (Leary and Edwards), and this layer therefore forms the outer lining of the subdural space. The subdural space is usually described as being a potential space only, but Penfield ('24), by freezing the heads of dogs, apparently demonstrated that there is in this space an appreciable amount of clear, yellow fluid that probably normally prevents intimate contact between the arachnoid and dura. Penfield and Norcross suggested that it is displacement of this fluid locally, thus allowing close contact between arachnoid and dura, that is responsible for posttraumatic headache. The origin of the subdural fluid is unknown.

DURAL SEPTA

In certain locations the dura, instead of remaining closely applied to the inner surface of the skull, projects inward to form septa that partially divide the cranial cavity. The largest and most important of these are the falx cerebri and the tentorium cerebelli (Figs. 1-8 and 1-9).

Falx Cerebri

The falx cerebri is a longitudinally directed septum, passing downward from the cranial vault in the midsagittal plane between the cerebral hemispheres; this large sickle-shaped fold is attached anteriorly to the crista galli and posteriorly blends with the upper surface of the tentorium cerebelli. Where it is continuous close to the midline with the dura of the vault of the skull, it helps to enclose the supe-

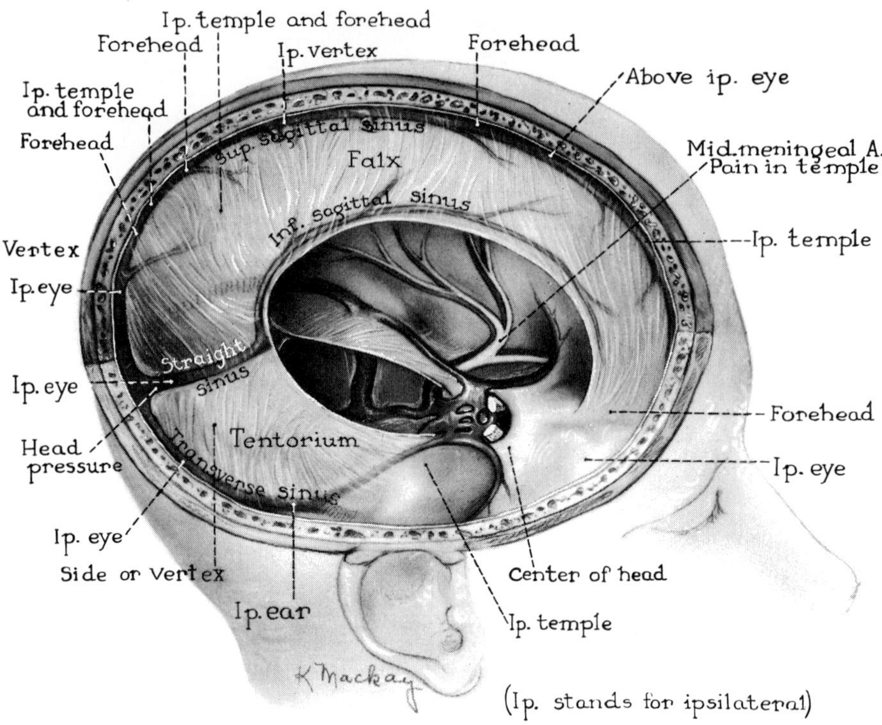

Fig. 1-8. The cranial dura, some of the venous sinuses, and the areas of the head to which pain has been said by the patient to be referred when the dura is stimulated. (Baker GS, Adson AW: Minn Med 26:282, 1943)

rior sagittal sinus; in its free margin, above the corpus callosum, it encloses the inferior sagittal sinus; finally, at its attachment to the tentorium cerebelli, it helps to form the walls of the straight sinus.

Tentorium Cerebelli

The tentorium cerebelli is placed approximately transversely, its center being, however, much higher than its sides (Fig. 1-9), so that it fits fairly snugly between the posterior portions of the two cerebral hemispheres and over the somewhat convex upper surface of the cerebellum. The line of attachment of the tentorium cerebelli to the skull extends backward from the posterior clinoid processes along the superior borders of the petrous portions of the two temporal bones ("petrous ridges") and backward and medially along the occipital bone at the grooves for the transverse sinuses; thus the tentorium largely separates the middle from the posterior cranial fossa. The superior petrosal and transverse sinuses, from before backward, lie at its line of attachment to the periosteal dura, and, as already mentioned, the straight sinus lies at the junction of its raised center and the falx cerebri; anteriorly, the tentorium cerebelli is entered, close to the posterior clinoid processes, by the trochlear nerves.

The anterior edge of the tentorium cerebelli presents a deep concavity, the tentorial notch (incisure) for accommodation of the brain stem as it passes into the posterior cranial fossa. Sunderland ('58) described variations in the anatomy at the tentorial notch and discussed injuries to the nerves, the cranial vessels, and the brain that may result from herniation through the notch as a result of increased supratentorial pressure.

In addition to the above-mentioned folds, there are others—the small falx cerebelli, the diaphragma sellae, and the arrangement of the dura about the trigeminal ganglion—

which are also of some importance and will be discussed in connection with the regions of the base of the skull in which they occur.

VENOUS SINUSES

The more constant dural venous sinuses are the superior and inferior sagittal, the transverse and sigmoid (lateral), the straight, the occipital, the superior and inferior petrosal, the cavernous, and the sphenoparietal sinuses. Of these, the sagittal, straight, and occipital sinuses are unpaired, lying approximately in the midplane of the head, while the others are paired.

Superior Sagittal Sinus

The superior sagittal sinus lies on the inner surface of the vault of the skull, where it forms a slight groove in the midsagittal plane. It is approximately triangular in section, its upper wall being formed by the periosteum–dura as this passes across the midline, its two inferolateral walls being formed by the two laminae of origin of the falx cerebri. Occasional trabeculae (chords of Willis) cross the sinus, but are not numerous enough to affect its blood flow. At its anterior end, where the falx is attached to the crista galli, the superior sagittal sinus is small, but it enlarges rapidly as it runs posteriorly. Anteriorly it receives communications from the frontal diploic veins, and it may be connected to the veins of the nose by an emissary vein passing through the foramen cecum; this connection, which can allow nasal infections to be transmitted to the sinus, is said to be regularly present in the fetus, but apparently is only sometimes maintained during adult life. The sinus may also connect by emissary veins with the angular veins on the face and, farther back, through the parietal bone, is regularly connected by such veins to the superficial temporal and occipital veins of the scalp. At the internal occipital protuberance the superior sagittal sinus enters into the formation of the confluence of the sinuses (torcular Herophili), where it usually passes into the right transverse sinus.

Laterally along the course of the superior sagittal sinus are irregular expansions that form the lateral lacunae, lying like the sinus itself largely within the dura but receiving the projecting arachnoid villi (Pacchionian granulations); the villi are believed to serve as the chief drainage point for return of cerebrospinal fluid to the bloodstream. The lacunae are said to form after birth and to be reduced after middle age to a single pair, one on each side of the sinus. They develop in otherwise tiny meningeal veins that empty into the sinus.

The position of the superior saggital sinus, its enlargement as it runs posteriorly, and the occurrence of the lacunae in connection with it must be taken into account in trephine op-

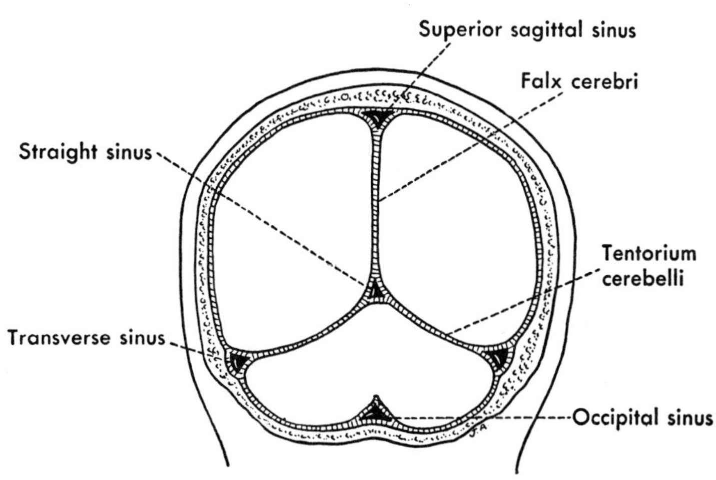

Fig. 1-9. The dura and venous sinuses in a frontal section through the posterior part of the skull.

erations upon the calvaria. The sinus lies in the midline except close to its termination, not far from the external occipital protuberance (torus), where it often deviates somewhat to one side, usually the right. However, the lacunae so increase its width that in trephining over the posterior half of the parietal region, where the lacunae are largest and most constant, the medial edge of the trephine should be not much closer than 2 cm or more (about an inch) to the midline.

The superior cerebral veins, from the convexity of the cerebral hemispheres, empty into the superior sagittal sinus; they penetrate its walls obliquely, usually running forward so that they empty against the flow of blood within the sinus itself. Stopford ('30) suggested that this direction of entrance helps maintain the circulation when the pressure of the cerebrospinal fluid is raised, ensuring sufficient pressure in these thin-walled vessels to prevent their collapse, as they cross the subarachnoid space, by increases in the pressure of the cerebrospinal fluid.

Ligation and resection of the anterior portion of the superior sagittal sinus has been carried out frequently, but in the past more posterior portions were resected successfully only after they had been slowly occluded by growth of a tumor. Bonnal and Brotch, however, have reported reconstruction of a resected portion of the sinus with a dural or vein graft, anastomosing to collaterals of the vein graft any cortical veins involved.

Inferior Sagittal Sinus

The inferior sagittal sinus runs in the free edge of the falx cerebri, where it receives small veins from the medial surfaces of the hemispheres and from the inferior part of the frontal region; it is small throughout its course, and at the junction of the falx and tentorium joins the much larger great cerebral vein (of Galen) to form the straight sinus. The straight sinus, in turn, runs posteriorly along the attachment of falx and tentorium to enter a transverse sinus or the confluence of the sinuses (torcular), the main stream of its blood typically going to the transverse sinus opposite that receiving the blood from the superior sagittal sinus. Among the specimens examined by Saxena and co-workers, the straight sinus bifurcated and joined both transverse sinuses in 50%; only a very few were found to end in the confluence.

Occipital Sinus

The occipital sinus (Fig. 1-10) usually begins as right and left parts about the margins of the foramen magnum; these portions are often known as the marginal sinuses and they are continuous with the internal vertebral venous plexus. The two sinuses usually unite at a variable distance above the margin of the foramen magnum to form a single sinus that runs upward in the base of the falx cerebelli to join the confluence. Through plexuses accompanying the hypoglossal nerves the occipital sinus communicates with the internal jugular vein or the inferior petrosal sinus outside the skull.

Transverse and Sigmoid Sinuses

The transverse sinuses begin at the confluence of the sinuses, at the internal occipital protuberance, and extend laterally to become continuous with the sigmoid sinuses. (The terms *transverse sinus* and *lateral sinus* have also been used to denote the combined transverse and sigmoid sinuses.) The transverse sinus lies in the dura at the attachment of the tentorium cerebelli to the occipital bone. The sigmoid sinus then runs downward in the posterior cranial fossa, grooving the petrous part of the temporal bone as it does so, and finally turning slightly forward in the floor of the fossa to reach the jugular foramen through which it makes its exit.

The transverse sinus receives superior cerebellar veins and may receive inferior cerebral ones. The sigmoid sinus receives variable veins from the cerebellum and brain stem, and it communicates with the occipital sinus and, through mastoid and condylar emissary veins, with the occipital and vertebral veins. The superior petrosal sinus enters it just as it begins to curve downward on the temporal bone.

Both transverse sinuses are typically large; the right is usually larger. The sigmoid sinus

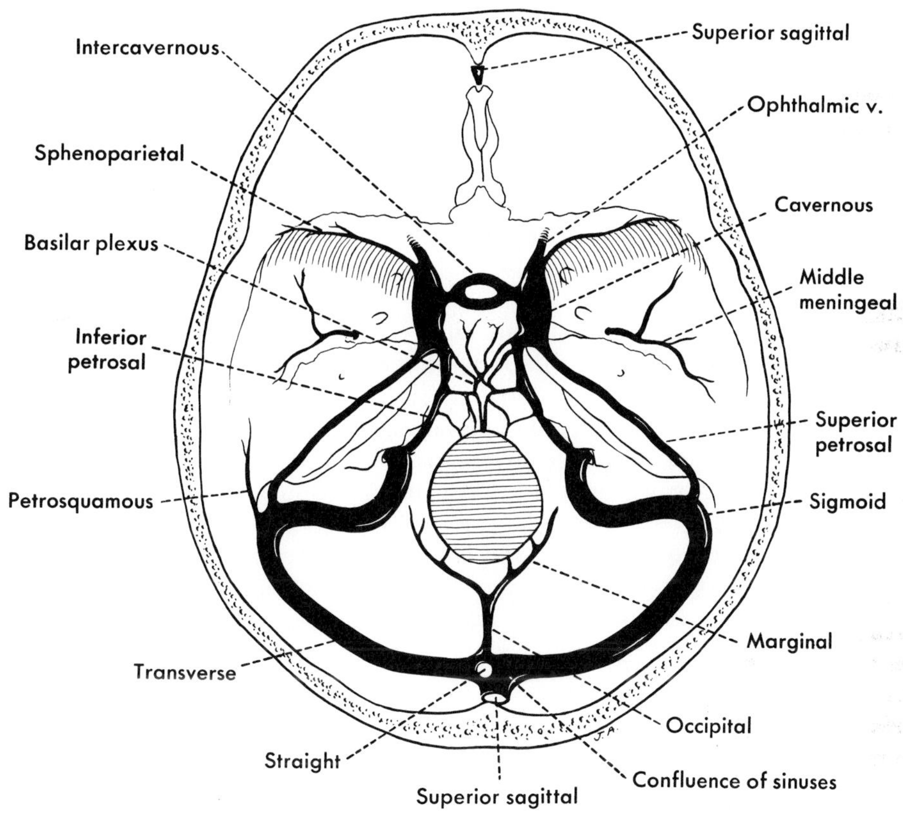

Fig. 1-10. The venous sinuses at the base of the skull.

ends by passing through the posterior part of the jugular foramen, outside of which it expands to form the superior bulb of the internal jugular vein. As it passes through the jugular foramen it is separated from the inferior petrosal sinus by the ninth to eleventh cranial nerves.

Superior Petrosal Sinuses

The superior petrosal sinuses leave the cavernous sinuses, on either side of the sella turcica, to run along the attached margins of the tentorium cerebelli (therefore between middle and posterior cranial fossae) and empty into the lateral sinuses at the beginning of their sigmoid portions. Usually the superior petrosal sinus passes above the mouth of the trigeminal cave, but Coates found this to be true in only 30 of 55 cases; in 12 the sinus passed below the trigeminal roots, and in 13 one portion passed above and one below the roots.

The superior petrosal sinus receives inferior cerebral and superior cerebellar veins and veins from the brain stem, including one that runs on or close to the roots of the trigeminal nerve in the posterior cranial fossa (the "petrosal vein" of Dandy).

Inferior Petrosal Sinuses

The inferior petrosal sinuses, like the superior, originate from the cavernous sinuses and somewhat parallel the superior sinuses; each runs backward along the petrooccipital suture, in the posterior cranial fossa, to pass through the anterior compartment of the jugular foramen and empty into the superior bulb of the internal jugular vein. The inferior petrosal sinus receives veins from the lower surface of the cerebellum, from the brain stem and, usually, from the veins from the internal ear. As it leaves the cavernous sinus the inferior petrosal sinus passes beneath the petroclinoid or petrosphenoid ligament (of Gruber) in company with the abducens nerve; as it passes through the jugular foramen, it usually runs laterally between the ninth and tenth nerves to reach the internal jugular vein. According to Rand and Hanafee, it may join a vein from the deep cervical plexus to the internal jugular or drain into the plexus rather than directly into the internal jugular.

Cavernous Sinuses

The cavernous sinuses (Fig. 1-10) are large venous spaces on the sides of the body of the sphenoid bone; they are connected to each other in front of and behind the hypophysis by anterior and posterior intercavernous sinuses, the whole being therefore sometimes referred to as the *circular sinus.* A large anterior channel is not infrequent (Renn and Rhoton, Kaplan and co-workers) and may be the source of heavy bleeding in the transsphenoidal approach to the pituitary gland. Unlike the sinuses already described, which are typically open channels, the cavernous sinuses are much broken up by heavy trabeculae that cross them (their name is derived from the fact that they resemble cavernous or erectile tissue), and the flow of blood through them is therefore very much slowed. This, and the fact that the trabeculae markedly increase the endothelial surface, both contribute to the likelihood of thrombosis of the sinus as a result of its infection. Harris and Rhoton agreed with this description, although Parkinson ('64, '65, and other papers) described the sinus as being actually a plexus of veins. These reports describe the relationships in the cavernous sinus in detail, emphasizing the branches that arise from the internal carotid artery in the sinus.

In the dural wall of the cavernous sinus are the oculomotor and trochlear nerves above, and the ophthalmic division of the trigeminal below. The carotid artery and the abducens nerve run through it more deeply. The trochlear and ophthalmic nerves converge as they approach the superior orbital fissure and thus form two sides of a triangle; it is through this triangle, with its base the dorsum sellae, that Parkinson ('65, '73) has repaired fistulas between the internal carotid and the cavernous sinus. The cavernous sinuses receive the superficial middle cerebral vein and veins from the lower surfaces of the frontal lobes; into them also enter the ophthalmic veins, which communicate with the angular veins and the pterygoid plexuses and are thus one route by which blood from the face, jaws, and pharynx can reach the cavernous sinuses. The emissary plexus of the foramen ovale (or a sphenoid emissary vein), veins that may traverse the foramen rotundum, and small veins that run through the foramen lacerum also connect the cavernous sinus and the pterygoid and pharyngeal plexuses. Laterally each cavernous sinus receives the sphenoparietal sinus, which communicates with the middle meningeal vein. Each sinus communicates with the internal jugular vein or the pharyngeal plexus through the variably developed plexus about the internal carotid artery and through the superior and inferior petrosal veins by which it is drained and with which it is continuous posteriorly; the two sinuses also communicate posteriorly with each other through an ill-defined plexus of small channels, the basilar plexus. Because of its connections to the facial vein, the cavernous sinus may become thrombosed as a result of ill-advised handling of boils or infected pimples in the "danger area" of the face (see Chap. 6).

Basilar Plexus

The basilar plexus lies in the dura mater of the posterior cranial fossa over the clivus (the sloping body of the sphenoid and base of the occipital). Anterosuperiorly it communicates with the cavernous and inferior petrosal sinuses; posteroinferiorly, with the marginal sinuses and the anterior internal vertebral venous plexus. It receives small veins from the brain stem and the cerebellum.

The importance of the connections of the basilar plexus and occipital sinus to the internal vertebral venous plexuses ("Batson's veins") as an alternative to drainage through the internal jugular veins has been emphasized by Eckenhoff. He pointed out that even during expiration more cranial venous blood leaves by way of the vertebral plexuses than by way of the internal jugulars, and that during weightlifting, with its high intrathoracic pressure, all blood leaves through the plexuses.

Sphenoparietal and Other Sinuses

The sphenoparietal sinuses arise from one of the meningeal veins, usually the middle meningeal, and run downward on the lesser wings of the sphenoid bone to empty each into the corresponding cavernous sinus, passing above the oculomotor nerve in the roof of the sinus to do so. It is in these sinuses that the anterior temporal diploic veins usually end.

In addition to the venous channels already described in the dura, small sinuses likewise accompany the meningeal arteries and are referred to as the meningeal veins. These vessels usually communicate with the diploic veins and with the various larger venous sinuses; two veins accompany the middle meningeal artery, and in addition to their usual communications with the cavernous sinus (through the sphenoparietal) pass through the foramen spinosum with the artery to empty into the pterygoid plexus.

Of the other sinuses that may sometimes be present, the petrosquamous, although it is present only occasionally, is the most constant. It lies in the groove between the anterolateral surface of the petrous and squamous portions of the temporal bone and posteriorly opens into the superior petrosal sinus or the beginning of the sigmoid sinus; it may communicate anteriorly with a deep temporal vein outside the skull. Other venous sinuses that may be present as variations from the normal are discussed later in this chapter.

MAJOR VARIATIONS

As a consequence of their development from a widespread venous network (Streeter), the cranial sinuses show numerous variations. Occasionally, as in the case described by Werden, multiple venous anomalies may be extreme and produce neurological symptoms, but other, more common, variations are also of clinical importance. Probably the most frequent are variations at the confluence of the sinuses (torcular), often associated with variations in the transverse and sigmoid sinuses. These have been analyzed by a number of workers.

In regard to the *sizes* of the vessels, Gibbs and Gibbs measured the cross-sectional area of those emptying into the confluence and found them to be extremely variable, even relative to the cross-sectional area of other vessels at the same point. Their average values, however, were as follows: for the superior sagittal sinus, 20 mm^2; for the straight sinus, 15 mm^2; for the occipital sinus, 7 mm^2; for the right lateral sinus, 30 mm^2; and for the left, 24 mm^2.

The varying disposition of the sinuses at the confluence, and the condition of the lateral sinuses, have been investigated both by dissection and from studies of roentgenograms or of dried bones. Although the cranial markings do not always provide precise data on the confluence and the sizes of the transverse sinuses, Woodhall and Seeds found that good routine roentgenograms of the occipital region were usually sufficient for clinical purposes; in a study of 100 such roentgenograms they found about the same percentage of types of variation as Woodhall found in a study of specimens by dissection.

Woodhall divided the types of confluence into five, of which the four chief types are shown in Figure 1-11. These were a common pool type (9%) in which straight and superior sagittal sinuses met and both lateral sinuses drained this junction; a plexiform type (56%) which, although divisible into several subvarieties, was characterized by several channels that provided adequate communications between the two sides; an ipsilateral type (31%) in which the sagittal sinus ran to one side and the straight sinus to the opposite side, although there was sometimes a foramen between the two; a unilateral type (4%) in which both the sagittal and straight sinuses entered the same transverse sinus and there was absence of the opposite one; and a fifth type which he termed the occipital, overlapping with some of the other types but characterized by the presence of a large single or paired occipital sinuses with large marginal sinuses.

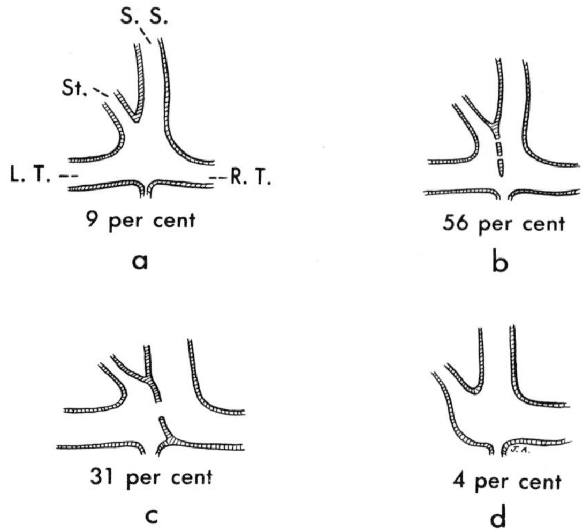

Fig. 1-11. Variations at the confluence of the sinuses. (*a*). Common pool type; (*b*) plexiform type; (*c*) ipsilateral type; and (*d*) unilateral type. (*S.S.*) Superior sagittal sinus; (*St.*) straight sinus; and (*L.T.*) and (*R.T.*), left and right transverse sinuses, respectively. Percentages are those given by Woodhall (Arch Surg 33:297, 1936)

Variations in the transverse sinuses are only in part correlated with variations at the confluence. According to Woodhall, the transverse sinuses are usually of equal size when the confluence is of the common pool type; in the plexiform type the drainage varies, both sagittal and straight sinuses draining about equally into both sides, or one draining predominantly to one side, the other to the other side. In the majority of cases of unequal drainage the sagittal sinus drained to the right, the straight sinus to the left. This tendency for the superior sagittal sinus to drain into the right transverse sinus and the straight sinus into the left one, when there is unequal drainage, is seen also in the ipsilateral type; among Woodhall's 31 cases this pattern occurred in 26, the reverse in only 5. This unequal drainage does not, however, necessarily affect the sizes of the transverse sinuses, for they were reported as being equal in size in 15, and the right as being only slightly larger in 6 more, while the left was the larger in 3 cases. In the unilateral type the right transverse sinus was missing in 3 cases, the left in 1.

Of the variations in the region of the confluence, the most important one clinically is that in which a transverse or sigmoid sinus is much reduced or missing (Fig. 1-12). In this condition the Tobey–Ayer test may give misleading results, since occlusion of the internal jugular on the side of the normal sinus will give the expected rise in intracranial pressure, while occlusion on the side of the defective sinus will fail to produce a rise and therefore may be interpreted as indicating a thrombosis of the sinus on that side.

According to Waltner, variations of the transverse and sigmoid sinuses are independent of each other, since they develop from separate anlagen, but the sigmoid sinus shows greater constancy and fewer variations than does the transverse one. Some that have been described are shown in Figure 1-13. Hoople stated that sometimes most of the blood from the transverse sinus leaves the skull through the mastoid foramen, the jugular foramen then being markedly contracted and the sigmoid sinus reduced or absent; he also reported a case in which the right transverse and sigmoid sinuses were small, and the sigmoid ended blindly at the jugular foramen, its drainage passing into sinuses of the middle cranial fossa.

The most striking variation among other sinuses is the atresia of the anterior end of the superior sagittal sinus reported by Kaplan and Browder. Of less interest is the occasional absence of the inferior sagittal sinus (Saxena and co-workers) and the high incidence of apparent absence of the occipital sinus (Das and Hasan). Variations in the occipital sinus that may be important in surgical approaches to the posterior fossa are described with that fossa.

Doubling of one of the major sinuses is occasionally found; Woodhall found two cases of doubling of the superior sagittal sinus for a portion of its length, and Hoople noted that the transverse portion of the lateral sinus is sometimes similarly divided into separate channels. The straight sinus also may be partially or wholly doubled (Browder and co-workers; Saxena and co-workers). In the latter case, one channel represented the great cerebral vein; the other, the inferior sagittal sinus.

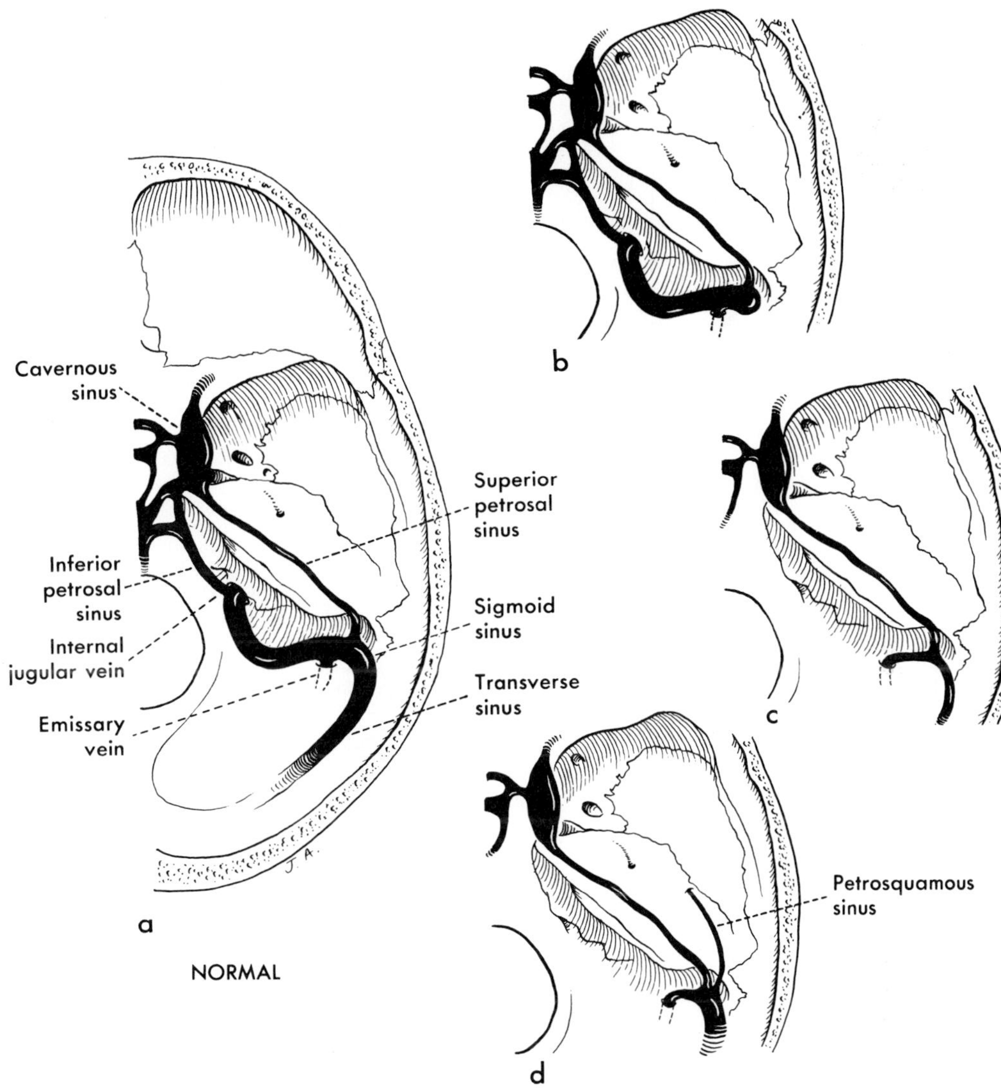

Fig. 1-12. Variations in the transverse and sigmoid sinuses, with the normal arrangement (*a*), for comparison. In (*b*) the transverse sinus is lacking. In (*c*) and (*d*) the major part of the sigmoid sinus is lacking, and the small transverse sinus in (*c*), the normal-sized one in (*d*), ends in the mastoid emissary vein shown in (*a*). (After Waltner JG: Arch Otolaryngol 39:307, 1944)

FLOW OF BLOOD IN THE SINUSES

The usual anatomical disposition of the sinuses is such that one would expect practically all the blood reaching the intracranial contents to make its exit by way of the large transverse and sigmoid sinuses into the internal jugular veins. Because the straight sinus, formed by the great cerebral vein and the inferior sagittal sinus, receives blood bilaterally from the cerebral hemispheres and the more deeply placed structures of the brain and the large superior sagittal sinus also receives blood from both hemispheres, one would also expect rather thorough mixing of the blood entering by the paired carotid vessels supplying the two sides. Finally, in many cases one would also expect the blood from the superior sagittal sinus to be directed into one of the

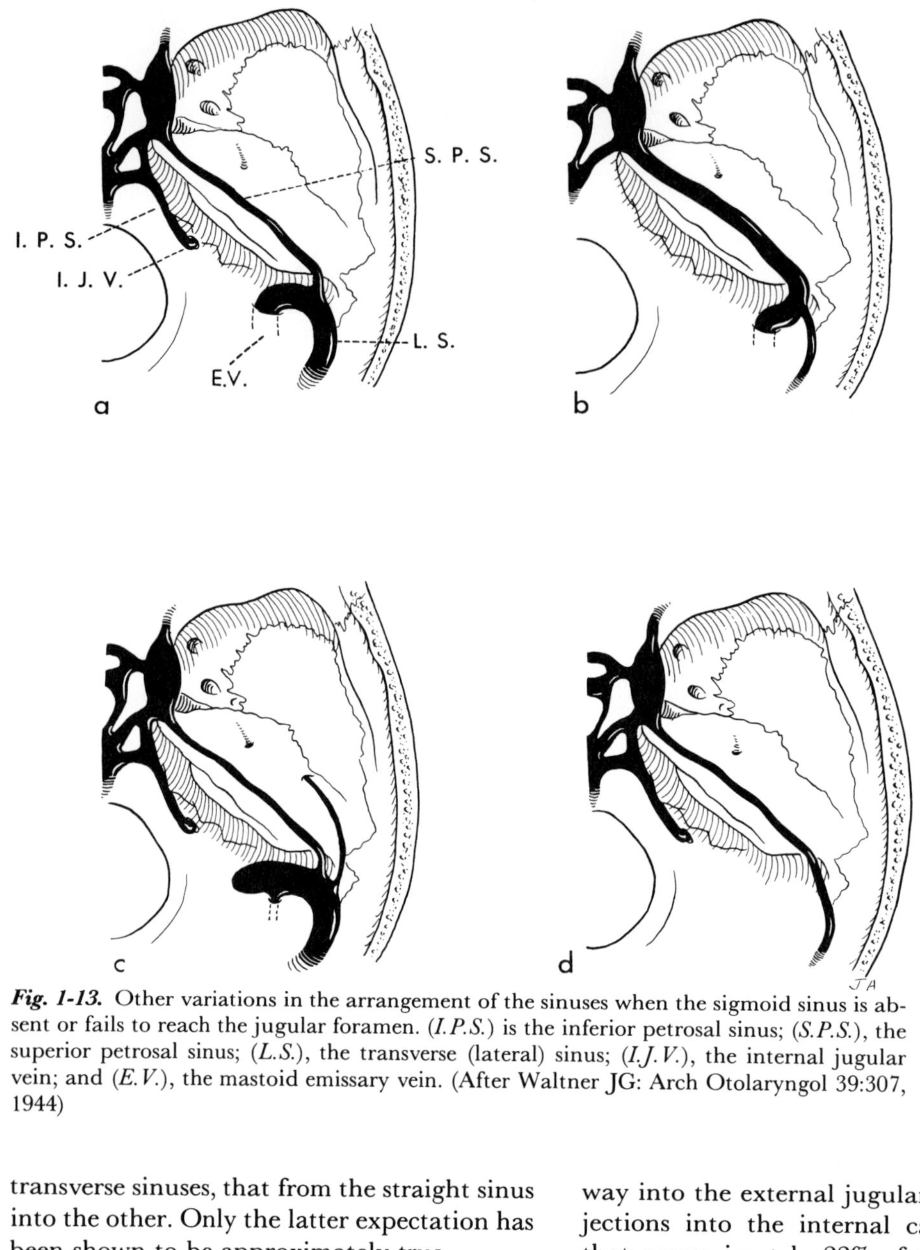

Fig. 1-13. Other variations in the arrangement of the sinuses when the sigmoid sinus is absent or fails to reach the jugular foramen. (*I.P.S.*) is the inferior petrosal sinus; (*S.P.S.*), the superior petrosal sinus; (*L.S.*), the transverse (lateral) sinus; (*I.J.V.*), the internal jugular vein; and (*E.V.*), the mastoid emissary vein. (After Waltner JG: Arch Otolaryngol 39:307, 1944)

transverse sinuses, that from the straight sinus into the other. Only the latter expectation has been shown to be approximately true.

Gibbs and Gibbs, through perfusion experiments, have apparently substantiated the general belief that at the confluence the greater part of the flow from the superior sagittal sinus is directed into the right transverse sinus and that from the straight sinus into the left. In regard to the pathways of exit of blood from the cranium, however, Shenkin and associates have found that a considerable proportion of internal carotid blood makes its way into the external jugular system; dye injections into the internal carotid indicated that approximately 22% of the blood in the external jugular vein is derived from this vessel. Obviously, the rather insignificant emissary veins are of more importance than they would appear to be. Since these authors found that only about 2.7% of the blood of the internal jugular vein is derived from extracerebral sources, it is apparent that the flow of blood through the emissary veins is as a rule predominantly or entirely outward.

Shenkin and co-workers have apparently

shown also that there is a surprisingly incomplete mixing, on the venous side, of the blood derived from the two carotids; their experiments indicated that, of the carotid blood in the internal jugular, about 66% is derived from the ipsilateral internal carotid, and only 34% from the contralateral one.

ANOMALOUS ACCESSORY SINUSES

The most commonly occurring anomalous sinus is probably the *petrosquamous* (Fig. 1-12*d*); it usually begins anteriorly as a communication with the deep temporal and middle meningeal veins, runs along the angle between the petrous and squamous portions of the temporal bone, and empties into either the transverse sinus or the posterior portion of the superior petrosal sinus. It is present in approximately 50% of heads and is about three times as frequent unilaterally as bilaterally. The sinus may receive small veins from the middle ear through the petrosquamous fissure.

The *accessory sinus of Kelch* is described as passing backward from the superior orbital fissure, where it usually communicates with some of the vessels of the orbit, to cross the upper border of the petrous portion of the temporal bone and empty either into the superior petrosal or the transverse sinus.

The *ophthalmopetrous sinus of Hyrtl* is probably identical with the sinus of Kelch, since it is described as passing backward from the superior orbital fissure over the inner surface of the great wing of the sphenoid and the anterior surface of the temporal bone to terminate in the transverse sinus; it is said sometimes to leave the skull anteriorly through the foramen ovale, and sometimes to communicate with the middle meningeal veins.

The *accessory sinus of Verga* is a communication between the cavernous sinus or the ophthalmic vein anteriorly and the transverse sinus posteriorly and may be another variation of those described above.

Occasionally, one of these accessory sinuses is of importance in the transmission of an infection. Thus Stewart reported a case in which Kelch's accessory sinus became infected, apparently during the course of an operation on the middle ear, and noted that the petrosquamous sinus has been said to be important in the passage of infection from the middle ear to the brain.

Stewart also described an additional accessory "sinus," which he termed the petrooccipital, although it probably should not be called a sinus since it lay for the most part outside the cranial cavity. According to Stewart it ran from the carotid venous plexus (at the point of communication of this with the cavernous sinus) to the jugular bulb, leaving the skull through an opening in the petrooccipital synchrondrosis to run a largely extracranial course along the inferior surface of this joint.

MENINGEAL ARTERIES

Although there are a variable number of arteries running in the dura and supplying it and the adjacent bone, most of them are small and confined to the base of the skull. They include meningeal branches of the ethmoidal arteries, branches of the ophthalmic, to the floor of the anterior cranial fossa; twigs of the sphenopalatine artery (from the maxillary—internal maxillary) to the same region; tiny branches from the internal carotid that are distributed largely in the neighborhood of the cavernous sinus and the trigeminal ganglion, and a longer one that supplies the tentorium cerebelli and the lower part of the falx cerebri; the accessory middle meningeal artery to the region of the ganglion, when it enters the skull (through the foramen ovale) at all; branches of the ascending pharyngeal artery to the floor of both middle and posterior cranial fossae; and branches from the vertebral and occipital arteries to the posterior cranial fossa. These are considered in the descriptions of the cranial fossas to which they are distributed. By far the largest of the meningeal arteries, and the most important one surgically, is the middle meningeal artery.

Gross and Savitsky pointed out that, as would be expected from its size, the middle meningeal artery is the one most commonly involved in extradural hemorrhage, and therefore the subtemporal exposure of the middle cranial fossa is usually adequate to reveal such hemorrhages. They emphasized, however, that on rare occasions an extradural hemorrhage may be limited to the anterior cranial fossa, usually resulting here from hemorrhage from the meningeal branch of the anterior ethmoidal (or from this latter vessel itself?); they reported that extradural hemorrhages in the posterior cranial fossa are extremely rare. They also pointed out that extradural hemorrhages may occur without a fracture of the skull, and stated their belief that most extradural hemorrhages result from injury to the meningeal arteries and not from tears in the intracranial venous sinuses. Reichert and Morrissey agreed also that extradural venous hemorrhage is rare as compared with subdural venous hemorrhage or extradural arterial hemorrhage; they suggested, however, that a number of extradural hemorrhages that are supposed to arise from torn middle meningeal arteries actually arise from torn meningeal veins or sinuses.

MIDDLE MENINGEAL ARTERY

The middle meningeal artery (see Fig. 1-8) supplies most of the supratentorial dura, with the exception of the floor of the anterior cranial fossa, and even sends some branches into the posterior cranial fossa; it therefore spreads widely in the dura and is of surgical importance from the standpoint of its possible involvement in fractures of the skull (see the preceding paragraph). This artery, a branch of the maxillary (formerly internal maxillary, from the external carotid) enters the middle cranial fossa through the foramen spinosum, which lies just posterolateral to the foramen ovale; very rarely, it enters through the latter.

Shortly after its entrance into the skull the middle meningeal artery divides into two major branches, frontal and parietal or anterior and posterior (Fig. 1-14); while some authors describe three branches, their so-called middle branch, although sometimes originating from the main vessel at its bifurcation, commonly arises from either of the other branches. The frontal branch of the middle meningeal artery is distributed to the dura over the convexity of the brain in both anterior and middle cranial fossae, while the parietal is distributed to the remainder of the supratentorial portion of the dura, with a few twigs going also to the dura of the posterior fossa.

Although the larger branches of the middle meningeal artery lie in the dura, they are so superficially located in it that, with their accompanying veins, they produce grooves on the interior of the cranium, and the smaller branches are distributed more extensively to the bone than to the relatively avascular dura. The middle meningeal artery anastomoses with arteries of the opposite side, with anterior and posterior meningeal arteries, with a meningeal branch of the internal carotid and, through foramina in the bone, with the deep temporal arteries. It usually has also an anastomotic branch with the lacrimal artery in the orbit; typically small, this vessel may be larger and form the stem of origin of the lacrimal artery or even of the ophthalmic artery itself; perhaps very rarely, as may have been true in the case described by Low, this anastomotic branch may form the origin of an anomalous middle meningeal artery.

Course

The *foramen spinosum* is located in approximately the deepest part of the concavity of the floor of the middle cranial fossa, fairly close to the anteroposterior levels of the posterior root of the zygomatic arch. The middle meningeal artery may branch almost immediately after entering the cranium or may run for some distance laterally and anteriorly before branching (Fig. 1-15). Chandler and Derezinski, in a study of 1200 half skulls, found no significant difference in the frequency of branching close to the foramen spinosum, near the level of the upper border of the greater wing of the sphenoid, and between these two points. If the branching is far lateral, it is situated at about the level of the upper border of the zygomatic arch and at about the middle of the length of this arch; in

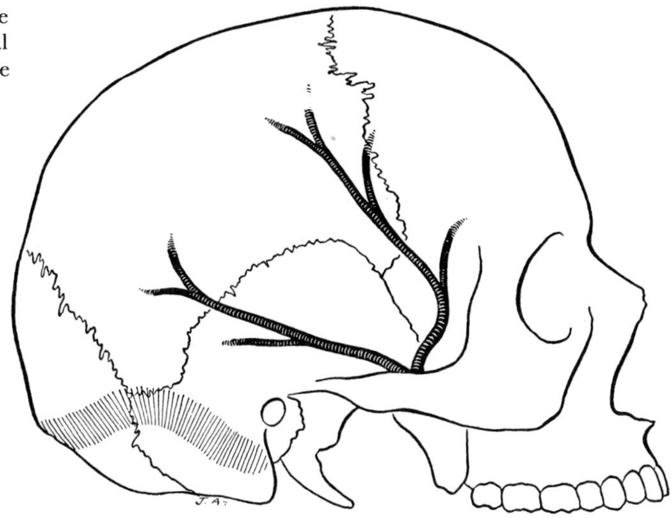

Fig. 1-14. Approximate course of the middle meningeal artery in relation to the cranial vault. The transverse and sigmoid sinuses are also indicated.

this case the parietal branch of the artery arches upward and backward about 1 cm above the zygomatic arch and the supramastoid crest. If the middle meningeal artery branches close to the foramen spinosum, the parietal branch does not come so far forward and appears above the level of the zygomatic arch only toward its posterior end.

The *frontal branch* of the artery, whether it arises from close to the foramen spinosum or at the level of the middle of the zygomatic process, describes a broad curve with its concavity directed backward; the foremost part of this curve, and the point at which the artery can be fairly constantly located from externally on the skull, is the pterion, or region of junction of the frontal, sphenoid, parietal and temporal bones. This is covered by the temporal muscle, but can be located approximately by measuring about 2.5 cm (a thumb's breadth) behind the zygomatic process of the frontal bone and about 3.5 cm (two fingerbreadths) above the zygomatic arch. Chandler and Derezinski found the frontal branch absent in only 9 of 1200 sides and the parietal absent in only 7.

The frontal branch of the middle meningeal artery in more than 50% of cases (677 of 1200, according to Chandler and Derezinski) runs in a canal in the bone, rather than simply in a groove, for part of its length and is most apt to do so in the region of the pterion.

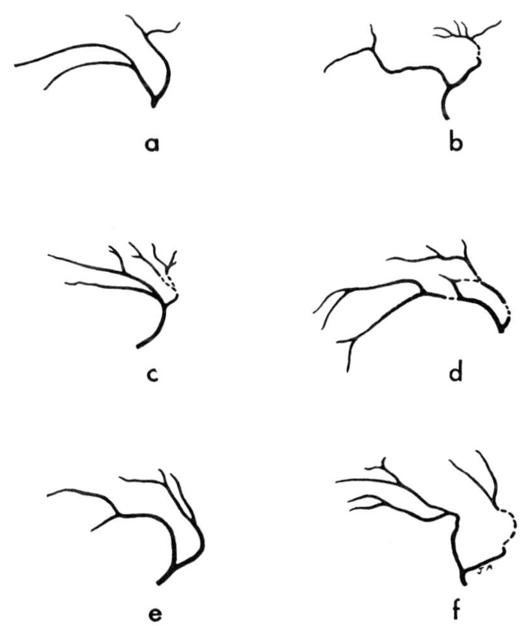

Fig. 1-15. Variations in the branching of the middle meningeal artery. The *broken lines* indicate that those parts of the vessels coursed in a canal in the bone. (After Chandler SB, Derezinski CF: Anat Rec 62:309, 1935)

For this reason, separation of the bone from the dura at this particular level may often lead to brisk bleeding from a torn middle meningeal. The approach for ligation of this branch or the artery as a whole is over the middle of the zygomatic arch; if the artery sought is not found at this level, the dura lateral and inferior to the temporal lobe must be elevated; the frontal branch of the artery, in such a case, may be found by following the dura medially, but if the main stem of the vessel is to be located, the dura must be followed not only medially but also posteriorly.

INNERVATION OF THE DURA OF THE CRANIAL VAULT

The vasomotor nerve supply of the dura mater is probably derived from plexuses accompanying the vessels, especially the middle meningeal artery, derived largely from the superior cervical ganglion. The sensory supply to the dura of the cranial vault is apparently sparse, in contrast to that of most of the cranial floor.

Although the dura of the middle cranial fossa is supplied usually by direct twigs from all three divisions of the trigeminal nerve, these are largely distributed at the base of the skull and to the tentorium or a limited area above this; the nerves to the dural vault run primarily with the middle meningeal artery and are limited to its immediate neighborhood (Fig. 1-16). They consist of fibers arising from both the maxillary and mandibular divisions of the trigeminal, which largely accompany the middle meningeal artery and its branches. Tentorial branches from the ophthalmic nerve may reach the dura of the parietal region, however, according to Penfield and McNaughton; these workers found suggestive evidence that all the meningeal branches of the fifth nerve, regardless of their anatomical origin, are derived actually from the ophthalmic. The nerves are described in more detail in connection with the middle cranial fossa.

Anatomical investigations on the innervation of the dura of the cranial vault seem to have been confirmed by clinical studies. Penfield and McNaughton, from stimulations of the dura at operation, concluded that all the supratentorial dura is innervated exclusively by the ipsilateral fifth cranial nerve. They were able to elicit only pain from the cranial dura, regardless of the type of stimulus used. They also found that most of the dural con-

Fig. 1-16. Nerves of the cranial vault and the posterior cranial fossa. See also Fig. 1-43. (After Penfield W, McNaughton F: Arch Neurol Psychiatr 44:43, 1940)

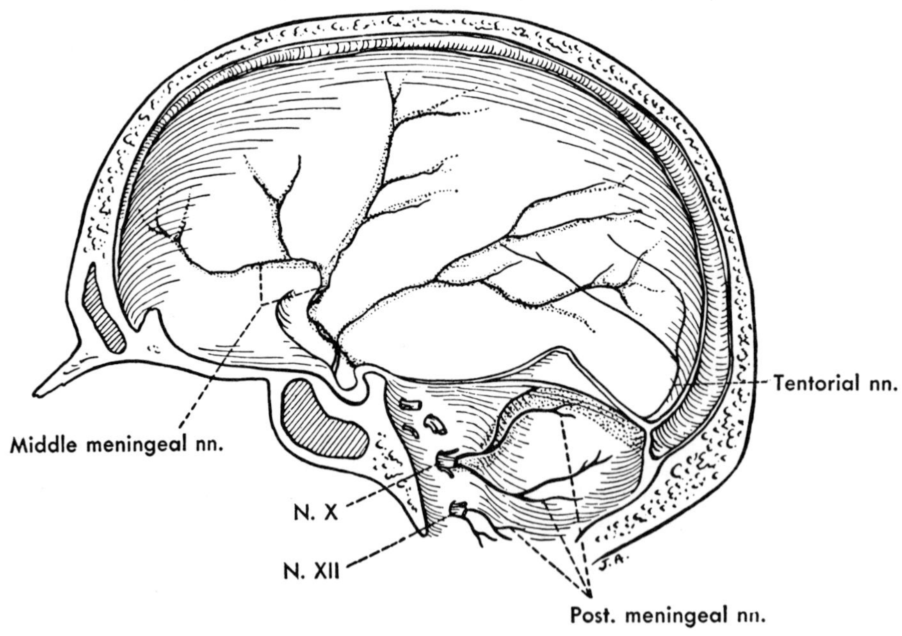

vexity is insensitive and that the sensitive points upon it coincide largely with the meningeal arteries, except for occasional points where nerve pass across the dura on their way to vascular channels.

Although typical findings in regard to referred pain from the dura are shown in Figure 1-8, Wirth and Van Buren were unable to find, in a study made with implanted electrodes, any correlation between the area to which pain was referred and the placement of the electrodes.

Ray and Wolff, in their more extensive study of the sensitivity of cranial structures, concluded as did Penfield and McNaughton that the dura between the branches of the middle meningeal artery is largely insensitive to pain; they reported that, except along the course of this artery and along the margins of the dural sinuses, the supratentorial dura is sensitive to stimulation of any type only at the base of the skull.

LEPTOMENINGES, SUBARACHNOID SPACE, AND VENTRICLES

The leptomeninges consist of the arachnoid and the pia, connected to each other by the arachnoid trabeculae and enclosing the subarachnoid space. The inner lining cells of the arachnoid, resembling mesothelial or endothelial cells, are continuous over the archnoid trabeculae with the similar cells forming the outer surface of the pia; arachnoid and pia actually represent a splitting of a single layer, the leptomeninx, and the subarachnoid space is really then an intraleptomeningeal space, just as the peritoneal cavity is an intraperitoneal space.

Measurements of the total amount of cerebrospinal fluid in the adult show wide variations, but the average amount is often quoted at about 135 to 140 ml. Of this, perhaps slightly more than half is in the spinal subarachnoid space; somewhat less than 20 ml, but again with wide variation, in the four ventricles (Bull); and the remainder in the cranial subarachnoid spaces.

Cisterns

Since the pia is intimately attached to the surface of the brain and therefore follows its form, while the arachnoid follows rather closely the conformation of the dura, the slit-like subarachnoid space is exaggerated in certain areas; these enlargements are known as cisterns. Because the cisterns are so interconnected that they have few definite boundaries, there has been little agreement as to how many there are and how they should be named. The NA recognizes three unpaired cisterns—the cerebellomedullary, the chiasmatic, and the interpeduncular (Fig. 1-17)—and the paired cisterns of the lateral cerebral fossae. The importance of pneumoencephalographic studies has, however, led to more detailed descriptions of the loci in which injected air may accumulate, most of them located around the brain stem. Yasargil and co-workers listed the contents of 11 cisterns that they considered of chief surgical importance.

Ventrally, the *chiasmatic cistern* lies in front of the optic chiasma, above the sella turcica; caudal to and continuous with it is the *interpeduncular cistern* between the cerebral peduncles; caudal to that is the *pontine cistern* ("Hilton's waterbed"), and caudal to that the *medullary cistern* (Fig. 1-18). The chiasmatic, interpeduncular, and pontine are sometimes grouped into the *basal cistern.* These ventrally placed cisterns have lateral extensions that have also received special names (for instance, Wilson). The lateral extensions from the chiasmatic cisterns are the *cisterns of the lateral cerebral fossae,* or of the Sylvian fissures, at the front end of each lateral (Sylvian) sulcus between the temporal pole and the lower surface of the frontal lobe. The lateral extensions from the interpenduncular cisterns are the *crural cisterns,* extending dorsally around the cerebral peduncles. Those from the pontine cistern, sometimes called its lateral recesses, expand dorsally to form the *cerebellopontine angle cisterns* in the angles between the caudal border of the pons, the medulla, and the cerebellum; it is into these cisterns that the lateral recesses of the fourth ventricle open by way of the lateral apertures (of Luschka).

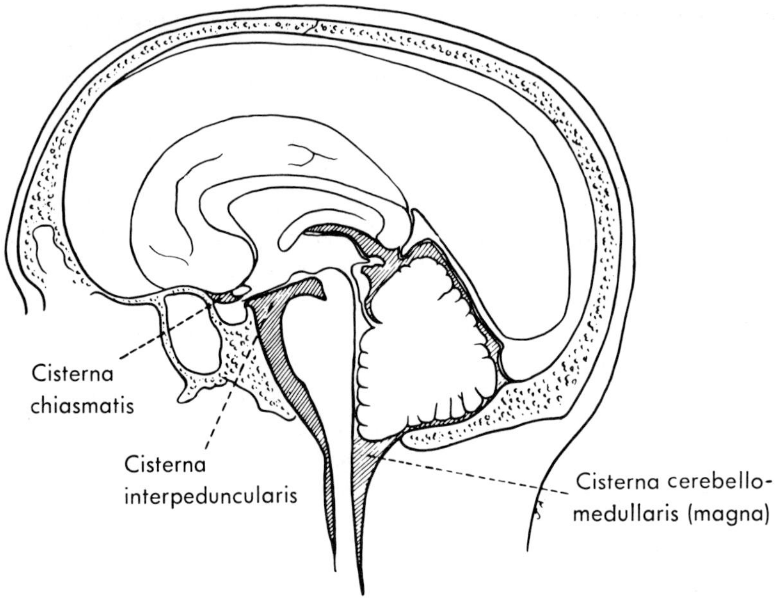

Fig. 1-17. The chief unpaired subarachnoid cisterns. Those labeled are recognized in the official NA terminology.

The lateral extensions of the medullary cistern pass around the sides of the medulla to become continuous with the *cerebellomedullary cistern,* better known as the *cisterna magna.* This dorsally situated cistern, in the angle between the lower surface of the cerebellum and the dorsal surface of the medulla, communicates with the median aperture (of Magendie) of the fourth ventricle. It extends highest in the midline, where it occupies the cerebellar vallecula, the notch between the cerebellar hemispheres. Introduction of a needle into it to record the pressure here in comparison with that at the lumbar level, or to secure cerebrospinal fluid from this area, has become a common practice; the tap is done through the foramen magnum but care must always be exercised not to thrust the needle against the medulla, with its numerous vital centers and pathways.

Other dorsally situated cisterns are also commonly described. The space between the upper surface of the cerebellum and the tentorium has been called the *superior cerebellar cistern* (and air largely over the vermis between the hemispheres has been said to be in the cerebellosagittal cistern). Anterior to the superior cerebellar cistern is the relatively commodious space dorsal to the midbrain colliculi, and between those, the splenium of the corpus callosum, and the cerebellum. Its older name is the *cistern of the great cerebral vein,* because it houses that vein; it has also been called *quadrigeminal cistern,* superior cistern, and internal cerebral cistern. Anterolateral extensions from it, around the sides of the colliculi, form the *ambient cisterns,* and Wilson describes "wings of the ambient cistern" as extending forward around the pulvinars of the thalamus to become continuous with the crural cisterns anterolaterally, and with each other across the caudal part of the roof of the third ventricle where they essentially form a forward extension of the quadrigeminal cistern. Air may also come to occupy the space at the bottom of the longitudinal cerebral fissure, therefore along the upper aspect of the corpus callosum; this space has been called the *pericallosal cistern.* It communicates caudally with the quadrigeminal cistern; rostrally, it extends over the genu and rostrum of the corpus callosum and then, as the *cistern of the lamina terminalis,* over the front of that lamina. The latter cistern is therefore a connection between the pericallosal and chiasmatic cisterns.

The leptomeninges and the subarachnoid space are invaginated along the blood vessels into the nervous system, where the arachnoid lies in contact with the outer wall of the blood vessel to form an inner lining to the perivascular space (the continuation of the subarachnoid space), and the pia forms the outer lining of this space (Fig. 1-19). Patek emphasized that the perivascular spaces are lined only in part by the pia and arachnoid, and in part by membranes composed of the glia itself; the arachnoid, according to Patek, extends along the blood vessels to the capillaries, while the pia extends about arteries for only a short distance, but about veins down to the smaller venules.

The subarachnoid space does not extend around the hypophysis, although it does encircle the infundibular stalk and is in contact with the upper surface of the hypophysis; below this the dura is fused to the outer portion of the gland.

Meningeal Relations of Nerves

As the cranial nerves make their exits through the pia, the arachnoid, and the dura, each of these elements contributes to the connective tissue of the nerve root; as the roots traverse the subarachnoid cavity they are covered by pia, and as they pierce the arachnoid they re-

Fig. 1-18. Roentgenogram of the subarachnoid cisterns at the base of the brain. The *arrow* indicates radiopaque material that has spread dorsolaterally into the cerebellopontine angle. Below, the material has spread downward along the spinal cord; above, it has reached the more horizontally-lying chiasmatic cistern. (Courtesy Dr. David G. Pugh)

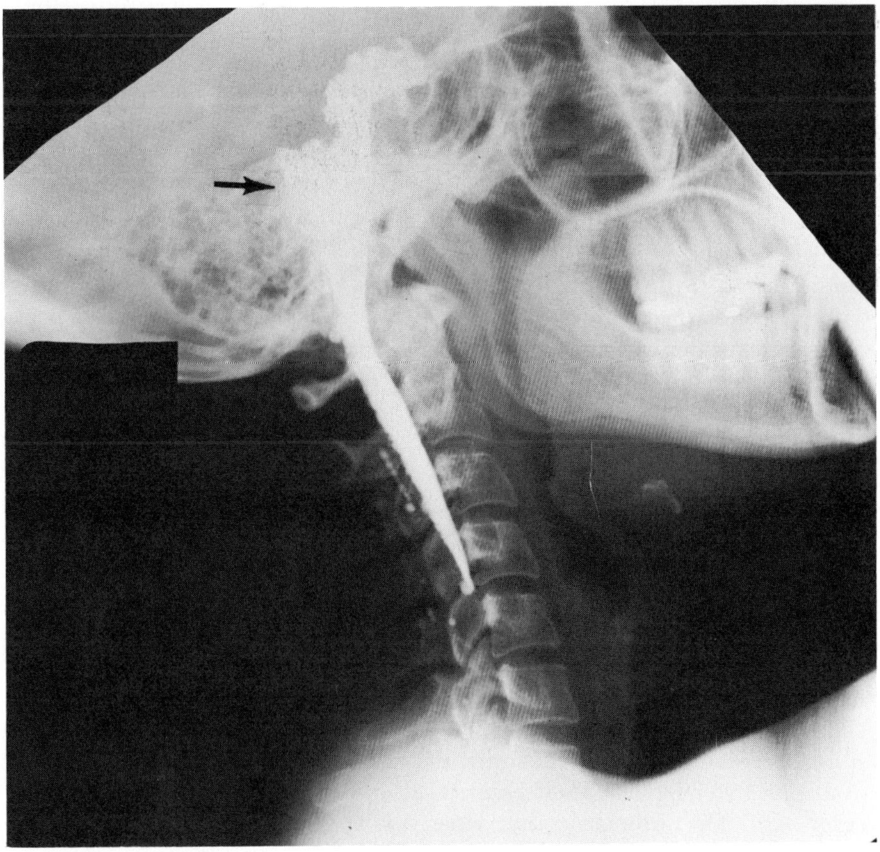

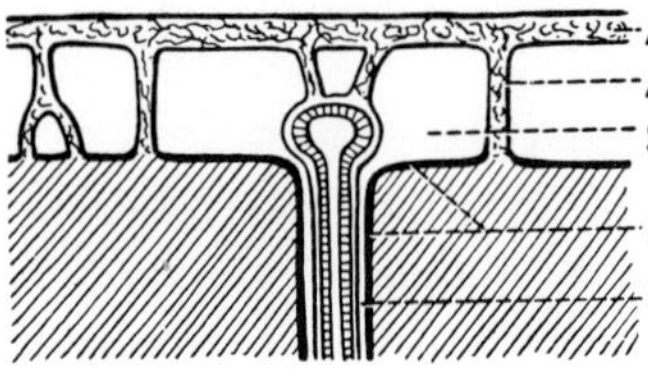

Fig. 1-19. Relation of the leptomeninges and the subarachnoid space to the perivascular spaces.

ceive connective tissue from the outer layer of this (the cellular lining of the subarachnoid cavity being reflected at this point from arachnoid to pia). Similarly, as the nerves penetrate the dura, the connective tissue of this element fuses with the connective tissue of the definitive nerve.

For the most part, the cranial nerves pierce the arachnoid and the dura in quick succession and do not carry out with them portions of either the subarachnoid or the subdural spaces; the optic nerve, however, carries the meninges extracranially and is surrounded by a continuation of the subarachnoid space as far forward as the eyeball, while the facial and eighth (vestibulocochlear, acoustic) nerves are similarly surrounded by arachnoid and dura to the end of the internal acoustic meatus; sometimes (Perlman and Lindsay) the subarachnoid space is prolonged about the facial nerve farther than this, extending even to the geniculate ganglion. Finally, there is an extension of the subarachnoid space around the roots of the fifth nerve, as they pass across the petrous ridge; it expands in the dura of the middle cranial fossa to form the trigeminal (Meckel's) cave.

Neuroglia in Nerve Roots

Although the pia covers the filaments of origin of the nerves throughout their courses through the subarachnoid space, the neuroglia can as a rule be traced for only a very short distance along the nerve roots as they leave the central nervous system. Tarlov ('37a, b) has shown that for most cranial and spinal nerves the glia extends only a fraction of a millimeter beyond their external origins (Fig. 1-20). The optic nerve is an exception to this, but is really a tract of the brain rather than a true nerve and therefore contains neuroglia throughout its entire length.

The outstanding exception to the distribution of glia along only a very short segment of a cranial nerve is the eighth nerve, which typically has glia extending for some 6 to 8 mm along its course; this undoubtedly explains the far greater occurrence of glial tumors on the eighth than on other cranial nerves. Tarlov noted that occasional patches of glia may be found farther out in a nerve than they normally appear.

Ventricles

The ventricular system (Figs. 1-21 to 1-23), an important source of cerebrospinal fluid, apparently contains on the average slightly less than 20 ml of it, but there are wide variations in its size (Bull). The two lateral ventricles, in the cerebral hemispheres, are the largest parts, and communicate with the slitlike third ventricle through the interventricular foramina (of Monro). Anatomically, each ventricle is divided simply into an anterior horn, in front of the interventricular foramen, a pars centralis or body, extending backward to the confluence with the other two horns (the trigone), a posterior horn running backward into the occipital region, and an inferior horn curving downward and forward into the temporal lobe. For radiologic purposes, however, it is customary to subdivide the lateral ventricle into numbered parts as follows: part 1 is the portion of the anterior horn that turns downward and laterally in front of the head of the caudate nucleus; part 2 is the remainder of the anterior horn; part 3 is the highest portion of the pars centralis or body ; part 4 is the trigone, also called the atrial region; part 5 is the posterior horn, subject to much variation in length; and part 6 is the inferior horn. The choroid plexus, invaginated into the thin medial wall of the ventricle, is particularly well developed in the inferior horn where the anterior choroidal artery supplies it, but curves upward and forward in the trigone, here bulging to form the glomus of the trigone, and in the central part continues through that to reach the interventricular foramen and pass into the third ventricle.

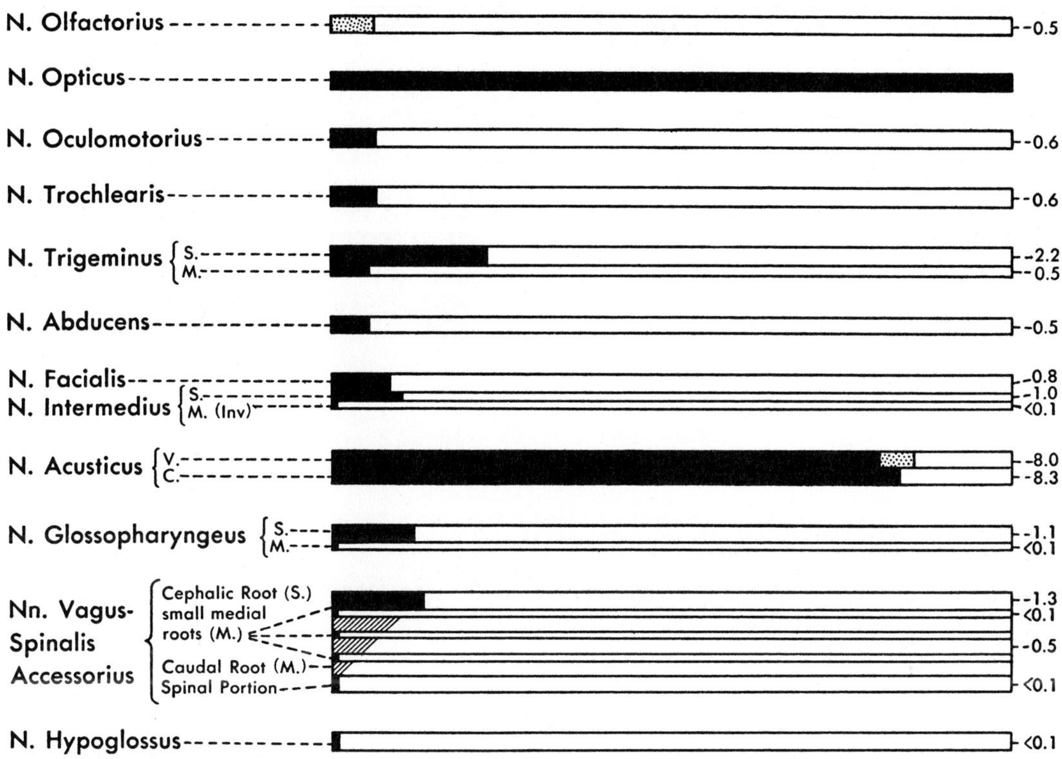

Fig. 1-20. The variable extensions of the pia along the roots of cranial nerves. The lengths of the glial segments are given in millimeters, the sign < indicating that the measurement is usually less than that given. *Stippling* on the bars for the olfactory and vestibular nerves indicates common variations in the extent of the glia along these nerves; *hatching* on the vagus-accessory complex indicates a gradual decrease in the lengths of the glial segments of the successive rootlets from above downward. (Tarlov IM: Arch Neurol Psychiatr 37:1338, 1937)

Torkildsen has discussed the distances between the ventricles and the surfaces of the brain.

The third ventricle has a particularly thin wall, the lamina terminalis, above the point at which the optic chiasma contributes to the floor. The floor is also relatively thin between the optic chiasma and the infundibulum, or base of the hypophyseal stalk. The roof (tela choroidea) of the third ventricle, with its paired choroid plexuses, is extremely thin; the internal cerebral veins pierce it and unite under the splenium (posterior end) of the corpus callosum to form the great cerebral vein (of Galen). The third ventricle communicates through the cerebral (Sylvian) aqueduct, in the midbrain, with the fourth ventricle.

The cerebral aqueduct, somewhat flared at both ends, varies much in diameter; Flyger and Hjelmquist found its average minimal cross-sectional area to be 1.5 mm^2, but some were as small as 0.4 mm^2. Such sizes may make passage of a catheter for relief of noncommunicating hydrocephalus particularly hazardous, and various shunt operations have been preferred by many workers (*e.g.,* Robertson and co-workers; Kempe and Blaylock). Keucher and Mealy reported that ventriculoperitoneal shunts were more satisfactory than ventriculoatrial ones in cases of infantile hydrocephalus. Scarff maintained that the older procedure of puncture of the lamina terminalis gives better results, and Elvidge reported success in placing a catheter in the aqueducts of patients ranging in age from 5 weeks to 46 years. Bilateral choroid plexectomy fails to

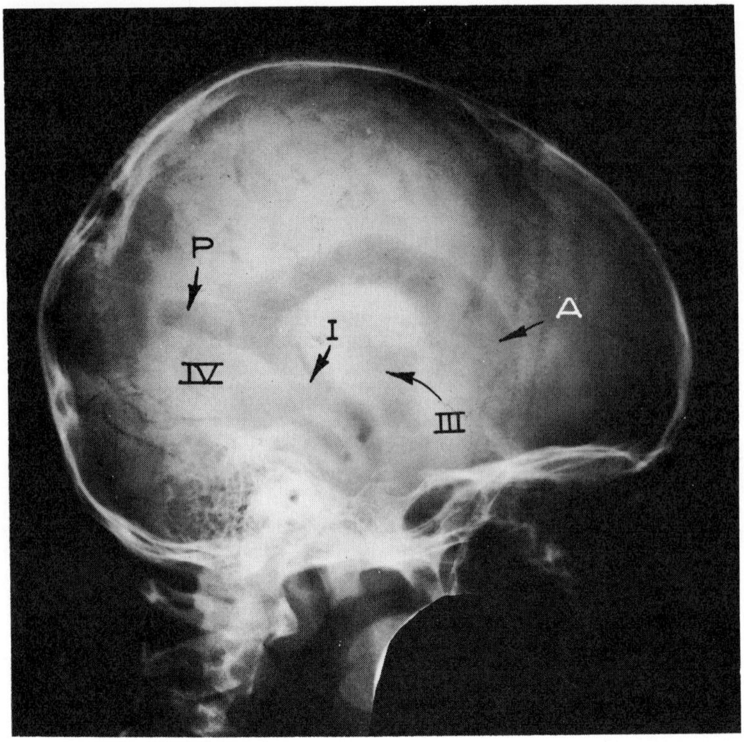

Fig. 1-21. Ventriculogram in which the ventricular system has been fairly well filled with air. (*A*), (*P*), and (*I*), the anterior, posterior, and inferior horns of the lateral ventricle (some of the lateral ventricle of the other side is also visible); (*III*) and (*IV*), the third and fourth ventricles, more clearly seen in Fig. 1-23. (Courtesy Dr. David G. Pugh)

Fig. 1-22. Ventriculogram showing particularly well the anterior horns and bodies (pars centralis) of the lateral ventricles and the third ventricle. There is some air in the inferior horns and over the convexities of the cerebral hemispheres. (Courtesy Dr. Colin B. Holman)

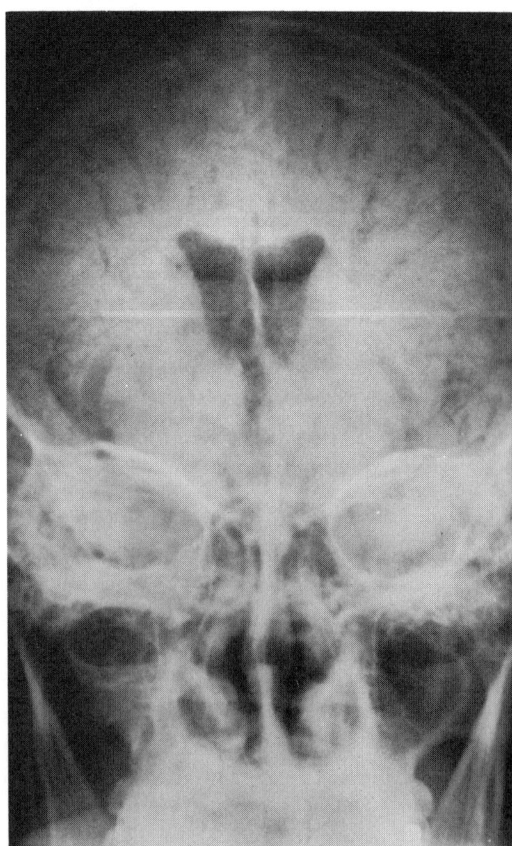

produce the expected results of markedly reducing the formation of cerebrospinal fluid (Milhorst; Milhorst and co-workers); thus, although Edvinsson and co-workers have described the plexus as being innervated by fibers from the superior cervical ganglion, and Owman and West have reported that superior cervical ganglionectomy reduced the formation of fluid at least briefly, denervation is not likely to have a favorable clinical outcome.

The fourth ventricle, with a thin roof that bears paired choroid plexuses, has marked lateral recesses that project laterally between cerebellum and brain stem and open, close to the ninth nerve, into the subarachnoid space (by way of the lateral apertures of the fourth ventricle, or foramina of Luschka). In the lowest part of the roof of the ventricle is the third opening into the subarachnoid space,

A

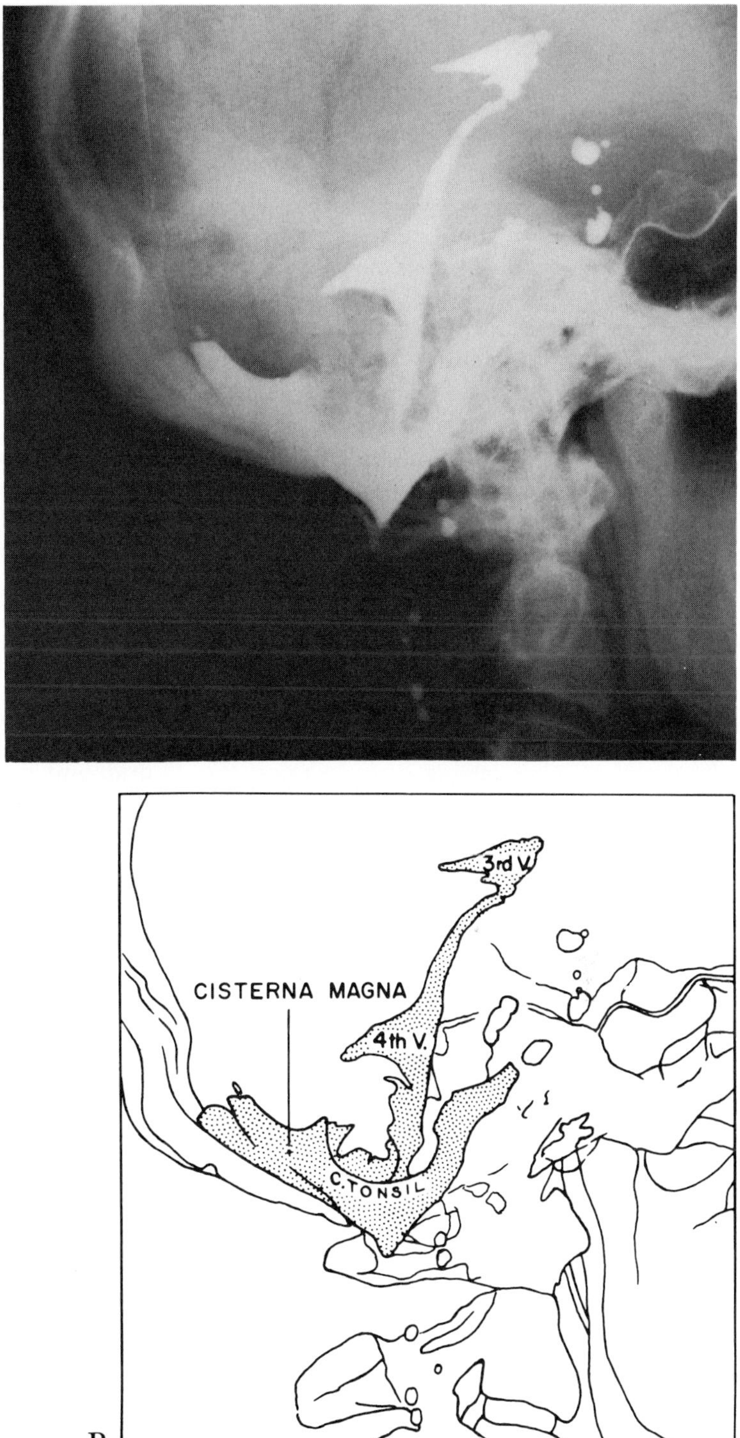

B

Fig. 1-23. (*A*) Roentgenogram and (*B*) drawing showing the third and fourth ventricles and the connecting aqueduct, with radiopaque material also in the cerebellomedullary cistern (cisterna magna) and around the cerebellar tonsil. (Courtesy Dr. Colin B. Holman) the median aperture of the fourth ventricle (foramen of Magendie).

CRANIOCEREBRAL RELATIONS

The relations between various parts of the brain and certain markings or measured points on the skull are discussed in surgical texts, and this question will therefore be discussed here only in a very general way. However, since different skulls vary somewhat in the details of their anatomy, and there is also a very considerable variation in the morphology of the cerebral hemispheres themselves, no craniocerebral relations given here can be regarded as accurate and exact; rather, they are only approximations that usually suffice for surgical purposes.

The cerebral hemispheres are entirely supratentorial, the frontal portion of the brain being lodged in the anterior cranial fossa, the temporal lobes fitting into the smooth lateral expansions of the middle cranial fossa, and the occipital lobes resting directly upon the tentorium. The level of attachment of the tentorium, and therefore the lowermost part of the occipital lobe, lies about 2.5 cm (an inch) above the plane through the upper border of the external acoustic meatus and the lower margin of the orbit; the lower surface of the frontal lobe, since it rests upon the roof of the orbit, corresponds approximately to the level of the upper margin of the orbit; the temporal lobes, in the deeper anterior portion of the middle cranial fossa, descend to the level of about the upper border of the zygomatic arch, therefore almost to the line between the lower border of the orbit and the upper border of the external acoustic meatus. The midbrain lies at the tentorial notch, and the pons, the medulla, and the cerebellum lie in the posterior cranial fossa; most of the cranial nerves therefore arise within the posterior cranial fossa, and the majority of them enter the dura or make their exits from the skull through the walls or floor of this fossa. The third to the sixth, however, traverse also the middle cranial fossa.

The superior sagittal sinus has already been discussed, as has the middle meningeal artery. The course of the transverse sinus can be approximated by a line, arching slighly upward, from the upper portion of the external occipital protuberance to the posterior aspect of the upper part of the root of the ear; the sigmoid sinus descends approximately along the posterior aspect of the root of the ear to the level of the lower border of the meatus and then curves forward to the level of the posterior border of the meatus, which approximates the anteroposterior position of the jugular foramen.

Sulci, Gyri, and Lobes

The pterion overlies approximately the stem of the *lateral* (*Sylvian*) *sulcus,* the chief or posterior limb of which runs backward more or less parallel to but some 4 to 5 cm above the zygomatic arch until it reaches the anteroposterior level of the mastoid process, after which it turns farther upward (Fig. 1-24).

Superiorly, the *central* or *Rolandic sulcus or fissure* begins slightly (about 1 cm) behind the midpoint of a line over the convexity of the skull between the nasofrontal suture (nasion) and the highest point of the greater occipital protuberance (inion or torus); from this midline point the central sulcus extends downward and forward at an angle usually somewhat less than 70° to end, of course, at the level of the lateral fissure. The lower end of the central sulcus is approximately vertically above the temporomandibular joint. Below the lateral sulcus and slightly behind the level of the lower end of the central sulcus (therefore at approximately the anteroposterior level of the external meatus) is the transverse temporal gyrus, the area for hearing.

The *parietooccipital sulcus* corresponds approximately to a short line drawn at right angles to the sagittal suture about 1 cm above or anterior to the lambda (the point of intersection between the sagittal and the lambdoid—occipitoparietal—suture, usually palpable as a depression). The calcarine area (center for sight) lies in the occipital lobe just above the level of the external occipital protuberance.

These lines suffice to indicate the general

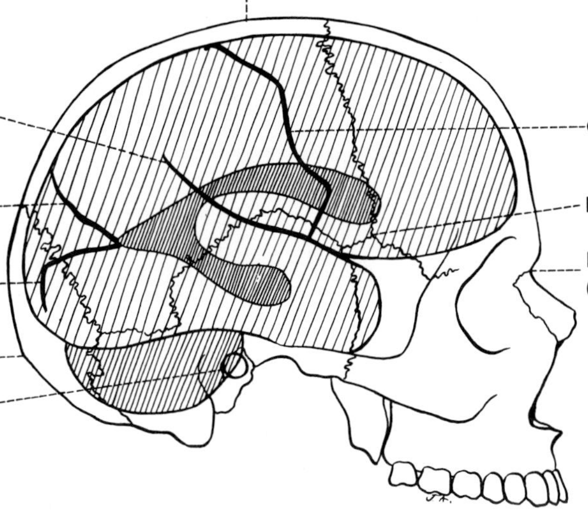

Fig. 1-24. Some approximate relations of the brain to the skull. Although these vary, this figure is fairly typical; the location of the sulci was recorded by roentgenograms after placing lead strips in the fissures. The lateral ventricle and the cerebellum are more heavily *shaded* than the cerebral hemispheres.

positions of the lobes of the brain, for the frontal lobe lies above the roof of the orbit and the anterior part of the lateral sulcus and extends backward to the central sulcus; the temporal lobe lies below the lateral sulcus and extends downward to about the level of the zygomatic arch, continuing backward approximately to a line connecting the parieto-occipital sulcus with the back of the root of the external ear; the occipital lobe lies behind the latter line, while the parietal lobe lies between the upper part of this line and the central sulcus.

It must be emphasized again that relationships between points on the skull and the fissures and gyri of the cerebral hemispheres are really quite variable. For instance, a line directed vertically upward from the external acoustic meatus typically crosses the central sulcus, but sometimes crosses close to the base of this sulcus and sometimes considerably higher, depending upon the position and the inclination of the sulcus. Rowland and Mettler, noting that the coronal suture was frequently used as a landmark in lobotomy, investigated the relationships of this suture and of the Freeman–Watts landmark to the cerebral hemisphere in a number of cadavers. Even among only 6 skulls chosen at random they found marked variations in the inclination of this suture, for it formed an angle of anywhere from 69° to 86° with the plane through the external acoustic meatus and the lower rim of the orbit; investigating the Freeman–Watts landmark—3 cm behind the lateral rim of the orbit and 6 cm above the zygomatic arch—they found this may overlie any one of the three major subdivisions of the inferior frontal gyrus. Observations such as these could probably be repeated with any chosen landmark on the skull, and serve to emphasize the variablity of craniocerebral relations.

Deeper Structures of Brain

The deeper structures of the brain also vary somewhat in their relations to the skull; Symington, among the early workers, pointed out a few average levels for certain structures (Fig. 1-25). According to his findings, the genu of the corpus callosum usually extends forward to the level of the lower end of the coronal suture. The highest part of the body of the lateral ventricle is said to lie about 7.5

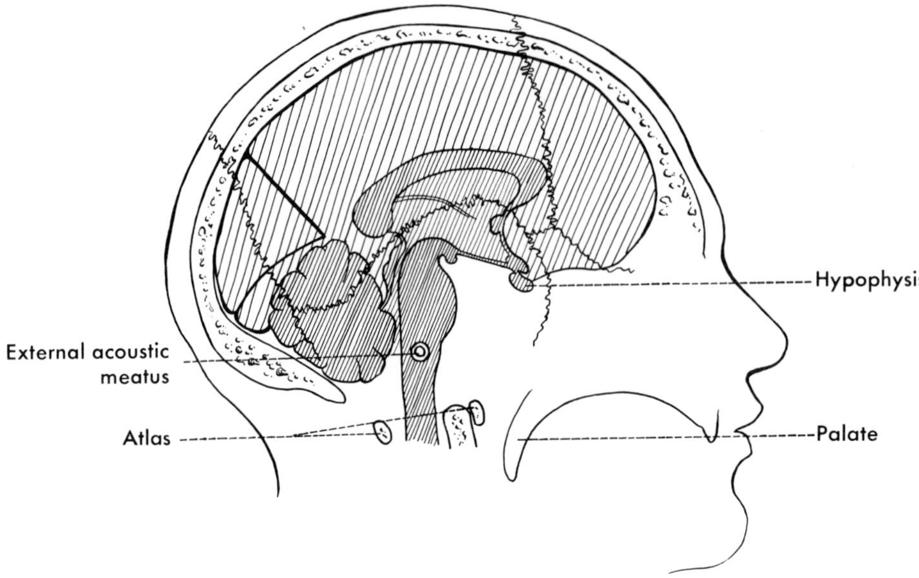

Fig. 1-25. Some approximate relations of the brain (as seen in sagittal section) to the skull. The brain stem, corpus callosum, and cerebellum are more heavily *shaded;* only the parietooccipital and calcarine sulci of the cerebral cortex are shown.

cm (3 inches) above the roof of the external acoustic meatus, therefore about 5 cm from the top of the head; the anterior horn, like the callosum, reaches approximately as far forward as the coronal suture. The posterior horn is quite variable in its extent, although it is nearer the surface than any other part of the ventricle; Symington found it usually located about 2.5 cm behind and 5 cm above the external acoustic meatus. It is customary to tap the lateral ventricle through its posterior horn, after a hole has been drilled in the upper posterior part of the parietal bone; the horn lies at a depth of about 6 to 7 cm from the surface. The posterior horn may, however, be absent.

The line of junction between midbrain and pons is said to be marked by a line about 2.5 cm above the top of the external acoustic meatus, and the junction of pons and medulla by a line at about the level of the roof of the meatus, therefore at about the same level as the vault of the nasal part of the pharynx. The foramen magnum, and the junction of spinal cord and medulla, is at about the level of the hard palate. That useful landmark, the pterion, marks approximately the level of the anterior perforated substance and therefore of the optic chiasma and of the floor of the diencephalon.

BLOOD SUPPLY OF THE BRAIN

ARTERIES

The blood supply of the brain is derived from the vertebral and the internal carotid arteries (Fig. 1-26). The details of the relationships of these vessels and their branches to the base of the brain and to the cranial nerves are given in connection with the descriptions of the three cranial fossae; their general courses only are given here, therefore.

Vertebral Arteries

The two vertebral arteries, after leaving the transverse foramina of the atlas, run posteriorly and medially on the upper surface of the posterior arch of this bone to pierce the posterior atlantooccipital membrane and the spinal dura and lie against the lateral side of the upper portion of the cord or lower end of the medulla, anterior to the spinal rootlets of the accessory nerve. They run upward

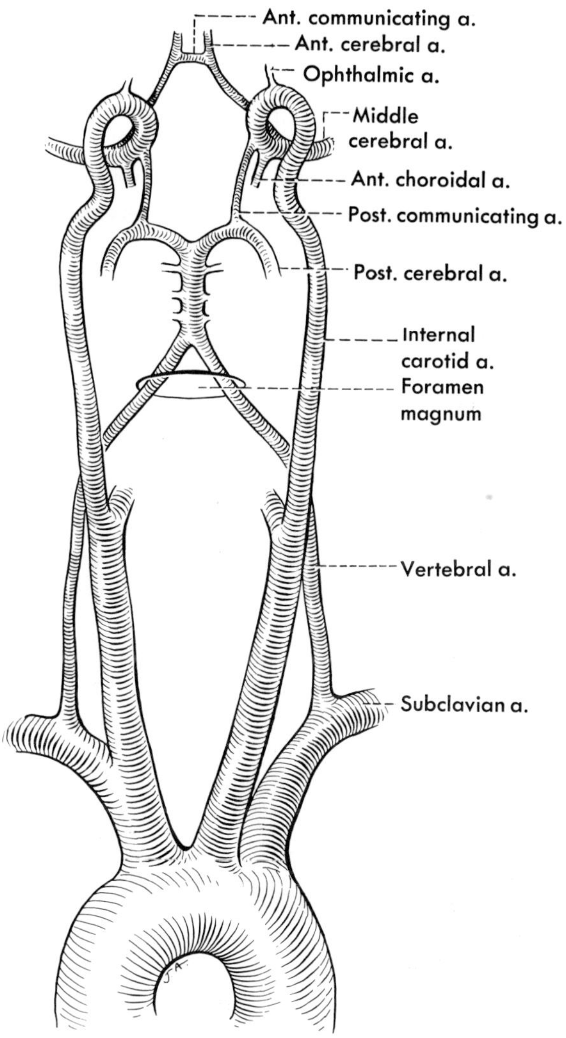

Fig. 1-26. Double origin of the blood supply to the brain, and the formation of the circulus arteriosus cerebri.

through the foramen magnum on the ventral side of the medulla and unite at about the caudal end of the pons to form the basilar artery. Although, in the neck, each artery is accompanied by a vertebral venous plexus, this plexus does not enter the cranial cavity with it but remains outside the dura and normally receives, through emissary veins, only a very tiny proportion of the blood borne intracranially by the artery.

Before their union each vertebral artery typically gives off a small meningeal branch, the anterior spinal artery to the cord, and sometimes, although not usually (Stopford, '16a), the posterior spinal artery. As the vertebrals pass alongside the medulla they give off a variable number of small branches to this, and close to their terminations the larger posterior inferior cerebellar arteries to the cerebellum and the posterolateral portions of the medulla (Fig. 1-27). The posterior inferior cerebellar artery normally runs upward along the dorsolateral surface of the medulla and then loops dorsally and caudally to descend along the medulla and inferior surface of the cerebellum; before leaving the medulla and branching on the inferior surface of the cerebellum it usually gives rise to the posterior spinal artery.

Basilar Artery

The basilar artery, formed by the junction of the two vertebrals at about the lower border of the pons, gives off in addition to a number of small pontine branches three or four main branches—the labyrinthine (internal auditory) artery, sometimes, and the anterior inferior cerebellar artery, the superior cerebellar, and the posterior cerebral; all these vessels are

Fig. 1-27. Schema of the vertebral and basilar arteries and their branches. Some variations of the vertebrals are shown in Fig. 1-30. The more common origins of the posterior spinal and labyrinthine (internal auditory) arteries, from the posterior inferior and anterior inferior cerebellar arteries, respectively, are shown on the left side of the figure; the less common origins, from the vertebral and basilar, respectively, are shown on the right. Any of the vessels shown here may be asymmetrical in size, or more or less tortuous.

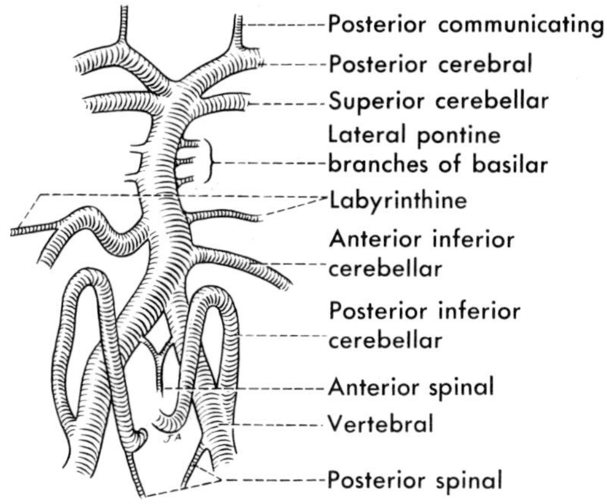

paired, and the posterior cerebrals are the terminal branches formed by the bifurcation of the basilar artery at the upper border of the pons.

The *labyrinthine artery* is a small vessel, passing on each side into the internal acoustic meatus between the facial and the eighth nerve; it was once described as a branch of the basilar artery, but is now recognized as arising from the anterior inferior cerebellar artery in the majority of cases. Mazzoni, for instance, found no origin from the basilar in 100 sides (and found more than 1 artery in 49%). The labyrinthine artery is further discussed in Chapter 3.

The *anterior inferior cerebellar arteries* usually originate from the basilar soon after this vessel is formed, but may arise from the vertebral; each passes laterally and backward around the pons or medulla to the rostral part of the inferior surface of the cerebellum, often looping laterally in contact with the seventh and eighth nerves.

The *superior cerebellar arteries* are given off from the upper end of the basilar shortly before the ending of this vessel; they pass backward and downward around the cerebral peduncles or upper border of the pons to reach the superior surface of the cerebellum.

The *posterior cerebral arteries* are the terminal branches of the basilar and pass laterally, where they lie at first parallel to the superior cerebellar arteries; close to their origins they are normally united to the internal carotids of their respective sides by the posterior communicating arteries. The posterior cerebral artery supplies most of the temporal lobe of the brain, except for the superior temporal gyrus, and most of the occipital lobe, thus including the important calcarine area.

The finer distribution of the vessels described briefly above, and the neurological symptoms resulting from their occlusion, cannot be entered into here. The former has been discussed in both its anatomical and clinical aspects by Stopford ('16b, '17); more recent discussions of some of the latter include that of Adams of a rare case of occlusion of the anterior inferior cerebellar artery, those of Merritt and Finland and of Anderson and coworkers on the more common occlusion of the posterior inferior cerebellar artery, of Kubik and Adams on occlusion of the basilar, and of Davison on occlusion of the anterior spinal artery.

Stopford ('17) stated that occlusion of the vertebral artery is rare, but when it does occur it is more common on the left side. He also emphasized the variation in distribution of the posterior inferior cerebellar artery, with the resultant variation in symptoms resulting from its occlusion, and stated that it, of all the branches of the vertebral or basilar, is anatomically most prone to occlusion by emboli, a fact borne out by clinical experience. It may also be pointed out that the posterior cerebral artery supplies not only cortex but also the medial and posterior portions of the thalamus, and gives off one or more posterior choroidal rami to the choroid plexus of the lateral ventricle and branches to the cerebellar peduncles and the corpora quadrigemina.

The vertebral arteries and their branches therefore supply an extensive area, including an upper portion of the spinal cord, the entire brain stem from the posterior portion of the thalamus downward, the entire cerebellum, and a large part of the posteroinferior portion of the cerebral cortex. It should be noted that while the cerebellar arteries anastomose freely with each other on the cortex, the bulbar arteries are essentially end arteries.

Internal Carotid Arteries

The internal carotid arteries, the other great arterial trunks to the brain, are largely limited in their distribution to the cerebrum itself.

According to the studies of Roberts and coworkers with an electromagnetic flow meter, about nine tenths of the blood supply to the brain is through the two internal carotids and one tenth is through the vertebrals. Apparently, normal movements of the head and neck will alter the distribution of blood, for turning the head away from the side being measured reduced the blood flow in that internal carotid. Complete obstruction of any one of the four vessels typically increases flow in the others, common carotid obstruction having, as would be expected, a much greater effect than vertebral obstruction. Surprisingly, these workers also reported that only in about 50% of apparently normal people did the external carotid deliver blood to the internal carotid after acute occlusion of the common carotid; in the remainder, blood flow in the internal carotid was reversed, blood from the cranial cavity being delivered to the external carotid system. According to these findings, ligation of the common carotid may thus decrease the blood supply to the brain more than would ligation of the internal carotid, producing a situation somewhat similar to the "subclavian steal" syndrome mentioned in Chapter 9.

The cranial course and chief branches of the internal carotid as seen in an arteriogram are shown in Figure 1-28. As the internal carotid artery passes upward through the carotid canal in the petrous portion of the temporal bone it lies in close relation to (anteromedial to) the middle ear cavity and the cochlea; it then turns sharply forward to run anteriorly and medially in the horizontal portion of the canal, at the end of which it enters the cranial cavity by turning upward upon the lateral surface of the body of the sphenoid bone, then turning again forward and proceeding along the lateral surface of the body through the cavernous sinus. Turning once again upward it leaves the cavernous sinus by piercing the dura medial to the anterior clinoid process. The two vessels converge somewhat in their course along the sphenoid; emerging from the dura, each vessel turns again caudally and divides, at the level of the anterior end of the lateral cerebral sulcus, into its two terminal branches, anterior and middle cerebrals. The portion of the artery lateral to and above the sphenoid bone is frequently referred to as the carotid siphon.

In the cavernous sinus, the artery is in immediate contact with the abducens nerve, which crosses its lateral surface, but the oculomotor, the trochlear, and the ophthalmic branch of the fifth nerve lie more laterally, being in the roof and lateral wall of the sinus. An internal carotid venous plexus accompanies the artery through the carotid canal and forms a communication between the cavernous sinus and either the pharyngeal plexus or the internal jugular vein. The carotid artery is also accompanied by a nerve plexus, derived from the superior cervical ganglion; in addition to accompanying the branches of the artery, this plexus also gives rise to the deep petrosal nerve (to the pterygopalatine ganglion) and a sympathetic ramus to the ciliary ganglion.

Within the carotid canal the carotid artery gives off small caroticotympanic branches to the middle ear cavity, and in about the last centimeter of its course (Paullus and co-workers) gives rise to a Vidian artery in perhaps a third of cases, and, more rarely, to a periosteal branch. While it lies in the cavernous sinus it gives off, besides the inferior hypophyseal artery, various branches to the dura mater and the trigeminal ganglion. Some of these branches anastomose with those of the opposite side while still in the cavernous sinus, and the meningeal branches typically anastomose with meningeal branches of the ophthalmic and vertebral arteries and with the middle meningeal and accessory meningeal. It is the importance of these anastomoses in maintaining an arteriovenous fistula in the cavernous sinus after trapping procedures on the internal carotid that led Parkinson ('64, '65), and later Harris and Rhoton, to their detailed studies of the cavernous sinus.

The first major intracranial branch of the internal carotid is the *ophthalmic artery.* This vessel usually arises just as the carotid perforates the dura at the medial surface of the anterior clinoid process, and therefore just posterior to the optic canal (or it may arise before the carotid has left the cavernous sinus). It enters the optic canal below the optic nerve, but within the dural sheath surrounding the nerve.

After giving off the ophthalmic artery the carotid curves upward and backward to divide into its two terminal branches, the small anterior cerebral and the large middle cerebral artery, this division taking place approximately lateral to the optic chiasma. Just before dividing, each carotid sends a posterior communicating artery to join the posterior

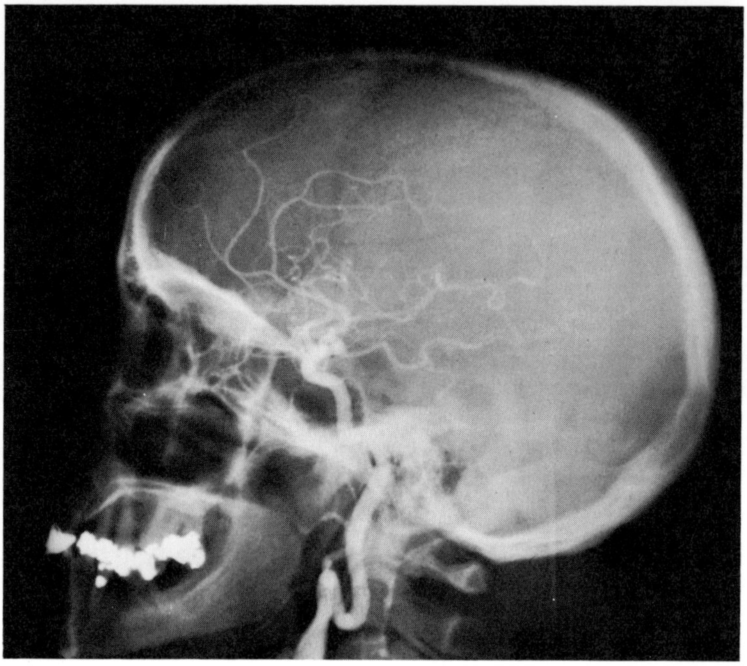

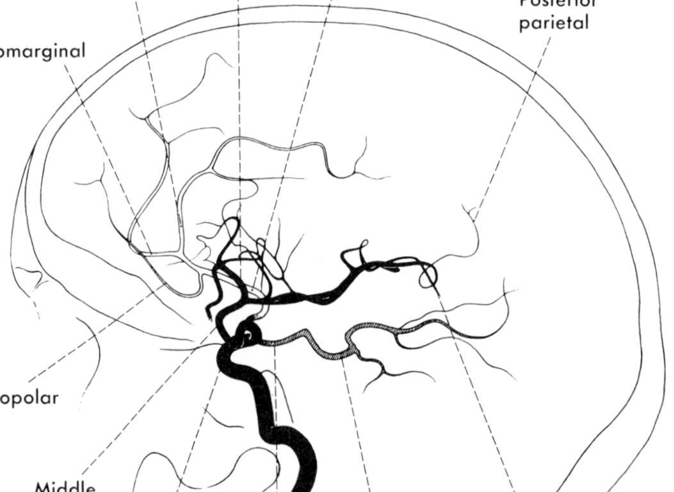

Fig. 1-28. Lateral view of an arteriogram of the internal carotid. (Courtesy Dr. Colin B. Holman), with an accompanying diagram.

cerebral artery of the same side and thus help form the circulus arteriosus (circle of Willis), and usually sends the anterior choroidal artery into the inferior cornu of the lateral ventricle. The anterior choroidal artery may arise from the middle cerebral.

The *anterior cerebral arteries* pass forward and medially, above the optic nerves, chiasma, or tracts, to the longitudinal cerebral fissure (between the hemispheres), where they are connected by a short trunk, the anterior communicating artery. Dunker and Harris have described the origin and distribution of various small branches, mostly perforating ones including the recurrent artery of Huebner, from the anterior cerebral and anterior communicating vessels. In the longitudinal fissure the two vessels lie close together and pass around the genu of the corpus callosum to be distributed to much of the medial side of the two cerebral hemispheres, anastomosing here with the posterior cerebral arteries at about the junction of parietal and occipital lobes.

The *middle cerebral artery,* anatomically the continuation of the internal carotid, passes laterally and upward over the insula; here it divides into cortical branches, most of which emerge through the lateral cerebral fissure to be distributed to the lateral surfaces of the frontal and parietal lobes of the cerebral hemisphere and to the superior portion of the temporal lobe. In its course the middle cerebral artery gives off perforating or striate arteries to the corpus striatum and internal capsule. Still other perforating arteries arise from the anterior choroidal and the posterior communicating. The distribution of these and other deep branches of the carotid system has been reviewed by Abbie, by Rubinstein, and by Finley; Beevor described and beautifully illustrated the distribution of the three cerebral arteries.

Although the *cortical branches* of a cerebral artery have little anastomosis with each other as they spread over the surface of the cerebral hemisphere, those of one cerebral artery do anastomose with those of another where their areas of distribution are adjacent; thus the anterior cerebral, for instance, may be a source of collateral circulation to an occluded branch of the middle cerebral. However, the deep group of vessels, to subcortical structures and the brain stem, are similar to the bulbar vessels of the vertebral system in that they are essentially end arteries, with no effective anastomoses.

Circulus Arteriosus Cerebri

The circulus arteriosus (circle of Willis) is normally formed by the origin of the two posterior cerebral arteries from the basilar, the junction of these to the internal carotids by the posterior communicating arteries, and by the anterior cerebrals and their anastomosis, the anterior communicating artery (Figs. 1-26 and 1-29). Although it thus typically forms anastomotic channels by which blood can conceivably pass freely from any one vessel to all branches of the circle, Rogers pointed out that there is little evidence that any appreciable amount of blood passes even from one internal carotid into that of the opposite side or, as a rule, into the posterior cerebral; injections of radiopaque materials into the carotid artery of one side usually outline that carotid system only, very little passing into other channels. He expressed the belief, therefore, that the arterial circle probably does not permit mingling of the various bloodsteams from the main vessels and should not be regarded as a distributor or equalizing station for a normal cerebral blood supply. Rather, it provides a bypass that may be capable of opening up should one of the main channels become obstructed, and its value is largely potential only. Shenkin and co-workers also emphasized that the connections of the circle are potential rather than normally used ones. By use of dye injections they showed that there was little mixing of carotid blood on the arterial side and even incomplete mixing on the venous side.

Woodhall and co-workers found blood-pressure drops in the internal carotid of 28% to 79% following its proximal occlusion in the neck, indicating the great individual variations in the anastomotic supply to the carotid circulation. Svien and Hollenhorst reported cases in which ligation of the internal carotid artery was followed by a significantly lowered

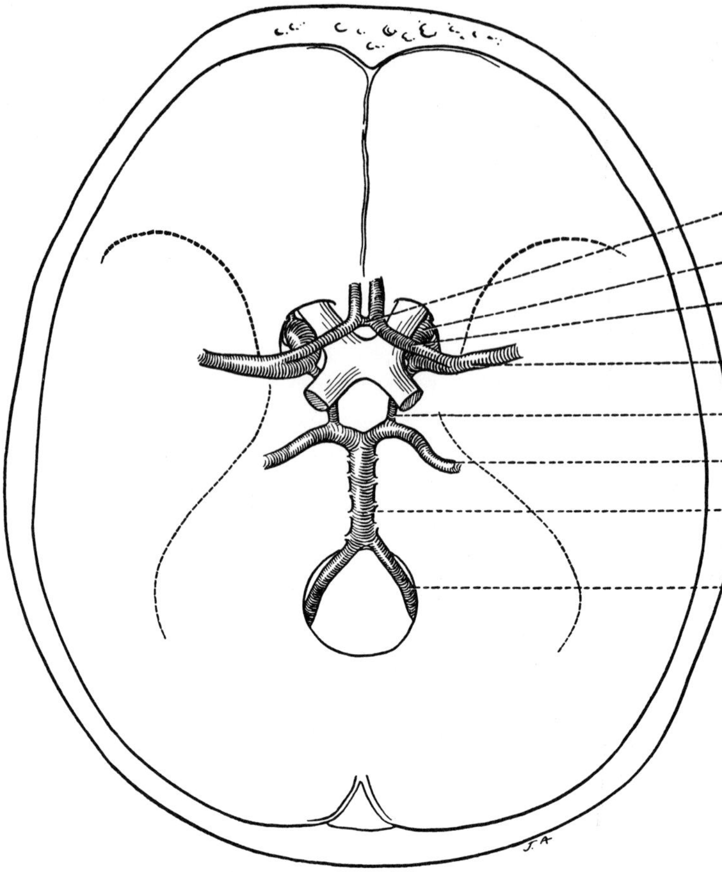

Fig. 1-29. Relation of the arterial circle of the brain to the base of the skull.

pressure in the ipsilateral retinal artery for as long as 10 years. In ligation of the common carotid, the situation is different, since the circulation through the circulus arteriosus can be supplemented by that through the ipsilateral external carotid. Heyman and coworkers believed this accounts for approximately one fourth of the residual pressure in the internal carotid; even with this, however, they found a significant reduction of pressure persisting for as long as 6 to 19 months. It has already been mentioned that immediately following occlusion of the common carotid artery the collateral circulation through branches of the external carotid may be less effective than that through the circulus arteriosus, so that the internal carotid may contribute blood to, rather than derive blood from, the external carotid.

Mount and Taveras and Lang and coworkers have studied angiographically the collateral circulation of the cerebral hemispheres and have emphasized especially the rather numerous, although typically small, anastomoses between the external and internal carotids.

Variations

Variations of the vertebrals, the basilar, or their branches are the rule rather than the exception, and variations in the sizes of the vessels participating in the arterial circle are very common.

The *vertebral arteries* are usually (92%—Stopford, '16a) unequal, the left being larger in somewhat more than half these cases (Fig. 1-30, a ventral view). Stopford found an excessively small right vertebral artery in 5 of

150 brains examined, and 1 instance in which both vessels were minute. McCullough reported two cases in which the left vertebral artery ended intracranially without joining the basilar, one with a double basilar artery, and one in which that artery was plexiform.

Other variations in the formation, course, and branching of the basilar artery have been recorded in detail by Stopford ('16a) and others. Stopford found absence of the posterior inferior cerebellar artery bilaterally in 3% and absence on the right in 15% and on the left in 6%; doubling of the anterior inferior cerebellar artery once, and absence bilaterally once and on the left twice (among 150 brains); doubling of the superior cerebellar artery on both sides in 3%, on the left only in 16% and on the right only in 12%.

Persistent *carotid–basilar anastomosis* is used most commonly to describe a branch that passes from the internal carotid in the cavernous sinus to the upper end of the basilar, but since it is only one of the vessels that supplied the basilar during development and may persist, the better terminology is *persistent trigeminal artery.* This has been reported a number of times, both in dissecting-room material (Sunderland, '48) and from angiographic studies (in between 0.1% and 0.2%, according to Parkinson and Shields). In this condition, the internal carotid artery divides within the cavernous sinus, one branch proceeding as the normal internal carotid, the other running caudally to join the basilar. The position of the anastomotic branch is variable; according to Schaerer it passed between the two posterior clinoid processes in 7 of 16 reported cases and lateral to the process in 9. It is the latter position that brings it in especially close contact with the roots of the trigeminal nerve. Binet and Young reported on a patient in whom it was present bilaterally, and Merry and Jamison reported on one in whom it had produced facial pain and diplopia.

The other two types of carotid–basilar anastomosis are much rarer. In *persistent hypoglossal artery* the vessel enters the skull through the hypoglossal canal to form the vertebral. Samra and co-workers apparently reported the first case in which the artery arose from the external carotid; in the eight previously reported cases that they found it arose from the internal carotid. In *persistent preatlantal artery* the contribution to the basilar

Fig. 1-30. Some variations in the vertebral arteries. In (*b*) a high and in (*c*) a low origin of the posterior inferior cerebellar artery is indicated. Note also the variations in the origin of the anterior spinal artery. (After Stopford JSB: J Anat 50:131, 1916)

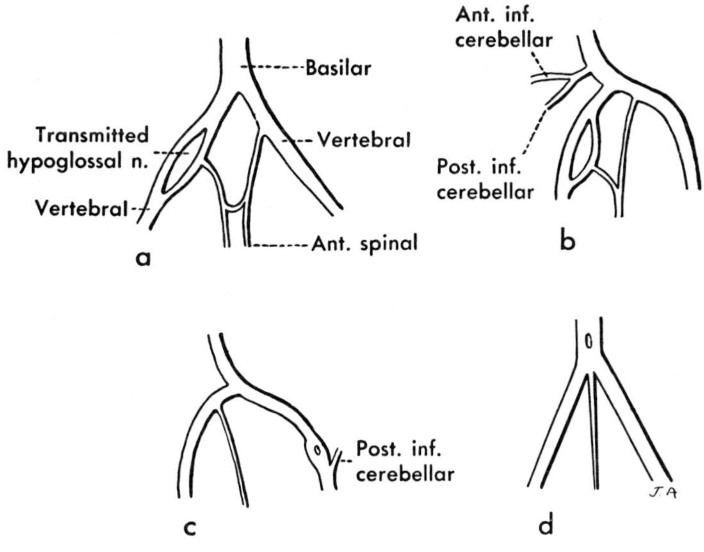

may arise from either the internal or external carotid (Rao and Seith); it is distinguished by the fact that it runs in front of the atlas, thereafter turning dorsally between atlas and skull to enter the foramen magnum as a vertebral artery. Flynn reported on a patient in whom a vertebral arose from the external carotid but took its usual course through the upper transverse foramina.

Although the internal carotid is almost constant in its course and branches, Primam and Christie reported one that gave off no intracranial branches but continued into the optic canal as the ophthalamic artery.

Fawcett and Blachford, Windle, Blackburn ('07, '10), and Stopford ('16a) have reported upon the *arterial circle* and its branches in a total of approximately 1500 brains; they agree that the most common defect in the circle is absence of one or both posterior communicating arteries, the exact figure varying with the author (*e.g.*, 7% in Stopford's series of 150, 3.9% in Fawcett and Blachford's series of 700, and 12.5% in Windle's series of 200).

Another relatively common major variation is a major or entire origin of the posterior cerebral artery from the carotid, by way of an "enlarged" posterior communicating artery (Fig. 1-31). The occurrence of this origin

Fig. 1-31. The two origins of the posterior cerebral artery, the less common one being shown on the right. The *dotted lines* show the parts of the arterial circle that may be very small or entirely lacking. (*I.C.*), the internal carotid artery; (*P.C.*) the posterior cerebral artery.

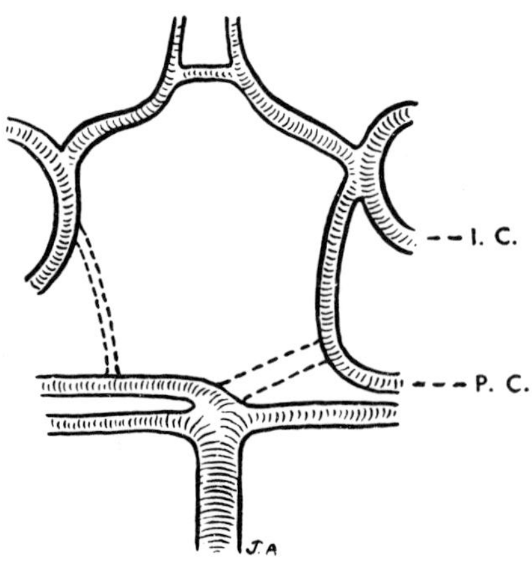

averaged, in the series already cited, approximately 5% on the right side, slightly less on the left, and about 2% bilaterally, but Sunderland ('48) found a carotid origin of the posterior cerebral bilaterally in 6% of 100 cases, and unilaterally in about 13% each of right and left sides; Vare and Bansal found such an origin bilaterally in 5.7% of 175 specimens and in 13.7% on the right and 5.7% on the left. Actually, this origin represents a retention of the embryonic condition, since the posterior cerebral is originally a branch of the internal carotid that subsequently transfers to the basilar. Apparently no major variations occur in the middle cerebral artery. The anterior cerebrals may be supplemented by a third, a so-called middle, anterior cerebral (Stopford found this in 6%, Blackburn in about 10%, Windle in about 4.5%, or one may, rarely, be absent, or the two may be united for a part or much of their course (Windle.) The anterior communicating artery is occasionally, but rarely, absent; more often it is double (7% to 9%) or even triple (Fig. 1-32). It is, of course, absent as a vessel when there is fusion between the two anterior cerebrals.

Although congenital defects in the arterial circle are thus common, they are compensated for by other departures from normal; for instance, if one anterior cerebral is hypoplastic at its origin or is absent, an enlarged anterior communicating artery supplies blood to the defective one (Puchades–Orts and coworkers). These workers concluded that compensation is always adequate to maintain normal function of the brain.

Aneurysms

The arterial circle and its branches are of importance from the standpoint of aneurysms. McKissock and co-authors found 29% of 599 aneurysms on the proximal part of the anterior cerebral or on the anterior communicating arteries, 28% on the internal carotid or posterior communicating, and 8% at the carotid bifurcation; thus approximately two-thirds were on or close to the circle. Carotid ligation, either of the common or the internal artery, has often been used to reduce the

Fig. 1-32. Variations in the anterior communicating artery. (After Stopford JSB: J Anat 50:131, 1916)

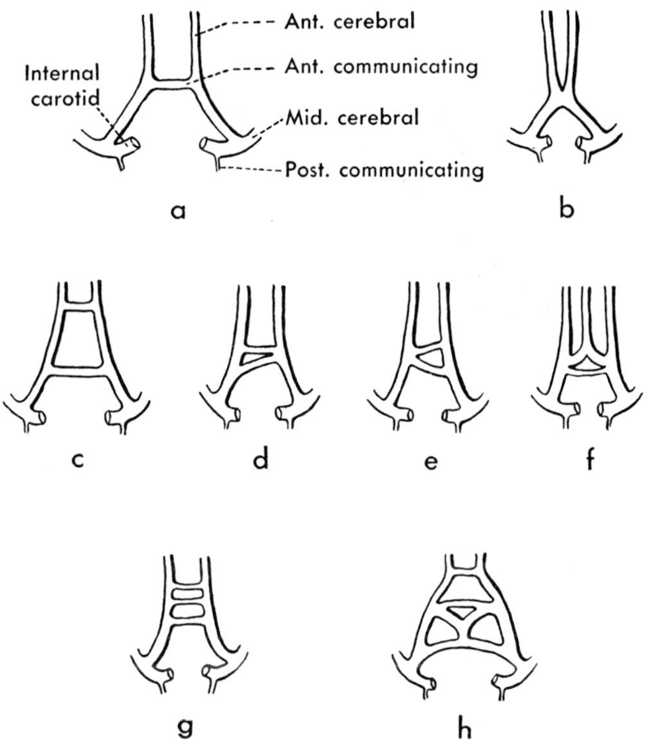

pressure when there are aneurysms of the carotid system. Although there is controversy as to its value, Heyman and co-workers found that common carotid occlusion may produce a relatively prolonged period of lowered pressure in the internal carotid.

Aneurysms of the arterial circle are particularly likely to cause palsies of the third nerve or may affect the visual pathway or other nerves to the eye muscles, in addition to pressure on the substance of the brain itself. Because the usual treatment of aneurysms of the internal carotid artery in the cavernous sinus (by trapping them between two occlusions of the artery, one intracranially and the other in the neck) may fail, since the circulation can be maintained by the anastomotic connections of the intracavernous carotid, Parkinson ('65) devleoped his technique for dealing with them in the sinus.

Innervation

As they enter the skull, the arteries bring with them nerve plexuses which follow the smaller branches of the vessels into the pia and into the nervous system itself (Penfield, '32; Clark). Sato and Suzuki found vessels at the base of the skull as small as 20 μm that possessed an innervation, but over the convexity of the brain the smallest ones receiving nerve fibers varied from 100 to 300 μm. The majority of the nerve fibers are probably vasoconstrictor or vasodilator, but Clark found apparent sensory endings in the adventitia of vessels, and Levine and Wolff expressed the belief that they have shown that the pial blood vessels possess an afferent as well as a motor supply and that this afferent supply may be important in the regulation of the cerebral circulation. Kimmel ('59) reported that the intracranial part of the vertebral artery is accompanied by sensory fibers derived from the second and third cervical nerves and sympathetic fibers derived from the superior cervical ganglion by way of the ramus communicans with the first cervical nerve. Apparently, however, there are no pain fibers among the afferent fibers on the smaller vessels, as Ray and Wolff were unable to obtain pain by stimulation of these vessels of the pia;

only the larger blood vessels were found to contain pain fibers.

The causes of *headache* are not completely understood, but following especially the work of Wolff and collaborators (*e.g.,* Schumacher, Ray, and Wolff; Schumacher and Wolff; Kunkle, Ray, and Wolff) it is generally granted that a major cause is distention of cerebral arteries or of these and extracranial arteries of the head. Although the distention of branches of the external carotid is stated by Ray and Wolff to be responsible chiefly for migraine headache and the headache associated with arterial hypertension, distention of intracranial arteries is said to cause the headache produced by various chemical agents, such as histamine, and headache associated with septicemia and fever. Thus the larger arteries in the cranial cavity may play an important part in the production of symptoms. Kunkle and colleagues ('48) expressed the belief also that clinical headaches related to hunger, recent trauma of the head, or emotional tension are due to distention of intracranial or extracranial arteries; they found that headaches of these types could be eliminated by exposure to a favorable increased gravity produced by the human centrifuge.

The *motor innervation* to the pial and cerebral arterioles is believed to be a double one, from both sympathetic and parasympathetic systems. Pool and co-workers and Stavraky stated that stimulation of the cervical sympathetic trunk produces constriction of the pial arteries that averages about a 10% decrease in diameter. According to Pool and co-workers, the constriction is confined to the ipsilateral side, there being no change in the contralateral side. The parasympathetic supply to the pial vessels is said to be derived at least in part from the facial nerve, as stimulation of this at the geniculate ganglion produces vasodilation (Forbes and associates, '37a). The vasodilator fibers of the seventh nerve to the pial vessels are believed to reach their destination by way of the greater (superficial) petrosal nerve and the internal carotid plexus, probably with a synapse upon cells located in this plexus (Chorobski and Penfield). Forbes and colleagues, '37b, and Cobb and Finesinger also described vasodilation of the pial vessels consequent to vagal stimulation, but suggested that this is a reflex involving the facial nerve on the efferent side. Fog ('39a) adduced evidence that the effect of the vasomotor nerves on the blood vessels is confined entirely to the larger arteries of the pia and that there is none on the arterioles; he stated ('39b) that although the pial arterioles respond to a fall in blood pressure by relaxing and to a rise by contracting, these reactions occur after bilateral section of the cervical sympathetic trunk and bilateral section of various vasosensory nerves. Florey was likewise unable to obtain evidence of a nervous control over the vessels to the cerebral cortex.

Despite the apparent demonstration of vasoconstrictor and vasodilator fibers to the vessels within the cranium, the relative physiological importance of these is subject to some doubt. Forbes and Cobb stated that the regulation of the blood flow within the brain and the meninges is largely controlled by chemical agents, especially carbon dioxide; they reported that cerebral vasoconstrictor nerves are only about one tenth as effective on the vessels in the pia as are vasoconstrictor nerves in the skin, and that the vasoconstrictor nerves present are distributed unequally to different parts of the brain. Later, Forbes stated that although the volume of blood flow through the brain is regulated largely by the arteriolar pressure, there are many other factors involved too; he also pointed out that although the blood vessels are particularly sensitive to slight variations in the tension of carbon dioxide, the resulting dilation of the arterioles may either increase the flow through the tissues of the brain or, if it is general, may cause such a large fall in blood pressure that there is an actual decrease in cerebral blood flow.

The ease with which cerebral blood flow can be modified by extracerebral factors is indicated by the experiments of Brown and coworkers. They found that stimulation of the cervical sympathetic trunk in the monkey, while producing inconstant constriction of the intracranial part of the internal carotid and of the middle and anterior cerebrals, produced such constriction of the external carotid and its branches that flow in the internal carotid increased, and there was even an intracranial shunt to the other side through the circulus arteriosus. Thus sympathetic activity may help to maintain or increase, rather than decrease, intracranial flow. Sympathectomy apparently has no appreciable effect on cerebral blood flow, but inhalation of 5% carbon dioxide increases the flow materially (Shackelford and Hegedus).

VEINS

The venous drainage of the brain differs from that of most parts of the body in that the very thin-walled veins, containing no valves, do not at all correspond to the arteries and do not, for the most part, accompany them. The veins resemble the arteries in only one essential respect—that the cerebral vessels may be divided into two chief groups, a cortical group and a deep group.

Cortical Veins

The veins of the pallium or cerebral cortex are usually divided into two groups, a superior and an inferior (Fig. 1-33). The *superior cerebral veins* communicate freely with each other, but do not form a constant pattern; they are usually described as forming some eight to a dozen trunks that empty into the superior sagittal sinus, but according to Merwarth usually unite into four principal trunks: a frontal, a precentral, a postcentral or Rolandic, and an occipital. Of these, the postcentral or Rolandic is usually the largest and drains the precentral and postcentral gyri. Merwarth has described the syndrome resulting from occlusion of this vein, the symptoms being especially marked in the lower extremity because the vein drains especially the upper portions of the precentral and postcentral gyri. As already noted, the veins usually enter the sinus obliquely and in the direction opposite to that in which the blood flows within the sinus.

The *inferior cerebral veins* drain into the cavernous sinus and into the inferior sagittal, superior petrosal, and transverse sinuses; a vein from the occipital lobe usually also joins the termination of the great cerebral vein, and a vein from the inferior surface of the frontal lobe usually empties into the superior ophthalmic vein, but occasionally opens into the superior petrosal sinus. This small vessel is known as the ophthalmomeningeal vein.

One of the larger superficial cerebral veins runs in the lateral fissure to end in the cavernous sinus, and is termed the *superficial middle cerebral* (or Sylvian) vein. This vein is united posteriorly to the inferior cerebral veins in the posterior temporal and occipital regions, the connection being usually described as the inferior anastomotic vein (of Labbé), but according to O'Connell being represented typically by two vessels. The superficial middle cerebral vein usually has also a large anastomosis with one or more of the superior cerebral veins, this anastomotic channel being known as the superior anastomotic vein (of Trolard). The superior anastomotic vein is, however, only one of several (according to O'Connell, usually at least three) vessels that unite the superior and inferior cerebral veins.

The *deep middle cerebral vein* drains the cortex of the insula and operculum and runs medially and ventrally to emerge at the medial side of the lateral sulcus, thus close to the cerebral peduncle and optic chiasma. Here it is joined by the anterior cerebral vein, which accompanies the proximal portion of the anterior cerebral artery; this junction forms the *basal vein* (of Rosenthal). This vein is also joined by one or more striate veins, and thereafter runs dorsally and caudally around the cerebral peduncle and across the colliculi to end in the great cerebral vein (or occasionally, it is said, in the superior petrosal sinus). It typically receives the deep veins from the inferior horn of the lateral ventricle and from the midbrain, although some of the latter may join the great cerebral vein directly.

The veins on the superior surface of the cerebellum join the great cerebral vein, the straight sinus, and the transverse and superior petrosal sinuses; those on the inferior surface of the cerebellum join the inferior petrosal, transverse, and occipital sinuses.

Deep Cerebral and Other Veins

Except for the deep venous drainage into the basal vein, just recounted, the deep structures of the cerebrum and of the brain stem above the region of the upper border of the pons drain into paired internal cerebral veins, which in turn unite to form the great cerebral vein (of Galen) that empties into the anterior end of the straight sinus. Since this venous drainage includes also the choroid plexuses, it has been suggested that obstruction of the great cerebral vein may be one cause of hy-

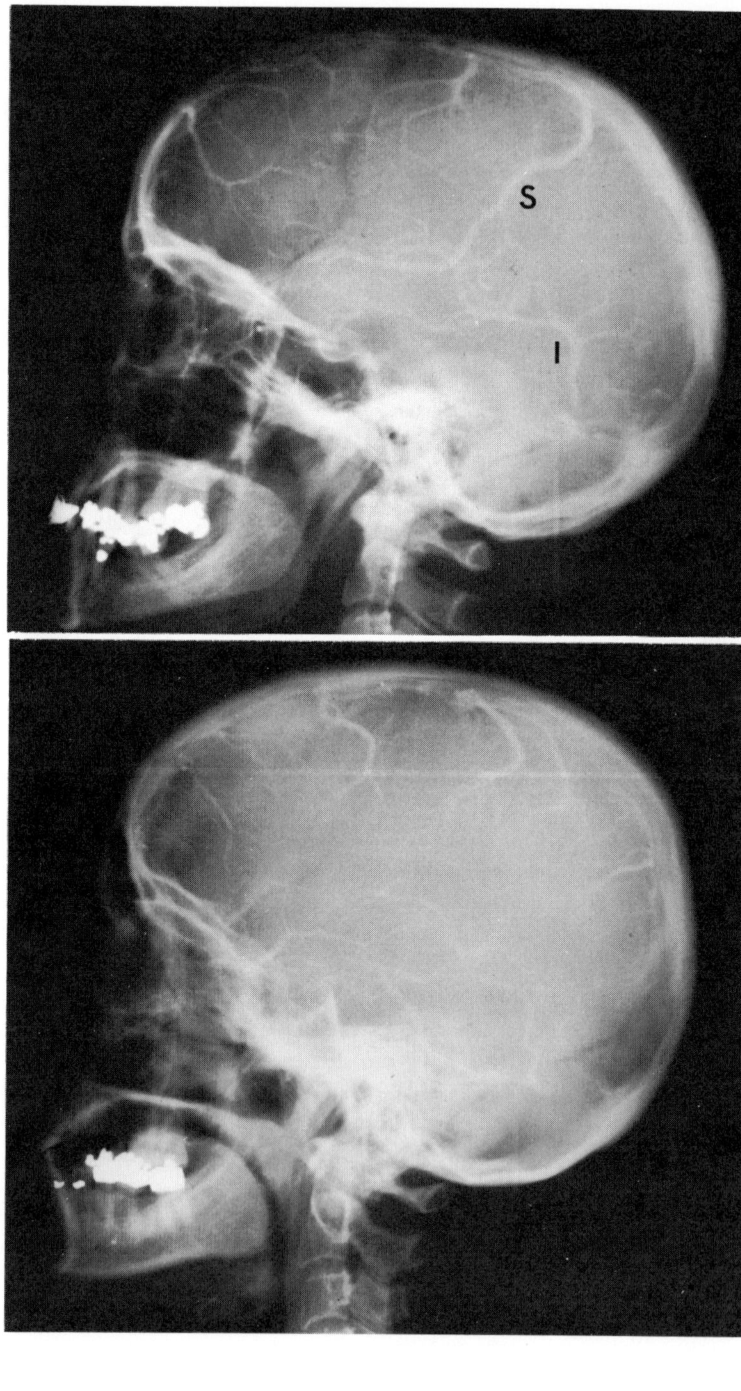

A

drocephalus; however, Hammoch and co-workers were unable to produce hydrocephalus by operative occlusion of the great cerebral vein in either the monkey or the dog, although the choroidal vessels, among others, were dilated.

Each internal cerebral vein is formed by the junction of a thalamostriate vein (formerly vena terminalis) and a choroid vein. The most prominent vein joining the thalamostriate is the vein of the septum pellucidum (septal vein); thereafter the thalamostriate turns sharply medially and backward through the interventricular foramen (Fig. 1-33) to join the choroid vein in the roof of the third ventricle. Lin and co-workers have emphasized the importance of angiographic examination of this turn, or "venous angle," in diagnosing deep-lying cerebral lesions.

Schlesinger found in the monkey large intracerebral anastomotic venous channels in the white matter, which establish connections between the tributaries of the great cerebral vein and the veins on the surface of the hemispheres; apparently, therefore, the deep and superficial veins of the cerebrum are united by functional collateral channels (in sharp contrast to the arterial system), and drainage of the great cerebral vein and of the large veins on the surface of the brain overlaps to a very considerable extent. Schlesinger found, however, no anastomotic veins in the basal ganglia, which seemed to be drained primarily or exclusively by the great cerebral vein; occlusion of this vein, in his experiments, produced no hyperemia in the choroid plexus and no hydrocephalus, but did produce numerous hemorrhages into the head of the caudate nucleus.

The veins from the lower part of the brain stem—that is, from the pons and medulla—end in the inferior petrosal and transverse sinuses. According to Schlesinger, however, even the pontine vessels can be injected through the communications between the great cerebral vein and the veins on the surface of the cortex.

Fig. 1-33A. Two venograms, the upper showing particularly well the superior (*S*) and inferior (*I*) anastomotic veins, the lower showing most of the superficial and deep cerebral veins. (Courtesy Dr. Colin B. Holman) For identification of the vessels in *A,* bottom, see *B;* here dural sinuses are *cross-hatched;* superficial veins, *cross-lined;* deep veins, *black.*

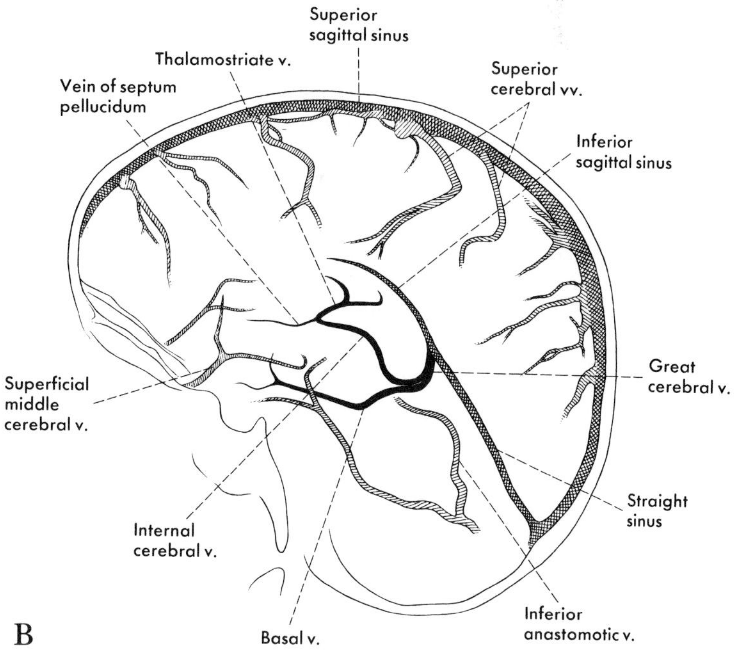

B

Anterior Cranial Fossa

The anterior portions of the frontal lobes of the brain, and the associated olfactory nerves, bulbs, and tracts, lie in the anterior cranial fossa, while the optic chiasma and hypophysis, and the anterior ends of the temporal lobes, are immediately adjacent to it and may be approached through it.

The cranial vault of this fossa is largely formed by the frontal bone, hence the approach to it is usually through this bone. Sagittal incisions on the forehead are avoided, when possible, for cosmetic reasons, but in the scalp they lie between the major nerves and blood vessels. Curving the incision laterally necessarily interrupts the supraorbital nerves and denervates some scalp to approximately the interauricular line, but allows the skin flap to be turned downward over the forehead, thus preserving the innervation of the latter.

In "turning" a frontal bone flap, it is desirable to avoid the frontal sinus if possible, in order to prevent possible infection from this source; because the sinus varies much in its upward extension into the frontal bone, its position should be ascertained by roentgenograms previous to operation. The bone flap is usually somewhat C shaped, with its base in the anterior temporal region at the thin squamous portion of the temporal bone, where it is most easily fractured. Nutrition of the bone flap may be maintained by leaving it attached to the temporal muscle, but this blood supply (through the deep temporal arteries) may be a source of troublesome bleeding and has been found to be unnecessary for healing of the replaced bone. "Free" (*i.e.,* detached) bone flaps are preferred by many surgeons.

After the bone flap is reflected (a bilateral approach is sometimes preferred for access both to the region of the cribriform plate and to the optic chiasma and hypophysis), the dura may be incised directly, but if structures close to the floor of the anterior cranial fossa are sought, the dura may be easily separated from the wall and most of the floor of the fossa. This approach allows the dura to be retracted with the frontal lobe, thus lessening the possibility of damage to the latter. The firmer attachments of the dura are at the cribriform plate, toward the midline, and posteriorly along the sphenoid ridge (between anterior and middle cranial fossae); after it has been elevated from the inner surface of the cranium to these points, the dura must be incised to gain access to the olfactory bulb, optic chiasma, or hypophysis.

FLOOR OF ANTERIOR CRANIAL FOSSA

The floor of the anterior cranial fossa (see Fig. 1-42) consists largely of the orbital portion of the frontal bone; medially, the cribriform plate (lamina cribrosa) of the ethmoid, which has the olfactory bulbs in contact with it, forms a small portion of the floor, and posteriorly the sphenoid contributes. The lamina cribrosa bears the crista galli in the midline, and at its lateral borders the openings of the ethmoid canals into the anterior cranial fossa—the posterior canal at the angle where frontal, ethmoid, and sphenoid bones meet, the anterior at about the middle of the length of the cribriform plate. Between the crista galli and the frontal crest, both of which give attachment to the falx cerebri, is the foramen cecum. The part of the body of the sphenoid contributing to the anterior cranial fossa is the jugum sphenoidale, bounded posteriorly by the chiasmatic sulcus; extending laterally from the jugum on each side are the lesser wings of the sphenoid, which form the major part of the sharp posterior margin (sphenoid ridge) of the anterior fossa. Each lesser wing gives rise to a prominent anterior clinoid process that projects posteriorly lateral to the optic canal and carotid sulcus.

Although it is correct anatomically to regard the chiasmatic sulcus and the sella turcica as being part of the middle cranial fossa, access to this region is most easily obtained through the frontal approach: the optic chiasma and the upper surface of the hypophysis lie at the level of the floor of the anterior fossa, but considerably above the lower surface of the temporal lobe and therefore above the lateral parts of the floor of the middle fossa. From the surgical standpoint the sella can be considered a part of the anterior fossa, the middle cranial fossa being therefore, from this standpoint, a paired structure.

The optic canals open between the lesser wings and the body of the sphenoid bone and are connected to each other by the chiasmatic sulcus (rarely actually occupied by the optic chiasma). A small eminence, the tuberculum sellae, separates the chiasmatic sulcus from the deeper hypophyseal fossa, and the sella turcica is completed posteriorly by the dorusm sellae with its more or less well-developed posterior clinoid processes. Behind the optic canals, on the sides of the tuberculum and under cover of the anterior clinoid processes, are the upper ends of the carotid sulci; these are sometimes bordered posteriorly by a middle clinoid process, which may even meet the anterior clinoid process and form a caroticoclinoid foramen (canal) here.

Fractures

The floor of most of the anterior cranial fossa is particularly delicate because of its invasion by ethmoid and frontal sinuses and its close relationship to the orbit and to the nose. Fractures of it may injure the olfactory bulb or tract, which lie in close contact with the cribriform plate; if they are accompanied by tearing of the meninges they may produce hemorrhage from the ethmoid branches of the ophthalmic artery, or their meningeal branches, which run along the lateral border of the cribriform plate; similarly, complete fractures of the floor of the anterior cranial fossa with tearing of the meninges may produce bleeding into the frontal sinus or directly into the nose or, even in the absence of bleeding, an escape of cerebrospinal fluid into either of these (cerebrospinal rhinorrhea). According to Adson ('41a), from 2% to 5% of fractures of the skull produce this latter result, but most of them heal spontaneously. Fractures across the floor of the anterior or middle cranial fossae extending to the region of the optic canals may also produce lesions of the visual pathway, either through hemorrhage about the optic nerve in the canal or by directly affecting this nerve.

Defects

The floor of the anterior cranial fossa may present a congenital defect, through which a basal *encephalocele* or *meningocele* protrudes. According to Anderson, basal encephaloceles or meningoceles typically pass either through the superior orbital (sphenoidal) fissure into the orbit (or on through the orbit into the pterygopalatine fossa) or protrude into the nose or nasopharynx through the ethmoid or sphenoid bones or through the suture between them. The latter he has termed the "sphenopharyngeal" type though it might in some cases be more properly termed "ethmonasal." Mood, reviewing some 20 cases of anterior cerebral herniation, classified the majority as being nasofrontal ones, but found some, including the case he described, to be nasoethmoidal and one to be nasolacrimal. The distinctions here do not seem to be of particular importance. According to Anderson, however, basal meningoceles are more commonly sphenopharyngeal, and the presence of these in the nasal cavity has been shown to be sometimes responsible for the occurrence of "spontaneous" rhinorrhea, that is, not associated with fracture.

Spontaneous Rhinorrhea

Spontaneous rhinorrhea has been demonstrated in several persons to be due to protrusion of an arachnoid diverticulum into the nasal cavity through a defect in the ethmoid bone (Fig. 1-34); thus Adson ('41a) described a case in which at operation there was a 2 mm opening in the cribriform plate through which the meninges passed into the nasal cavity, paralleling a filament of the olfactory nerve. Love and Gay reported two similar cases, with small defects, in adults, and a large bilateral defect with herniation of a part of the frontal lobe in a child. It seems to be generally accepted that cerebrospinal rhinorrhea in the absence of fracture of the skull is usually due to defects in the cribriform plate with protrusion and thinning of the meninges.

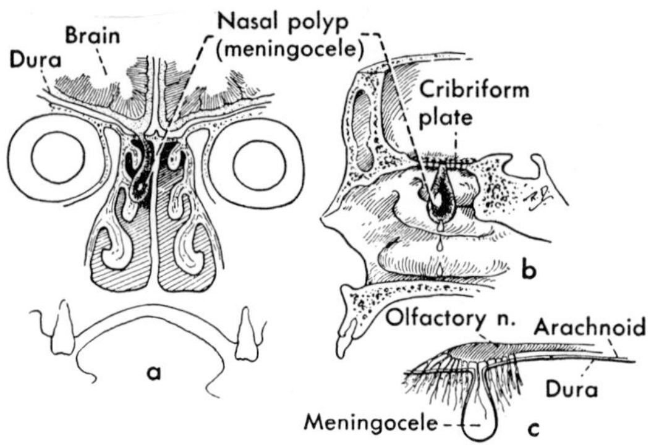

Fig. 1-34. Views of the cribriform region showing the meningeal sac in a case of cerebrospinal rhinorrhea. (Adson AW, Uihlein A: Arch Surg 58:623, 1949)

In the adult cases of Love and Gay, the rhinorrhea was precipitated by nasal infection—thus, although the condition is apparently congenital, symptoms may not occur, or may not be sufficiently marked, until some trauma precipitates them. It is generally recognized that meningitis is the most serious potential complication of this condition. The accepted method of repair is an intracranial approach (Adson '41a; Love and Gay), a piece of muscle often being placed over the defect of the cribriform plate as a whole. Adson found the bilateral approach, with mobilization of the dura beneath both frontal lobes, most satisfactory.

Arteries and Nerves

Laterally and superiorly, the dura of the anterior cranial fossa and its covering vault is supplied by the frontal branch of the middle meningeal artery. The floor of the fossa is supplied by twigs from the anterior and posterior ethmoidal arteries, branches of the ophthalmic which run in the dura on the lateral border of the cribriform plate. A branch from the anterior ethmoidal may be large enough to be called the anterior meningeal artery. The meningeal branches of the ethmoidal arteries are reinforced by twigs of the sphenopalatine artery (from the maxillary), which ascend through the cribriform plate to anastomose with them.

Ray and Wolff found the dural floor of the anterior cranial fossa to be inconsistently sensitive to pressure, but uniformly sensitive to faradic stimulation; such stimulation produced pain that became localized in the general region of the homolateral eye (apparently, therefore, involving the ophthalmic division of the fifth nerve). The dura close to the midline was found to be especially sensitive, as were portions of the arteries in the floor of the anterior cranial fossa; 2 cm above the floor, however, the dura became insensitive. The innervation of the dura of the floor of the anterior cranial fossa is largely from the anterior ethmoidal (anterior nasal) nerve from the nasociliary branch of the ophthalmic, which like the artery enters the cranial cavity from the orbit and runs anteriorly along the lateral border of the cribriform plate to leave the cranial cavity and enter the nose; fine twigs also accompany the branches of the middle meningeal artery.

Venous Sinuses

The venous sinuses encountered in the anterior cranial fossa are small and easily avoided. The origin of the *superior sagittal sinus* lies in the attachment of the falx cerebri to the dura of the anterior wall and vault of the anterior fossa, while the small *inferior sagittal sinus* runs in the free edge of the falx. Toward its anterior end, where it attaches to the crista galli, the falx is often perforated; it can, if desired, be cut away from the crista without special precaution, since the inferior sagittal sinus at this level is practically nonexistent, and the superior sagittal sinus lies anterior to the crista.

At the posterior boundary of the anterior cranial fossa is the *sphenoparietal sinus,* running along the sphenoid ridge; this is a slender connection between the middle meningeal vein and the cavernous sinus, and would be encountered only if an attempt were made to cut the dura away from its close attachment to the ridge.

The *cavernous sinuses* lie lateral to the sella turcica and are therefore in the operative field in anterolateral approaches to the optic nerves and the hypophysis; their relations are of more importance in connection with the middle cranial fossa, however, and they are therefore described with this. Anterior and posterior to the hypophyseal stalk the cavernous sinuses are connected by anastomotic channels (anterior and posterior intercavernous sinuses, or the circular sinus) in the diaphragma sellae.

Olfactory Bulb

By elevation of the frontal lobe, the olfactory bulb and tract may be examined throughout their length; the olfactory nerves themselves are 15 or 20 delicate filaments on either side, which arise from bipolar cells situated in the mucous membrane, pass through the cribriform plate of the ethmoid, and enter into the inferior surface of the olfactory bulb. The olfactory pathway is often interfered with in the frontal approach to the cranial contents, either accidentally or purposefully; although olfaction is not under ordinary circumstances a very important function in man, and the number of olfactory fibers present seems to decrease steadily after birth, bilateral interruption of the olfactory tracts should be avoided. There are no blood vessels of importance related to the olfactory nerves. The olfactory system is, of course, subject to damage from tumors pressing upon it.

Associated with the olfactory system is the *nervus terminalis* (rarely seen even by anatomists) from the nose; regarded as both sensory and autonomic, its functional significance is unknown. Intracranially, the nervus terminalis runs parallel to the olfactory bulb (Brookover). Between the two cerebral hemispheres are the anterior cerebral arteries; these, and the anterior communicating artery that usually connects them, are most closely related to the region of the optic chiasma and hypophysis and are therefore discussed in connection with this.

CHIASMAL AND HYPOPHYSEAL REGION

Since the upper border of the sella turcica lies almost in the plane of the floor of the anterior cranial fossa and considerably above that of the floor of the lateral portions of the middle cranial fossa, it is most easily approached by the frontal route; the importance of the structures concentrated in this immediate area—the optic nerves, chiasma and tracts, the internal carotid arteries, the hypophysis and the hypothalamus—make the anatomical relations here exceedingly important.

Most anteriorly, and also most laterally, the optic nerves emerge from the optic canals to be directed posteriorly and medially to join in the optic chiasma (Fig. 1-35). Immediately posterior to the optic canals, and under the backward projecting hook of the anterior clinoid processes, the internal carotid artery of each side emerges from the cavernous sinus on the side of the sella turcica and occupies the lateral angle of the optic chiasma (between the optic nerve and the optic tract); here it proceeds to give off its branches.

Behind the optic chiasma and between the diverging optic tracts is the *hypothalamus,* connected to the hypophysis by the slender hypophyseal stalk which passes downward in the midline behind the optic chiasma. The hypophysis occupies the sella turcica, its upper surface more or less covered by the dura of the diaphragma sellae. The depths of the sella are relatively inaccessible from the cranial cavity, because the anterior and posterior bony walls dip sharply below the general level of the cranial floor, and its lateral walls are the two cavernous sinuses. These transmit the internal carotid artery, the third, fourth, and sixth cranial nerves, and a portion of the fifth. Posteriorly, the dorsum sellae separates this region from the posterior cranial fossa, the floor of which slopes sharply downward in contrast to the approximate horizontal disposition of this middle part of the middle cranial fossa.

As the frontal lobe is elevated, the *olfactory tract* may be seen just a little lateral to the midline; it is attached anteriorly, through the olfactory bulb and nerves, to the cribriform plate, while posteriorly it attaches to the inferior surface of the frontal lobe of the brain just in front of the anterior perforated substance. Inferior to the attachment of the olfactory tract to the brain is the optic nerve,

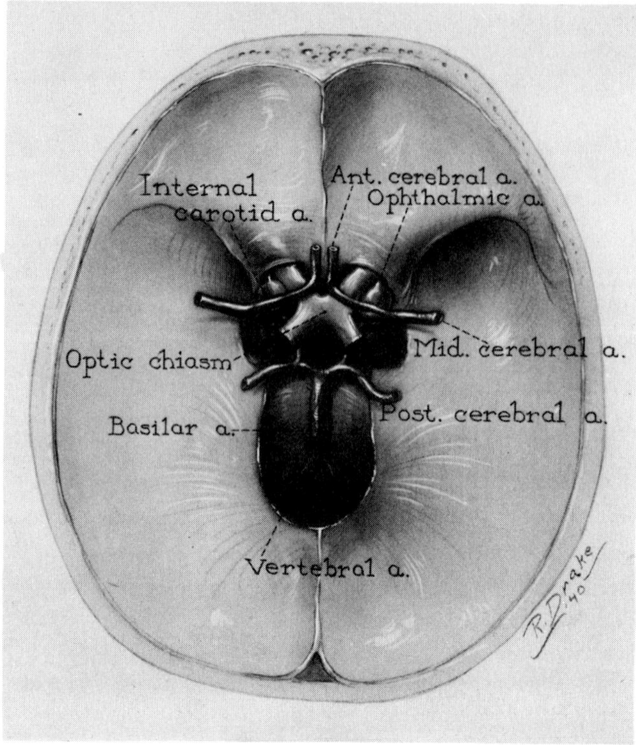

Fig. 1-35. The optic chiasma and the arterial circle. The cavernous sinus and all the soft structures related to it except the internal carotid artery have been removed. (Craig WMcK. In Bancroft FW (ed): Surgical Treatment of the Nervous System, pp 176–203. Philadelphia, JB Lippincott, 1946)

which may be traced from the optic canal to the optic chiasma.

Lateral to the optic chiasma and posterolateral to the nerve, lying against the anterior perforated substance, is the *terminal portion of the internal carotid artery* and the first part of its major continuation, the middle cerebral artery; the latter runs laterally across the perforated substance, to which it is held by the numerous small arteries that it gives off into this substance, thereafter disappearing into the medial aspect of the lateral cerebral (Sylvian) fissure, between the frontal and temporal lobes of the brain.

The *anterior cerebral artery* arises from the anterior wall of the internal carotid and passes medially and forward, lying inferior to the base of the olfactory tract, but crossing above the optic nerve or the optic chiasma; in the midline, between the bases of the olfactory tracts, the two anterior cerebral arteries come close together, are usually connected by the anterior communicating artery, and disappear as they turn upward between the two cerebral hemispheres.

The *posterior communicating artery* leaves the posterior surface of the internal carotid, proximal to the level of origin of the anterior cerebral, and passes posteriorly and somewhat medially about the hypothalamus, crossing below the optic tract, to form a posterolateral segment of the arterial circle of the brain. The *anterior choroidal artery* usually arises from the posterior aspect of the internal carotid or from the first part of the middle cerebral, lateral to the origin of the posterior communicating artery; it typically at first lies lateral to the posterior communicating artery and then diverges farther laterally to disappear into the medial surface of the temporal lobe.

With the exception of the internal carotid artery (which curves upward and backward from its entrance through the dura—medial to the anterior clinoid processes and immediately posteromedial to the entrance of the optic nerves into the cranial cavity—to its attachment to the brain at the anterior perforated substance, just lateral to the optic chiasma), the cerebral vessels of this region lie entirely in close contact with the inferior sur-

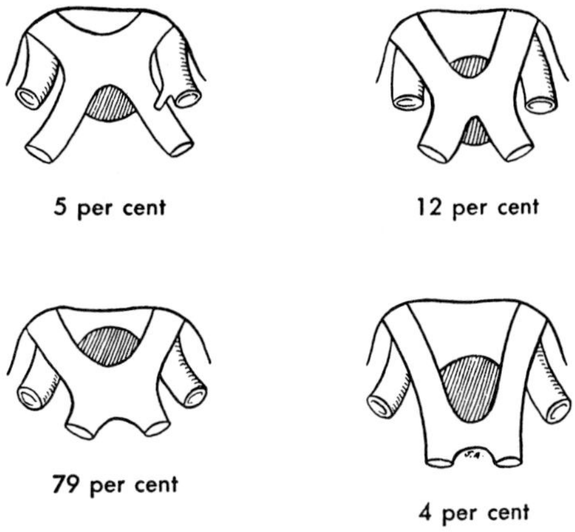

Fig. 1-36. Variations in the relation between the optic chiasma and the hypophysis. (Hollinshead WH: Surg Clin North Am 32:1115, 1952, after de Schweinitz GE: Trans Ophthalmol Soc UK 43:12, 1923)

face of the brain and are necessarily elevated with the frontal lobe. So also is the proximal end of the olfactory tract. However, the optic nerves and the internal carotid arteries necessarily extend across the subarachnoid space from the base of the skull to the base of the brain, and limit the space available for an anterolateral approach to the hypophysis. If, as sometimes occurs, the intracranial portions of the optic nerves are particularly short and the optic chiasma lies well forward toward the so-called chiasmatic groove, an anterior approach, between the optic nerves and below the chiasma, is impossible, and the approach must be a more lateral one, posterolateral to the optic nerve and either between this and the internal carotid or behind this vessel. If, on the other hand, the optic nerves are long intracranially, the chiasma can be lifted with the base of the brain, and the upper surface of the hypophysis inspected through the gap enclosed by the anterior lip of the sella, the two optic nerves, and the chiasma. In such cases the hypophyseal stalk can be seen passing downward, or downward and forward, in the midline, as it runs from the infundibulum, just behind the optic chiasma, to the central aperture in the diaphragma sellae.

In the more lateral approach to the hypophysis across the sphenoid ridge, it should be remembered that the entire lateral dural wall of the sella is occupied by the cavernous sinus and the structures that traverse this sinus.

OPTIC NERVES AND OPTIC CHIASMA

As the optic nerves enter the cranial cavity through the optic canals their dural and arachnoid sheaths, which extend about them to the eyeball, become continuous with these layers in the cranial cavity; the nerves then run through the chiasmatic cistern to the chiasma. The segments of the nerves within the cranial cavity may be relatively long or very short, for the position of the optic chiasma varies. Bergland and co-workers and Renn and Rhoton reported slightly different findings from those shown in Figure 1-36. Dividing the positions of the chiasma into only three groups, prefixed (largely in front of the diaphragma), normal (over the diaphragma), and postfixed, Bergland and co-workers found a prefixed position in 9%, a normal one in 80%, and a postfixed one in 11%. Renn and Rhoton's figures, for a smaller number of cases, were 10%, 75%, and 15%, respectively.

Defects in the visual fields produced by enlargements of the hypophysis depend, therefore, not only upon the constant arrangement of the fibers within the optic nerve and chiasma, but also upon the variable relationship between the hypophysis and chiasma; enlargements of the hypophysis may, depending upon their positions and on the positions of the optic chiasma, affect different portions of the nerves, chiasma, or tracts.

Localization of Fibers in Optic Nerve

Within the optic nerve as it leaves the orbit the nerve fibers arising from the various quadrants of the retina are arranged almost exactly in the relative positions of these parts of the retina—that is, the fibers originating from the temporal half of each retina (representing the nasal portion of the visual field) occupy the lateral part of the nerve, while those from the nasal half of the retina (representing the temporal field) occupy the medial portion of the nerve. Similarly, fibers from the

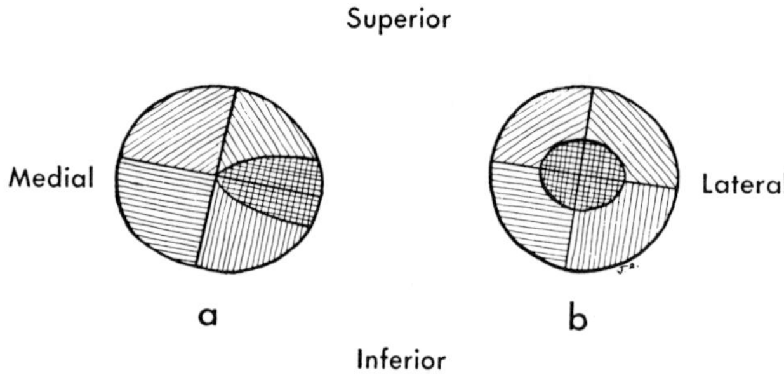

Fig. 1-37. Positions of the fibers from the macula and the four quadrants of the peripheral retina in the optic nerve. (*a*), Close to the eyeball; (*b*), well behind the eyeball.

upper half of the retina (representing the lower visual field) lie uppermost in the nerve, while those from the lower half of the retina (representing the upper portion of the visual field) lie in the lower part of the nerve. Finally, macular fibers, mixed with extramacular ones, occupy a large central area in the nerve (Fig. 1-37).

Details of the fiber arrangement in the optic chiasma and of the blood supply of the optic nerve and chiasma are given in Chapter 2. For present purposes, it should suffice to recall that the fibers arising from the temporal half of each retina, thus representing the nasal visual fields, enter the ipsilateral optic tract, while those from the nasal halves, representing the temporal fields, cross in the chiasma to enter the contralateral tract. Thus it comes about that the optic tract, like other afferent tracts to the thalamus, conducts impulses originating from the opposite side of the body (Fig. 1-38). As a result of the position of the fibers in the chiasma, lateral pressure upon it may first produce a defect in the nasal portion of the visual field of the eye on the affected side, while midline pressure may be expected to produce some type of bitemporal visual field disturbance.

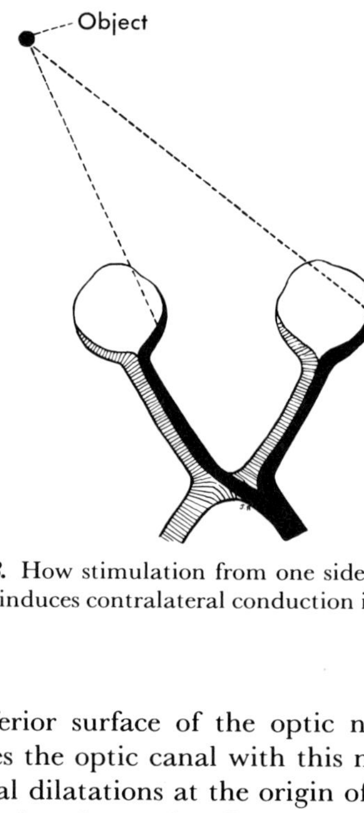

Fig. 1-38. How stimulation from one side of the visual field induces contralateral conduction in the optic tract.

Vascular Relations of Optic Nerve

The immediate vascular relations of the optic nerves, chiasma, and tracts in the region of the sella are to branches of the internal carotid artery. Since this vessel lies lateral to the chiasma in the angle formed by the converging optic nerves and the diverging optic tracts (Figs. 1-35 and 1-39), aneurysms of it may affect the optic nerves, the chiasma, or the tracts. The ophthalmic artery comes immediately into relationship with the inferior or latero-inferior surface of the optic nerve and traverses the optic canal with this nerve; aneurysmal dilatations at the origin of this vessel may therefore early affect the optic nerve. The anterior cerebral arteries pass forward and medially just above the optic nerves, the chiasma, or the optic tracts, and pressure from one of these arteries may therefore produce various defects in the visual fields. Finally, the posterior communicating artery as it runs backward from the carotid crosses beneath the optic tract.

INTERNAL CAROTID ARTERY AND BRANCHES

The general course of the internal carotid and its branches has already been decribed, as have its important relations to the hypophyseal region and the optic system. To summarize, the internal carotid artery leaves the dura and the cavernous sinus at the medial border of the anterior clinoid process, and turns backward and upward to reach the lateral angle between the optic nerve and optic tract. Its exact relation to the optic nerve, chiasma, and tract obviously varies with the position of the chiasma.

The first branch of the carotid is the *ophthalmic,* which lies below the optic nerve and may or may not be visible lateral to it; this branch arises just as or barely before the carotid penetrates the dura. As the carotid crosses the chiasmatic cistern it has no branches, but as it reaches its definitive position against the base of the brain it gives rise to the *posterior communicating artery.* This passes posteromedially along the base of the brain, crossing inferior to the optic tract.

Before it divides into its terminal branches, the anterior and middle cerebral arteries, the internal carotid commonly gives off the *anterior choroidal artery,* which parallels but lies lateral to the posterior communicating artery. Less frequently, the anterior choroidal arises from the middle cerebral, and occasionally it arises from the posterior communicating.

The *middle cerebral artery* passes laterally into the medial aspect of the lateral cerebral fissure, while the *anterior cerebral artery* swings medially and anteriorly, dorsal to the optic nerve, chiasma, or tract, and (above or anterior to the optic chiasma) is typically united to its fellow of the opposite side by the anterior communicating artery.

It is important to note that the anterior cerebral, the middle cerebral, and the posterior communicating arteries are all attached to the brain in this region by numerous delicate perforating branches; those from the middle cerebral are particularly important, as some of them (striate arteries) form the chief supply to the motor portion of the internal capsule.

HYPOTHALAMUS

Lying chiefly above and behind the optic chiasma, in the region bounded anteriorly and anterolaterally by the chiasma and the optic tracts, and laterally and posteriorly by the posterior communicating and posterior cerebral arteries, is the hypothalamus. The infundibulum, a downwardly directed projection of the floor of the hypothalamus, lies just behind the optic chiasma and gives rise to the hypophyseal stalk connecting the hypothalamus to the posterior lobe of the hypophysis. The floor of the third ventricle is especially thin in the midline above the optic chiasma, where it is formed by the lamina terminalis. Numerous small vessels from all the elements forming the cerebral arterial circle converge toward the hypothalamus.

Because of the importance of the hypothalamus in the control of metabolic and endocrine functions, great care must be exercised in working in this region; despite such care it is not uncommon, for instance, for an alarming hyperthermia (due to a disturbance of the heat-regulating mechanism located in the hypothalamus) to develop following operations in this neighborhood.

HYPOPHYSIS

The hypophysis, or pituitary gland, normally lies entirely in the sella turcica (Fig. 1-40). Although it is connected to the hypothalamus by the hypophyseal stalk, it is largely separated from the other structures in this location by the diaphragma sellae, a projecting shelf of dura over the sella. The diaphragma

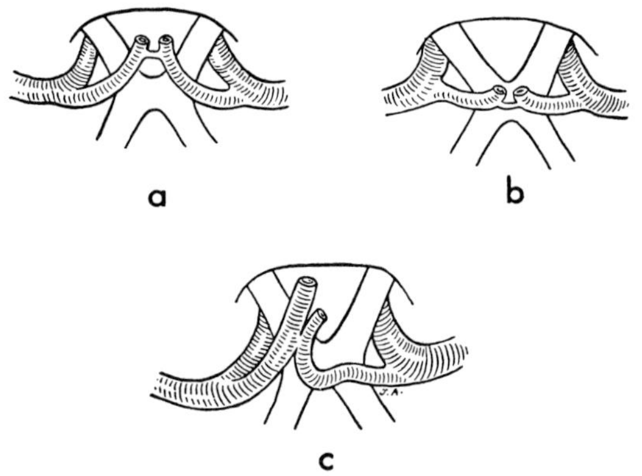

Fig. 1-39. Variations in the relation between the anterior cerebral arteries and the optic chiasma. (After de Schweinitz GE: Trans Ophthalmol Soc UK 43:12, 1923)

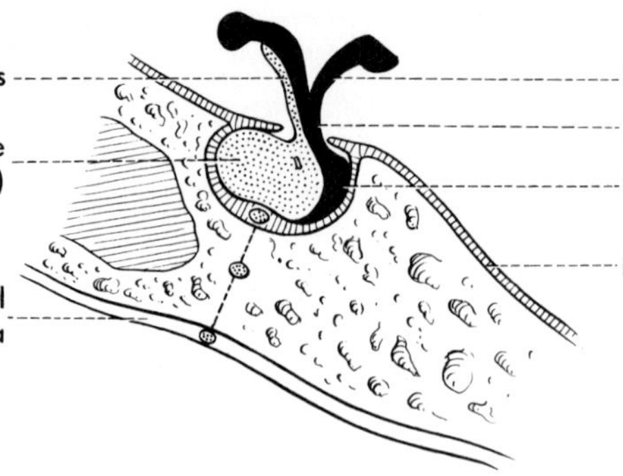

Fig. 1-40. Schema of a sagittal section of the hypophysis and of the pathway of migration of the anterior lobe to the sella turcica. Various locations in which anterior lobe tissue might be left during this migration are indicated. (Hollinshead WH: Surg Clin North Am 32:1115, 1952)

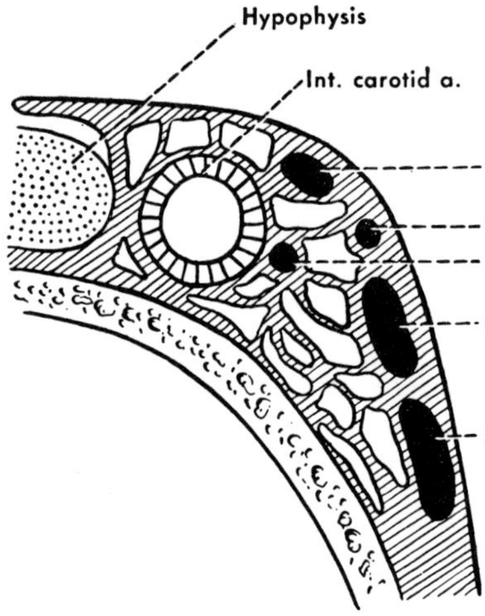

Fig. 1-41. Hypophysis and cavernous sinus in a frontal section. (Hollinshead WH: Surg Clin North Am 32:1115, 1952)

varies markedly both in its strength and in the size of the central aperture that transmits the hypophyseal stalk; this presumably influences in part the direction in which hypophyseal tumors expand and, of course, affects the relative safety of transsphenoidal approaches to the gland. The varying relationship between the hypophysis and the optic chiasma has already been mentioned. The subarachnoid cavity extends through the foramen in the diaphragma and expands over the upper surface of the hypophysis; except on its upper surface, however, the outer portion of the gland is fused to the dura, so that the gland is held in place both by the diaphragma and its dural attachments.

Within the sella turcica the hypophysis is related laterally to the cavernous sinuses, which are also connected to each other, anterior and posterior to the hypophysis, by intercavernous sinuses of varying size. The cavernous sinuses and their contents are discussed in more detail in connection with the middle cranial fossa; it need only be mentioned here that since these sinuses form the lateral walls of the sella turcica and since the anterior and posterior walls of the sella are bony, the intracranial approach to the hypophysis must be from above—a lateral approach through the dural wall of the sella is impossible because of the presence of the cavernous sinus and the structures that traverse it (the internal carotid artery and the third, fourth, sixth, and part of the fifth cranial nerves—Fig. 1-41). The internal carotid artery lies immediately adjacent to the hypophysis and may even compress it (Bergland and co-workers). Tumors of the hypophysis expanding toward the cavernous sinus therefore tend to displace this portion of the carotid, which may therefore transmit its pulsations to the gland. Such tumors may also exert pressure upon the cranial nerves in the sinus; those most frequently affected are the third and fourth, the sixth sometimes being spared, apparently because it lies on the internal carotid and is therefore not subject to direct pressure from a hypophyseal tumor.

Bergland and co-authors described the anatomy of the sellar region in great detail, and Ray regarded the transfrontal approach as preferable in a total hypophysectomy. Renn and Rhoton also reported on the anatomy in detail and enumerated, for both the transfrontal and transsphenoidal routes, variations that might prove troublesome. Adams and co-workers preferred transsphenoidal cryohypophysectomy for acromegalic patients to avoid destroying all the gland; Landolt reported "large areas" of regenerated tissue in patients dying 252 to 353 days after transsphenoidal coagulation. Svien and Litzow developed a transantral–transsphenoidal approach for large tumors essentially confined to the sella, and Konrad and Canalis found a distinct advantage in a transpalatal approach to the hypophysis and other deep-lying midline structures. Rand and Hanafee described cavernous sinus venography as being very helpful in placing a cryoprobe stereotactically.

Vascular Supply

The vascular supply of the hypophysis, as described by Xuereb and co-workers, is through two sets of arteries: superior and inferior hypophyseal arteries. The superior hypophyseal arteries arise from the internal carotid just after it has left the cavernous sinus, therefore close to the origin of the ophthalmic; there may be one on each side, dividing into an anterior and a posterior branch, but these branches may arise separately from the carotid, and Stanfield found this to be most common. The inferior hypophyseal arteries, also one on each side, arise from the internal carotid while it lies in the cavernous sinus and reach the sides of the hypophysis. Both sets of arteries also supply twigs to adjacent dura, and the superior ones help supply the optic chiasma and the infundibular region of the hypothalamus. Insofar as the hypophysis is concerned, both sets of vessels supply the hypophyseal stalk and anastomose with each other in the sella; the inferior vessels also supply the posterior lobe or neurohypophysis. The anterior lobe or adenohypophysis receives most of or all its blood from the hypophyseoportal system of veins. Xuereb and colleagues found no arterial supply to this lobe; McConnell described a supply from the inferior hypophyseal arteries to a small portion of the lobe in front of the stalk; and Stanfield found a small and variable arteriolar supply from the anastomotic circle around the hypophysis. Section of the stalk above the sella interrupts the upper part of the hypophyseoportal system and descending branches of the superior hypophyseal arteries that supply capillaries in the lower part of the stalk, but it leaves intact the supply to the lower part from the inferior hypophyseal arteries. Since the hypophyseoportal system originates in part from the lower end of the stalk, only a part of the blood supply to the sinusoids of the anterior lobe is interrupted by suprasellar section.

The *venous drainage* from the capillaries of the posterior lobe and the sinusoids of the anterior lobe is laterally into the cavernous sinuses, upward into the intercavernous sinuses, and downward into a subhypophyseal sinus. The hypophyseoportal system of veins was first described as collecting blood *from* the hypophysis and its stalk.

Innervation

Sympathetic nerve fibers reach the anterior lobe of the hypophysis from the carotid plexus by accompanying the arterial supply, but have not been shown to be of importance for the activity of the gland. The numerous nerve fibers constituting the hypophyseal stalk end primarily in the posterior lobe. The importance of the nerve supply to the posterior lobe in the control of water balance was demonstrated by Fisher and co-workers; the relationship between damage to the stalk or the cells of origin of its fibers (the supraopticohypophyseal system) and the onset of diabetes insipidus is now well established, as is the fact that the anterior lobe is controlled not by the nerve fibers in the hypophyseal stalk but by neurohumoral substances produced in the hypothalamus and reaching the anterior lobe through the hypophyseoportal veins.

Aberrant Tissue

Since the anterior lobe of the hypophysis (and the pars intermedia) develop from the diverticulum of the buccal epithelium known as Rathke's pouch, these portions of the gland are originally connected to the roof of the mouth. With increase in depth of the buccal pouch, this attachment becomes shifted posteriorly until it lies at the caudal border of the nasal septum, and the stalk passes upward through the tissue that eventually forms the body of the sphenoid bone. As the hypophysis becomes detached from the pharyngeal wall, it leaves an apparently constant mass, the pharyngeal hypophysis, in this wall (Melchionna and Moore). Anterior lobe tissue has also been reported in the body of the sphenoid bone (*e.g.,* by Suchannek), between the sella turcica and the pharynx (see Fig. 1-40). *Craniopharyngiomas,* thought to be developed from remnants of this stalk, are, however, according to Zeitlin and Oldberg, more commonly found above the sella turcica than below it—perhaps owing to the necessity for more room for these usually cystic tumors. Tumors of the hypophyseal stalk proper—that is, of the stalk of the posterior lobe, connecting this to the brain—also occur; their symptomatology has been discussed by Beckmann and Kubie.

Remnants of the hypophysis within the body of the sphenoid bone are frequently described as lying in the course of the craniopharyngeal canal, a canalicular defect sometimes found in the bone; apparently this canal does not, however, represent the course of migration of the gland, but only roughly parallels that course (Arey). There is therefore no correlation between the presence of the canal and the presence of intrasphenoid remnants of the hypophysis.

Middle Cranial Fossa

In order to avoid the auriculotemporal nerve and superficial temporal artery, incisions of the scalp for an approach to the middle cranial fossa may begin above the ear and curve upward and forward along the superior temporal line, that is, at about the uppermost origin of the temporalis muscle from the side of the skull. The anterior end of the incision can curve downward, as desired, behind the orbit. The origin of the temporal muscle from the parietal and temporal bones can then be elevated and reflected downward. As in the case of frontal bone flaps, temporal bone flaps have been left attached to the muscle; detachment has the advantages of providing more room and, since a bone flap separated from its vascular connections does not bleed, lessening the hazard of a postoperative subdural hematoma.

In turning the bone flap for a temporal approach to the cranial contents, the course of the middle meningeal artery should be borne in mind. Briefly repeated, the frontal branch makes a broad curve upward, with its concavity directed backward; the most forward point of this curve, and the point at which the artery can be most easily located, is the pterion. If the middle meningeal artery branches late after entering the cranial cavity, this branching occurs at about the level of the upper border of the zygomatic arch and fairly close to the middle of this arch; the parietal branch of the middle meningeal artery then arches upward and backward and almost parallels the zygomatic arch, lying approximately a fingerbreadth above it.

FLOOR OF MIDDLE CRANIAL FOSSA

The chiasmal grooves and the sella turcica, discussed in this volume with the anterior cranial fossa, form the small middle part of the middle cranial fossa (Fig. 1-42). Each expanded lateral part, concave for the anterior part of the temporal lobe, is bounded anteriorly by the "sphenoid ridge" and posteriorly by the upper border of the petrous part ("petrous ridge") of the temporal bone. The greater wing of the sphenoid bone anteriorly and the anterior surface of the pars petrosa posteriorly form most of the floor, with a part of the squama of the temporal bone lateral to and between them. The greater wing of the sphenoid, projecting lat-

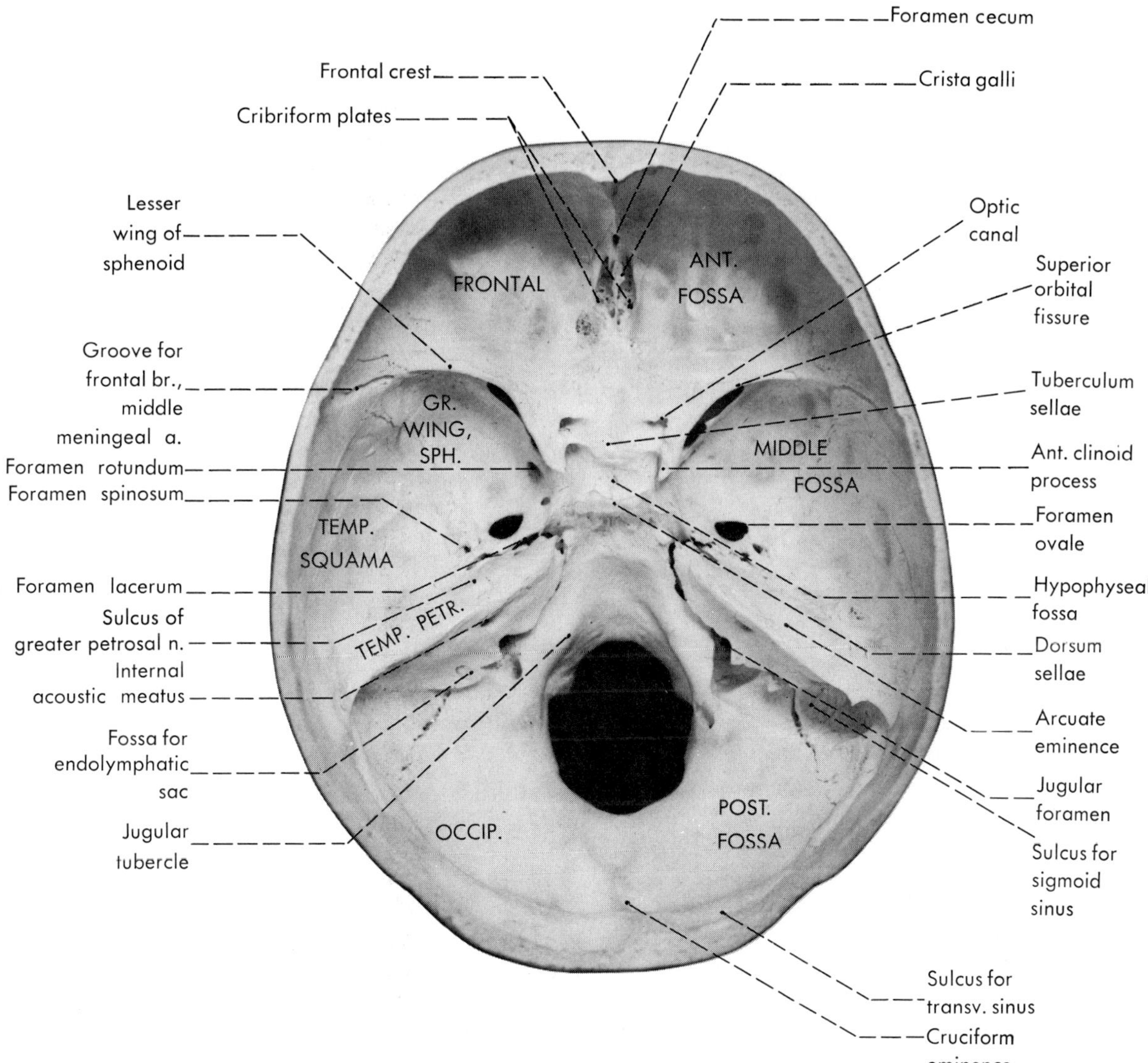

Fig. 1-42. Interior of the base of the skull. *GR. WING, SPH.* is the greater wing of the sphenoid; *TEMP. SQUAMA* and *TEMP. PETR.*, the squamous and petrous portions of the temporal bone; and *OCCIP.*, the occipital bone.

erally from the body, is separated from the lesser wing by the superior orbital fissure; posteromedially to it, the body of the sphenoid and the petrous part of the temporal bone are separated from each other by the foramen lacerum. From just above the foramen lacerum, on the side of the body of the sphenoid, the carotid sulcus curves forward and upward toward the anterior clinoid process. Behind the medial end of the superior orbital fissure, between the greater wing and the body of the sphenoid, is the foramen rotundum for the maxillary nerve; posterolateral to that, in the greater wing lateral to the foramen lacerum, is the foramen ovale for the mandibular nerve; just posterolateral to that is the foramen spinosum for the middle meningeal artery. Anteromedial to the foramen ovale there may or may not be a sphenoid emissary foramen (foramen of Vesalius), which transmits a vein of the same name.

Other than the grooves for the middle

meningeal artery and its branches, the squama of the temporal bone has no features of particular interest in the middle fossa. At its anterior end, the petrous part of the temporal presents the anterior opening of the carotid canal, and there are several markings of interest on the anterior surface of the bone. Close to its anterior end is the trigeminal impression for the trigeminal ganglion, extending downward to the roof of the carotid canal, which is here formed of very thin bone or of membrane only. The flattened lateral part of the petrous temporal is the thin tegmen tympani, roofing the tympanic cavity and mastoid antrum. Anteromedial to the tegmen is the more or less well-defined arcuate eminence, over the anterior (superior) semicircular canal; more anteromedially still, and lower on the surface of the petrous ridge, is the small hiatus of the canal of the greater petrosal nerve (formerly hiatus of the facial canal), and leading from it toward the foramen lacerum is the groove for the greater petrosal nerve. (Thirty percent of specimens in the series of Rhoton, Pulec, and associates had no canal for the greater petrosal nerve; either the nerve and at least part of the geniculate ganglion, or the nerve alone, lacked any bony covering.) At about the same level, but farther down the petrous ridge, there may or may not be recognizable the tiny hiatus of the canal of the lesser petrosal nerve, and a groove for this nerve running toward the suture between sphenoid and temporal bones just posteromedial to the foramen spinosum (the lesser petrosal nerve may leave the skull here, continue forward to leave through the foramen ovale, or leave through a tiny unnamed foramen just behind the foramen ovale).

The upper border of each petrous ridge is marked by a sulcus for the superior petrosal sinus; the attachment of the tentorium cerebelli here separates the lateral parts of the middle fossa from the posterior fossa, and behind the petrous ridges the temporal lobes rest on the tentorium.

The floor of the middle cranial fossa is distinctly thicker than that of the anterior fossa, but is nevertheless weakened by its numerous foramina. A large proportion of *fractures* of the base of the skull traverse the middle fossa. Fractures that extend across the body of the sphenoid bone may injure the internal carotid artery, or they may allow communication between the subarachnoid cavity and the sphenoid sinus, with discharge of cerebrospinal fluid through the nose and ascending infection from the nose to the meninges. The middle meningeal artery is particularly subject to damage in fractures of the middle fossa. If the fracture extends into or through the petrous portion of the temporal bone, facial paralysis and deafness are common sequelae. Fractures through the tegmen tympani offer the same possibilities for the escape of blood or cerebrospinal fluid into the mastoid and middle ear cavities and of infection of the meninges from these cavities, as exists in fractures involving the paranasal sinuses; if a fracture is so severe as to produce a tear of the tympanic membrane or to extend into the roof of the external acoustic meatus, blood and cerebrospinal fluid may also escape directly through the ear canal. Obviously, bleeding from the ear or nose, or both, in cranial injuries may be a sign of severe damage to the skull.

Attachment of Dura

The dura of the middle cranial fossa is especially strongly attached about the numerous foramina of this fossa—the superior orbital fissure, the foramen rotundum and the formen ovale, and the foramen lacerum in particular. Anteriorly, it is attached to the sphenoid ridge; medially, where it lies on the side of the sella turcica, it forms the walls of the cavernous sinus and must not be stripped from the bone; posteromedially, it is attached to the petrous ridge, which also receives the attachment of the tentorium cerebelli.

MENINGEAL VESSELS AND NERVES

Middle Meningeal Artery

The middle meningeal artery enters the middle cranial fossa through the foramen spinosum. From this position it runs forward and laterally in the dura; although it supplies

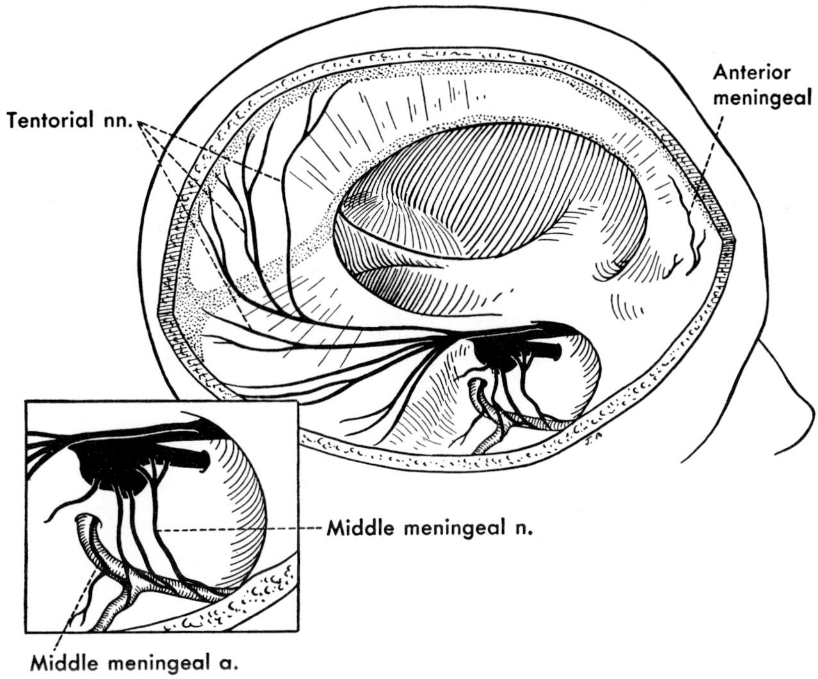

Fig. 1-43. Nerves of the dura of the middle cranial fossa. (After Penfield W, McNaughton F: Arch Neurol Psychiatr 44:43, 1940)

twigs to the dura of the floor of the fossa, its distribution is largely to the calvaria, and it has already been discussed earlier in this chapter. While it lies in the dura of the floor of the middle fossa, the middle meningeal artery supplies a petrosal branch to the facial nerve and tympanic cavity; this accompanies the greater (superficial) petrosal nerve into the petrous bone, to anastomose with the stylomastoid branch of the posterior auricular artery. The middle meningeal also gives off a small superior tympanic artery to the tensor tympani muscle and the middle ear cavity.

Frequently there is an accessory middle meningeal artery, derived from the middle meningeal or directly from the (internal) maxillary; its intracranial branch, if any, is usually small and enters through the foramen ovale to supply the dura in the immediate vicinity of the trigeminal ganglion and the ganglion itself. In the absence of the accessory meningeal, small branches from the middle meningeal artery supply this same area; other branches to the dura of the middle cranial fossa include tiny branches from the internal carotid artery, distributed largely in the neighborhood of the cavernous sinus and the trigeminal ganglion, and twigs of the ascending pharyngeal artery that pass through the foramen lacerum.

Innervation

The dura of the middle cranial fossa, including the upper surface of the tentorium, is supplied entirely by branches of the trigeminal nerve. All three divisions of the nerve may apparently participate in this, although Penfield and McNaughton suggested that the ophthalmic division of the ganglion may contain the cells of origin of all the trigeminal meningeal branches. In addition to variable twigs from the second and third divisions of the nerve to the dura close to the ganglion, both these divisions typically send branches to the middle meningeal artery to be distributed along it (Fig. 1-43). The meningeal branch from the mandibular, once called the nervus spinosus, typically arises outside the skull to join the middle meningeal artery here, give a tiny communicating branch to the otic ganglion, and re-enter the skull along the artery through the foramen spinosum; a part or all of it may join the artery intracranially, however. The meningeal branch (recurrent branch, middle meningeal nerve) of the maxillary, sometimes double or triple (Grzybowski), joins the plexus on the anterior branch, or sometimes both branches, of the middle meningeal artery intracranially.

The largest meningeal branch of the ophthalmic nerve is frequently called the tentorial nerve or nerves (of Arnold). It runs a recurrent course across the trigeminal ganglion close to the trochlear nerve and is distributed to the tentorium, the posterior third of the falx cerebri, and dura in the parietooccipital region; Penfield and McNaughton, whose paper should be consulted for further details on the anatomy of the dural nerves, found the tentorial nerve to be constantly present.

Penfield and McNaughton, and Ray and Wolff, have agreed that most of the dural convexity is insensitive, except where the dura contains the meningeal arteries. Similarly, the falx cerebri is said to be largely insensitive except along the margins of the superior sagittal sinus, but the floor of the middle cranial fossa and the upper surface of the tentorium are generally and uniformly sensitive to pain (although the lower surface of the tentorium is not, except close to the venous sinuses).

VENOUS SINUSES

The venous sinuses in or adjacent to the middle cranial fossa include the small sinuses (meningeal veins) accompanying the middle meningeal artery and the sphenoparietal, cavernous, and superior petrosal sinuses. There are typically two venous channels with the middle meningeal artery; they communicate with the cavernous sinus through the sphenoparietal sinus and also pass through the foramen spinosum with the artery to empty into the pterygoid plexus.

The *sphenoparietal sinus* forms a communication between the meningeal veins and the cavernous sinus, running downward and medially on the sphenoidal ridge to empty into the cavernous sinus above the oculomotor nerve as it lies in the wall of the sinus. The sphenoparietal sinus usually also receives the anterior temporal diploic veins.

The *cavernous sinuses* lie on the sides of the body of the sphenoid bone, therefore on the sides of the hypophyseal fossa and just anterior and medial to the trigeminal ganglion. The oculomotor and trochlear nerves, the ophthalmic division of the trigeminal, and sometimes the upper portion of the maxillary division of this nerve run in the dural wall of the sinus and, more deeply, the carotid artery and the abducens nerve traverse it. Posteriorly, it lies against the anterosuperior edge of the trigeminal ganglion and gives off the superior petrosal sinus, which runs backward along the petrous ridge close to the root of this ganglion. Its connections are described in more detail in the introductory section of this chapter.

The *superior petrosal sinus* lies along the petrous ridge in the attached margin of the tentorium cerebelli; it forms a communication between the cavernous sinus and the upper end of the sigmoid sinus. It is typically fairly small; in dividing the tentorium cerebelli from its attachment to the petrous ridge, in order to approach from above a laterally lying tumor in the posterior fossa, care need only be taken not to section the tentorium at its very attachment to the petrous ridge; section within a few millimeters of the ridge is usually safe, however (although the trochlear nerve on the lower surface of the tentorium should be avoided). An incision into the tentorium should not, of course, encroach upon the transverse sinus and, if extended posteriorly, must be curved so as to parallel the anterior border of this sinus as it lies in the attachment of the tentorium to the occipital bone.

The roots of the trigeminal nerve also pass across the petrous ridge and are therefore closely related to the superior petrosal sinus; the latter more commonly passes above the trigeminal roots, but many pass below them or divide so that one portion passes above and one below the roots.

In addition to the sphenoparietal, cavernous, and superior petrosal sinuses, which essentially border the middle cranial fossa, other sinuses, usually small, may pass across the concavity of the fossa. These include the petrosquamous sinus, present on at least one side in approximately 50% of persons, the ophthalmopetrous sinus of Hyrtl, and the accessory sinuses of Kelch and of Verga. The latter three are relatively rare and are probably variations of the same fundamental pattern. All four of these sinuses have already been discussed.

The chiasmal and hypophyseal regions may be approached through the temporal or subtemporal route, but can be less well exposed than through the frontal approach, because of the marked concavity of the floor of the middle fossa. The cavernous sinuses cover the sides of the hypophyseal fossa, and the internal carotid arteries emerge from the cavernous sinuses to lie lateral to the optic chiasma; anterior and posterior to the chiasma the anterior cerebral and posterior communicating arteries pass respectively anteriorly and posteriorly, the former running above the optic nerves and the latter below the optic tracts; the middle cerebral artery, the continuation of the internal carotid, passes laterally and slightly upward into the medial side of the lateral cerebral fissure.

CRANIAL NERVES TRAVERSING THE MIDDLE FOSSA

MOTOR NERVES TO THE ORBIT

All three of the motor nerves to the orbit pass through the cavernous sinus and therefore through the middle cranial fossa (Fig. 1-44); their relationships within the sinus to each other have already been mentioned and are described also in connection with the orbit. The nerves are so closely grouped here that a lesion—for instance, an aneurysm of the internal carotid or an adenoma of the hypophysis—may effect all three nerves, causing a complete ophthalmoplegia; the abducens nerve is sometimes spared in hypophyseal enlargements, at least temporarily. The distribution of these nerves also is described in connection with the orbit; in brief, they to-

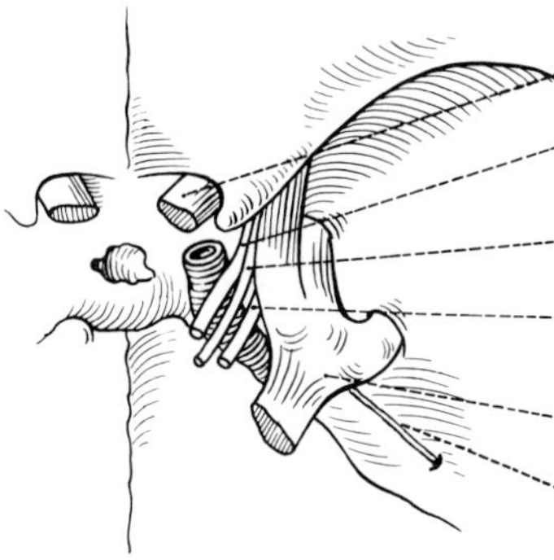

Fig. 1-44. Nerves to the orbit, and the trigeminal ganglion, in the middle cranial fossa.

gether supply all of the voluntary muscles of the eyeball, the levator palpebrae superioris, and two sets of involuntary muscle within the eyeball, the sphincter pupillae and the ciliary muscle.

The entrance of the abducens nerve into the cavernous sinus cannot be seen from the middle cranial fossa, since this nerve first runs intradurally in the floor of the posterior cranial fossa. The oculomotor nerve enters the dura mater just anterolateral to the posterior clinoid process, passing between this process and the most anterior end of the free edge of the tentorium cerebelli. The trochlear nerve enters the dura just beneath the free border of the tentorium cerebelli (therefore technically in the posterior cranial fossa), only a little behind the posterior clinoid process. It would not be easily injured here in cutting the tentorium from the petrous pyramid unless the tentorium were cut very close to the bone (which should be avoided also for the sake of the superior petrosal sinus); however, in dividing the tentorium, it would probably be safer to identify the trochlear nerve and cut the free edge of the tentorium behind the entrance of this nerve. The tendency of the delicate trochlear never to cling to the lower surface of the tentorium may nevertheless lead to injury to the nerve in extending a tentorial incision posteriorly, unless the nerve is first separated.

The three nerves to the eye muscles are also discussed in connection with the posterior cranial fossa, in which they originate.

TRIGEMINAL NERVE

The trigeminal, or fifth cranial, nerve is not only the most important nerve of the middle cranial fossa, but also the largest of all cranial nerves except the optic. It is especially important clinically because of its involvement in major trigeminal neuralgia or tic douloureux, an accepted treatment of which is section of the sensory root of the fifth nerve, most commonly carried out through the middle cranial fossa.

The large sensory root (once portio major) and the small motor root (portio minor, actually motor and afferent) of the fifth nerve arise from the lateral pontine region, where the fibers pass through the brachium pontis. It is the sensory root that bears the large trigeminal (semilunar or Gasserian) ganglion, the equivalent of the dorsal root ganglion of a spinal nerve, from which the three great branches of the nerve arise; the motor root blends neither with the sensory root nor with the ganglion, but passes deep to the ganglion to accompany the mandibular division and join this just outside the foramen ovale.

Janetta and Ley and co-workers have reported sparing the accessory rootlets of the trigeminal nerve, which lie between the motor and chief sensory roots, in sectioning the latter. Janetta found light touch well preserved and the corneal reflex normal after this operation, while Ley and co-workers reported partial preservation of sensation and the corneal reflex. These rootlets thus seem to contain the same types of sensory fibers found in the chief sensory roots; Pelletier and co-workers, investigating them in the monkey, found that they also contain fibers of the same modalities as those found in the motor root. Pelletier and co-workers and Saunders and Sachs described the accessory rootlets as joining the motor rootlets, while Gudmundsson and co-workers, although regarding them as largely motor, said they join the sensory root.

Trigeminal Ganglion

The cells of the trigeminal ganglion give rise to fibers of general sensation that form the ophthalmic and maxillary branches of the nerve and the major portion of the mixed mandibular nerve; these fibers are distributed peripherally to skin and mucous membrane of the head, while their central branches form the sensory root. The motor root contains voluntary motor fibers to and afferents from the muscles of mastication and a few others derived from the mandibular arch; its distribution corresponds mostly to that of the motor distribution of the mandibular nerve.

It may be remembered that although the fifth nerve in its peripheral distribution is intimately connected with all four of the larger parasympathetic ganglia in the head, its roots contain no preganglionic fibers for these ganglia; similarly, although the lingual branch of the fifth nerve contains fibers for the taste buds of the anterior two thirds of the tongue, it is generally granted that these fibers enter the brain stem not with the fifth nerve, but with the seventh.

Although the fifth nerve arises in the posterior cranial fossa, its roots extend anterolaterally and cross the petrous pyramid, so that the ganglion lies in the middle cranial fossa, on the sloping anterolateral surface of the petrous pyramid. Because of the slope of this surface of the petrous pyramid, the trigeminal ganglion, closely applied to it, has somewhat complicated relations to the major planes of the body (Fig. 1-44; and see Fig. 1-47). Thus its upper edge, almost horizontal, is also situated somewhat medially; moreover, since the petrous pyramids of each side converge toward each other, the upper edge of the ganglion, when traced forward, also inclines medially. Similarly, the posterior edge of the ganglion extends both laterally and downward from the root, and if it extends slightly forward at the same time it becomes an inferolateral rather than a strictly posterior border; the curved distal edge of the ganglion extends from a more posterior, inferior, and lateral position to a more anterior, superior, and medial one.

Mira and co-workers reported finding cells similar to those of the trigeminal ganglion on both the motor and sensory roots of the nerve; Young suggested that the presence of 300 to 1000 unmyelinated fibers in the motor root may be one explanation for failure to obtain relief of pain by section of the sensory root.

Peripheral Branches of Trigeminal

From the ganglion the peripheral branches of the fifth nerve make their exits through three separate foramina of the skull. The *ophthalmic division* runs through the lateral wall of the cavernous sinus, in which position it gives off its major meningeal branch, receives sympathetic fibers from the carotid plexus (by way of the abducens nerve), and divides into three branches that then enter the orbit through the superior orbital (sphenoid) fissure; the details of this are described in connection with the orbit and the face.

The *maxillary division* of the fifth nerve runs along the lower portion of the lateral wall of the cavernous sinus, its relationship to the sinus varying with the size of the latter; it may be below the sinus, or the sinus may extend medial to and even below it. It leaves the skull by the way of the foramen rotundum to enter the pterygopalatine (sphenomaxillary) fossa; its branches of distribution to the cheek, nose, palate, and upper jaw are described in connection with these structures.

The large sensory portion of the *mandibular nerve,* derived from the trigeminal ganglion, passes through the foramen ovale with the motor root, which lies deep (medial) to it; immediately outside the skull they join to form a mixed nerve.

Trigeminal (Meckel's) Cave

The relationship of the roots and ganglion of the fifth nerve to the arachnoid and dura is of importance in the surgical approach to the ganglion and its sensory root. Within the posterior cranial fossa the motor and sensory roots of the ganglion pass freely through the large subarachnoid space, but as they cross the petrous ridge they enter a diverticulum of the dura and arachnoid, the trigeminal or Meckel's cave, that extends onto the sloping anterolateral surface of the petrous ridge, thus into the middle cranial fossa. Dura and arachnoid, and consequently cerebrospinal fluid, thus surround the roots of the fifth nerve in the middle fossa (Fig. 1-45), but peripherally the meninges fuse with the connective tissue of the ganglion. The meninges of the posterior cranial fossa are therefore evaginated out along the roots and ganglion of the fifth nerve in a fashion analogous to the evagination of the spinal meninges about the spinal nerves. The subarachnoid space surrounds the nerve roots as far as the ganglion; close to the junction of root and ganglion it is obliterated on the anterolateral (outer) surface by fusion of the meninges to the ganglion, but on the posteromedial (deep) surface, along the mandibular part, the subarachnoid space extends over about the proximal two thirds of the ganglion. This relationship is of special importance from the standpoint of attempted injections of the ganglion, since fluid delivered inadvertently into this space spreads readily backward toward the brain stem.

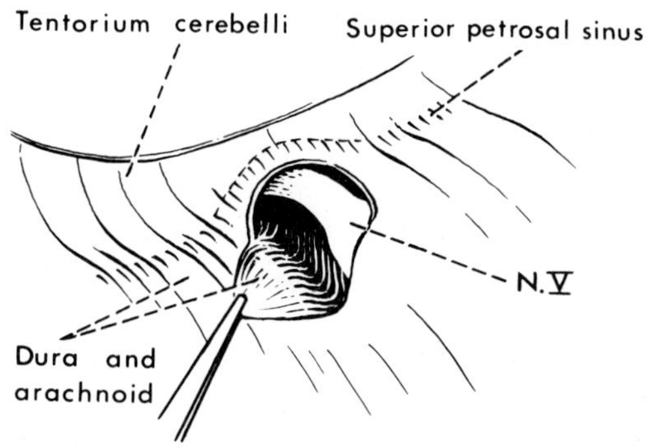

Fig. 1-45. The trigeminal (Meckel's) cave, a blind extension of the subarachnoid space of the posterior cranial fossa into the middle fossa.

The oval mouth to the cave, transmitting the roots of the trigeminal nerve, usually lies below the superior petrosal sinus, for the latter normally runs in the base of the tentorium above the entrance to the cave. The sinus may, however (in more than 40%?), pass below the mouth and therefore against the petrous ridge or split to pass both above and below (Coates). Within the cave the trigeminal ganglion is on the lateral surface of the anterior part of the petrous portion of the temporal bone, and its mandibular part is separated from the internal carotid artery only by dura and the thin bone or membrane roofing the anterior end of the carotid canal.

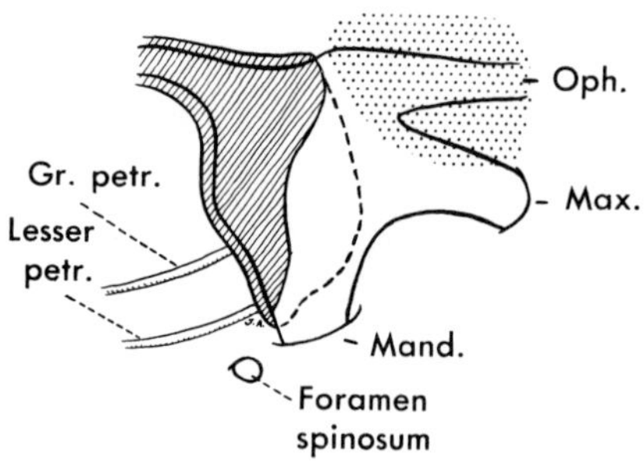

Fig. 1-46. Some approximate relations of the trigeminal ganglion. The heavily *shaded* area indicates the extension of the subarachnoid cavity of the trigeminal cave on the outer surface of the roots and ganglion; the *broken line* on the ganglion, its more distal extension deep to the ganglion; the *stippled* area, the cavernous sinus.

The upper edge of the ganglion has a variable contact with the lateral wall of the cavenous sinus (Fig. 1-46), and in the temporal approach to the sensory root this fact is taken into consideration; the dura along the upper and medial portion of the ganglion is avoided, and it is the dura along the posterior border that is incised to expose the sensory root. Above the ganglion, in the dural wall of the sinus, run the trochlear and oculomotor nerves.

TRIGEMINAL NEURALGIA

The cause or causes of major trigeminal neuralgia (tic douloureux) are not understood; it has been suggested that the precipitating cause may lie within the central nervous system, but it might be noted that Dandy ('34), in reporting upon his observations in 215 operations for this disease, stated that he found some type of gross anatomical involvement of the nerve root in 60% of these patients. Haines and co-workers compared the vascular relations of 40 nerves from cadavers known to have been free of facial pain with a similar number of nerves of patients explored because of such pain; 14 of the nerves from cadavers showed contact between the nerve and a blood vessel, but the nerve was compressed or distorted in only 4. Among patients, however, 34 nerves (85%) had contact with and 31 (77.5%) were compressed by an adjacent vessel—usually the superior cerebellar artery but sometimes the anterior inferior cerebellar or a vein. Pressure from other sources has also been suggested. Thus, Taarnhøj thought that the dura over the ganglion and nerve root might be responsible, and advocated decompression; Kerr ('63a) suggested that pressure by the internal carotid artery, when the bony roof of the carotid canal deep to the ganglion is deficient, may be a cause.

Regardless of the real causative factor of trigeminal neuralgia, the paroxysmal attacks of this painful and disabling disease are ordinarily precipitated by stimulation of pain fibers from a so-called trigger zone, and the attacks may be alleviated or abolished by interruption of the pain fibers in this nerve. This has been attempted by injection of the ganglion with alcohol, extirpation of the ganglion, section of the sensory root, section of the pain fibers as they course caudally in the medulla, decompression, and, more recently, "compression" or deliberate slight injury to the ganglion, and thermocoagulation. Of these, section of the root has been the most common procedure.

Trigeminal neuralgia involves apparently only the pain fibers of the fifth nerve and most frequently arises in the zones of the mandibular or maxillary divisions of this nerve, being relatively rare in the first division; Grant reported that in a series of 100 cases of major trigeminal neuralgia the pain was primarily in the ophthalmic division in only 6. Because of these facts, the arrangement of the fibers in the sensory root has been investigated from the standpoint of the possibility of differential section of them by one of the following routes: close to the ganglion by the temporal approach, close to the brain stem by a "cerebellar" or posterior fossa approach (Fig. 1-47), or by a subtemporal transtentorial one. Most workers agree that the fibers in the sensory root close to the ganglion are arranged in the same order as the major branches of the nerve—that is, the fibers of the sensory root representing the mandibular division of the nerve lie laterally, posteriorly, and inferiorly, those representing the ophthalmic lie medially and superiorly, while those of the maxillary lie between the two.

Although a topographic arrangement of

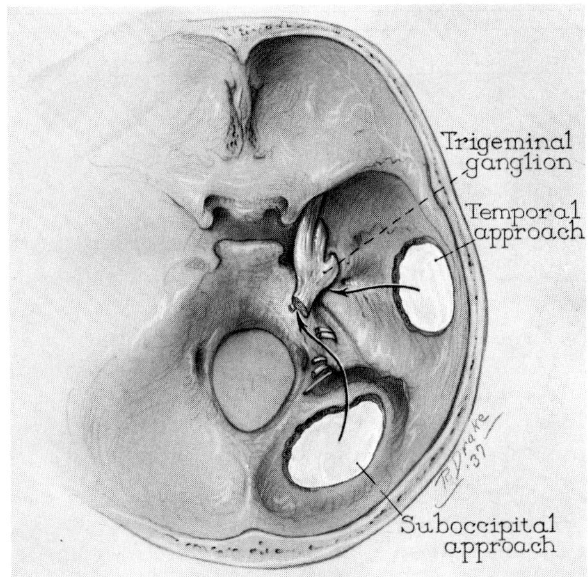

Fig. 1-47. The two approaches to the sensory root of the trigeminal nerve. The small motor root is also shown. (Craig WMcK, Parkinson D: In Oxford Medicine, Vol 6, pp 747–780. New York City, Oxford University Press, 1957)

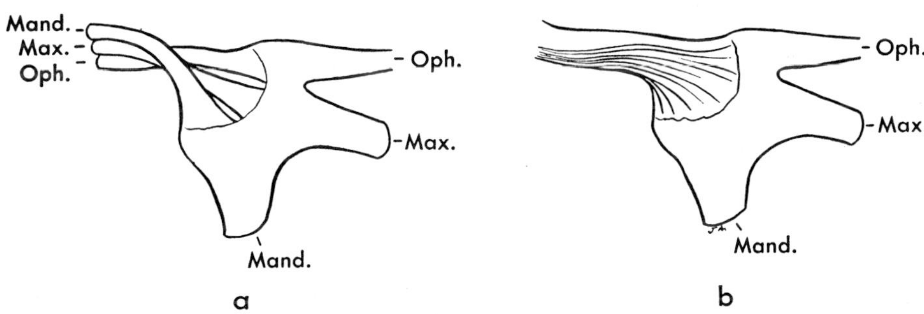

Fig. 1-48. Two concepts of the rearrangement of fibers said to occur in the sensory root of the trigeminal nerve between the ganglion and the brain stem. In (*a*) the three divisions are seen to be rotated upon each other, as in Fig. 1-49; in (*b*) pain fibers from all three divisions are shown concentrated in the posterior part of the root at the pons.

the sensory fibers in the root of the trigeminal nerve adjacent to the ganglion has been widely accepted, much less agreement has existed as to their arrangement between this point and the brain stem. Although, according to Grant, the sensory root of the fifth nerve is rarely more than 2 cm long, it has been both claimed and denied that (1) extensive intermingling of the fibers of the three major divisions of the trigeminal occurs proximal to the ganglion, (2) there is not a great deal of intermingling, but there is a definite rotation of the root between the ganglion and the pons (Fig. 1-48*a*), and (3) there is a grouping, at the pons, of the pain fibers from all three divisions (Fig. 1-48*b*).

In view of all the relatively recent investigations of this subject, which are in general agreement, there seems to be no need for an extensive review of the older literature. In brief, Davis and Haven were among the earlier workers who reported that the three components of the nerve root are relatively distinct as they near the pons, and undergo a definite rotation before reaching it (Fig. 1-49). Kerr ('63b), working with both the cat and the monkey, confirmed both statements; Gudmundsson and co-workers, studying the trigeminal root of human beings, also confirmed them. As for the supposed sorting out of pain fibers, apparently first suggested by Dandy, the only support it has received in re-

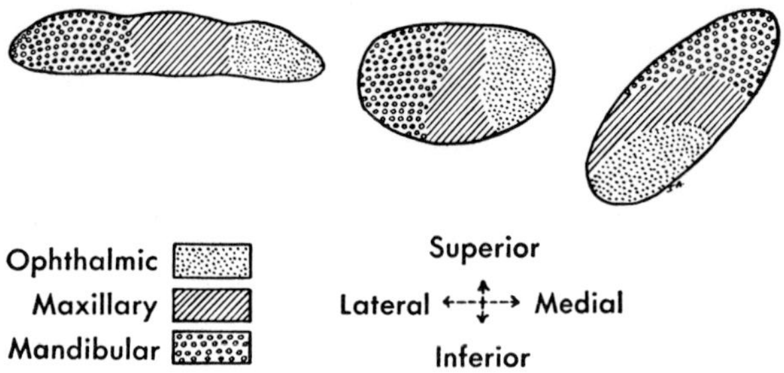

Fig. 1-49. Torsion of the sensory root of the trigeminal nerve between the ganglion and the pons, as seen in cross sections close to the ganglion, at the middle of the root, and close to the pons. The positions of the nerve roots are given in relation to the longitudinal axis of the pons and the medulla. (After Davis L, Haven HA: Arch Neurol Psychiatr 29:1, 1933)

cent years is from Hitselberger and Pulec and Pulec and Hitselberger, who reported that they had abolished pain but spared light touch by sectioning only the posterior part of the nerve. Most neurosurgeons have not found this to be true, and according to Pelletier and co-workers Dandy himself came to doubt it. Davis and Haven found no evidence of such sorting in the cat, nor did Kerr ('63b) in either cat or monkey; Emmons and Rhoton and Pelletier and co-workers, working with monkeys but using different methods, also denied this report. Lende and Poulos extended the finding to the trigeminal ganglion, reporting that there is in the mandibular part (again in the monkey) no segregation of cells according to function.

Davis and Haven based their diagram of the extent of the rotation on their experiments on the cat, while Gudmundsson and co-workers worked exclusively with the nerve roots of human beings. The latter group found wide variation among 50 specimens in the amount of rotation between ganglion and pons: usually between 40° and 50°, but occasionally as low as 10° or as high as 80°. The apparent discrepancy between Davis and Haven's statement that the mandibular division fibers are largely superior to the ophthalmic ones and Gudmundsson and co-workers' contention that they are largely caudolateral (varying from almost directly caudal to almost directly lateral) is due to the fact that the former workers used the pons and medulla as a longitudinal axis and the latter used the longitudinal axis of the body.

Temporal Approach to Sensory Root

The temporal approach to the sensory root for section of it in trigeminal neuralgia is usually made largely extradurally, by elevation of the dura from the floor of the middle cranial fossa. In this approach the foramen spinosum and the middle meningeal artery are found by following the concavity of the middle cranial fossa to its deepest part; in order to retract the dura here the middle meningeal artery must be divided. After the dura is freed at the foramen spinosum by section of the artery, the foramen ovale and the mandibular branch of the fifth nerve may be found just anteromedial to the foramen spinosum (Fig. 1-50). The posterior border of the mandibular nerve and of the trigeminal ganglion leads medially and upward to the sensory root, and can be followed by carefully incising the enveloping dura about the posterior edge of the mandibular nerve and ganglion (this, or more commonly the adjacent arachnoid, is sometimes called "dura propria") and freeing it slightly from these structures. As the subarachnoid space of the trigeminal cave is entered cerebrospinal fluid is released, and the sensory root of the nerve is exposed. If the incision of the dura is carried upward to the tentorium cerebelli, as in a decompression operation, the superior petrosal sinus may be encountered. The motor root is not visible; although at the origin of the fifth nerve from the brain stem it lies somewhat anterior to the sensory root, it crosses obliquely below this root in order to reach the mandibular nerve, and as it enters the trigeminal cave lies deep to (medial to and beneath) the sensory root. In sectioning the sensory root, great care should be taken to identify and spare the motor root in order to avoid paralysis of the muscles of mastication on that side; sometimes, however, this is not possible, and the motor root also is sectioned.

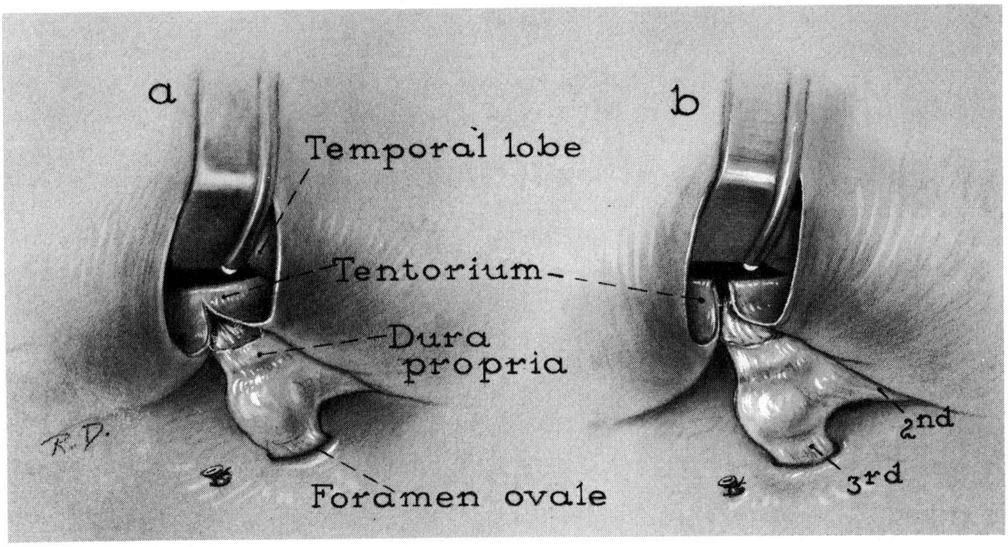

Fig. 1-50. Decompression of the trigeminal roots. The dura has here been dissected off the mandibular nerve and the posterior part of the ganglion, and the dura and arachnoid (the latter identified as "dura propria") have been incised over the roots. (Dodge HW Jr, Love, JG: J Am Geriatr Soc 2:467, 1954)

A rare anomaly is the presence of a persistent trigeminal artery deep to the ganglion and sensory root.

Since the subarachnoid space extends very little, if at all, over the anterolateral or outer surface of the trigeminal ganglion, the dura is largely fused to this surface. Lockhart found that the dura is, however, easily freed from the ganglion and advocated an intradural approach, with incision of the dura over the lateral aspect of the ganglion. On its posteromedial or deep surface, and especially along its posterolateral or mandibular border, the ganglion is in contact with the subarachnoid space in at least its proximal two thirds and is fused with the dura only in its distal third; thus the extradural approach allows freer access to the subarachnoid cavity as it surrounds the roots of the nerve, and carries less danger of inadvertently opening the cavernous sinus.

If the dura along the posterior border of the trigeminal ganglion is elevated from the base of the skull, the greater petrosal nerve is necessarily elevated also; this nerve passes forward and somewhat medially to disappear beneath the deep surface of the trigeminal ganglion about halfway between the foramen ovale and the petrous ridge, and leaves the middle fossa through the foramen lacerum.

A transient paralysis of the facial nerve sometimes occurs following section of the sensory root of the trigeminal nerve by the temporal route; this usually has been thought to be due to damage to the facial nerve at its geniculum (external genu) through traction upon the greater petrosal nerve; Grant expressed the belief that it might be due either to this or to rupture of the petrosal vein (Dandy, '29), which leaves the cerebellum and crosses the subarachnoid space to join the superior petrosal sinus, usually anterior to the root of the fifth nerve, but sometimes upon it. It is difficult for the present writer to believe that tension upon the dura containing the greater petrosal nerve can usually be transmitted through the bony canal of this nerve to the main stem of the facial with sufficient force to injure the latter nerve. However, Dandy ('29) noted that the geniculate ganglion of the facial nerve may lack a bony covering and hence be in direct contact with the dura, and this is not particularly uncommon (Rhoton, Pulec and co-workers). Such exposure of the ganglion, or a very short canal of the greater petrosal nerve, may thus explain some cases of facial paralysis. Ischemia following division of the middle meningeal artery, whose petrosal branch follows the greater petrosal nerve toward the ganglion and helps supply the facial nerve, is an unlikely possibility in view of the overlapping blood supply of this nerve (*e.g.*, Blunt), and it seems that a truly satisfactory explanation of the paralysis has not yet been advanced.

The anesthesia accompanying section of the sensory root of the fifth nerve is to some patients very annoying, and if it involves the first division of the nerve, care has to be taken to prevent ulceration of the cornea. Taarnhøj reported that trigeminal neuralgia could be abolished by simply removing the dura from over the posterior part of the trigeminal ganglion and sectioning the dural roof of the trigeminal cave and the attached edge of the tentorium—thus decompressing this part of the ganglion and the sensory root as it crosses the petrous ridge. Although several workers at first had apparent success with this operation, the relief from pain has very frequently proved to be temporary only, and many surgeons have therefore abandoned it.

Shelden and colleagues concluded that the success of decompression depended upon inadvertent damage to the ganglion, and they described purposeful slight damage to the ganglion with a blunt instrument (compression) as a more reliable method than decompression in abolishing pain, yet produces relatively little general sensory loss. This technique has proved so effective that some surgeons have employed it as a primary procedure, to be supplemented by root section in case of recurrence. Stender and Grumme concluded that reoperation may be required for about a third of the patients, but with a marked postoperative sensory defect, recurrence of pain should occur in only about 20%.

Pertinent anatomical points to be considered in incising and reflecting the dura over the trigeminal ganglion and roots are the possibility of opening the cavernous sinus if the reflection is carried too far forward; the probability of encountering the superior petrosal sinus if the incision is carried over the petrous ridge; and, in the latter instance, the possibility that the trochlear nerve, which enters the tentorium cerebelli anterior to the entrance to the cave, may cling to the lower surface of the tentorium and be injured during sectioning of that.

Cerebellar Approach and Section of Spinal Tract

Hyndman regarded the cerebellar approach to the root of the fifth nerve as more dependable than the temporal, but more difficult technically; this approach is briefly discussed later. The pain of trigeminal neuralgia can also be abolished by section of the spinal tract of the fifth nerve in the medulla, as originally described by Sjöqvist and more recently discussed by Grant and colleagues, Grant and Weinberger, Weinberger and Grant, and Falconer. This operation is not so commonly attempted as section of the root and is undoubtedly more difficult technically, but it offers the advanteage of differential section of the pain fibers of the fifth nerve without causing anesthesia. Trigeminal tractotomy may also produce analgesia over the areas of distribution of the seventh, ninth, and tenth nerves (Brodal) and therefore can be used in neuralgias of these nerves.

Other Techniques

Other approaches for sectioning the trigeminal root close to the pons include the retrolabyrinthine one used by Pulec and Hitselberger and the subtemporal transtentorial approach used by Janetta.

Sweet and Wepsic, and Tew and Mayfield, noting that alcohol injections produce a high incidence of anthesia over the first division, inserted an electrode into the ganglion through the foramen ovale; they were thus able to destroy pain fibers while leaving intact many fibers for touch.

Posterior Cranial Fossa

The posterior cranial fossa, roofed by the tentorium cerebelli, contains the midbrain, pontine region, medulla, and cerebellum, and therefore the attachments to the brain stem of all except the first two cranial nerves. Of these, the third is the only one not entering the dura in the posterior fossa; instead, it enters the dura of the middle fossa just anterolateral to the posterior clinoid process.

Although they arise in the posterior cranial fossa, the third, fourth, and sixth nerves all traverse the cavernous sinus and leave the cranial cavity, to enter the orbit, through the superior orbital fissure. The fifth nerve likewise, although arising in the posterior cranial fossa, does not leave the cranium through this fossa; its roots pass anterolaterally, beneath the edge of the tentorium cerebelli and across the petrous ridge, to enter the middle cranial fossa, in which the trigeminal ganglion lies and through the walls of which the three major divisions of the fifth nerve leave the cranial cavity.

The posterior cranial fossa is usually approached posteriorly, through the occipital bone; this approach is referred to as the suboccipital or cerebellar approach. The upper surface of the cerebellum, the midbrain, and some of the higher cranial nerves in the fossa may also be approached by the temporal route—after elevation of the temporal and occipital lobes from the upper surface of the tentorium cerebelli, the tentorium may be cut close to its attachment to the petrous ridge and reflected.

The skin of the occipital region is supplied primarily by the greater occipital nerve, supplemented by the third occipital nerve and more laterally, toward the mastoid region, by the lesser occipital; the remainder of the skin over the mastoid region and behind the ear is supplied by the great auricular and the auricular branches of the seventh, ninth, and tenth nerves. Transverse incisions, or incisions that curve laterally from the midline, necessarily transect at least the greater occipital nerve, which supplies the scalp about as far as the interauricular line. Adson ('41b) sometimes used a straight incision some 3 cm from and almost paralleling the midline, therefore somewhat between the distributions of the greater and lesser occipital nerves.

OCCIPITAL MUSCULATURE

The occipital belly of the occipitofrontalis muscle arises from the supreme nuchal line on the occipital bone, thus a little above the level of the external occipital protuberance. Here, therefore, the galea aponeurotica is firmly attached to the skull.

Laterally, upon the superior nuchal line and the posterolateral aspect of the mastoid process, the *sternocleidomastoid* attaches (Fig. 1-51). Deep to it on the superior nuchal line is the attachment of the *splenius capitis.* Deep to the lateral border of the splenius capitis, attaching not to the occipital bone but only to the mastoid process, is the *longissimus capitis* muscle.

Medially, on either side of the midline, the *trapezius* muscle typically arises from the skull along the superior nuchal line; sometimes the trapezius fails to reach the skull. In any event, there is a gap, of variable size, beteween the trapezius and the medial or posterior border of the sternocleidomastoid. Through this gap (the upper part of the posterior triangle) the occipital artery usually becomes subcutaneous and turns upward superficial to the trapezius. The greater occipital nerve appears medial to the artery and crosses it superficially to run lateral to it into the scalp.

Deep to the trapezius is the heavy *semispinalis capitis,* attached to the skull between the superior and inferior nuchal lines. The greater occipital nerve emerges through it, and the ligamentum nuchae separates the two muscles. Somewhat under cover of the semispinalis, attached to the lateral portion of the inferior nuchal line, is the *obliquus capitis superior* (the obliquus inferior does not reach the skull, but inserts on the transverse process of the atlas).

In its course upward and backward the occipital artery may pass either superficial or deep to the longissimus capitis, but regularly lies superficial to the upper end of the obliquus superior, or entirely cranial to it, and passes around the lateral border of the semispinalis capitis. It runs almost transversely across the superficial surface of the

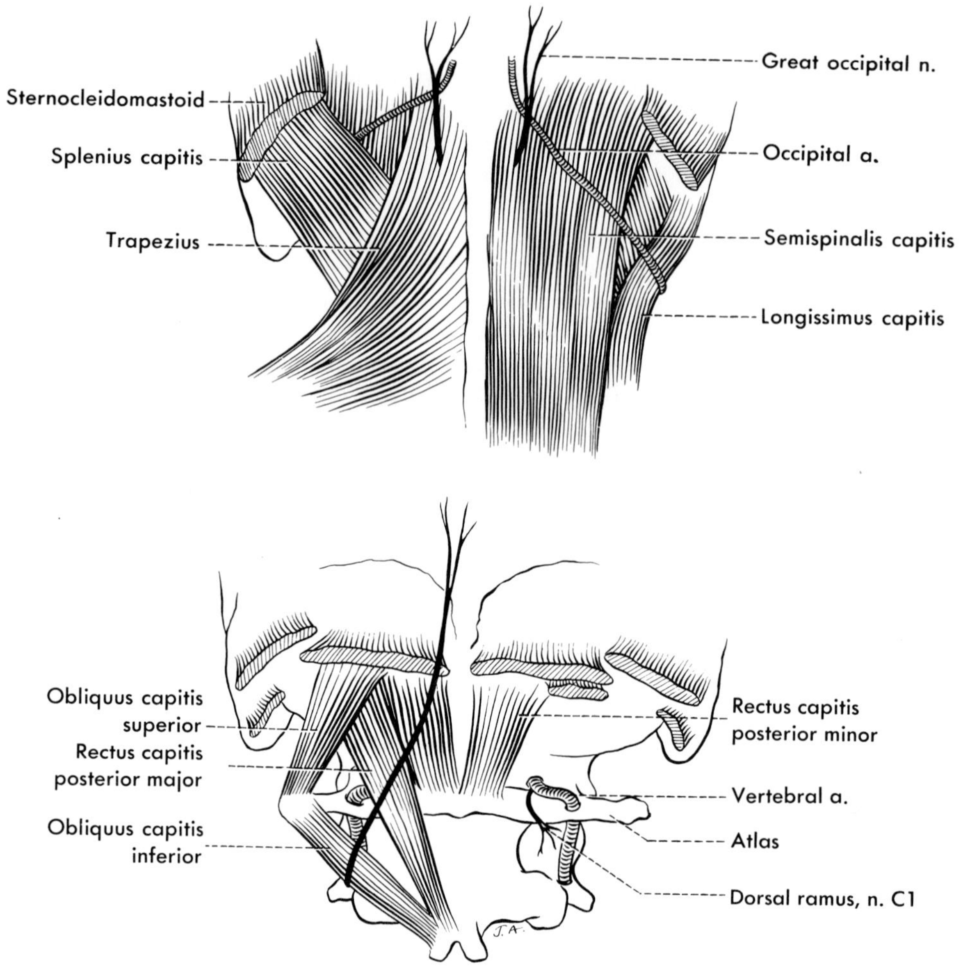

Fig. 1-51. The chief muscles, nerves, and vessels of the occipital region.

semispinalis, just below the attachment of the muscle to the bone, until it makes its turn upward into the scalp.

Below the inferior nuchal line are the attachments of the two other suboccipital muscles that reach the skull—the *rectus capitis posterior minor* and the *rectus capitis posterior major.* The rectus capitis posterior minor attaches on either side of the midline under cover of the semispinalis capitis and the medial edge of the rectus capitis posterior major. The rectus capitis posterior major attaches more laterally, extending to the attachment of the obliquus superior and with the two obliques forms the sides of the suboccipital triangle. In this triangle are the suboccipital (the dorsal ramus of the first cervical) nerve and a portion of the vertebral artery.

It is usually necessary to reflect most of these muscles from their attachments to the skull, in the suboccipital approach, and the more medial muscles must be reflected if an exposure through the posterior arch of the atlas is planned. If the scalp is not reflected from these muscles, and the latter are detached at their insertions on the bone, the base of the skull and the posterior arch of the atlas can be uncovered without sacrificing the nerve and blood supply to the scalp.

The posterior atlantooccipital membrane stretches from the upper border of the posterior arch of the atlas to the posterior border of the foramen magnum and is pierced by both the vertebral artery and the suboccipital nerve. The vertebral artery here at first runs medially and backward in the groove on the upper border of the posterior arch of the atlas, being covered by the oblique muscles, and then turns forward to enter the vertebral canal. The suboccipital nerve typically emerges between the artery and the arch of the atlas, but many occasionally pierce the atlantooccipital membrane above the artery.

CRANIAL FLOOR AND DURA

The central part of the floor of the posterior fossa, the clivus, is made up of the sphenoid and occipital bones and slopes backward and downward from the dorsum sellae to the foramen magnum in the occipital bone (see Fig. 1-42). The hypoglossal canals open above and lateral to the foramen magnum, and farther laterally the jugular foramen lies between the occipital bone and the petrous part of the temporal. Leading to the anterior border of the foramen is the sulcus for the inferior petrosal sinus, and to the posterolateral border the sulcus for the sigmoid sinus. On the posterior surface of the pars petrosa, above the sigmoid sulcus at about the point where it is crossed by the suture between temporal and occipital bones, is a smooth flattened or slightly concave area that accommodates the endolymphatic sac; at the upper border of this area, opening downward, is a small slit representing the external aperture of the vestibular aqueduct. This aqueduct transmits the endolymphatic duct and, according to most reports, a vein from the internal ear. Farther forward on this surface of the pars petrosa is the opening (porus) of the internal acoustic meatus. A slight depression, the subarcuate fossa, just above and posterior to the porus is usually inconspicuous unless it is marked by one or more prominent foramina for veins and a tiny branch of the labyrinthine artery.

The tentorium cerebelli, attached to the superior border of the pars petrosa and to the edges of the sulcus for the transverse sinuses, forms the roof of the posterior cranial fossa.

The floor is frequently regarded as being especially sturdy, but is actually very thin in places, and fractures of it are apt to be overlooked although they may have serious consequences. Even severe fractures of the posterior cranial fossa may not be easily recognized (unlike those of the middle and anterior cranial fossae, which are likely to lead to discharge through the nose and ear respectively), both because of the mass of muscle covering this region and because blood and fluid in the posterior cranial fossa has no ready passageway into which it can escape. Sometimes, however, there is ecchymosis over the mastoid process (Battle's sign). Fractures of the body of the sphenoid bone may cause paralysis of the abducens nerve, which was a relatively long course of contact with this bone.

The dura of the posterior cranial fossa is firmly attached at the foramen magnum, at which level it splits into its definitive two layers—the periosteal layer that is reflected onto the outer surface of the skull and the dural layer proper that is continued downward as the dura about the spinal cord. It is also strongly attached to the basioccipital region, the basal portion of the sphenoid bone, and the jugular foramen. It is not, however, as firmly attached as one might suppose about the venous sinuses, and it is possible to dissect the occipital bone away from the posterior surface of the transverse sinus without injuring this vessel, just as in mastoidectomy the sigmoid sinus may be freed of bone.

The dura generally follows the conformation of the skull, but extending downward from the lower surface of the tentorium cerebelli at its posterior attachment to the midline is a usually small sickle-shaped fold, the falx cerebelli; this protrudes slightly forward between the two cerebellar hemispheres and contains the occipital sinus. Hasan and Das found the falx duplicated in 76 of their 100 specimens, and there were more than two in four specimens.

Arteries

The dura of the posterior cranial fossa, below the tentorium, receives some branches from the middle meningeal artery but is supplied largely by variable small branches from the ascending pharyngeal, the vertebral and the occipital arteries, none of which is of surgical importance. A fairly constant posterior meningeal artery from the ascending pharyngeal enters the cranium through the jugular foramen; smaller twigs of this artery may pass through the hypoglossal canal. The usually tiny meningeal branches from the occipital artery may variously enter through the jugular foramen, the hypoglossal canal, the condylar canal, or the mastoid and parietal foramina, thus frequently accompanying emissary veins.

The artery from the mastoid branch of the occipital, traversing the skull with the mastoid emissary vein, is frequently the only one of these vessels to attain considerable size; it supplies the more lateral and posterior portion of the dura, entering at the posterior margin of the sigmoid sinus.

The meningeal branch of the vertebral artery usually arises from that vessel as it pierces the dura mater and anastomoses with other branches in the posterior fossa. The arteries to the dura of the posterior fossa vary considerably in size, and any one of the several vessels in this region may predominate.

Innervation of the Dura

Ray and Wolff found the dura on the lower surface of the tentorium to be insensitive to pain unless the venous sinuses were approached, but the dura of the floor of the posterior fossa, on the other hand, was found to be uniformly sensitive like the floors of the other fossae. In one of their patients, pain produced by stimulation of the more lateral portion of the dura of the floor of the posterior fossa, and referred to an area behind the ear, was abolished by section of the ninth and tenth cranial nerves. Pain from the more medial part of the dura usually was referred to a point low in the back of the head and was abolished in one patient by section of the posterior roots of the first three cervical nerves.

The two major nerves to the dura of the posterior cranial fossa are the meningeal (recurrent) branches of the vagus and hypoglossal nerves (Fig. 1-52). Since the hypoglossal nerve contains neither sensory nor autonomic fibers in its roots, the fibers of its meningeal branch must be derived through its connections with other nerves. The meningeal branch of the hypoglossal is believed to contain fibers derived both from the superior cervical ganglion and from upper cervical nerves. Although the meningeal branch of the vagus has generally been regarded as truly vagal, Kimmel ('61) found no vagal fibers in it in fetuses; like the hypoglossal branch, it was composed exclusively of fibers from the first two cervical nerves and the superior cer-

Fig. 1-52. The nerves of the posterior cranial fossa and cranial vault. (After Penfield W, McNaughton F: Arch Neurol Psychiatr 44:43, 1940)

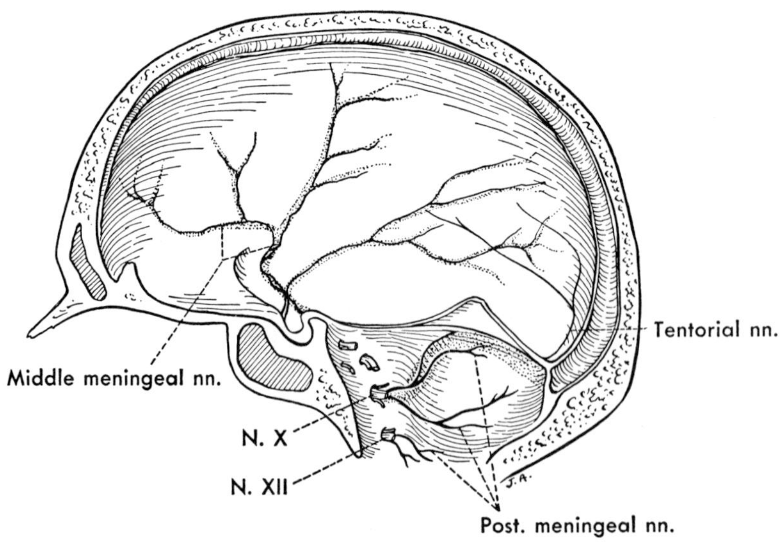

vical ganglion. Kimmel also found meningeal twigs from the first three cervical nerves, especially C2, that ascended through the foramen magnum and were distributed to the dura of the clivus.

VENOUS SINUSES OF POSTERIOR CRANIAL FOSSA

The posterior cranial fossa is largely bounded by sinuses that lie at the attachment of the tentorium cerebelli; the sigmoid sinuses, the inferior petrosal sinuses, the occipital sinus, and the basilar plexus all lie within the posterior cranial fossa. Knowlege of the exact positions of the sinuses related to the posterior cranial fossa is obviously useful, but unfortunately the more important of these tend to vary somewhat, and marked anomalies are sometimes encountered.

The *transverse sinuses* extend from the confluence of the sinuses (torcular), close to the internal occipital protuberance, to the posterior ends of the petrous ridges; each sinus lies between the layers of the tentorium cerebelli where this attaches to the dura. The sigmoid sinus, the continuation of the transverse, curves downward in a deep groove on the posteromedial aspect of the petrous ridge, to run forward and medially along the junction of the petrous ridge and the occipital bone and empty through the jugular foramen. As the sinus passes through the jugular foramen it is usually separated from the inferior petrosal sinus by the emerging rootlets of the ninth to eleventh cranial nerves.

The usual relations of the lateral sinuses to external landmarks on the skull have been described previously; however, it should be noted that there are variations in this regard. As the superior sagittal sinus reaches the confluence, it frequently does not remain in the midline but deviates to the right; the right transverse sinus is usually not only larger than the left transverse sinus, but its upper border may also lie almost 1 cm higher, while the left sinus may dip slightly downward in the posterior fossa, especially near the midline. Further, the internal occipital protuberance, near which the confluence lies, does not correspond exactly in position to the external occipital protuberance, and the relation of the external occipital protuberance to the beginning of the transverse sinus varies. Thus, although the transverse sinus is usually said to be marked on the external surface of the skull by a slightly curved line running from above the external occipital protuberance to the posterior border of the ear, the lower border of the sinus may actually lie 2 cm or more below the level of the external occipital protuberance. Removal of bone from the posterior cranial fossa must therefore be done carefully; the upper limit at which the transverse sinus will be uncovered cannot be surely known beforehand. The variations at the confluence of the sinuses, already discussed, do not as a rule markedly affect the carrying out of surgical procedures in the posterior fossa.

The *superior petrosal sinus* has already been described as running along the edge of the petrous ridge, in the attachment of the tentorium cerebelli to this ridge; although it lies at the boundary between the posterior and middle cranial fossae, it is usually not of surgical importance; however, as already mentioned, it will frequently have to be divided in the operation of decompression of the trigeminal ganglion and sensory root.

Each of the paired *inferior petrosal sinuses* leaves the posterior aspect of the cavernous sinus and runs along the groove between the petrous temporal laterally and the basal portions of the sphenoid and occipital bones medially—that is, on the lateral aspect of the floor of the narrower upper portion of the posterior cranial fossa—and swings laterally to pass through the anterior compartment of the jugular foramen and empty, typically, into the superior bulb of the internal jugular vein. After its exit at the jugular foramen, the inferior petrosal sinus usually lies between the ninth and tenth cranial nerves. The two inferior petrosal sinuses communicate with each other across the floor of the posterior cranial fossa through the ill-defined basilar plexus.

The usually small *occipital sinus* runs in the sagittal plane in the attachment of the falx cerebelli, from the foramen magnum to the

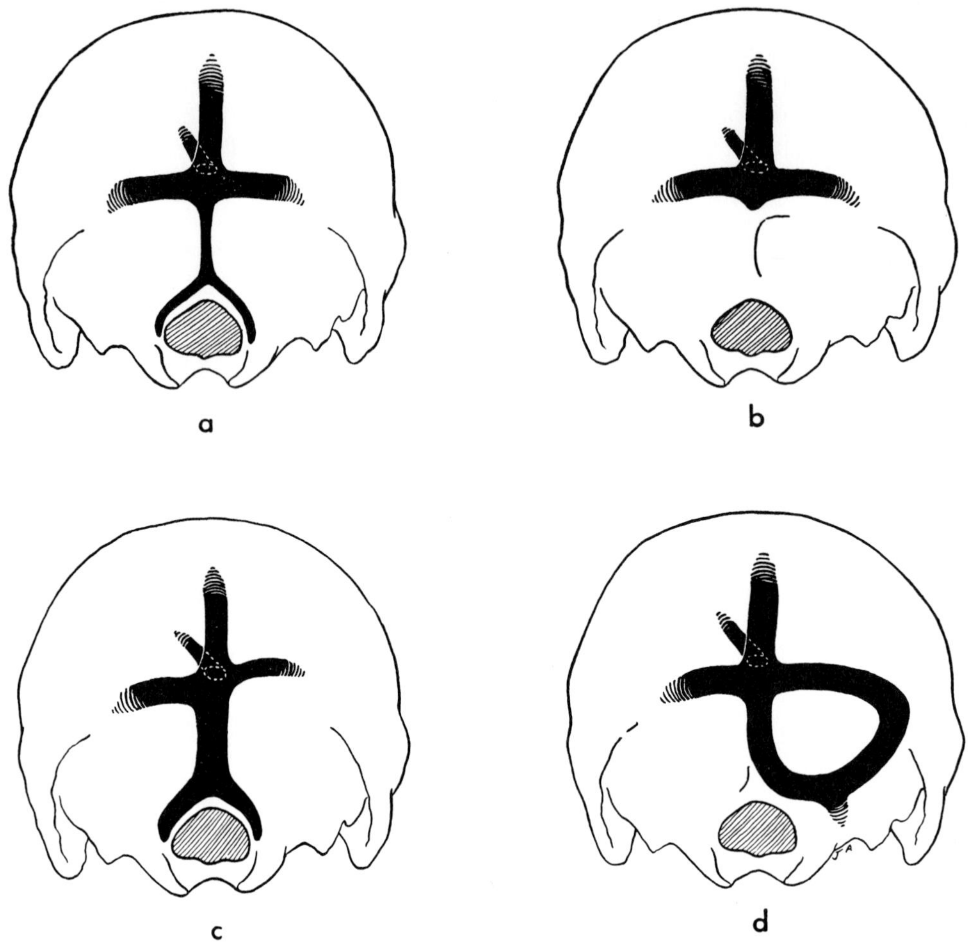

Fig. 1-53. Variations in the occipital sinus.

confluence of the sinuses. It varies markedly in its size, however, and may have an aberrant course (Fig. 1-53); it is therefore always a potential source of difficulty in posterior approaches to the fossa. Although this sinus can usually be ligated and divided with ease in a bilateral or midline approach to the contents of the posterior cranial fossa, it is sometimes quite large and may even replace one of the sigmoid sinuses. (Das and Hasan found a single midline sinus in only 35% of 200 specimens and no sinus in 35.5%.) It may deviate to one side to join the sigmoid sinus as this passes through the jugular foramen; if such a deviated occipital sinus is large (and this writer has seen one 5 mm in diameter) it may offer a definite hazard in approaches to the posterior fossa.

CRANIAL NERVES AND ARTERIAL RELATIONS IN THE POSTERIOR FOSSA

ACCESSORY, VAGUS, AND GLOSSOPHARYNGEAL NERVES

Reflection of the dura of the posterior fossa allows examination of the cerebellum and of the pontine and medullary portions of the brain stem with the roots of the cranial nerves arising laterally from these (Fig. 1-54). The cerebellar vermis marks the midline, while the hemispheres largely fill the posterior and lateral, therefore the greater, portions of the posterior fossa and must be retracted to reveal the roots of the cranial nerves and the brain stem itself. Elevation of the cerebellum in the midline allows examination of the caudal portion of the roof of the fourth ventricle, with its contained choroid plexus, and, at the inferior end of this, the median aperture of the fourth ventricle (foramen of Magendie). The boundaries of the fourth ventricle as seen from this aspect are the tubercles of the posterior column nuclei and, farther upward, the inferior cerebellar peduncle (restiform body). The posterior inferior cerebellar artery, descending along the lateral aspect of the medulla, curves onto the posterior surface of the medulla at about the level of the apex of the fourth ventricle (obex) and usually runs again upward between the cerebellum and the posterolateral aspect of the medulla before branching on the inferior surface of the cerebellum.

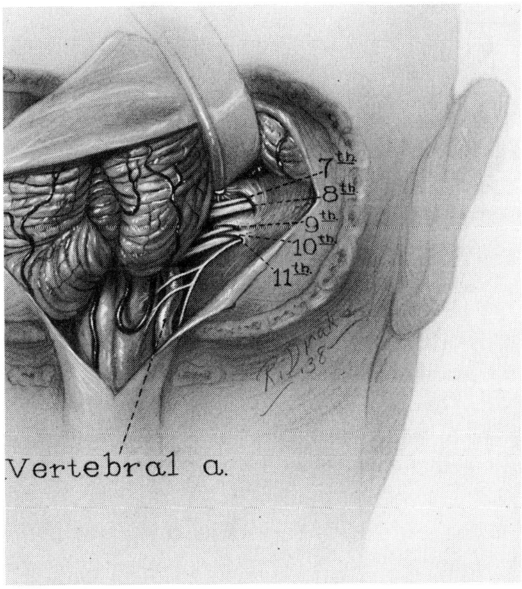

Fig. 1-54. The seventh through the eleventh cranial nerves as they leave the posterior cranial fossa. An enlarged vertebral artery here bulges laterally against them. (Craig WMcK: Am Heart J 17:40, 1939)

If the cerebellar hemisphere is lifted, the roots of most of the more caudal cranial nerves can be examined from their origin to their disappearance into the foramina by which they leave the cranial cavity. At the foramen magnum the uppermost denticulation of the *denticulate ligament* attaches to the dura and is crossed dorsally by the ascending spinal root of the accessory nerve; ventral or anterior to both the denticulate ligament and the spinal root of the accessory nerve, and at first usually lateral to the nerve, the vertebral artery runs upward and somewhat medially, disappearing anterior to the medulla as it converges with its fellow of the opposite side (Fig. 1-55).

As the spinal root of the accessory runs upward, diverging somewhat laterally toward the jugular foramen, it receives from the side of the medulla additional rootlets, the cranial (bulbar) rootlets of the accessory nerve. Farther upward, arising from the side of the medulla in line with the cranial rootlets of the accessory, are the rootlets of origin of the vagus, and above this the one or more rootlets of the glossopharyngeal nerve.

All these nerve roots arise along the same line on the posterolateral aspect of the medulla and largely just dorsal to the olive; they

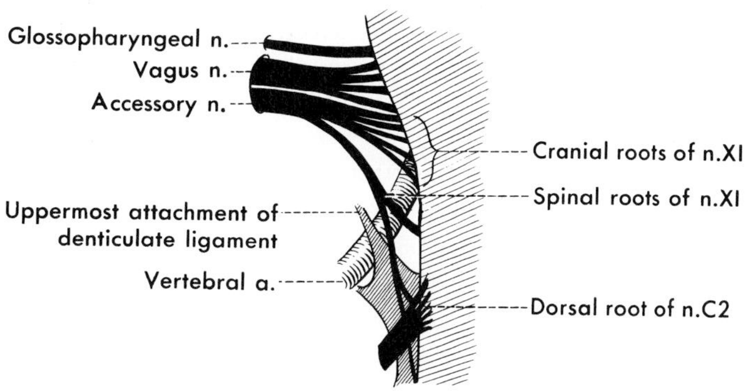

Fig. 1-55. Roots of origin of the ninth, tenth, and eleventh cranial nerves in a posterior view.

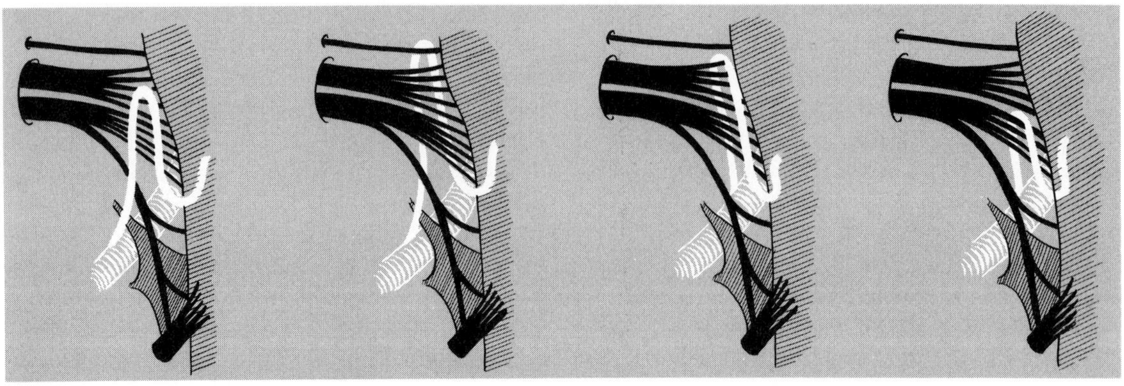

Fig. 1-56. Some variations in the relationship between the posterior inferior cerebellar artery and the roots of the ninth, tenth, and eleventh cranial nerves.

form a series of rootlets that cannot be definitely assigned to any of the three cranial nerves to which they contribute until they are traced far enough laterally toward the jugular foramen to see the formation of the definitive nerves.

Some of the roots of the ninth, tenth, and eleventh nerves are necessarily closely related to the *posterior inferior cerebellar artery.* This artery arises from the vertebral, sometimes soon after that vessel has entered the foramen magnum and sometimes close to the formation of the basilar artery, therefore near the caudal border of the pons. Unless its origin is fairly high, the vessel usually ascends alongside the medulla and in this part may lie either anterior or posterior to the roots of the ninth, tenth, and eleventh nerves (Fig. 1-56); at the end of this ascending course the artery loops dorsally and caudally to descend along the side of the medulla and the inferior surface of the cerebellum, and this loop may lie above the origin of the ninth nerve, pass between the ninth and tenth, between the rootlets of the tenth, or between the rootlets of the tenth and eleventh. The most that can be said, therefore, concerning the relationship of the posterior inferior cerebellar to these nerve roots is that its loop, ascending and descending alongside the medulla, may lie largely ventral to the nerves, may lie largely dorsal to them, or may lie partly ventral and partly dorsal.

It has long been recognized that many of the *voluntary motor fibers* distributed in the vagus—that is, especially the fibers to the muscles of the pharynx and larynx—reach this nerve by way of the cranial rootlets and internal ramus (accessory branch to vagus) of the accessory nerve. The fibers of the cranial root arise from vagal nuclei and are distributed with the vagus, being associated with the accessory for only a few millimeters—they join the spinal root of the accessory nerve within the cranial cavity, but as the accessory and vagus nerves pass through the jugular foramen the spinal and cranial portions of the accessory nerve separate again (Fig. 1-57); the cranial root forms the internal ramus of the accessory and joins the vagus, while the spinal root forms the external ramus of the accessory, distributed to the sternocleidomastoid and trapezius muscles. The cranial rootlets of the accessory nerve are therefore more properly regarded as inferior vagal rootlets.

Accessory Nerve

The spinal portion alone of the accessory nerve thus corresponds to the extracranial accessory nerve; although this nerve is joined outside the skull by afferent fibers from cervical nerves (for proprioception, but apparently not for pain, according to Windle and DeLozier), the spinal portion of the nerve is entirely voluntary motor and is distributed exclusively to the sternocleidomastoid and trapezius muscles. It can be sectioned intracranially, as for torticollis, below the level at which it is joined by the cranial root, to produce an isolated paralysis of these two muscles.

This part of the nerve arises by a series of rootlets from the lateral aspect of the spinal cord; among 40 nerves investigated Holl found the lowermost rootlets to vary from the level of C3 to that of C7, but more commonly lying at C5 or C6. The fibers after their emergence turn and run upward between the dorsal and ventral roots of the spinal nerves, being joined at successively higher levels by rootlets from these levels (and, in 68% of the 50 sides investigated by Ouaknine and Nathan, by some or all of the dorsal root of the first cervical nerve). Passing upward in this position in the subarachnoid cavity, and lying dorsal to the denticulate ligament, the spinal portion of the accessory nerve crosses dorsal (posterior) to the vertebral artery, to enter the cranial cavity through the foramen magnum. Alongside the medulla it usually lies lateral to the posterior inferior cerebellar artery, although varying in its anteroposterior relationship to this artery, and is joined here by the cranial root of the nerve.

Vagus Nerve

The vagus arises from the lateral surface of the medulla, between the inferior cerebellar peduncle and the olive, as a series of rootlets in line with the more cranially placed roots of the glossopharyngeal nerve and the more caudally placed ones of the cranial portion of the accessory nerve. Foley and DuBois found that the sensory fibers in the vagus nerve (of the cat) constitute from 65% to 80% of the total fibers, while the motor fibers number from 20% to 35% of the total. Many of the voluntary motor fibers in the vagus, however, as already stated, leave the brain stem by way of the cranial root of the accessory rather than by way of the vagus itself, and Tarlov ('40) has described the main cephalic portion of the vagus–accessory complex as primarily sensory. Delicate motor roots lie ventromedial to the large cephalic sensory roots, but the caudal portion of the vagus–cranial accessory complex is chiefly motor.

In a subsequent paper Tarlov ('42) reported that after the section of the cephalic third of these rootlets no motor changes in the voluntary muscles could be demonstrated, and sensation of the posterior pharyngeal wall and soft palate appeared normal; there was, however, absence of sensitivity to pain over the side of the epiglottis and diminished sensitivity to pain in the lower portion of the pharynx and in the larynx. Subsequent histological examination of the sectioned nerve roots again indicated that the uppermost roots were largely (about 95%) sensory, although the more caudally sectioned ones contained greater numbers of motor fibers. (Uihlein and co-workers found evidence that the upper sensory roots of the vagus are distributed to most of the posterior pharyngeal wall, rather than being limited to the lower part of the pharynx as usually described.)

From their origins the roots of the vagus nerve run laterally and slightly anteriorly to converge to form the trunk of the vagus and pass through the anterior portion of the jugular foramen, separate from the ninth nerve

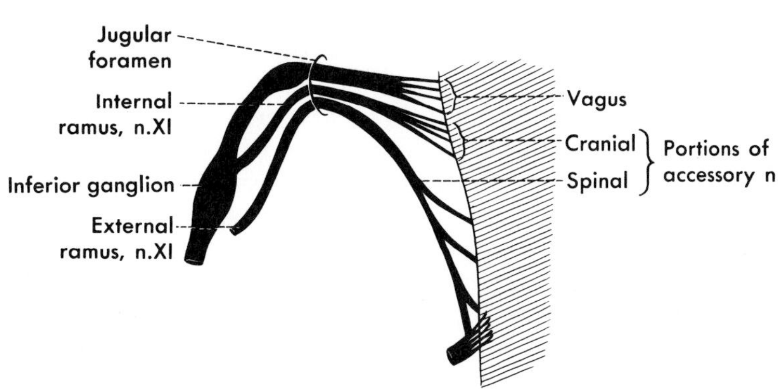

Fig. 1-57. The grouping of the cranial (bulbar) rootlets of the accessory nerve as the internal ramus that joins the vagus, and the continuation of the spinal rootlets as the external ramus, or accessory nerve of gross anatomy.

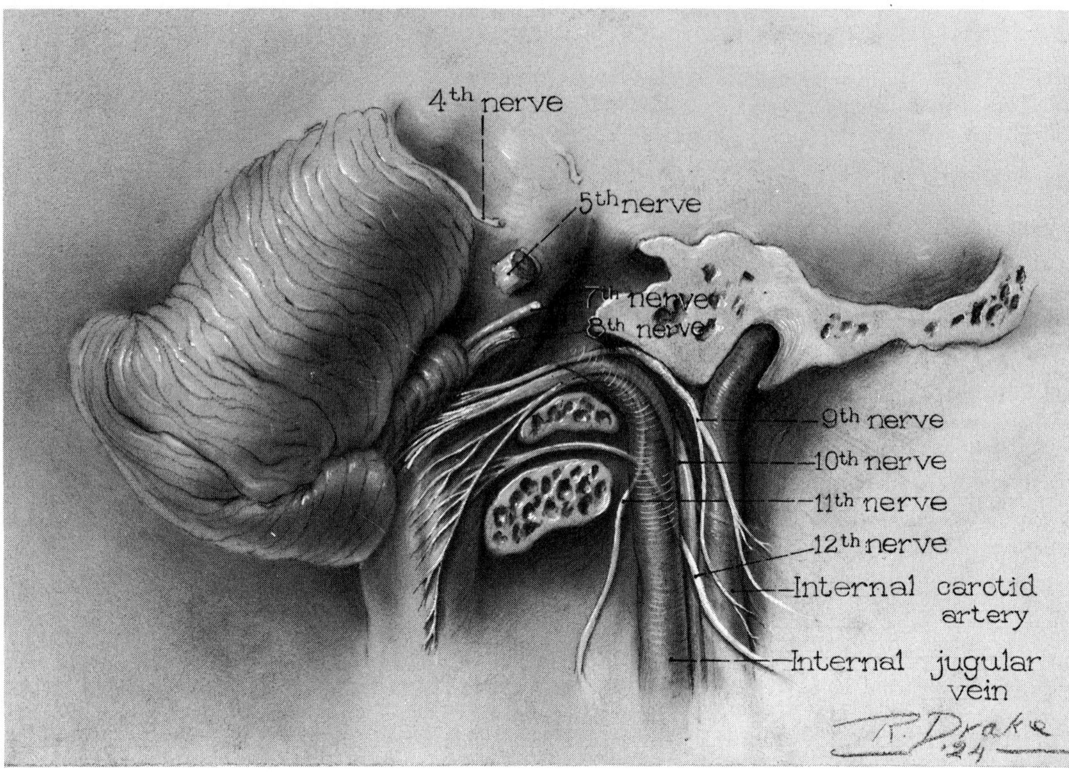

Fig. 1-58. The ninth through the eleventh nerves in the jugular fossa. (Adson AW: Arch Neurol Psychiatr 12:487, 1924)

but closely paralleled by the accessory nerve (Fig. 1-58); the sigmoid sinus passes through the posterior part of the jugular foramen, behind the accessory nerve, to form the superior bulb of the internal jugular vein, and the inferior petrosal sinus passes through the most anterior portion of the jugular foramen. As the vagus passes through the jugular foramen it bears its small superior (once jugular) ganglion; from this arise both the auricular and meningeal branches of the vagus. (Rhoton and Buza found the upper pole of the superior ganglion visible intracranially in 14% of their 50 specimens.) The former branch enters the mastoid canaliculus in the lateral wall of the jugular fossa; the latter runs recurrently through the jugular foramen to be distributed to dura of the posterior cranial fossa.

Glossopharyngeal Nerve

The glossopharyngeal nerve, like the vagus, arises from the lateral aspect of the medulla just caudal to the pons and in the groove between the olive and the inferior cerebellar peduncle. According to Tarlov ('40) it usually presents two roots, a large sensory one and a small ventromedial motor one. These roots form the most superior or cephalic ones of the glossopharyngeal-vagus-accessory complex, represented by the entire series of rootlets in this position. From its origin the glossopharyngeal nerve runs laterally and slightly anteriorly in company with the vagus and accessory nerves and in close contact with the lateral recess of the fourth ventricle; it usually comes in contact with the posterior inferior cerebellar artery and makes its exit through the anterior part of the jugular foramen. Rhoton and Buza found its small superior ganglion visible intracranially in 32% of the specimens they examined. Just outside the foramen the inferior petrosal sinus, commonly the most anterior structure in the foramen, usually passes between the ninth and tenth nerves to enter the superior bulb of the internal jugular vein; however, it varies both in its course (Rhoton and Buza) and its ending (Rand and Hanafee).

According to Partridge the nerve typically lies in a groove in the anterior part of the jugular foramen and is separated from the other structures in the foramen by a band of fibrous tissue that, in 25% of the sides that he investigated, had been converted to osseous tissue. Thus in these cases the glossopharyngeal nerve passed through a bony canal of its own. Regardless of this, the separation between the ninth and tenth nerves as they pierce the dura at the jugular foramen facilitates intracranial section of the ninth nerve without damage to the tenth.

The glossopharyngeal nerve is largely a sensory one, although it contains some motor fibers; according to Foley and Sackett these number about 10% of the total (in the cat). The motor fibers are of two types: voluntary ones to the only voluntary muscle, the stylopharyngeus, supplied by the ninth nerve, and preganglionic parasympathetic fibers to the otic ganglion, for the innervation of the parotid gland. The sensory fibers of the ninth nerve are distributed to the tympanic cavity and the auditory (eustachian) tube, to at least the tonsillar region of the pharynx and the posterior third of the tongue, and to a limited region of skin of the external acoustic meatus and ear.

The sensory distribution of the glossopharyngeal nerve accounts for the distribution of the pain of *glossopharyngeal neuralgia,* which is commonly induced by swallowing and usually starts in the tonsillar region or the base of the tongue, but may extend to the ear or even begin in the ear. The pain is of the paroxysmal type, resembling trigeminal neuralgia, from which it can be distinguished by its distribution and the location of its trigger zone. The present accepted method of treatment is intracranial section of the root of the ninth nerve through a unilateral suboccipital craniotomy (Fig. 1-59). In this operation advantage is usually taken of the fact that, as the ninth and tenth nerves pierce the dura, they do so distinctly separated from each other. In some instances, however, the dura at the jugular foramen and about the ninth and tenth nerve passes so deeply into the foramen that the actual points at which these nerves penetrate the dura cannot be seen. In such a case it may be difficult to determine exactly which rootlets belong to the glossopharyngeal nerve; since, however, the upper vagal rootlets are largely sensory, one or more of these may be divided without untoward damage and indeed are often purposely divided.

The glossopharyngeal nerve has been sectioned extracranially for glossopharyngeal neuralgia, but the surgical approach here is difficult because of the close relations of the nerve to the carotid and internal jugular vessels; even though it is possible to section it high enough to interrupt it before it has given off its pharyngeal branch or branches—although they may come off high, only a little outside the jugular fossa—it is impossible to interrupt the branch to the ear by extracranial section, for this arises from the nerve as it traverses the jugular fossa. In such a case as that reported by Erickson, in which the pain was entirely limited to the tympanic branch of the ninth, extracranial section of the nerve would, of course, completely fail to alleviate the symptoms.

Section of the ninth nerve is usually described as producing loss of pain and touch over the fauces, the tonsils, and the pharyngeal wall from the larynx to the auditory (eustachian) tube; loss of the gag reflex; and loss of taste, pain, and touch over the posterior third of the tongue (Dandy, '27; Erickson; and others). However, Uihlein and co-workers, reviewing their cases of section of the nerve for glossopharyngeal neuralgia, found no effect on pharyngeal sensation, other than the relief of pain, in the majority of patients; they attributed loss of touch and of the gag reflex to purposeful section or inadvertent injury to the upper sensory rootlets of the vagus. Their conclusion that the ninth nerve may conduct primarily or only pain from the posterior wall of the oral pharynx may be questionable, but their analysis certainly indicates that the vagus participates in the innervation of this region; even pain fibers from here, they

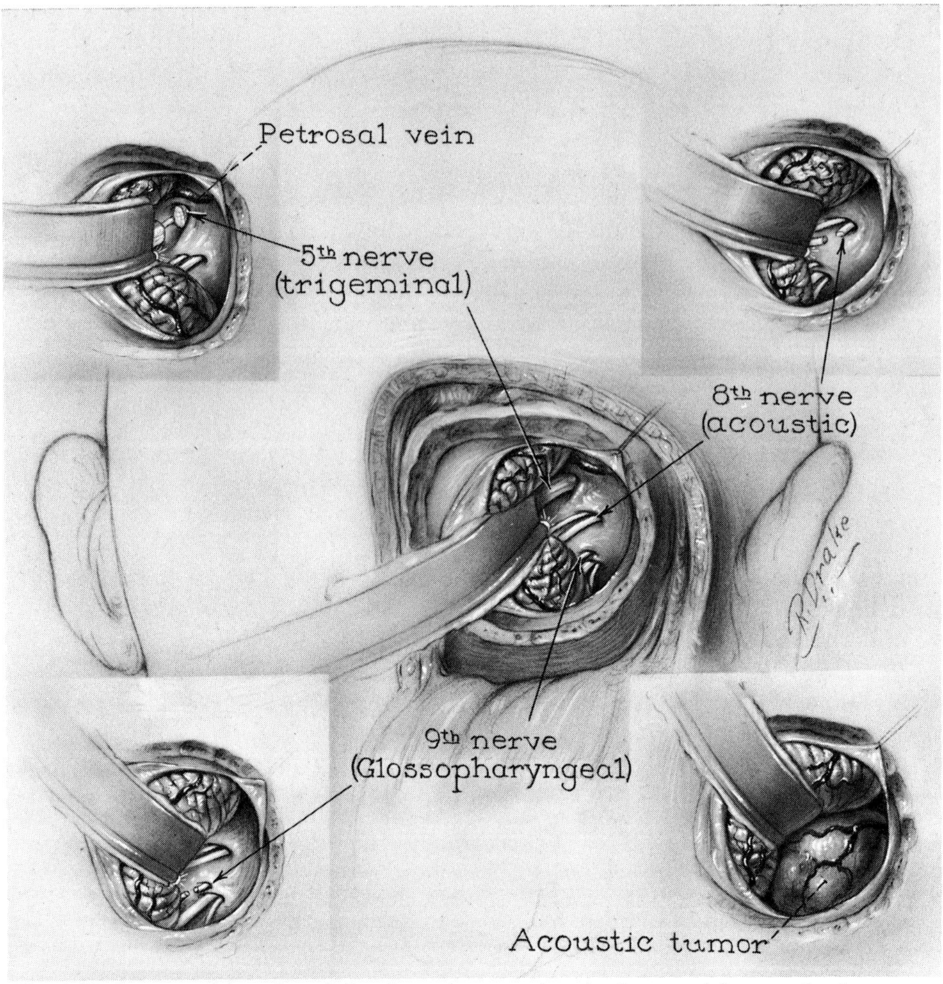

Fig. 1-59. Intracranial section of the glossopharyngeal and other cranial nerves in the posterior fossa.

found, may enter through upper tenth rootlets rather than through the ninth, so that section of these rootlets is sometimes necessary to relieve the pain of "glossopharyngeal" neuralgia.

FACIAL AND EIGHTH NERVES

The seventh (facial) and eighth (vestibulocochlear, previously acoustic or auditory) cranial nerves attach to the lateral surface of the brain stem close to the caudal border of the pons (Fig. 1-60). These two nerves are closely related during their course through the subarachnoid space and the internal acoustic meatus. The eighth nerve lies dorsal and slightly caudal to the facial, which is therefore located somewhat more ventrally and slightly above the eighth. From the cerebellar approach, therefore, the facial nerve may not be visible, being entirely hidden by the eighth. As the nerves leave the medulla they lie close against the cerebellum.

The eighth nerve consists of two portions, which may or may not be grossly identifiable since they frequently form a single bundle; these are the cochlear (auditory) and vestibular parts. The cochlear part constitutes the more dorsal and caudal portion of the nerve and passes dorsally over the inferior cerebellar peduncle (restiform body), while the vestibular part forms the more superior and anterior portion of the nerve and passes into the side of the medulla deep to the peduncle.

The facial nerve also has two roots; the voluntary motor root is large and is situated anterior and cranial to the mixed sensory and parasympathetic root (nervus intermedius, nerve of Wrisberg or glossopalatine nerve), which is a rather small filament lying be tween the eighth nerve and the voluntary portion of the facial nerve, or bound to the eighth nerve for a variable distance, often well into the internal meatus (Rhoton, Kobayaski, and Hollinshead); it may join the facial through small twigs instead of a single stem. These conditions make it impossible to section the nerve in the posterior fossa in some patients, and difficult to be sure that the entire nerve is sectioned in others. In cases of geniculate neuralgia, Pulec has exposed the facial nerve and the geniculate ganglion through the middle fossa and sectioned the nervus intermedius, removed the geniculate ganglion without apparent damage to the motor fibers, or done both.

The seventh and eighth nerves pass laterally and slightly upward to enter the internal acoustic meatus; as they pass across the subarachnoid space and enter the meatus the voluntary portion of the facial tends to lie most anteriorly, the eighth nerve most pos-

Fig. 1-60. Some relations of the seventh and eighth cranial nerves. In (*a*) is shown, somewhat exaggerated spatially, the disposition of the four roots of these nerves at the brain stem and in (*b*), also exaggeratedly spread apart, the relations of the three peripherally distinct components (cochlear and vestibular being ordinarily not separable here) as they might be seen from above. In (*c*), (*d*), and (*e*) are shown variable relations between these nerves and the anterior inferior cerebellar and labyrinthine (internal auditory) arteries as they might appear in a cerebellar approach, from which aspect the eighth nerve commonly hides the seventh.

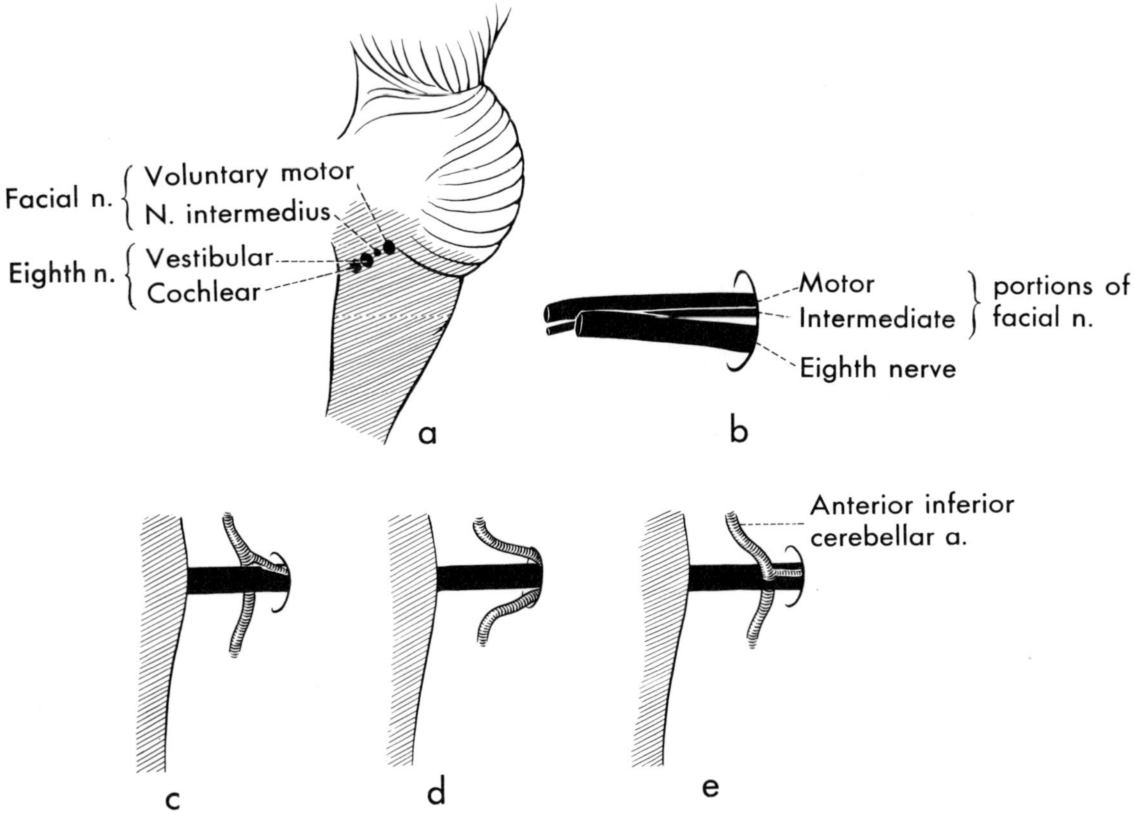

teriorly, with the nervus intermedius between them. Within the meatus the nervus intermedius typically joins the motor portion of the facial, and the common trunk crosses the anterior surface of the eighth nerve to lie above it at the distal end of the meatus.

As the seventh and eighth nerves pass laterally, they come in close but varying relationship to the *anterior inferior cerebellar artery,* which may lie either dorsal or ventral to them, loop around them, or pass between any two of the three divisions (motor portion of the facial, nervus intermedius, eighth nerve— Fig. 1-60*c, d,* and *e*) as noted by Stopford ('16a) and others. Stopford found the artery lying more commonly ventral to the nerves, beteween them and the pons. The artery often loops laterally along these nerves, extending to the entrance of the meatus or even into the meatus. The varying relation of the anterior inferior cerebellar artery is of particular importance in operations at the cerebellopontine angle; Atkinson warned that occlusion of it may involve grave danger to the patient, especially if the posterior inferior cerebellar artery is small.

The seventh and eighth nerves are accompanied into the meatus by the *labyrinthine* (internal auditory) *artery,* which frequently insinuates itself between them; this artery may be a branch of the basilar artery, but is most commonly from the anterior inferior cerebellar artery. It may be doubled.

Distribution of Facial Nerve

The voluntary motor fibers of the facial, constituting the bulk of the nerve, are distributed to all the facial or minetic muscles, including the buccinator, and to the platysma, the stylohyoid, the posterior belly of the digastric, and the tiny stapedius. The nervus intermedius root of the facial nerve contains preganglionic parasympathetic fibers destined for the pterygopalatine (sphenopalatine) and submandibular (submaxillary) ganglia and therefore especially for the glands and blood vessels of the nose and mouth.

This root of the nerve also contains the sensory fibers derived from cells of the geniculate ganglion (usually located in the petrous portion of the temporal bone, at the geniculum or external genu of the nerve, but partially or completely uncovered by bone, thus in contact with the dura, in 15% of cases according to Rhoton and Hall). Some of these fibers are for taste from the anterior two-thirds of the tongue, while others are possibly proprioceptive; still others are concerned with cutaneous sensation and with pain. The further relations of this nerve in the skull are described in connection with the ear, with which it is closely associated in a part of its course, and its branching and distribution are discussed in connection with the face, nose, and tongue.

Distribution of Eighth Nerve

The fibers of both parts of the eighth nerve are distributed to the internal ear, but the two divisions have markedly different functions, since the cochlear part is distributed to the cochlea, while the vestibular part is distributed primarily to the semicircular ducts and the utriculus and sacculus. As discussed in connection with the ear, both divisions of the nerve have been shown to contain, in their final distribution, efferent fibers that arise in the brain stem and leave it through the vestibular division. Section of vestibular fibers close to the brain stem has been carried out for Ménière's disease (endolymphatic hydrops); in such a case the more anterior and superior fibers of the eighth nerve have been cut, and this has been done with relatively little clinical evidence of damage to acoustic functions. Unless the two divisions are rather clearly delineated from each other, which commonly is not true, it is impossible to demonstrate anatomically how much of the nerve is composed of vestibular fibers, and differential section of the vestibular with sparing of the cochlear can be only approximate. Many otolaryngologists favor other methods of dealing with Ménière's disease.

TRIGEMINAL NERVE

The fifth (trigeminal) cranial nerve, like the seventh through the eleventh, attaches laterally to the brain stem, where its fibers penetrate the pons. It arises by two roots, a large sensory and a small motor one (portio major and portio minor). The motor root lies above the sensory one, but in contact with it; as the roots run anterolaterally, the motor root is applied to the upper and anteromedial surface of the sensory root, and is therefore visible with difficulty or not at all from the cerebellar approach. Both roots, closely applied together, pass across the petrous ridge into the narrow mouth of the trigeminal (Meckel's) cave; the further relationships here, in the middle cranial fossa, have been described in a preceding section.

Section of the trigeminal root for neuralgia of this nerve has been carried out through an approach through the posterior fossa, as described by Dandy ('29), and some surgeons have preferred this approach; it presents certain obvious difficulties, especially that of avoiding injury to other cranial nerves in reaching the origin of the fifth. Dandy regarded this as much the superior approach, stating that it obviated the unpleasant sequelae—including facial paralysis—that sometimes occurred after section of the root by the temporal approach. The relationship of motor and sensory roots of this nerve at the brain stem is apparently constant and well known, although it may not be obvious to the surgeon working in this region; the arrangement of the sensory fibers of the various peripheral divisions of the nerve has already been discussed, as has the finding, first reported by Dandy, that even complete section of the sensory root close to the pons does not abolish all sensation on the face, because of the presence of the aberrant sensory rootlets.

Stopford ('16a) called attention to an almost constant and particularly large pontine branch of the basilar artery that runs laterally across the pons to pass out along the ventral surface of the sensory root of the trigeminal nerve. This artery could apparently be a source of troublesome bleeding in section of the root in the posterior fossa. A venous connection to the superior petrosal sinus rather regularly passes along the lateral surface of the nerve, according to Sunderland ('48), and may also be a source of hemorrhage. Presumably this is the vein described by Dandy ('29) as the petrosal vein; he stated that it commonly lay anterior to the nerve root, to which it could serve as a guide, but varied much in size and position (Fig. 1-61). He found no difficulty in occluding and dividing it when it lay against the nerve.

Dandy ('34) described a branch of the superior cerebellar artery as lying either beneath the trigeminal root (that is, between it and the brain stem) or along its lateral surface; he stated that in 30.7% of 215 patients with trigeminal neuralgia on whom he had operated this artery had in some way affected the nerve, either lifting or bending the sensory root or leaving a mark upon it. Larger arteries may also be related to the root in the posterior fossa: Sunderland ('48) described the superior cerebellar, the posterior inferior cerebellar, and even the basilar artery as being sometimes in contact with the trigeminal root, and in three instances found a persistent trigeminal artery accompanying this root. For further discussion see the section on trigeminal neuralgia.

TROCHLEAR NERVE

The trochlear nerve has the longest intracranial course, and the longest course through the subarachnoid cavity, of any of the cranial nerves. It is also the only one arising from the dorsal surface of the brain stem. From its external origin just caudal to the inferior colliculi the threadlike trochlear nerve passes laterally, downward, and forward in the transverse fissure between cerebral and cerebellar hemispheres. It is often described as lying parallel to and between the superior cerebellar and the posterior cerebral arteries, but is actually, of course, separated from the latter by the tentorium cerebelli; it passes lateral to the cerebral peduncles, lying against the lower surface of the tentorium, and enters this dural shelf close to its free border, only a little behind the posterior clinoid process. In its subarachnoid course this nerve is conceivably subject to compression against the tentorium by the superior cerebellar artery; in its further course through the cavernous sinus it is brought in close relation to the other nerves

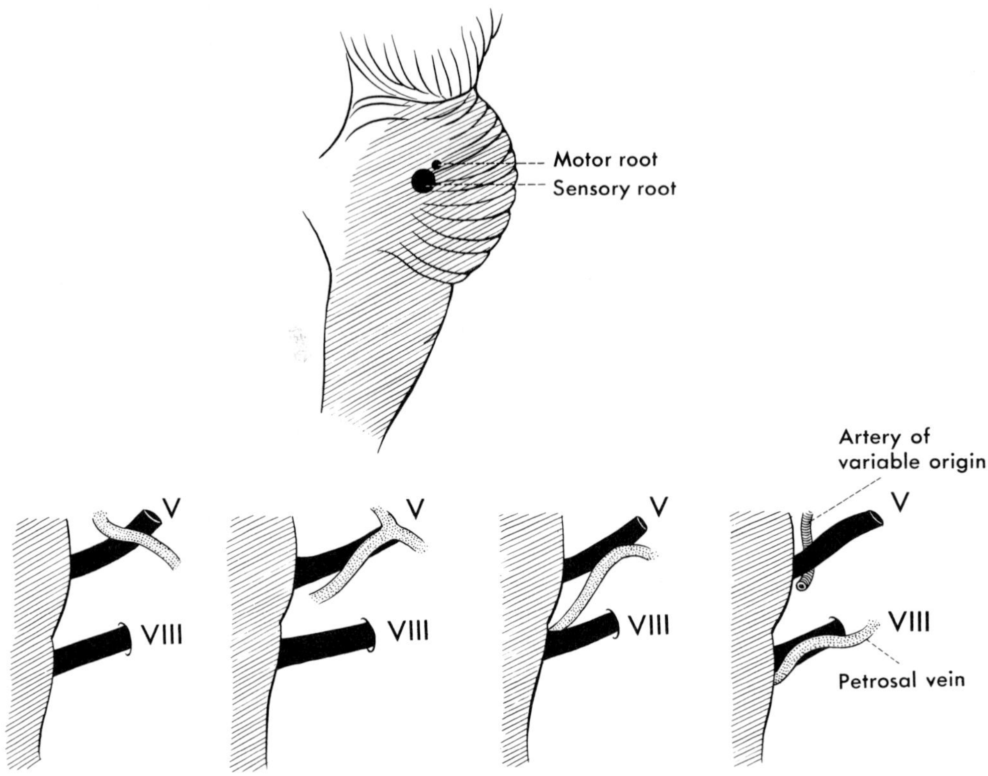

Fig. 1-61. Some relations of the fifth nerve roots in the posterior cranial fossa. In the lateral view of the brain stem the relation between the sensory and motor roots is shown; the remaining sketches, after Dandy (Arch Surg 18:687, 1929), illustrate varying relationships between the fifth nerve root and the petrosal vein, with fifth and eighth nerves identified; in the last sketch an artery is shown related to the fifth nerve (see text).

entering the orbit and is here subject to damage from the same sources as these nerves. This nerve supplies only the superior oblique muscle.

HYPOGLOSSAL, ABDUCENS, AND OCULOMOTOR NERVES

The three nerves arising ventrally from the brain stem, between this and the base of the skull in the posterior fossa, are the twelfth, sixth, and third (hypoglossal, abducens, and oculomotor).

Hypoglossal Nerve

The hypoglossal nerve arises from the ventral surface of the medulla, between the olive and the pyramid, its rootlets being in line with the anterior roots of the spinal nerves. From their origin the approximately dozen filaments of the hypoglossal nerve, which are entirely voluntary motor in function, unite to form two bundles that pass forward and laterad to traverse the hypoglossal canal (anterior condyloid foramen). In or just external to the canal these two trunks unite to form the main hypoglossal trunk; they usually pierce the dura separately and may traverse separate bony canals. Lillie found no sign of subdivision of the bony hypoglossal canal in only about 50% of the specimens that he examined, while in some 34% to 39% there was a spur partially dividing the canal, and in 11% to 17% there was complete division of the canal into two parts.

The two converging vertebral arteries usually lie medial and ventral to the hypoglossal nerves, and if they are tortuous and deviated laterally may produce pressure upon these nerves; occasionally, filaments of the nerve pass through an island in the vessel. The relationship of the posterior inferior cerebellar artery to the hypoglossal nerve varies considerably; as already noted in connection with the ninth, tenth, and eleventh nerves, this artery may lie largely dorsal to these nerves, or it may lie in part ventral to the nerves and therefore between them and the hypoglossal. The artery may arise from the vertebral either above or below the level of origin of the nerve from the brain stem and, in the former case especially, have no intimate relation to the nerve. It might be recalled that there is normally a venous plexus accompanying the hypoglossal nerve through the hypoglossal canal and connecting the marginal sinus with the vertebral vein; in those cases in which the occipital sinus is much enlarged, the venous connection through the hypoglossal canal may also be of considerable size.

Abducens Nerve

The abducens nerve, like the trochlear supplying only one muscle in the orbit (the lateral rectus), emerges from the medulla at the posterior edge of the pons, just lateral to the pyramid and therefore not far from the ventral midline. Close to its origin it is crossed by the anterior inferior cerebellar artery (Fig. 1-62); according to Stopford ('16a) the artery lies ventral to the nerve in more than four fifths of instances, but otherwise crosses dorsal to it or may even pass through the roots of the nerve. When the artery lies dorsal to the nerve it may compress the latter against the sphenoid bone. The labyrinthine artery may have a similar relationship to the nerve in that minority of instances in which it arises from the basilar. Cushing, reporting compression of the sixth nerve by the anterior inferior cerebellar and labyrinthine arteries, expressed his belief that constriction by an artery, due to displacement of the brain by a tumor or through edema, is one of the most common reasons for isolated paralysis of the sixth nerve. As is well known, such a paralysis of the sixth nerve is not uncommon in intracranial disturbances and has little localizing value.

From its origin at the brain stem the abducens nerve passes forward and slightly laterally to enter the cranial dura on the side of the base of the sphenoid bone; it then continues upward and forward in the dura to pass over a groove in the petrous ridge just

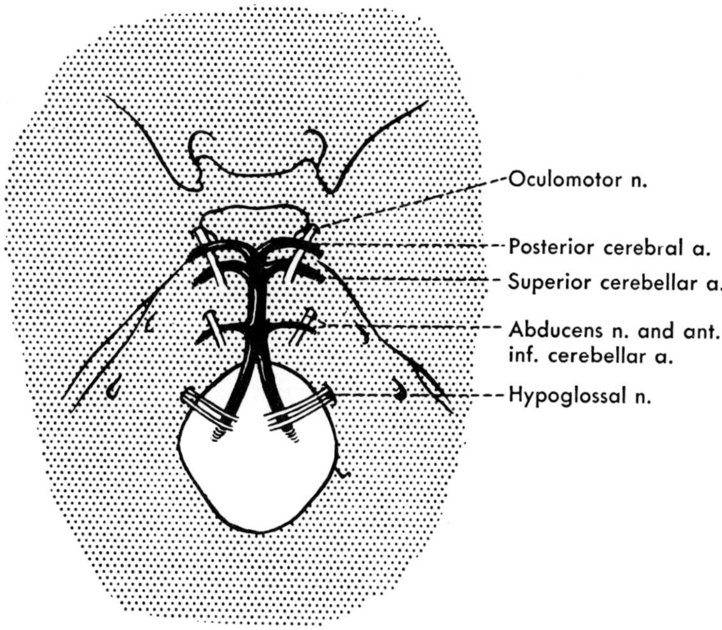

Fig. 1-62. Relationships between the arteries and the nerves at the base of the brain. The more common relation between the abducens nerve and the anterior inferior cerebellar artery is shown on the left; a less common one, on the right.

posterior to the base of the posterior clinoid process and enter the cavernous sinus. This groove is converted into a canal (Dorello's canal) by the petroclinoid or petrosphenoid ligament (of Gruber), and contains also the inferior petrosal sinus, which usually lies medial to the nerve but may (Houser) lie above it. (Nathan and colleagues reported that when the nerve is double only one of the two passes through the canal; the other penetrates or passes above the ligament.) In this canal the abducens nerve is adjacent to the tip of the petrous apex, and therefore exposed to infection from petrous tip cells, and may be relatively close to the sphenoid sinus (less than 3 mm in 15.7% of cases studied by Houser). These relationships may also account for isolated paralysis of the sixth nerve. Further, any downward shift of the brain stem would necessarily quickly produce tension upon the sixth nerve, with possible injury to it.

These relationships, its fixed position in the dura of the cranial floor—where it may be involved by skull fractures—and perhaps other anatomical factors probably account for the frequency of involvement of this nerve; the often quoted statement that paralysis of the sixth nerve is due to its long intracranial course is illogical, for the sixth has a short subarachnoid course, and the trochlear nerve, very rarely paralyzed alone, has both a much longer subarachnoid and total intracranial course than has the sixth. While it is in the cavernous sinus the sixth nerve receives fibers from the internal carotid plexus, which travel with it for a short distance and then join the ophthalmic branch of the trigeminal (see Chap. 2).

Oculomotor Nerve

The oculomotor nerves arise at the upper border of the pons, emerging between or through the more medial fibers of the two cerebral peduncles. Unlike the other two motor nerves to the orbit the oculomotor nerve contains, in addition to its voluntary motor fibers (distributed to the levator palpebrae, superior, medial and inferior recti, and the inferior oblique), preganglionic parasympathetic fibers; these supply, through the ciliary ganglion, the ciliary muscle and the sphincter pupillae.

Close to its origin the oculomotor nerve typically passes between the posterior cerebral and the superior cerebellar arteries; in 50% of instances, according to Stopford ('16a) the posterior cerebral artery lay in actual contact with the root of the nerve. Each nerve runs forward through the interpeduncular cistern, crosses the attachment of the tentorium cerebelli to the posterior clinoid process (this attachment may or may not form a fairly sharp ridge against which the nerve can be injured), and then, proceeding lateral to the posterior clinoid process, pierces the dura to traverse the lateral wall of the cavernous sinus and enters the orbit through the superior orbital fissure.

Lying as it does usually between the posterior cerebral and the superior cerebellar arteries, the oculomotor nerve may be compressed by either, or by aneurysms of the anterior part of the basilar artery; as the nerve passes upward and forward it is also in close contact with the posterior communicating artery and is therefore potentially subject to compression by aneurysms of that vessel.

ARTERIES

The general courses and connections of the arteries in the posterior cranial fossa have been described in connection with the blood supply of the brain; their variable relations to the cranial nerves of the posterior fossa have been discussed in connection with these nerves.

Briefly summarized, the two *vertebral arteries* (Fig. 1-63) pass alongside the anteroventral aspect of the medulla, anterior or ventral to the roots of origin of the ninth, tenth and eleventh cranial nerves; as they converge toward their point of junction to form the basilar artery, they necesarily pass close to the hypoglossal nerves, which are more commonly affected by aneurysms of these vessels than are the more laterally lying nerves.

The *posterior inferior cerebellar artery* usually arises from the vertebral, but varies in the level of its origin and many even arise from

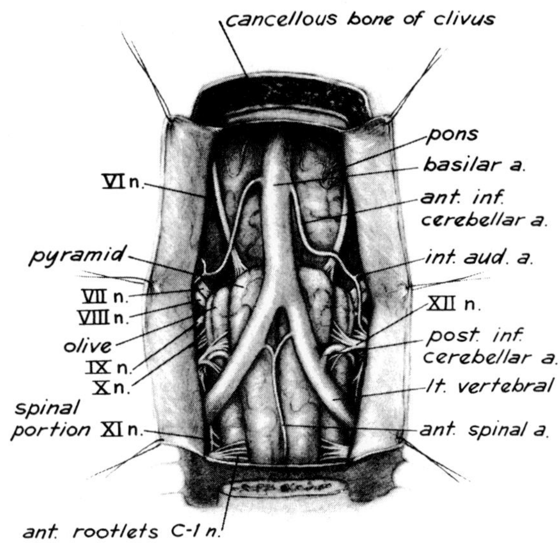

Fig. 1-63. Exposure of the vertebral and basilar arteries by an approach through the clivus. (Courtesy Dr. George C. Stevenson)

the basilar; it typically runs cranially to loop dorsally and caudally and in so doing is in close relation to the roots of the ninth, tenth, and eleventh nerves, but this relationship varies, as already mentioned. Finally, it usually turns upward, between the medulla and the cerebellum, before passing onto the latter.

The *basilar artery* is typically formed by the union of the two vertebrals at about the lower border of the pons, but may have a higher or a lower origin; although it should run in the midline, it is not unusual to find it deviated markedly to one side.

The *anterior inferior cerebellar artery* usually arises from the basilar, but may arise from the vertebral; it passes typically either above or below the sixth nerve, and in close relation to it, and as it passes toward the inferior surface of the cerebellum comes also into close relationship with the seventh and eighth nerves. It commonly loops laterally with these nerves, so that it lies adjacent to them at the internal acoustic meatus. It may lie above or below the nerves, or between them.

The *labyrinthine artery,* if it arises from the basilar, runs laterally with the seventh and eighth nerves; if it arises from the anterior inferior cerebellar artery, as it does more commonly, it usually arises from the loop on this vessel and may therefore leave it in the internal meatus.

The *superior cerebellar artery,* from the upper end of the basilar, passes laterally, backward, and downward around the pons; close to its origin it usually passes just below (caudal to) the oculomotor nerve, and farther laterally crosses the trochlear nerve. The *posterior cerebral artery* also passes laterally, paralleling the superior cerebellar; it usually lies above the oculomotor nerve, so that the roots of this nerve pass between the posterior cerebral and superior cerebellar vessels. The *posterior communicating arteries,* from the internal carotids, run on the sides of the hypothalamic region to join the posterior cerebrals and complete the arterial circle of the brain.

In addition to these larger branches, there are multiple small branches to the brain stem. Attention might be called again to the fact that an artery of appreciable size may be associated with the sensory root of the trigeminal nerve in the posterior fossa.

References

ABBIE AA: The clinical significance of the anterior choroidal artery. Brain 56:233, 1933

ADAMS JE, SEYMOUR RJ, EARLL JM, TUCK M, SPARKS LL, FORSHAM PH: Transphenoidal cryohypophysectomy in acromegaly. Clinical and endocrinological evaluation. J Neurosurg 28:100, 1968

ADAMS RD: Occlusion of the anterior inferior cerebellar artery. Arch Neurol Psychiatr 49:765, 1943

ADSON AW: Cerebrospinal rhinorrhea: Surgical repair of craniosinus fistula. Ann Surg 114:697, 1941a

ADSON AW: A straight lateral incision for unilateral suboccipital craniotomy. Surg Gynecol Obstet 72:99, 1941b

ANDERSON AG, LOCKHART RD, SOUTER WC: Lateral syndrome of the medulla: Occlusion of the posterior inferior cerebellar artery; vascular lesions of the hindbrain. Brain 54:460, 1931

ANDERSON FM: Intranasal (sphenopharyngeal) encephalocele: A report of a case with intracranial repair and a review of the subject. Arch Otolaryngol 46:644, 1947

AREY LB: The craniopharyngeal canal reviewed and reinterpreted. Anat Rec 106:1, 1950

ATKINSON WJ: The anterior inferior cerebellar artery: Its variations, pontine distribution, and significance in the surgery of cerebello-pontine angle tumours. J Neurol Neurosurg Psychiatr 12:137, 1949

BECKMANN JW, KUBIE LS: A clinical study of twenty-one cases of tumour of the hypophyseal stalk. Brain 52:127, 1929

BEEVOR CE: On the distribution of the different arteries supplying the human brain. Philos Trans R Soc Lond [Biol] 200:1, 1909

BERGLAND RM, RAY BS, TORACK RM: Anatomical variations in the pituitary gland and adjacent structures in 225 human autopsy cases. J Neurosurg 28:93, 1968

BINET EF, YOUNG RF: Bilateral persistent trigeminal arteries. Case report. J Neurosurg 47:619, 1977

BLACKBURN IW: Anomalies of the encephalic arteries among the insane: A study of the arteries at the base of the encephalon in two hundred and twenty consecutive cases of mental disease, with special reference to anomalies of the circle of Willis. J Comp Neurol 17:493, 1907

BLACKBURN IW: On the median anterior cerebral artery as found among the insane. J Comp Neurol 20:185, 1910

BLUNT MJ: The blood supply of the facial nerve. J Anat 88:520, 1954

BONNAL J, BROTCH J: Surgery of the superior sagittal sinus in parasagittal meningeomas. J Neurosurg 48:935, 1978

BRODAL A: Central course of afferent fibers for pain in facial glossopharyngeal and vagus nerves: Clinical observations. Arch Neurol Psychiatr 57:292, 1947

BROOKOVER C: The nervus terminalis in adult man. J Comp Neurol 24:131, 1914

BROWDER J, KAPLAN AH, KRIEGER AJ: Anatomical features of the straight sinus and its tributaries. Clinical correlations. J Neurosurg 44:55, 1976

BROWNE KM, WARNER WA, WALKER, AE: Cerebral arterial shunt in the monkey. J Neurosurg 12:634, 1955

BULL JWD: The Robert Wartenberg Memorial Lecture: The volume of the cerebral ventricles. Neurology 11:1, 1961

CHANDLER SB, DEREZINSKI CF: The variations of the middle meningeal artery within the middle cranial fossa. Anat Rec 62:309, 1935

CHOROBSKI J, PENFIELD W: Cerebral vasodilator nerves and their pathway from the medulla oblongata: With observations on the pial and intracerebral vascular plexus. Arch Neurol Psychiatr 28:1257, 1932

CLARK SL: Innervation of the choroid plexuses and the blood vessels within the central nervous system. J Comp Neurol 60:21, 1934

COATES AE: A note on the superior petrosal sinus and its relation to the sensory root of the trigeminal nerve. J Anat 68:428, 1934

COBB S, FINESINGER JE: Cerebral circulation. XIX. The vagal pathway of the vasodilator impulses. Arch Neurol Psychiatr 28:1243, 1932

CUSHING H: Strangulation of the nervi abducentes by lateral branches of the basilar artery in cases of brain tumour. Brain 33:204, 1910

DANDY WE: Glossopharyngeal neuralgia (tic douloureux): Its diagnosis and treatment. Arch Surg 15:198, 1927

DANDY WE: An operation for the cure of tic douloureux: Partial section of the sensory root at the pons. Arch Surg 18:687, 1929

DANDY WE: Concerning the cause of trigeminal neuralgia. Am J Surg 24:447, 1934

DAS AC, HASAN M: The occipital sinus. J Neurosurg 33:307, 1970

DAVIS L, HAVEN HA: Surgical anatomy of the sensory root of the trigeminal nerve. Arch Neurol Psychiatr 29:1, 1933

DAVISON C: Syndrome of the anterior spinal artery of the medulla oblongata. Arch Neurol Psychiatr 37:91, 1937

DUNKER RO, HARRIS AB: Surgical anatomy of the proximal anterior cerebral artery. J Neurosurg 44:359, 1976

ECKENHOFF JE: The physiologic significance of the vertebral venous plexus. Surg Gynecol Obstet 131:72, 1970

EDVINSSON L, NIELSEN KC, OWMAN CH, WEST KA: Adrenergic innervation of the choroid plexus. Am J Anat 139:299, 1974

ELVIDGE AR: Treatment of obstructive lesions of the aqueduct of Sylvius and the fourth ventricle by interventriculostomy. J Neurosurg 24:11, 1966

EMMONS WF, RHOTON AL JR: Subdivision of the trigeminal sensory root. Experimental study in the monkey. J Neurosurg 35:585, 1971

ERICKSON TC: Paroxysmal neuralgia of the tympanic branch of the glossopharyngeal nerve: Report of a case in which relief was obtained by intracranial section of the glossopharyngeal nerve. Arch Neurol Psychiatr 35:1070, 1936

FALCONER MA: Intramedullary trigeminal tractotomy and its place in the treatment of facial pain. J Neurol Neurosurg Psychiatr 12:297, 1949

FAWCETT E, BLACHFORD JV: The circle of Willis: An

examination of 700 specimens. J Anat Physiol 40:63, 1905

FINLEY KH: Angio-architecture of the substantia nigra and its pathogenic significance. Arch Neurol Psychiatr 36:118, 1936

FISHER C, INGRAM WR, HARE WK, RANSON SW: The degeneration of the supraoptico-hypophyseal system in diabetes insipidus. Anat Rec 63:29, 1935

FLOREY H: Microscopical observations on the circulation of the blood in the cerebral cortex. Brain 48:43, 1925

FLYGER G, HJELMQUIST U: Normal variations in the caliber of the human cerebral aqueduct. Anat Rec 127:151, 1957

FLYNN RE: External carotid origin of the dominant vertebral artery. Case report. J Neurosurg 29:300, 1968

FOG M: Cerebral circulation. I. Reaction of pial arteries to epinephrine by direct application and by intravenous injection. Arch Neurol Psychiatr 41:109, 1939a

FOG M: Cerebral circulation. II. Reaction of pial arteries to increase in blood pressure. Arch Neurol Psychiatr 41:260, 1939b

FOLEY JO, DUBOIS FS: Quantitative studies of the vagus nerve in the cat. I. The ratio of sensory to motor fibers. J Comp Neurol 67:49, 1937

FOLEY JO, SACKETT WW: On the number of cells and fibers in the glossopharyngeal nerve of the cat (abstr). Anat Rec 106:303, 1950

FORBES HS: Physiologic regulation of the cerebral circulation. Arch Neurol Psychiatr 43:804, 1940

FORBES HS, COBB S: Vasomotor control of cerebral vessels. Brain 61:221, 1938

FORBES HS, NASON GI, WORTMAN RC: Cerebral circulation. XLIV. Vasodilation in the pia following stimulation of the vagus, aortic and carotid sinus nerves. Arch Neurol Psychiatr 37:334, 1937a

FORBES HS, NASON GI, COBB S, WORTMAN RC: Cerebral circulation. XLV. Vasodilation in the pia following stimulation of the geniculate ganglion. Arch Neurol Psychiatr 37:776, 1937b

GIBBS EL, GIBBS FA: The cross section areas of the vessels that form the torcular and the manner in which flow is distributed to the right and to the left lateral sinus. Anat Rec 59:419, 1934

GRANT FC: Major trigeminal neuralgia. Am J Surg 27:430, 1935

GRANT FC, WEINBERGER LM: Experiences with intramedullary tractotomy. I. Relief of facial pain and summary of operative results. Arch Surg 42:681, 1941

GRANT FC, GROFF RA, LEWY FH: Section of the descending spinal root of the fifth cranial nerve. Arch Neurol Psychiatr 43:498, 1940

GROSS SW, SAVITSKY N: Extradural hemorrhage in the anterior cranial fossa. Ann Surg 116:821, 1942

GRZYBOWSKI J: L'innervation de la duremère crânienne chez l'homme. Arch Anat Histol Embryol 14:387, 1931

GUDMUNDSSON K, RHOTON AL JR, RUSHTON JG: Detailed anatomy of the intracranial portion of the trigeminal nerve. J Neurosurg 35:592, 1971

HAINES SJ, JANETTA PJ, ZORUB DS: Microvascular relations of the trigeminal nerve. An anatomical study and clinical correlation. J Neurosurg 52:381, 1980

HAMMOCK MK, MILHORST TH, EARLE K, DI CHIRO G: Vein of Galen ligation in the primate. Angiographic, gross, and light microscopic evaluation. J Neurosurg 34:77, 1971

HARRIS RS, ROTON AL JR: Anatomy of the cavernous sinus. A microsurgical study. J Neurosurg 45:169, 1976

HASAN M, DAS AC: A note on the falx cerebelli. Acta Anat 76:624, 1969

HEYMAN A, TINDALL GT, FINNEY WHM, WOODHALL B: Measurement of retinal artery and intracarotid pressures following carotid artery occlusion with the Crutchfield clamp. J Neurosurg 17:297, 1960

HITSELBERGER WE, PULEC JL: Trigeminal nerve (posterior root) retrolabyrinthine selective section. Operative procedure for intractable pain. Arch Otolaryngol 96:412, 1972

HOLL M: Ueber den Nervus accessorius Willisii. Arch Anat Physiol p. 491, 1878

HOOPLE GD: Anomaly of the lateral sinus: Report of a case. Ann Otol Rhinol Laryngol 45:1019, 1936

HOUSER KM: Anatomic relation of the sphenoid sinus to Dorello's canal: Abducens paralysis. Arch Otolaryngol 16:488, 1932

JANETTA PJ: Gross (mesoscopic) description of the human trigeminal nerve and ganglion. J Neurosurg 26:109, 1967

JEFFERSON G, STEWART D: On the veins of the diploë. Br J Surg 16:70, 1928

KAPLAN HA, BROWDER J: Atresia of the rostral superior sagittal sinus: Substitute parasagittal venous channels. J Neurosurg 38:602, 1973

KAPLAN HA, BROWDER J, KREIGER AJ: Intercavernous connections of the cavernous sinuses. The superior and inferior circular sinuses. J Neurosurg 45:166, 1976

KEMPE LG, BLAYLOCK R: Ventriculolymphatic shunt. J Neurosurg 47:86, 1977

KERR FWL: The etiology of trigeminal neuralgia. Arch Neurol 8:15, 1963a

KERR FWL: The divisional organization of afferent fibres of the trigeminal nerve. Brain 86:721, 1963b

KEUCHER TR, MEALY J: Long term results after ventriculoatrial and ventriculoperitoneal shunting for infantile hydrocephalus. J Neurosurg 50:179, 1979

KIMMEL DL: The cervical sympathetic rami and the vertebral plexus in the human fetus. J Comp Neurol 112:141, 1959

KIMMEL DL: Innervation of spinal dura mater and dura mater of the posterior cranial fossa. Neurology 11:800, 1961

KONRAD HR, CANALIS RFG: Transpalatal approach to

deeper cranial structures. Laryngoscope 87:165, 1977

KUBIK CS, ADAMS RD: Occlusion of the basilar artery: A clinical and pathological study. Brain 69:73, 1946

KUNKLE EC, LUND DW, MAHER PJ: Studies on headache: Analysis of vascular mechanisms in headache by use of the human centrifuge, with observations on pain perception under increased positive G. Arch Neurol Psychiatr 60:253, 1948

KUNKLE EC, RAY BS, WOLFF HG: Experimental studies on headache: Analysis of the headache associated with changes in intracranial pressure. Arch Neurol Psychiatr 49:323, 1943

LANDOLT AM: Regeneration of the human pituitary. J Neurosurg 39:35, 1973

LANG EK, HAHN EC, LUROS TJ: Arteriographic demonstration of external-internal carotid anastomoses and their correlation to RISA circulation studies. Radiology 83:632, 1964

LEARY T, EDWARDS EA: The subdural space and its linings. Arch Neurol Psychiatr 29:691, 1933

LENDE RA, POULOS DA: Functional localization in the trigeminal ganglion in the monkey. J Neurosurg 32:336, 1970

LEVINE M, WOLFF H G: Cerebral circulation: Afferent impulses from the blood vessels of the pia. Arch Neurol Psychiatr 28:140, 1932

LEWY FH, GRANT FC: Physiopathologic and pathoanatomic aspects of major trigeminal neuralgia. Arch Neurol Psychiatr 40:1126, 1938

LEY A, MONTSERRAT L, BACCI F, LEY A JR: Clinical and electrophysiological studies on sensory conduction mediated by the accessory rootlets of the human trigeminal nerve. J Neurosurg 42:513, 1975

LILLIE RD: Variations of the canalis hypoglossi. Anat Rec 13:131, 1917

LIN PM, MOKROHISKY JF, STAFFER HM, SCOTT M: The importance of the deep cerebral veins in cerebral angiography with special emphasis on the orientation of the foramen of Monro through the visualization of the "venous angle" of the brain. J Neurosurg 12:256, 1955

LOCKHART RD: The dural relations of the gasserian ganglion with reference to a new method of surgical approach. J Anat 62:105, 1927

LOVE JG, GAY JR: Spontaneous cerebrospinal rhinorrhea: Successful surgical treatment. Arch Otolaryngol 46:40, 1947

LOW FN: An anomalous middle meningeal artery. Anat Rec 95:347, 1946

MAZZONI A: Internal auditory canal: Arterial relations at the porus acusticus. Ann Otol Rhinol Laryngol 78:797, 1969

MCCONNELL EM: The arterial supply of the human hypophysis. Anat Rec 115:175, 1953

MCCULLOUGH AW: Some anomalies of the cerebral arterial circle (of Willis) and related vessels. Anat Rec 142:537, 1962

MC KISSOCK W, PAINE KWE, WALSH LS: An analysis of the results of treatment of ruptured intracranial aneurysms: Report of 772 consecutive cases. J Neurosurg 17:762, 1960

MATAS R: Aneurysms of the circle of Willis. A discussion of Dr. Dandy's intracranial occlusion of the internal carotid for aneurysms of the circle of Willis, with supplementary remarks. Ann Surg 107:660, 1938

MELCHIONNA RN, MOORE RA: The pharyngeal pituitary gland. Am J Pathol 14:763, 1938

MERRITT H, FINLAND M: Vascular lesions of the hindbrain: Lateral medullary syndrome. Brain 53:290, 1930

MERRY GS, JAMISON KG: Operative approach to persistent trigeminal artery producing facial pain and diplopia. Case report. J Neurosurg 47:613, 1977

MERWARTH HR: The syndrome of the rolandic vein: Hemiplegia of venous origin. Am J Surg 56:526, 1942

MILHORST TH: Failure of choroid plexectomy as treatment for hydrocephalus. Surg Gynecol Obstet 139:505, 1974

MILHORST TH, HAMMOCK MK, CHIEN T, DAVIS DA: Normal rate of cerebrospinal fluid formation five years after bilateral choroid plexectomy. Case report. J Neurosurg 44:735, 1976

MIRA KM, ELNAGA IA, EL-SHERIF H: Nerve cells in the intracranial part of the trigeminal nerve of man and dog. J Neurosurg 34:643, 1971

MOOD GF: Congenital anterior herniations of brain. Ann Otol Rhinol Laryngol 47:391, 1938

MOUNT LA, TRAVERAS JM: Angiographic demonstration of the collateral circulation of the cerebral hemispheres. Arch Neurol Psychiatr 78:235, 1957

NATHAN H, OUAKNINE G, KOSARY IZ: The abducens nerve. Anatomical variations in its course. J Neurosurg 41:561 1974

O'CONNELL JEA: Some observations on the cerebral veins. Brain 57:484, 1934

OUAKNINE G, NATHAN H: Anastomotic connections between the eleventh nerve and the posterior root of the first cervical nerve in humans. J Neurosurg 83:189, 1973

OWMAN C, WEST KA: Effect of superior cervical ganglionectomy on experimentally induced intracranial hypertension. Brain Res 18:469, 1970

PARKINSON D: Collateral circulation of cavernous carotid artery: Anatomy. Can J Surg 7:251, 1964

PARKINSON D: A surgical approach to the cavernous portion of the carotid artery. Anatomical studies and case report. J Neurosurg 23:474, 1965

PARKINSON D: Carotid cavernous fistula: Direct repair with preservation of the carotid artery. Technical note. J Neurosurg 38:99, 1973

PARKINSON D, SHIELDS CB: Persistent trigeminal artery: Its relationship to the normal branches of the cavernous carotid. J Neurosurg 40:244, 1974

PARTRIDGE EJ: The relations of the glossopharyngeal

nerve at its exit from the cranial cavity. J Anat 52:332, 1918

PATEK PR: The perivascular spaces of the mammalian brain. Anat Rec 88:1, 1944

PAULLUS WS, PAIT TG, RHOTON AL JR: Microsurgical exposure of the petrous portion of the carotid artery. J Neurosurg 47:713, 1977

PELLETIER VA, POULOS DA, LENDE RA: Functional localization in the trigeminal root. J Neurosurg 40:504, 1974

PENFIELD W: The cranial subdural space: A method of study. Anat Rec 28:173, 1924

PENFIELD W: Intracranial vascular nerves. Arch Neurol Psychiatr 27:30, 1932

PENFIELD W, MCNAUGHTON F: Dural headache and innervation of the dura mater. Arch Neurol Psychiatr 44:43, 1940

PENFIELD W, NORCROSS NC: Subdural traction and posttraumatic headache: Study of pathology and therapeusis. Arch Neurol Psychiatr 36:75, 1936

PERLMAN HB, LINDSAY JR: Relation of the internal ear spaces to the meninges. Arch Otolaryngol 29:12, 1939

POOL JL, FORBES HS, NASON GI: Cerebral circulation. XXXII. Effect of stimulation of the sympathetic nerve on the pial vessels in the isolated head. Arch Neurol Psychiatr 32:915, 1934

PRIMAN J, CHRISTIE, DH: A case of abnormal internal carotid artery and associated vascular anomalies. Anat Rec 134:87, 1959

PUCHADES-ORTS A, NOMBELA-GOMEZ M, ORTAÑO-PACHECO G: Variation in form of circle of Willis: Some anatomical and embryological considerations. Anat Rec 185:119, 1976

PULEC JL: Geniculate neuralgia: Diagnosis and surgical management. Laryngoscope 86:955, 1976

PULEC JL, HITSELBERGER WE: Trigeminal neuralgia; Retrolabyrinthine selective posterior root section. Laryngoscope 87:1861, 1977

RAND RW, HANAFEE WN: Cavernous sinus venography and stereotaxic cryohypophysectomy. J Neurosurg 26:521, 1967

RAO TS, SEITH PK: Persistent preatlantal artery with carotid-vertebral anastomosis. J Neurosurg 43:499, 1975

RAY BS: Intracranial hypophysectomy. J Neurosurg 28:180, 1968

RAY BS, WOLFF HG: Experimental studies on headache: Pain-sensitive structures of the head and their significance in headache. Arch Surg 41:813, 1940

REICHERT FL, MORRISSEY EJ: Extradural venous hemorrhage. Ann Surg 113:204, 1941

RENN WH, RHOTON AL JR: Microsurgical anatomy of the sellar region. J Neurosurg 43:288, 1975

RHOTON AL JR, BUZA R: Microsurgical anatomy of the jugular foramen. J Neurosurg 42:541, 1975

RHOTON AL JR, KOBAYASHI S, HOLLINSHEAD WH: Nervus intermedius. J Neurosurg 29:609, 1968

RHOTON AL JR, PULEC JL, HALL GM, BOYD AS JR: Absence of bone over the geniculate ganglion. J Neurosurg 28:48, 1968

ROBERTS B, HARDESTY WH, HOLLING HE, REIVICH M, TOOLE JF: Studies on extracranial cerebral blood flow. Surgery 56:826, 1964

ROBERTSON JT, SCHICK RW, MORGAN F, MATSON DD: Accurate placement of ventriculo-atrial shunt for hydrocephalus under electrocardiographic control. J Neurosurg 18:255, 1961

ROGERS L: The function of the circulus arteriosus of Willis. Brain 70:171, 1947

ROWLAND LP, METTLER FA: Relation between the coronal suture and cerebrum. J Comp Neurol 89:21, 1948

RUBENSTEIN HS: Relation of circulus arteriosus to hypothalamus and internal capsule. Arch Neurol Psychiatr 52:526, 1944

SAMRA K, SCOVILLE WB, YAGHMAI M: Anastomosis of carotid and basilar arteries. Persistent primitive trigeminal artery and hypoglossal artery: Report of two cases. J Neurosurg 30:622, 1969

SATO S, SUZUKI J: Anatomical mapping of the cerebral nervi vasorum in the human brain. J Neurosurg 43:559, 1975

SAUNDERS RL, SACHS E JR: Relation of the accessory rootlets of the trigeminal nerve to its motor root. A microsurgical autopsy study. J Neurosurg 33:317, 1970

SAXENA RC, BEG MAQ, DAS AC: The straight sinus. J Neurosurg 41:724, 1974

SCARF JE: Evaluation of treatment of hydrocephalus. Arch Neurol 14:382, 1966

SCHAERER JP: A case of carotid basilar anastomosis with multiple associated cerebrovascular anomalies. J Neurosurg 12:62, 1955

SCHLESINGER B: The venous drainage of the brain, with special reference to the galenic system. Brain 62:274, 1939

SCHUMACHER GA, WOLFF HG: Experimental studies on headache. A. Contrast of histamine headache with the headache of migraine and that associated with hypertension. B. Contrast of vascular mechanisms in preheadache and in headache phenomena of migraine. Arch Neurol Psychiatr 45:199, 1941

SCHUMACHER GA, RAY BS, WOLFF HG: Experimental studies on headache: Further analysis of histamine headache and its pain pathways. Arch Neurol Psychiatr 44:701, 1940

SHACKELFORD RT, HEGEDUS SA: Factors affecting cerebral blood flow—experimental review: sympathectomy, hypothermia CO_2 inhalation and pavarine. Ann Surg 163:771, 1966

SHELDON CH, PUDENZ RH, FRESHWATER DB, CRUE BL: Compression rather than decompression for trigeminal neuralgia. J Neurosurg 12:123, 1955

SHENKIN HA, HARMEL MH, KETY SS: Dynamic anatomy of the cerebral circulation. Arch Neurol Psychiatr 60:240, 1948

SJÖQVIST O: Studies on pain conduction in the

trigeminal nerve: A contribution to the surgical treatment of facial pain. Acta Psychiatr Neurol (Suppl) 17, 1938

STANFIELD JP: The blood supply of the human pituitary gland. J Anat 94:257, 1960

STAVRAKY GW: Response of cerebral blood vessels to electric stimulation of the thalamus and hypothalamic regions. Arch Neurol Psychiatr 35:1002, 1936

STENDER A, GRUMME T: Late results of gangliolysis as a treatment for trigeminal neuralgia. J Neurosurg 31:21, 1969

STEWART JP: The persistence of fetal blood sinuses and their relation to the middle ear spaces. Arch Otolaryngol 10:266, 1929

STOPFORD JSB: The arteries of the pons and medulla oblongata. J Anat 50:131, 1916a

STOPFORD JSB: The arteries of the pons and medulla oblongata. Part II. The precise distribution of the arteries supplying the medulla and pons. J Anat 50:255, 1916b

STOPFORD JSB: The arteries of the pons and medulla oblongata. Part III. The clinical application of Parts I and II. J Anat 51:250, 1917

STOPFORD JSB: The functional significance of the arrangement of the cerebral and cerebellar veins. J Aant 64:257, 1930

STREETER GL: The development of the venous sinuses of the dura mater in the human embryo. Am J Anat 18:145, 1915

SUCHANNEK: Ein Fall von Persistenz des Hypophysenganges. Anat Anz 2:520, 1887

SUNDERLAND S: Neurovascular relations and anomalies at the base of the brain. J Neurol Neurosurg Psychiatr n.s.11:243, 1948

SUNDERLAND S: The tentorial notch and complications produced by herniation of the brain through that aperture. Br J Surg 45:422, 1958

SVIEN HJ, HOLLENHORST RW: Pressure in retinal arteries after ligation or occlusion of the carotid artery. Proc Mayo Clin 31:684, 1956

SVIEN JH, LITZOW TJ: Removal of certain hypophyseal tumors by the transantral–sphenoid route. J Neurosurg 23:603, 1965

SWEET WH, WEPSIC JG: Controlled thermocoagulation of trigeminal ganglion and rootlets for differential destruction of pain fibers. Part 1. Trigeminal Neuralgia. J Neurosurg 40:143, 1974

SYMINGTON J: Observations on the relations of the deeper parts of the brain to the surface. J Anat Physiol 37:241, 1903

TAARNHØJ P: Decompression of the trigeminal root and the posterior part of the ganglion as treatment in trigeminal neuralgia: Preliminary communication. J Neurosurg 9:288, 1952

TARLOV IM: Structure of the nerve root. I. Nature of the junction between the central and the peripheral nervous system. Arch Neurol Psychiatr 37:555, 1937a

TARLOV IM: Structure of the nerve root. II. Differentiation of sensory from motor roots; observations on identification of function in roots of mixed cranial nerves. Arch Neurol Psychiatr 37:1338, 1937b

TARLOV IM: Sensory and motor roots of the glossopharyngeal nerve and the vagus–spinal accessory complex. Arch Neurol Psychiatr 44:1018, 1940

TARLOV IM: Section of the cephalic third of the vagus–spinal accessory complex: Clinical and histologic results. Arch Neurol Psychiatr 47:141, 1942

TEW JM, MAYFIELD FH: Trigeminal neuralgia: A new surgical approach. (Percutaneous coagulation of the trigeminal nerve.) Laryngoscope 83:1096, 1973

UIHLEIN A, LOVE JG, CORBIN KB: Intracranial section of the glossopharyngeal nerve: Sensory changes observed postoperatively. Arch Neurol Psychiatr 74:320, 1955

VARE AM, BANSAL PC: Arterial pattern at the base of the human brain. J Anat Soc India 19:71, 1970

WALTNER JG: Anatomic variations of the lateral and sigmoid sinuses. Arch Otolaryngol 39:307, 1944

WEINBERGER LM, GRANT FC: Experiences with intramedullary tractotomy. III. Studies in sensation. Arch Neurol Psychiatr 48:355, 1942

WERDEN DH: Intracranial venous anomaly. Am J Surg 29:115, 1935

WILSON MCC: The Anatomical Foundation of Neuroradiology of the Brain. Boston, Little, Brown & Co, 1963

WINDLE BCA: On the arteries forming the circle of Willis. J Anat Physiol 22:289, 1888

WINDLE WF, DELOZIER LC: The absence of painful sensation in the cat during stimulation of the spinal accessory nerve. J Comp Neurol 54:97, 1932

WIRTH FP, VAN BUREN JM: Referral of pain from dural stimulation in man. J Neurosurg 34:680, 1971

WOODHALL B: Variations of the cranial venous sinuses in the region of the torcular Herophili. Arch Surg 33:297, 1936

WOODHALL B, ODOM GL, BLOOR BM, GOLDEN J: Direct measurement of intravascular pressure in components of the circle of Willis: A contribution to the surgery of congenital cerebral aneurysms and vascular anomalies of the brain. Ann Surg 135:911, 1952

WOODHALL B, SEEDS AE: Cranial venous sinuses: Correlation between skull markings and roentgenograms of the occipital bone. Arch Surg 33:867, 1936

XUEREB GP, PRICHARD MML, DANIEL PM: The arterial supply and venous drainage of the human hypophysis cerebri. Q J Exp Physiol 39:199, 1954

YASARGIL MG, KASDAGLIS K, JAIN KK, WEBER H–P: Anatomical observations of the subarachnoid cisterns of the brain during surgery. J Neurosurg 44:298, 1976

YOUNG RE: Unmyelinated fibers in the trigeminal motor root. Possible relationship to the results of trigeminal rhizotomy. J Neurosurg 49:538, 1979

Chapter 2

THE EYELIDS, ORBIT, AND EYEBALL

The external muscles, nerves, and vessels about the orbit are described in the chapter on the face; only those points of particular interest in connection with the orbit will be described here.

The Lids

The structure of the upper and lower lids is similar, although the upper lid is more movable and presents certain features not found in the lower. The lids are essentially reduplicated folds of skin, the inner layer of which is modified to form the conjunctiva. When the lids are open the interval between them is known as the palpebral fissure (rima); this is limited medially and laterally at the angles or canthi by the medial and lateral palpebral commissures (Fig. 2-1). The medial commissure, unlike the lateral one, does not rest directly against the eyeball but is separated from it by a small space, the lacrimal lake. The fleshy, yellowish projection within the lacrimal lake is the lacrimal caruncle, and the thin semilunar fold extending laterally from this is the plica semilunaris. The caruncle contains large modified sweat and sebaceous glands; the plica semilunaris is a vestige of the movable third lid, or nictitating membrane, of lower animals.

The eyelashes (cilia), or hairs on the free margin of each lid, are arranged in two or three rows that extend from the lateral canthus to about the level of the plica semilunaris. Associated with the hair follicles of the cilia are sebaceous glands (of Zeis) and rudimentary sweat glands (of Moll), the latter sometimes opening to the skin surface but frequently opening into the sebaceous glands or their ducts or into a hair follicle. A stye is due to blockage of either type of gland.

At about the level of the lateral border of the plica semilunaris, and always medial to the rows of cilia, each lid margin presents, when slightly everted, a posteriorly directed lacrimal papilla bearing a central opening or lacrimal punctum, these elevations and openings representing the origins of the lacrimal canaliculi, which drain the tears from the lacrimal lake. Close to the posterior edge of the free margin of the lids may be seen a row of minute apertures representing the opening of the tarsal glands (of Meibom), the glands themselves being visible as yellowish lines on the inner surface of the everted eyelid. A chalazion is due to blockage of this type of gland. A grayish line or a slight groove sometimes visible between the lashes and the openings of the tarsal glands represents the junction of the two fundamental portions of the lid, the skin and muscle on the one hand, and the tarsus (the plate of closely packed tarsal glands) and conjunctiva on the other, and indicates a plane along which the lid may be split into anterior and posterior portions with minimal scarring.

Each eyelid consists, from without inward, of skin, subcutaneous tissue, voluntary muscle, orbital septum and tarsus, smooth muscle and conjunctiva (Fig. 2-2); in addition, the more freely movable upper lid receives the insertion of the levator palpebrae superioris muscle. The skin of the eyelids is thin, and its texture is difficult to match with skin grafts from other areas of the body. It is attached by loose connective tissue to the underlying muscle; hence, in contrast to most areas of the face, relatively large quantities of fluid may accumulate subcutaneously in this loose connective tissue.

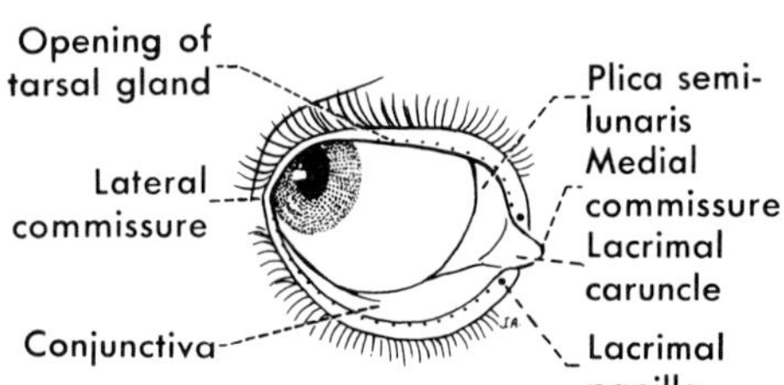

Fig. 2-1. Palpebral fissure with both eyelids everted.

Orbicularis Oculi

The part of the orbicularis oculi in the lids (*palpebral portion*) is composed of rather thin bundles that are loosely held together by delicate connective tissue. This portion of the orbicularis arises largely from the medial palpebral ligament and inserts (Fig. 2-3) into the lateral palpebral raphe (not the lateral palpebral ligament), but a part of it closest to the margins of the lids arises from the posterior lacrimal crest (behind the lacrimal sac) and is known, at its origin only, as the pars lacrimalis, the tensor tarsi, or the muscle of Horner (see Fig. 2-10). In addition to this deep origin, Jones ('57) described some of the more peripheral ("preseptal") fibers of the palpebral portion in each lid as having an origin from the fascia over the lacrimal sac. Some particularly delicate muscle fibers at the free edge of the lid have been separately designated as the ciliary muscle of Riolan, a complication of no practical usefulness.

Paralysis of the orbicularis abolishes tight closure of the eye, allows the lower lid to drop away from the eyeball (ectropion), and results in improper drainage of tears with conse-

Fig. 2-2. Structure of the upper eyelid. Opinions differ on whether tendinous fibers of the levator extend through the orbital septum, as shown here.

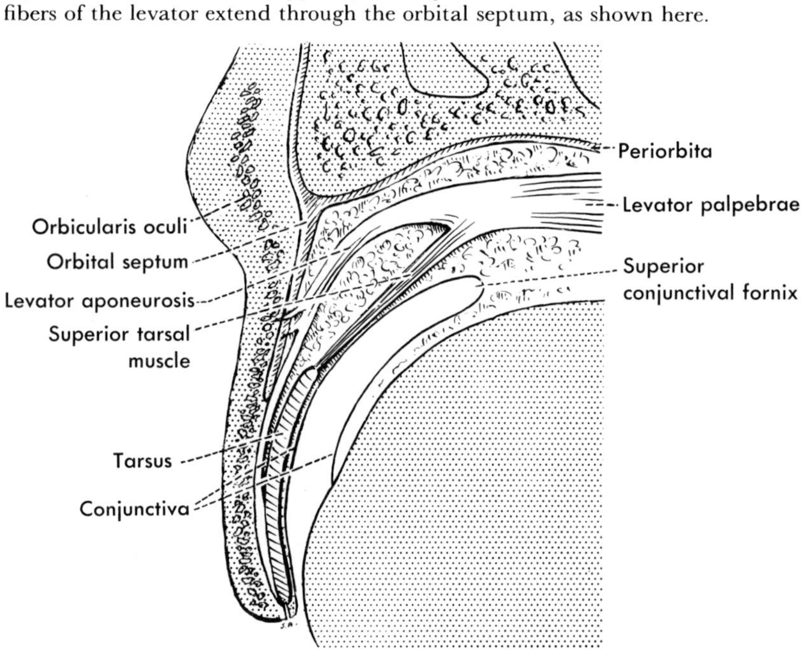

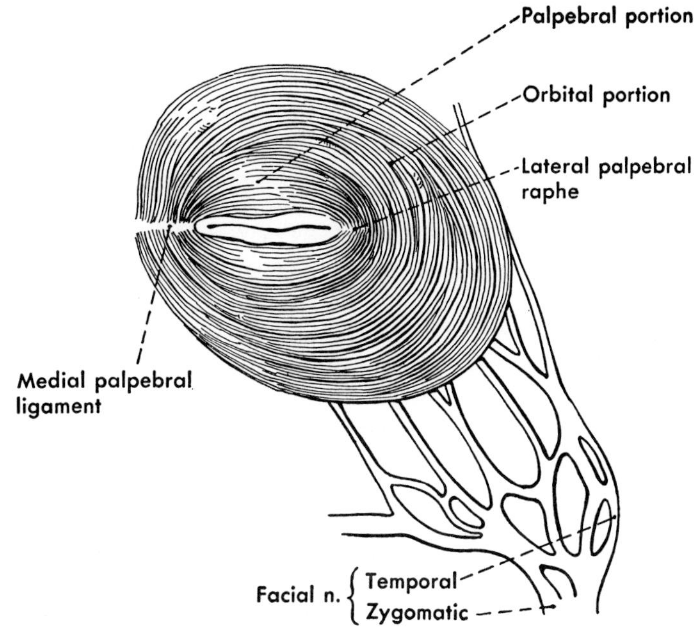

Fig. 2-3. Orbicularis oculi from the front. For the relations of the pars lacrimalis see Fig. 2-10.

quent spillage over the lower lid (epiphora). Induced paresis of the orbicularis oculi by injection of a local anesthetic into the fibers of the muscle at their insertion into the lateral raphe (technique of Van Lint) or by injection about the parotid plexus of the facial nerve (technique of O'Brien) has become a popular procedure in many intraocular operations carried out under local anesthesia; this paresis forestalls the squeezing action of the eyelids upon the eyeball, with its tendency to produce loss of part of the vitreous body.

In order to alleviate blepharospasm, McCabe and Boles found it necessary to section all the upper branches of the facial nerve, including that to the upper lip; they stated that the nerve supply to the lower lip maintains an adequate sphincter about the mouth.

Behind the orbicularis oculi is another layer of loose connective tissue, in which the chief blood vessels and nerves of the lid run. It is through this layer that the lid may be most easily split. Behind this loose connective tissue is an area of denser tissue, formed in the lower lid by the inferior tarsus (tarsal plate) and the orbital septum and in the upper lid by the orbital septum, the deeper lamina of the levator palpebrae, and the superior tarsus.

Orbital Septum

The orbital septum (palpebral fascia) extends into the upper and lower lids from approximately the margins of the bony orbit, where it is continuous with the periosteum of the orbit (periorbita) and the periosteum of the outer surface of the skull. From this peripheral origin it extends as a membranous sheet into the eyelids. The orbital septum is not actually divided into upper and lower parts but is a continuous membrane; however, it is often convenient to speak of it in terms of its association with the lids; hence the upper and lower segments are sometimes referred to individually as the superior and inferior orbital septa.

The superior orbital septum is notably thicker at its origin laterally, but both parts of the septum extend as thin sheets towad the free margins of the lids. The inferior septum loses its entity by fusing with the connective tissue of the front of the inferior tarsus. The superior septum overlaps the front of the superior tarsus and fuses with that medially and laterally; centrally, its relations are disputed, for it has been said to be separated from the tarsus by the deep lamina of the levator palpebrae superioris, to be united to the tarsus by fibers that pass between the fibers of the levator, and to end as an entity above the tarsus, the continuation downward on the front of the tarsus being described as formed chiefly by fibers derived from the levator (see the description of that muscle).

The orbital septum separates the connective tissue spaces of the lids from those of the orbit, and with expansions from the fascial sheaths of the ocular muscles restrains the orbital fat and forms a sort of fire screen that tends to prevent infection of the lid from passing into the orbit, and vice versa. Superiorly, it is attached to the orbital rim except medially, where it is partially interrupted by emerging vessels and nerves. Medially, the attachment extends downward at first along the posterior lacrimal crest, behind the pars lacrimalis muscle, thus excluding this muscle and the upper part of the lacrimal sac from the orbit proper; the attachment then crosses the fascia over about the middle of the lacrimal sac to continue downward along the anterior lacrimal crest below the medial palpebral ligament to the inferior rim of the orbit. The attachment follows this rim almost to the lateral border of the orbit and then swings slightly forward to lie in front of the orbital rim; it maintains this anterior position through about the lower half of the lateral orbital wall and then recedes to the orbital rim and follows this to the superior orbital rim. In its lateral part the septum lies behind the lateral raphe of the orbicularis oculi, but in front of the lateral palpebral ligament (whereas it lies behind the medial palpebral ligament).

Fig. 2-4. The tarsi and their attachments.

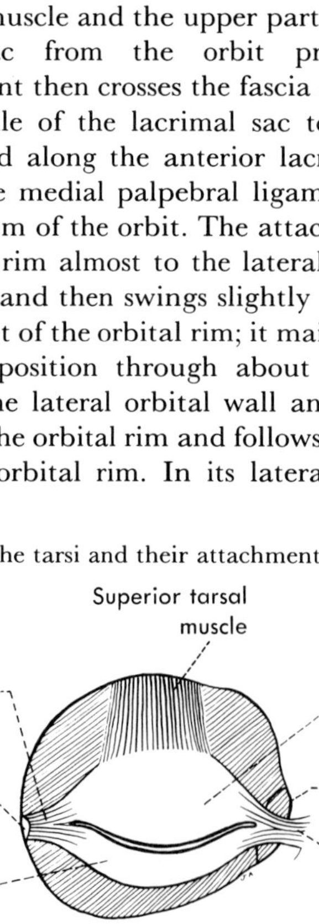

Tarsi

The tarsi or tarsal plates consist of dense connective tissue in which are embedded large sebaceous glands, the tarsal or Meibomian glands; the tarsi (Fig. 2-4) are the main structural support of the lids. The edge of each tarsus adjacent to the free border of the lid parallels this, while the deeper border of each (that of the upper much more so than that of the lower) is curved so that the tarsi are somewhat semilunar in shape. They are also, of course, curved to conform to the outer surface of the eyeball. The superior tarsus is considerably larger than the inferior one, the greatest height of the superior being reported as between 8 and 12 mm, and that of the inferior between 5 and 7 mm. Wesley and co-workers, although agreeing with these measurements for the superior tarsus, found the inferior one usually no higher than 4 mm. To forestall either inversion or eversion of the eyelid as the result of scar tissue formation in the tarsus, incisions into this are usually made in the frontal or vertical plane.

Medial Palpebral Ligament

The tarsi are anchored medially and laterally by the medial and lateral palpebral (canthal, tarsal) ligaments. The medial palpebral ligament, the stronger of the two, has an indefinite upper border but a lower border that is thick and prominent. This ligament is attached laterally to the medial angles of the two tarsi and extends medially, adjacent to the lacrimal canaliculi, to pass in front of the lacrimal sac and attach to the frontal process of the maxilla on and medial to the anterior lacrimal crest. As a landmark for locating the lacrimal sac, the medial palpebral ligament is quite important. This ligament can be demonstrated in the living person by drawing the lids laterally, so as to tense it, at which time the lower border is easily palpable. The ligament is frequently divided, or cut from its insertion into the periosteum of the anterior lacrimal crest, in order to gain access to the lacrimal sac in surgical procedures around this area. Callahan regarded suture of a divided or avulsed ligament to the periosteum as unsatisfactory, and described a method for anchoring it firmly to bone.

In addition to its attachment to the anterior lacrimal crest, the medial palpebral ligament is also said, by many authors, to send a weaker continuation lateral to and behind the lacrimal sac to attach to the posterior lacrimal crest, so that the nasal attachment is often described as U shaped. Other investigators do not recognize the existence of a posterior part of the medial palpebral ligament, or describe it as being no more than delicate fascia on the anterior surface of the pars lacrimalis muscle. The majority of the fibers of the palpebral portion of the orbicularis oculi are attached firmly to the anterior surface of the medial palpebral ligament, thus hiding much of that; the fibers of the pars lacrimalis are sometimes described as related to the posterior surface of the posterior lamina, and sometimes as arising from the anterior surface of this lamina, indicating again the differences of opinion concerning this part of the ligament. Jones ('57) described it as a strong band extending between the anterior and posterior lacrimal crests and reinforcing the fascia over the lacrimal sac. He disagreed also with most authorities by maintaining that the medial palpebral ligament does not extend from the anterior lacrimal crest to the tarsi and that it has no attachment to the latter; he regarded the free lower border extending from the lacrimal crest as being not ligament but the tendon of origin of part of the orbicularis.

Lateral Palpebral Ligament

The lateral palpebral ligament is frequently confused with the lateral palpebral raphe. The latter, however, represents merely an interlacing of muscle fibers at the lateral corner of the lid; it lies in front of the orbital septum, to which it is partly fused. Since it is not especially attached to the tarsi, it can hardly be regarded as a ligament of these. The true lateral palpebral ligament is completely separate from the raphe, since it lies behind both the orbital septum and the strong lateral horn of the levator aponeurosis. It consists of a band of connective tissue that attaches medially to the lateral angles of both tarsi, and laterally to the zygomatic bone at the marginal (orbital) tubercle. At the tubercle the ligament is fused to the aponeurosis of the levator and to the check ligament of the lateral rectus muscle, but more medially it is separated from the check ligament by the lower edge of the palpebral portion of the lacrimal gland.

Probably because of its fusions with the other structures attaching to the marginal tubercle and also because it is by no means as dense as the medial ligament, considerable variation exists in descriptions and illustrations of this ligament. It may be distinct only at and just lateral to its attachment to the tarsi.

Levator Palpebrae Superioris

The upper lid, but not the lower, contains in addition to the tarsus and orbital septum a third element, the tendon of insertion, or aponeurosis, of the levator palpebrae superioris. The aponeurosis of the levator at first descends almost vertically in the lid behind the oribital septum, but accounts of its attachment to the lid vary. Its main insertion is usually described as being through fine tendinous fibers that pass through the orbital septum and thereafter between fascicles of the orbicularis oculi to reach the skin of the lid, and these have been called its superficial lamina. Isaksson could not identify it either by dissection or in histologic sections. There is, however, a definite deep lamina of levator tissue that descends in front of the tarsus and then attaches to it. It is variously described as passing through, below, or blending with the lower part of the orbital septum. Isaksson described the aponeurosis proper as ending in a thickening and then fusing with the orbital septum. According to his account, the only attachment to the lid is through a loose lamella of connective tissue derived from this fusion of levator and septum; the lamella extends downward in the lid and exchanges fibers with both the tarsus behind it and the connective tissue of the orbicularis in front of it, but with no tendinous attachment to either. Anderson described an anterior part as attaching to the lid crease and sending septa in a diffuse pattern to the entire pretarsal portion of the orbicularis and skin of the lid. Regardless of the details, it is through its attachment to the lid that the levator produces the superior palpebral fold, and in paralysis of this muscle the fold disappears.

The aponeurotic portion behind the orbital septum is much wider than the muscle from which it is derived, and its medial and lateral expansions are known as horns (or cornua). The lateral horn is prominent and deeply indents the anterior portion of the lacrimal gland to divide it into thin palpebral and thick orbital portions; its attachment on the orbital wall is to the marginal (orbital) tubercle, already mentioned as the point of attachment of the lateral palpebral ligament. The weaker medial horn of the levator aponeurosis blends with the orbital septum and the medial check ligament and, according to Anderson, also attaches to the posterior lacrimal crest. From the upper surface of the levator as it passes into the lid a reflection of its fascia (called by clinicians its check ligament) can usually be demonstrated passing upward behind the superior orbital septum to attach to the orbital rim. From the standpoint of the surgical correction of blepharotosis, the levator palpebrae superioris is the most important muscle of the eyelid. Jones and co-workers reported that the most frequent defect of the levator is a dehiscence in its aponeurosis, but that sometimes upper and lower parts are completely separate.

The *superior tarsal or superior palpebral muscle* (of Müller), a thin sheet of smooth muscle, is firmly attached to the inferior surface of the levator and passes downward behind the aponeurotic part; according to Isaksson, additional smooth muscle is contributed to its posterior surface by the bulbar sheath (fascia bulbi, Tenon's capsule). The insertion of this smooth muscle is into the upper border of the tarsus. Although it is attached to the levator and has a function similar to this, it should not be forgotten that it is smooth muscle and is innervated by the sympathetic nervous system. Although this muscle probably plays no part in normal active elevation of the eyelid, its paralysis, as in Horner's syndrome, leads to drooping of the lid.

Lower Lid

In the lower lid the place of the levator palpebrae is taken by fascia from the inferior rectus, which attaches in part to the lower border of the inferior tarsus but also sends fibers into the orbicularis in front of the tarsus and through that into the subcutaneous tissue of the lid (Jones , '60; Hargiss). These attachments enable the inferior rectus to retract the lower lid, and just as the levator aponeurosis is repaired for blepharoptosis, so can this tissue be repaired to correct entropion, as apparently first proposed by Jones ('60). Arising from the posterior surface of the fascia is a layer of smooth muscle, described as either inserting into the tarsus or (Whitnall, '21a) passing toward it but failing to reach it and inserting again into the fascia from which it arose. This *inferior tarsal muscle* (also of Müller) can be seen in living persons through the conjunctiva.

Conjunctiva

The conjunctiva forms the posterior layer of the eyelid. It begins at the free margin of the lid as a modification of the skin, is reflected upward or downward, as the case may be, to form the posterior surface of the lid, and from here is reflected onto the eyeball; that portion of the conjunctiva reflected from lid to bulb is sometimes termed the *fornix* (cul-de-sac), but the entire space between lid and eyeball is also referred to as the fornix. The superior fornix (in either sense) reaches to the orbital margin, and the inferior fornix almost does this; laterally the two fornices extend deeply, to just behind the equator of the globe, but toward the medial side a fornix is largely nonexistent because of the presence of the caruncle and the plica semilunaris. Caution is required in surgical manipulations of the lateral part of the superior fornix, lest the orifices of the ducts of the lacrimal gland be injured.

The conjunctiva is intimately attached to the posterior surfaces of the two tarsi and is almost inseparable from them; above the superior tarsus the conjunctiva lies relatively loosely on the smooth muscle of the lid, although it is so delicate that it may be diffcult to separate surgically. The superior fornix is described as having attached to it an expansion from the fascia of the levator palpebrae and superior rectus muscles, while the inferior fornix is said to receive an expansion from the fascia of the inferior rectus muscle; thus the fornices tend to move with movements of the lids and of the eyeball.

As the conjunctiva is reflected at the fornix onto the bulb it lies loosely on the anterior portion of the bulbar sheath (vagina bulbi; fascia bulbi or Tenon's capsule), and subconjunctival vessels lie in this position; at about 3 mm from the cornea it becomes more closely applied to the sheath and subsequently to the sclera and cornea, from both of which, however, it remains separable surgically until it fuses with the cornea. The line along which fusion of the conjunctiva to the cornea occurs is the conjunctival anulus or ring, often known also as the conjunctival limbus (limbus meaning edge or border). It is about 1 mm anterior to the true or corneal limbus (the junction of cornea and sclera; see Fig. 2-21). The "limbus" formed by the insertion of the bulbar sheath into the sclera lies about 1.5 mm posterior to the corneal limbus; the area between the limbus of the bulbar sheath and that of the conjunctiva is of surgical importance as the site of various fistulizing procedures for glaucoma.

FASCIAL SPACES IN THE LIDS

Attention has already been called to the fact that the skin is only loosely attached to the underlying orbicularis oculi muscle, thus presenting here a fascial space in which fluid can readily accumulate. This space is largely limited to the orbit by the more dense connective tissue attaching the skin to the musculature over the remainder of the face and forehead. Behind the muscle, also, is an area of loose

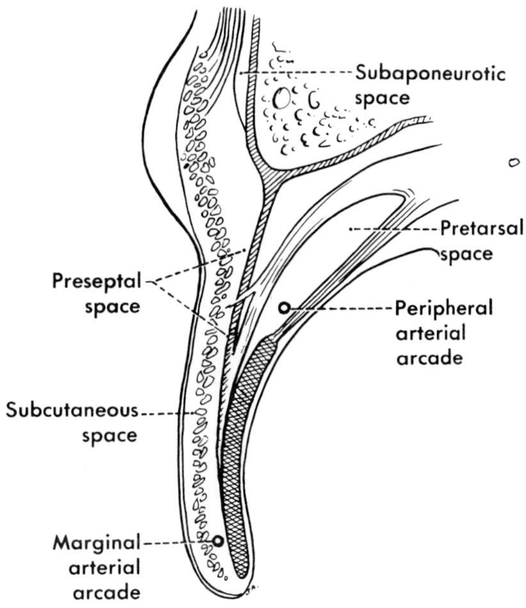

Fig. 2-5. Fascial spaces of the upper lid.

connective tissue, which is traversed by the chief blood vessels and nerves of the lid. In the lower lid this space is a closed one and is known as the preseptal space.

In the *upper lid* (Fig. 2-5) two spaces are described. The *pretarsal space* lies behind the aponeurosis of the levator palpebrae and in front of the superior tarsal muscle and an upper part of the superior tarsus. It contains the peripheral arterial arcade and laterally, extending downward behind the levator, the palpebral portion of the lacrimal gland. It is bounded above by the origin of the superior tarsal muscle from the levator palpebrae, and below by the attachment of the levator to the front of the tarsus.

The other space in the upper lid is the *preseptal space,* bounded in front by the orbicularis, behind by the septum, and behind and below by the levator and other fibers on the front of the tarsus. It contains loose connective tissue and, in its lower portion the marginal arterial arcade. The preseptal space is often described as being continuous above, deep to the frontalis muscle, with the subaponeurotic or danger space of the scalp; Whitnall, following Charpy, states, however, that these two spaces are separated by the strong attachment of the galea to the orbital margin and that injections into this space do not rise above the eyebrows, but rather pass back into the pretarsal space or pass outward among the fiber bundles of the orbicularis. This was observed in a case in which the preseptal space and the orbicularis muscle were thoroughly infiltrated and distended by injected latex, but none passed above the eyebrow into the subaponeurotic space. Clinically, according to Wolff, blood and pus may pass between the two spaces; this may occur, of course, along the branches of the ophthalmic artery and nerve which round the upper margin of the orbit to lie at first deep to the galea.

NERVES AND VESSELS OF THE LIDS

NERVES

The orbicularis oculi muscle is supplied by the facial nerve (temporal and zygomatic branches, and upper buccal according to McCabe and Boles) branches of which enter the lateral portion of the muscle both above and below the lateral raphe, while some, probably including the supply from upper buccal branches, run deep to the lower border of the muscle to join the infraorbital nerve and enter the muscle here. It might be noted, however, that the levator palpebrae, the other voluntary muscle of the lid, is an orbital muscle and is innervated by the oculomotor nerve. The smooth muscle in the lids is supplied through the superior cervical sympathetic ganglion.

Fig. 2-6. The cutaneous nerves related to the eyelids.

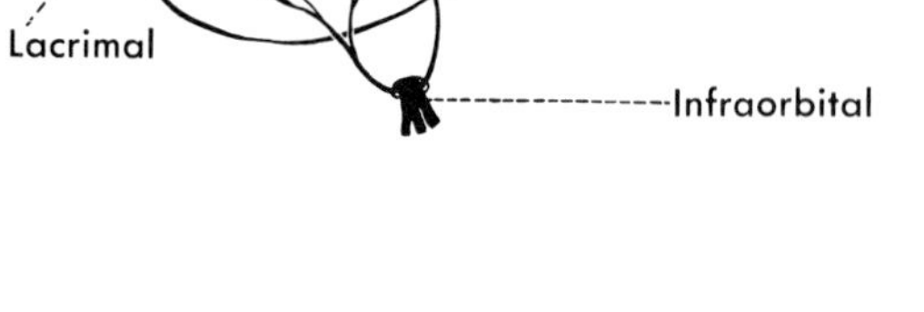

Paralysis of the facial nerve results in an inability to close the lids tightly, and the lower lid tends to drop away from the eyeball and become everted. Paralysis of the oculomotor nerve produces a ptosis of the lid and inability to open the eye voluntarily; interruption of the sympathetic supply to the superior tarsal muscle, most frequently seen as a result of interruption of the cervical sympathetic trunk (Horner's syndrome), interferes with the maintenance of a normal elevation of the upper lid, producing a constant ptosis of the lid.

In contrast to the ptosis produced by levator paralysis, however, paralysis of the superior tarsal muscle does not produce obliteration of the superior palpebral fold. Mahoney and Sheehan have shown on experimental animals that the superior tarsal muscle may completely overcome the ptosis of levator paralysis during moments of excitement.

The cutaneous nerves of the eyelids are those branches of the ophthalmic and maxillary nerves that appear subcutaneously to supply skin of the face and of the lids themselves (Fig. 2-6). The cutaneous branches of the ophthalmic nerve traverse the orbit and therefore necessarily penetrate the orbital septum.

The supraorbital nerve, the largest of these, leaves the orbit by the supraorbital notch or foramen and turns up to supply skin of the forehead; at the same time, however, it gives off minute branches that supply the larger portion of the upper lid. Medial to the supraorbital the smaller supratrochlear nerve also rounds the orbital rim to supply the forehead, and similarly supplies a more medial portion of the upper lid. Laterally, the upper lid near the lateral angle is supplied by twigs from the lacrimal nerve. The small infratrochlear nerve, another branch of the ophthalmic, supplies skin of both upper and lower lids at the medial angle (inner canthus) and along the side of the nose, while the major portion of the lower lid is supplied through upward running branches of the infraorbital nerve (the terminal portion of the

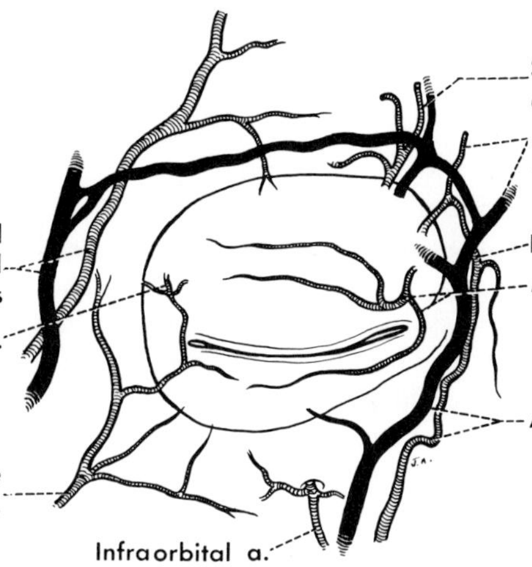

Fig. 2-7. The chief blood vessels about the orbital margin and in the lids; the veins of the lids are not shown.

maxillary). Skin lateral to the lateral angle is likewise supplied by the maxillary nerve, through its zygomaticofacial and zygomaticotemporal branches.

VESSELS

Arteries

Passing upward between the medial corner of the eye and the side of the nose is the terminal (angular) branch of the facial (external maxillary) artery (variable in size, and sometimes it or the vein is missing); usually accompanying it is the angular vein, the upper part of the facial (anterior facial) vein, which anastomoses above with veins of the forehead and communicates through the orbital septum with the superior ophthalmic vein (Fig. 2-7; see also Fig. 2-46). The vein lies lateral to the artery and is often visible through the skin, but in any case it is important to visualize the position of these vessels in an external approach to the lacrimal sac, since their cut ends tend to retract and make hemostasis difficult. It is preferable to avoid them, if at all possible, in incisions in this area. Inferior to the orbit are branches of the transverse facial artery, while lateral to and above it run branches of the superficial temporal artery. All these vessels usually help supply the skin of the lids and anastomose with branches of the palpebral arterial arcades, but the chief supply to the lids is from the ophthalmic artery itself.

The larger branches of the ophthalmic artery are concentrated in the upper nasal quadrant of the orbit and pierce this portion of the orbital septum; hence, in operations around the orbit, the vessels in this position are usually the most difficult to ligate and section. The supraorbital artery arises from the ophthalmic within the orbit and accompanies the nerve of the same name through the supraorbital notch or foramen; as it turns upward deep to the frontalis muscle it gives twigs to the upper eyelid. More medially, the supratrochlear (frontal) artery, one of the terminal branches of the ophthalmic, also pierces the orbital septum to turn upward on the forehead with the supratrochlear nerve. Close to the supratrochlear artery, and immediately above the strong part of the medial palpebral ligament, the dorsal nasal artery (the other terminal branch of the ophthalmic) penetrates the orbital septum to supply skin at the base and side of the nose, give off filaments to the lacrimal sac, and anastomose with the angular artery.

The chief vessels of the upper and lower lids are the superior and inferior medial palpebal arteries, which arise either by a common stem or individually from the ophthalmic or one of its two terminal branches. The *superior medial palpebral artery* passes above the medial palpebral ligament and usually divides to form two trunks that run laterally in the eyelid as the peripheral and marginal arterial arcades. The peripheral arcade runs behind the aponeurosis of the levator palpebrae, between this and the superior tarsal muscle, while the marginal one lies on the front of the tarsus. The arcades of the upper lid may be completed laterally by lateral palpebral branches from the lacrimal artery, and they give off branches that supply the deeper structures of the lid and most of the conjunctiva. These vessels usually anastomose also with the supraorbital artery and with the zygomaticoorbital branch of the superficial temporal.

The *inferior medial palpebral artery* passes downward behind the medial palpebral ligament and then turns laterally to run transversely in the connective tissue on the front of the inferior tarsus. The arch in the lower lid may likewise be completed laterally by a twig from the lacrimal artery, and usually communicates inferiorly with the transverse facial artery. A secondary or peripheral arch may be present, but is inconstant in the lower lid.

The arterial supply of the conjunctiva of the upper lid is largely from the peripheral arterial arcade, but is supplemented toward the free margin of the lid by a supply from the marginal arcade. These branches supply not only the conjunctiva of the lid but also that of the fornix and most of the bulbar conjunctiva, and are said to show little anastomosis with each other. The ascending branches from the peripheral arcade penetrate the superior tarsal muscle and pass toward and around the edge of the fornix, to form posterior conjunctival vessels that descend behind the bulbar conjunctiva; these vessels are prominent in acute conjunctivitis. About 4 mm from the cornea they anastomose with the anterior conjunctival branches of the anterior ciliary arteries, forming vessels that supply the pericorneal portion of the conjunctiva through a series of anastomosing branches. The vessels of the pericorneal arcade become prominent in intraocular inflammations, such as iritis, of the anterior part of the eyeball. The branches to the conjunctiva of the inferior fornix are similarly distributed, but often are necessarily derived entirely from the marginal arcade.

Veins

The veins of the eyelids are larger and more numerous than the arteries, and drain into the ophthalmic and the angular veins medially and into the superficial temporal laterally. A small venous plexus that lies between the aponeurosis of the levator and the superior tarsal muscle accompanies the peripheral arterial arcade of the upper lid; it connects with the veins of the levator and of the superior rectus, which are in turn tributaries of the ophthalmic vein.

The larger veins about the lids vary considerably, but usually the supratrochlear communicates with the supraorbital and ends below by becoming continuous with the angular (the two sending a communication through the lid to join the superior ophthalmic). The most prominent component of the supraorbital vein is frequently a transverse branch that runs above the orbit to anastomose with the superficial temporal vein; the supraorbital vein usually communicates with the supratrochlear before passing through the supraorbital foramen or notch (where it receives the venous drainage from the frontal sinus) into the orbit.

Lymphatics

The lymphatics of the lids are, like the veins, divisible into pretarsal and posttarsal plexuses that are, however, connected to each other by anastomosing channels; the posttarsal vessels drain the conjunctiva and the Meibomian glands, while the pretarsal or superficial vessels drain the skin and adjacent layers. According to Burch the superficial lymphatics parallel the lid margins rather than run perpendicular to them as usually described. Both plexuses drain into the same nodes, those from most of the upper lid, from the lateral angle, and from a lateral portion of the lower lid (that is, the majority of the vessels) draining into superficial parotid (preauricular) lymph nodes, while those from the medial angle of the eye and the medial portion of the lower lid drain into submandibular (submaxillary) nodes (Most). In many severe types of conjunctivitis there is swelling of the superficial parotid nodes indicating spread of the infection by way of lymphatics; their enlargement may be the first evidence of spread of malignancy from the eyelids. Enlargement of submandibular nodes as the result of ocular disease is rarely seen.

The Lacrimal Apparatus

The lacrimal apparatus consists of a secretory portion, the lacrimal gland and its ducts, and of a drainage apparatus, the lacrimal lake, the lacrimal canaliculi and sac, and the nasolacrimal duct.

SECRETORY APPARATUS

The lacrimal gland is partially divided into two portions (Fig. 2-8) by the lateral horn of the aponeurosis of the levator palpebrae; that is, the gland is folded about a posterior border of the horn in such a fashion that one part lies above the aponeurosis, while another part extends behiind the aponeurosis into the lid. The larger orbital portion of the gland ("superior lacrimal gland") lies in a shallow fossa on the frontal bone and is in contact anteriorly with the orbital septum, which separates it from other structures of the lid. Behind the free posterior edge of the lateral horn of the levator aponeurosis it is continuous with the palpebral portion of the gland. The smaller and thinner palpebral portion (palpebral process, "inferior lacrimal gland") extends into the lateral part of the upper lid to lie against the conjunctiva, through which it may be seen when the lid is everted; farther medially, it lies in front of the lateral portion of the superior tarsal muscle. Its position should be kept in mind during surgical procedures upon the upper lid.

The lacrimal gland is supported and maintained in position by the aponeurosis of the levator and by the lateral rectus muscle, on the upper surfaces of which it rests. Some accounts of the fascia of the orbit describe the lacrimal gland as surrounded by a capsule derived from a splitting of the periorbita or as

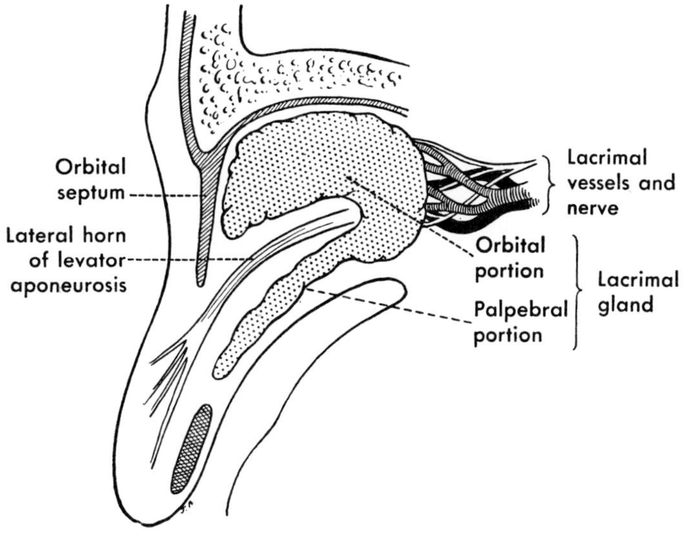

Fig. 2-8. The lacrimal gland in a sagittal section of the front of the orbit.

held in place by septa from the periorbita, but the gland has no very definite capsule and is normally only weakly attached to the periorbita. In malignant diseases of the gland it becomes firmly attached to the periorbita, and this in turn becomes more firmly attached to the adjacent bone. This may account for the pain and tenderness evoked in this condition by palpation of the orbital rim. It also accounts for the early invasion of bone by malignancies of the lacrimal gland and for the difficulty of eradicating them even by radical surgery.

The *nerves and vessels* to the lacrimal gland enter it at its undivided posterior border and consist of numerous branches of the lacrimal artery and nerve. In orbitotomy by the lateral approach it is well to avoid these vessels, if possible, for they bleed briskly and tend to retract into the gland substance. A branch of the infraorbital artery may also sometimes supply the gland, extending upward from the inferior orbital fissure to reach its lower border. The branches of the lacrimal nerve proper to the gland are sensory, but this nerve is usually joined by fibers from the pterygopalatine (sphenopalatine) ganglion that are secretory to the gland. The venous drainage of the gland is into the ophthalmic vein; the lymphatics, probably very superficial, are said to join the conjunctival lymphatics and hence empty into parotid nodes.

Fig. 2-9. The lacrimal drainage system.

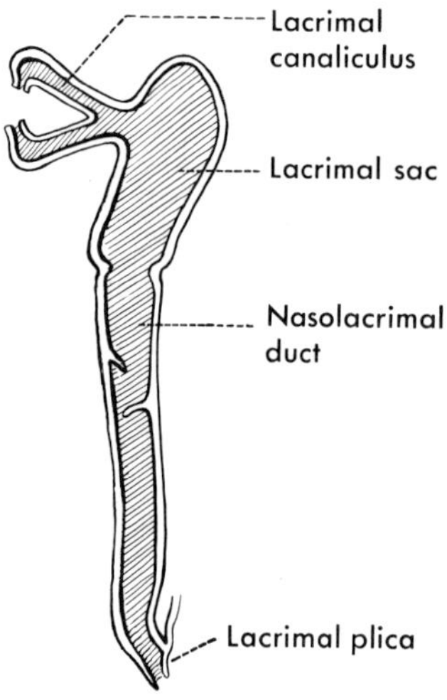

There are approximately 12 *ducts* (excretory ductules) of the lacrimal gland; those which come from the orbital part of the gland run through or close to the posterior part of the palpebral portion and are joined by ducts from this portion, although some of the palpebral ducts usually open independently. Removal of the palpebral portion may therefore destroy the drainage of the entire gland, but apparently does not always do so. The ducts are delicate and open for the most part into the outer part of the superior fornix, but one or more may also open into the inferior fornix. In addition to the lacrimal gland itself small accessory lacrimal glands (glands of Krause) lie adjacent to the conjunctiva and are regarded as a continuation of the palpebral portion of the gland. These smaller glands are normally sufficient to keep the cornea moist, but if neither they nor the lacrimal gland are functional the ensuing xerophthalmia may be severe enough to warrant transposition of the parotid duct to the lower conjunctival sac (Pierce and colleagues).

LACRIMAL DRAINAGE

Movements of the eyelid distribute the secreted tears over the surface of the eyeball, and any excess tends to accumulate in the lacrimal lake. The *lacrimal canaliculi* (lacrimal ducts) drain the lacrimal lake. The lacrimal canaliculi begin at the puncta, which open just lateral to the lacrimal lake, the orifice of the upper punctum facing downward and backward and that of the lower upward and backward. From the puncta the two canaliculi pass at first almost vertically away from the margins of the lids, but after about 2 mm turn sharply medially, the upper canaliculus then running medially and downward, the lower one medially and upward, so that they converge toward the lacrimal sac (Fig. 2-9). The canaliculi are largest at their bends, where each shows an ampulla; they are approximately 8 mm long. After piercing the lacrimal fascia, the two canals enter the lacrimal sac either close together or by a common stem, often at a small dilation of the sac, the sinus of Maier. In their courses the canaliculi lie for the most part behind the medial palpebral ligament and are surrounded by fibers of the pars lacrimalis (tensor tarsi, Horner's muscle) of the orbicularis.

The *lacrimal sac* lies in a fossa on the anteromedial wall of the bony orbit, but actually only partly within the orbital cavity, since the orbital septum attaches in front of its lower part only. At the posterior lacrimal crest the periosteum splits into two layers that surround the sac and converge against the anterior lacrimal crest, these portions of periosteum being referred to as the lacrimal fascia (Fig. 2-10). Loose connective tissue in which is embedded a venous plexus lies between the lacrimal sac and the lacrimal fascia; because of this plexus, operations upon the sac may be accompanied by considerable bleeding, even though the angular vein has been avoided. Anteriorly, the lacrimal fascia is in contact with the medial palpebral ligament, while posteriorly it is in contact with the pars lacrimalis (and when the ligament is described as U shaped the fascia is then believed to be in contact also with the posteriorly reflected limb of the ligament).

The lacrimal sac is approximately 12 mm long, and its upper end is covered by the medial palpebral ligament; in consequence, abscesses within the sac typically appear below the medial palpebral ligament, where incision should therefore be made for drainage. In approaches to the sac the presence of the angular vein must always be borne in mind. This vessel typically lies close to the side of the nose, although lateral to the corresponding artery (see Fig. 2-7) and therefore almost directly anterior to the lacrimal sac; hence the approach to the sac is usually made between the angular vessels and the medial angle of the eye. The medial wall of the lacrimal sac roughly corresponds to the most anterior part of the middle nasal meatus and that portion of the nasal cavity (a part of the atrium) lying just anterior and extending inferior to the middle concha on the nose, a fact worth remembering in attempting the intranasal approach to the sac or in searching for the opening of a dacryocystorhinostomy.

Since the lacrimal canaliculi open into the lacrimal sac approximately 2 to 3 mm below

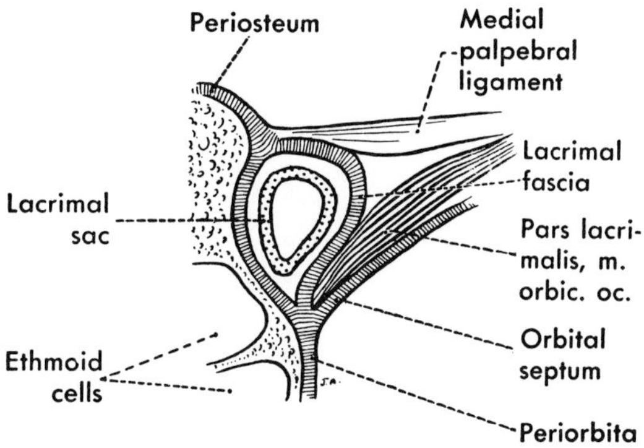

Fig. 2-10. Relations of the lacrimal sac in a horizontal section; right side, from above.

its apex, the sac ends blindly above in a fornix or fundus; below, it continues as the nasolacrimal duct.

It has long been argued that during closure of the lids the canaliculi are shortened and their ampullae compressed by the pars lacrimalis muscle fibers that surround them, so that blinking exerts a pumping action that helps move tears into the lacrimal sac. Jones ('57) regarded this as only a part of the "lacrimal pump" mechanism, adding the production of negative pressure in the lacrimal sac as the other component; he described muscle fibers of the preseptal portion of the orbicularis as arising from the lacrimal fascia and therefore pulling on this fascia during contraction of the muscle. The report (Kuribayashi) that the lacrimal plica (see below) moves with opening and shutting of the lids seems to confirm that these movements produce changes of pressure in the lacrimal sac.

The *nasolacrimal duct,* the connection between the lacrimal sac and the nasal cavity, lies within the bony nasolacrimal canal which is largely directly downward but slightly backward and laterally, lying in the thin bony wall between the maxillary sinus and the nasal cavity. It may be of the diameter of the sac itself, but the junction of sac and duct is typically narrowed; the two may even unite side to side rather than end to end (Fig. 2-11), thus making impossible probing of this system without injury to its walls. Within the nasolacrimal canal the duct is closely fused to the periosteum, in contrast to the separation between the lacrimal sac and its surrounding

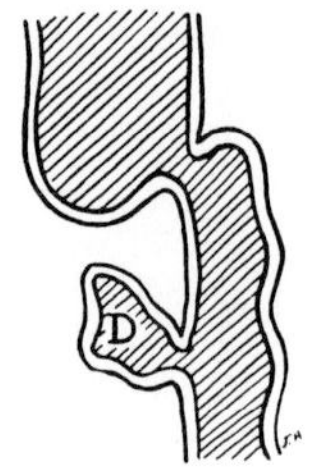

Fig. 2-11. Side-to-side junction of the lacrimal sac and nasolacrimal duct, making probing of the passage without tearing its mucous membrane impossible. (*D*), a diverticulum of the duct. (After Schaeffer JP: The Nose, Paranasal Passageways, and Olfactory Organ in Man. Philadelphia, Blakiston, 1920)

periosteum, but its walls contain a venous plexus that is connected above to that about the lacrimal sac and below to that of the nasal mucous membrane. The nasolacrimal duct terminates by passing through the nasal mucous membrane, usually fairly obliquely, to open into the inferior nasal meatus. This opening is usually about a fourth of the way back from the front border of the meatus and commonly high, just below the attached border of the inferior concha. Schaeffer ('12a) pointed out that the nasolacrimal duct develops by a canalization of a solid epithelial cord, and this development undoubtedly explains the not unusual presence of valve-like folds of mucous membrane in this duct. The most constant of these folds, which apparently do not function as valves, is a flap of mucous membrane, the plica lacrimalis or "valve" of Hasner, at the opening into the nasal cavity.

Schaeffer ('12b) and others have pointed out that there is no invariably typical form for the opening of the nasolacrimal duct, but that there are several normal types (see Fig. 4-13). Thus the ostium may be rather indefinite and slitlike and guarded by a fold of mucous membrane, may pass almost directly through the mucous membrane of the nose, may empty into a deep groove, or even open on a raised projection; occasionally also there may be more than one opening. Geddes reported a case in which the nasolacrimnal duct opened into the middle instead of the inferior meatus.

Operations to establish a new communication between the lacrimal sac and the nasal cavity in cases of obstruction of the nasolacrimal duct (dacryostenosis) are designed to take advantage of the findings of Whitnall on the relations of the sac to the ethmoidal air cells. In 100 orbits he examined, Whitnall ('21a) found some contact between one or more anterior ethmoid cells and the upper half of the fossa of the lacrimal sac in every case, and in more than half of these the whole medial wall of the upper part of the fossa was directly related to these cells; on the other hand, he found that approximately the lower half of the fossa was in every instance directly related to the middle meatus of the nose. Obviously, the formation of new communications should therefore take place in the lower part of the fossa and preferably in the posterior part of this through the thin lacrimal bone. Kinosian opened the lacrimal sac without displacing it and made the opening into the nose at the junction of the sac and the nasolacrimal duct.

The *blood supply* of the lacrimal sac and nasolacrimal duct is from the medial palpebal, dorsal nasal, angular, infraorbital and sphenopalatine arteries; the veins, similarly, drain both toward the orbit and toward the nose. The *nerve supply* is largely from the infratrochlear branch of the ophthalmic, with twigs to the duct probably also coming from the anterior superior alveolar branch of the maxillary nerve.

The Bony Orbit

The orbital cavities are, for convenience, usually described as pyramidal in shape, although they obviously depart considerably from this. Under this analogy, the orbital rim represents the base of the pyramid, while the apex is at the optic canal (foramen); the four sides of the pyramid are then the roof, the medial wall, the floor, and the lateral wall (Fig. 2-12). There are, however, several important deviations from this roughly pyramidal shape. The first is that the widest portion of the orbit is not at the base (the rim of the orbit) but lies about 1.5 cm behind this,

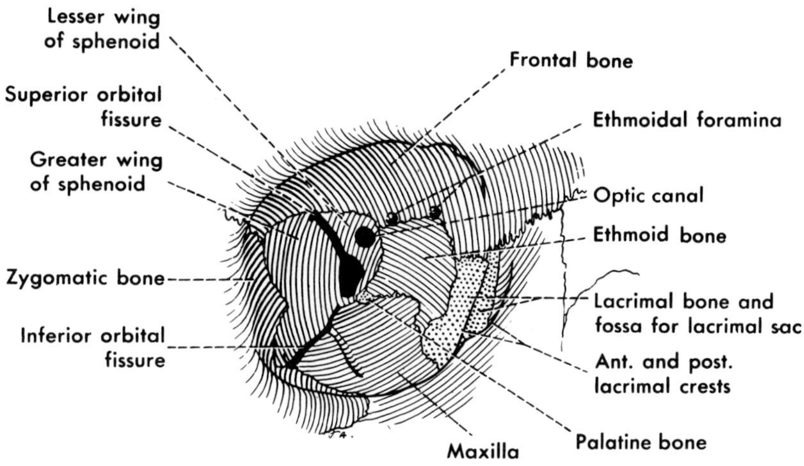

Fig. 2-12. The right bony orbit, somewhat from the lateral side.

where the lacrimal gland and the greatest circumference of the eyeball are. The second deviation pertains to the medial wall, which, unlike the other walls, is not particularly triangular, but more nearly quadrilateral. It is also important to note that the apex of the orbital pyramid does not lie directly behind the middle of the base of the orbit, but decidedly to the medial side of this; the medial walls of the two orbital cavities are approximately parallel to each other, and therefore the apex (the optic canal) lies almost directly posteriorly along the medial wall of the orbit. It is important to remember this point if the optic nerve is to be sectioned well behind the eyeball, as is required, for instance, in enucleation for retinoblastoma.

Ronthal and co-workers studied 48 orbits of prepared skulls (imported from India) and reported various measurements along each of the four walls; for the medial wall, for instance, they reported the distances from the anterior lacrimal crest to the anterior and posterior ethmoidal foramina and to the optic canal, and the distances between these points.

Medial Wall of Orbit

The medial wall of the orbit is formed above by a part of the orbital process of the frontal bone, which also forms the major portion of the roof of the orbit; the remainder of the medial wall (the larger part) is formed by the lacrimal, ethmoid, and sphenoid bones, arranged anteroposteriorly in that order. Of these elements, the orbital lamina of the ethmoid (formerly lamina papyracea, because of its extreme thinness) forms the larger and more important component. It is a common route through which sinus infections produce an orbital cellulitis, since it separates the orbit from the ethmoid air cells, and it is so thin that the outlines of these cells as their walls attach to the lamina may usually be seen easily in dried skulls. Ethmoid air cells typically invade also the lacrimal bone and may likewise invade the frontal, or the frontal sinus may extend backward in the upper portion of the medial wall and in the roof of the orbit. Behind the ethmoid is a portion of the body of the sphenoid bone, where this forms the medial portion of the optic canal; the sphenoid sinus may here be in close relation to the orbit. At the upper border of the ethmoid, usually on the frontoethmoidal suture, the anterior and posterior ethmoidal foramina can be seen. Anteriorly, the lacrimal bone and the frontal process of the maxilla lodge the fossa for the lacrimal sac; as already stated, the upper portion of this fossa is regularly in contact with anterior ethmoid cells, while the lower portion regularly shares a common wall with the middle meatus of the nose.

Floor of Orbit

The floor of the orbit slopes up on its medial side to join the medial wall; hence no sharp

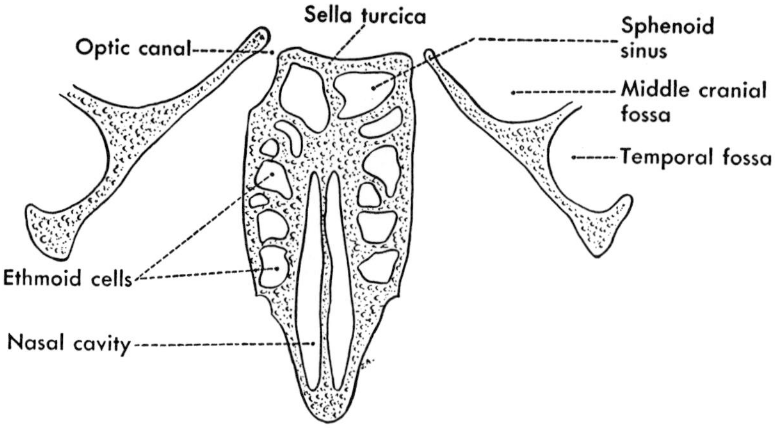

Fig. 2-13. Relations of the bony orbits as seen in horizontal section.

line can be drawn between these two walls. Laterally and posteriorly the inferior orbital (sphenomaxillary) fissure may be considered as separating the floor from the lateral wall. The floor is formed primarily by the orbital process of the maxilla, supplemented anterolaterally, in front of the inferior orbital fissure, by the orbital process of the zygomatic, and posteriorly by a small portion of the palatine. Extending forward in the floor of the orbit from about the middle of the inferior orbital fissure is the infraorbital groove, which transmits the infraorbital nerve and vessels to the infraorbital canal.

Lateral Wall of Orbit

The larger portion of the lateral wall of the orbit is the greater wing of the sphenoid bone; anterior to this, at and behind the orbital rim, is the frontal process of the zygomatic; it is along the suture between sphenoid and zygomatic bones that the bone flap is fractured in lateral orbitotomies. Posteriorly, at the lateral rim of the optic canal, is a portion of the lesser wing of the sphenoid bone, these bones together separating the orbital cavity from both the temporal and the middle cranial fossae (Fig. 2-13). A slight projection, the marginal (zygomatic) tubercle, or orbital tubercle of Whitnall, is usually present just behind the orbital margin and represents the area of attachment of the lateral horn of the levator aponeurosis, the lateral palpebral ligament, and the lateral check ligament, which attach in that order, from before backwards (although usually fused at their attachments here). This tubercle was identified by Whitnall ('21a) in 95% of 2000 skulls.

The greater and lesser wings of the sphenoid are separated by the superior orbital (sphenoidal) fissure, which is also prolonged anteriorly as a partial separation between the lateral wall and the roof of the orbit. The superior orbital fissure leads from the cranial cavity (middle cranial fossa) into the orbit; the inferior orbital fissure forms a communication in its posterior part from the pterygopalatine (sphenomaxillary) fossa to the orbit, while anteriorly it joins the orbit and the infratemporal fossa. Anterior to the superior orbital fissure, where the lateral wall blends with the roof of the orbit, this wall is formed by a small portion of the frontal bone.

Roof of Orbit

The roof of the orbit is formed largely by the orbital process of the frontal bone, with the lesser wing of the sphenoid contributing a minor portion posteriorly. Anteriorly, the roof presents in its lateral part a slight impression, the fossa for the lacrimal gland, and medially there may be either a more pronounced but much smaller impression, the trochlear fovea, or a small trochlear spine. The roof of the orbit is thin and lies between the orbit and the anterior cranial fossa above and the orbit and the temporal lobe of the brain (middle cranial fossa) posterolaterally. The anterior portion of the roof is commonly occupied by a supraorbital extension of the frontal sinus that varies considerably in size. The medial portion of the roof is a favorite route by which mucoceles invade the orbit; it is closely related to ethmoidal cells that may also spread laterally some distance in the roof. At the orbital rim the roof of the orbit presents medially a notch, the supraorbital notch, which is frequently converted into a foramen by ossification of the ligament crossing it. There may be a more medial notch or foramen, the frontal notch, for the medial branch (once frontal nerve) of the supraorbital.

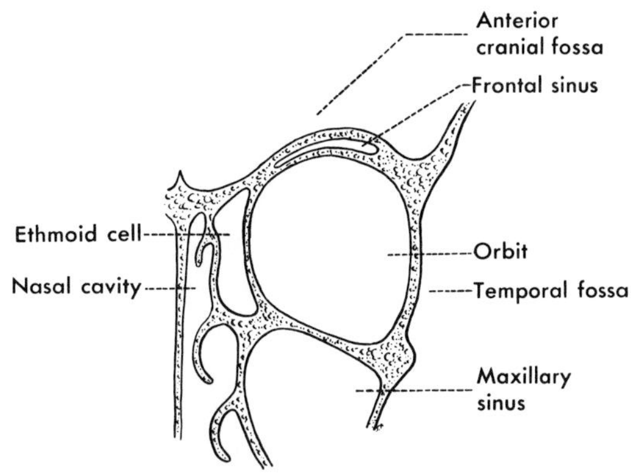

Fig. 2-14. Relations of the bony orbit as seen in frontal section.

Orbital Relations

The important relation of the floor of the orbit is to the maxillary sinus, or antrum of Highmore, which lies beneath almost the entire floor; the bone between the orbit and maxillary sinus may be as thin as 0.5 mm or less, and therefore tumors of this sinus may easily invade the orbit. Conversely, "blowout fractures" of the orbit involving this thin bone allow soft tissues of the orbit to be displaced into the sinus. The relationship of much of the lateral wall of the orbit is to the temporal fossa, occupied by the powerful temporalis muscle; historically, this is the route through which optic nerve tumors were removed and, more recently, lateral decompressions of the orbit through this wall have been advocated for exophthalmos. Posteriorly, a rather thin portion of the lateral wall separates the orbital cavity from the middle cranial fossa and the temporal lobe of the brain.

Since the orbit on three of its four sides comes in close relation to the paranasal sinuses (Fig. 2-14), it may be affected by infection or other pathological processes within these sinuses; Porter postulated that a large majority of orbital infections, perhaps as many as 75%, originate from infected sinuses. As would be judged by the thinness of the intervening wall, an ethmoiditis is one of the common causes of general infections of the orbital contents. Also, since the walls of the orbit are thin, sharp instruments can readily penetrate these walls and cause abnormal communications between the orbit and the various paranasal sinuses or, more important, enter the anterior or middle cranial fossae and damage the brain. The superior orbital fissure is sometimes the route of an encephalocele or meningocele, which may remain within the orbit or may leave it, by the inferior orbital fissure, to protrude into the pterygopalatine fossa.

Optic Canal

The optic canal (foramen), some 4–10 mm long, is the rounded passageway at the apex of the orbit through which the optic nerve and the opththalmic artery pass from an intracranial to an intraorbital position; it is formed medially by the body and laterally by the lesser wing of the sphenoid bone (see Fig. 2-12). The walls of this canal thus have an important but varying relationship to the sphenoid sinus and to posterior ethmoid sinuses, depending upon the extent to which these sinuses may have invaded the lesser wing and the anterolateral aspect of the body of the sphenoid. The medial wall of the optic canal is rather regularly adjacent to the sphenoid sinus, unless this is particularly poorly developed or partially replaced by an ethmoid cell; Van Alyea, in a study of 100 sphenoid sinuses, found the wall of the canal actually projecting into the sinus in 40, and Dixon found it projecting deeply enough to be almost completely surrounded in 7% of 1600 skulls. Whitnall ('21a) quoted another worker as having found the sphenoid sinus projecting into the lesser wing of the sphenoid, above the optic canal, in about one-third of skulls examined, and refers to the fact that the canal may be completely surrounded by the sphenoid sinus, so that it is a very thin-walled tube passing through the sinus. A similar relationship to a posterior ethmoid cell was described by Goodyear, who stated that he had seen complete blindness follow surgical opening of the sphenoid sinus and of posterior ethmoid cells. Less extensive growth of a posterior ethmoid cell may bring it into relation with the optic canal only medially or, through invasion of the bone between the canal and the superior orbital fissure, inferiorly and laterally; Van Alyea found the canal projecting into an ethmoid cell in 5% of 100 canals examined. In any case, the bony wall between the cavities of the canal and sinus may be quite thin (0.5 mm or less), or may even present dehiscences.

The Periorbita and the Orbital Fascia

PERIORBITA

The periosteum of the orbit is usually referred to as the periorbita, but differs from periosteum on the outer surface of the skull only in that it is, for the most part, rather loosely attached (as is much of the periosteal portion of the dura intracranially), so that it may be fairly easily lifted from the bone by accumulations of blood or pus. Where the orbit communicates with the cranial cavity—that is, at the optic canal, the superior orbital fissure, and the anterior and posterior ethmoidal canals—the periorbita is of course continuous with the periosteum on the inner surface of the skull and therefore with the dura (Fig. 2-15); in these places, also, the periosteum is rather tightly attached to the bone, as it is also at the sutures and deep to the lacrimal gland. At the margin of the orbit the periorbita thickens and becomes continuous with the periosteum on the external surface of the skull; here also it gives rise to sheets of tissue that are reflected into the eyelids, like curtains, to form the orbital septum, already described in connection with the lids. At the posterior lacrimal crest the periorbita splits to surround the lacrimal sac and form the so-called lacrimal fascia, which at the nasolacrimal canal continues downward as the periosteum of this canal. It is sometimes stated that a thin sheet of periorbita passes on the lower surface of the lacrimal gland, and the gland is then described as being enclosed by the periorbita. This would appear to be a particularly strained description of the periorbita, since in fact the gland can be very easily separated from its bed against the orbital wall.

Posteriorly, around the optic canal and a medial portion of the superior orbital fissure, the periosteum is thickened as it gives rise to the tendons of origin of the four rectus muscles, which are here fused to form a common tendon, the anulus tendineus communis (anulus or tendon of Zinn; see Fig, 2-27). The portion of the anulus arising from the rim of the optic canal represents a point of fusion between tendons of origin of the muscles, periorbita, and the dural sheath about the optic nerve. At the anulus, also, the periorbita is sometimes described as being continuous with the fascial sheaths on the muscles; since contiguous connective tissue layers generally are continuous with each other this seems entirely reasonable, but the fascial sheaths of the muscles are so thin posteriorly as to be practically nondemonstrable.

Over the inferior orbital fissure the gap in the periorbita is bridged by connective tissue that contains a quantity of smooth muscle, the *orbitalis muscle* (inferior orbital, of Müller). It is this muscle that has been held to cause, by its contraction, the proptosis of exophthalmic goiter, although it is doubtful that this theory has any proponents today. Ruedemann, for instance, pointed out that a very considerable increase in the orbital contents, amounting to apporoximately 1600 mm^3, is

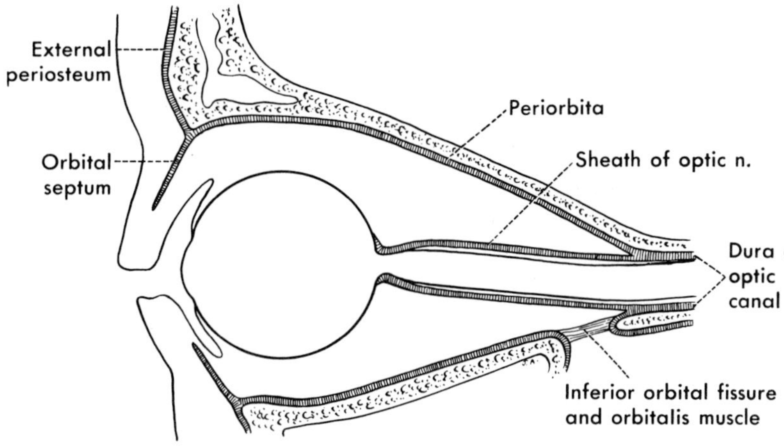

Fig. 2-15. Periorbita and dural sheath of the optic nerve in a vertical section of the orbit along the optic nerve.

necessary to move the eyeball forward 1 mm. It is impossible to imagine this muscle, slung as it is across the narrow infraorbital fissure, causing any appreciable decrease in the orbital space by its contraction; nor is the alternative idea, that it can produce marked congestion of the veins, much more attractive. However, Van Buren and co-workers suggested that posterior venous obstruction by a tumor contributes importantly to exophthalmos and that the anterior collateral pathways are inadequate under these conditions. It is interesting that unilateral exophthalmos is not necessarily due to an orbital tumor, but may be associated with hyperthyroidism (*e.g.*, Hinton).

FASCIA OF THE ORBIT

Orbital fat fills the portions of the orbit not occupied by other structures, but in addition to the fibrous connective tissue permeating this fat there are certain rather definite layers of fascia within the orbit. Each of the muscles of the orbit has a fascial capsule or sleeve; as already noted, this is a histological rather than a gross anatomical feature posteriorly, but anteriorly a thick layer of fascia appears about each muscle as it approaches the eyeball. In the latter location the sheaths of the four rectus muscles gradually spread laterally so as to come in contact with each other and form an intermuscular membrane that separates the fat within the muscle cone from that between the cone and the periosteum. The intermuscular membrane thus helps to delimit the space within the muscle cone for a short distance behind the eyeball; in consequence, explorations of the orbital contents outside the cone may fail to reveal tumors or other masses (including pus, according to Wolff) within the cone.

BULBAR SHEATH

The bulbar sheath (vagina bulbi, fascia bulbi, or Tenon's capsule; Figs. 2-16 and 2-18) is a definite fibrous layer intervening between the orbital fat and the eyeball; it is attached firmly to the sclera around the point of entrance of the optic nerve posteriorly and likewise firmly attached anteriorly to the sclera about 1.5 mm posterior to the limbus (the sclerocorneal junction). Between these two rings of attachment the inner surface of the sheath is attached to the sclera only by fine but numerous trabeculae that bridge the underlying space (episcleral space, intravaginal or intervaginal space, interfascial space of Tenon). When these trabeculae are broken by dissection, the inner surface of the sheath presents a smooth, almost serous appearance. The outer surface is less distinct, since it re-

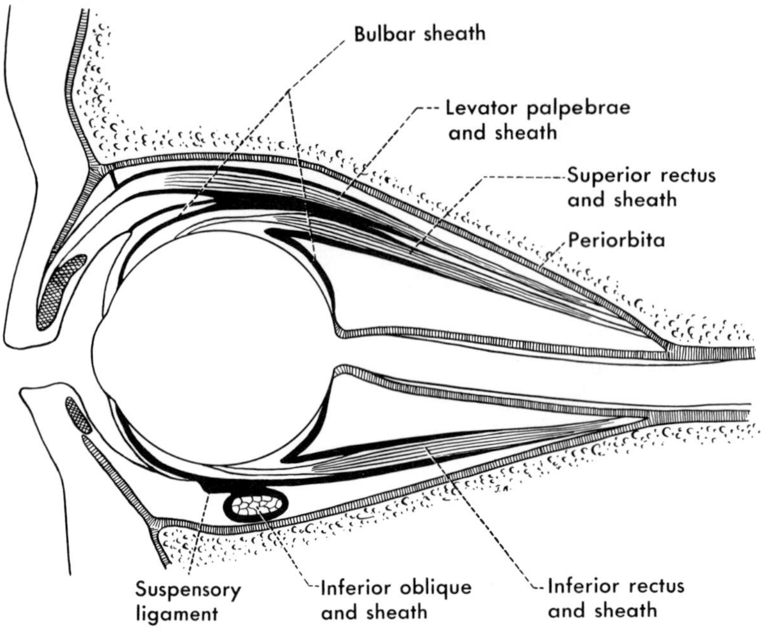

Fig. 2-16. Fascia of the orbit in a vertical section.

ceives the heavier trabeculae from the surrounding orbital fat. Around the optic nerve the bulbar sheath is pierced by the ciliary nerves and vessels, and in each of its four posterior quadrants by a vorticose vein. All six muscles moving the eyeball also enter the sheath. Although it was once held that the eyeball moved fairly freely within the bulbar sheath, it is now generally accepted that relatively little movement can take place between the bulb and the sheath and that the two move together upon a bed of surrounding fat. When the eyeball is removed the bulbar sheath can serve as a socket for the prosthetic bulb used to replace it; it has the additional advantage, in such cases, of moving with the contractions of the orbital muscles since, as described in the following paragraph, it is continuous with their sheaths.

The fascial sheaths of all six muscles of the eyeball, well developed in the more anterior part of the orbit, become continuous with the bulbar sheath as the muscles pierce the sheath to run between it and the eyeball to their insertion on the latter (see Figs 2-16 and 2-18). Because of this continuity, the bulbar sheath can be regarded as being formed by fusion of the fascial sheaths of the muscles, or the fascial muscle sheaths can be regarded as extensions of the bulbar sheath along the muscles. Thus when the eyeball has been removed from the capsule, the cut ends of the muscles can be seen protruding from individual sleeves of tissue that continue back around the muscles.

SUSPENSORY LIGAMENT

Although the suspensory ligament (of Lockwood) is sometimes described and illustrated as if it were an entity apart from the bulbar sheath and the sheaths of the muscles that contribute to it, it is really a thickened lower part of the bulbar sheath. Thus its middle and lowest portion consists of thick fascia contributed to the bulbar sheath by the inferior rectus and inferior oblique muscles as they cross each other below the eyeball; its ends, contributed to by the fascias of the medial and lateral rectus muscles, are attached to the orbital wall by the expansions of these fascias.

The suspensory "ligament" is therefore a hammocklike part of the bulbar sheath (Fig. 2-17) that, through its strong attachments to the walls of the orbit, supports the eyeball and resists anteroposterior displacement of both the sheath and the bulb. The intimate relation of the dense connective tissue of the ligament to the inferior oblique and inferior rectus muscles makes recessions and advancements of the tendons of these muscles more difficult, and Dyer regarded tenotomy of the inferior oblique as preferable to recession.

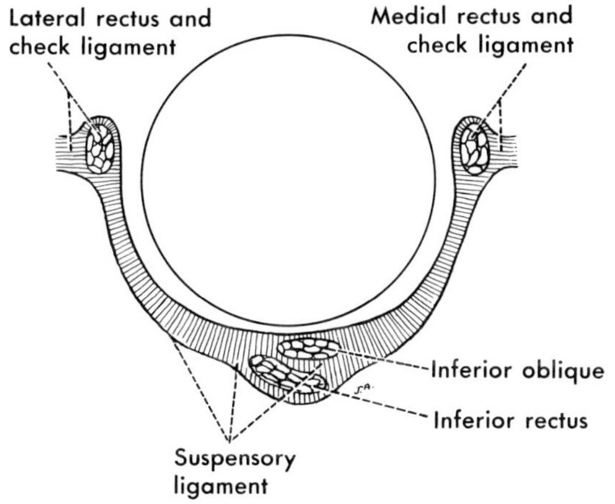

Fig. 2-17. The suspensory ligament.

OTHER EXPANSIONS OF THE MUSCULAR FASCIAS

In addition to their continuity with the bulbar sheath, the fascial sleeves of the muscles of the eyeball also send out certain other expansions, which are, for the most part, not so well marked. The most pronounced of these are from the fascias of the medial and lateral rectus muscles, which extend to attach to the adjacent walls of the orbit; because they apparently limit the action of the muscles they are named the medial and lateral check ligaments (Fig. 2-18). Posteriorly, where they are bordered by orbital fat, these ligaments are well defined, but anteriorly and inferiorly they are continuous with other connective tissue at the angles of the eye.

The *medial check ligament* is an expansion of the fascia on the medial surface of the medial rectus muscle, which forms a broad sheet attaching the muscle sheath to the orbital wall just behind the posterior lacrimal crest; anteriorly it is fused with the medial attachment of the orbital septum and the medial horn of the levator palpebrae (but not to the medial palpebral ligament, since the pars lacrimalis

Fig. 2-18. Fascia of the orbit in horizontal section.

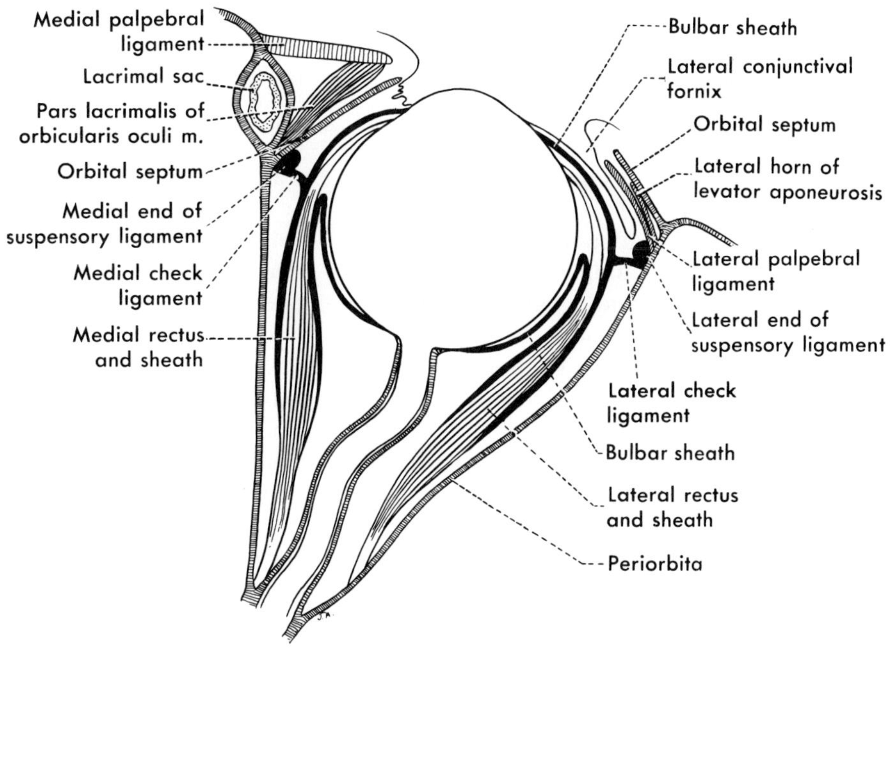

of the orbicularis, and the lacrimal canaliculi and sac, intervene). The *lateral check ligament* (lacertus of the lateral rectus), stronger than the medial, is an expansion from the lateral surface of the lateral rectus sheath to the marginal (orbital) tubercle on the zygomatic bone. At its lateral attachment it is fused anteriorly to the lateral palpebral ligament, and through this to the lateral horn of the levator. An anterior portion of each of these ligaments is sometimes described as forming the ends of the suspensory ligament but cannot, in fact, be separated from the remainder of the check ligaments. Whitnall ('21a) noted that these ligaments may occasionally contain aberrant strands of voluntary muscle and that several investigators have described smooth muscle in them. The major function of the check ligaments is not to limit the actions of the muscles with which they are associated, but to stabilize the eyeball and prevent it from being pulled backward by the contraction of any of the rectus muscles or forward by either of the obliques.

Less marked superior and inferior check ligaments are sometimes also referred to, but are often ill defined. A thickening of fascia on the upper surface of the levator palpebrae superioris has been described by Whitnall ('21a) and others as extending transversely below the superior margin of the orbit and functioning as the superior check ligament, or check ligament of the levator. Whitnall described it as attaching medially to the trochlea and helping to bridge the supraorbital notch. Anderson and Dixon regarded the horns of the levator aponeurosis as being the chief check ligament and, while granting that Whitnall's ligament could have a subsidiary function as a check ligament, stressed that its real importance is its contribution to supporting the lid. They warned that it should not be cut during operations on the lid.

A thin expansion from the sheath of the inferior rectus is usually described as passing into the lower lid to insert in a manner similar to the attachment of the levator in the upper lid, and therefore as helping to pull the lower lid downward as the eye moves down; likewise, most of the muscles are said to send thin fascial expansions to attach to the cul-de-sac of the conjunctival fornices, but none of these finer expansions is easily demonstrable grossly. An inferior check ligament, from the inferior rectus, is not usually described, but the fusion of the fascia about this muscle with that of the inferior oblique, and its continuity with the suspensory ligament, apparently should afford a definite check to extreme downward movements of the eye produced by the inferior rectus.

The fascial sheath about the superior oblique muscle is continuous with the U-shaped cartilage that forms the pulley for this muscle and with the ligaments that attach the cartilage to the wall of the orbit. The fusion of the fascia of the inferior oblique to that of the inferior rectus has already been referred to.

The expansions of the sheaths of the muscles to the walls of the orbit tend to aid the orbital septum in preventing herniation of orbital fat into the eyelids.

The Eyeball

A detailed description of the anatomy of the eyeball (bulbus oculi) is beyond this text's scope; for this, consult the various ophthalmological texts or detailed anatomies devoted to this subject. In this chapter it is possible to deal only with a few major features; the application of these, and of others, to surgery of the bulb has been discussed by Fralick.

The eyeball departs slightly from a perfect sphere, its horizontal and vertical diameters being about 23.5 mm and its anteroposterior one, because of the protrusion of the cornea, about 24 mm. Its most anterior part lies close to a line drawn between the upper and lower orbital margins, but because of the indentation of the lateral margin of the orbit approximately a third of the sphere projects in front of this margin. The eyeball occupies half or more of the total depth of the orbit which, as measured from the inferior orbital margin to the external end of the optic canal is some 44–50 mm.

The eyeball is enclosed largely by the bul-

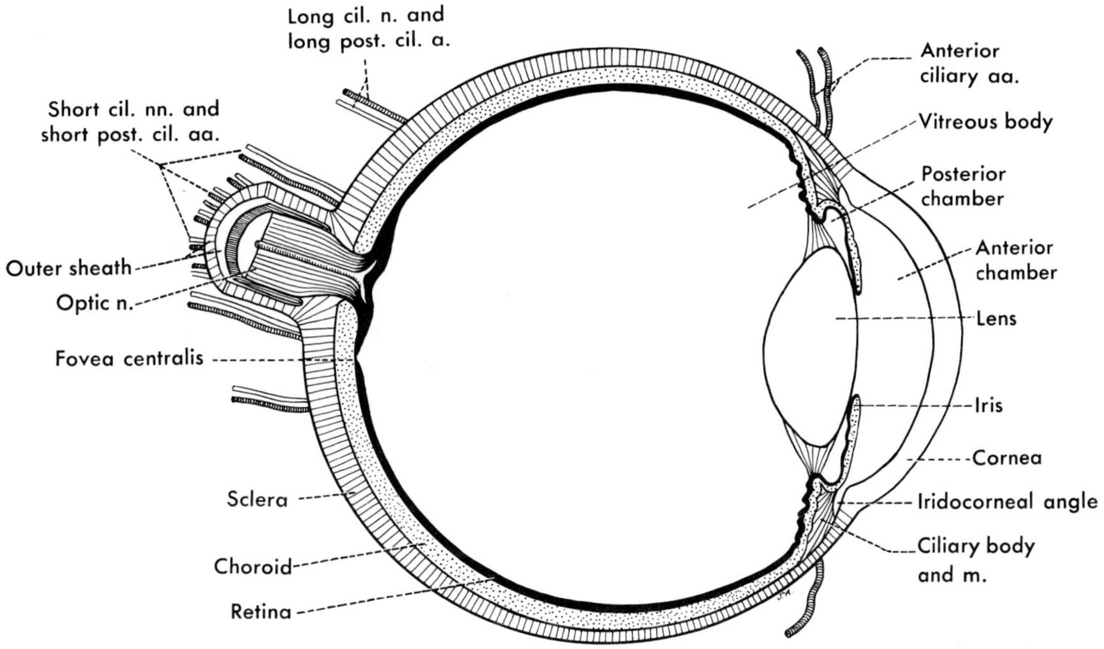

Fig. 2-19. General structure of the eyeball in a horizontal section.

bar sheath (capsule of Tenon, fascia bulbi), which surrounds it from the point of entrance of the optic nerve almost to the sclerocorneal junction; it is acted upon by six voluntary muscles that pierce the capsule to insert into the outer layer of the eyeball, the sclera.

In the major portion of its circumference the eyeball consists of three layers or *tunics,* which are, from without inward, the sclera (fibrous tunic or connective tissue layer), the choroid (vascular tunic), and the retina (internal tunic) or neural layer (Fig. 2-19). Toward the front of the eyeball the retina becomes rudimentary and with the choroid extends across the cavity of the eyeball, in front of the lens, to form the colored iris with its adjustable central aperture or pupil. In front of the iris and pupil the sclera is modified to form the cornea.

CORNEA AND SCLERA

Approximately the posterior five sixths of the fibrous tunic of the eyeball is the white and opaque sclera; the remainder is transparent cornea. The region of transition between the sclera and cornea is the corneal limbus, marked externally by a slight groove, the scleral sulcus. Wiener stated that the limbus is the weakest part of the eyeball, and rupture of the globe, which allows the lens to protrude against the conjunctiva, may therefore occur here.

Cornea

The cornea is more sharply curved than is the remainder of the eyeball and therefore looks like a segment of a smaller transparent sphere placed upon a larger opaque one. Refraction of light takes place in the cornea, and it is largely distortions of its curvature that give rise to the condition known as *astigmatism.* Aside from such distortions, the cornea is still not a perfect segment of a sphere, since it is said to be somewhat more flattened peripherally than centrally. The cornea is thinnest at its center and thickest at the periphery, varying according to Wiener from 0.8 mm at the center to 1.2 mm peripherally; this author has also stated that the cornea becomes thinner with advancing years. Von Bahr reported markedly different findings in measurements on the living eye—an average thickness of 0.565 mm, with a normal variation of 0.46 mm–0.67 mm and no sex or age difference. The anterior surface of the cornea is covered with an epithelium that is continuous peripherally with the conjunctiva. The small blood vessels supplying the cornea are limited to the region of the limbus, so that the cornea is largely avascular, but it is supplied abundantly with sensory nerves derived mostly from the ciliary nerves.

Sclera

The sclera receives the attachments of the muscles of the eyeball and is perforated posteriorly by the fibers of the optic nerve, the area of perforation being the lamina cribrosa. Wiener pointed out that the sclera is about 0.6 mm thick at the limbus and is thinnest just behind the insertions of the recti upon the eyeball, where it is 0.3 mm or less. Behind this point it increases again in thickness and reaches 0.6 mm at the equator and 1 mm posteriorly; at the lamina cribrosa it becomes very thin. Since tendon recessions or advancements offer obvious advantages and are apparently now usually preferrred to tenotomies (*e.g.*, Peter), the sclera's varying thickness is of considerable practical importance. Obviously, slight recessions of rectus tendons involve sewing them to the thinnest part of the sclera.

The *lamina cribrosa* is the thin portion of sclera pierced by numerous holes for the axons forming the optic nerve; it is situated not in the strict longitudinal axis of the eyeball, but above and medial to its posterior pole. Around the periphery of the lamina cribrosa the sclera is also pierced by the ciliary nerves and the short posterior ciliary arteries; slightly farther away on the medial and lateral sides of the attachment of the optic nerve the long posterior ciliary arteries pass through the sclera, and still more peripherally, in each of the four posterior quadrants of the eyeball, the venae vorticosae leave the sclera. Anteriorly, the anterior ciliary arteries enter the sclera at the attachments of the four rectus muscles.

Within the eyeball just behind the sclerocorneal junction (Fig. 2-20) lies a ring-shaped canal, the sinus venosus sclerae (sclerovenous sinus, canal of Schlemm); this endothelially lined vessel usually contains no blood, for it drains aqueous humor from the anterior

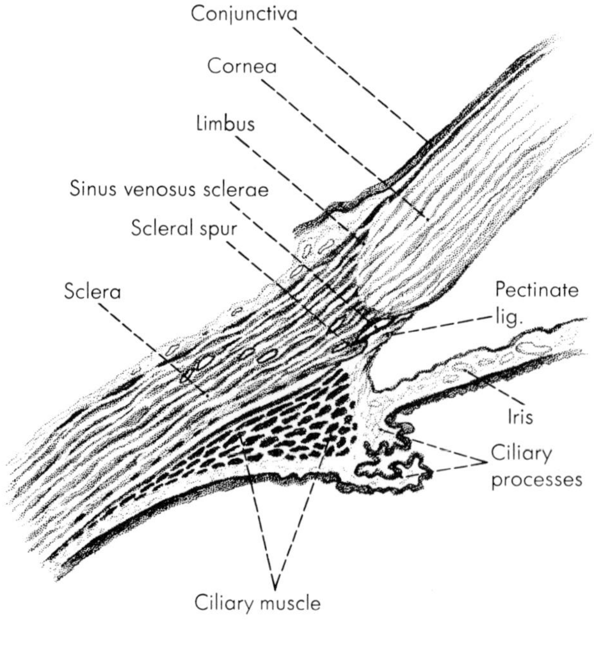

Fig. 2-20. The iridocorneal angle.

chamber of the eye. The sinus is drained by veins that open into it, and in the cat more than 99% of the aqueous humor is said to then drain into the adjacent intrascleral plexus (Bill); in man, veins draining aqueous humor vary in their origin, arising directly from the canal, arising indirectly through several communications with it, or arising from anastomoses between deep and superficial scleral venous plexuses (Ashton). Between the sinus venosus sclerae and the anterior chamber of the eye is the pectinate ligament (of the iridocorneal angle), consisting of fibers passing from the inner surface of the cornea to the base of the iris and containing interconnecting spaces (spaces of the iridocorneal angle, spaces of Fontana). These spaces open into the sinus on one side and the anterior chamber of the eye on the other and are usually referred to as the "trabecular meshwork," or network. The pectinate ligament is said to be responsible for about 75% of the normal resistance to filtration of the aqueous humor (Grant), and pathologic changes in it might therefore be expected to alter this resistance. Maul and co-workers found such changes in a 5-month old child with congenital glaucoma: the trabecular network was covered by mesodermal tissue, and the trabeculae were thickened and contained an increased amount of collagen fibers.

Just behind the pectinate ligament is a slight inward projection of the sclera, the scleral roll or scleral spur, which affords attachment to the ciliary muscle.

Sclerocorneal Junction

The sclerocorneal junction, sharply defined macroscopically although said to appear microscopically as an area of gradual transition, is the true or corneal limbus. The term *limbus* is sometimes used in a broader sense to include a strip almost 3 mm wide about the sclerocorneal junction, extending from the anterior fusion of the bulbar sheath with the sclera to the conjunctival ring (also called conjunctival limbus) on the cornea (Fig. 2-21); thus the bulbar sheath and the conjunctiva are both sometimes said to attach to the limbus. Since the area between these attachments is of particular surgical importance in operative procedures for glaucoma, confusion can be avoided by defining these attachments as the limbus or anulus of the bulbar sheath and the conjunctival anulus or limbus, respectively, and restricting the unqualified term to the macroscopic sclerocorneal junction.

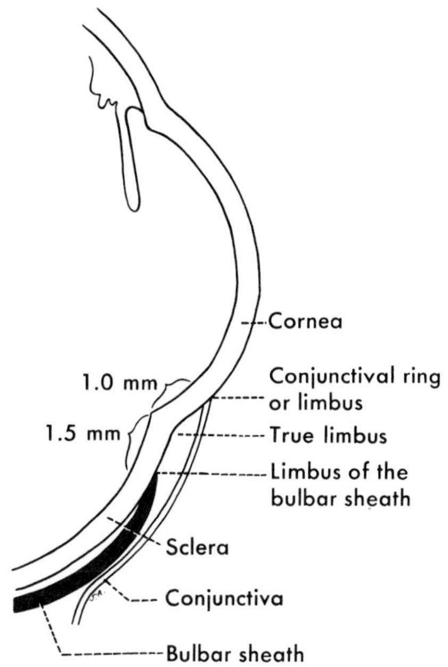

Fig. 2-21. Subdivisions of the limbus and the approximate distances between these parts. In the upper half of the figure is indicated the fashion in which the sclera participates in the formation of a part of the wall of the anterior chamber.

VASCULAR TUNIC

The middle, vascular, layer of the eyeball, often referred to as the *uvea* or uveal tract, consists of three continuous parts; from behind forward, they are the choroid, the ciliary body, and the iris. The choroid is firmly attached to the sclera in a ring about the optic nerve, and the ciliary body is attached firmly to the sclera at and behind the sclerocorneal junction; between these two areas of attachment, however, only very delicate connective tissues unites the two layers, and the resulting potential space between them is known as the perichoroidal space; it transmits the ciliary nerves and the long posterior ciliary arteries from their entrance posteriorly in the eyeball to the anteriorly situated ciliary body.

Choroid

The choroid coat, forming approximately the posterior two thirds of the uvea, is deficient posteriorly over the lamina cribrosa, to allow for the exit of the fibers forming the optic nerve, and anteriorly blends with the ciliary body. It consists largely of a plexus of vessels, the finer vessels lying in its interior part, adjacent to the retina, and the larger vessels lying in its outer portion. The innermost layer of vessels, the choroidocapillaris (choriocapillaris) nourishes the outer portion of the retina; it is separated from this by a basement membrane, the lamina basalis or Bruch's membrane. The outer surface of the choroid, the suprachoroid lamina (lamina fusca), contains pigmented cells, which are also found in the connective tissue about the larger vessels; the inner part of the choroid is generally devoid of pigment.

Ciliary Body

Anterior to the choroid the ciliary body represents a thickened middle portion of the uveal tract; the ciliary body and the choroid are continuous with each other at a wavy line known as the ora serrata. Like the choroid, the outer surface of the ciliary body conforms to the curvature of the sclera, but as it is traced forward it thickens internally so that in cross section it is triangular. Toward its anterior border the ciliary body is continuous centrally with the iris; on its inner surface, from the ora serrata forward, it and the ciliary part of the retina are fused; thus when retina and choroid are separated from each other, the retina tears at the ora serrata. Beyond this line, which lies abcout 8.5 mm behind the limbus, the retina is devoid of nerve cells and is represented only by layers of epithelium that, as the pars ciliaris and pars iridica retinae, form the inner surface of the ciliary body and the posterior surface of the iris.

On its inner surface, for about 2 mm behind the iris, the ciliary body presents many prominent longitudinally running ridges, the *ciliary processes,* to which are attached the zonular or suspensory fibers of the lens. This part of the ciliary body is the corona ciliaris and is the most vascular region of the entire eyeball, since the ciliary processes consist essentially of venous plexuses continued forward from the choroid. It is through the more functionally inert posterior portion of the ciliary body, the ciliary ring (orbiculus ciliaris, also called pars plana), an area some 3 to 4 mm wide in which neither vessels nor muscle fibers are plentiful, that incisions are often made for the removal of intraocular foreign bodies.

The bulk of the ciliary body is composed of the ciliary muscle (Fig. 2-20). Stieve described three sets of fibers in this muscle: meridional, radial, and equatorial or circular (the last have also been known as a muscle of Müller). Both the meridional and radial fibers take origin from the scleral spur; the meridional fibers insert into connective tissue on the scleral side of the ciliary body, particularly at the ora serrata, some of the fibers attaching into the sclera; the radial fibers curve inward to attach along the surface of the ciliary body adjacent to the vitreous body; and the circular fibers lie in the angle between the surfaces of the ciliary body that abut against the vitreous body and the anterior chamber, respectively.

Lens and Accommodation

The transparent lens (crystalline lens) is a biconvex body about 9–10 mm in diameter and 4–5 mm in anteroposterior thickness, the latter varying with focusing of the eye on near and far objects. Although it does not belong developmentally with the vascular tunic, since it arises from superficial ectoderm closely related to the primitive retina or optic cup, the lens becomes closely associated topographically with the ciliary body, and its increase in curvature during accommodation is made possible by the activity of the ciliary muscle.

The lens lies behind the iris, the central portion of its anterior surface being related to the pupil and through this to the anterior chamber of the eye; the peripheral portion of this surface forms a part of the posterior wall of the posterior chamber of the eye (see Fig. 2-19). The posterior surface of the lens, more highly convex than its anterior surface, lies in a concavity of the anterior surface of the vitreous body, a small retrolental space separating the two. In this position the lens is suspended by a complex series of fibers, collectively known as the ciliary zonule (of Zinn), the suspensory ligament, or the suspensory apparatus, individually referred to as zonular or suspensory fibers; these fibers attach centrally to the capsule of the lens, peripherally to the ciliary body, including both the ciliary processes and the ciliary ring or pars plana. Some of these fibers cross each other as they pass to their attachments upon the anterior and posterior surfaces of the lens, but the anterior and posterior fibers are connected by interstitial material that converts them into membranes (*e.g.,* Troncoso). Under the electron microscope, the fibers are said to be apparently identical with those that form the stroma of the vitreous body (Fine and Tousimis). The space between these two lamellae and the periphery of the lens is known as the zonular "spaces," also often identified as Petit's canal. According to Troncoso, and others, however, Petit's canal lies between the posterior lamella of the zonule and the vitreous body, and the perilenticular or circumlental space is known eponymically as "Hanover's space."

The lens is of complex structure and cannot be described in detail here. The elongated fibers of which it is largely composed have in part a laminated arrangement, but none of them completely encircles the lens; the areas of attachment of the ends of fibers to the ends of others form the lens sutures (stars, rays) visible upon inspection with a slit-lamp. New lens fibers are added to the periphery of the lens until at least adult life; this continuing growth gradually transforms the almost spheroid lens of the fetus to the flatter biconvex lens of the adult. The older, more centrally placed fibers form the denser nucleus, while the more superficially placed fibers form the softer cortex of the lens.

Adjacent to the anterior surface of the cortex is the lens epithelium (anterior epithelium). There is no corresponding posterior epithelium, for the cells once forming such a layer are completely converted into fibers. The highly elastic *capsule* is the outermost layer of the lens, and although varying in its thickness in different locations, forms a complete envelope around the other fibrous elements.

Accommodation is an increase in the curvature of the lens that focuses the eye for near vision. It results from a combination of two factors, contraction of the ciliary muscle and the high elasticity of the lens capsule. The soft cortex can be molded by the pressure exerted upon it by the very elastic capsule, but during the resting condition, when the ciliary muscle is relaxed, the capsule itself is under radial tension because of the pull of the zonule (suspensory ligament) upon it. The lens is then held in a more flattened state through this pull by the zonule and perhaps also because of its contact with vitreous body posteriorly. Contraction of the ciliary muscle relieves the tension upon the lens, which then becomes more rounded through the inherent elasticity of its capsule. Thus, contraction of the ciliary muscle produces relaxation of the lens and accommodation for near vision; paralysis of the ciliary muscle incident to injury to the oculomotor nerve abolishes the power of accommodation.

The exact method by which the ciliary muscle releases the tension upon the lens has long been in dispute, particularly as to whether the ciliary body can be displaced forward to narrow the circle from which the lens is suspended, but Fincham is reported to have recorded such a movement. According to Sieve, it is the radial fibers that should draw the inner surfaces of the ciliary body forward, at the same time increasing the bulk of the angle between the long and short limbs of the body; the circular fibers, of course, would simply bring the angle inward; and Stieve regarded the meridional fibers as pulling inward on the sclera at the same time that the recti compressed it from outside, thus both moving the ciliary body inward and elongating the optical axis of the eye. Although it has usually been assumed that loss of elasticity on the part of the lens capsule or increased density of the cortex can account entirely for the hypermetropia typical of advancing age, Stieve reported that degenerative changes in the ciliary muscle occur also.

Iris

Toward the anterior edge of the ciliary body the vascular tunic or uvea once again changes its character and abruptly changes its direction to become the major portion of the iris. This latter is arranged transversely across the eyeball in the frontal plane and presents a central diaphragmatic opening, the pupil. Beyond the attachment of the iris the ciliary body is continued forward to the sclerocorneal junction, to which it is firmly attached. The root or attachment of the iris therefore lies distinctly behind the sclerocorneal junction. The lens lies on the anterior surface of the vitreous body; the iris, in turn, divides the space in front of the lens into a small posterior chamber between iris and lens and a larger anterior chamber between iris and cornea, both occupied by aqueous humor.

The iris is usually believed to contain both circular and radially arranged muscle fibers, which are designated as the *sphincter pupillae* and the *dilator pupillae* muscles; the existence of radially arranged muscle fibers to form a dilator pupillae has been doubted, but the fibers are now recognized as being myoepithelial cells, not typical smooth muscle. Richardson has described them in the rabbit; Van Alpen, recalling that the action of the sympathetic system in dilating the eye has been said to be largely a vascular effect, studied monkey eyes that were frozen while highly midriatic, and protested that the changes observed could result only from muscular action. However, the sphincter pupillae is undoubtedly much stronger than the dilator, and many variations in the size of the pupil are due to variations in the contraction of the sphincter, rather than to activity of the dilator. The innervation of the sphincter pupillae is through the oculomotor nerve and the ciliary ganglion; active dilation of the pupil is controlled by fibers from the superior cervical ganglion by way of the carotid plexus and the ciliary nerves. According to Korczyn and co-workers, the more miotic pupil often found in aged persons is due to lowered sympathetic activity.

As already noted, the retina is continued over the inner surface of the ciliary body and the posterior surface of the iris. It is the pigment in the pars iridica retinae that is responsible for blue eye color; additional darkness in the color of the iris depends on the pigment present in the stroma anterior to this layer. The epithelium of the ciliary body is believed to be responsible for the formation of the aqueous humor.

Blood vessels constitute a considerable portion of the bulk of the iris and form two circular channels from which numerous radial vessels are given off. At the base of the iris, actually in the ciliary body, is the major vascular circle of the iris (circulus arteriosus iridis major) formed by anastomoses of the long posterior ciliary arteries with each other and with the anterior ciliary arteries. Radially directed branches from this circle run toward the free margin of the iris, close to which they anastomose to form the circulus arteriosus iridis minor. The minor arterial circle is paralleled by a corresponding venous circle, but the major arterial circle is not.

RETINA

Classic descriptions of the optic part of the retina usually include ten layers, but there are only three fundamental layers of nerve cells; these are, from without inward, the layer of rods and cones, the layer of bipolar cells, and the layer of ganglion cells (now respectively named the neuroepithelial stratum, the ganglionic stratum of the retina, and the ganglionic stratum of the optic nerve, and collectively grouped as the cerebral stratum). External to the layer of rods and cones is the pigmented epithelial layer of the retina (stratum pigmenti), and internal to the ganglionic layer of the optic nerve are the nerve fibers that compose this nerve, usually known as the optic stratum. Light entering the retina must

Fig. 2-22. The conversion of the optic vesicle into the retina. The mantle layer is shaded; the rods and cones develop in the deeper part of this, next to the cavity of the optic cup and the pigmented layer of the retina.

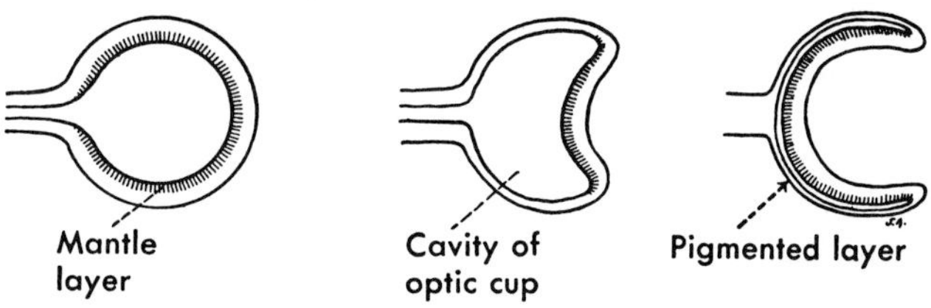

therefore pass through all other layers of this structure, with the exception of the pigmented layer, in order to reach the layer of rods and cones; this circumstance, which may seem unphysiologic, is brought about through the method of development of the retina (Fig. 2-22).

The anlage of the retina develops as an outgrowth, the optic vesicle, from the central nervous system, and like other typical parts of the central nervous system has its mantle or cellular layer located centrally, about the lumen, while its marginal layer, in which fibers will develop, lies peripherally. When the outer wall of the optic vesicle is invaginated to form the optic cup, the relationship between mantle and marginal layers is not changed, of course, so the marginal layer, even though it forms the innermost layer of the retina, is still between the mantle layer and the external source of light. Thus the rods and cones develop adjacent to the original central cavity, which separated the neural retina from the pigmented epithelial layer, and it is the optic stratum and its associated vessels that is first presented to light entering the eyeball.

The method of development of the retina also has another consequence of practical importance. The outer, epithelial, layer of the retina becomes firmly attached to the choroidocapillaris, the two sharing a basement membrane; however, the original lumen of the optic vesicle is still represented in the retina by a potential slit between the layer of rods and cones and the epithelial layer, and it is through this area of weakness that separation of the retina occurs.

At the *ora serrata* the retina becomes thinner, the neuroglia increases, and the retina ceases as a functional layer, although it is prolonged forward over the ciliary body and the posterior surface of the iris as a two-layered epithelium. Posteriorly, at the *fovea centralis,* the retina becomes markedly thin because of the absence of ganglion cells and most of the nerve fibers, and the rods (primarily for night vision) gradually disappear to be replaced entirely by cones (primarily for day and color vision); thus this portion of the retina is most suited for acute vision in the light-adapted eye. Surrounding the fovea, in the macula (lutea), the ganglion cell layer of the optic nerve is particularly thick. However, where the optic nerve fibers converge to form the optic disk or nerve head (optic papilla), all other layers of the retina disappear, so that the disk is formed entirely of nerve fibers and hence is insensitive to light; its presence is responsible for the blind spot in the visual field. From the optic disk the nerve fibers pass through the lamina cribrosa, become myelinated, and form the optic nerve.

AQUEOUS HUMOR; VITREOUS BODY

The *aqueous humor* occupies the anterior and posterior chambers of the eye and is formed largely or entirely by the vascular ciliary body. According to Bárány's experiments on the rabbit, ultrafiltration plays a minimal role in the formation of the aqueous humor, which is formed entirely or primarily by secretion. The drainage of the aqueous humor, as already stated, is through the spaces of the iridocorneal angle ("trabecular meshwork") into the scleral venous sinus and hence into the anterior ciliary veins. Increased pressure of the aqueous humor is designated as glaucoma.

The viscid *vitreous body* (often called vitreous humor) lies behind the lens and against the inner surface of the retina. It is formed in the embryo during the stage in which the developing lens receives a branch (hyaloid artery) from the ophthalmic artery; the proximal portion of this branch forms the central artery of the retina, but the distal portion degenerates and leaves an empty canal, the hyaloid (Cloquet's) canal, through the vitreous. The anterior expanded end of this canal forms a retrolental space that may be observed in slit-lamp ophthalmoscopy (Mann). According to Mann, however, the hyaloid canal is usually displaced inferiorly in the adult, so that it no longer runs in the median horizontal plane. Since the vitreous body, unlike the aqueous humor, is not replaced, its loss during intraocular manipulations is to be avoided.

The vitreous body is usually described as consisting of stroma or framework and a more fluid vitreous humor. According to Fine and Tousimis, the stroma consists of interconnecting laminae formed by fine filaments.

BLOOD VESSELS

Central Artery

The blood supply of the eyeball is largely from the short and long posterior ciliary arteries and from the anterior ciliary arteries, but the retina is also supplied by the central artery of the retina. This artery, with its accompanying vein, lies in the center of the distal centimeter or more of the optic nerve and passes with this nerve through the lamina cribrosa, appearing thus at the center of the optic disk (Fig. 2-23). From this point the branches of the vessels pass over the retina, embedded in the inner layer of the stratum opticum, adjacent to the internal limiting membrane. The arteries here are distinctly smaller in caliber than the veins, but both sets of vessels are so placed as to interfere with the path of light; neither is distributed directly to the region of the fovea, however, but rather encircles it, with branches proceeding somewhat radially toward it. In its course the central artery of the retina gives off branches into

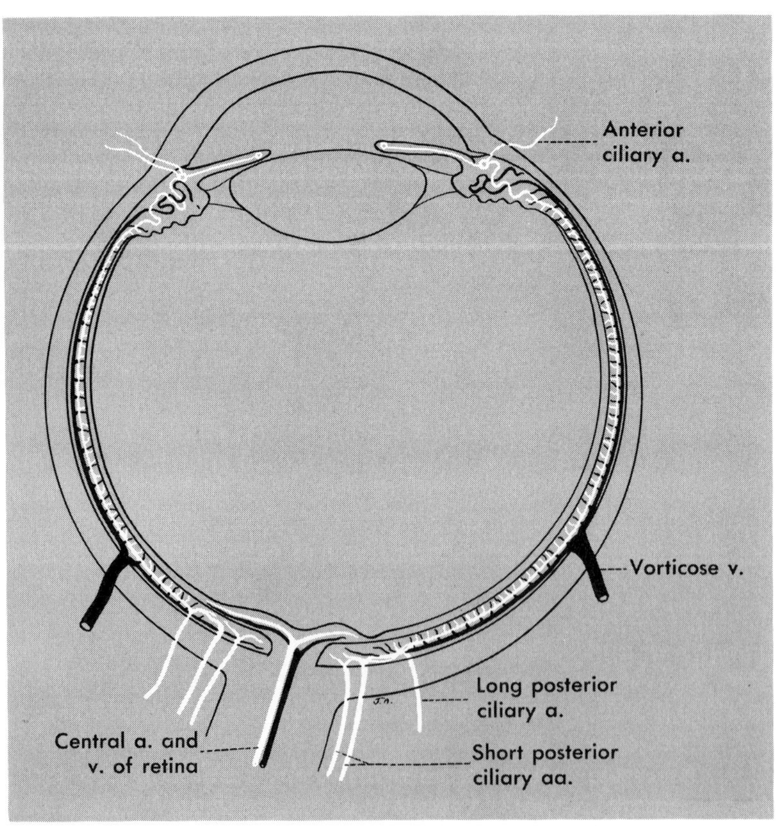

Fig. 2-23. Chief vessels of the eyeball. The details of their distribution within the eyeball are not shown here (see text).

the inner layers, but the outer layer, of rods and cones, is devoid of vessels and depends largely for its nourishment upon its close relationship to the exceedingly vascular choroidocapillaris of the choroid.

Other Arteries of Eyeball

The arteries to the choroid and the iris are the short and long posterior ciliary and the anterior ciliary arteries. Minute twigs from these also reach the sclera, which is, however, almost avascular.

The short posterior ciliary arteries, arising from the ophthalmic, penetrate the sclera in a ring about the optic nerve and immediately branch to supply the choroid; the two long posterior ciliary arteries enter the sclera on the medial and lateral sides of the optic nerve, run forward between the choroid and sclera to the region of the ciliary body, and here pierce the ciliary muscle to anastomose with each other and with the anterior ciliary arteries to form the major arterial circle of the iris. Recurrent branches from the long ciliary arteries, the major arterial circle, or the anterior ciliary arteries supply the choroid as far back as the equator, where they meet branches of the short posterior ciliary arteries. Here the two systems anastomose through the networks of the choroid, or by direct continuity (Wybar, '54).

The anterior ciliary arteries arise from the muscular branches of the ophthalmic artery to the rectus muscles; according to Wolff the lateral rectus muscle usually furnishes only one anterior ciliary artery, while the remaining three muscles furnish two each. These arteries supply twigs to the sclera and the conjunctiva and pierce the sclera at the attachment of the tendons of the muscles, subsequently joining with the long posterior ciliaries in forming the greater arterial circle of the iris. As already described, the lesser arterial circle is formed by anastomosis of branches given off from the greater circle.

The two circulations to the eyeball, that of the central artery and that of the ciliary arteries, are completely separate except for very small anastomoses in the pia of the distal part of the optic nerve and similar small ones in the sclera about the nerve head (Wybar, '56). Otherwise, the central artery and its branches are end arteries; they neither anastomose with each other nor with the anomalous cilioretinal artery (reported by Justice and Lehmann as present in 49.5% of 1000 patients) that a short ciliary vessel may send to the retina. Brown and Shields found that of 107 patients with central artery occlusion, 26% had some macular sparing because of a patent cilioretinal artery; these patients regained a variable amount of visual acuity, depending on the distribution of the artery. According to Scarlett, occlusion of the central artery is usually brought about by a combination of arteriosclerosis and a thrombus, and is more likely to occur in the neighborhood of the lamina cribrosa; he stated that digital massage is sometimes useful in restoring partial vision after such an occlusion. Since the outer portion of the retina, the layer of rods and cones, is primarily dependent for its nutrition upon contact with the choroidocapillaris, separation of the retina from the choroidocapillaris leads to nonfunction of the receptive layer of the retina; death of this layer does not immediately follow detachment, however, because of diffusion from the periphery of the detached area and from the inner, vascular, portion of the retina.

Venous Drainage

The venous drainage from the inner portion of the retina is by means of the central vein, which accompanies the central artery through much of its course and usually ends independently in the cavernous sinus, although it normally communicates with the superior ophthalmic vein and may end by joining this.

Almost all the drainage of the choroid is by means of four veins which do not accompany the ciliary arteries but pierce the sclera some considerable distance from the optic nerve; one of these veins appears in each of the four posterior quadrants of the eyeball, being formed by the convergence of veins in a radial fashion; from the manner in which they are formed, they are known as the *vorticose veins* or venae vorticosae; after penetrating the sclera they pass through the bulbar sheath, and in the intraorbital fat the two superior vorticose veins join the superior ophthalmic vein (or the lateral one may join the lacrimal), while the inferior ones join the inferior ophthalmic vein. Anterior ciliary veins are said to parallel the arteries of the same name; some authors also describe true posterior ciliary veins, but apparently these are minute and inconstant, and the venae vorticosae represent them functionally as the drainage of the choroid.

No *lymph vessels* have been demonstrated in the eyeball or indeed elsewhere within the orbit except for the lacrimal gland, derived from the modified skin of the conjunctiva. (Wolter and Rosenberg suggested that there were probably others in the orbit from which they removed an apparently normal lymph node.) However, Behrman regarded a perivascular space around the central artery of the retina, said to open into the orbit, as a lymphatic space draining the eyeball, and attributed papilledema to obstruction of this space.

The Muscles of the Orbit

The seven voluntary muscles of the orbit (Fig. 2-24; see Figs. 2-36 and 2-37) are the levator palpebrae superioris, the superior, inferior, medial, and lateral rectus muscles, and the superior and inferior oblique muscles. The smooth musculature of the orbit includes the orbitalis muscle, the superior and inferior tarsal muscles, and the ciliary and iridial musculature within the eyeball, all of which have already been described.

The voluntary muscles within the orbit are supplied by the third, fourth, and sixth cranial nerves: the superior oblique is supplied by the fourth nerve, the lateral rectus by the sixth, and the remainder by the third. The tarsal and orbital muscles (of Müller) are supplied by sympathetic fibers derived from the carotid plexus and hence from the superior cervical ganglion; so likewise is the dilator pupillae; the sphincter pupillae and the ciliary muscle are supplied by parasympathetic fibers through the third or oculomotor nerve.

LEVATOR PALPEBRAE

The levator palpebrae superioris arises from the lesser wing of the sphenoid bone just above and in front of the optic canal and inserts into the upper lid; although it is not among the muscles forming the anulus tendineus, its tendon of origin blends with that of the superior rectus at the periosteal attachments. The levator receives its nerve supply from the superior division of the oculomotor nerve, which first supplies the superior rectus muscle and then penetrates the medial portion of that muscle, or passes around its edge, to reach the levator. Paralysis of the levator produces a ptosis of the lid and, since the tonus of the muscle is responsible for the superior palpebral fold, a disappearance of this fold.

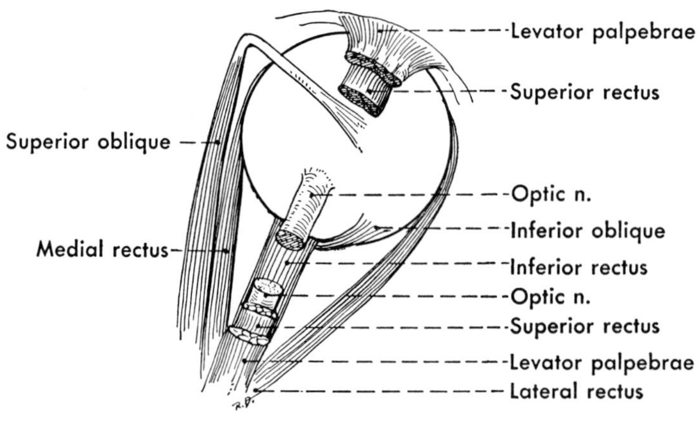

Fig. 2-24. Muscle of the eyeball.

From its origin the muscle extends forward above the superior rectus, lying in close contact with this; the muscle fibers end in a strong tendon which, after passing the equator of the eyeball, expands medially and laterally to form an aponeurosis (see Figs. 2-2, 2-24, and 2-36) that attaches to the orbital walls before continuing into the lid. Of these expansions, usually known as horns or cornua, the stronger lateral one deeply indents the front of the lacrimal gland, dividing it into palpebral and orbital parts, to attach to the marginal (zygomatic, orbital) tubercle between the lateral muscular raphe and the lateral palpebral ligament. The medial horn can be traced with more difficulty to an attachment behind the medial palpebral ligament. Just before the muscle changes to tendon it gives rise on its deep surface to the superior tarsal (Müller's) muscle. The attachment of the levator into the lid, and the fascial expansions from its sheath, have already been described.

The sympathetic fibers from the carotid plexus to the superior tarsal muscle are said by Whitnall ('21a) probably to reach the lid by way of the oculomotor branch to the levator, but not everyone agrees that the oculomotor receives a communicatin from the carotid plexus. Alternative routes would be by way of branches of the ophthalmic nerve or by way of the ophthalmic artery and its branches. Collin and co-workers thought the latter more probable for at least the last part of their course, since in the monkey they found degeneration of some nerve endings in the muscle both when they severed a blood and possible nerve supply through the levator and when they severed part of the blood supply from the lid. Paralysis of this muscle, like paralysis of the levator itself, produces a ptosis of the lid but, in contrast to paralysis of the levator, no change in the superior palpebral fold.

Through much of its course the levator palpebrae is in direct contact below with the superior rectus muscle. The trochlear nerve crosses above the proximal end of the levator to reach the medially situated superior oblique muscle; the frontal nerve runs above the muscle for a considerable distance in gradually crossing from lateral to medial sides of the orbit; the supraorbital artery appears on the medial side of the muscle abut a third of the distance from origin to insertion and is thereafter in relation to the upper medial border of the muscle as far as the supraorbital notch, lying medial to the supraorbital nerve.

SUPERIOR OBLIQUE

The superior oblique muscle, the longest and most slender of the muscles of the orbit, arises by a thin tendon above and medial to the optic canal, just outside the anulus. The muscle is fusiform rather than ribbonlike and becomes tendinous some distance before reaching the trochlea (Fig. 2-25). The trochlea itself is a small piece of U-shaped cartilage attached by ligaments to the fovea or spine on the frontal bone; the trochlea and its ligaments are continuous both proximally and distally with the heavy connective tissue sheath of the muscle, of which the trochlea and trochlear ligaments can be regarded as a special development. Because of the necessity of free movement of the tendon over the trochlea, the space about the tendon is particularly well developed here and is lined by a synovial tendon sheath.

Fig. 2-25. Superior and inferior oblique muscles from behind and laterally. (After Fink WH: Surgery of the Oblique Muscles of the Eye. St. Louis, CV Mosby, 1951)

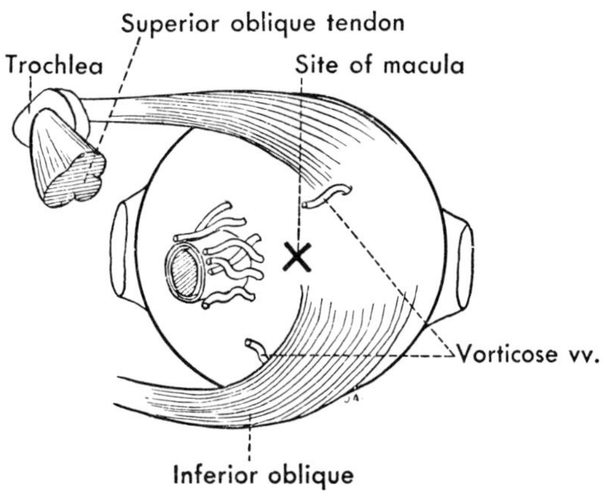

The superior oblique is usually described as consisting of straight and reflected portions, the bend at the trochlea constituting the junction of these. At the trochlea the tendon of the superior oblique muscle is reflected sharply backward, outward, and downward (at an angle of about 55°) toward the upper surface of the eyeball, where it pierces the bulbar sheath, passes between the superior rectus muscle and the eyeball, and spreads out in a fan-shaped insertion on the postero-superior quadrant of the eyeball. The line of insertion into the eyeball is typically concave medially and forward and is largely lateral to the longitudinal axis of the eyeball. As the tendon of the muscle pierces the bulbar sheath, this sheath and the particularly thick fascia over the reflected tendon blend.

The superior oblique muscle is innervated by the trochlear or fourth cranial nerve, which runs along its lateral border for approximately 1 cm before entering the muscle. The actions of this muscle are discussed with those of the other muscles of the eyeball.

The superior oblique is in almost direct contact with the periorbita of the roof of the medial wall of the orbit, while inferior to it are the main stem of the ophthalmic artery, after the supraorbital branch has been given off, and the nasociliary nerve; consequently, the ethmoidal branches of this artery and nerve commonly pass transversely below the muscle to reach the ethmoidal canals. The supratrochlear nerve lies above and lateral to the muscle for most of the course of these two structures, but crosses above the reflected tendon just lateral to the trochlea.

Fig. 2-26. Origins of the orbital muscles at the apex of the orbit. Here they are artificially separated from each other.

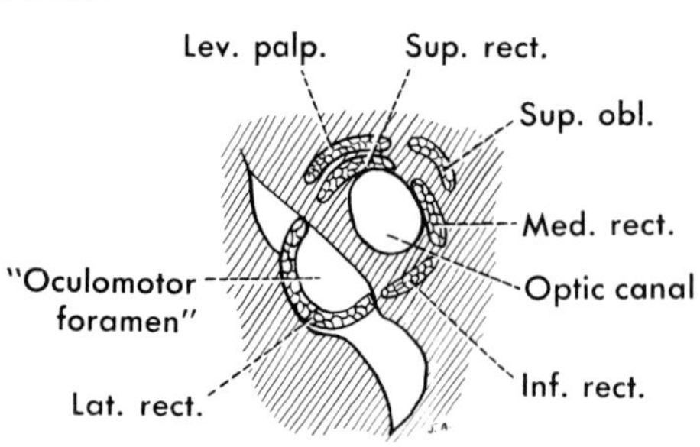

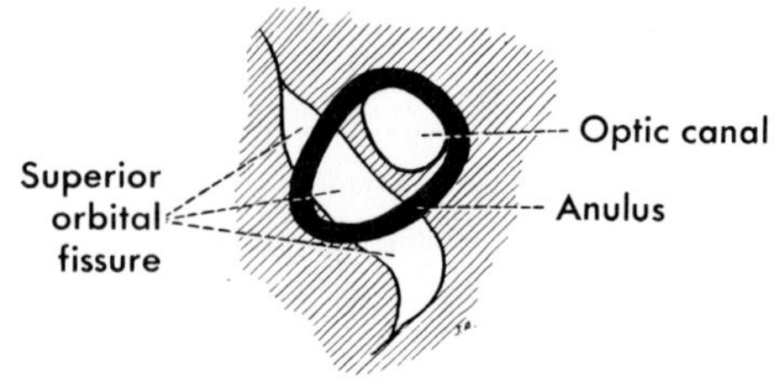

Fig. 2-27. Relation of the anulus tendineus (of Zinn) to the optic canal and superior orbital fissure; the latter extends higher and farther laterally than is apparent from this angle. Right orbit, viewed from the front.

RECTUS MUSCLES

The four rectus muscles arise from a thickening of the periosteum at the apex of the orbit, where they are placed so closely together that their tendons of origin are first fused with each other to form a common annular tendon, the anulus tendineus communis or tendon of Zinn. This tendon surrounds the orbital end of the optic canal and a portion of the superior orbital fissure, and is somewhat oval in shape (Figs. 2-26 and 2-27). The medial portion of the oval, which is both wider and higher than the lateral, surrounds the optic canal, which it immediately borders above and medially; below it swings somewhat away from the rim of the canal. The lateral portion of the anulus extends across the lower portion of the superior orbital fissure, enclosing a portion of this fissure (the so-called oculomotor foramen) within the anulus. From this common origin the recti diverge toward their attachments to the eyeball, thus forming a muscle cone that partially separates the orbital contents into two portions, that lying outside the cone and that lying within it. The actions of the four recti will be discussed in connection with the actions of the orbital muscles as a whole.

SUPERIOR RECTUS

The superior rectus arises from the medial portion of the upper part of the common tendon, where this borders the upper rim of the external end of the optic canal; it is here firmly attached both to the periosteum at this opening and to the sheath of the optic nerve. From its origin the muscle runs forward and somewhat laterally, covered above by the levator, at an angle of about 27° to the anteroposterior axis of the eyeball; in its distal portion it, like the other muscles of the eyeball, is provided with a heavy fascial sheath that blends with the fascial sheath of the eyeball at the point at which the muscle penetrates the latter. Within the bulbar sheath the superior rectus passes above the tendon of insertion of the superior oblique, to insert into the sclera on a slightly curved, obliquely placed line. The most medial point of this insertion is anterior to the most lateral part, but its midpoint lies about 7.7 mm behind the sclerocorneal junction.

As quoted by Wolff, the sheaths of the superior rectus and the levator palpebrae are attached by a definite band, in which a bursa may be found, this band insuring the synergistic action of the two muscles. It is usually also stated that fusion of fascia on the lower surface of the levator and the upper surface of the rectus gives rise to a slip that attaches to the superior fornix of the conjunctiva. The superior rectus muscle is supplied by the superior branch of the oculomotor nerve, which enters the inferior (deep) surface of the muscle; the muscle is usually pierced on its medial side by the continuation of this branch to the levator palpebrae.

The upper surface of the muscle is in contact with the levator palpebrae from its origin to the bulbar sheath; on its deep or inferior surface the muscle borders orbital fat, embedded in which lie the ophthalmic artery, the nasociliary nerve, and tributaries of the superior ophthalmic vein.

LATERAL RECTUS

The lateral or external rectus muscle arises from the lateral portion of the common tendon, as this portion bridges the superior orbital fissure, and from the attachments of the tendon to the greater and lesser wings of the sphenoid across the fissure. The anulus here is on the curve of the lateral end of its oval and from above runs laterally across the fissure to attach to the greater wing, upon which it curves down, then medially once more across the fissure to attach again to the lesser wing. Thus this portion of the anulus completely cuts off a small segment of the fissure, and it is from this part that the lateral rectus arises. The oculomotor, abducens and nasociliary nerves, usually a sympathetic branch from the carotid plexus, and sometimes the ophthalmic vein all enter or leave the orbit through the part of the superior orbital fissure that is cut off from the more lateral part by the origin of the lateral rectus muscle; this part is sometimes referred to as the oculomotor foramen (see Fig. 2-26).

Because the attachment of the muscle is curved or bowed around the structures passing medial to it, the lateral rectus is sometimes described as possessing two heads, a superior and an inferior; similarly, the oculomotor nerve and the other structures listed here are sometimes described as entering the orbit between the two heads of the lateral rectus muscle. In reality, the potential gap where the muscle bridges the upper border of the oculomotor foramen is closed by connective tissue.

From its origin, the lateral rectus muscle passes forward in fairly close contact with the lateral wall of the orbit, pierces the bulbar sheath (to which its sheath of course contributes) and inserts into the sclera about 7 mm behind the limbus, on an approximately vertical line. As already described, the sheath of the lateral rectus muscle sends also an expansin to the lateral wall of the orbit to attach to the marginal (orbital) tubercle and form the lateral check ligament (lacertus of the lateral rectus); inferiorly, as the sheath contributes to the bulbar sheath, it also helps form the suspensory ligament of Lockwood. The lateral rectus is supplied by the sixth or abducens nerve, which traverses the superior orbital fissure within the anulus, passes upward and laterally between the limbs of the Y formed by the upper and lower divisions of the oculomotor nerve, and enters the lateral rectus muscle on its medial surface, almost at its middle.

Relations at Orbital Fissure

At its origin, bridging the superior orbital fissure, the lateral rectus is in close relation with most of the structures entering the orbit. Above the upper portion of the tendon of origin the trochlear, frontal, and lacrimal nerves enter the orbit, thus lying immediately outside the muscle cone, and usually the ophthalmic vein lies here too. Most of these structures therefore enter the orbit through the most lateral portion of the medial, wider, part of the superior orbital fissure and lie just above and lateral to the tendon of origin of the muscle. The anastomotic branch between the lacrimal artery and the frontal (anterior) branch of the middle meningeal artery lies considerably more laterally, however, usually entering through the extreme lateral part of the fissure or a special foramen just lateral to this.

Within the anulus, and therefore in contact with the inner surface of the lateral rectus, pass, in this order from above downward, the upper division of the oculomotor nerve, the nasociliary nerve and a branch from the sympathetic (if this latter is an independent branch), the lower division of the oculomotor nerve, then the abducens and, sometimes, the ophthalmic vein (Fig. 2-28). All these structures are crowded within such a small space that the actual order of their arrangement is really of little practical importance except to the dissector. Below the origin of the muscle, through the most medial and inferior portion of the superior orbital fissure, the inferior ophthalmic vein may occasionally leave the orbit. (The ophthalmic veins may leave the orbit separately or together and, as indicated above, leave through various portions of the superior orbital fissure.

In addition to the abducens nerve, close against the medial side of the muscle in its posterior third or more, the ophthalmic artery and the ciliary ganglion lie between the muscle and the optic nerve about 1 cm forward from the origin of the muscle; above the muscle through much of its length are the lacrimal artery and nerve, and more anteriorly the lacrimal gland rests upon the upper and lateral surfaces of the muscle. Below and medially, the branch of the oculomotor nerve to the inferior oblique muscle passes forward in the orbit.

MEDIAL RECTUS

The medial or internal rectus, the largest of the muscles of the eyeball, has an extensive origin from the common tendinous ring me-

Fig. 2-28. General arrangement of the structures entering and leaving the back of the orbit. The ophthalmic vein is shown above the anulus, but it or a part of it may pass through or below this.

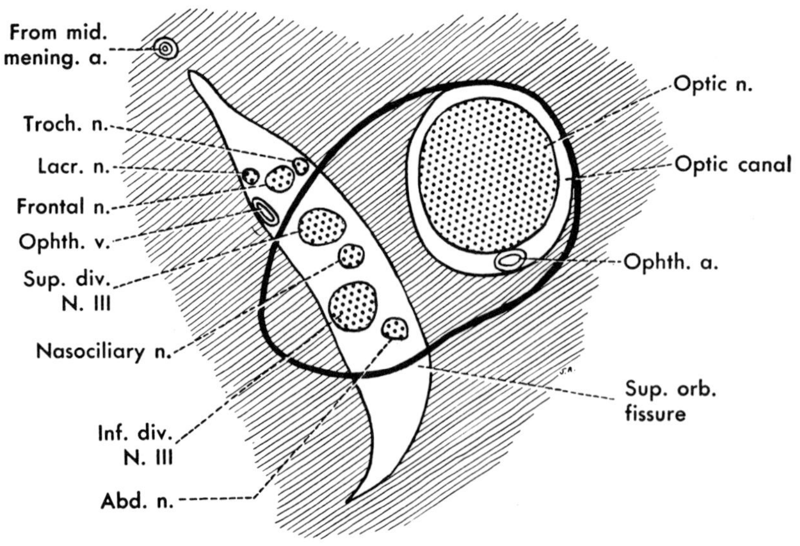

dial to and below the optic canal. Since it arises at the border of the canal the medial rectus is, like the superior, also closely attached to the sheath of the optic nerve; hence contraction of either of these muscles is more likely to cause pain in retrobulbar neuritis than is contraction of other ocular muscles.

From its origin the medial rectus passes forward parallel to the medial wall of the orbit to penetrate the bulbar sheath and attach by a tendon to the sclera, the line of attachment being straight and vertical and also closer to the cornea (5.5 mm) than that of any of the other rectus muscles. The fascial sheath of the muscle contributes, of course, to the bulbar sheath, forms the medial check ligament (wider but less strong than the lateral check ligament), and contributes to the medial portion of the suspensory ligament.

On its medial aspect the medial rectus is closely related to the medial wall of the orbit, being separated from the lamina orbitalis (lamina papyracea) of the ethmoid bone by only a small amount of orbital fat; the superior oblique lies above it, and between the two muscles are branches of the ophthalmic artery and the nasociliary nerve, especially the ethmoidal branches of these. The muscle is innervated by a nerve from the inferior branch of the oculomotor, which enters it at about the junction of its proximal and middle thirds.

INFERIOR RECTUS

The inferior rectus arises from the anular tendon below the optic canal and passes forward and somewhat outward (at an angle of about 27°) along the floor of the orbit, paralleling the superior rectus in this regard; after penetrating the bulbar sheath it attaches to the sclera by a broad tendon, the line of attachment of which is curved with its convexity forward; like the attachment of the superior rectus, this line of attachment is also obliquely placed in respect to the sclerocorneal junction, the nasal end being farther forward than is the temporal end, but the insertion averaging about 6.5 mm from the corneal limbus.

The inferior rectus is innervated by a branch from the inferior division of the oculomotor nerve, which enters the upper surface of the muscle.

Through most of its course the muscle is separated from the floor of the orbit, and therefore from the roof of the maxillary sinus and the infraorbital nerves and vessels, by a moderate amount of orbital fat. Lateral to the muscle, and also above its level, runs the branch of the inferior division of the oculomotor nerve to the inferior oblique muscle. As the muscle attains the eyeball it contributes to the bulbar sheath and passes above the inferior oblique muscle, at which point the sheaths of both muscles blend to contribute to the suspensory ligament. The anterior portion of the inferior rectus is so intimately associated with the inferior oblique through common fascial connections that only minimal recessions or resections of the tendon of this rectus muscle are possible.

INFERIOR OBLIQUE

The inferior oblique muscle is the only voluntary muscle within the orbit that does not arise from its apex; the muscle arises from the floor of the orbit just posterior to the orbital rim and lateral to the upper aperture of the nasolacrimal canal, and may arise also in part from the lacrimal fascia over the lacrimal sac. From this origin the muscle passes laterally, backward and upward, following the curve of the eyeball, to pass between the inferior rectus muscle and the floor of the orbit and reach the posterolateral aspect of the eyeball where it inserts into the sclera under cover of the lateral rectus muscle (Fig. 2-29). The line of attachment of the tendon to the sclera is convex above and laterally, its upper portion lying at about the midhorizontal plane of the eyeball, while its lateral and lower portion lies distinctly below this plane. The posteromedial edge of the insertion of this muscle is an external landmark for locating approximately the macula of the eye.

The inferior oblique muscle is supplied by the inferior division of the oculomotor nerve; a long branch specifically for this muscle (but usually close to its origin giving rise to the

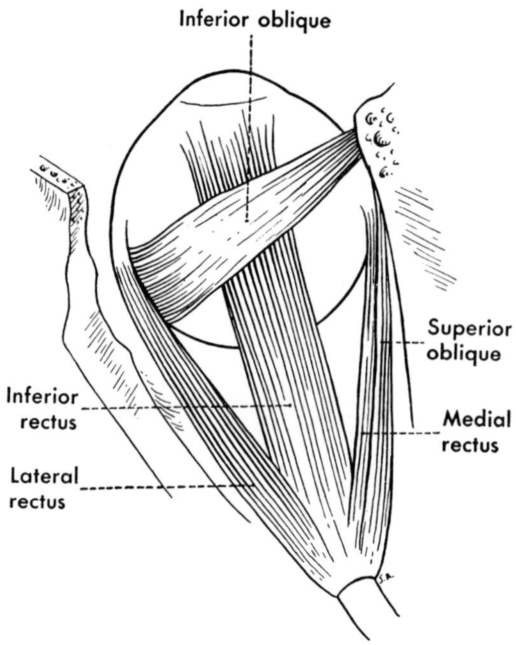

Fig. 2-29. The right inferior oblique muscle from below. See also Figs. 2-24 and 2-25. (After Fink WH: Surgery of the Oblique Muscles of the Eye. St. Louis CV Mosby, 1951)

oculomotor root of the ciliary ganglion) passes forward through the orbit. It has been said to enter the muscle on its upper surface, near the junction of its anterior and middle thirds. This is, however, variable, for the nerve may also enter the inferior surface of the muscle or go directly into its posterior border.

The sheath of the inferior oblique muscle blends with that of the inferior rectus in contributing to the suspensory ligament and to the bulbar sheath.

TENDONS OF THE BULBAR MUSCLES

The six extrinsic muscles of the eyeball vary considerably in the length of their tendons of insertion; the superior oblique has the longest tendon of all, extending from well behind the trochlea to the sclera, while the inferior oblique has the shortest tendinous insertion, which sometimes appears to be actually muscular rather than tendinous. The four recti vary among themselves both in the length of their tendons (usually quoted as averaging 8.8 mm for the lateral rectus, 5.8 mm for the superior and 5.5 mm for the inferior, and 3.7 mm for the medial, thus generally a little longer than shown in Fig. 2-30) and in the relation of their lines of attachment to the sclerocorneal junction; the superior rectus is attached farthest behind the sclerocorneal junction (almost 8 mm), the medial rectus closest to the cornea (about 5.5 mm), and the lateral rectus is slightly farther from the cornea than is the inferior one (Fig. 2-30). Thus, beginning with the superior rectus and proceeding laterally around the eyeball, the attachments of the recti lie progressively farther forward. This fact, and the general differences in lengths of the various tendons, are the important points, and absolute measurements should not be taken seriously, since they necessarily vary with the individual. It should be pointed out also that, as would be expected, there is some variation in the lines of attachment of any one muscle on different bulbs—these may be more or less curved than usual or vary in their obliquity, as well as in their

Fig. 2-30. Attachments of the rectus muscles to the eyeball, with the approximate length (in millimeters) of their tendons, the distance between the tendons, and the distance of the attachments from the limbus, shown on the right eyeball. (After Duke-Elder WS: Text-book of Ophthalmology, Vol I. St. Louis, CV Mosby, 1938)

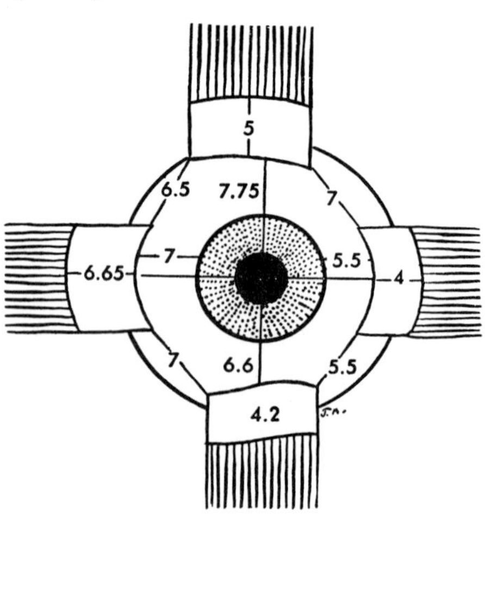

distance from the cornea. The oblique muscles vary considerably in their attachments to the bulb, as emphasized especially by Fink (Figs. 2-31 and 2-32).

It is worth emphasizing that as the muscles enter the bulbar sheath, they are so tightly bound to their fascial sheaths that they retract very little after their tendons are cut. It is this, plus the fact that the tendons apparently may reattach themselves to the sclera, that has led some workers to prefer simple tenotomy in place of certain more difficult recessions (Dyer, and others).

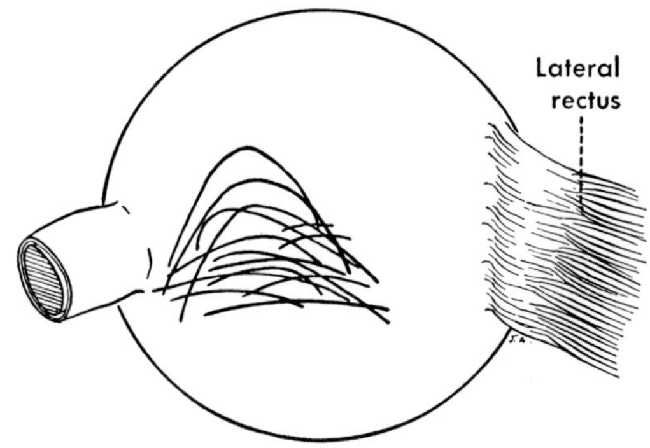

Fig. 2-32. Variations in the attachment of the inferior oblique muscle to the eyeball. (After Fink WH: Surgery of the Oblique Muscles of the Eye. St. Louis, CV Mosby, 1951)

ACTIONS OF THE BULBAR MUSCLES

The movements of the eyeball are defined in terms of the direction in which the pupil is moved. Thus, "elevation" is directing the pupil upward and "depression" is directing it downward; "abduction" is directing the pupil laterally and "adduction" is directing it medially. Finally, the eyeball may (under abnormal conditions only, according to Jampel) also be rotated around its longitudinal (sagittal) axis, and these movements are defined according to the direction in which the upper part of the pupil moves. Thus "inward rotation" or "intorsion" is that movement in which the upper part of the pupil is rotated medially or nasally; "external rotation" or "extorsion" is, correspondingly, a rotation of the upper part of the pupil in the lateral or temporal direction.

Fig. 2-31. Variations in the attachment of the superior oblique muscle to the eyeball. (After Fink WH: Surgery of the Oblique Muscles of the Eye. St. Louis, CV Mosby, 1951)

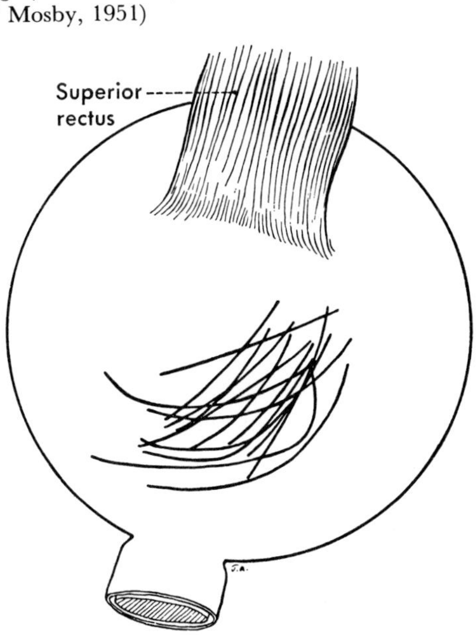

In considering the actions of the two obliques and of the superior and inferior recti, one must distinguish between their active movement as seen and tested clinically on the intact or partially paralyzed eye and the slight and relatively passive action that they exert in balancing each other. Both sets of actions can be deduced from the anatomy of the muscles; in clinical practice, however, one rarely observes the action of a single eye muscle, and one naturally tests in the direction of the strongest movement. It is this latter, therefore, that is of prime importance.

The influence of the vestibular apparatus on movements of the eye is discussed in Chapter 3.

Action of Superior Oblique

The chief action of the superior oblique is to turn the pupil downward. Since the line of pull of this muscle is from the trochlea to the posterolateral aspect of the eyeball, adduction, which turns this aspect of the eyeball farther laterally, increases the distance between trochlea and insertion and therefore increases the strength of the downward movement. Abduction, on the other hand, decreases the distance between trochlea and insertion and therefore decreases the strength of a downward movement; marked abduction of the eye abolishes the depressor action of the superior oblique. Hence this muscle is always tested by having the patient attempt to look downward *while the pupil is directed nasally.*

When acting alone, the superior oblique will also produce intorsion, a movement that can also be seen when the oculomotor nerve is paralyzed. When, however, it acts with the inferior rectus in depressing the pupil, the latter muscle prevents the rotational movement (Jampel).

Action of Inferior Oblique

The chief action of the inferior oblique is to elevate the pupil, or turn it upward. Since the line of pull of this muscle, like that of the superior oblique, is from an anteromedial position on the orbital wall to a posterolateral one on the eyeball, the strength of the movement of this muscle also is increased by adduction and decreased or abolished by abduction. Hence both obliques are tested by requesting the appropriate upward and downward movements *with the eye in the adducted position.*

The inferior oblique, acting alone, will produce extorsion. Just as the rotatory action of the superior oblique is canceled by the simultaneous action of the inferior rectus, however, so is that of the inferior oblique canceled, on elevation of the pupil, by the action of the superior rectus.

Balancing Actions of Oblique Muscles

Both obliques have the additional action of abduction, which is primarily concerned with counterbalancing opposite tendencies upon the part of the superior and inferior recti. Thus, the superior oblique tends to produce slight abduction, while the inferior rectus with which it may act in depression tends to produce slight adduction. Since the tendency to abduction markedly lowers the effectiveness of the superior oblique in directing the pupil downward, isolated action of this muscle is largely ineffective in regard to both movements simultaneously, as can be noted in paralysis of the lateral rectus muscle.

The same statement holds true in regard to the inferior oblique, except that its abduction counterbalances the tendency to adduction on the part of the superior rectus. Thus the two obliques share the function of abduction. Therefore, for both muscles adduction increases and abduction decreases their contribution to vertical movements.

Action of Recti

Of the four rectus muscles, the medial and lateral direct the pupil only medially and laterally; since they are usually symmetrically attached to the eyeball, they have no subsidiary actions. The superior rectus chiefly directs the pupil upward, and the inferior rectus likewise chiefly directs the pupil downward. Like the obliques, these recti really run somewhat obliquely (in regard to the sagittal plane) and insert obliquely upon the eyeball, and both tend to produce adduction. It is usually said that they also tend to rotate, the superior rectus producing internal rotation and the inferior rectus producing external rotation. However, Jampel found that although they prevent the rotation that the obliques would otherwise produce, they themselves do not rotate. Abduction increases, and adduction decreases, the power of movement of the superior and inferior recti; thus these two muscles are always tested by upward and downward movements *while the eye is well abducted.* In this position the obliques are so shortened that they have no action, and the recti, which are correspondingly lengthened, are the only muscles that can produce these movements.

Similarly, in the adducted position used for testing the obliques, these are the only muscles that can produce upward and downward movements, since the action of the superior and inferior recti is obviated by the adduction. Thus it is possible to test inferior oblique and superior rectus (elevators), and superior oblique and inferior rectus (depressors), independently, although both elements composing a pair have exactly the same chief action.

In addition to their actions on the eyeball, the superior rectus presumably aids the levator in elevation of the upper lid and of the conjunctival fornix, through the attachment of its fascial sheath to that of the levator, while the inferior rectus is said to depress the lower lid and conjunctiva.

VARIATIONS

Slight variations in the attachments of the muscles of the eyeball are probably fairly common, but major anomalies are at least infrequent. Whitnall ('21b) recorded variations in a number of muscles and noted that there may be complete absence of a muscle, absence of the medial rectus being sometimes a cause of divergent strabismus and that of the lateral rectus the cause of some cases of medial strabismus. The levator palpebrae and the superior rectus, associated in action, also are closely associated in development and therefore tend to be defective simultaneously, but not always. Diamond and co-workers operated upon 12 patients with craniofacial dysostosis to correct strabismus; among the 5 who had muscular anomalies, the superior rectus was absent bilaterally in 3, but the levator palpebrae was present in all. (The fourth patient did not have a right inferior rectus, and the left medial rectus was so tight that it limited abduction to the midline; the fifth patient lacked a superior oblique muscle or tendon on one side and a tendon on the other.) Insertion of the levator into the trochlea, forming a tensor trochleae, is probably the most common anomalous development of the muscles of the orbit and has been reported a number of times. The levator may also have a transverse belly, and in a case reported by Whitnall ('21b) had a slip inserted into the lacrimal gland. MacEwen described a separate levator of the fornix and other anomalous muscles in an orbit, including an attachment to the lacrimal gland; he also described a case of continuity between the superior rectus and superior oblique muscles at their insertions. MacEwen described doubling of the inferior oblique. Whitnall ('11) described a case of a retractor bulbi, a remnant of a cone of muscle attached to the eyeball within the cone of the recti in some mammals, but not frequently found in man; this writer saw such a case during his first dissection of the human orbit, but has not seen one since. Posey reviewed much of the early literature upon anomalies of the extrinsic bulbar muscles and reported cases that he himself observed.

Nerves

OPTIC NERVE

Since the retina is embryologically an outgrowth from the diencephalon, the optic nerves are not, strictly speaking, peripheral nerves but tracts connecting two portions of the brain, and they show this in their morphology. Thus, they, unlike any of the other cranial or spinal nerves, contain neuroglia throughout their entire lengths and are surrounded, from the lamina cribrosa of the sclera inward, by the typical three layers of cerebral meninges and by a subarachnoid space.

The fibers forming the optic nerve arise from the ganglionic cells which form the innermost cellular layer of the retina and converge upon a point situated slightly above and internal to the longitudinal axis of the eyeball, where they form the optic disk (blind spot, papilla of the optic nerve, nerve head) and leave the eyeball through the lamina cribrosa. As they emerge from this plate they acquire myelin sheaths and become grouped together to form the optic nerve. Close behind the eyeball, the fibers from the macula are said to lie superficially on the lateral side of the nerve, but proximally they run deeper into the nerve so that they come to occupy the center of it (Fig. 2-33). Hoyt and Luis ('62), studying the effects of retinal lesions made by the photocoagulator in monkeys, have emphasized that they are not segregated here, but are intermingled with fibers from the peripheral parts of the retina; the latter form a

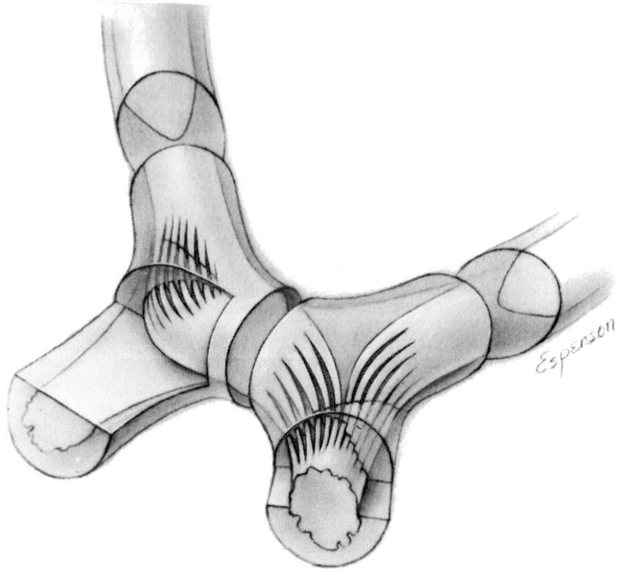

Fig. 2-33. The macular projection. As the fibers enter the optic chiasma they are separated by superior quadrant extramacular fibers of the same side passing through them. The crossed fibers occupy a large central part of the chiasma; as they leave it to become concentrated in the dorsal part of the optic tract they are again separated, this time by extramacular fibers of the opposite nerve. (Hoyt WF, Luis O: Arch Ophthalmol 70:69, 1963)

relatively thin peripheral ring in the optic nerve, but in addition are apparently distributed throughout the large central core occupied also by macular fibers.

The optic nerve occupies a central position in the orbit, running obliquely from the back of the eyeball to the optic canal; after traversing this it converges toward the opposite optic nerve intracranially to form the optic chiasma. The gross anatomy of the optic chiasma is discussed in Chapter 1.

The details of the crossing of fibers in the chiasma, as worked out Hoyt and Luis ('63) in the monkey (Figs 2-33 to 2-35), differ somewhat from the various arrangements suggested on the basis of clinical studies. Rather than macular fibers crossing posteriorly in the chiasma, they found them throughout all the chiasma except for small anteroinferior and posteroinferior parts, which were occupied largely or exclusively by inferior quadrant and superior quadrant extramacular projections, respectively; thus macular and extramacular fibers are mixed in the optic chiasma just as they are in the optic nerve. Although most inferior quadrant fibers were found to cross anteroinferiorly in the chiasma and most superior quadrant fibers to cross posteriorly, some of each were mingled with the main group of the other (see Figs. 2-34 and 2-35). Further, just as the macular fibers in the chiasma become concentrated in the superior portion of the optic tract, so do the paramacular fibers become concentrated adjacent to the macular ones. This involves little rearrangement of the fibers from the inferior retinal quadrant (Fig. 2-34), but a complete rearrangement of the superior quadrant fibers; thus fibers from the peripheral parts of the retina run downward to reach the lower medial part of the optic tract (Fig. 2-35), and the paramacular fibers run upward, eventually occupying a dorsomedial position in the tract.

Sheaths

As the optic nerves are traced forward toward the eyeball from the optic chiasma, they lie at first in the subarachnoid space at the base of the brain, and as each nerve enters and traverses the optic canal it is accompanied by pia, arachnoid, and dura.

The pia on the optic nerve is closely applied to that nerve, just as pia elsewhere is closely applied to the central nervous system. The arachnoid forms a looser sleeve, separated from the pia by the subarachnoid space, but united to it by subarachnoid trabeculae that traverse the space. The arachnoid is closely adjacent to the dura, separated from it only by the potential subdural space. In the

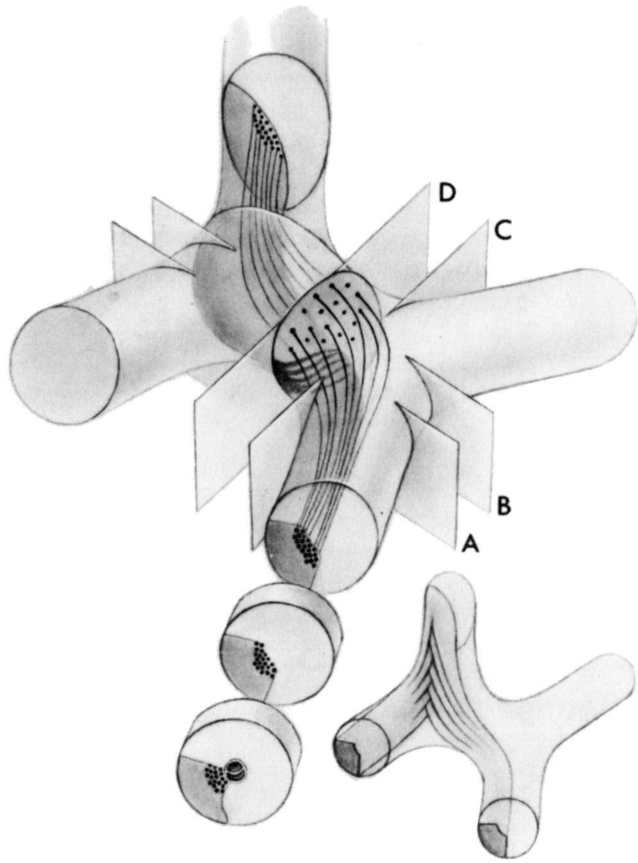

Fig. 2-34. Crossing of the extramacular inferior nasal fibers in the optic chiasma. Note that most of the fibers cross anteroinferiorly, but those closest to the macula are widely scattered through the posterodorsal part of the chiasma. The smaller insert (bottom right) shows how the fibers from the crossed and uncrossed inferior quadrants interlock as they go into the optic tract. The lettered planes in the larger figure indicate the positions of photomicrographs. (Hoyt WF, Luis O: Arch Ophthalmol 70:69, 1963)

optic canal, the dura is like cranial dura elsewhere, consisting of an inner component that is the dura proper and an outer component that represents periosteum. At the anterior end of the optic canal these components separate, the periosteal part continuing as periorbita, the truly dural part continuing as the outermost and most obvious sheath about the optic nerve. Above and medial to the optic canal the parts of the anulus tendineus formed by the superior and medial recti arise in the angle between dura proper and periorbita. The inferior rectus and the lateral rectus muscles are not attached to the margin of the optic canal, however, for the inferior rectus attaches both below and in large part lateral to it, while the lateral rectus bridges the medial end of the superior orbital fissure. The line of attachment of the anulus thus corresponds only in part to the point of separation of the nerve sheath and the periorbita, and the inferior and lateral rectus muscles are attached only to periorbita, not to the nerve sheath. Pia, arachnoid, and dura continue forward about the optic nerve throughout its intraorbital course, ending only at the cribriform lamina of the eyeball, where pia and arachnoid become continuous with each other, and the dura fuses with the sclera and the bulbar sheath.

The intravaginal space of the NA is the subarachnoid space, and the external and internal sheaths of the optic nerve are dura–arachnoid and pia, respectively.

Vascular Supply

The optic nerve in the orbit and optic canal is supplied by tiny branches from the central artery of the retina, from the posterior ciliary arteries, and from the ophthalmic directly; the ophthalmic also contributes to the blood supply within the cranial cavity, as does a su-

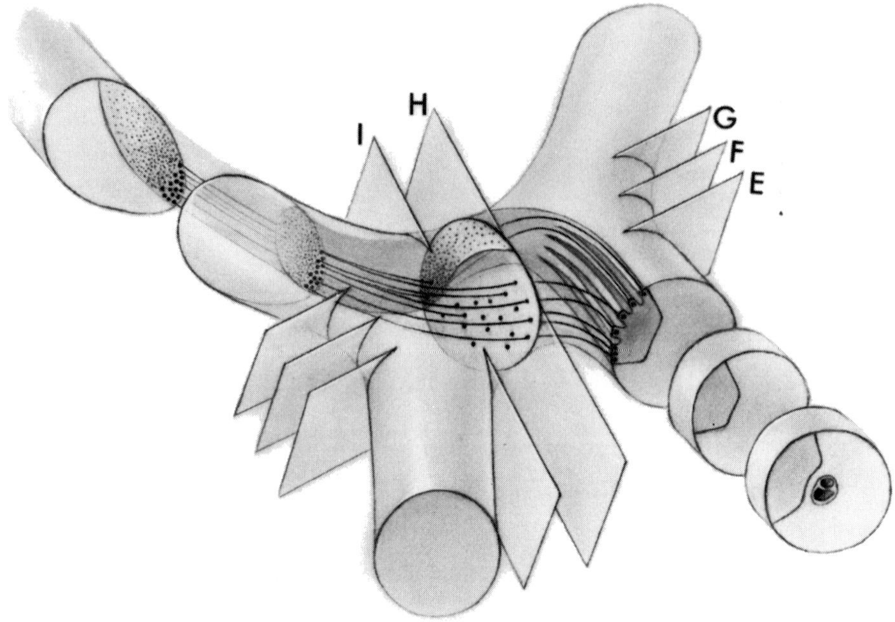

Fig. 2-35. Crossing of the extramacular superior nasal fibers in the optic chiasma. Although most of these fibers cross posteriorly, the more peripheral ones are scattered through the anterior part of the chiasma. (Hoyt WF, Luis O: Arch Ophthalmol 70:69, 1963)

perior hypophyseal artery (Steel and Blunt). The blood supply of the chiasma is from numerous sources, including superior hypophyseal, anterior and posterior communicating, anterior cerebral, and internal carotid arteries (Steele and Blunt; Dawson). Bergland and Ray found the branches from the anterior cerebral going to the upper surfaces of the optic nerves and tracts, but to the lateral parts only of the chiasma, while the remaining vessels went to the inferior surfaces of the nerves, chiasma, and tracts. They suggested that the bitemporal field defects produced by hypophyseal tumors are caused not by direct pressure but by ischemia resutling from distortion of these inferior vessels.

Relations

The relations of the optic chiasma are discussed in the chapter on the cranium. As the optic nerve traverses the optic canal it is accompanied by the ophthalmic artery; this arises from the internal carotid artery just as (or sometimes just before) the latter vessel penetrates the dura and thus lies below the optic nerve. As the artery runs forward beneath the optic nerve it gradually sinks into the dural floor of the canal and, at the distal end of the canal, emerges into the orbit, lying here at first inferior and somewhat lateral to the optic nerve.

Within the orbit the optic nerve is at first closely related to the superior and medial rectus muscles especially and, to a lesser extent, to the inferior rectus, but not at all closely to the lateral rectus. Close to the apex of the orbit the otic nerve is normally (85%) crossed above by the ophthalmic artery, which turns upward and medially after having given off its lacrimal branch (Fig. 2-36); only slightly beyond this, it is crossed above, in a lateral to medial position, by the nasociliary nerve (Fig. 2-37), and still farther distally the medial branch of the superior ophthalmic vein crosses in the fat between the superior rectus and the nerve. The trochlear nerve, passing from the apex of the orbit to the medial side, is separated from the optic nerve by the levator palpebrae and superior rectus muscles; the oculomotor nerve, however, is in closer relation at the apex, since it lies only slightly lateral to the optic nerve, separated from it by the ophthalmic artery. The superior and inferior divisions of the oculomotor nerve then separate to run above and below the optic nerve.

Not far distal to this, and still well back toward the apex of the orbit, the ciliary ganglion lies against the lateral surface of the optic nerve, between it and the lateral rectus muscle; from this point forward the optic nerve is surrounded by numerous ciliary nerves and arteries that approximately parallel its course to reach the eyball. During about the distal centimeter or so of its course the optic nerve contains the central artery and vein of the retina, more or less in its middle; proximal to the point of entrance of the artery through the dural sheath the vessel lies in contact with the sheath for some distance along its lower surface, mingled with the ciliary vessels.

Fig. 2-36. Dissections of the orbit. On the left, bone and periorbita have been removed. On the right, the more superficially lying nerves and vessels and part of the levator palpebrae have been removed, and the proximal part of the nasociliary nerve in the orbit has been moved laterally and anteriorly to bring it into view.

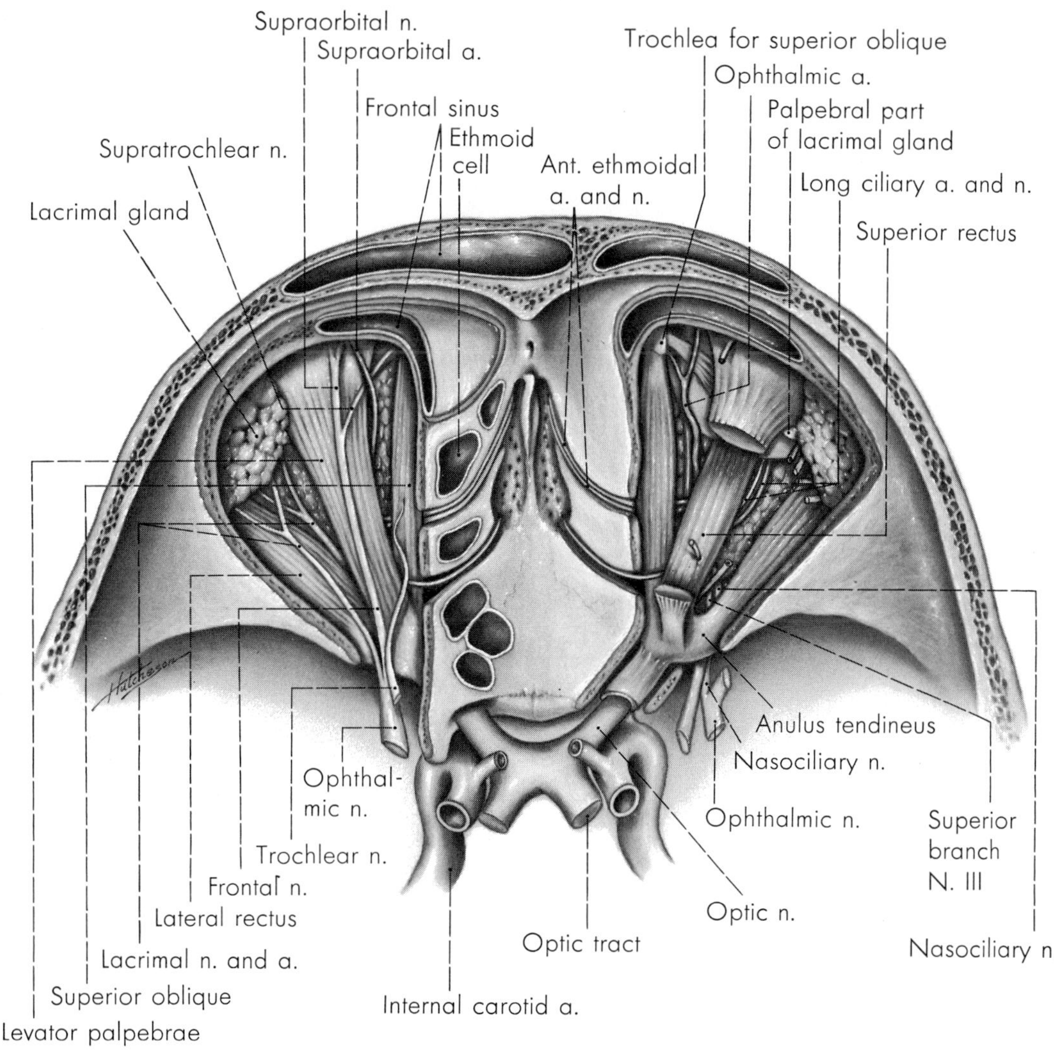

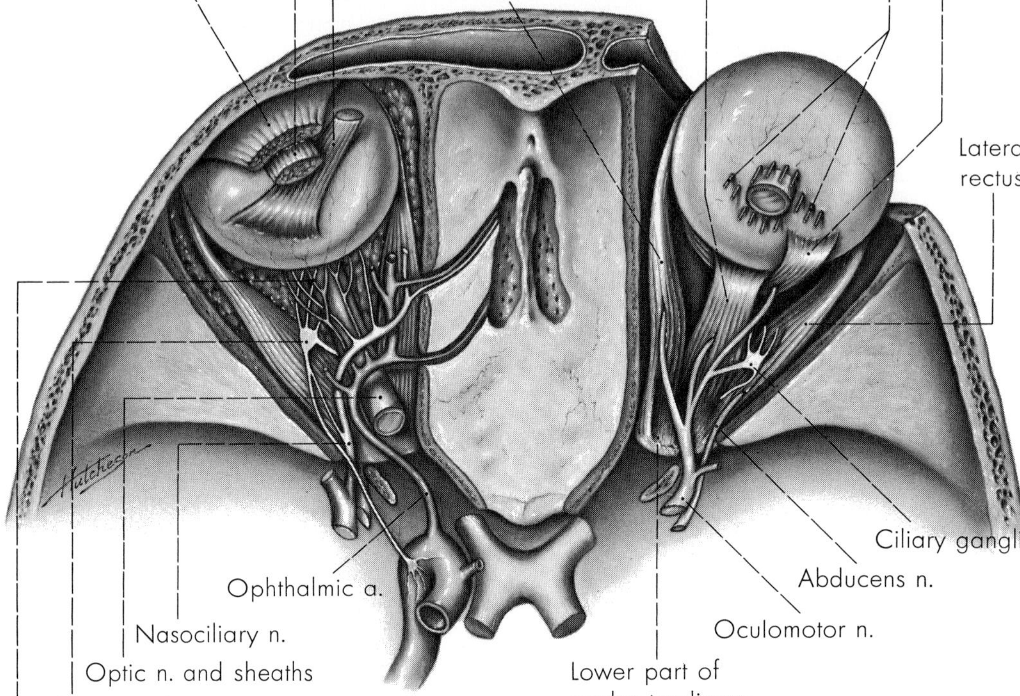

Fig. 2-37. Further dissections of the orbit. On the left, the levator palpebrae, superior rectus, and superior oblique have largely been removed, and a segment has been taken from the optic nerve. On the right, most of the nerves and vessels have been removed to show the abducens nerve and the inferior branch of the oculomotor.

In head injuries in which the optic nerve is injured indirectly, it is usually the intracranial portion of the nerve or that within the optic canal which is so affected, according to Turner. He said that hemorrhage into the subarachnoid space about the nerve is not a factor, but that intraneural hemorrhage or thrombosis is the usual cause of damage in such cases. Within the orbit the nerve is protected both by the surrounding soft tissues and by its laxity. It is approximately 5 mm longer intraorbitally than the distance (some 2–2.5 cm) between the back of the eyeball and the optic canal. Physiologically, this excess length allows unimpeded movement of the globe; surgically, it allows a considerable forward displacement of the globe, in retrobulbar operations, before undue tension is exerted on the nerve.

Because of the close association of the optic canal with the sphenoid sinus and posterior ethmoid cells, and the thinness of the bony wall between the nerve and the mucosa of the sinus, the optic nerve has been said to be vulnerable to infection from these sources. However, many workers now believe that there is practically never a causal relationship between sinusitis and optic neuritis.

MOTOR NERVES OF THE ORBIT

The three motor nerves to the voluntary muscles of the orbit all enter the orbit through the superior orbital fissure (see Fig. 2-28), but while the oculomotor and abducens nerves enter through that portion (the "oculomotor foramen") cut off by the tendinous ring, and therefore lie within the muscle cone, the trochlear nerve enters above the ring and throughout its course lies outside the muscle cone. The trochlear and abducens nerves supply only one muscle each, the superior oblique and lateral rectus muscles, respectively, while the oculomotor supplies the remainder. The origins and intracranial courses of these nerves have already been described in the preceding chapter and need only be summarized here. Their orbital courses and distributions (Figs. 2-36 and 2-37) are discussed in more detail.

The nerves to the muscles of the eyeball are remarkable for their relatively large sizes in proportion to the sizes of the muscles that they innervate. The abducens nerve, for instance, is said to be the largest nerve in the body in this relationship. Counts and estimates of the total numbers of nerve fibers in the motor nerves to the eyeball, and of the muscle fibers supplied by them, indicate a ratio of approximately one nerve fiber to every three muscle fibers. This necessarily includes sensory fibers in the nerves, since these have never been eliminated experimentally, but even if they are numerous it indicates a high proportion of motor nerve fibers to muscle fibers and therefore a very small motor unit; this in turn, of course, indicates the delicate control of eye movements, probably matched by no other muscles of the body.

OCULOMOTOR NERVE

The oculomotor nerve arises from the base of the midbrain, passes between the posterior cerebral and the superior cerebellar arteries and, running forward and downward through the subarachnoid space, passes lateral to the posterior clinoid process to reach the lateral side of the sella turcica and penetrate the dura not far from the posterior clinoid process; having done so, it comes to lie in the lateral wall of the cavernous sinus (Fig. 2-38). It is at first the most superior of the several nerves running in the lateral wall of the sinus, for the fourth nerve and the upper two divisions of the fifth nerve lie below it here. At the anterior part of the sinus the oculomotor nerve separates into a small superior and a large inferior division, just before it passes through the superior orbital fissure. At about this point the nerve is crossed on its lateral side by the trochlear nerve, which then passes above it to lie more medially; just distal to this crossing by the trochlear, the oculomotor nerve is also crossed by the ophthalmic division of the fifth, which runs from below upward lateral to the oculomotor and usually here divides into its three major branches.

The trochlear nerve and the frontal and lacrimal branches of the ophthalmic pass above the anulus and are therefore separated from the oculomotor by the lateral rectus muscle, but the nasociliary branch of the ophthalmic passes medially to lie between the two branches of the oculomotor as it enters the orbit through the anulus. At the anterior end of the cavernous sinus the oculomotor nerve also comes in closer relationship to the adbucens, which had been lying more deeply in the sinus; the oculomotor nerve at first lies above the abducens, but as the nerves pass through the anulus the abducens runs upward and then turns laterally between the two divisions of the oculomotor nerve to reach the lateral rectus muscle.

Course Within Orbit

Within the orbit the *superior divison* of the oculomotor nerve passes upward and medially above the optic nerve, paralleling but proximal to the nasociliary nerve, and enters the superior rectus muscle at the junction of about its middle and posterior thirds; the branch to the levator palpebrae continues into that muscle either through or around the medial border of the superior rectus.

The larger *inferior division* of the oculomotor

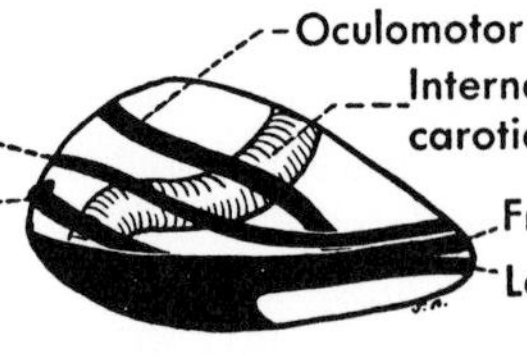

Fig. 2-38. Arrangement in the cavernous sinus of the nerves of the orbit that traverse this, as seen from above and laterally. The nasociliary branch of the ophthalmic lies medial to the other branches of this nerve and is not shown.

divides into three branches to supply medial and inferior recti and the inferior oblique. The branch to the medial rectus passes below the optic nerve to enter the muscle on its lateral side at about the junction of middle and posterior thirds; that to the inferior rectus runs forward to enter this muscle, usually on its upper surface, in about the corresponding position; the nerve to the inferior oblique is of necessity much longer and runs forward close to the floor of the orbit, lateral to and slightly above the inferior rectus muscle.

Close to its origin the nerve to the inferior oblique gives rise to the short but stout oculomotor root of the ciliary ganglion (Fig. 2-39), containing the parasympathetic fibers of the oculomotor nerve; details of this connection will be found in a following section. The oculomotor nerve thus innervates all the voluntary muscles of the orbit except the superior oblique and the lateral rectus and also, through its branch to the ciliary ganglion, innervates the sphincter pupillae and the ciliary muscle.

Intracranial Injury

In its intracranial course the oculomotor nerve is subject to damage through sclerosis or aneurysm of the several arteries related to it—particularly the posterior cerebral and the superior cerebellar, but also the basilar, the posterior communicating, or the internal carotid in the cavernous sinus. Paralysis of this nerve results in the pupil of the eye being turned laterally, as a result of the unopposed actions of the superior oblique and lateral rectus muscles; the downward movement usually imparted by the superior oblique is minimized in this case by the abduction of the eye.

An individual so affected is unable to look downward, upward, or inward beyond the midline with the affected eye, but if the superior oblique is still intact the eye can be intorted. (Movement of a paralyzed eye to about the midline, owing to reciprocal innervation of the extraocular muscles, has been often affirmed and sometimes denied; however, it is useless to study late cases of paralysis to demonstrate such reciprocal innervation, since by that time the unparalyzed muscles have contracted too far to allow proper relaxation.) Also, because of its parasympathetic fibers, paralysis of the oculomotor nerve produces a somewhat dilated pupil which is "fixed," reacting neither to light nor accommodation.

TROCHLEAR NERVE

The fourth or trochlear nerve supplies only the superior oblique muscle and is the longest and most slender of the three eye-muscle nerves. Arising from the roof of the brain just behind the mesencephalon, it runs forward below the tentorium cerebelli and penetrates the inferior surface of the tentorium a little behind the posterior clinoid process. It then runs forward intradurally into the lateral wall of the cavernous sinus. In this position the oculomotor lies above and slightly medial to it, while the ophthalmic division of the fifth lies below and slightly lateral to it.

As the trochlear nerve leaves the cavernous sinus and enters the superior orbital fissure (above the anulus), it crosses the oculomotor nerve, passing lateral to and then above it and, as it enters the orbit, bending sharply medially. It courses above the levator palpebrae and enters the upper border of the superior oblique muscle, usually by several branches that are distributed over the muscle for almost 1 cm of its length.

The trochlear nerve may be subjected to pressure from the same vessels as the oculomotor. Because of their higher positions in the cavernous sinus, both oculomotor and trochlear nerves are more likely to be affected by enlargements of the hypophysis than is the abducens nerve, which lies in a lower position and is also protected by the presence of the internal carotid artery.

The greatest limitation of movement produced by paralysis of the trochlear nerve is seen when the patient is asked to look downward with the eye adducted; in this position the action of the inferior rectus is minimal, while that of the superior oblique should be maximal. The nerve is not often paralyzed alone, however.

ABDUCENS NERVE

The abducens nerve, like the trochlear, supplies a single muscle, the lateral rectus. It leaves the base of the brain stem at the lower border of the pons and passes forward and slightly laterally to enter the cranial dura over the clivus and then run upward and forward within the dura covering this. It passes over the most anterior end of the petrous portion of the temporal bone (petrous tip), usually crossing superficial to the inferior petrosal sinus to lie lateral to this, and passing with the sinus through "Dorello's canal." This canal is the foramen formed by the attachment of the petrosphenoidal or petroclinoid ligament (of Gruber) across the notch of the petrosphenoid joint.

Nathan and co-workers found this description to be true in 86.5% of 62 dissections. In the remainder, the nerve either split into two trunks in the subarachnoid space or originated in two trunks; in these, one of the trunks passed through the canal, the other penetrated the ligament or passed above it. The nerve has been described as double all the way to the lateral rectus, but these workers found no example of this.

The abducens makes a sharp, almost right-angle, turn over the petrous tip to enter the cavernous sinus and lie against the lateral aspect of the internal carotid artery (not, therefore, in the lateral wall of the sinus). It crosses the vertical portion of the carotid artery in the cavernous sinus and more anteriorly lies approximately parallel to the horizontal portion of this vessel; as it enters the superior orbital fissure it lies at first below the two divisions of the oculomotor nerve, but immediately passes upward and laterally to run between these two divisions and reach the medial surface of the lateral rectus muscle.

Intracranial Injury

The well-known vulnerability of the abducens nerve, evidenced by the fact that an isolated paralysis of the lateral rectus (resulting in an internal strabismus) is a relatively common concomitant of intracranial lesions, especially those accompanied by increased intracranial pressure, is probably due to several different factors. These have been for the most part discussed in the preceding chapter.

Fig. 2-39. Schema of the distribution of the oculomotor nerve.

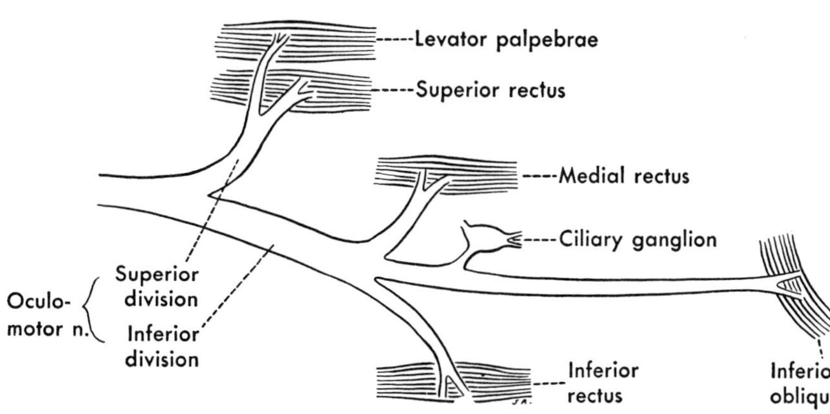

At the point at which the abducens lies in Dorello's canal, it may be subject to injury as a result of infection of the air spaces: if the petrous tip cells are well developed, the nerve may be involved in middle ear infections extending to these, and since it may lie relatively close to the sphenoid sinus it may be affected by disease here. In a series of cases in which Houser measured the distance between the mucosa of the sphenoid sinus and the abducens nerve, he found a distance of 2 mm or less in 5.2% of sinuses and 10.5% of heads and a distance of 3 mm or less in 10.5% of sinuses and 15.7% of heads.

Finally, as pointed out by Bucy and Weaver, the abducens nucleus is so located that it is subject to pressure from a cerebellar abscess; in 3 out of 7 cases of such abscess they found paralysis of lateral conjugate movement toward the side of the abscess, due apparently to implication of both the abducens nucleus and the medial longitudinal fasciculus (or the center for movement?). Obviously, a number of factors may account for injury to this nerve.

Fig. 2-40. Locations of muscle spindles in the inferior rectus muscle. The nerves in the cross sections are *shaded* and in the fourth section up are easily confused with the spindles; comparison with the surface view of the muscle at this level shows that it contains few spindles, however. (After Cooper, S, Daniel, PM: Brain 72:1, 1949)

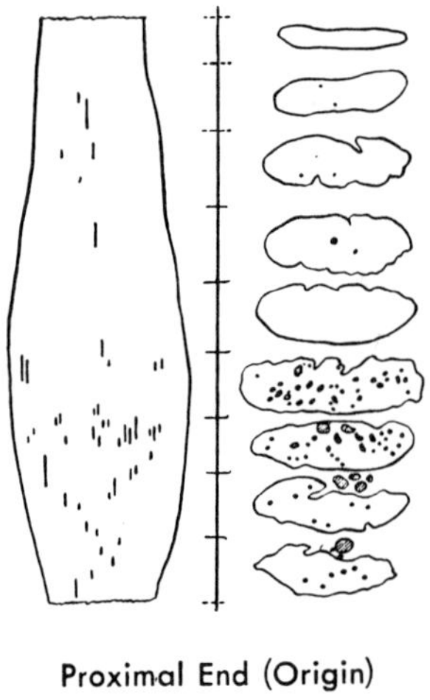

SENSORY FIBERS IN THE EYE MUSCLE NERVES

Nerve endings of sensory type at the musculotendinous junctions of the eye muscles, and other apparently sensory endings, have been described. However, Cooper and Daniel were the first to describe typical muscle spindles in human eye muscles (Fig. 2-40), and other workers have since confirmed their findings. There are also apparently numerous nerve fibers for pain in the orbital muscles, since they are very sensitive to stretching; for example, a muscle anesthetized with procaine so that it may be cut without pain may still produce exquisite pain when it is plulled upon. There can be no doubt, therefore, that the "motor" eye-muscle nerves contain afferent fibers, at least in their peripheral course, but there are still no uncontroverted data on the cells of origin of these afferent fibers or the routes by which they enter the brain stem.

Nothing is known concerning the pain pathway from the orbital muscles, although Abd-El-Malek's production of a rise in blood pressure upon stimulating the central ends of the severed motor nerves to the orbit can be interpreted as indicating that pain fibers enter the brain stem with these nerves.

It is generally admitted that the occasional ganglion cells that may be found along the nerves to the ocular muscles are too inconsistent and too few to account for the sensory fibers in them, and it has been suggested that the cells lie in the mesencephalic nucleus of the fifth nerve, known to be the origin of proprioceptive fibers to the muscles of mastication. Cooper and associates ('53) supported this concept, suggesting from their findings in the goat that although the cell bodies are few and scattered, they do form a part of the mesencephalic nucleus. Tarkhan expressed the belief that he could trace, histologically, larger numbers of nerve fibers from cells of the mesencephalic nucleus into the third and fourth nerves. Corbin expressed doubt that there were any degenerating myelinated fibers in these two nerves and was sure that there were none in the sixth and none running from the fifth to the motor nerves, following destruction of the mesencephalic nucleus. Corbin and Harrison concluded that the sensory cell bodies concerned with the third nerve are in the oculomotor nucleus itself, and course with the motor fibers of that nerve. On the other hand, Cooper and coworkers ('55) said that in the goat the fifth nerve may send a branch directly to an eye muscle; they recalled the numerous reports of small connections between the fifth nerve and the motor nerves of the orbit and suggested that it is through these connections to the fifth nerve, thence by way of the latter nerve, that the afferent nerve fibers enter the brain stem. Winckler also argued that in man the afferent fibers join the ophthalmic nerve in the cavernous sinus. However, Bach-Y-Rita and Murata have apparently shown that in the cat the abducens nerve contains proprioceptive fibers at its origin from the brain stem.

Exactly what part the various proprioceptive endings in the eye muscles play in ocular movements is unknown. They surely must help guide ocular movements, may play a part in the perception of depth and space, and perhaps may even be involved in the initiation of the accommodation accompanying convergence.

OTHER CONNECTIONS OF THE EYE MUSCLE NERVES

In addition to the fine connections, just mentioned, between the ophthalmic nerve and the eye-muscle nerves in the cavernous sinus, it is also possible that these nerves receive sympathetic fibers from the internal carotid plexus while they are closely related to that. Both the fourth and sixth nerves have been said to receive filaments from the plexus, although Koch stated that only the sixth nerve contains unmyelinated fibers; and if the superior tarsal muscle is innervated through sympathetic fibers that reach it through the oculomotor nerve, as Whitnall indicated, they also probably join the nerve in the cavernous sinus.

VARIATIONS

Variations in the nerves to the muscles of the orbit are apparently rare, but several have been described including occasional penetration of a muscle by a branch of a nerve on its way to some other distribution, communications between the third and sixth nerves, and such anomalous distribution of the nerves as partial supply of the superior oblique from the oculomotor, or replacement of the innervation of the lateral rectus by the oculomotor with absence of the abducens nerve.

OPHTHALMIC NERVE

All three branches of the ophthalmic nerve, a purely sensory branch of the trigeminal, traverse the orbit and have a cutaneous distribution to the face, but only two of these, the lacrimal and the nasociliary, are important in the innervation of the orbital contents. After leaving the trigeminal ganglion the ophthalmic nerve passes forward in the lateral wall of the cavernous sinus, between the maxillary nerve below and the trochlear and oculomotor nerves above; just before it enters the superior orbital fissure the ophthalmic nerve divides into its three main branches (the nasociliary, the lacrimal, and frontal), the nasociliary usually being the first to be given off and the remainder of the nerve then dividing to form the frontal and lacrimal (Fig. 2-41). At about this point the frontal and lacrimal branches cross lateral to the oculomotor nerve to lie above it.

All three of these branches of the ophthalmic then enter the orbit through the superior orbital fissure, but the frontal and lacrimal do so above the common anular tendon and thus lie outside of the muscle cone (see Fig. 2-36), while the nasociliary passes through the anulus. Since the larger terminal branches of both the ophthalmic nerve and the ophthalmic artery leave the orbit through the superior medial portion of the orbital septum, it is in operations in and around the upper nasal quadrant of the orbit that one must be especially alert for these structures.

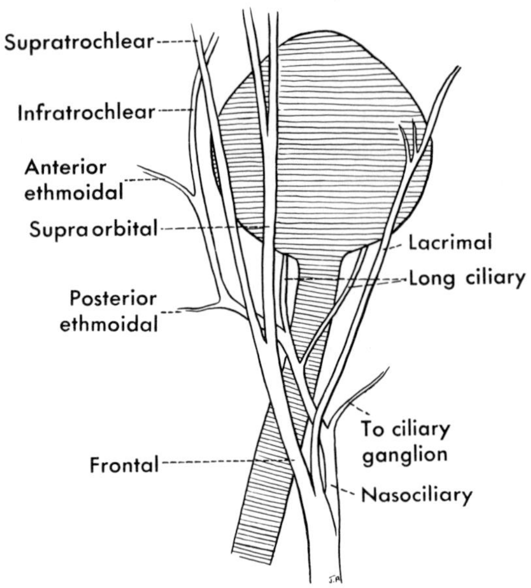

Fig. 2-41. Course and branches of the ophthalmic nerve in the orbit.

Lacrimal Branch

The lacrimal branch of the ophthalmic passes through the superior orbital fissure above the upper portion of the lateral rectus muscle and in this position lies lateral to the frontal and trochlear nerves, and above and medial to the ophthalmic vein. It runs forward along the lateral or upper border of the lateral rectus muscle and in the distal two thirds of this course is joined by the lacrimal artery, the two running together to reach the lacrimal gland. Like the artery, branches of the lacrimal nerve are continued beyond the gland to supply the conjunctiva and, usually, the skin of the lateral corner of the eyelids. Through the anastomosis that the lacrimal nerve usually receives from the zygomaticotemporal branch of the maxillary nerve, before this leaves the orbit, the lacrimal nerve receives secretory fibers for the lacrimal gland.

It seems to be generally agreed that, as reviewed by Ruskin, the preganglionic parasympathetic fibers concerned with tear secretion run at first with the facial nerve and leave this at the geniculate ganglion to form a part of the greater (superficial) petrosal nerve, by which they reach the pterygopalatine (sphenopalatine) ganglion. Postganglionic fibers originating from this ganglion then join the maxillary nerve, leave it by way of the zygomatic and the zygomaticotemporal, and reach the lacrimal gland through the anastomosis with the lacrimal nerve. Ruskin noted that sympathetic fibers from the deep petrosal nerve are also distributed to the lacrimal gland (and apparently others reach it along the lacrimal artery), but whether they are secretory or purely vasomotor is not known. He also stated that blocking of the pterygopalatine ganglion does not abolish tear secretion entirely, but does markedly reduce it.

Whitwell reported that the branch for the lacrimal gland from the zygomaticotemporal nerve may enter the gland directly, in which case it enters a little more laterally than does the lacrimal nerve. He quoted others as having observed direct branches from the ganglion to the gland, entering the orbit through the inferior orbital fissure.

Frontal Branch

The frontal nerve, the largest branch of the ophthalmic, enters the orbit through the superior orbital fissure above the lateral rectus muscle where it lies between the trochlear and the lacrimal nerves, the trochlear being above and medial, and the lacrimal inferior and lateral. The frontal nerve passes forward above the levator palpebrae, thus close to the roof of the orbit, to divide into its two terminal branches, the supratrochlear and the supraorbital nerves.

Of these the *supraorbital* is by far the larger and continues the direction of the frontal nerve to pass through the supraorbital notch or foramen and be distributed to the skin of the forehead and of the upper eyelid. As it passes through the supraorbital notch it supplies a twig to the frontal sinus through the aperture for the frontal diploic vein. The supraorbital nerve may divide into medial and lateral branches before it crosses the orbital rim, and the medial branch has been called, confusingly, the frontal nerve.

The *supratrochlear nerve* runs forward and medially; it passes above the trochlea of the superior oblique muscle, usually communicating at about this point with the infratrochlear nerve, and then pierces the orbital septum with the supratrochlear (frontal) artery to curve upward with this artery and supply skin and conjunctiva of the upper lid, and skin on the lower medial portion of the forehead. At their points of exit from the orbit both supraorbital and supratrochlear nerves lie deep to the orbicularis and frontalis muscles, and their terminal cutaneous branches pierce these muscles.

Nasociliary Branch

The nasociliary (nasal) nerve arises from the medial and inferior portion of the ophthalmic; it is intermediate in size between the lacrimal and the frontal and is usually the first of the three branches to be given off. It enters the orbit within the anulus tendineus communis, being therefore the only branch of the ophthalmic to lie within the muscle cone; lying first lateral to the oculomotor nerve, between this and the remainder of the ophthalmic, it passes medially between the two divisions of the former nerve close to the back of the orbit. It at first lies lateral and superior to the optic nerve, but as the ophthalmic artery crosses above the optic nerve to attain a more medial position the nasociliary nerve does likewise. It passes between the superior rectus and the optic nerve and then runs along the medial wall of the orbit between the superior oblique and medial rectus muscles.

Before or after the nasociliary nerve enters the orbit it gives off a filament (ramus communicans) to the ciliary ganglion, this ramus commonly being known as the long or sensory root of the ganglion. Within the orbit this branch runs along the lateral surface of the optic nerve to reach the upper and posterior part of the ganglion; as the nasociliary crosses the optic nerve it gives off the posterior ethmoidal nerve and the long ciliary nerves.

The *posterior ethmoidal nerve,* which is often lacking (about two thirds of the time, as quoted by Whitnall), runs transversely across the orbit to enter the posterior ethmoidal foramen with the corresponding branch of the ophthalmic artery and supplies posterior ethmoid cells and the sphenoid sinus.

There are typically two *long ciliary nerves,* which run forward, usually one along the medial side of and one above the optic nerve, to pierce the sclera medial and lateral to the nerve and be distributed to the eyeball. The long ciliary nerves are largely sensory, but are thought to contain also most of the sympathetic fibers for pupillodilation. After giving off these branches and passing forward along the medial wall of the orbit, the nasociliary divides into its two terminal branches, the infratrochlear and the anterior ethmoidal.

The *infratrochlear nerve* continues the direction of its parent branch to pass below the pulley of the superior oblique, pierce the orbital septum and the orbicularis oculi, and be distributed to the medial portion of the upper and lower lids, to the lacrimal sac and to the upper part of the lateral aspect of the nose.

The *anterior ethmoidal nerve,* a large and constant branch, leaves the parent stem at almost a right angle to pass through the anterior ethmoidal foramen in company with the corresponding branch of the ophthalmic artery. Through this foramen it gains access to the anterior cranial fossa, where it runs forward along the cribriform plate of the ethmoid bone; it then enters the nasal cavity through the slitlike ethmoidal fissure at the side of the crista galli, and within the nose divides into internal and external nasal branches.

The internal nasal branches are distributed to the upper part of the nose; the external nasal branch runs downward on the inner surface of the nasal bone, passes between the nasal bone and the lateral nasal cartilage, and supplies skin over the lower part of the dorsum of the nose. As the anterior ethmoidal nerve and artery leave the orbit, they are at the level of the cribriform plate and therefore serve as a landmark for this; they thus indicate the division between that part of the medial orbital wall adjacent to the anterior cranial fossa and that part adjacent to the nasal cavity.

MAXILLARY NERVE

The major portion of the maxillary nerve, the second division of the trigeminal, also has a close relationship to the orbit. This branch leaves the skull by way of the foramen rotundum to enter the pterygopalatine (sphenomaxillary) fossa, where it has suspended from it the pterygopalatine (formerly sphenopalatine) ganglion; through the ganglion it gives off branches to the nose, palate, and pharynx, and the posterior superior alveolar nerves, and it also receives fibers from the ganglion. (The connections between nerve and ganglion are known as the pterygopalatine nerves.) The zygomatic nerve usually also arises in the fossa to run forward parallel to the continuation of the maxillary, hereafter known as the infraorbital.

The infraorbital and zygomatic nerves leave the pterygopalatine fossa by the inferior orbital fissure and attain the infraorbital sulcus in the floor of the orbit; they are separated from the orbital contents by the orbitalis muscle, which bridges the fissure and sulcus. While it lies in the sulcus the *infraorbital nerve* gives off the middle superior alveolar branch (when this is present) and becomes separated from the zygomatic nerve as this latter diverges laterally. The infraorbital continues forward to enter the infraorbital canal, within which it gives off anterior superior alveolar branches, and then divides into terminal cutaneous branches as it emerges from the infraorbital foramen. This foramen may often be located, for purposes of local anesthesia, by dropping a vertical line from the supraorbital foramen or notch (when that is palpable); the infraorbital foramen usually lies on or just lateral to this line.

The infraorbital nerve is accompanied in its course by the infraorbital artery, one of the terminal branches of the maxillary; in the pterygopalatine fossa the artery arises lateral to the nerve, but during its infraorbital course it may cross or pierce the nerve and thus emerge medial to it. In the floor of the orbit the infraorbital nerve is also in the roof of the maxillary sinus, and the position of the canal may be marked by a ridge on this roof; sometimes the bone below the nerve is deficient, the nerve then lying in contact with the mucous membrane of the sinus.

The *zygomatic branch* of the maxillary nerve, leaving the infraorbital nerve soon after the two have passed through the inferior orbital fissure, enters the orbit and runs anteriorly and laterally; low on the lateral wall it divides into its two terminal branches, the zygomaticofacial and zygomaticotemporal nerves, which leave the orbit by minute foramina to become cutaneous. While it is within the orbit the zygomaticotemporal nerve typically gives off a branch to the lacrimal gland. This branch, containing secretory fibers from the pterygopalatine ganglion, usually joins the lacrimal nerve (Fig. 2-42) but, as already noted, may proceed independently to the lacrimal gland.

Fig. 2-42. Schema of the innervation of the lacrimal gland.

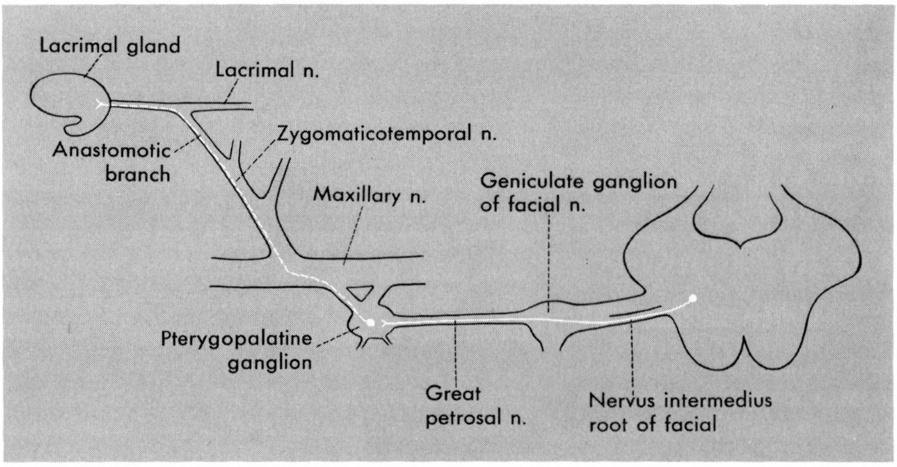

AUTONOMIC NERVES OF ORBIT

CILIARY GANGLION

Associated with both the oculomotor and the nasociliary nerves is the ciliary ganglion, the parasympathetic ganglion connected with the eyeball. This small ganglion, usually about the size of the head of an ordinary pin, but according to Devos and Marcelle varying in size from about 1 × 1 mm to 3 × 4 mm, lies well back in the orbit (3–4 cm on the slant from the superolateral orbital rim, or about 1 cm from the back of the orbit) between the optic nerve and the lateral rectus muscle and is closely associated with the ophthalmic artery or its branches as these lie also lateral to the nerve. It is said to be supplied by a special artery, from a posterior ciliary or other branch. The ganglion typically receives three incoming connections: the ramus communicans with the nasociliary nerve (also called the long, superior, or sensory root); an oculomotor (short, inferior, or motor) root; and a sympathetic ramus (intermediate or sympathetic root) from the carotid plexus (Fig. 2-43). Devos and Marcelle, however, found more than three roots in 35% of the specimens that they dissected, owing usually to doubling of the oculomotor root.

In spite of its several roots the ciliary ganglion is a purely parasympathetic one, which like the other parasympathetic ganglia in the head is traversed by sensory and sympathetic fibers so that its branches of distribution (short ciliary nerves) contain both these and the fibers from the ganglion. Scattered ganglion cells or small ganglia may be found along the course of the short ciliary nerves, and ganglion cells also occur along the nerves within the eyeball, but beyond the ciliary ganglion most of the parasympathetic fibers are postganglionic.

Nasociliary Connection

The ramus communicans with the nasociliary nerve (long or sensory root of the ciliary ganglion) arises from the nasociliary nerve either before or after it enters the orbit. It may contain sympathetic fibers derived from the carotid plexus by the nasociliary nerve or may be joined by the sympathetic ramus to the ciliary ganglion, but otherwise consists of sensory fibers; it enters the ciliary ganglion at its posterosuperior angle. The fibers of this root do not, however, end in the ganglion; since they are sensory or postganglionic sympathetic ones, they traverse the ganglion without synapse and help to form the short ciliary nerves arising from it.

Oculomotor Root

The oculomotor root of the ganglion consists of preganglionic fibers that arise from the nucleus of Edinger–Westphal, close to the voluntary nucleus of the oculomotor nerve; they leave the brain stem with this nerve to make synapse in the ganglion. Kerr and Hollowell have shown that the majority of fibers concerned with both pupillary constriction and accommodation lie at first very superficially in the nerve, thus confirming the findings of Sunderland and Hughes that the parasympathetic fibers are so placed in the intracranial portion of the nerve as to be particularly vulnerable to pressure from above or medially. The fibers then run in the inferior division of the nerve; they separate from the branch to the inferior oblique muscle, to form the true (short, motor) root of the ganglion. It has been described as absent only when the ganglion was located directly on the nerve to the inferior oblique, and in the real sense is therefore apparently never absent. Devos and Marcelle found it to be double in about a third of the specimens they dissected, and occasionally triple.

The postganglionic fibers derived from the cells of the ciliary ganglion join the sympathetic and sensory fibers that traverse the ganglion, to form the short ciliary nerves. Within the eyeball they are distributed to at least the sphincter pupillae and the ciliary muscle; hence, activity of this system produces constriction of the pupil and accommodation of the lens for near vision.

It is usually assumed that the preganglionic fibers to the sphincter pupillae synapse almost entirely in the ciliary ganglion, as do other

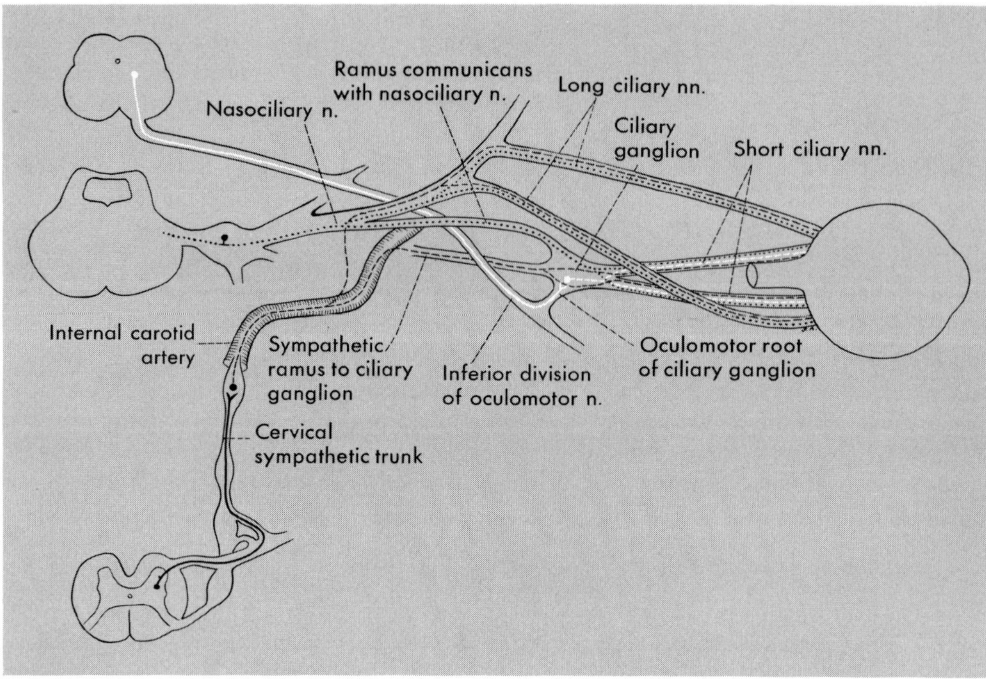

Fig. 2-43. The ciliary ganglion and ciliary nerves. Parasympathetic fibers are *white,* sympathetic ones *black,* the preganglionic fibers being indicated in each case by *solid lines* and the postganglionic ones by *broken lines.* Sensory fibers are indicated by the *black dotted lines.*

parasympathetic fibers to the smooth muscle of the eyeball. However, Nathan and Turner adduced evidence which they thought indicated that only the motor pathway concerned with the light reflex synapsed in the ciliary ganglion, and that the pathway for accommodation bypassed that. Although they suggested that the fibers for accommodation might synapse in episcleral ganglia (and in light of later knowledge they might also synapse on intrabulbar ganglion cells), they have not explained the anatomy of such a bypass; they quoted evidence that pupillary constriction may accompany accommodation after removal of the ciliary ganglion and reported the occurrence of Argyll Robertson pupils after peripheral nerve lesions, but failed to find any pathway by which oculomotor fibers to the eyeball bypass the ganglion.

Sympathetic Ramus

The classic description of the sympathetic ramus to the ciliary ganglion is that it leaves the internal carotid artery in the cavernous sinus, enters the orbit through the anulus tendineus just below the nasociliary nerve, and joins the posterior border of the ganglion between the upper and lower roots. This connection is the most variable of the three in its gross anatomy and is often not seen in dissections; Devos and Marcelle expressed the belief that it is nevertheless constantly present. Since it arises from fibers of the carotid plexus, it may in some cases join the ophthalmic nerve directly and hence be contained in the nasociliary ramus to the ganglion; it may join this at a greater or lesser distance from the ganglion. In contrast to the usual description, Johnston and Parkinson, and Parkinson and colleagues found, in 80 dissections and three specimens examined by serial section, no instance in which sympathetic fibers ran directly from the carotid plexus to the ciliary ganglion; instead, they found fibers from the plexus first joining the abducens nerve and then leaving that to join the ophthalmic nerve while still in the cavernous sinus. Sunderland and Hughes had observed similar connections earlier, but did not regard them as the sympathetic root of the ciliary ganglion.

Devos and Marcelle also reported that the sympathetic root may not arise within the cavernous sinus at all, but instead may leave the carotid plexus later. They found it passing through the optic canal medial and inferior to the ophthalmic artery; they said also that the fibers representing it may form a part of the plexus about the ophthalmic artery, passing to the ganglion where these two are in close apposition; in this case, obviously, the sympathetic ramus as a gross anatomical entity is lacking, but the fibers are present nevertheless. Like the fibers from the nasociliary nerve, those of the sympathetic ramus merely traverse the ganglion and make no functional connections with it; upon leaving the ganglion they too help form the short ciliary nerves. The sympathetic fibers in the short ciliary nerves are usually said to be primarily vasomotor, in contrast to the pupillodilator sympathetic fibers in the long ciliary nerves.

Short Ciliary Nerves

From the anterior end of the ciliary ganglion arise from 4 to 12 short ciliary nerves; these pass forward above and below the optic nerve, mingled with the short ciliary arteries, and diverge slightly so that nerves and arteries practically surround the optic nerve at the eyeball and pierce the sclera in a circle about this nerve. The short ciliary nerves may have delicate anastomoses with each other and with the long ciliaries.

From the above statements, it should be apparent that the long ciliary nerves contain sensory and sympathetic fibers; the short ciliary nerves, sensory, sympathetic, and parasympathetic ones. The sensory fibers of both sets of nerves are derived from sensory cells located in the trigeminal ganglion; the fact that they may traverse the ciliary ganglion on their way to the eyeball therefore has no functional significance.

SYMPATHETIC FIBERS

The sympathetic fibers to the eyeball constrict the blood vessels and dilate the pupil; they are derived from the superior cervical ganglion or perhaps also from smaller ganglia that are known to be located in the sympathetic plexus on the carotid artery as it lies in the cavernous sinus (Gellért) and are therefore already postganglionic fibers as they enter the orbit. Although many of them traverse the ciliary ganglion, their connection with this is purely adventitious.

Origin and Course

The preganglionic sympathetic fibers for the innervation of the orbit are usually derived from upper thoracic segments of the spinal cord. Insofar as the pupil is concerned, there is considerable variation in the segmental origin of these fibers; some apparently always leave the spinal cord by way of the first thoracic nerve, and all may do so, but pupillary dilation can sometimes also be produced by stimulation of all the ventral spinal thoracic roots as low as T4 and occasionally as high as C8 (Ray and co-workers). Whether these observations hold only for the pupil or are true also of the sympathetic innervation of all orbital smooth muscle is not known; according to Hyndman and Wolkin the highest origin of sympathetic fibers to the skin of the face (presumably including that of the lids) is the second thoracic nerve.

The postganglionic sympathetic fibers to the orbit from the superior cervical ganglion follow the internal carotid artery into the skull, and most of them leave the internal carotid plexus either by joining the ophthalmic nerve or by way of the sympathetic ramus to the ciliary ganglion, perhaps with still others traveling along delicate plexuses around the ophthalmic artery and its branches. Many of the details are not yet known, however.

There is clear evidence that in some animals pupillodilator fibers do not follow the internal carotid plexus into the cavernous sinus, but instead leave that plexus to pass through the middle ear cavity before joining the trigeminal ganglion or its ophthalmic division, and there is indirect evidence that at least some of the pupillodilator fibers in man may take the same course (*e.g.,* Barlow and Root). Further, the course of sympathetic fibers to the smooth muscle of the lids is not known. It has already been mentioned that the arteries rather than the third nerve may transmit sympathetic fibers to the superior tarsal muscle; and the sixth nerve is said to be the only one of the three motor nerves that contains unmyelinated fibers, expected of postganglionic sympathetic neurons. Sunderland and Hughes described the sympathetic twig to the sixth nerve, which does contain apparent sympathetic fibers, as running with this nerve for only a short distance and then leaving it to join the ophthalmic and maxillary nerves, from which sympathetic fibers might reach both lids and the floor of the orbit. The orbitalis muscle in the floor of the orbit is usually said, however, to be supplied by orbital twigs from the pterygopalatine (sphenopalatine) ganglion that enter its inferior surface (bringing in sympathetic fibers from the deep petrosal nerve).

Sympathetic fibers to the lacrimal gland, whether vasomotor or otherwise, are believed to reach it both through a plexus around the lacrimal artery and through the branch which the zygomaticotemporal nerve sends to the gland; sympathetic fibers leave the internal carotid plexus through the deep petrosal nerve, which joins the pterygopalatine ganglion and can thus send fibers into the branches of the maxillary nerve.

Of clinical importance is the innervation of the superior tarsal muscle by the sympathetic, since interruptions of the cervical sympathetic trunk produce both a ptosis of the lid and a myosis of the pupil, two of the characteristics of Horner's syndrome. Evidence of paralysis of the inferior tarsal muscle has apparently not been reported.

Interruption of the cervical sympathetic trunk has been said also to produce an enophthalmos, due to paralysis of the orbitalis muscle, but this is apparently a false impression caused by the ptosis that does occur. Likewise, stimulation of the orbitalis, through the cervical sympathetic trunk or otherwise, was once seriously considered as a cause of exophthalmos. Naffziger pointed out that cervical sympathetic stimulation does not produce exophthalmos, and it is now rather generally agreed that exophthalmic goiter produces its effects through edematous and degenerative changes in the orbital contents.

Distribution Within the Eye

The nerves entering the eyeball are the so-called nerve of Tiedemann, a plexus around the central artery of the retina; the short ciliary nerves; the long ciliary nerves; and perhaps, although never demonstrated, fibers derived from the occasional episcleral ganglia.

The nerve of Tiedemann is presumably a sympathetic vasomotor plexus distributed to the central artery of the retina and its branches. Nothing is known about either the central or peripheral connections of the episceral ganglia. The long ciliary nerves, as already noted, apparently contain both sensory and sympathetic fibers (the latter said to include many of the pupillodilator fibers); the short ciliary nerves, sensory, sympathetic, and parasympathetic fibers.

After they enter the eyeball, the short ciliary nerves are said (Beatie and Stilwell) to proceed directly to the ciliary muscle and iris (in the rabbit), which, if there is not a species difference, casts some doubt upon the statement that in man the sympathetic fibers in these nerves are primarily vasoconstrictor. On the other hand, the long ciliary nerves are said in the rabbit to give off fibers that follow arteries into the choroid and thereafter proceed to the region of the ciliary body. Wolter ('60a) found a rich nerve supply to the human choroid, consisting both of perivascular and intervascular fibers. Branches of both the long and the short ciliary nerves run forward in the perichoroidal (suprachoroidal) space and form a plexus in the region of the ciliary body. From this plexus branches pass to the ciliary body and iris, while others penetrate the sclera just behind the limbus to supply the cornea. Stimulation of the cervical sympathetic trunk in the rabbit has been reported to decrease the rate of formation of the aqueous humor, apparently by decreasing the rate of blood flow through the ciliary processes (Langham and Rosenthal).

The nerves to the iris presumably consist of

sensory fibers, sympathetic fibers for the dilator pupillae, and parasympathetic fibers for the sphincter pupillae. By electron microscopy, Richardson found consistent differences between the nerve endings on these two muscles, in agreement with what has usually been described as their different innervations. Ehinger described adrenergic fibers, mostly not associated with blood vessels, in both the ciliary muscle and the sphincter pupillae, and Laties and Jacobowitz described the fibers in the ciliary muscle and the sphincter pupillae as being mostly cholinergic but partly adrenergic and those in the dilator pupillae as being mostly adrenergic but partly cholinergic. The implication that this indicates a double innervation—sympathetic and parasympathetic—for the smooth muscle within the eyeball rests upon the assumptions that the techniques used actually do demonstrate the functional characteristics of the fibers. Still to be explained is the physiologic significance of a second innervation.

The pectinate ligament (trabecular meshwork) has a particularly rich innervation, shown experimentally to consist of fibers from the ophthalmic nerve, the superior cervical ganglion, and the ciliary ganglion (Holland and co-workers). The function of this innervation is not known. The sympathetic part of it, however, apparently does not significantly influence the rate of drainage of the aqueous humor (Langham and Rosenthal). Finally, Wolter ('60b) apparently demonstrated centrifugal or motor fibers in the human optic nerve and described them as ending on blood vessels of the optic nerve and retina; Brindlay and Hamasaki were unable to demonstrate such fibers in the cat.

Blood Vessels

OPHTHALMIC ARTERY

The ophthalmic artery has typically small anastomoses inside the orbit with the middle meningeal and maxillary (internal maxillary) arteries, from the external carotid, and variable larger ones in the face and scalp with other branches of the external carotid.

The ophthalmic artery normally arises from the internal carotid just as this vessel leaves the caverous sinus, therefore just medial to the anterior clinoid process, but occasionally it arises in the cavernous sinus. The artery lies below the optic nerve and runs forward within its dural-arachnoid sheath; it may also be at first slightly medial to the nerve, but if so proceeds slightly laterally, gradually pierces the sheath, and emerges from the optic canal to lie outside the sheath of, and lateral and inferior to, the optic nerve (Figs. 2-37 and 2-44). After its entrance into the orbit the ophthalmic artery lies between the optic nerve and the lateral rectus muscle; in this position it gives off numerous small branches that include muscular ones, posterior ciliary arteries, and the central artery of the retina.

Occasionally, the ophthalmic artery enters the proximal end of the optic canal through a small separate foramen placed below that for the optic nerve.

Fig. 2-44. Course and branches of the ophthalmic artery in the orbit. The variable muscular branches are not shown.

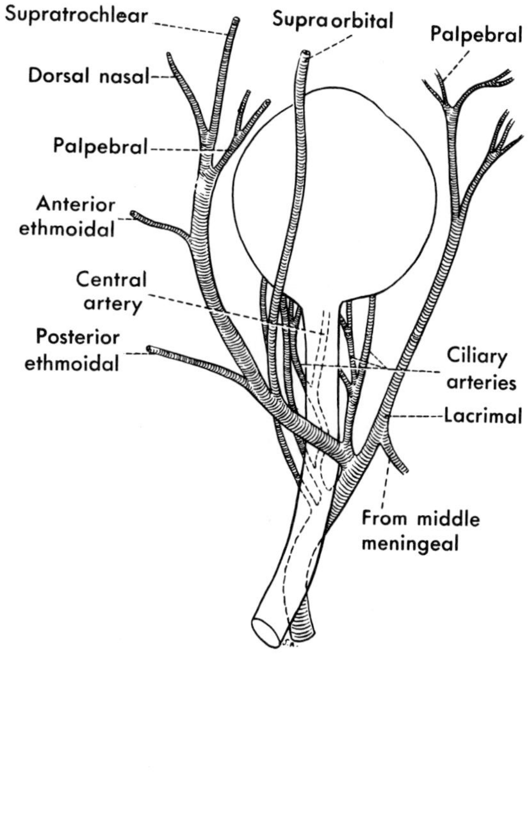

Posterior Ciliary Arteries

The posterior ciliary arteries are divided into two sets, short and long. Between 10 and 20 short ciliary arteries reach the eyeball, but they arise from fewer stems. They gradually surround the optic nerve to become intermingled with the short ciliary nerves; running forward with these they pierce the sclera close to the entrance of the optic nerve. The long posterior ciliary arteries, typically two, also parallel the optic nerve and pierce the sclera medial and lateral to it, but outside the circle of the short ciliaries. The intrabulbar distribution of these arteries has already been described.

Central Artery of Retina

The central artery of the retina arises in company with the posterior ciliary arteries from the ophthalmic while this lies inferolateral to the optic nerve, and may arise in common with one; it runs forward beneath the optic nerve and some 10–15 mm behind the eyeball turns upward to penetrate the dura and arachnoid, pass through the subarachnoid space in a short oblique course, and then turn abruptly upward to penetrate the optic nerve and pass (usually) to the center of this. During this course it gives off tiny twigs to the meninges about the nerve. The course of this artery within the optic nerve and the eyeball has already been noted.

After giving off these multiple small branches in the posterior part of the orbit, the ophthalmic artery typically turns upward and sharply medially to cross above the optic nerve, thus usually passing between the optic nerve and the superior rectus muscle; in about 15% of instances the artery crosses below the nerve. Before or as it crosses the nerve it gives off the lacrimal artery, which runs along the lateral wall of the orbit, and after crossing the nerve gives off the supraorbital artery, which accompanies the corresponding branch of the frontal nerve to the forehead. The main stem of the ophthalmic then runs along the medial wall between superior oblique and medial rectus muscles. In this position it is paralleled by the nasociliary nerve and like this nerve usually gives off branches, the posterior and anterior ethmoidal arteries, which leave the orbit through the similarly named canals. As it passes below the trochlea of the superior oblique muscle the ophthalmic artery gives off the medial palpebral branches to the lid and then divides into its two terminal branches, the supratrochlear (frontal) and dorsal nasal arteries.

Lacrimal Artery

The lacrimal artery usually arises close to the apex of the orbit before or as the ophthalmic crosses the optic nerve; it runs forward along the lateral walk of the orbit, at first within the muscle cone but later leaving this to accompany the lacrimal nerve in the distal part of its course and lie above the lateral rectus muscle; it ends largely in the lacrimal gland. In its course the lacrimal artery supplies small branches to other structures in the orbit and beyond the lacrimal gland sends twigs (lateral palpebral arteries) to the lateral angle of the eye. These twigs may complete the palpebral arterial arches. The lacrimal artery is usually the origin of the muscular branch to the lateral rectus muscle and may give rise to some of the short posterior ciliary arteries.

The lacrimal artery receives from the frontal branch of the middle meningeal artery an anastomotic channel (the meningeal's ramus anastomoticus with the lacrimal artery, also called orbital branch of the meningeal, meningeal branch of the lacrimal, and sometimes accessory ophthalmic). This enters the orbit either through the upper corner of the superior orbital fissure or through a separate foramen close by and varies markedly in size. When it is large, it obviously reinforces the lacrimal artery, and it may replace this artery, giving rise to the usual branches of that vessel. More rarely, it completely replaces the entire ophthalmic artery, giving off the usual branches of this artery and following that vessel's usual intraorbital course (Fig. 2-45). Bansal reported such a case, with partial replacement on the other side of the same body; he found previous reports of only four cases of complete replacement, but others have surely gone unreported; Gillilan found one among ten well-injected heads that she investigated, and this author encountered one, never reported, while making a filmed dissection that many ophthalmologists have seen.

Ethmoidal Arteries

The posterior ethmoidal artery leaves the ophthalmic well back in the orbit and passes between the superior oblique and medial rectus muscles (sometimes above the superior oblique) to enter the posterior ethmoidal foramen. The anterior ethmoidal artery, more constant and typically larger than the posterior, leaves the orbit a variable distance farther forward by entering the anterior ethmoidal foramen. Both vessels pass between ethmoid cells to reach the anterior cranial fossa and thereafter turn down into the nasal cavity. Their variations and relations are of particular importance if they are a source of serious nasal bleeding, and are discussed in connection with the blood supply of the nose (see Chap. 3).

Medial Palpebral Arteries

The superior and inferior medial palpebral arteries arise from the ophthalmic by separate stems or by a common trunk at about the level of the trochlea, penetrate the orbital septum (the superior above, the inferior below the medial palpebral ligament), and run across the lids to form the palpebral arches; twigs from those vessels also supply the conjunctiva, the lacrimal sac, and the nasolacrimal duct.

Terminal Branches

The terminal branches of the ophthalmic artery are the *supratrochlear* (frontal) and *dorsal nasal* (nasal) *arteries,* and their common stem has been referred to as the frontonasal artery. These vessels usually separate from each other behind the orbital septum to penetrate this individually, but may divide after penetrating the septum. The supratrochlear artery passes upward over the superior margin of the orbit to supply the forehead, while the dorsal nasal artery passes above and then downward in front of the medial palpebral ligament to

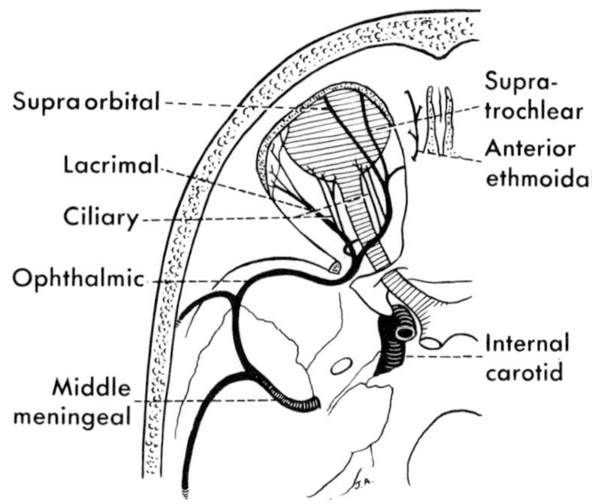

Fig. 2-45. Anomalous origin of the ophthalmic artery from the middle meningeal. (After Harvey JC, Howard LM: Anat Rec 92:87, 1945)

descend along the side of the nose, which it helps to supply, giving off also twigs to the lacrimal sac and usually anastomosing with the angular artery.

The anastomosis between the dorsal nasal artery and the angular or a nasal branch of that vessel is the most direct and often the largest of the collateral channels to the orbit, but it varies in size and may be lacking; its place may or may not be taken by a similar anastomosis between the supraorbital and the angular. In addition, the branches of the ophthalmic in the lids, the forehead, and the scalp anastomose especially with the transverse facial, the frontal branch of the superficial temporal, and with other vessels supplying the scalp, both from the external carotid and the ophthalmic of the opposite side. Gillilan reported on the various anastomoses in some detail and quoted evidence that about 90% of people have enough collateral circulation to prevent permanent blindness following occlusion of the internal carotid or the ophthalmic artery.

Variations in the branches of the ophthalmic artery are frequent, but of no particular importance clinically. In addition to those already cited, they consist primarily in the origin of several arteries from a common stem or in the separate origin of these from the ophthalmic artery.

INFRAORBITAL ARTERY

Also in relation to the orbit is the infraorbital artery, a terminal branch of the maxillary (formerly internal maxillary). This artery arises from the maxillary in the pterygopalatine fossa, where it is therefore in close relation to the maxillary nerve. It accompanies the continuation of the maxillary nerve, the infraorbital, through the inferior orbital fissure and the infraorbital groove and canal in the floor of the orbit.

It usually lies at first lateral to the nerve, to cross and lie medial to it as they approach the orbital rim, but may lie medial to the nerve even as they enter the fissure. In the floor of the orbit the infraorbital artery usually gives off small branches that help supply the orbital fat and the inferior rectus and inferior oblique muscles and may also reach the lacrimal gland.

VEINS OF THE ORBIT

The superior and inferior ophthalmic veins, the two major vessels draining the orbit, may both have anastomoses with the veins of the face; those of the superior vein are regularly so well developed that it is described as originating from these anastomoses (Fig. 2-46). These communications are of particular importance because neither the ophthalmic nor the facial veins contain valves, and therefore the ophthalmic may readily conduct infected thrombi from a superficial position in the face, with resultant cavernous sinus thrombosis. The flow of blood from the orbit is probably normally both to the cavernous sinus and to the facial vein (Whitnall). The communications between the angular and ophthalmic veins at the medial corner of the eye are also of obvious importance to the surgeon, since they may be of considerable size.

At the upper medial corner of the eye the supratrochlear (frontal) vein and the laterally lying supraorbital vein (a chief tributary of which often parallels the upper orbital rim) unite to form the angular vein. Typically, before this union occurs the supraorbital vein sends a branch through the supraorbital notch, where it receives the frontal diploic vein, and close to its upper end the angular vein usually sends a branch through the upper part of the orbital septum. These branches, both of which have been called the nasofrontal vein, typically unite to form the beginning of the superior ophthalmic vein; or the branch from the supraorbital may join the superior ophthalmic well back in the orbit, after the branch from the angular has received tributaries that drain the orbital contents.

Fig. 2-46. General disposition of the ophthalmic veins and their chief connections.

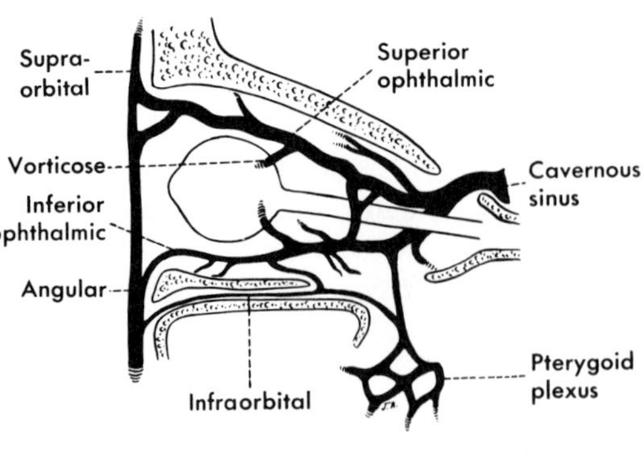

Superior Ophthalmic Vein

The superior ophthalmic vein is the larger of the two ophthalmic veins and, accompanying the ophthalmic artery above the optic nerve and below the superior rectus muscle and surrounded by orbital fat, receives most of the branches corresponding to the branches of this artery. Thus it receives the anterior and posterior ethmoidal veins, the lacrimal vein, and most of the muscular branches. It also communicates with the central vein of the retina. The superior ophthalmic vein also receives the two venae vorticosae from the upper half of the eyeball. As described in connection with the eyeball, the venae vorticosae represent posterior ciliary veins, but differ from the corresponding arteries in being less numerous and in leaving the eyeball at some distance from the optic nerve.

As the superior ophthalmic vein nears the

apex of the oribit it usually receives the inferior ophthalmic vein, so that the two veins open together into the cavernous sinus; more commonly the united veins pass through the superior orbital fissure above the lateral rectus muscle and therefore outside the anulus (usually lying here lateral to and below the nerves that pass through this portion of the superior orbital fissure) but the position of the common vein is somewhat variable and it may pass through the anulus or pass through the lower portion of the superior orbital fissure below the anulus. Also, the superior and inferior ophthalmic veins may empty separately into the cavernous sinus and may then have different relations to the anulus as they leave the orbit.

Central Vein of Retina

The central vein of the retina accompanies the central artery through the subarachnoid space and is here subject to the effects of increased pressure from the cerebrospinal fluid during elevation of the general intracranial pressure. Paton and Holmes pointed out many years ago that papilledema or "choked disk" associated with increased intracranial pressure may be largely due to a rise in venous pressure sufficient to maintain circulation through the central vein, and Gibbs demonstrated in the dog and cat that when cerebrospinal pressure was increased above the venous pressure of the nerve head, the venous pressure also rose and remained some 2–4 mm of mercury above the cerebrospinal pressure. Edema and swelling of the optic nerve within its tough dural sheath can also compress the vein, but Behrman denied that venous compression causes papilledema. He argued that swelling of the optic nerve compresses a perivascular channel around the central artery of the retina and that it is occlusion of this channel (not yet demonstrated in man) that causes the swelling of the nerve head.

The central vein of the retina is often described as ending by joining the superior ophthalmic vein at the apex of the orbit, but according to both Whitnall ('21a) and Wolff it usually ends independently in the cavernous sinus, although regularly communicating with the superior ophthalmic vein. As pointed out by Whitnall, this communication with the superior ophthalmic provides a collateral channel from the retina in cases of thrombosis of the cavernous sinus.

Inferior Ophthalmic Vein

The inferior ophthalmic vein lies below the optic nerve and begins by the confluence of small vessels on the floor of the orbit; these in turn may communicate with the facial vein across the orbital rim. It passes backward along the floor of the orbit in the orbital fat, usually having communications with the superior ophthalmic vein, and receives the two inferior vorticose veins and branches from the lower muscles of the orbit. As it passes across the inferior orbital fissure it communicates with the pterygoid plexus. It usually ends, as already noted, by joining the superior ophthalmic, but if it fails to do so it may pass alone through or below the anulus.

Infraorbital Vein

The infraorbital vein accompanies the infraorbital nerve and and artery in the floor of the orbit; it communicates anteriorly with the angular vein and posteriorly with the pterygoid plexus. Through the inferior orbital fissure it usually receives a tributary that helps drain structures in the floor of the orbit and communicates with the inferior ophthalmic vein.

References

ABD-ÉL-MALEK S: On the presence of sensory fibres in the ocular nerves. J Anat 72:524, 1938

ANDERSON RL: The levator aponeurosis. Attachments and their clinical significance. Arch Ophthalmol 95:1437, 1977

ANDERSON RL, DIXON RS: The role of Whitnall's ligament in ptosis surgery. Arch Ophthalmol 97:708, 1979

ASHTON N: Anatomical study of Schlemm's canal and aqueous veins by means of neoprene casts. Part II. Aqueous veins (continued). Br J Ophthalmol 36:265, 1952

BACH-Y-RITA P, MURATA K: Extraocular propriocep-

tive responses in the VI nerve of the cat. Q J Exp Physiol 49:408, 1964
VON BAHR G: Measurements of the thickness of the cornea. Acta Ophthalmol 26:247, 1948
BANSAL PC: A case of abnormal ophthalmic arteries. J Anat Soc India 10:41, 1961
BÁRÁNY, EH: The relative importance of ultrafiltration and secretion in the formation of aqueous humour as revealed by the influence of arterial blood pressure on the osmotic pressure of the aqueous. Acta Physiol Scand 13:81, 1947
BARLOW CM, ROOT WS: The ocular sympathetic path between the superior cervical ganglion and the orbit in the cat. J Comp Neurol 91:195, 1949
BEATIE JC, STILWELL DL JR: Innervation of the eye. Anat Rec 141:45, 1961
BEHRMAN S: Pathology of papilloedema. Brain 89:1, 1966
BERGLAND R, RAY BS: The arterial supply of the human optic chiasm. J Neurosurg 31:327, 1969
BILL A: Aspects of the drainage of aqueous humor in cats. Arch Ophthalmol 67:148, 1962
BRINDLEY GS, HAMASKI DI: Histological evidence against the view that the cat's optic nerve contains centrifugal fibres. J Physiol 184:444, 1966
BROWN GC, SHIELDS JA: Cilioretinal arteries and retinal arterial occlusion. Arch Ophthalmol 97:84, 1979
BUCY PC, WEAVER TA JR: Paralysis of conjugate lateral movement of the eyes in association with cerebellar abscess. Arch Surg 42:839, 1941
BURCH GE: Superficial lymphatics of human eyelids observed by injection in vivo. Anat Rec 73:443, 1939
CALLAHAN A: Secondary reattachment of the medial canthal ligament. Arch Ophthalmol 70:240, 1963
COLLIN JRO, BEARD C, WOOD I: Terminal course of nerve supply to Müller's muscle in the rhesus monkey and its clinical significance. Am J Ophthalmol 87:234, 1979
COOPER S, DANIEL PM: Muscle spindles in human extrinsic eye muscles. Brain 72:1, 1949
COOPER S, DANIEL PM, WHITTERIDGE D: Nerve impulses in the brain stem of the goat. Short latency responses obtained by stretching the extrinsic eye muscles and the jaw muscles. J Physiol 120:471, 1953
COOPER S, DANIEL PM, WHITTERIDGE D: Muscle spindles and other sensory endings in the extrinsic eye muscles; the physiology and anatomy of these receptors and of their connexions with the brainstem. Brain 78:564, 1955
CORBIN KB: Observations on the peripheral distribution of fibers arising in the mesencephalic nucleus of the fifth cranial nerve. J Comp Neurol 73:153, 1940
CORBIN KB, HARRISON F: Further attempts to trace the origin of afferent nerves to the extrinsic eye muscles. J Comp Neurol 77:187, 1942
DAWSON BH: The blood vessels of the human optic chiasma and their relation to those of the hypophysis and hypothalamus. Brain 81:207, 1958
DEVOS L, MARCELLE R: Variations morphologiques du ganglion et des nerfs ciliaries. Arch Anat Histol Embroyol 27:277, 1939
DIAMOND GR, KATOWITZ JA, WHITAKER LA, QUINN GE, SCHAFFER DB: Variations in extraocular muscle number and structure in craniofacial dysostosis. Am J Ophthalmol 90:418, 1980
DIXON FW: A comparative study of the sphenoid sinus: A study of 1600 skulls. Ann Otol Rhinol Laryngol 46:687, 1937
DYER JA: Tenotomy of the inferior oblique muscle at its scleral insertion: An easy and effective approach. Arch Ophthalmol 68:176, 1962
EHINGER B: Connections between adrenergic nerves and other tissue components of the eye. Acta Physiol Scand 67:57, 1966
FINCHAM E: Discussion on ophthalmic photography. Proc R Soc Med 42:451, 1949
FINE BS, TOUSIMIS AJ: The structure of the vitreous body and the suspensory ligaments of the lens. Arch Ophthalmol 65:95, 1961
FINK WH: Surgery of the Oblique Muscles of the Eye. St. Louis, C. V. Mosby, 1951
FRALICK FB: Surgical anatomy of the eye. Surg Gynecol Obstet 74:589, 1942
GEDDES AC: An abnormal nasal duct. Anat Anz 37:5, 1910
GELLÉRT A: Ganglia of the internal carotid plexus. J Anat 68:318, 1934
GIBBS FA: Relationship between the pressure in the veins on the nerve head and the cerebrospinal fluid pressure. Arch Neurol Psychiatr 35:292, 1936
GILLILAN LA: The collateral circulation of the human orbit. Arch Ophthalmol 65:684, 1961
GOODYEAR HM: Ophthalmic conditions referable to diseases of the paranasal sinuses. Arch Otolaryngol 48:202, 1948
GRANT WM: Experimental aqueous perfusion in enucleated human eyes. Arch Ophthalmol 69:783, 1963
HARGISS JL: Inferior aponeurosis vs orbital septum tucking for senile entropion. Arch Ophthalmol 89:210, 1973
HINTON JW: Unilateral exophthalmos. Ann Surg 100:1184, 1934
HOLLAND MG, VON SALLMANN L, COLLINS EM: A study of the innervation of the chamber angle. Part II. The origin of trabecular axons revealed by degeneration experiments. Am J Ophthalmol 44:206, 1957
HOUSER KM: Anatomic relation of the sphenoid sinus to Dorello's canal: Abducens paralysis Arch Otolaryngol 16:488, 1932
HOYT WF, LUIS O: Visual fiber anatomy in the infrageniculate pathway of the primate: Uncrossed and crossed retinal quadrant fiber projections studied with Nauta silver stain. Arch Ophthalmol 68:94, 1962

HOYT WF, LUIS O: The primate chiasm: Details of visual fiber organization studied by silver impregnation techniques. Arch Ophthalmol 70:69, 1963

HYNDMAN OR, WOLKIN J: Sympathectomy of the upper extremity: Evidence that only the second dorsal ganglion need be removed for complete sympathectomy. Arch Surg 45:145, 1942

ISAKSSON I: Studies on congenital genuine blepharoptosis: Morphological and functional investigation of the upper eyelid. Acta Ophthalmol (Suppl) 72, 1962

JAMPEL RS: Ocular torsion and the function of the vertical extraocular muscles. Am J Ophthalmol 79:292, 1975

JOHNSTON JA, PARKINSON D: Intracranial sympathetic pathways associated with the sixth cranial nerve. J Neurosurg 40:236, 1974

JONES LT: Epiphora. II. Its relation to the anatomic structures and surgery of the medial canthal region. Am J Ophthalmol 43:203, 1957

JONES LT: The anatomy of the lower eyelid and its relation to the cause and cure of entropion. Am J Ophthalmol 49:29, 1960

JONES LT, QUICKERT MH, WOBIG JL: The cure of ptosis by aponeurotic repair. Arch Ophthalmol 93:629, 1975

JUSTICE J JR, LEHMANN RP: Cilioretinal arteries. A study based on review of stereofundus photographs and fluorescein angiographic findings. Arch Ophthalmol 94:1355, 1976

KERR FWL, HOLLOWELL OW: Location of pupillomotor and accommodation fibers in the oculomotor nerve: Experimental observations on paralytic midriasis. J Neurol Neurosurg Psychiatr 27:473, 1964

KINOSIAN HJ: A new technique for dacrocystorhinostomy. Arch Ophthalmol 70:33, 1963

KOCH SL: The structure of the third, fourth, fifth, sixth, ninth, eleventh and twelfth cranial nerves. J Comp Neurol 26:541, 1916

KORCZYN AD, LAOR N, NEMET P: Sympathetic pupillary tone in old age. Arch Ophthalmol 94:1905, 1976

KURIBAYASHI Y: Observation of the opening of nasolacrymal duct. Report II. Japn J Ophthalmol 1:96, 1957

LANGHAM ME, ROSENTHAL AR: Role of cervical sympathetic nerve in regulating intraocular pressure and circulation. Am J Physiol 210:786, 1966

LATIES AM, JACOBOWITZ D: A comparative study of the autonomic innervation of the eye in monkey, cat, and rabbit. Anat Rec 156:383, 1966

MACEWEN EM: Orbital variations. Anat Rec 46:223, 1930

MAHONEY W, SHEEHAN D: Experimental ptosis in primates. Arch Neurol Psychiatr 35:99, 1936

MANN IC: The relations of the hyaloid canal in the foetus and in the adult. J Anat 62:290, 1928

MAUL E, STROZZI L, MUÑOZ C, REYES C: The outflow pathway in congenital glaucoma. Am J Ophthalmol 89:667, 1980

MCCABE BF, BOLES R: Surgical treatment of essential blepharospasm. Ann Otol Rhinol Laryngol 81:611, 1972

MOST A: Ueber die Lymphgefässe und die regionaren Lymphdrüsen der Bindehaut und der Lidder des Auges. Arch Anat Physiol, p 96, 1905

NAFFZIGER HC: Progressive exophthalmos associated with disorders of the thyroid gland. Ann Surg 108:529, 1938

NATHAN H, OUAKNINE G, KOSARY IZ: The abducens nerve. Anatomical variations in its course. J Neurosurg 41:561, 1974

NATHAN PW, TURNER JWA: The efferent pathway for pupillary contraction. Brain 65:343, 1942

PARKINSON D, JOHNSTON J, CHANDHURI A: Sympathetic connections to the fifth and sixth cranial nerves. Anat Rec 191:221, 1978

PATON L, HOLMES G: The pathology of papilloedema: A histological study of sixty eyes. Brain 33:389, 1910

PETER LC: Present status of tendon transplantation of the ocular muscles. Am J Surg n.s. 42:30, 1938

PIERCE MK, GOLDBERG JL, BROOKS CJ: A direct approach for transposition of the parotid duct: A description of an intra-oral approach to the inferior conjunctival cul-de-sac. Arch Ophthalmol 64:566, 1960

PORTER CT: The etiology and treatment of orbital infection. Ann Otol Rhinol Laryngol 41:1136, 1932

POSEY WC: Concerning some gross structural anomalies of the muscles of the eye and its adnexa. Trans Am Acad Ophthalmol 28:243, 1923

RAY BS, HINSEY JC, GEOHEGAN WA: Observations on the distribution of the sympathetic fibers to the pupil and upper extremity as determined by stimulation of the anterior roots in man. Ann Surg 118:647, 1943

RICHARDSON KC: The fine structure of the albino rabbit iris with special reference to the identification of adrenergic and cholinergic nerves and nerve endings in its intrinsic muscles. Am J Anat 114:173, 1964

RONTHAL E, RONTHAL M, GUILFORD FT: Surgical anatomy of the orbit. Ann Otol Rhinol Laryngol 88:382, 1979

RUEDEMANN AD: Exophthalmos. Ann Otol Rhinol Laryngol 50:1064, 1941

RUSKIN SL: Control of tearing by blocking the nasal ganglion. Arch Ophthalmol n.s. 4:208, 1930

SCARLETT H: Occlusion of the central artery of the retina. Ann Surg 101:318, 1935

SCHAEFFER JP: The genesis and development of the nasolacrimal passages in man. Am J Anat 13:1, 1912a

SCHAEFFER JP: Types of ostia nasolacrimalia in man and their genetic significance. Am J Anat 13:183, 1912b

STEELE EJ, BLUNT MJ: The blood supply of the optic nerve and chiasma in man. J Anat 90:486, 1956

STIEVE R: Über den Bau des menschlichen Ciliarmuskels, seine Veränderungen während des Lebens und seine Bedeutung für die Akkommodation. Anat Anz 97:69, 1949

SUNDERLAND S, HUGHES ESR: The pupillo-constrictor pathway and the nerves to the ocular muscles in man. Brain 69:301, 1946

TARKHAN AA: The innervation of the extrinsic ocular muscles. J Anat 68:293, 1934

TRONCOSO MU: Microanatomy of the eye with the slitlamp microscope. II. Comparative anatomy of the ciliary body, zonula, and related structures in mammalia. Am J Ophthalmol 25:1, 1942

TURNER JWA: Indirect injuries of the optic nerve. Brain 66:140, 1943

VAIL HH: Retrobulbar optic neuritis originating in the nasal sinuses: A new method of demonstrating the relation between the sphenoid sinus and the optic nerve. Arch Otolaryngol 13:846, 1931

VAN ALPEN GWHM: The structural changes in miosis and midriasis of monkey eye. Arch Ophthalmol 69:802, 1963

VAN ALYEA OE: Sphenoid sinus: Anatomic study, with consideration of the clinical significance of the structural characteristics of the sphenoid sinus. Arch Otolaryngol 34:225, 1941

VAN BUREN JM, POPPEN JL, HORRAX G: Unilateral exophthalmos. A consideration of symptom pathogenesis. Brain 80:139, 1957

WESLEY RE, MCCORD CD, JONES NA: Height of the tarsus of the lower eyelid. Am J Ophthalmol 90:102, 1980

WHITNALL SE: An instance of the retractor bulbi muscle in man. J Anat Physiol 46:36, 1911

WHITNALL SE:: The Anatomy of the Human Orbit and Accessory Organs of Vision. New York City, Oxford, 1921a

WHITNALL SE: Some abnormal muscles of the orbit. Anat Rec 21:143, 1921b

WHITWELL J: Denervation of the lacrimal gland. Br J Ophthalmol 42:518, 1958

WIENER M: Applied anatomy in eye surgery. Surg Gynecol Obstet 84:777, 1947

WINCKLER G: L'innervation proprioceptive des muscles extrinsèques du globe oculaire chez l'homme. Comptes Rendus Assoc Anat 43:848, 1956

WOLFF E: The Anatomy of the Eye and Orbit: Including the Central Connections, Development, and Comparative Anatomy of the Visual Apparatus, 3rd ed. Philadelphia, Blakiston, 1948

WOLTER JR: Nerves of the normal human choroid. Arch Ophthalmol 64:120, 1960a

WOLTER JR: Regenerative potentialities of the centrifugal fibers of the human optic nerve. Arch Ophthalmol 64:697, 1960b

WOLTER JR, ROSENBERG RJ: Ectopic lymph node of the orbit simulating a tumor of the lacrimal gland. Am J Ophthalmol 83:909, 1977

WYBAR KC: A study of the choroidal circulation of the eye in man. J Anat 88:94, 1954

WYBAR KC: Anastomoses between the retinal and ciliary arterial circulation. Br J Ophthalmol 40:65, 1956

Chapter 3
THE EAR

The ear is conveniently divided for purposes of discussion into three portions (Fig. 3-1): (1) the external ear, consisting of the auricle or pinna, the external acoustic meatus, and the tympanic membrane at the medial, blind, end of the meatus; (2) the middle ear cavity with its contained bones, its connection with the pharynx by way of the auditory or eustachian tube, and its expansions in the form of air cells in the temporal bone; and (3) the internal ear, embedded in the petrous portion of the temporal bone and consisting of a membranous (otic) labyrinth that lies within the particularly dense bone (otic capsule) that is excavated to form the osseous or periotic labyrinth. The internal ear is the sense organ, both for hearing and balance; the external and middle ears represent simply accessory sound-conducting apparatus for the auditory part of the internal ear.

The External Ear

AURICLE AND MEATUS

Auricle

The skeleton of the auricle consists of a single plate of elastic cartilage that conforms closely to the shape of the ear as seen in the living (Fig. 3-2), except for the lobule, which is obviously devoid of skeletal support. The cartilage of the auricle is continuous with that of the external meatus. It is attached to the bony skull by three ligaments, of which the anterior attaches the helix and tragus to the zygomatic process; the superior, the spine of the helix to the superior margin of the bony external meatus; and the posterior, the medial surface (eminence) of the concha to the mastoid process. The auricle is also anchored both through its continuity with the cartilage of the meatus and through the skin and extrinsic muscles. Perhaps the most important ligament is the fibrous band stretching between the cartilage of the tragus and that of the helix to complete the front of the external meatus.

The auricle consists essentially of the elastic cartilaginous plate and its covering skin, the latter being more loosely attached on the medial or posteromedial aspect of the ear than it is on the lateral aspect. Some six or more small and rudimentary intrinsic muscles have been described in connection with the cartilage of the external ear, but they are of no apparent importance. The extrinsic muscles of the ear—anterior, superior, and posterior auricular muscles—are distinctly better developed than the intrinsic and, although usually functionless in human beings, are sometimes subject to voluntary control as in "wiggling" the ears. Both extrinsic and intrinsic muscles of the ear are supplied by the facial nerve.

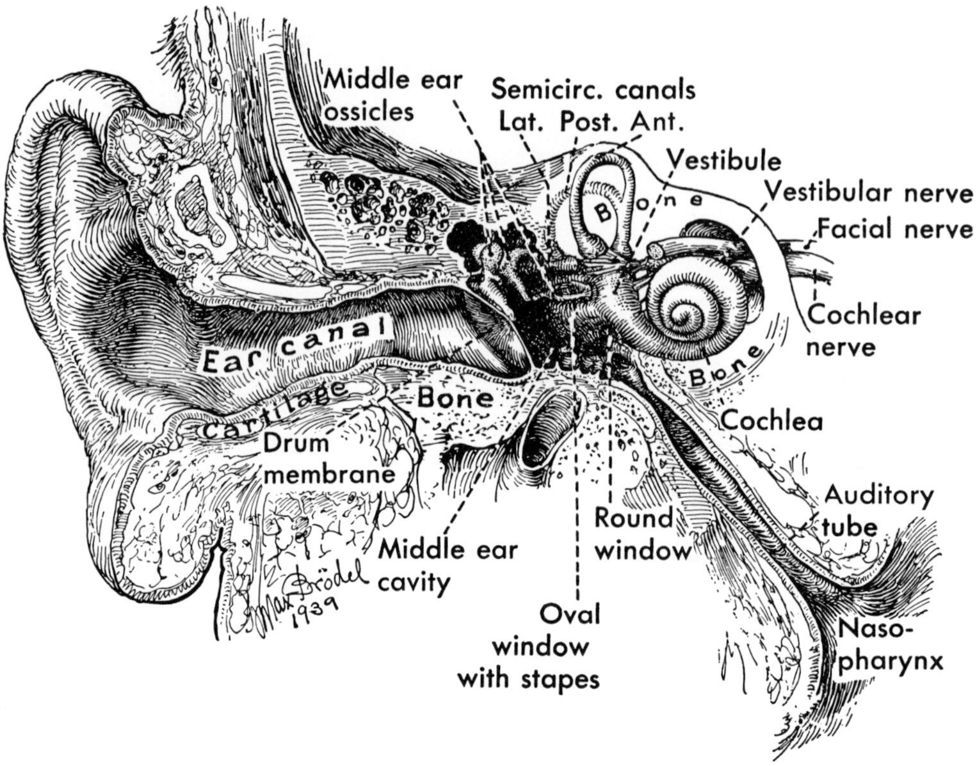

Fig. 3-1. Diagram of the external, middle, and internal ears. (Brödel M: Three Unpublished Drawings of the Anatomy of the Human Ear. Philadelphia, WB Saunders, 1946)

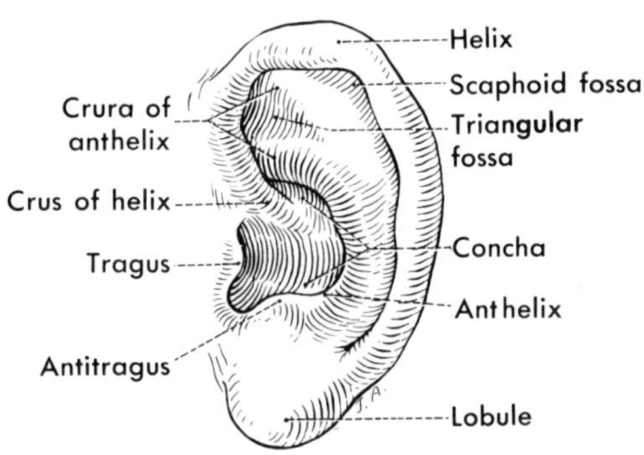

Fig. 3-2. Parts of the external ear.

Meatus

The external acoustic (auditory) meatus, or auditory canal, extends from the concha to the tympanic membrane. The lateral portion of its wall consists of cartilage and connective tissue, the medial portion of bone. Laterally, the cartilage of the external meatus is continuous with that of the body of the auricle through a narrow cartilaginous band, the isthmus, and in turn bears the cartilage of the tragus; thus the cartilage of the tragus is continuous with that of the auricle proper only through the meatus, and a gap (the incisura terminalis, or anterior or tragohelicine incisure) exists anterosuperiorly between the cartilage of the helix and that of the tragus (Fig. 3-3). The incision for the endaural approach to the ear is made between tragus and helix and thus avoids cutting cartilage. Medially, the meatal cartilage is firmly attached to the outer lip of the bony meatus (Fig. 3-4). The cartilage itself forms a trough, open superiorly and posteriorly, where the squamous part of the temporal bone completes the tube, and dense connective tissue binds the parts together.

The *cartilaginous part* of the meatus constitutes somewhat less than half its total length. Inconstant fissures (fissures of Santorini) filled with connective tissue occur in the cartilage and may transmit infections from the meatus to the parotid and superficial mastoid regions, or vice versa. The cartilaginous meatus is slightly concave anteriorly, but can be somewhat straightened, so that the insertion of a speculum is made easier, by drawing the auricle posteriorly and upward.

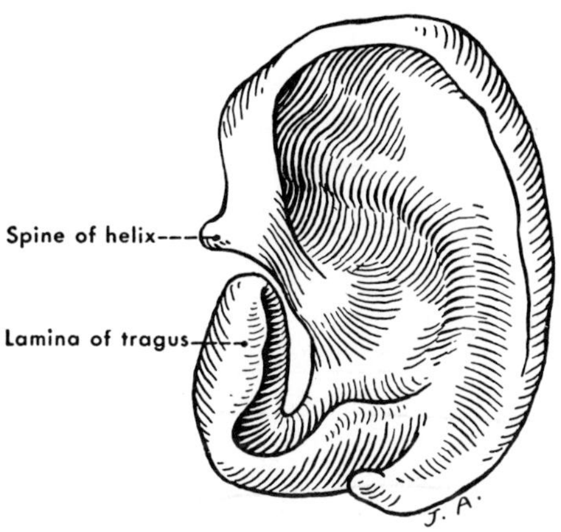

Fig. 3-3. Cartilage of the left ear, lateral view, showing the gap between the cartilaginous support of the helix and tragus.

The *bony meatus,* a tunnel into the temporal bone, is in turn slightly concave posteriorly, so that the meatus as a whole is slightly S shaped. In the child up to the age of about 4 years, and sometimes in the adult, the bony meatus is deficient anteroinferiorly, and through this gap (foramen of Huschke) infections may pass between the adjacent parotid gland and the meatus. The entire meatus is lined by a reflection inward from the skin.

The anterior, inferior, and lower posterior parts of the bony wall of the external meatus are formed by the C-shaped tympanic part of the temporal bone, developed from the narrow anulus tympanicus of the fetus. However, this anulus was never complete above or posterosuperiorly, and therefore the wall of the adult meatus is completed by the squamous and petrous parts of the temporal bone. In the

Fig. 3-4. Cartilage of the left ear and its relation to the external acoustic meatus: (*a*), from the front; (*b*), from behind; (*c*), from above in a horizontal section through the meatus; (*d*), the relationship between cartilage and bone in the planes in (*a*).

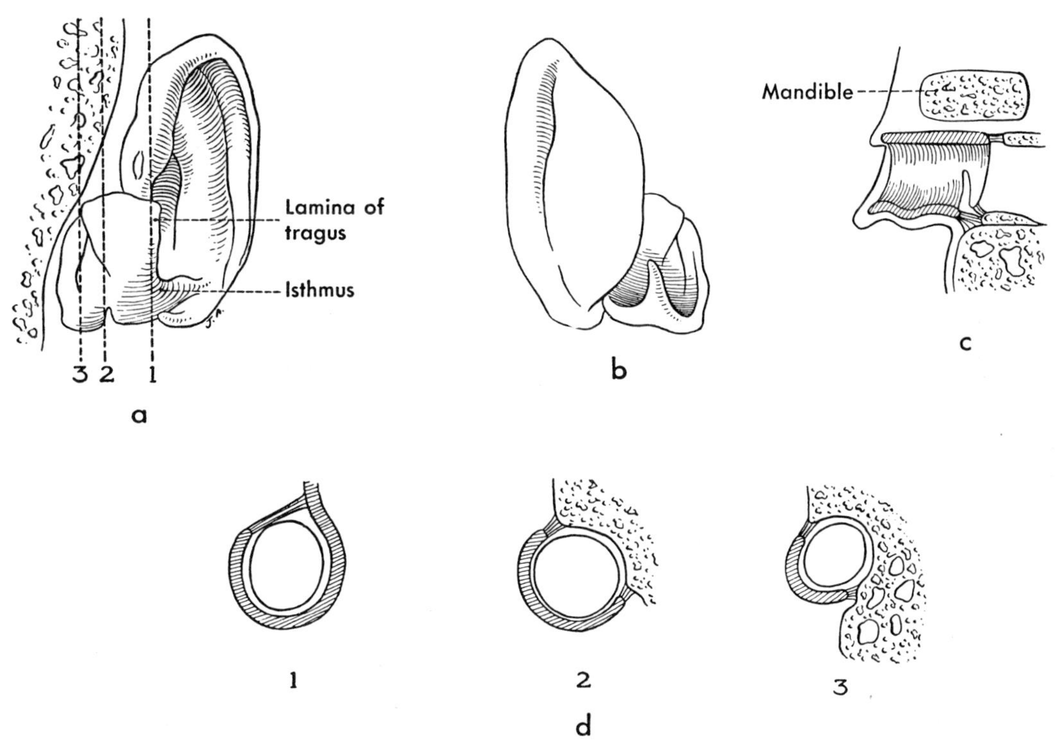

infant, before the anulus has grown much and before the mastoid process has developed, the bony canal is very short, and it and the outer surface of the tympanic membrane are directed more inferiorly than laterally. The foramen of Huschke results from a lag in the growth of the part of the tympanic ring that should form the floor, with the anterior and posterior walls subsequently coming together but not fusing completely.

In the adult, because the tympanic membrane still retains some of its original downward and inward slope, the posterosuperior wall of the external meatus is only about 25 mm long, while the anteroinferior wall is about 6 mm longer. The anterior and inferior walls of the meatus are related closely to the parotid gland; the anterior wall of the bony meatus is also closely related to the condyle of the mandible, but the more important relations of the meatus are that of the posterior wall to the mastoid cells and that of the medial portion of the superior wall to the epitympanic recess—these air cavities being separated from the meatus by relatively thin bone. Lateral to the epitympanic recess thicker bone, which, however, often contains extensions of the mastoid air cells, intervenes between the meatus and the middle cranial fossa.

VESSELS AND NERVES

The auricle has a rather abundant blood supply from the posterior auricular artery and from small auricular rami of the superficial temporal (branches of the external carotid); however, since the vessels are necessarily superficially located and are not embedded in any appreciable amount of fat, the ear is particularly subject to frostbite despite its vascularity.

The blood supply to the external acoustic meatus is the same as that to the auricle, but in addition the meatus also receives twigs from the *deep auricular artery*, a branch from the first part of the maxillary (internal maxillary). The deep auricular artery passes through the parotid gland and behind the capsule of the temporomandibular joint to penetrate either the cartilaginous or bony part of the meatus and supply the deeper part of the meatus and the external surface of the tympanic membrane.

The *veins* from the auricle and meatus drain into the superficial temporal and posterior auricular veins; the former joins the retromandibular (formerly posterior facial) vein, which usually divides and joins both jugular veins; the latter joins the external jugular but sometimes connects also with the sigmoid sinus through the mastoid emissary vein.

The *lymphatics* of the auricle and meatus join and empty into neighboring lymph nodes, that is, parotid ones in front of the ear, the superficial cervical nodes along the external jugular vein below the ear, and the retroauricular (mastoid) nodes behind it. Because of drainage of the meatus to the retroauricular nodes, enlargement of these has in the past caused confusion between the diagnoses of external otitis and acute mastoiditis.

The external ear represents a region of juncture between skin originally over the branchial region and postbranchial skin, and this is reflected in its *nerve supply* (Fig. 3-5). Because of the usual overlap between the sensory distribution of adjacent cutaneous nerves, and probably also because of a good deal of variation in the contribution that the rudimentary cutaneous branches of the seventh and ninth nerves make to the innervation of the external ear, the exact distribution of the various components to the auricle and the external meatus is not known. However, the approximate distribution is as follows.

The auriculotemporal branch of the mandibular division of the trigeminal nerve is distributed to only the most anterior limb of the helix and a portion of its crus, and to the tragus; it also supplies the anterior and upper walls of the external acoustic meatus and a corresponding portion of the external surface of the tympanic membrance (Cushing). In some of the individuals studied by Cushing, however, the deeper (medial) portion of the posterior wall, and the entire drum membrane, seemed to be supplied by the fifth (that is, auriculotemporal) nerve.

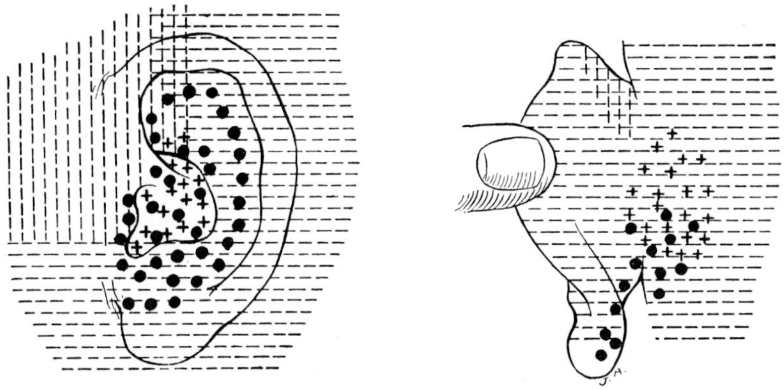

Fig. 3-5. Cutaneous innervation of the ear on the lateral and posterior aspects. The former is adapted from Cushing (Bull Johns Hopkins Hosp 15:213, 1904); the latter, from Graves and Edwards (Arch Otolaryngol 39:359, 1944). The outlines of the areas of distribution undoubtedly vary with the individual, and so do the zones of overlap. *Vertical broken lines* indicate the trigeminal field (auriculotemporal); *horizontal lines,* the cervical field (C2 and C3). The area of distribution of the ninth and tenth cranial nerves is indicated by *crosses;* the probable distribution of the seventh, by *circles.*

The greater portion of the posterior or medial surface of the auricle, and the more posterior portion of the lateral surface (including most of the helix, and the anthelix and lobule, but not dipping into the concha proper) is supplied by cutaneous branches of the cervical plexus, primarily the great auricular nerve from C2 and C3. The skin over most of the mastoid region is supplied by the lesser occipital nerve, also receiving fibers from the second and third cervical nerves through the cervical plexus. The lesser occipital also supplies the skin on the upper medial portion of the auricle, overlapping here with the nerve supply from the great auricular.

The concavity of the concha, between the area supplied by the fifth cranial and that supplied by cervical nerves, is supplied by fibers of the tenth and ninth nerves and apparently also (Hunt and others) of the seventh nerve. The tenth or ninth and tenth nerves, and the facial nerve, also supply the posterior portion of the external meatus and usually the posterior portion of the external surface of the drum membrane; according to Hunt, these nerves also are distributed, individually or together, to a small area of skin on the posteromedial aspect of the auricle and adjacent mastoid region.

These nerves reach the ear through the *auricular branch* of the vagus (nerve of Arnold). This arises from the superior (jugular) ganglion and enters the temporal bone through a foramen in the lateral wall of the jugular fossa, traverses the tiny mastoid canaliculus, and emerges from the skull through the tympanomastoid fissure (it crosses the facial canal and has been said to emerge sometimes through the stylomastoid foramen; Guild, '53, described it or part of it as sometimes running at first outside the skull to enter the stylomastoid foramen). It is usually joined by a communicating branch from the ninth (often called auricular branch, but not to be confused with the tympanic branch of this nerve, although they may arise together). While it is in the temporal bone the auricular branch of the tenth may receive a filament from the seventh nerve in the facial (Fallopian) canal, and after it emerges it typically receives a communication from the posterior auricular branch of the facial; this branch contains most of the sensory fibers of the facial that are distributed here.

Although it is impossible anatomically to differentiate between the distributions of the seventh, ninth, and tenth nerves to skin, Hunt's observations on the location of the eruptions in herpes zoster of their ganglia have indicated that all three nerves participate in the innervation thus outlined, although there may be considerable variation from individual to individual.

TYMPANIC MEMBRANE

The tympanic or drum membrane, sometimes also referred to as the ear drum or tympanum (confusingly, since these terms are more properly applied to the middle ear as a whole), is a fibrous sheet interposed between the external acoustic meatus and the middle ear cavity; it is covered on its lateral side by a thin layer of skin continuous with that lining the meatus and on its medial surface by mucosa continuous with that lining other surfaces of the middle ear cavity. The connective tissue interposed between these two layers consists largely of radiating fibers attached to the manubrium of the malleus, but these are reinforced peripherally by circular fibers, which are thickened at the margin of the drum membrane to form a fibrocartilaginous ring (anulus fibrocartilagineus) that attaches the membrane to the tympanic sulcus of the temporal bone. In a limited upper portion, where the pars flaccida occurs, these connective tissue fibers are wanting; in this area also the tympanic portion of the temporal bone bearing the sulcus (sometimes referred to as the anulus) is deficient, and in this gap (the *tympanic incisure,* or notch of Rivinus) the membrane attaches to the squamous portion of the temporal bone.

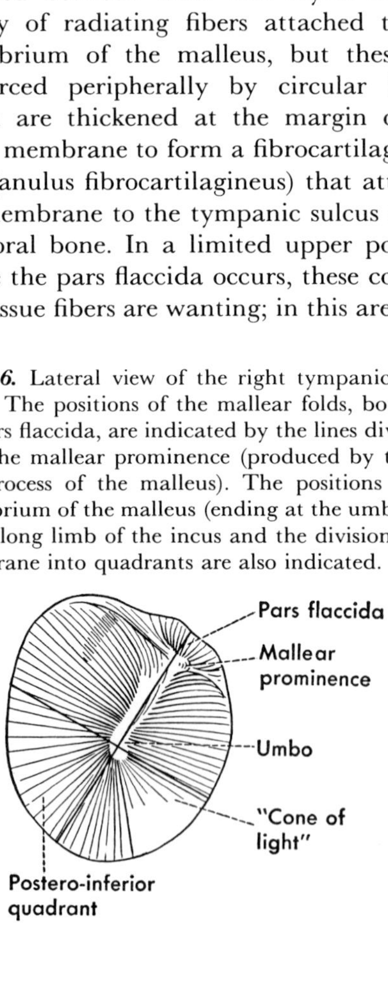

Fig. 3-6. Lateral view of the right tympanic membrane. The positions of the mallear folds, bounding the pars flaccida, are indicated by the lines diverging from the mallear prominence (produced by the lateral process of the malleus). The positions of the manubrium of the malleus (ending at the umbo) and of the long limb of the incus and the division of the membrane into quadrants are also indicated.

The tympanic membrane (Fig. 3-6) is set obliquely at the end of the external acoustic meatus, so that it slopes medially both from above downward and from behind forward. Its outer aspect is slightly concave, the center of this concavity being known as the *umbo;* this marks the strong attachment of the manubrium of the malleus to the tympanic membrane. In the anterosuperior portion of the membrane the lateral process of the malleus is also attached to it, and from this point of attachment the anterior and posterior mallear (formerly malleolar) folds pass to the cartilaginous anulus and separate the pars flaccida (Shrapnell's membrane) from the pars tensa. The *cone of light,* seen in otoscopic examination of the tympanic membrane, is a bright area of reflection that extends downward and forward from the umbo.

Vessels and Nerves

The vessels and nerves of the tympanic membrane are, as might be expected, derived in part from those supplying the external meatus and in part from those supplying the middle ear cavity, and may thus be divided into external and internal branches. Most of the arterial twigs to the external surface of the membrane are from a small peripheral vascular ring formed by the deep auricular branch of the maxillary (internal maxillary) and enter the membrane radially from below, in front, and behind; the largest artery (the descending or manubrial) is derived also from the deep auricular, but descends, sometimes duplicated, from above, at first slightly posterior to and then on the manubrium mallei (Fig. 3-7). The arterial twigs to the mucous surface are from the anterior tympanic branch of the maxillary artery and from the stylomastoid branch of the posterior auricular; branches of these two vessels form an inner peripheral vascular ring and an inner descending artery (Nager and Nager).

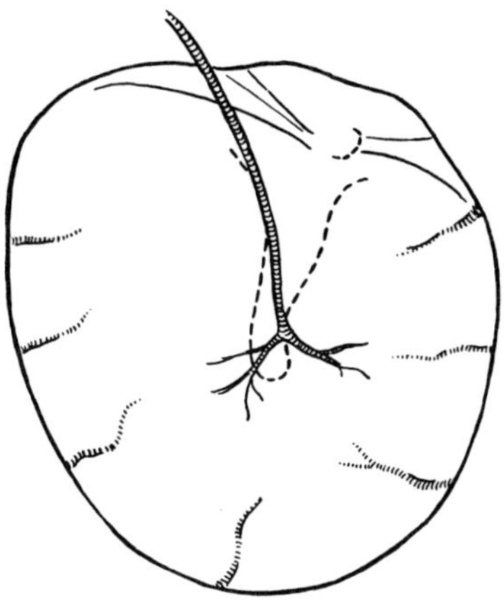

Fig. 3-7. Arteries of the external surface of the right tympanic membrane.

The innervation of the external surface of the drum membrane, as already noted, is usually described as being from the auriculotemporal branch of the fifth and the auricular branch of the tenth (or seventh, ninth, and tenth) nerves, thus corresponding to the innervation of the external meatus (Fig. 3-8); the innervation of its inner surface is believed to be from the tympanic plexus, therefore largely from the ninth nerve, with probably some participation also by the seventh. Bosman found branches from the chorda tympani distributed to the inner surface of both the pars flaccida and pars tensa.

The majority of the nerve fibers in the human tympanic membrane descend from posterosuperiorly across the flaccid portion toward the umbo, accompanying the principal artery of the membrane, and lying just behind the level of the manubrium of the malleus (Wilson; Fig. 3-9). The pars flaccida has a particularly rich innervation; the pars tensa, a poorer one. While numerous small branches enter the membrane around its periphery, most of these are derived from the nerves of the meatus, and relatively few enter it from the tympanic cavity. Wilson was unable to distinguish the origin of the nerves to the external surface in people, but had previously found that after section of the auriculotemporal nerve in the monkey almost all the fibers on the membrane disappeared; the few fibers still remaining, which crossed the flaccid portion and were irregularly distributed over the tense portion, were presumably derived from the seventh-ninth-tenth complex of the auricular branch of the vagus.

The fact that its chief arterial and nerve supply enter the tympanic membrane from above is one of several factors influencing the site of incision of it (myringotomy) in middle ear infections. A preferable location for this is the posteroinferior quadrant. Since the distance between the tympanic membrane and the inner wall of the middle ear cavity may be no more than 2–4 mm, the relations of the membrane to structures in the middle ear cavity must also be considered. Druss pointed out that its anterosuperior quadrant is closely related to the tympanic opening of the auditory (eustachian) tube, its anteroinferior quadrant to the carotid canal (the bony wall of which may be deficient), its posterosuperior quadrant to the long process of the incus and to the stapes and oval window, and its posteroinferior quadrant to the promontory.

The chief danger in incisions of the posteroinferior quadrant of the membrane, but a relatively remote one according to Druss (although cases have been reported in the older literature), is that of wounding the internal

Fig. 3-8. Distribution of sensory nerves to the meatus, tympanic membrane, and middle ear. The *numbers* refer to the corresponding cranial nerves. (After Graves GO, Edwards LF: Arch Otolaryngol 39:359, 1944)

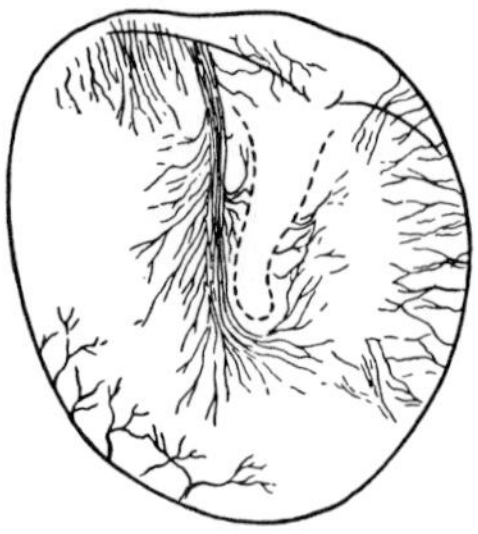

Fig. 3-9. Nerves of the external surface of the right tympanic membrane. (After Wilson JG: Am J Anat 11:101, 1911)

jugular vein (jugular bulb); the bony roof over this vein occasionally protrudes into the lower portion of the middle ear cavity, and there are sometimes dehiscences in the bone between the vein and the middle ear cavity.

In performing a myringotomy it is customary to turn the point of the knife upward, as close to the medial surface of the tympanic membrane as possible, in order to open Prussak's space, which lies medial to the pars flaccida; this has poor or no natural gravity drainage and therefore may continue to harbor infection and pus in spite of the incision in the drum.

VARIATIONS AND ANOMALIES

Wolff ('33) pointed out that in the 6-month fetus the tympanic membrane is almost horizontal and lies below rather than lateral to the otic capsule; the subsequent shift to the lateral and more nearly vertical position may be more or less complete, thus accounting for the variations in slope of the tympanic membrane commonly encountered. Retraction and calcification of the tympanic membrane are mentioned in connection with malformations of the middle ear.

Congenital atresia of the external acoustic meatus has, according to Richards, been estimated to occur in about 1:2000 patients suffering from aural disease and to be bilateral in about 1:6 of these. Cohen and Fox found a higher percentage in their own practice, about 1:1000 patients, but the incidence in the general population must be very low; males were found to be more frequently affected than females. Maldevelopment of the middle ear and its associated bones is commonly associated with astresia of the external meatus, but the internal ear may or may not be affected, and deafness may be solely of the conduction type (Richards).

Cohen and Fox suggested that the unilateral condition is perhaps best left alone, but that in bilateral atresia operation should be done in early life; they and Pattee have outlined operations for this condition. It should be remembered that in distorted and anomalous formations of the ear the facial nerve may also have an anomalous course.

Calcification and ossification of the external ear have been occasionally reported and have been discussed by Scherrer. "Supernumerary external ears," which usually do not particularly resemble a normal ear and are often therefore termed *"preauricular appendages"* or *"preauricular nodules,"* also occur occasionally anterior to the normal ear (Fig. 3-10); Wood–Jones and I-Chuan quoted a finding of 33 such cases in a series of 50,000 children. Costello and Shepard pointed out that the nodules have been mistaken for fibromas, and that the anomaly is usually solitary and unilateral but may be multiple. They stated that the literature indicates that the condition may be inherited and is more frequent in males than females.

Preauricular cysts, fossae, sinuses, and fistulas

Fig. 3-10. Distribution of (*a*) preauricular fistulas and (*b*) appendages. (After Wood-Jones F, I-Chuan W: J Anat 68:525, 1934)

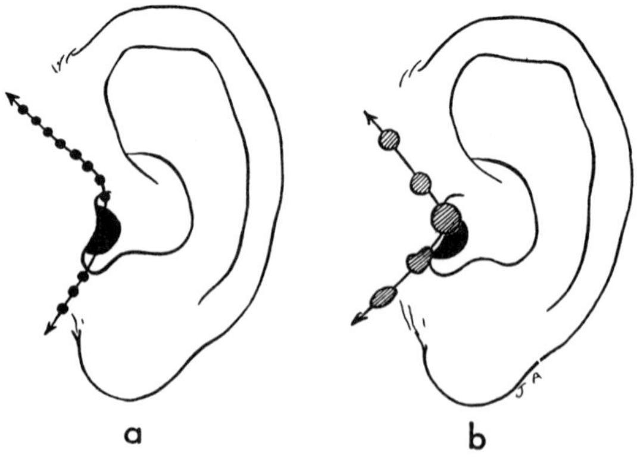

also occur in the same area in which nodules are found and, like these, have a low incidence: Wood–Jones and I-chuan quoted an occurrence of about 2:1000 births, but Pastore and Erich noted that 12 such cases were reported among 2000 children in one study and 48 cases among 3200 in another, and that the incidence among Orientals has been said to be as much as 4% to 6%.

Most of the so-called fistulas are apparently really blindly ending sinuses (Becker and Brunschwig) and are of no clinical importance unless they become infected. They are congenital and apparently hereditary (Pastore and Erich) and, depending upon one's interpretation of the development of the ear, are sometimes regarded as remnants of the first branchial cleft (Wood–Jones and I-Chuan; Becker and Brunschwig) or as due to incomplete fusion of the tubercles described by His as contributing to the auricle (*e.g.*, Congdon and colleagues). Congdon and his co-workers found that although about 90% of juxta-auricular fossae (of a series of 470) were close to the anterior border of the helix, the remainder were in various locations on the ear itself or even behind; they expressed the belief that these variations in position could not be compatible with a development from the first branchial cleft. Indeed, there is now general agreement (see Chap. 9) that although remnants of the ventral part of the first branchial cleft are sometimes found, they are not identical with the cysts and sinuses discussed here.

The Middle Ear

The middle ear cavity consists of a central air-filled space, the tympanic cavity; its connection to the pharynx, the auditory tube; and its extensions into neighboring parts of the temporal bone. These extensions, outgrowths from the central cavity, are likewise normally filled with air and are known as air cells.

TYMPANIC CAVITY

The tympanic cavity (cavum tympani, tympanum proper, middle ear cavity) lies in the temporal bone, between the tympanic membrane and parts of the tympanic and squamous portions of the temporal bone laterally, and the petrous portion of the temporal bone, surrounding the internal ear, medially. It communicates anteriorly with the pharynx by way of the auditory tube and posteriorly with the mastoid air cells by way of the aditus and mastoid antrum (Figs. 3-11 and 3-12). In it lie, covered by mucous membrane, the three bones of the middle ear cavity, their ligaments, and the tendons of their associated muscles and through it (also covered by mucous membrane) runs the chorda tampani from the facial nerve.

Subdivisions or Pouches

Since the structures in and traversing the middle ear cavity are all covered by reflec-

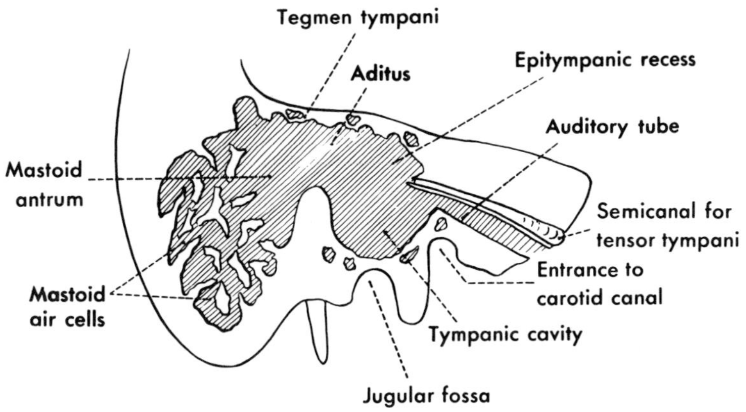

Fig. 3-11. Schema of the tympanic cavity and its associated air spaces.

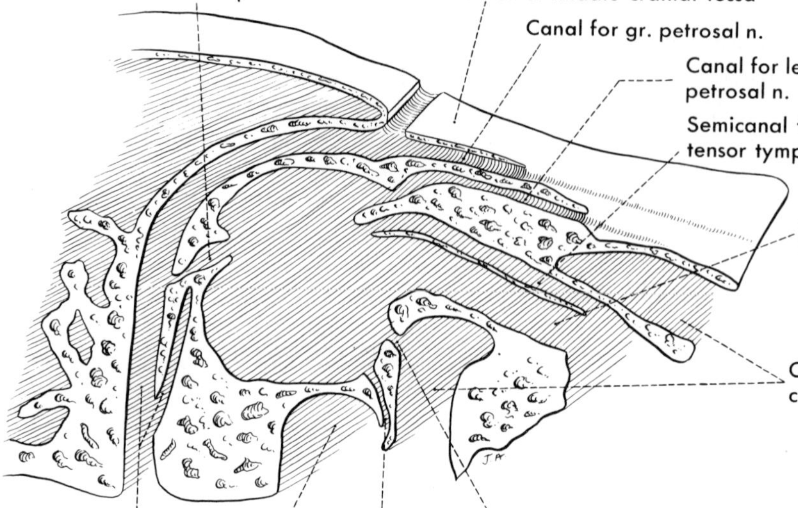

Fig. 3-12. Some relationships of the tympanic cavity. All those indicated do not actually lie in the same plane, but are at varying depths from the lateral wall.

tions of mucous membrane, they and the folds of mucosa associated with them tend to partially subdivide the tympanic cavity into various "pouches." Most of these communicate widely with the rest of the cavity and with each other and are of little import: for instance, the anterior pouch of the malleus is simply that portion of the tympanic cavity lying between the head of the malleus, the anterior tympanic wall, and the superior and anterior mallear ligaments; similarly, the superior and inferior "pouches" of the incus lie above and below the short process of the incus (Fig. 3-13).

The one subdivision of clinical importance is the pouch of Prussak, or upper tympanic recess. This is bounded laterally by the pars flaccida of the tympanic membrane, medially by the neck of the malleus, above by the lateral mallear ligament, and below by the lateral process of the malleus (Fig. 3-14). The mucosal folds reflected upon the ligament and the process may convert this space into one with only relatively small anterior and posterior openings and therefore no good gravity drainage. (The pouches of von Tröltsch, or anterior and posterior recesses of the tympanic membrane, are of historical interest only; they lie between the tympanic membrane and the anterior and posterior mallear folds, respectively, open downward, and therefore do not tend to harbor pus.)

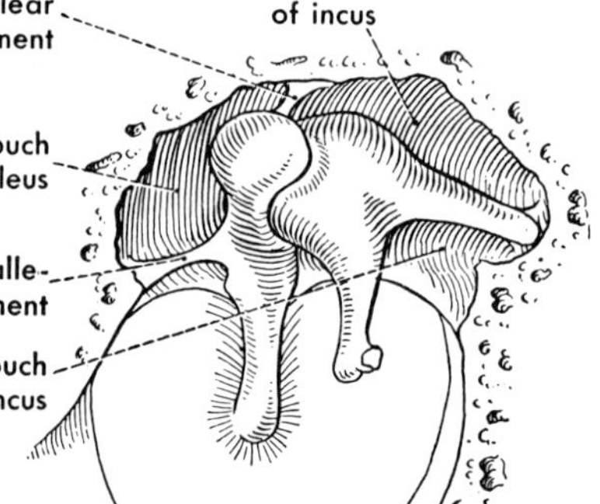

Fig. 3-13. Malleus and incus from the medial side, to illustrate the larger "pouches" associated with them.

Lateral Wall

The tympanic cavity extends well above the level of the tympanic membrane as the epitympanic recess (attic); in it lie the head of the malleus and the body and short process of the incus. The epitympanic recess projects laterally above the external acoustic meatus, and it is this portion of the tympanic cavity that has as its lateral wall a part of the squamous portion of the temporal bone; below this level the lateral wall of the cavity is composed of the tympanic membrane and the pars tympanica of the temporal bone to which the membrane is attached.

Roof

The roof of the tympanic cavity, that is, of the epitympanic recess, is a thin plate of bone, the *tegmen tympani,* which separates the middle ear cavity from the cranial cavity; prolonged backward it separates the mastoid antrum from the cranial cavity, and prolonged forward it forms the roof of the canal for the tensor tympani muscle. In children the unossified petrosquamous suture in the tegmen tympani may allow the direct passage of infection from the middle ear to the meninges of the middle cranial fossa, and even in the adult veins from the middle ear perforate this suture to end in the petrosquamous sinus (present in about 50% of cases) and the superior petrosal sinus and thus may transmit infection directly to the cranial venous sinuses.

Floor

The floor of the middle ear cavity, lying slightly below the level of that of the meatus, is usually a thin plate of bone separating the cavity from the internal jugular vein; if the superior bulb of the vein is particularly small, the floor may be even as much as 1 cm thick and may contain air cells (hypotympanic cells) between the middle ear cavity and the vein; if the bulb is large, it bulges upward into the middle ear, and the wall is thus very thin and may present dehiscences (Maybaum and Goldman). The major portion of the epitympanic recess has, of course, no floor, since it is simply an upward extension of the main portion of the tympanic cavity, but the laterally projecting portion has as its floor the roof of the external meatus. In this also there may be air cells (cells of Kirchner). The hypotympanic cells in the floor of the tympanum may extend also below the labyrinth to lie adjacent to "petrous apex" cells; in this case, they afford a good surgical approach to the more anterior cells of the pars petrosa.

Posterior Wall

The posterior wall of the tympanic cavity is deficient above where it opens through the narrow aditus ad antrum into a wider cavity, the *mastoid* (tympanic) *antrum;* below the aditus a relatively thin bony wall intervenes between the tympanic cavity and the antrum with its connecting mastoid air cells. From this posterior wall the *pyramidal eminence* projects, with an aperture at its apex transmitting the tendon of the stapedius muscle; the groove above the eminence is the *sinus posterior.* Above and behind the base of the pyramidal eminence the facial nerve, in the medial wall of the tympanic cavity, curves downward to change its course from an al-

Fig. 3-14. Some "pouches" related to the tympanic membrane in frontal section through the ear. (*a*), Through the malleus; (*b*), posterior to the malleus.

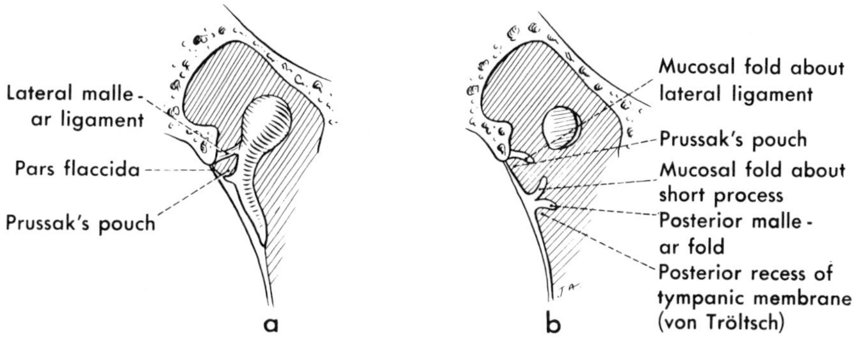

most horizontal to an almost vertical one. The chorda tympani, arising from the facial, then enters the tympanic cavity through the canaliculus of the chorda (once iter posterius) in the posterior wall.

Anterior Wall

The lower part of the anterior wall of the tympanic cavity, part of the petrous apex, consists of a thin plate of bone, which may be incomplete or may contain air cells, between this cavity and the carotid canal housing the internal carotid artery. In the Ramadier approach to "petrous apex" cells, the exact position of the carotid canal is of major importance; the technique involves apicotomy through the anteromedial aspect of the tympanic wall between the carotid canal and the promontory, and the space between these two varies considerably. The upper part of the wall, like the corresponding part of the posterior wall, is deficient, since the canal containing the tensor tympani muscle opens here and immediately below this is the tympanic orifice of the auditory tube. Through the auditory tube the tympanic cavity, the mastoid antrum, and the mastoid and other air cells connected with the middle ear cavity are in continuity with the nasopharynx.

Wigand and Trillisch described a niche in the bone above the semicanal for the tensor tympani and suggested it be called the sinus epitympanicus. They said it was frequently present, and that it could both harbor hidden disease and serve as a route to preepitympanic structures.

Medial Wall

The medial wall of the tympanic cavity is the petrous portion (once "pyramid") of the temporal bone surrounding the internal ear and here separating the cavities of the middle and internal ears. It presents on its surface, more or less clearly, several markings of particular importance (Fig. 3-15). Posteriorly and above, mostly in the aditus, is the broad prominence produced by the anterior end of the lateral semicircular canal; below this, and also extending more anteriorly, is the less marked *prominence of the facial* (Fallopian) *canal,* produced by the horizontal portion of the facial nerve as it passes in the relatively thin bony wall between internal and middle ears. The bone overlying the facial nerve, especially anteriorly, is usually quite thin; it may even be lacking, so that the nerve is covered only by mucosa (Anson). Anterior to the prominence of the facial canal the hollow *cochleariform process,* transmitting the tendon of the tensor tympani muscle, projects posteriorly along the medial wall from the junction of this with the anterior one; its apex, at which the tensor tympani muscle turns sharply laterally, is the landmark for the position of the turn (geniculum or external genu) between the anterolaterally and the posteriorly directed horizontal portions of the facial nerve.

Immediately below the facial canal is a depression, the *fossula fenestrae vestibuli,* or "stapes niche," which contains the vestibular fenestra (oval window), closed by the base of the stapes. Below the vestibular window is the rounded bulge of the *promontory,* formed by the basal turn of the cochlea; the tympanic nerve plexus lies upon the promontory, covered by its mucous membrane. Below the back part of the promontory the *cochlear fossula* or round window niche leads to the round window or fenestra cochleae; this window is closed by a fibrous membrane, the *secondary tympanic membrane.* Depending upon the bulge of the promontory, the orifice of the cochlear fossula may be readily visible or may be largely hidden by the promontory. The round window itself, at the anterosuperior end of the fossula, is visible with difficulty, if at all, from the middle ear cavity. Anson has pointed out, however, that the secondary tympanic membrane can be ruptured by incautious packing or too forceful aspiration, as well as by ill-advised introduction of an instrument into the fossula.

Behind the promontory, medial and inferior to the pyramidal eminence and immediately below the ridge of the facial canal, is a depression, the *sinus tympani* (recessus tympanicus subcanalis Fallopii); the depth to which this extends into the petrous portion of the temporal bone varies greatly (Saito and

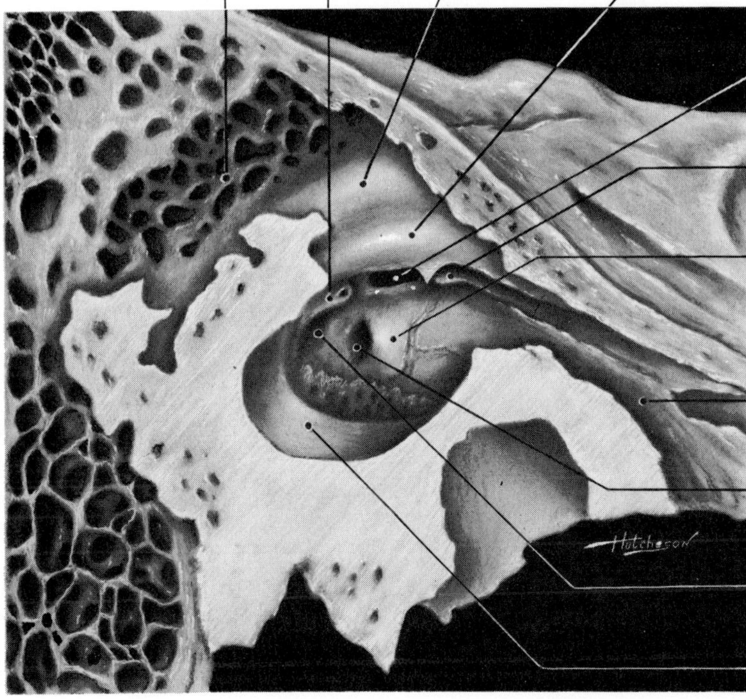

Fig. 3-15. Medial wall of the tympanic cavity.

co-workers measured depths of 0.61 to 5.87 mm and volumes of 0.43 to 84.72 mm^3). It may harbor infections, and when it extends deeply enough to lie close to the ampullary end of the posterior canal and the posterior end of the lateral canal it may transmit infections to these. Its opening, which lies approximately midhorizontally between the oval and round windows, may look forward rather than laterally and therefore be inconspicuous.

Bones of Middle Ear

Of the three bones of the middle ear, the *malleus* is closely attached to the tympanic membrane by both its manubrium (handle) and its lateral process, while its head projects somewhat above the membrane into the epitympanic recess to articulate with the body of the incus (Figs 3-16 and 3-17). The anterior process ("long" process, processus gracilis) of the malleus is a slender spicule extending obliquely downward from the neck toward the tympanosquamous fissure; in the infant it may reach this fissure but in the adult its distal part is converted into connective tissue that forms the anterior ligament of the malleus, and the bony process itself is short. (Through the tympanosquamous fissure the anterior ligament of the malleus may be continuous with the sphenomandibular ligament; both represent remains of Meckel's cartilage.) The malleus is also attached to the tympanic wall by a superior and a lateral mallear ligament. The superior ligament descends almost straight down from the roof of the epitympanic recess. The malleus may be fused to the tegmen tympani at the site of this ligament, thereby causing conductive hearing loss that may be confused preoperatively with otosclerosis. The broad but short lateral liga-

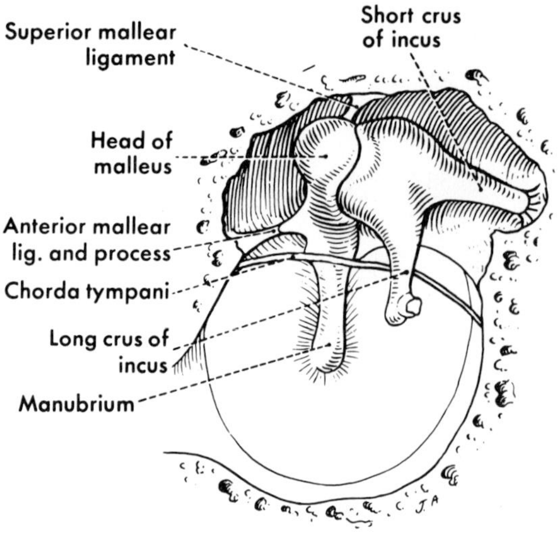

Fig. 3-16. The malleus, incus, and drum membrane seen from the medial side.

ment attaches the neck of the bone to the margin of the tympanic notch.

The body of the *incus,* fitted against the head of the malleus, lies also in the epitympanic recess, and the short limb of this bone rests in a depression, the fossa of the incus, situated in the posterior wall of the tympanic cavity below the aditus ad antrum. The long limb of the incus descends parallel and slightly posteromedial to the manubrium of the malleus, but at its lower end turns medially at almost right angles to articulate with the stapes. The incus is held in place by a posterior ligament attaching to its short limb and by a superior ligament that descends from above to attach to the body. The incudomallear and incudostapedial joints are synovial (diarthroidal) ones; the knoblike expansion of the long crus of the incus at the incudomallear joint is the lenticular process.

From its articulation with the incus, the *stapes* passes almost horizontally to the oval window. The periphery of the base or footplate of the stapes is covered with cartilage, as is also the rim of the vestibular window, and the two are united by an annular ligament; this is a ring of elastic fibers that allows movement of the stapes and yet seals the potential space between its footplate and the edges of the oval window. Cartilage also persists on the vestibular surface of the base of the stapes and may in spots form the entire thickness (as little as 0.0425 mm) of the base (Anson).

In addition to their ligaments, the stapes and the manubrium of the malleus have muscles attached to them. The *stapedius muscle* lies within the hollow of the pyramidal eminence in the posterior wall of the tympanic cavity; its tendon, emerging through the opening at the apex of the eminence, turns slightly inferiorly to run anteriorly and somewhat medially and insert on the posterior aspect of the head of the stapes. This muscle diminishes the excursion of the base of the stapes by its reflex contraction, which has been assumed to protect the internal ear from excessive sound; Liden and co-workers, however, regarded its protective action as being rather ineffective, saying it contracts somewhat slowly in response to noise and maintains maximal tension for only a few seconds. They found that it responds most to low-frequency sounds and is little affected by the high-frequency sounds that most damage the ear. From a study of patients, they concluded that an important function is to filter out low tones in background noise and thus improve discrimination for the higher frequencies used in speech. Mahoney and co-workers paralyzed both it and the tensor tympani and found that this

Fig. 3-17. Middle ear cavity in diagrammatic frontal section. The stapes has here been rotated through approximately 90° in order to show its crura.

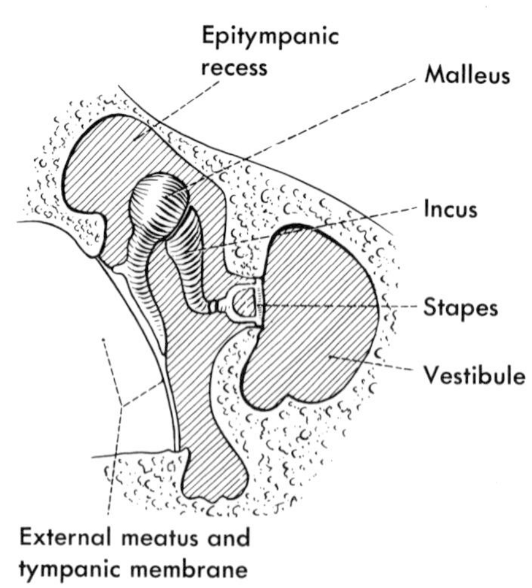

interfered with the understanding of speech at higher decibels with or without background noise.

The stapedius muscle is innervated by a branch of the facial nerve, which leaves this nerve within the facial canal at about the level of the muscle.

The *tensor tympani muscle,* like the stapedius, enters the tympanic cavity through a bony canal, but is much larger than the latter. The muscle is usually described as arising in part outside the skull, from the cartilaginous portion of the auditory tube and an adjacent portion of the sphenoid bone, and in part from the bony canal (termed the *semicanal for the tensor tympani*) in which it lies; this is situated directly above the bony auditory tube, a very thin wall of bone separating the lumina of the two canals. According to Lupin, however, the tensor tympani is primarily a continuation of the tensor veli palatini (and Kameres has described the two muscles as acting together), with only a few fibers from the sphenoid spine. Rood and Doyle agreed with this. Upon its emergence at the tip of the cochleariform process the tendon of the tensor tympani turns sharply laterally across the tympanic cavity to attach to the manubrium of the malleus close to the neck. The muscle is innervated by a branch of the mandibular nerve which usually traverses the otic ganglion.

The tensor tympani draws the manubrium medially and thus tightens the tympanic membrane. Like the action of the stapedium, this has been assumed to be primarily a protection against excessive noise. Kameres pointed out that the muscle responds not only to noise but also to swallowing and electric stimulation of the tongue and that the tensor veli palatini also reacts to these stimuli; the simultaneous action of these two muscles, he reasoned, must "pump" air from the tympanic cavity into the auditory tube, the tensor veli opening the ostium of the tube and the inward movement of the tympanic membrane forcing air toward the pharynx and thus helping to open the isthmus.

It might be pointed out that the stapes represents a portion of the hyoid arch that has been enclosed within the middle ear, and its muscle, the stapedius, is therefore innervated by the nerve of the hyoid arch—the facial nerve. On the other hand (except for the manubrium mallei and the long crus of the incus, which are largely of second arch origin; Hanson and co-workers), the malleus and incus represent a backward-projecting portion of the original cartilage of the lower jaw, and the muscle of the malleus, the tensor tympani, is therefore innervated by the nerve to the musculature of the lower jaw—the mandibular. The course of the cartilage from the jaw to the middle ear cavity is evidenced in the adult by the anterior ligament of the malleus, extending through the tympanosquamous fissure (which on the base of the skull appears as a lateral continuation of the perhaps better known petrotympanic fissure) to the sphenoid spine, and the sphenomandibular ligament extending from the spine to the mandible. Both these ligaments represent degenerated remains of the mandibular (Meckel's) cartilage, the original skeleton of the lower jaw from which the incus and malleus are largely derived.

The blood supply of the bones of the middle ear has been described somewhat differently by Hamburger and Wersäll and by Anson and Winch. Both pairs of workers described communications between the mucosal and intraosseous vessels, but Hamburger and Wersäll described definite nutrient vessels for the malleus and incus, whereas Anson and Winch regarded the communications with mucosal vessels as being the source of the blood supply. They differed also in regard to the stapes. Anson and Winch stated that blood vessels are prominent only in the footplate, while Hamburger and Wersäll found vessels primarily in the head, saying that crura and footplate may be completely devoid of vessels. Lindeman also described the footplate as having the more prominent vessels. (Hamburger and Wersäll found no haversian systems in any of these bones, while Lindeman found both malleus and incus to be definitely composed of compact or haversian bone and believed the stapes probably has the thickest bone.)

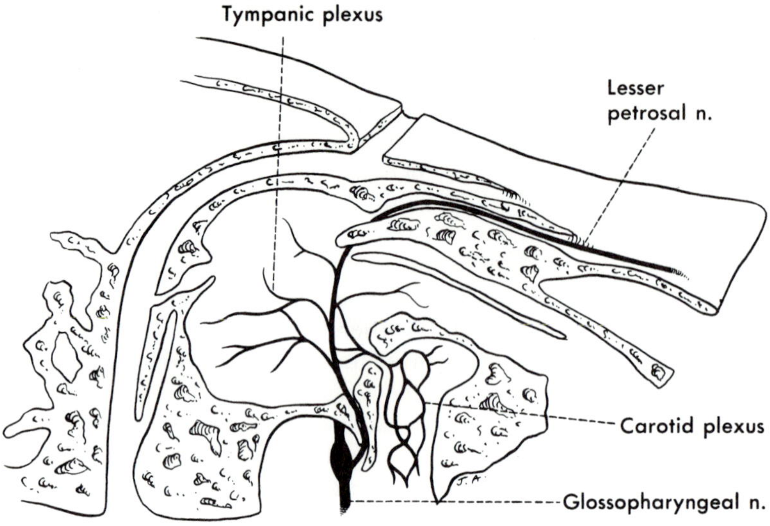

Fig. 3-18. Formation of the tympanic plexus and the lesser petrosal nerve.

NERVES AND VESSELS

Tympanic Plexus

The nerves of the middle ear cavity (Fig. 3-18) are represented by the tympanic plexus, which lies on the promontory deep to the mucosa in grooves or even actual canals (Frenckner) in the bone. This plexus is formed chiefly by the tympanic branch (nerve of Jacobson) of the ninth cranial nerve, but is reinforced by one or more caroticotympanic nerves derived from the internal carotid plexus. The tympanic branch of the ninth nerve supplies sensory fibers to the mucosa of the middle ear, while the sympathetic fibers are presumably vasomotor. It has already been mentioned that sympathetic fibers to the dilator pupillae may pass through the middle ear cavity.

The sympathetic filaments from the carotid plexus enter the tympanic cavity inferiorly, penetrating the thin bone between the middle ear cavity and the carotid canal in company with the caroticotympanic branches of the internal carotid artery. The tympanic branch of the glossopharyngeal nerve, from the inferior (petrosal) ganglion, traverses the tympanic canaliculus in the thin lamina of bone between the carotid canal and the jugular fossa; most commonly, it emerges on the promontory just anterior to the lower border of the fossula of the round window. Hunt suggested that sensory branches from the facial nerve also may join the tympanic plexus to be distributed to the middle ear cavity, reaching this by connections between the facial and the lesser (superficial) petrosal nerve, which arises from the plexus. Vidic and Yound found a connection from the facial nerve, the geniculate ganglion, or the greater petrosal in all their dissections, but Vidic later reported that this communicating branch consists of vagal fibers, probably preganglionic ones, that leave the auricular branch of the vagus and run recurrently in the facial nerve to reach the lesser petrosal nerve.

From the tympanic plexus branches are given off to the tympanic cavity, to the mastoid air cells and to the auditory tube; a connection to the cochlear nerve has been suspected but not proved. From the upper border of the plexus arises the *lesser petrosal nerve* (lesser superficial petrosal nerve). This nerve is formed primarily by preganglionic fibers of the ninth nerve that traverse the plexus and upon leaving it run through a canal in the petrous portion of the temporal bone (medial to the canal of the tensor tympani) to enter the middle cranial fossa. Here the lesser petrosal parallels but lies lateral to the greater (superficial) petrosal nerve, with which it may exchange branches, and then, like this latter, leaves the middle cranial cavity. The point of exit of the lesser petrosal nerve varies but is through or close to the foramen ovale; immediately outside the foramen ovale the nerve ends in the otic ganglion. The fibers in this nerve are described as preganglionic ones for the innervation of the parotid gland.

Tumors said to be histologically identical with carotid body tumors and usually termed "glomus jugulare" tumors, apparently arise from cell groups associated with the nerves to the ear. Guild ('53) reported such cells groups in the normal ear associated with the tympanic branch of the ninth nerve on the promontory and beyond, and also along the auricular branch of the vagus—thus not always adjacent to the jugular bulb, where the tissue of this type related to the ear was first found, and hence obtained its name. Parkinson reported what was apparently the second case of secretory glomus jugulare tumor: after its removal, the tissue assayed was found to contain more than 150 times the amount of noradrenalin than does the normal adrenal medulla. Simpson and co-workers embolized glomus jugulare tumors immediately before operation to minimize bleeding.

The chorda tympani traverses the middle ear cavity and gives branches to the tympanic membrane (Bosman); its course is described in connection with that of the facial nerve.

Arteries

Several arteries supply the mucosa and adjacent bone of the tympanic cavity (Fig. 3-19), most of them derived from the external carotid. The *anterior tympanic artery,* a small branch of the maxillary (internal maxillary) enters through the tympanosquamous fissure (the fissure that transmits the anterior ligament of the malleus, lateral to but continuous with the petrotympanic fissure through which the chorda tympani makes its exit from the skull), or its branches enter through several canals adjacent to the fissure (Nager and Nager); it is distributed to the more anterior part of the middle ear cavity, including a portion of the tympanic membrane. It and the other branches to the middle ear anastomose freely. The *inferior tympanic artery,* derived from the ascending pharyngeal, accompanies the tympanic branch of the ninth nerve through the tympanic canaliculus to enter the cavity in-

Fig. 3-19. Schema of the arteries of the tympanic cavity.

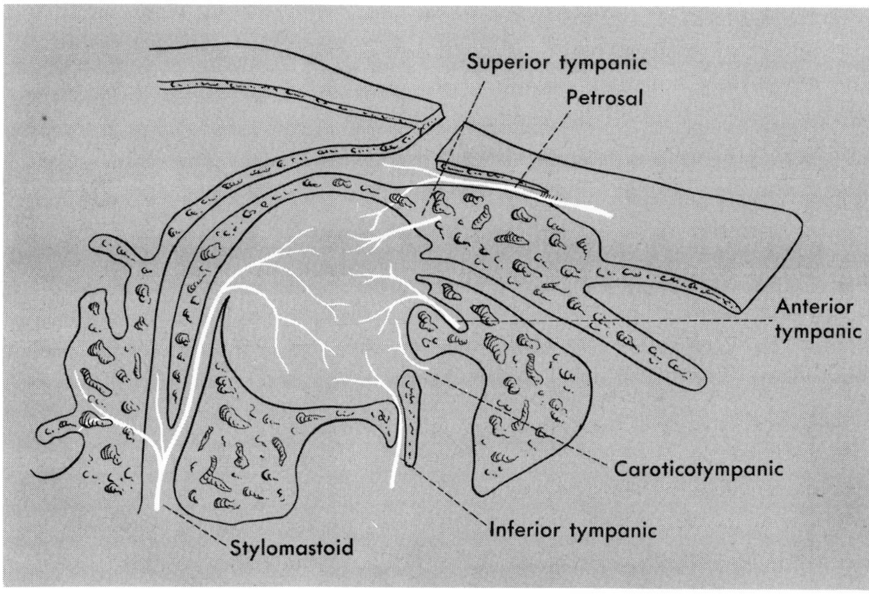

feriorly and anastomose with the other vessels. The *stylomastoid artery* from the posterior auricular (or the occipital) follows the facial nerve retrogradely within the facial canal, helps to supply that nerve and the mastoid cells, and sends a branch, the *posterior tympanic artery,* into the tympanic cavity through the canaliculus of the chorda tympani. This branch has been said to supply the posterior part of the tympanic membrane, the promontory and much of the rest of the medial wall, but Nager and Nager found it very small, contributing primarily to the inner peripheral vascular ring about the drum membrane.

The *middle meningeal artery* (from the maxillary) furnishes two branches to the middle ear, a petrosal (superficial petrosal), which enters the facial canal by way of the hiatus through which the greater petrosal nerve emerges, and a superior tympanic artery, which enters the middle ear through the petrosquamous fissure.

Finally, the internal carotid artery supplies one or two *caroticotympanic branches,* which accompany the nerves of the same name, arising from the carotid artery within the canal and entering the middle ear cavity to supply its anterior wall.

Nager and Nager described the distribution of these arteries in detail, including their distribution to the ossicles, and Anson and associates ('62) described and illustrated particularly those vessels closely related to the stapes.

Veins and Lymphatics

The veins of the middle ear roughly parallel the arteries and empty into the pterygoid plexus of veins or into the superior petrosal sinus. The lymphatics are usually said to end in the retropharyngeal and parotid lymph nodes.

AUDITORY TUBE

The auditory, pharyngotympanic, or eustachian tube extends from its tympanic ostium high on the anterior wall of the tympanic cavity to a pharyngeal ostium situated just posterior to the inferior nasal concha. In the adult the tympanic ostium is approximately 2 to 2.5 cm higher than is the pharyngeal one, and the tube thus runs downward, medially, and anteriorly toward the pharynx. It is, however, not straight, but slightly S shaped. In the infant the tube is shorter, relatively wider, and more nearly horizontal than in the adult and therefore affords an easier pathway for infections ascending from the pharynx to the tympanic cavity; on the other hand, however, the tube also opens relatively and actually lower in the tympanic cavity, so that it affords better drainage from this cavity. The length of the entire tube in the adult varies from 31 to 38 mm (Graves and Edwards); of this the lateral or tympanic third has a bony wall and is therefore referred to as the osseous portion, while the anteromedial two-thirds has a cartilaginous and connective tissue wall and is referred to as the cartilaginous portion of the tube.

Osseous Canal

The osseous canal, the semicanal of the auditory tube, is separated by a very thin lamina of bone from the similar canal of the tensor tympani muscle lying above it, the two semicanals together forming the *musculotubal canal;* medially it is related to the lateral aspect of the carotid canal, and inferiorly a lamina of bone of varying thickness, which may contain air cells, separates it from the jugular fossa. The osseous canal is widest at its tympanic orifice and gradually narrows throughout its length; its anteromedial end forms the narrowest part, or isthmus, of the entire tube. At the isthmus the cartilage of the tube is firmly attached to the bony tube (and may extend posteriorly as a part of the roof of this), and from this point to the pharynx the tube gradually widens again. The cartilaginous and bony tubes meet at an obtuse angle.

Cartilaginous Canal

Unlike the bony tube, which is complete and consequently remains permanently open, the cartilaginous portion of the tube is not completely surrounded by cartilage and in the resting condition is converted into a closed, slitlike cavity. The cartilage of the tube is essentially hooklike in section, being applied to the entire medial wall and turning over the superior aspect to descent slightly on the lateral wall; fibrous connective tissue (the lamina membranacea or salpingopharyngeal fascia) completes the lateral wall of the tube. In addition to the twisted sheet of cartilage forming the main support of the tube, accessory cartilages may occur about the pharyngeal orifice; Proctor described the cartilage as being formed from several different centers and presenting sliding joints that facilitate movement of the tube.

Most accounts agree that the tensor veli palatini, through its attachment to the lateral wall of the tube, is the chief opener of the tube, pulling downward on its wall (*e.g.*, Ross, Proctor). Rood and Doyle called the medial part of the muscle, that attaching to the tube, the dilator tubae muscle. Ross added that the portion of the muscle arising lateral to the tube is also important, particularly in children, since according to his findings it protects the tube from pressure exerted by the medial pterygoid muscle. (Kamares, as already mentioned, regarded the tensor tympani as also helping to open the tube, the inward movement of the eardrum membrane producing a flow of air through the isthmus.)

Some workers have held that the levator veli palatini also helps to open the tube. Proctor agreed that it may do so by pulling outward on its fibrous wall as it elevates the cartilage. However, Misurya reported that in dogs the levator, although it dilated the lumen of the isthmus, narrowed the lumen of the ostium, and the pressure in the auditory tube remained unchanged. Cantekin and coworkers reported that stimulation of both the levator and the medial pterygoid caused constriction of the tube, while stimulation of the tensor veli palatini dilated it.

Except when it is open, the pharyngeal end of the auditory tube strongly resists passage from the pharynx to the tympanic cavity. Thus, pressure in an air chamber can be raised as much as 130 mm of mercury without forcing the auditory tube open (Kobrak, '59); the painful pressure has to be equalized by swallowing. Passage from tympanic cavity to pharynx is much easier, however, since lowering the atmospheric pressure by 30 mm of mercury will produce flow of air from the tympanic cavity. Abnormal patency, with its symptoms of autophony and uncomfortable fullness in the ear, can be treated by injections of polytetrafluoro-ethylene (Teflon) if a patient's gain in weight proves ineffective (Pulec). Crary and co-workers reported that this treatment much improves the psychological state of persons suffering from abnormal patency.

The tube is lined throughout its length with a low columnar ciliated epithelium, except at its pharyngeal end where there is pseudostratified epithelium containing many goblet cells. In its cartilaginous portion it contains also mucous glands; it is often described as containing lymphoid (adenoid) tissue—"Gerlach's tubal tonsil"—near its pharyngeal end, but Eggston and Wolff found such tissue in only 1 of 47 auditory tubes examined. According to Wolff ('34a), the air cells opening into the osseous portion of the tube are not originally outgrowths from this tube, but rather of the tympanic cavity itself; only secondarily do they come to empty into the tube.

Blood Supply

The blood supply to the auditory tube is through twigs from the pharyngeal branches of both the (internal) maxillary and ascending pharyngeal arteries, from the ascending palatine branch of the facial (external maxillary) artery, the superior tympanic branch of the middle meningeal, and the artery of the pterygoid canal (a branch of the sphenopalatine from the maxillary). The veins drain into the pterygoid venous plexus. The lympahtics drain into the retropharyngeal and deep cervical nodes; the several courses described for them have been summarized by Graves and Edwards.

The blood vessels of the tube apparently have a neural control similar to that of the vessels of the nose. Nathanson and Jackson reported that stimulation of the nerve of the pterygoid canal (Vidian nerve) in the dog increased the perfusion pressure of the tube, indicating vasodilation; section of the cervical sympathetic trunk also increased the pressure, indicating a tonic effect by that system on the vessels.

Innervation

The sensory supply to the auditory tube is largely by way of the tympanic plexus and therefore through the ninth nerve (see Fig. 3-8), but Bosman traced twigs from the chorda tympani and the mandibular nerve to its lateral wall. According to Graves and Edwards, the reports of various workers indicate that the pharyngeal orifice is probably innervated through the ninth nerve in about 50% of cases, but through the maxillary portion of the trigeminal (through the pharyngeal branch of the pterygopalatine ganglion) in the other 50%.

The relation of the tube to infections of the middle ear cavity is obvious, since it forms an easy pathway for ascending infections from the pharynx. The close association between the pharyngeal orifice and the pharyngeal recess (fossa of Rosenmüller), which contains the lymphoid tissue of the pharyngeal tonsil or adenoid, renders the tube particularly susceptible to occlusion from overgrowth of this tissue.

CONDUCTION OF SOUND THROUGH THE MIDDLE EAR

The literature upon this subject and its application to problems of hearing is now enormous. Detailed discussions of many aspects and an extensive bibliography may be found in Kobrak's book ('59).

Direct observations have shown that the greatest movement of the tympanic membrane in response to sound is not at the umbo, as might be thought from the shape of the membrane, but below, fairly close to the lower attachment to the meatus. The vibrations of the membrane necessarily set up vibrations both of the air in the tympanic cavity and of the ossicular chain, and various views of the relative importance of these two routes have been held.

Conduction by Air

By this is meant the impingement of air waves on the secondary tympanic (cochlear or round window) membrane (see a following section) with consequent vibration of this and of the endolymph on its cochlear side. It has been argued that this is an important route, if not for all tones at least for high ones, but it is now generally conceded that it contributes little to hearing in the normal ear. Its inefficiency, as compared with the ossicular route, has been demonstrated by observations on perforations of the tympanic membrane in human beings and by experiments on animals.

It is now well known that while small perforations of the tympanic membrane may affect primarily hearing of lower tones, large ones, even though they improve the efficiency of the air route, produce loss of all tones. The ossicular route is thus more effective than the air route for high tones as well as low ones, and tympanic membrane prostheses can be expected to improve hearing in all tonal ranges.

The relative roles of air and ossicular conduction were analyzed by Wever and coworkers ('48a) in experimental animals. They found that breaking the ossicular chain at the incudostapedial joint, thus allowing the vibrations of the tympanic membrane to affect the inner ear through the air route only, produced a great loss of sensitivity (60 decibels in the middle tonal frequencies and only a little less for lower and higher frequencies). Removal of the membrane and ossicles, which through their mass impede the transmission of airborne energy to the middle ear cavity, moderated the loss; wider opening of the middle ear to the outside by removal of bone decreased the loss still more, until it amounted to only about 28 decibels. The great loss in acuity following interruption of the ossicular chain indicates the inefficiency of the air route through the drum membrane; the persistent decreased acuity of about 28 decibels represents the efficiency of the ossicular chain over the best airborne pathway and is recognized as approximately the conductive deficit expected after a fenestration operation.

Conduction by the Ossicular Chain

The ossicular chain has been regarded as a lever system that converts the vibrations of the tympanic membrane into intensified movements of the stapes against the perilymph. Wever and colleagues ('48a) interpreted their results to indicate not so much a lever mechanism as a "hydraulic mechanism," through which the vibrations of the air can impart their energy more efficiently to the heavier perilymph; in this transmission the important factor is the ratio of the area of the drum membrane to that of the stapedial footplate. As summarized by Lawrence, the ratio is such that in the cat about 54% of the vibrations from the air reached the perilymph—and the energy imparted to the perilymph by the drum-ossicle route is at least 1000 times that of the energy of vibrations entering through the round window.

The concept that the efficiency of the ossicular chain depends primarily upon the ratio of the vibratory area of the tympanic membrane to that of the stapedial footplate (about 17:1) is now accepted, but the lever mechanism of the ossicular chain, with an efficiency of about 1.3:1, increases that efficiency to about 22:1 (Goodhill); this is said to approximate 25 to 28 decibels in the speech range and represents, as has already been noted, the unrecoverable loss when the air route is substituted for the ossicular one. (According to Goodhill, this expected loss may be partially offset in fenestration operations by a transformer gain through the tympanomeatal membrane, but even then the loss is never less than 15 decibels.) On the other hand, the lever mechanism of the ossicular route adds so relatively little to the transmission of sound that marked improvement in hearing can be obtained by a rod prosthesis between the tympanic membrane and the stapes.

Movement of the incus is said to lag behind that of the malleus because of laxity of the incudomallear joint, and the amplitude of movement of the long process of the incus is less than that of the malleus (Kobrak, '59). Further, the usual movement of the stapes is primarily at its upper border and anterior end, with the posterior end and lower border moving very little; thus otosclerosis that involves only the anterior end of the base of the stapes, which is often the first part involved, may seriously decrease hearing. With loud tones, however, the movement of the stapes is said to change to a rocking one around its long axis, one edge moving inward and the other outward. This, resulting in less displacement of fluid, is apparently a protective mechanism for the inner ear; the change to this type of movement is presumably brought about by contraction of the stapedius muscle.

The actions of the stapedius and tensor tympani muscles are usually interpreted as being largely protective, although not necessarily entirely so. (As noted in the descriptions of these muscles, the stapedius has been described as functioning to filter out lower tones primarily, and thus improve the understanding of speech; the tensor has been described as helping the passage of air downward through the auditory tube.) In man, contraction of the stapedius actually moves the stapes only if the ossicular chain is broken (Kobrak, '59), but its isotonic contraction dampens stapedial movement; such contraction is generally a reflex contraction, with its afferent pathway in the eighth nerve only (Lindsay and coauthors); thus if there is hearing with bone conduction only, the reflex cannot be provoked by air conduction. Goodhill, noting that the stapedius tendon is sometimes cut in operations on the stapes, suggested that long-range studies of the results of these relatively new operations will be necessary to show whether lack of the protective function of the stapedius will damage the cochlea. Liden and co-workers apparently did not investigate that, but did advise that the stapedius tendon be spared, if possible, to preserve discrimination of higher tones.

Contraction of the tensor tympani has been said to protect the ear against loud sounds by diminishing the excursion of the malleus and tympanic membrane, to increase the efficiency of transmission through the ossicular chain or to maintain the membrane and ossicles in a state that favors transmission, to decrease the transmission, or to change the natural frequency of vibration of the bones so that some tones are favored over other ones. The evidence is not entirely clear. Kobrak

('59) stated that both the tympanic muscles may somewhat improve the transmission of sound in the middle tonal ranges; Crowe and co-workers reported that tension on the tensor tympani tendon (of the cat) tightened the ossicular chain and impaired the transmission of all low tones, while section of it relaxed the ossicular chain and impaired the transmission of higher tones; Wever and Bray reported that tension on the tendon in the cat produced diminution of hearing of all sounds, as recorded from the cochlear microphonics, except those of high intensity, which showed many variations. They concluded that contraction of the muscle both reduces sound conduction as a whole and raises the natural frequency of vibration of the chain.

Round Window

The secondary tympanic or round window membrane, only about 2 mm^2 in area, has been observed to bulge outward as the stapes moves inward, and vice versa; its inward movement is greater than its outer (Kobrak, '49). The function of the round window or cochlear fenestra may therefore be largely one of protecting the cochlea from excess pressure, but it seems probable that its functions are not completely understood.

Kobrak ('49), Crowe and associates, and others have stated that slight pressure on the round window membrane may somewhat improve the acuity of hearing (by slightly increasing the pressure in the cochlea?). Hughson ('40) used periosteal grafts in man to immobilize the round window and reported that they produced an increase in the acuity of hearing. (This report has apparently never been confirmed, and the operation is no longer done.) Wever and Lawrence, in contrast, reported that attempts to block the round window in the cat produced a reduction rather than an increase in the response to tones. Sullivan and Hodges reasoned that movement of the round window membrane increases the acuity of hearing and that deafness may be expected to result when it is immobilized; they suggested that deafness after a fenestration, for example, may be caused by scar tissue filling in the round window niche. W. F. House ('59) quoted evidence that complete blockage of both windows in man usually produces severe loss of hearing by both air and bone conduction. He therefore advocated removal of otosclerotic bone from the round window niche when that was involved, although he concluded that the exact clinical effect of closure of the round window was unknown. Tonndorf and Tabor reported that in the cat closure of the round window alone had a relatively small effect on hearing by air conduction (the ossicular chain) and that closure of both windows had no more effect on bone conduction than closure of the oval window alone, unless the cochlear aqueduct (perilymphatic duct) was also occluded.

The air route to the round window, although apparently unimportant in the normally functioning ear, may become important when the ossicular chain is broken, and Wever and co-workers ('48a) reported that in the cat the cochlea can be stimulated almost as readily by way of the round window as by way of the oval window when the middle ear mechanism is absent. It is also said, however, that when the tympanic cavity is open hearing can be improved by placing moist cotton in the round window niche (Kobrak, '59); perhaps this results from interfering with the sound waves reaching the round window so that those reaching the oval window have a greater effect. Simultaneous arrival of sound waves at both windows may also be the explanation for the loss of hearing if the tympanic membrane is perforated after a successful fenestration. According to Zoellner (Kobrak, '59), there must be a difference in impedance between the two windows for good hearing.

Changes of Pressure

Wever and colleagues ('48b) reported that although small changes in the pressure in the middle ear may slightly augment hearing in the range of about 2000 cycles, the general effect of either increased or decreased pressure is to stiffen the tympanic membrane and therefore decrease hearing (as recorded by cochlear microphonics), particularly affecting the lower tones. Kobrak ('59) noted that with obstruction of the auditory tube fluid may be

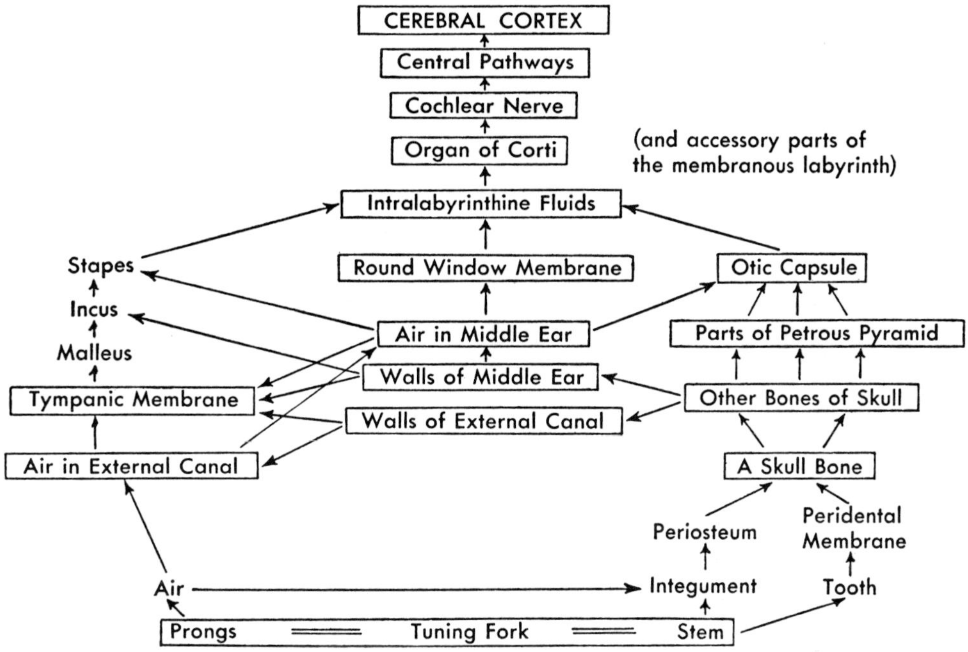

Fig. 3-20. Conduction of sound to and through the ear. (After Guild SR: Ann Otol Rhinol Laryngol 45:736, 1936)

expected to accumulate in the middle ear as a result of the lowered oxygen pressure there; the increased mass of this fluid then particularly affects the transmission of high tones. Earlier, Loch had considered decreased pressure itself to affect particularly high tones.

Increased or decreased pressure in the middle ear cavity may alter hearing not only by its effect on the tympanic membrane, but also perhaps by altering the mobility of the round window or the pressure in the cochlea. Wever and colleagues suggested that there was some effect on the internal ear, and Kobrak ('35) reported that in the rabbit pressure changes in the middle ear cavity affect also the pressure in the labyrinth.

Bone Conduction

In contrast to hearing by air conduction, hearing by bone conduction is largely through an osseous pathway rather than primarily through the osseotympanic one (Fig. 3-20). As summarized by Hirsch, low frequency (primarily below 800 cycles per second) sinusoidal vibration applied to the skull tends to move it as a whole, so that it vibrates toward and away from the less rigid ossicular chain, thereby essentially producing movement of the stapes at the oval window. This type of conduction is known as *inertia bone conduction.* It may also involve air conduction, however, for vibration of the skull moves it against the mandible, and this transmits vibrations to the cartilaginous portion of the external acoustic meatus and hence to the air in that meatus; thus if the meatus is occluded at the concha, the intensity of the sound is increased by movements of the tympanic membrane and the osseous chain.

As the vibratory rate is increased to some 800 to 1600 cycles per second, inertia bone conduction becomes less important and is largely replaced by *compression bone conduction;* in this process the skull is distorted rather than vibrating as a whole, so that parts of it, including the petrous part of the temporal bone, are alternately compressed and expanded. Perilymphatic pressure is most easily adjusted at the round window; hence pressure in the scalae alternately shifts toward and away form the scala tympani in response to compression and expansion of the otic capsule.

PNEUMATIZATION OF THE TEMPORAL BONE

The mastoid (tympanic) antrum communicates through the aditus with the posterior portion of the epitympanic recess; it extends somewhat lateral to the tympanic cavity and communicates in turn with the mastoid air cells. Since it is a relatively large cavity, it is the key to surgery of this region; both the mastoid air cells and the middle ear cavity proper can be approached from it.

The older surgical approach to the mastoid antrum is just behind the posterosuperior quadrant of the external meatus, through the so-called suprameatal triangle. In the adult, in whom the bony markings are best developed, the superior boundary of this triangle is the continuation backward above the meatus of the root of the zygomatic arch; its anterior border is the posterosuperior segment of the bony external meatus, and its posteroinferior border is an imaginary line joining these two parts. In the endaural approach to the antrum, in which the incision is made between the tragus and helix and the skin is reflected from the bony meatus, the antrum is entered through its anterior wall, from the meatus. In opening the antrum or the mastoid air cells in the postaural approach the sigmoid sinus and the facial nerve are in particular danger, but can best be avoided by making the original opening as close to the angle of union between the root of the zygomatic arch and the posterior wall of the meatus as possible and carefully enlarging downward and backward from this. From the lateral aspect, the facial nerve lies approximately on the line of the groove between the tympanic and mastoid parts of the temporal bone. In the endaural approach, the attachment of the tympanic membrane serves as a landmark for the depth of the facial nerve from the external surface; the vertical portion of the nerve lies only slightly deeper than this.

Mastoid Process

The mastoid process is constantly pneumatized in the adult (though not in the infant) but varies very considerably in the extent of its pneumatization and in the arrangement of the air cells. These cells grow out not only from the antrum, but also from each other, so that they form complex interlocking chains of thin-walled cavities opening into each other (see Fig. 3-11). Although there are apparently always air cells in the region of the antrum, the mastoid process as a whole is usually described as being of one of several types: pneumatized, diploic (containing marrow rather than air cells), mixed (containing both marrow and air cells), or sclerotic (very dense bone). Tremble reported that about 80% of mastoids are well pneumatized by the age of 3 or 4 years, but that in 20% of persons normal pneumatization fails to appear.

The *terminology* describing the cell groups in the mastoid is complex and variable. The cell groups within the body of the mastoid may be along the posterior aspect of the facial canal (retrofacial cells of Broca), over the lateral surface of the sigmoid sinus, behind the sinus (marginal cells), or at the tip of the mastoid process (mastoid tip cells). Meltzer has considered them in three major groups: a superior group, an anteroinferior group, and a medioposterior group, according to their relations to the sigmoid portion of the lateral sinus. He pointed out that the position of the sigmoid sinus in relation to the mastoid process (a variable relationship, since the sinus shows variations in both its course and size) determines to what extent pneumatization by these various cell groups will occur, and causes definite variations in their development (Fig. 3-21). He called attention to the fact that when the sinus plate (the harder bone over the sinus) is in contact with the cortex of the squama, cells cannot develop posterior to the sinus and the medioposterior group is therefore missing; if the sinus plate is deep and close to the floor of the antrum and the "facial ridge" (bone about the facial canal), cells develop freely both anteroinferiorly and medioposteriorly; if it is in contact with the posterior wall of the external acoustic meatus and with the lateral

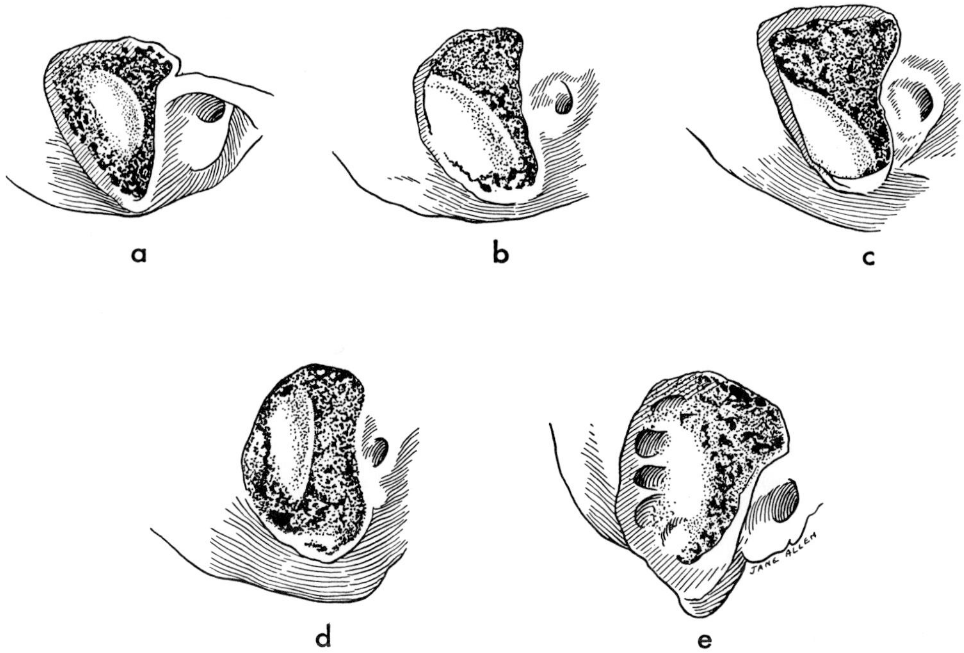

Fig. 3-21. Various relations of the sigmoid sinus in the mastoid process, indicating the fashion in which the size and placement of this sinus affect the development of cell groups in the mastoid. (After Meltzer PE: Arch Otolaryngol 19:326, 1934)

surface of the mastoid process then cells cannot develop either anteroinferiorly or medioposteriorly.

False Bottom

As the mastoid process develops from two different portions of the temporal bone (squamous and petrous), the cells growing downward from the antrum are, at least at first, divisible into two groups, squamous and petrous, separated from each other by the petrosquamous (petromastoid) suture. The suture itself is usually obliterated in adult life, though it may be marked on the surface of the mastoid process by a line at which the periosteum is firmly attached. Internally, the plane of the suture may divide the mastoid cells into a superficial and upper (squamous) group and a deeper (petrous) group that extends downward into the mastoid tip. These two cell groups may differ in the degree of their development; moreover, it sometimes happens that they are separated by an appreciably heavy plate of bone along the petrosquamous suture. If this is well developed, it is called the "false bottom" or Koerner's septum

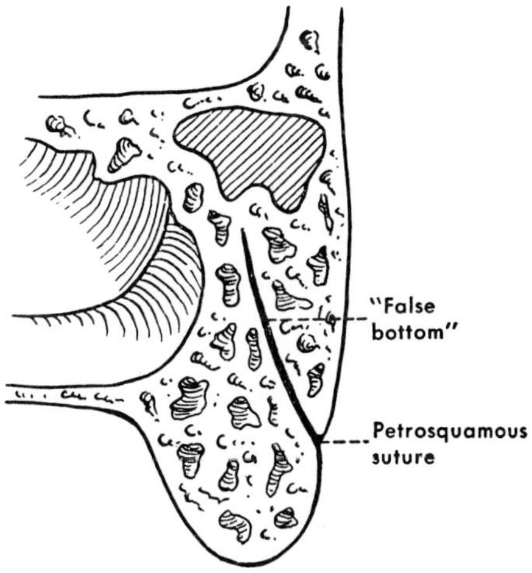

Fig. 3-22. Diagram of the "false bottom."

(Fig. 3-22) and is of special significance in operations upon the mastoid cells.

The postaural approach to the mastoid leads to the superficial and upper group of cells; if these are poorly developed, it may be difficult to reach the deeper cells through the more dense bone. Moreover, if the false bottom is well developed, it may mislead the inexperienced surgeon into believing that the cellular portion of the mastoid has been entirely explored, while in fact the cells deep to the false bottom are likely to be larger and better developed than are the superficial cells. It is the deep cells, also, that lead to the lowest cells in the mastoid, those at the tip.

Accessory Cell Groups from Mastoid

The air cells developed in connection with the mastoid process are not necessarily confined to this process, but may also invade other parts of the temporal bone to form accessory cells groups (Fig. 3-23). Thus cells from the mastoid may also extend below the labyrinth and into the floor of the tympanic cavity, between this and the jugular bulb, where they are usually referred to as *hypotympanic cells* or as the jugular bulb group. Similarly, accessory cells derived from the mastoid may also extend into the root or body of the zygomatic arch, into the squamous portion of the temporal bone, and into the floor of the middle cranial cavity as *epitympanic cells* (both above the labyrinth and, as cells of Kirchner, in the roof of the external meatus).

Among the few papers pertaining to radical mastoidectomy that have been written since that procedure became rather infrequent are those of Allam and of Williams ('66). Allam divided the cells into three major groups—associated with the mastoid process, the perilabyrinthine cells, and the petrous apex cells—and described the various cell tracts leading to each of these groups. A fourth group consisted of regions occasionally invaded by cells—the zygomatic and occipital bones, the squamous part of the temporal bone, and the styloid process. Williams emphasized the necessity for thorough exenteration of all infected parts and described both the anatomy and the techniques involved.

Petrous Apex

The new terminology describes the petrous portion of the temporal bone as including the mastoid process and discards the older term *petrous pyramid,* which included the labyrinth and the parts of the petrous temporal anterior to it. The term *petrous apex,* that part of the petrous temporal anterior to the labyrinth and the middle ear cavity, is still accepted, however. Since much of the work on pneumatization has been reported in terms of the petrous pyramid, the discussion that follows will employ that term as well as petrous apex. As already noted, some cells of the mastoid group may become related to the labyrinth

Fig. 3-23. The chief groups of air cells in the temporal bone. (After Tremble GE: Arch Otolaryngol 19:172, 1934)

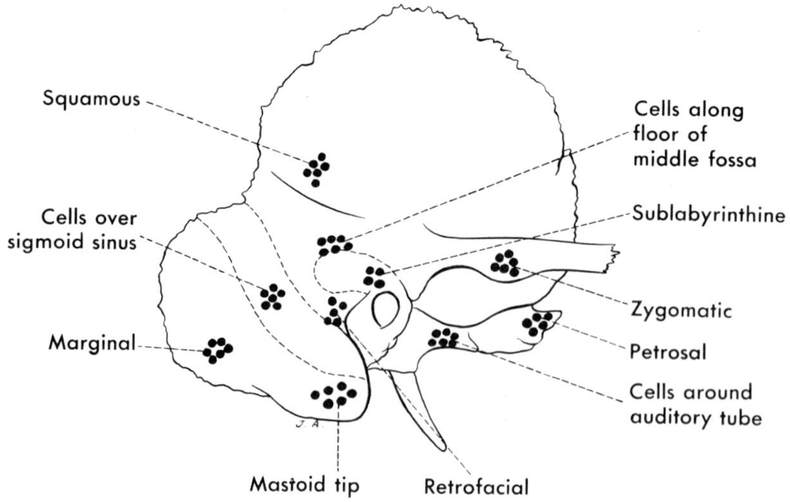

and thus invade the base of the petrous pyramid.

The petrous apex is also subject to pneumatization, particularly by outgrowths from the tympanic cavity. The cells extending into the petrous apex are necessarily related to the musculotubal and carotid canals, which largely fill the apex, and have been variously termed peritubal cells, petrous cells, angle cells, and cells of the petrous tip. They vary much, and it is not always clear whether the cells so described actually lie in the apex rather than the base of the pyramid.

Subtubal air cells are said to be present in 90% of children up to the age of 6 years and may or may not spread widely through the petrous apex. Hagens reported that approximately 34% of the temporal bones that he studied showed air cells in the petrous apex; Myerson and colleagues reported that only 11% of the bones they studied had well-pneumatized petrous pyramids and that none of these occurred in individuals under 15 years. As noted by Myerson's group, the cellular character of the mastoid does not necessarily indicate the extent of pneumatization of the pyramid.

Stelter and co-workers reviewed a number of investigations on the pneumatization of the petrous pyramid and reported the results of their own study of 126 bones. They described one group of cells as originating from the medial wall of the antrum above the ampulla of the anterior (superior) canal and extending medially around this canal—sometimes only around its anterior part, sometimes completely around it and also coming to lie above and medial to the cochlea. (They suggested that these supracochlear cells are what are usually called *apical* cells.) They described two groups of cells originating from the tympanic cavity, one at approximately the level of the tip of the cochleariform process, the other one posteriorly below the level of the semicircular canals. The first group extends medially between the cochlea and the tensor tympani and, when best developed, comes to lie in front of the cochlea and above, anterior, and posterior to the carotid canal. The second group extends below the semicircular canals and may also assume a postcarotid position. Finally, cells were found in infants arising from the auditory tube close to its tympanic end and extending forward below the tube. (Wolff, '34a, stated that the peritubal cells in the fetus are derivatives of the tympanic cavity and not of the auditory tube itself, but that secondarily they may open into the tube.)

The most evident pathway for *infection* of the petrous apex, in addition to those furnished by the pathways of the nerves and vessels of this region, is by direct extension from the middle ear cavity, and Lindsay ('45) regarded osteomyelitis of the apex as usually caused by this. Earlier workers stressed especially the possibility of extension from the mastoid along either an epitympanic or a hypotympanic route; while this is of course a possibility, it is lessened by the apparent fact that the mastoid cells are not, as once thought, continuous with the precochlear and carotid cells.

Farrior ('49) emphasized that the hypotympanic cells may come into contiguity with peritubal and petrous apex cells and should be explored in all radical mastoidectomies. He described a technique for complete exenteration of cells in the hypotympanum and about the carotid artery and auditory tube in a radical mastoidectomy; earlier ('42) he described a sublabyrinthine approach to the petrous apex, which can be developed, after a simple mastoidectomy, through the sublabyrinthine cells.

COURSE OF THE FACIAL NERVE THROUGH THE TEMPORAL BONE

FACIAL NERVE

The origin of the facial nerve is discussed in Chapter 1, and its more specific relations and distributions are described in connection with the external ear, the face, the nose, and the mandibular and lingual regions. The only concern here is the relationships of the nerve within the temporal bone, a point of considerable importance to the aural surgeon.

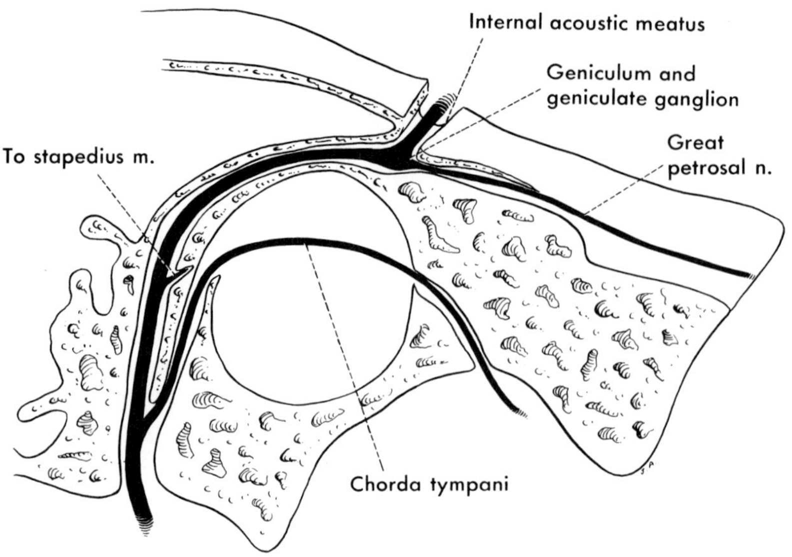

Fig. 3-24. The facial nerve in the temporal bone. The chorda tympani actually passes between the malleus and incus, not shown, in its course across the tympanic membrane.

The facial nerve traverses the internal acoustic meateus, within the petrous portion of the temporal bone, in company with the eighth nerve and the labyrinthine (internal auditory) artery; at the outer end of this canal the facial lies above the other elements and pierces the arachnmoid and dura to enter its own bony canal, the *facial canal* (aqueduct of Fallopius, Fallopian canal). Occasionally, the subarachnoid space may accompany the facial nerve into the canal, even as far as the geniculate ganglion (Perlman and Lindsay, '39). This canal continues for a short distance the lateral direction of the internal meatus, and the facial nerve thus comes to lie just above the base of the cochlea (therefore above and medial to the promontory), where it bears the geniculate ganglion (Fig. 3-24). From the geniculate ganglion is given off the greater petrosal (greater superficial petrosal) nerve, which passes anteriorly and slightly medially to enter the middle cranial fossa; there is also a twig from or from close to the ganglion to the tympanic plexus or the lesser petrosal nerve, and there may be a twig, the external (external superficial) petrosal nerve, from the ganglion to the plexus on the middle meningeal artery.

Immediately beyond the geniculate ganglion the facial nerve turns sharply laterally and posteriorly around the front of the vestibule of the internal ear, between this and the cochlea. The position of this turn (geniculum, external genu) can be located from the middle ear by the emerging tendon of the tensor tympani muscle, for the geniculum lies just deep to the tip of the cochleariform process. As it runs backward in the bone of the lateral wall of the vestibule, which is also the medial wall of the tympanic cavity, the facial nerve inclines slightly downward and laterally; here the bone about it forms a projection (the *prominence of the facial canal*) into the latter cavity, so that the course of the canal beneath the prominence of the lateral (horizontal) semicircular canal and above the vestibular or oval window can be seen upon inspection of the middle ear cavity. Its bulge may be great enough to partly cover the oval window and the base of the stapes. Anson warned that the bone covering the facial nerve here is always thin and may be entirely lacking in spots. In the posterior part of its short "horizontal" course the facial nerve, although it lies below the prominence of the lateral semicircular canal, may come to lie lateral and then posterior to the posterior end of this canal and also just lateral to and above the ampullary end of the posterior vertical canal.

In this position, behind the base of the pyramidal eminence, the facial nerve makes a rather broad curve downward to run almost vertically, but inclining slightly outward, through the mastoid process to the stylomastoid foramen; here the facial canal ends, and the nerve finally emerges from the skull. The position of the stylomastoid foramen is located in mastoidectomy by the anteromedial end of the digastric crest, the upward projection of cortical bone formed by the mastoid (digastric) incisure on the inferior surface of the mastoid process. As it passes behind the pyramidal eminence the nerve gives off a filament for the innervation of the stapedius muscle, and shortly before leaving the stylomastoid foramen it gives off the chorda tympani, the course of which is described later in this section.

As it runs downward through the mastoid process behind the tympanic cavity the facial nerve may have a straight course or may be slightly concave anteriorly; the course of the nerve in relation to the exterior of the skull corresponds approximately to the suture groove between the tympanic and mastoid portions of the temporal bone, though the nerve lies, of course, medial to this. Kettel ('46) quoted Batson as having found the nerve never lying posterior to this line, but sometimes slightly anterior to it; among 125 operations on the nerve within the canal, however, Kettel reported on three patients in whom there was a marked deviation from this course and the nerve lay posterior to its usual line. In two of these three cases the nerve emerged from the anterior aspect of the mastoid process rather than from between mastoid and styloid processes (Fig. 3-25).

More marked abnormalities in position have also been found. For instance, Basek reported three instances in which the nerve divided in the facial canal and its branches emerged through separate canals, Meyer and Crabtree reported finding three nerves that ran below the oval window and found reports of nine others, and Greisen found a facial

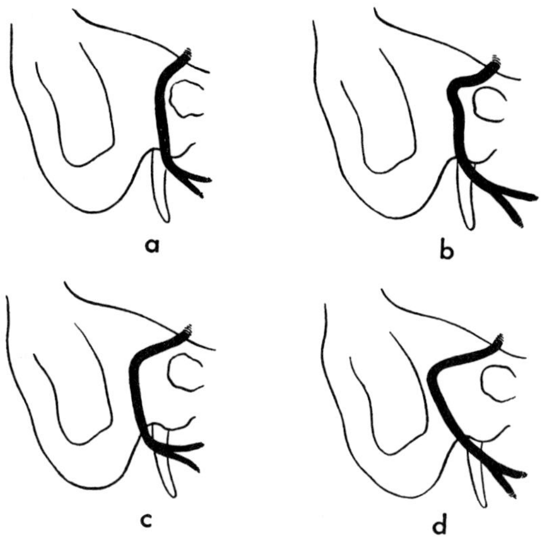

Fig. 3-25. Normal (*a*) and abnormal (*b, c,* and *d*) courses of the facial nerve. (After Kettel K: Arch Otolaryngol 44:406, 1946)

nerve lying just deep to the skin of the external auditory meatus some 2 to 3 mm lateral to the tympanic membrane. In the infant, in whom the mastoid process has not yet developed, the facial nerve lies superficially in the bone behind the external meatus, and upon its emergence through the stylomastoid foramen is almost subcutaneous, since it is not protected by a mastoid process.

The *blood supply* of the facial nerve has been reported in detail by Blunt, and the supply to that part of the nerve in the facial canal was described also by Nager and Nager. In the posterior cranial fossa the nerve receives twigs from the anterior inferior cerebellar artery; in the internal meatus from the labyrinthine (internal auditory); in the facial canal from both the stylomastoid and petrosal arteries, which meet and anastomose at the bend between the horizontal and vertical parts of the nerve; and in the parotid gland and on the face by twigs from the stylomastoid, posterior auricular or occipital, superficial temporal, and transverse facial arteries. There are apparently no anastomoses between the labyrinthine twigs and other vessels, but otherwise the vessels to the nerve anastomose freely with each other.

Injuries in the Temporal Bone

Because of its relationship to the middle ear cavity and the mastoid cells, the facial nerve during its intratemporal course is subject to damage both from infections of the middle ear and the mastoid and from operations upon the mastoid. Bunnell estimated that in not more than 7% of patients with facial paralysis is the paralysis due to infections of the middle ear cavity; the incidence of surgical injury is of course unknown. So-called idiopathic paralysis of the facial nerve within the canal also occurs, as described originally by Bell; it has been thought to be due to edema of the nerve with compression by the bony canal. Kettel ('47) concluded that the primary change is one of ischemia, due to malfunction of the vasa nervorum, and has reported on and discussed a number of patients in whom decompression of the nerve was carried out by opening the facial canal from the stylomastoid foramen to the lateral semicircular canal (Fig. 3-26). Lindeman found the normal nerve occupying no more than 25% to 40% of the space in the canal; Sade and coworkers incised the perineurium and epineurium in their decompressions, but expressed doubt that edema and compression cause the paralysis. Gussen arrived at the opposite conclusion from a study of the nerves of two patients who had been afflicted with facial palsy.

Jongkees stated that it is unnecessary to expose the horizontal part of the nerve, since the sheath here is composed of elastic fibers instead of the collagenous fibers found in the vertical part. However, Fisch and Essen exposed the nerve throughout its intratemporal course in 12 instances of facial palsy and found that the nerve was involved proximal to the geniculate ganglion—in the internal meatus and the first part of the facial canal—in 11 of these. Wolferman concluded that although decompression is the only treatment

Fig. 3-26. Decompression of the facial nerve. (Courtesy Dr. H. L. Williams)

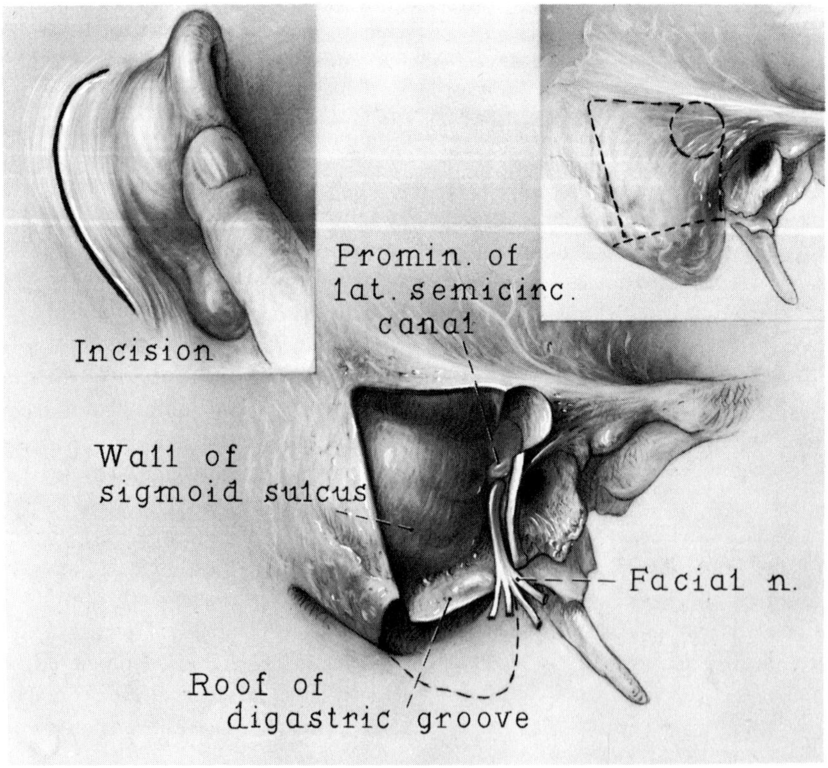

worth considering, it has not been, and probably cannot be, proved that it is effective. Greer and co-workers cast doubt on the value of sheath decompression, for they found that when used in normal cats it caused more damage than did bony decompression and actually slowed recovery of an experimentally produced edematous nerve.

The diagnosis of the level of injury in cases of lesions of the facial nerve, dependent upon the level at which various branches leave it, has been discussed by Bunnell and by Tschiassny; it should be mentioned, however, that the latter worker's conclusion that no taste fibers enter the brain stem through the seventh nerve is not generally accepted. In cases of destruction of a portion of the nerve in the facial canal, Ballance and Duel found nerve grafts preferable to displacement of the nerve and bringing its ends together for primary suture.

CHORDA TYMPANI

The chorda tympani is composed of sensory and preganglionic motor fibers, both of these running proximally in the nervus intermedius portion (pars intermedia, nerve of Wrisberg) of the facial nerve. Beyond the geniculate ganglion, the fibers destined for the chorda tympani are an intrinsic part of the facial trunk; Saito and co-workers described the shifting positon of its fibers in the nerve. At an average distance of 5.3 mm above the stylomastoid foramen, but varying from 1.5 mm below it to 10.9 mm above (Kullman and co-workers), the chorda tympani leaves the facial trunk and takes a recurrent course upward and forward in its canaliculus (once "iter chordae posterius"), in company with branches of the stylomastoid and petrosal vessels, to enter the tympanic cavity through its posterior wall. Within the cavity it lies at first close to the posterior border of the tympanic membrane, but diverges somewhat to pass between the malleus and the incus, as it crosses the upper aspect of the membrane; it leaves the tympanic cavity through a canal in the petrotympanic fissure (this part of its course was once known as the "iter chordae anterius"). It then joins the lingual nerve to be distributed to taste buds of the anterior two thirds of the tongue and to the submandibular (submaxillary) ganglion, through which postganglionic fibers reach the submandibular (submaxillary) and sublingual salivary glands. After its emergence from the skull it has a tiny ramus communicans with the otic ganglion; some workers believe that through this communication the facial nerve also supplies secretory fibers to the parotid gland, which is otherwise supplied by the ninth nerve.

Bosman studied the distribution of the chorda tympani in serial sections of ears from human, monkey, and baboon fetuses. He found fibers in common with those of the ninth and tenth nerves to skin behind the root of the pinna and to the posterior wall of the external auditory meatus; he also found twigs to the medial surface of the tympanic membrane and of the malleus and to the incudomallear joint, to perivascular plexuses and, with fibers of the mandibular nerve, to skin of the tragus, the anterior wall of the external auditory meatus, and the lateral wall of the auditory tube. (He also noted a connection to the middle meningeal plexus and the one to the otic ganglion.)

Despite its theoretical importance as indicated by the above statements, *injuries* to the chorda tympani are not necessarily associated with any marked symptoms. The chorda tympani may be destroyed by disease in chronic mastoiditis or removed in radical mastoidectomy; it is often divided in stapes operations (when the nerve is stretched but not divided a bitter taste may persist for several months), yet the patient does not necessarily complain of the loss of taste on the anterior two thirds of the tongue. This may occasionally be caused by the fact that the greater petrosal nerve, rather than the chorda tympani, contains the fibers for taste; often, however, the patient is simply not conscious of the loss. (Jeppson reported that after the chorda was severed, only 9% of the patients thought they had impaired taste, although testing revealed that all did.) Likewise, the expected reduction in salivation is apparently not marked, either because of autonomous activity on the part of the denervated glands or because, as suggested by several authors, the glands may likewise be innervated through the glossopharyngeal nerve.

In its course to the middle ear the chorda tympani is accompanied by a branch of the stylomastoid artery to the middle ear, and as it leaves the ear it is paralleled by the anterior tympanic artery. The nerve is subject to damage through anything affecting the intracranial portion of the facial nerve as a whole and may be injured by any injury to the upper part of the tympanic membrane. Like the other structures in the middle ear cavity, the chorda tympani is covered by a reflection of the mucous membrane of this cavity and therefore does not actually traverse this space.

ANOMALIES AND PATHOLOGY

Variations in the middle ear due to varying arrangements of air cells are numerous and are discussed briefly in the section on pneumatization of the temporal bone.

Atresia of the external acoustic meatus is usually accompanied by deformations of the middle ear cavity and of the ossicular chain, and the bones may be so deformed as to be hardly recognizable.

Other purely anatomic anomalies of the middle ear are also occasionally found. Walls described three specimens in which a muscle arising from the sphenoid spine had a tendon of insertion that passed through the petrotympanic fissure to join the anterior process of the malleus. Describing this under the name of the *laxator tympani muscle,* he noted that it has been previously described as the major laxator, but that it is now generally agreed that the so-called minor laxator is actually a ligament rather than a muscle.

Druss and Allen described a rare case of *fistula* communicating with the middle ear cavity; this was presumably a first branchial cleft derivative (Chap. 9). The auricle on the involved side was markedly abnormal; a fistula began on the upper part of the auricle close to the external acoustic meatus and opened into the neck immediately behind the lower margin of the auricle, but from this lower opening a second fistula extended medially and forward deep into the neck, and one part of it communicated with the lower part of the tympanic cavity. They found only one similar case previously reported.

Polvogt described a number of abnormalities of the tympanic membrane and other parts of the middle ear in persons who had, before death, been shown to have had apparently normal hearing. A number of these ears showed pathological variations of the membrane ranging from slight thickening to marked calcification and variations in its position ranging from moderate to extreme retraction, in the latter cases with the drum even adherent to the promontory. He found that in such ears the region of the oval window and the niche of the round window were seldom free from strands of embryonic tissue, these strands being sometimes small and delicate and sometimes broad and fibrous. In 16 of 63 ears the mucous membrane over the promontory was abnormally thickened, probably as a result of old infections. He also noted that in some cases the malleus and incus were composed of dense bone with few marrow spaces, while in others there were large marrow spaces with a thin bony cortex. None of these variations had apparently any effect upon hearing.

The glomus jugulare tumors of the middle ear have been discussed in the section on the tympanic plexus.

Otosclerosis

The most common and serious anatomicopathologic change of the middle ear is otosclerosis. This was said by Davenport and associates to occur probably in about 0.2% of the white population of the United States, but Guild ('44) pointed out that a sharp distinction should be made between clinical otosclerosis and histological otosclerosis; in a large series of ears which he investigated, the incidence of histological otosclerosis was 1 in 8 for white females and 1 in 15 for white males above the age of 5 years. From Guild's study it appears that race, sex, and age all influence the incidence of the disease. While Davenport and co-workers concluded that there was a genetic factor involved, Guild's study indicates the impossibility of drawing valid conclusions concerning this from purely clinical studies.

Wilson and Anson, among others, have described the histologic characteristics of otosclerosis, noting that the diseased bone is richly vascularized by large thin-walled vessels and usually restricted to certain portions of the capsule; they and others agree that the disease is not dependent on inflammatory changes.

Bast ('40), Anson and co-workers ('47), and others have emphasized that the otic capsule in the adult may not be completely ossified but may contain, rather, islands of cartilage, and have said that these islands, which may show calcification, are foci for the origin and spread of otosclerosis. Bast suggested that attempts to repair these bony defects later in life, when ossification of the capsule is normally completed, might be the initiating cause. These writers have especially stressed the relationship of the fissula ante fenestram to the onset of otosclerotic changes, since this opens usually close to the oval window and regularly contains cartilage at its vestibular end with islands of calcified cartilage scattered elsewhere within it. Guild, in his extensive study, found no support for this concept. Although extension of otosclerotic bone from the fissula ante fenestram to the adjacent oval window has apparently been shown to occur, clinical observations have indicated that the disease need not affect the anterior part of the window, as it should if it always spread from the fissula, but may occur irregularly at other parts of the window or in other parts of the ear (Goodhill).

Wittmaack concluded that venous stasis may be a factor in the onset of otosclerosis; upon producing such stasis in the hen he found that changes similar to those in human clinical otosclerosis occurred. Noting that otosclerotic changes are more common near the oval window, he pointed out that the chief drainage from this region is a vein running with the lesser petrosal nerve to join the pterygoid plexus. Lempert and Wolff have regularly found pathological changes in the blood vessels of the bones of otosclerotic ears and expressed the belief that these changes precede and probably initiate the changes in the bones; as further discussed by Wolff ('50), they have interpreted their observations to mean that otosclerosis has no specific etiologic agent, but is the result of damage to the peripheral blood supply, of whatever cause—chemical, vasomotor, or allergic.

Regardless of the etiology of the disease and its locus of origin, the clinical symptom, deafness, that results from the disease is generally accepted as being caused by fixation of the base of the stapes to the adjacent margin of the vestibular window; as a result of this, the stapes loses some or all of its mobility, and transmission through the ossicular chain becomes ineffective. The first really successful surgical attack on this disease was developed particularly by Lempert (*e.g.* '38, '48) and is commonly referred to as *fenestration,* since it involves making a new oval window (fenestra nov-ovalis) into the perilymphatic cavity. Basic steps are an endaural approach (reflecting skin lining the external meatus); removal of the incus and the head of the stapes; carefully removing bone from the anterior or ampullary end of the lateral semicircular canal until the perilymphatic space opens into the epitympanum; and sealing this opening with a viable membrane, commonly a skin flap from the external acoustic meatus. Experience has shown that best results are obtained when the opening is made over the anterior end of the lateral semicircular canal; the landmarks are also clearest here, and an opening so made in the bony labyrinth leads directly into the vestibule.

Lempert, Lindsay ('46), and others have emphasized the precautions that must be taken, particularly in removing all bony fragments and attaining a firm attachment of the flap covering the window, if closure of the window by new bone formation is to be prevented. It is also important that the round window open into a closed tympanic cavity and that there be no communication between this cavity and the new oval window; this prevents the cancelling effect of a sound wave arriving simultaneously at both windows.

The second type of operation for otosclerosis, pioneered in its modern form by Rosen ('52), was directed toward the stapes; although it has taken various forms, the primary effort was toward mobilization of the stapes without breaking the ossicular chain. Through an endaural approach, the stapes was visualized by appropriate removal of bone at the inner end of the meatus, and the sclerotic connection between the stapes and the window was fractured or otherwise disrupted. Techniques have improved and changed, and the operation has been discussed by many workers (*e.g.,* Rosen, '60; Kobrak, '59; Goodhill). Anson emphasized anatomical relations that may affect the outcome of attempted stapes mobilization; among these are the fragility of the base or footplate of the stapes, leading to its easy fracture; the short working distances that allow a tool passed only a little too deeply to impinge on important structures; and the superficial position of the facial nerve.

The third type of operation, again directed at the stapes, consists of removal of part or all of that bone with replacement of the missing part by a prosthetic device (Shea, H. P. House, Hough, and others). It was developed because the mobilized stapes became re-fused at the lines of fracture in a high percentage of cases; H. P. House reported serious regression of hearing in about 15% of patients during the first postoperative year and in about 10% per year thereafter.

Approach to the middle ear cavity is by elevation of skin and the fibrocartilaginous tympanic ring from the posterior and superior walls of the external acoustic meatus and removal of bone from the adjacent part of the pars tympanica ("tympanic ring") of the temporal bone. A variety of prostheses have been used successfully. These include vein, fat, fibrous tissue, and absorbable gelatin sponge (Gelfoam) to cover the oval window, with wire, polyethylene tubing, or the replaced head and crura of the stapes to transmit vibratory energy from the incus to the new cover of the oval window. (A variant of the technique employs a piston between the incus and a hole of 1 mm diameter drilled into the base of the stapes.) Belluci reported that a fat and wire prosthesis has improved hearing in 96% of afflicted persons.

The Internal Ear

The internal ear consists of two fundamental parts: complex cavities (perilymphatic or periotic spaces) hollowed out within the petrous portion of the temporal bone and referred to, with the surrounding bone, as the osseous labyrinth; and, within the osseous labyrinth, the delicate and slightly more complex membranous (otic) labyrinth.

EMBRYOLOGY

The membranous labyrinth is developed from a placodal thickening of the ectoderm, which subsequently sinks beneath the surface ectoderm to become a closed otic vesicle. By differential growth of this tiny vesicle and fusion of portions of its walls three semicircular ducts, a utricle and a saccule, a cochlear duct, an endolymphatic (otic) sac, and their various connections are formed (Fig. 3-27). The membranous labyrinth, being essentially tubular and saccular, contains a cavity termed the *endolymphatic space,* which in turn is filled with fluid, the endolymph or endolymphatic fluid.

Since the early development of the membranous labyrinth takes place in mesenchyme and subsequently even in the cartilage destined to form the petrous portion of the temporal bone, it is obvious that the membranous labyrinth must lie in cavities excavated from this mesenchyme or cartilage (later transformed to bone). The cavities of the temporal bone surrounding the membranous labyrinth are larger than necessary to accommodate the latter, and the excess space thus existing between the inner surface of the bony wall and the outer surface of the membranous labyrinth is the perilymphatic (periotic) space; this space is in most locations traversed by

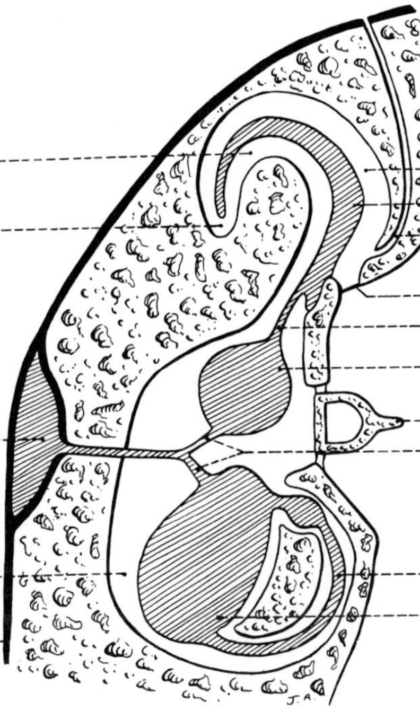

Fig. 3-27. Highly schematic representation of the perilymphatic and endolymphatic systems. The endolymphatic system is *shaded.*

numerous fibrous trabeculae, between which are interstices filled with fluid, termed the perilymph or perilymphatic fluid. In other locations, notably the cochlea, the trabeculae are reduced or absence and fluid alone occupies the perilymphatic space.

The perilymphatic spaces develop around the membranous labyrinth by a process of fusion of mesenchymal spaces to form larger ones, eventually almost completely surrounding the membranous portion; as the petrous portion of the temporal bone is represented in cartilage for some time before the mebranous labyrinth has completed its growth, further growth of this portion is allowed for by a constant dedifferentiation of the cartilage surrounding the membranous labyrinth and subsequent enlargement of the perilymphatic spaces following degeneration of this cartilage.

The essential steps in the formation and growth of the membranous labyrinth and the perilymphatic spaces have been described by Streeter ('17a and b); further details have been contributed by Foley. The development of the bony capsule has been analyzed especially by Bast; references on this subject will be found in his paper of 1942 and in the text of Bast and Anson.

OSSEOUS LABYRINTH

The shape of the bony labyrinth in the adult may be determined by casts of the perilymphatic cavity (periotic cistern) or, because the bone adjacent to the perilymphatic cavity is usually more dense than bone elsewhere, may be outlined by dissections to the surface of this denser bone. This procedure is especially accurate in the infant.

The cavity of the bony labyrinth is lined by a thin layer of internal periosteum, sometimes called endosteum, and most of it is crossed, as already mentioned, by a spongelike arrangement of delicate trabeculae somewhat similar to the arachnoid trabeculae of the subarachnoid space. According to Lempert and coworkers, this spongelike tissue is so dense that it prevents the free escape of perilymph from

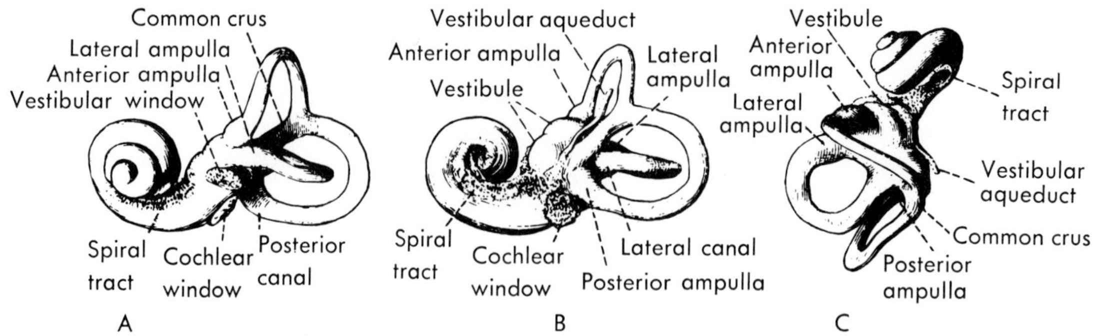

Fig. 3-28. Casts of the bony labyrinth (perilymphatic space). *A* is a lateral view of the left labyrinth, *B* a medial view of the right one, and *C* a view of the left one from above. (After Henle, from Hollinshead WH, Textbook of Anatomy, ed 3. New York, Harper & Row, 1974)

the trephine opening of a fenestration; however, they found this tissue to be lacking in the immediate neighborhood of the footplate of the stapes and throughout the bony cochlea (scala vestibuli and scala tympani). The periosteum and the trabeculae lying within the bony otic capsule can properly be referred to as the periotic labyrinth (Bast and Anson), a term that distinguishes between these soft tissues and the bone that surrounds them—either or both may be implied by the term "osseous labyrinth." The periotic labyrinth and the perilymphatic (periotic) fluid in turn surround the membranous (otic) labyrinth except in certain locations where the membranous labyrinth is fused to the periosteum lining the bony labyrinth.

Anson and Bast emphasized the very small size of the perilymphatic space, pointing out that a cast of the bony labyrinth displaces only 0.2 ml of water, or about 3 drops, and that part of the space represented by this volume is occupied by the membranous labyrinth and the soft tissues of the bony one; thus the amount of perilymph, much of it disposed in capillary form among trabeculae, must be very limited.

Vestibule

The central portion of the cavity of the bony labyrinth is the vestibule (Fig. 3-28), a relatively large ovoid perilymphatic space (about 4 mm in diameter) containing both the sacculus and the utriculus of the membranous labyrinth. In the floor of the bony vestibule may be seen the elliptical recess for the anterior end of the utriculus and anterior and slightly lateral to this the spherical recess for the sacculus. In the lateral wall of the vestibule is the vestibular fenestra or oval window, housing the base of the stapes. Through the stapes the perilymph of the vestibule receives the vibrations of the tympanic membrane and ossicular chain set up by sound waves reaching the tympanic membrane. On the medial wall and floor of the vestibule, where it abuts on the lateral end of the internal acoustic meatus, are minute openings for the entrance of the nerve branches to the vestibular portion of the ear.

Semicircular Canals

The vestibule is continuous with the three bony semicircular canals; the cavities within these are markedly larger than the ducts (membranous canals) that they contain. A portion of the anterior (superior) semicircular canal usually forms a ridge (arcuate eminence) on the floor of the middle cranial fossa (anterior surface of the petrous ridge). The lateral ("horizontal") semicircular canal projects, as already described, as a ridge on the medial wall of the aditus and epitympanic recess, and it is here, toward the anterior end of this canal, that the perilymphatic space is opened in the operation of fenestration. The cavities (perilymphatic spaces) of all three semicircular canals open into the vestibule at both their ends.

The anterior (superior) and posterior semicircular canals (sometimes referred to together as the vertical canals) are both arranged in the vertical plane, but their long axes are at approximate right angles to each other. The anterior canal is directed anterolaterally at an angle of about 45 ° to the midsagittal plane, and the posterior canal is directed posterolaterally at a corresponding angle. The posterior end of the anterior canal and the anterosuperior end of the posterior canal join together before entering the vestibule, to form a common crus. The lateral canal lies only approximately in the horizontal plane; its anterior end is higher than its posterior one, so that it makes an angle of about 30 ° with this plane; both its anterior and posterior ends open independently into the vestibule.

Cochlea

The perilymphatic cavity of the vestibule is also continuous with that of the other functional division of the inner ear, the cochlea (Figs. 3-27, 3-28, and 3-29). The bony cochlea, a part of the otic capsule, is named from its resemblance to a snail shell; it is a hollowed spiral of about two and three fourths turns which diminishes from a relatively broad base to a pointed cupula or apex. The base lies against the anteromedial surface of the vestibule and abuts also on the anterior surface of the lateral or blind end of the internal acoustic meatus. Part of the large basal coil of the cochlea projects into the middle ear cavity as the promontory. From the vestibule the cochlea is directed anterolaterally and slightly upward. A central core of bone, the *modiolus,* runs forward in the cochlea, but fails to reach the cupula; it is about this central core that the spiral channels (perilymphatic and endolymphatic) of the cochlea are arranged. A layer of bone, arranged in a spiral fashion, unites the modiolus to the peripheral wall of the bony cochlea and separates the successive spiral cavities from each other.

The modiolus is hollow, to accommodate the cochlear nerve; its base lies against the lateral end of the internal acoustic meatus, to which it transmits the nerve. From the outer surface of the modiolus, like the flanges of a screw, an *osseous spiral lamina* projects into the perilymphatic space and partially divides it into two parts. This is also hollow, and transmits the branches of the cochlear nerve to the end organ, the spiral organ (of Corti) of the membranous cochlear duct. Its cupular end, the hamulus, projects a little beyond the modiolus. The cochlear duct or scala media is attached to the osseous spiral lamina and extends from this cental core of bone across the perilymphatic space to attach to the outer wall of this space; thus the spiral lamina and the membranous cochlea divide the perilym-

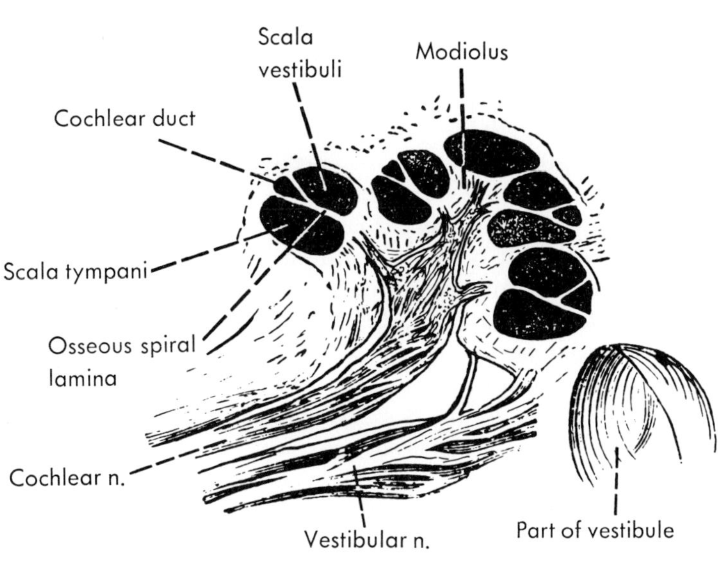

Fig. 3-29. A longitudinal section through the cochlea. (After Henle J: Handbuch der systematischen Anatomie des Menschen, Vol 2. Braunschweig, Vieweg und Sohn, 1868)

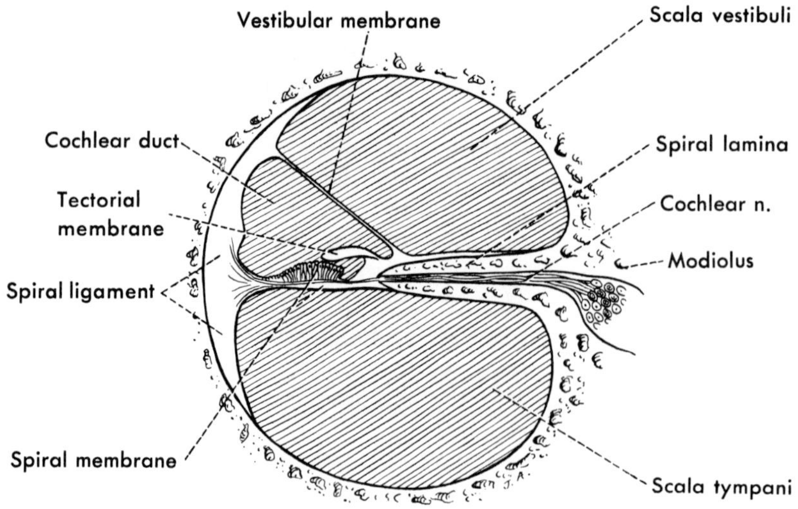

Fig. 3-30. Section through a single turn of the cochlea.

phatic space into its two portions, an ascending and a descending spiral.

The ascending spiral of the bony labyrinth is the *scala vestibuli,* which lies "above" the spiral lamina of the cochlea. (Since the cochlea points anteriorly, laterally, and slightly upward, the usual anatomical terms indicating position become hopelessly complicated in attempting to describe this spiral structure; hence it is customary to describe the cochlea as if it were sitting on its base with the cupula pointing upward. The true directions can be somewhat approximated if, in the present account, "anterior" is substituted for "above," and "posterior" is substituted for below.) Although the scala vestibuli is separated for most of its length from the descending spiral, the scala tympani, by the bony spiral lamina of the modiolus and by the cochlear duct (Fig. 3-30), both modiolus and duct end at the cupula without coming in contact with the apex of the bony wall; around these ends the scala vestibuli becomes continuous with the descending scala tympani, this communication being known as the helicotrema. The scala vestibuli begins at the base of the cochlea as a large opening from the vestibule; from the helicotrema the descending perilymphatic space, the *scala tympani,* runs backward and downward "below" the spiral osseous lamina and the cochlear duct to end blindly at its lower end, being provided here with a membranous window that seals off the perilymphatic space from the middle ear cavity.

Connections of Perilymphatic Spaces

The perilymphatic space of each osseous semicircular canal is continuous at both ends with the perilymphatic space of the vestibule, and this space is continuous widely with that of the scala vestibuli, which is then in turn continuous with that of the scala tympani at the helicotrema. All the perilymphatic spaces, therefore, open widely into each other. Although these perilymphatic spaces lie within the compact bone of the petrous portion of the temporal bone, this bone is deficient in certain locations and therefore presents regions of actual or potential communication between the perilymphatic space and other cavities. Thus, in addition to the oval and round windows which form potential communications between the middle and internal ears, the fissula (fissura) ante fenestram, and sometimes the fossula post fenestram, are defects in the bony wall between these two parts, while the vestibular and cochlear aqueducts, and the foramina for the nerves and vessels of the internal ear, are channels between the inner ear and the cranial cavity. Of these channels, all except possibly one are normally obstructed.

At the *oval window* the vestibule is, of course, sealed off from the middle ear cavity by the base of the stapes and its annular ligament, while the *round window* is similarly sealed off by the secondary tympanic membrane. The *fissula ante fenestram* and the *fossula post fenestram* are irregular clefts in the lateral wall of the internal ear, extending from the vestibule toward the tympanic cavity; the fissula usually extends completely through the bony lateral wall of the vestibule, while the fossula usually has a vestibular end only, but in about 25% of cases extends to the tympanic cavity. In any case, however, these fissures contain no marked extension of the perilymphatic space since they are obliterated by connective tissue (Anson and co-workers, '47).

The *vestibular aqueduct* (canaliculus) extends through the otic capsule from the vestibule to the posterior cranial fossa and transmits the endolymphatic (otic) duct and, according to most reports, an accompanying vein. The terminal dilation of the duct, the endolymphatic sac, ends blindly outside the dura; further, the duct and the vascular connective tissue that surrounds it so fill the aqueduct that there is no perilymphatic space here (Bast and Anson; Anson and associates, '68). Thus there is no free communication here from the perilymphatic space of the vestibule. Although Ogura and Clemis agreed that the perilymphatic space does not reach the dura, they said it seemed to extend to the proximal part of the endolymphatic sac; they described the artery and vein as lying in a separate canal approximately paralleling the aqueduct and being surrounded by a perilymphatic space that apparently reached the posterior fossa but not the subarachnoid space. Sando and co-workers said that the canaliculus (paravestibular canaliculus) containing the vessels merges with the aqueduct near the latter's cranial orifice and does not reach the posterior fossa.

The *cochlear aqueduct* (perilymphatic duct, cochlear canaliculus) opens at one end into the lower end of the scala tympani and, according to most descriptions, into the subarachnoid cavity at the other end. Traditionally, therefore, it has been regarded as the one free communication from the perilymphatic space (but see a following discussion of this duct). Incidentally, the vein of the cochlear aqueduct does not lie, as its name implies, in the same canal as the aqueduct, but in a separate parallel channel (Bast, '46; Anson).

Finally, the channels transmitting the nerves and vessels, although forming routes by which infections may reach the internal ear, are normally closed by the structures traversing them and therefore do not actually open into the perilymphatic space.

COMPOSITION AND CIRCULATION OF THE PERILYMPH

Comparisons between the composition of the perilymph, the cerebrospinal fluid, and the endolymph have influenced concepts concerning the origin and fate of the perilymph, although they have not solved these questions. Aldred and co-workers reported that the osmotic pressure of the perilymph and endolymph are usually identical but higher than that of the cerebrospinal fluid, thus suggesting that perilymph might be formed from endolymph by diffusion through the thin roof (vestibular or Reissner's membrane) of the cochlear duct. Later, Ledoux added that the index of refraction of perilymph is only slightly higher than that of endolymph, but much higher than that of cerebrospinal fluid, and that perilymph (of the cat) contains more protein than endolymph and far more than cerebrospinal fluid; he interpreted these findings as indicating that perilymph is cerebrospinal fluid in which proteins have become concentrated through the loss of water.

Apparently, accurate protein analysis on the minute amounts of labyrinthine fluids obtainable is difficult, for Vilstrup and Jensen said that in Acanthias (a shark) the perilymph contains less protein than does the endolymph, while Smith and associates, agreeing with Ledoux, found that in the guinea pig it may contain perhaps twice as much. Regardless of the findings in regard to total protein, however, various reports agree that in mammals there are similarities between cerebrospinal fluid and perilymph, but great differences between the latter and endolymph.

Thus, according to Smith and co-workers, the sodium, potassium, and chloride content of perilymph is similar to that of cerebrospinal fluid, but it differs markedly from that of endolymph. Vilstrup and Jensen noted that in Acanthias the perilymph is watery and clots soon after it is withdrawn, while the endolymph is so viscous that it will not mix with perilymph. For further details on the endolymph, see that section.

The similarities between perilymph and cerebrospinal fluid thus seem to support a common theory of the origin of the perilymph, namely, that it is derived from cerebrospinal fluid. The anatomical basis for such an origin has long been described as the opening of the cochlear aqueduct (perilymphatic duct) into both the subarachnoid and perilymphatic spaces; hence belief in this theory of origin depends in part upon one's concept of this duct. Descriptions of its anatomy have varied markedly, and observations and experiments upon interchange between the cerebrospinal fluid and the perilymph are also conflicting.

Altmann and Waltner, who had previously found evidence that in the rabbit (in which the cochlear aqueduct is apparently particularly wide) there is a strong current from the subarachnoid to the perilymphatic space, reported that there probably is no active flow through the narrower duct of monkeys. They concluded that such interchange as occurs between cerebrospinal fluid and perilymph is by diffusion, not active flow, and that the cerebrospinal fluid is not an essential source of perilymph. Ledoux seemed to agree on the activity of the circulation, reporting that fluorescein in the cerebrospinal fluid of the cat does not appear in the perilymph unless one opens the round window (thus draining off perilymph); he also said, however, that continued withdrawal of perilymph leads to a steady drop of the refractive index of perilymph toward that of cerebrospinal fluid. In contrast, Ritter and Lawrence were unable to demonstrate in human beings that withdrawal of perilymph after stapedectomy produced an inflow of cerebrospinal fluid into the perilymphatic space.

An increase in the cerebrospinal fluid's pressure does apparently produce flow into the labyrinth. Meyers found that in the monkey bilateral compression of the internal jugular veins for about 20 seconds produced a rise in pressure of the perilymph; he used the same technique to demonstrate poststapedectomy fistulas in patients' ears.

Altmann and Waltner favored the concept that perilymph originates largely from the perilymphatic blood vessels, which might explain its similarity to cerebrospinal fluid without involving free flow from one to the other; and Smith ('54) found the vascularity of the upper and lower ends of the spiral ligament (thus in both cochlear scalae) such as to suggest that perilymph may be both formed and absorbed there. Lindsay and co-workers ('52) found no changes in the threshold for hearing and no histological changes in two cats in which the cochlear aqueducts had been destroyed for 4 and 8 months, respectively, thus apparently indicating that the aqueduct is not essential to the formation of perilymph. In a later, similar experiment, Silverstein and co-workers analyzed the perilymph and found that although total protein at first rose, it eventually returned to normal. They concluded that perilymph (in the cat) is derived partly from the cerebrospinal fluid and partly from blood vessels. Crifo and Crifo went further; from their analysis of the amino acid content of the perilymph of the horse, they deduced that the perilymph must also be in part secreted.

In contrast to a dual or multiple origin, Schneider ('74a and b, '75) maintained, as had Schreiner, that perilymph is both produced and absorbed by vessels in the walls of the perilymphatic space. Schneider reported that in the guinea pig half the perilymph is replaced every 10 minutes; as measured by the decrease in this half-life, CO_2 was most effective in dilating the vessels, as it is also for cerebral vessels. Histamine increased the half-life.

Cochlear Aqueduct

Views concerning flow of cerebrospinal fluid through the cochlear aqueduct (perilymphatic or periotic duct) have already been presented. As indicated there, a common concept of the cochlear aqueduct has been that it forms an open channel between the subarachnoid space and the perilymphatic space (lower end of the scala tympani), but there has been doubt that this applies to primates. Waltner agreed with Mygind ('48) that any interchange between cerebrospinal fluid and perilymph must be by diffusion rather than active flow, but Mygind said this was so because of the dense meshwork of tissue that fills the aqueduct, while Waltner attributed it to the presence of a "barrier membrane of the cochlear aqueduct" that he described in the ears of animals and of fetal and adult man. According to Waltner's observations, this thin membrane completely occludes the aqueduct at its supposed opening into the scala tympani; only when the membrane is destroyed by pathologic processes, therefore, would interchange between cerebrospinal fluid and perilymph include nondiffusible substances such as erythrocytes. Ritter and Lawrence failed to find the barrier membrane (Kelemen and co-workers found only a few examples of it among 270 temporal bones studied in serial sections), but they and Anson and co-authors ('65) emphasized the small size of the aqueduct—according to Rask–Andersen and colleagues the mean diameter at its narrowest point is 0.14 mm—and, agreeing with Mygind, the fact that it is largely filled with connective tissue, although it is said to contain far more space for perilymph than the tissue around the endolymphatic duct (Anson and co-workers, '68). According to this, it is only through the tissue spaces, rather than through a channel such as is implied by the term "perilymphatic duct," that there can be continuity between the perilymphatic and the subarachnoid space. However, in another study of the human aqueduct, Palva and Dammert describe it as lined by dura and arachnoid and said interchange between cerebrospinal fluid and perilymph should be perfectly feasible. More recently, Wlodka studied the aqueduct in 50 specimens from each of five age groups. He found that all the aqueducts were patent in the fetal and newborn term specimens, but that with each succeeding group patency declined; in the group above 60 it was only 30%.

There is another possible role for the cochlear aqueduct, namely, that of allowing adjustments in the pressure of the perilymph; Kobrak ('59), for instance, regarded the aqueduct as the primitive pressure-regulating system, but suggested that in mammals, and particularly human beings, this function is largely taken over by the cochlear (round) window. Thus, the apparent failure of destruction of the cochlear aqueduct to cause permanent change in intralabyrinthine pressure (Lindsay, Schuknecht, and others) would seem to indicate that a connection of the aqueduct to the subarachnoid space is not essential to maintenance of a normal internal ear. However, the concept of the cochlear aqueduct as *one* part of a pressure-adjusting mechanism is supported by the finding of Tonndorf and Tabor that blocking both the aqueduct and the round window, but not the latter alone, decreases the cochlear response to bone conduction after the oval window is also blocked; they regarded the aqueduct and both windows as outlets for compressional forces set up in the perilymph by bone conduction.

MEMBRANOUS LABYRINTH

It is the growth of the membranous (otic) labyrinth that gives rise to the form of the osseous one, since the osseous labyrinth consists simply of periosteum-lined spaces surrounding and enclosing the membranous labyrinth. The differentiation of the membranous labyrinth leads to the formation of an anterior (superior), a posterior, and a lateral semicircular duct, pinched off from the central vesicle except at their ends; a spiral cochlear duct, an outgrowth from the central vesicle; and an endolymphatic duct and sac, likewise an outgrowth. Finally, the remainder of the vesicle, lying in the vestibule, becomes subdivided into two parts, utricle and saccule, united only by a narrow duct. The interconnecting spaces of the membranous labyrinth constitute the endolymphatic cavit (see Fig. 3-27).

Semicircular Ducts

The anterior or superior duct (membranous canal), like its surrounding bony canal, is directed anterolaterally, while the posterior duct is directed posterolaterally, and the two ducts form a laterally directed angle of approximately 90° between themselves. At this angle they, like the bony canals, join to form a common crus that leads into the utricle; from this common crus the anterior duct curves upward, while the posterior duct curves backward and then downward, so that the major portion of the anterior duct comes to lie largely above the level of the posterior one.

At their other ends, that is, at the anterior end of the anterior duct and the inferior end of the posterior one, are the membranous ampullae which contain the sensory endings of the ducts. The ampullary ends of the anterior and posterior semicircular ducts empty by short, wide continuations of the ducts into the utricle. The lateral semicircular duct sits only approximately, like the canal, in the horizontal plane, both of its ends being also connected with the utricle, and the anterior end being the ampulla. The semicircular ducts, much smaller than the osseous canals, are situated eccentrically in the perilymphatic space, so that they lie against the outer or convex surface of the lining periosteum of the canals, to which they are united by denser connective tissue than the trabeculae that elsewhere bridge the perilymphatic spaces of the bony canals.

Utricle

The utricle or utriculus is an elongated portion of the membranous labyrinth, larger in diameter than the semicircular ducts and receiving both ends of each semicircular duct (five in all, since the anterior and posterior ducts have a common opening). It and the saccule, another large part of the membranous labyrinth, lie in close association in the larger chamber of the perilymphatic space, the vestibule. On the inferior surface of the utricle and extending also slightly onto its lateral surface lies the *macula,* which is provided with sensory endings; from the utricle arises usually a utriculosaccular duct, which communicates with the endolymphatic duct and also connects the utriculus to the sacculus.

Saccule

The saccule is more rounded than the utricle and is situated somewhat anteromedial to the upper (anterior) end of the latter. It is continuous with the utricle through the "utriculosaccular duct" and with the cochlear duct (membranous cochlea) by the tiny *ductus reuniens* (canalis reuniens of Hensen). The macula of the saccule is an oval thickening on its lateral wall, containing the sensory nerve endings of this portion of the internal ear.

Utriculosaccular Duct

According to modern descriptions of the endolymphatic duct (see the following section), the so-called utriculosaccular duct consists of two ducts, a utricular and a saccular one; instead of meeting at an angle to form a single duct and giving off a correspondingly slender endolymphatic duct, each joins the dilated base of the endolymphatic duct. "Utriculosacular" is therefore a misnomer, since the two parts are connected only through the endolymphatic duct. Occasionally (14.1%, according to Bast, '37) the utricular duct is very short or lacking, so that the endolymphatic duct seems to arise from the utricle rather than being joined by both saccular and utricular ducts.

Endolymphatic Duct and Sac

The endolymphatic duct has traditionally been described as a slender duct extending medially through the petrous bone in the periosteum-lined channel of the vestibular aqueduct and, after emerging from the vestibular aqueduct through an aperture situated posterolateral to the opening of the internal acoustic meatus, expanding to form the endolymphatic sac. Anson and Wilson ('36) and Anson and Nesselrod, however, made reconstructions of the endolymphatic duct and sac of children and adults and found that the duct differs considerably from this. According to their description the first part of the endolymphatic duct is not a narrow duct but a sinuslike dilatation lying medial to the utriculus, with the narrow utricular duct joining it laterally and the saccular duct joining it anteriorly; this dilatation is said to attain sometimes a size almost equal to that of the utricle, but to be more elongated. Its wall was mostly smooth, but occasionally the distal portion was found to be plicated.

Following this dilated portion they found a constricted section (isthmus) and then a second sinuslike dilatation. This second dilatation they described as having extensive folds of epithelium containing cores of vascular connective tissue; beyond this there was another narrowing followed by a terminal dilatation.

Anson and Nesselrod called these dilatations (in order) Sinus I, Sinus II, and Sinus III; they concluded that Sinus II, with its concentration of vessels, was the site of greatest physiological activity of the endolymphatic duct and sac.

This description sounds somewhat complicated; the important part seems to be that the peripheral dilatation of the endolymphatic duct—the *endolymphatic sac*—may be divided histologically into two portions, a proximal (Sinus II), apparently physiologically active portion, and a more distal (Sinus III), less active or inactive portion.

Surgical approach to the endolymphatic sac, to provide a shunt for the relief of Ménière's disease, has been pioneered particularly by House (*e.g.,* '65). In the dissecting room, the endolymphatic sac can be found relatively easily in the dura on the posterior surface of the petrous temporal, about halfway between the internal acoustic meatus and the lateral wall of the skull. In the dried skull the external opening of the vestibular aqueduct, through which the endolymphatic sac protrudes, is usually slitlike and covered by an overlying ledge of bone (see Fig. 1-42). Variations in its position and in the size and position of that portion of the sac lying in the posterior cranial fossa are common. Shea and colleagues reported a number of measurements useful in locating it; they found the size of the entire sac to be very constant.

Cochlear Duct

The membranous cochlea (cochlear duct or scala media) is, as already mentioned, united to the sacculus by the ductus reuniens; it has a short blind end, the vestibular cecum, before it is joined by the ductus reuniens, and then spirals forward within the osseous cochlea to end blindly (as the cupular cecum) below the apex of this. Hardy ('38) measured the spiral organ of Corti (therefore essentially the membranous cochlea) from 68 cochleas of persons aged 10 weeks–85 years and found its average total length to be 31.52 mm, with a range of 25.26–34.45 mm. Of this length, the basal turn contributed an average of 57.9%, the middle one 29.4%, and the apical 12.8%. She found the number of turns to vary between 2½ and 2¾ in 59 ears, or 87% of the total; of these, nine had exactly 2½ turns. Eight had more than 2¾ but less than three, and one had only 2⅙ turns.

The cochlear duct is roughly triangular in cross section, the apex of the triangle being sharply pointed and attached to the osseous spiral lamina from the modiolus (Fig. 3-31); from this attachment the duct extends across the bony labyrinth, separating the scala vestibuli from the scala tympani except at the apex of the cochlea.

Thickened periosteum at the free edge of the osseous spiral lamina forms the limbus (limbus laminae spiralis ossei), which divides into a vestibular and a tympanic lip separated by a groove, the internal spiral sulcus. The fibrous *basilar membrane* (*lamina basilaris,* NA) stretches from the tympanic lip of the limbus to the projecting *basilar crest* of the spiral ligament; with the spiral organ (of Corti) it forms the *spiral membrane,* which is the "floor" or tympanic wall of the cochlear duct. The *spiral ligament* is a thickened, modified portion of the periosteum (endosteum) of the bony cochlea, which forms the outer, curved wall of the cochlea duct and adjacent parts of the scalae. The very thin *vestibular membrane* (vestibular wall, Reissner's membrane) forms the roof of the cochlear duct, separating its

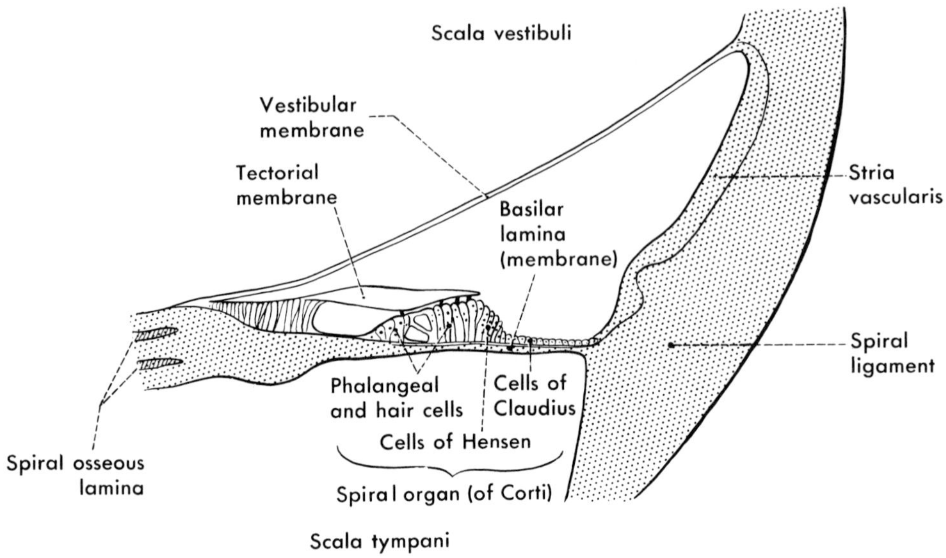

Fig. 3-31. Cross section of the cochlear duct.

cavity from that of the scala vestibuli, and passing from the upper surface of the limbus (the tectorial membrane arises from the vestibular lip of the limbus) to the upper part of the spiral ligament.

The *spiral organ* (of Corti), the sensory organ of hearing, lies upon the basilar membrane and consists of a thickened specialized epithelium; the tectorial membrane, a fibrogelatinous membrane with its base attached to the periosteum of the spiral osseous lamina, projects laterally to lie "upon" (really anterior to) the epithelial hair cells of the spiral organ, to which it is firmly attached by the stiff hairlike processes, or phalanges, of the supporting or phalangeal cells.

In the outer wall of the cochlear duct, and therefore on the inner surface of the spiral ligament, is the vascular area known as the *stria vascularis;* the stria is supplied by some 30 to 35 small arteries that originate in the modiolar region of the scala vestibuli and pass outward to the lateral wall of the osseous labyrinth (Belemer), and it is generally believed that it is the source of the endolymph (Guild, '27). Altmann and Waltner produced evidence that tissue spaces of the spiral ligament also serve as a site of absorption.

FINER STRUCTURE OF THE MEMBRANOUS LABYRINTH

Any tenable theory of the mechanism of vestibular function or of hearing must take into account the known structure of the parts concerned; therefore, a brief review of some of these facts and the ways in which they have been interpreted seems warranted. The following incomplete account takes into consideration recent observations and interpretations of the anatomy of the ear that have especially influenced physiologic concepts.

The membranous labyrinth, as already described, consists of epithelium-lined channels surrounded by connective tissue and largely suspended in the perilymphatic cavity. However, nerves and vessels must reach the membranous labyrinth, and the points of entrance are chiefly through areas of adhesion between the connective tissue of the membranous labyrinth and that of the periosteum lining the perilymphatic cavity. Thus the semicircular ducts are situated eccentrically in the perilymphatic space and are attached along their greater curvatures by tissue stronger than the trabeculae that elsewhere unite the membranous ducts and the bony canals. The attachments are especially strong at the ampullae, where the nerves enter. Similarly, the utriculus and sacculus are firmly attached to the floor and medial wall of the vestibule, through which their nerves reach them. Finally, the cochlear duct is attached to the spiral bony lamina, through which it receives its nerves, and its peripheral border is firmly fused with the periosteum of the osseous labyrinth.

Sensory Organs

The three basic divisions of the membranous labyrinth—the semicircular ducts, the utriculus and sacculus, and the cochlear duct—have an essentially similar structure: the sensory organ of each of these parts consists of a specialized thickened epithelium that is surmounted by and has attached to it a gelatinous mass. The specialized epithelium consists of at least two kinds of cells: supporting cells and hair cells or neuroepithelium; the latter, which have about them the endings of the eighth nerve, are provided with processes ("hairs") that project from the free edge of the cells and are embedded in or closely adjacent or attached to the gelatinous mass.

The advent of the scanning electron microscope has allowed more critical analysis of the hairs, or cilia. The typical neuroepithelial cell has numerous rows of hairs of varying length, called stereocilia, and a single kinetocilium or kinocilium, which is presumably moved toward or away from the cell, acting as a plunger on the cell membrane, by bending of the stereocilia. The exception to this seems to be the mammalian spiral organ, in which the kinetocilium is represented by a rudimentary structure called the basal body (Kimura; Smith, '68; and others).

The sensory endings in the ampullas of the ducts are the *cristae.* Each crista consists of thickened epithelium surmounted by a gelatinous *cupula.* The hairs of the neuroepithelial hair cells project into the base of the cupula. As shown first by Steinhausen and confirmed by Dohlman, the cupula so fills the ampulla that actual flow of fluid past it is impossible; because the cupula is gelatinous, however, it may be bent by the pressure of the endolymph, and this bending apparently stimulates the hair cells and therefore the nerve endings of the cristae. The physiology of the semicircular ducts is discussed in a following section.

The sensory areas of the utriculus and sacculus, the *maculae,* are identical in structure with each other and fundamentally similar also to the cristae of the semicircular canals. The epthelium consists of supporting and hair cells, and the hair cells have stiff, nonmotile projections that are embedded in a gelatinous statoconic or otolithic membrane. In the statoconic membrane are also embedded numerous crystalline bodies containing calcium carbonate (the statoconia, otoconia, or otoliths). It is usually stated that these lie close to the free surface of the membrane, but Ulrich has apparently shown that in the living ear (of the pike) there is an extensive gelatinous membrane covering them. It is only in the occurrence of statoconia that the maculae differ sharply from the other sensory areas of the ear.

A further detail of the cilia of the maculae seems to be that in both maculae there is a line of division, so that on each side of that line the kinetocilia face in different directions: in the utricle, the kinetocilia are on the side of the cell toward the line; in contrast, they are on the side away from the line in the saccule (*e.g.,* Lim, '71).

The sensory part of the cochlear duct, the *spiral organ* or organ of Corti, is, of course, arranged in a spiral like the duct itself and is supported by the spiral basilar membrane (see Fig. 3-31). It, also, consists of supporting cells and hair cells. The supporting cells are of several different types, but all contain stiffening fibrils within their cytoplasm, and their free edges expand to form a reticular (cuticular) membrane against which the gelatinous tectorial membrane rests. Possibly among the most important of the supporting cells are the phalangeal cells, arranged in two groups: the inner group is a single row of cells, while the outer (outer phalangeal cells, cells of Deiters) is said to form from three to five rows, depending upon the level of the cochlea investigated (fewer rows toward the base, more toward the apex). The inner row of phalangeal cells is associated with a single row of hair cells, while the outer rows of phalangeal cells alternate with rows of hair cells—each phalangeal cell apparently being closely associated with a hair cell. The phalangeal cells get their names from the shape of the stiff processes that project from the cells and contribute to the reticular (cuticular) membrane that covers the free surface of the organ. Mygind ('48) suggested that deformation of the phalanges by pressure of the tectorial membrane aids in deforming the hair cells and thus in initiating nerve impulses in response to hearing.

Peripheral to the outer phalangeal cells—that is, toward the attachment to the spiral ligament—are some tall cells (cells of Hensen) and still more peripherally shorter cells (of Claudius); their functions are unknown, but Mygind suggested that the cells of Hensen act as an elastic pillow that keeps the tectorial membrane firmly applied to the spiral organ.

Intercellular spaces among the cells of the spiral organ are apparently filled by intercellular substance, sometimes described as perilymph, sometimes believed to be different from both endolymph and perilymph. It is generally agreed that this substance cannot be endolymph, for the high potassium content of the latter is incompatible with the function of nerve fibers; the reticular (cuticular) membrane that covers the spiral organ keeps the endolymph from contact with all parts of the organ except the hairs that project through the membrane. The largest of these spaces runs the entire length of the spiral organ between the inner and outer rows of phalangeal and hair cells and is known as the tunnel, inner tunnel, or canal of Corti; this is bounded by special supporting cells, the inner and outer pillars (Corti's rods). The canal and its pillars together form Corti's arch. Mygind suggested that the spiral organ might make seesaw movements at Corti's arch, swinging on this somewhat as on a hinge.

The hair cells of the spiral organ have numerous (about 40–100 per cell) "hairs" projecting from their reticular (cuticular) surface. As discussed in connection with the tectorial membrane, there have been varying concepts concerning the relation of the hairs to that membrane, but the usual view has been that mechanical distortion of the hairs between the cells and the tectorial membrane is the trigger mechanism that activates the hair cells and their associated nerve endings. The inner hair cells are long, apparently less susceptible to damage than the outer cells, and also probably less sensitive to sound; the more sensitive outer cells, which are said to be responsible for the cochlear microphonics, are short and wedged between the apices of the phalangeal cells so that they do not reach the basilar membrane. (Discussion of many details of the structure and function of the spiral organ, with an extensive bibliography, can be found in the symposium edited by Rasmussen and Windle.)

Since the spiral organ is supported by the basilar membrane (basilar lamina) and in turn has the tectorial membrane attached to it, one or both of these would seem to be particularly important in the stimulation of this organ, and most of the theories of the mechanism of hearing have been based upon their supposed stucture. The most striking element in the *basilar membrane* is its fibers (basilar fibers, auditory strings) that pass from the bony spiral lamina to the basilar crest of the spiral ligament. The presence of some 24,000 of these fibers in the human ear, increasing in length from the basal to the apical coil, has been used as evidence for a resonator action on the part of the basilar membrane. Von Békésy's studies (summarized in '56, and recently in Rasmussen and Windle) have apparently completely discredited the idea of vibration of individual fibers, but have indicated that there is a continuous shift in the mechanical properties of the membrane from its stiff lower end to its soft upper end. According to his observations, the deformation of the basilar membrane by sound always begins at its lower end and travels upward toward the helicotrema (hence called a "traveling wave"); waves provoked by a certain frequency have a locus in the membrane at which they reach their maximum, and as the frequency is lowered the place of maximal displacement of the membrane advances toward the helicotrema.

The *tectorial membrane* is generally described as a gelatinous structure, but numerous fine fibers have been described in it also. Because of the fibers and because, like the basilar membrane, it increases in size from base to apex of the cochlea, the tectorial membrane has also been regarded as the vibrator that affects the hair cells. The nature of the attachment between the tectorial membrane and the hairs of the hair cells has been variously described; the hairs have been said to be continuous with the fibers in the membrane or imbedded in tiny canaliculi in the membrane. Kimura reported that only the tallest hairs of the outer hair cells are embedded, not deeply but firmly, in the tectorial membrane, and Lim ('72) agreed that none of the inner hair cells is attached.

COMPOSITION AND CIRCULATION OF THE ENDOLYMPH

Endolymph has a low protein content, usually reported to be about the same as that of cerebrospinal fluid, but differs in many respects from most body fluids. Its protein is entirely globulin instead of a mixture of globulin and albumin; its viscosity, said to resemble that of the vitreous body of the eye, is due to its high content of the mucopolysaccharide, hyaluronic acid (Vilstrup and Jensen). Even more striking is its content of sodium and potassium, similar to intracellular rather than extracellular fluid; thus endolymph contains more than 30 times as much potassium as does perilymph or cerebrospinal fluid but only about one tenth as much sodium (Smith and co-workers). Hinojosa and Rodriguez-Echandia suggested that certain epithelial cells of the stria vascularis produce this unusual ratio by reabsorption of material from the endolymph.

It is thus obvious from its composition that endolymph is not a filtrate from the bloodstream, and as early as 1908 Shambaugh described some of the epithelium of the stria vascularis in the cochlear duct as probably secretory in character. Following the work of Guild ('27) it is generally conceded that the chief source of endolymph is the stria vascularis; to this Dohlman (in Rasmussen and Windle) would add some of the epithelium of the ampullae of the semicircular ducts and of the maculae of the utricle and saccule, since he found radioactive sulfur rapidly excreted, apparently as a component of a mucopolysaccharide, in all these locations.

Guild described the endolymph as circulating downward toward the base of the cochlea, thence through the ductus reuniens into the sacculus, and thence into the endolymphatic duct and sac where it is reabsorbed; Dohlman agreed that, as indicated by his experiments with radioactive sulfur, the endolymphatic duct and sac are responsible for reabsorption of endolymph. Whether the endolymphatic sac is indispensable to reabsorption of endolymph and the maintenance of normal pressure in the endolymph seems to vary among species. It apparently is not in the monkey (Lindsay, '47) or the cat (Schuknecht and Kimura), but in the guinea pig its destruction leads to hydrops (Suh and Cody; Kimura and Schuknecht). The stria vascularis may, of course, be a locus for both origin and reabsorption of endolymph.

A valvelike folding of the wall at the junction of the utriculosaccular duct and the utricle, called the *utriculoendolymphatic valve,* has been described (*e.g.,* Bast, '28; Perlman and Lindsay, '36) and regarded as favoring the flow of endolymph into the utricle but impeding its outward flow; Wolff ('34b) found this fold in the late fetus to be usually small and probably not capable of functioning as a valve, but Bast ('34) described two ears in which rupture of the cochlear duct and sacculus had not produced collapse of the utricle and semicircular ducts, a circumstance he attributed to the endolymphatic valve. In view of the minute size of the utricular duct, and the evidence on viscosity of the endolymph, it seems improbable that, valve or no valve, there could be a sudden outpouring of utricular endolymph into a ruptured saccule.

Ménière's Disease

Subjective symptoms of dizziness and of tinnitus, which may eventually be followed by deafness, constitute a syndrome known as Ménière's disease. Its cause is not certainly known, but it is generally recognized that the chronic symptoms of Ménière's disease are accompanied by a dilation of the endolymphatic system. It has generally been believed, therefore, that the symptoms are due to excessive accumulation of endolymph, which exerts pressure upon the nerve endings; this disease is therefore also referred to as endolymphatic hydrops. The extent of the dilation of the endolymphatic duct system varies, but the cochlear duct is apparently always dilated and the sacculus and utriculus are usually affected, while the semicircular ducts remain free of dilation. Mygind ('66) suggested that edema of the hair cells, and perhaps also of the tectorial membrane, is responsible for the auditory symptoms.

Endolymphatic hydrops can result from decreased absorption, increased formation of endolymph, or both. Atkinson and many others have favored increased formation as the primary phenomenon, but it is also known that hydrops may be associated with a narrow or obstructed endolymphatic duct. With this in mind, Arenberg and colleagues studied biopsies of the endolymphatic sacs of six patients with hydrops and found in each loss of epithelium and perivascular fibrosis; they concluded that lack of absorption by the sac is one cause of hydrops.

Conjectures as to the possible mechanism of overproduction of endolymph have often focussed on changes in the vascular supply of the ear. Atkinson wrote that Ménière's disease is due to abnormalities in that; Williams ('47) suggested that autonomic dysfunction produces labyrinthine ischemia and increases capillary permeability; Choward took the opposite view, hypothesizing that parasympathetic fibers in the nervus intermedius may reach the ear through the vestibular nerve and through overactivity lead to abnormal vasodilation. Since successful medical treatment of the vertigo produced by hydrops often involves vasodilators, ischemia rather than vasodilation seems to be implicated. It is doubtful, however, if any marked changes in the ear's vascular supply can be brought about by autonomic action. Many years ago Rambo and co-workers failed to find any changes in the ear as a result of interruption of either the sympathetic or various possible parasympathetic pathways, and they questioned whether neural control over branches of the labyrinthine artery is any more effective than the weak effect on pial vessels.

More recent investigations indicate it is not; Clairmont and Jackson reported that sympathetic stimulation in dogs had no effect on blood flow in the temporal bone in most cases and never more than a small effect and that norepinephrine had no effect. Likewise, Suga and Snow's ('69a and b) pharmacologic experiments indicated that the sympathetic system can exert only a very weak influence over cochlear blood flow and that the parasympathetic probably exerts none.

Not only is the cause of endolymphatic hydrops unknown, but even the precise condition is in doubt; according to Alford and coworkers, it is not known whether the dilation is due to an increased volume of endolymph or increased osmotic pressure. Further, since perilymphatic fistulas may cause vertigo and hearing loss similar to that typical of Ménière's disease (Althaus and House), it may be that an apparent increased endolymphatic pressure is actually only relative, an imbalance between perilymphatic and endolymphatic pressures.

Since medical treatment to control the vertigo may not be successful and, in any event, does not stop the progress of the disease, a variety of surgical methods to at least control the vertigo have been developed. Older methods included partial or total labyrinthectomy and vestibular nerve section, the latter being the more common procedure for many years. W. F. House ('65) popularized drainage of the endolymphatic sac into the subarachnoid cavity; drainage into the mastoid cavity has also been used. A different approach to decompression is the tack operation of Cody and associates, designed to automatically and repetitively decompress the saccule.

Still another technique is ultrasonic irradiation of the vestibule, apparently best carried out through the round window (*e.g.,* Barnett and Kossoff; Tabb and co-workers). Irradiation and the various techniques of decompression are all designed to alleviate the vertigo and improve hearing; results have been thought both acceptable and unacceptable.

Maddox compared the results of simple decompression, endolymphatic–arachnoid shunt, and endolymphatic–mastoid shunt and found no difference; he noted that recurrence was frequently found if the patient was followed for a long time. Both he and Turner and colleagues, although reporting relief from vertigo in 55% and 60% of their patients, respectively, found little evidence that endolymphatic shunt can be expected to improve hearing significantly, and in 50% of Turner and colleagues' patients and 18% of Maddox's, hearing loss increased. In contrast, Arenberg and Spector reported that among those patients with a positive glycerol test, 59.3% had improved hearing and in 86 percent hearing was either improved or stabilized. Arenberg and Bayer suggested that the operation be done early in the course of the disease in order to preserve hearing (Maddox regarded the likelihood of recurrence as indicating it should not be done early), but said that if a negative glycerol test indicates that hearing loss is irreversible, vestibular nerve section is the best procedure. J. J. Shea and co-workers also found nerve section preferable when considerable hearing loss has occurred, but said that a subarachnoid–endolymphatic shunt prevents further hearing loss in patients who have no more than a slight loss and also tends to lessen or eliminate vertigo. House and Fraysse advised revision of the shunt if operation had relieved the symptoms for 3 months or more; they stated that revision controlled vertigo over a long term in 55% of patients and improved hearing in 26%.

Some workers do not favor endolymphatic drainage at all. Bryan and Bucy regarded all treatment of the endolymphatic system as inadequate for the control of vertigo and found vestibular nerve section preferable. Schuknecht and Alford and co-workers agreed; they challenged both the rationale and the effectiveness of drainage procedures and regarded vestibular nerve section as allowing preservation of any useful hearing present.

Elies and Plester found that patients with basilar impression may have symptoms similar to those of Ménière's disease. They used roentgenograms to establish a differential diagnosis.

VESTIBULAR FUNCTION

The sensory areas of the vestibular apparatus consist, as already described, of the cristae in the ampullae of the semicircular canals and the maculae of the utriculus and of the sacculus.

Semicircular Ducts

It seems to be generally granted that the semicircular ducts are stimulated by movement and not by position—that is, that they are organs responding only to acceleration and deceleration. In consequence, it is customary to speak of "flow" in a canal, the original concept being that movement of the head, carrying the semicircular ducts with it, would at first produce a streaming of the endolymph in the opposite direction because of the inertia of the endolymph; stimulation would then cease as the inertia was overcome, but with sudden cessation of movement it would begin again as the momentum of the endolymph continued to carry it through the duct and ampulla in the direction of turning. Although this simple concept is no longer accepted, it is still customary, for convenience, to speak of flow in a canal and to designate its direction as toward or away from the ampulla (ampullopetal or ampullofugal).

The concept that active flow of endolymph through the ampulla is necessary to stimulate the nerve endings was challenged by Morgan and co-workers, who pointed out that normal responses from the ducts may be obtained when circulation in them is prevented by sectioning, ligating, or plugging them; both Steinhausen and Dohlman showed by actual observation in animals that the gelatinous cupula completely occludes the ampulla, so that no fluid can pass through it. In consequence, the stimulus is now regarded as being a bending of the cupula in response to pressure. The common concept has been that this bending results in a push–pull or a shearing action on the hairs of the hair cells that are embedded in the cupula. However, it has been pointed out (Rasmussen and Windle) that since bending of the cupula through an arc of less than ½ ° may be an adequate stimulus, any movement of the hairs must be extremely minute; Dohlman (in Rasmussen and Windle), adopting the concept that the fundamental phenomena involve changes in electric potential of the hair cells, rejected both the idea that such tiny mechanical distortion of the hairs could produce a change in potential of the cells and the idea that the hair cells themselves are a source of the potential. He suggested instead that static electric forces are built up on the hairs by negatively charged ions in the endolymph that accumulate in tiny canaliculi surrounding the hairs and that distortion of the cupula produces discharge from the hair cells not by movement of the hairs, but by movement of fluid from the canaliculi. This ingenious concept is negated for the cochlea by the later finding that only the taller hairs of the outer hair cells are attached to the tectorial membrane and that the inner hair cells are not at all attached.

Mygind ('65, '66) suggested that the stimulus for the entire ear—ampullae, maculae, and spiral organ—is fundamentally the same and consists of an increased push on one part of the sense organ with therefore a pull or decreased push on the other part. He further suggested that the electrical discharge might be due to bending of large molecules of hyaluronic acid located between the apices of the cells.

"Ewald's second law" says that ampullopetal flow in the lateral canal, and ampullofugal flow in the other two, produces maximal stimulation of each. Whether this is strictly true in man is still being debated, but Szentagothai, moving the various cupulas by microinjection in dogs and cats and recording the activity of the ocular muscles, could produce contraction only through ampullopetal flow in the lateral canal and ampullofugal in the other two. Dohlman reported (in Rasmussen and Windle) that bending the cupula toward the utricle in the lateral canal and away from it in the other two decreased the electric potential in the ampulla, while bending in the opposite direction increased it.

It is generally accepted that in man movements of endolymph in either direction is an effective stimulus for eye movements, this being the theory on which caloric tests are based; in the lateral canal, ampullopetal flow is said to produce nystagmus with the slow phase to the opposite side, ampullofugal flow nystagmus with the slow phase to the same side (*e.g.,* Favill, Hyndman). Fluur stimulated the ampullary nerves of cats, recording from the ocular muscles and observing eye movements. He produced conjugate movements only when there was spontaneous activity in the muscles at the time of nerve stimulation, but the movements were those expected of maximal stimulation—to the opposite side for the lateral duct, upward for the anterior duct, downward for the posterior duct. (When he stimulated the left anterior and posterior ampullary nerves simultaneously, he produced conjugate counterclockwise rotation.)

Although Szentagothai apparently showed that in animals each duct controls primarily two muscles, and according to Favill's analysis each duct controls two muscles when flow is in one direction, two others (opposing muscles) when flow is in the opposite direction, it should be sufficient for clinical purposes to remember that the *slow* phase of vestibular-induced nystagmus is in the direction of flow; nystagmus is, of course, reported in terms of its fast phase, therefore would be opposite the direction of flow. Thus, forward flow in the right lateral duct and backward flow in the left one (thus circling to the left) should produce a slow phase to the left; forward (and upward) flow in one anterior duct and its parallel contralateral posterior duct should produce an upward slow phase; backward (and downward) flow in a posterior duct and its contralateral anterior duct should produce a downward slow phase. These are eseentially the movements analyzed by Favill for man and produced by Fluur in the cat. Further, from Fluur's observations, simultaneous stimulation of the anterior and posterior ducts of one side should produce a rotational nystagmus with slow phase to the opposite side—counterclockwise for the left labyrinth, clockwise for the right.

Utriculus and Sacculus

The maculae of the utricle and saccule have been regarded as responding only to position, not to acceleration or movement *per se;* that is, they have been regarded as organs of static equilibrium controlling posture and righting reflexes. It has also been said that they are stimulated not by pressure but by the pull of the statoconia (otoliths). It has also been argued that there is no proof that the sacculus (the hearing organ of fishes) has an equilibratory function in mammals. Newer evidence (*e.g.,* Bos and co-workers) has forced reconsideration of many of these views. Thus, linear acceleration has been shown to stimulate one or both maculae in rabbits, producing eye movements; and by analyzing the relation between posture and the effect of gravity on the statoconia in animals with one labyrinth destroyed, Bos and his co-workers produced evidence that the saccule shares a vestibular function with the utricle and that both push and pull by the statoconia are effective, with push being the more effective. In linear acceleration, however, the statoconia produce their greatest effect when they are neither above nor below the hair cells, but are hanging sideways, so that acceleration produces the greatest shearing, or push–pull, action. These observations seem to accord well with Mygind's ('65) concept that the fundamental stimulus to all parts of the ear is a combined push–pull one, and accord also with the later observations made with the scanning electron microscope.

COCHLEAR FUNCTION

Although there is disagreement on the meaning of some of the phenomena described, much progress has been made toward understanding the mechanisms involved in hearing. The subject can be reviewed only briefly here; discussions and references to many aspects of the problem are presented in Rasmussen and Windle and by Davis ('57), von Békésy ('56), and Mygind ('65, '66).

Movements of Perilymph

As already discussed in connection with the middle ear, movements of the perilymph are usually set up by movements of the stapes; thus inward movement of the stapes should displace perilymph from the vestibule upward in the scala vestibuli, where it can both exert pressure on the endolymph through the vestibular membrane and, traveling upward to the helicotrema and downward in the scala tympani, exert direct pressure on the basilar membrane and finally upon the round window membrane. That such a relatively massive displacement of the perilymph can occur seems obvious from observations that the round window membrane bulges outward when the stapes is pushed inward, but Pohlman long ago expressed doubt that it is necessary to normal hearing. Von Békésy's report (in Rasmussen and Windle) that movement of the basilar membrane, which presumably reflects movement of the perilymph, is confined to a basal portion of the cochlea when high tones are used as a stimulus and reaches the apex only when low tones are used, would seem to indicate that the route through the helicotrema is used only when low tones are producing the perilymphatic displacement; and although the movement of the basilar membrane varies with the intensity of the sound, being greater for louder ones, it may be almost infinitesimally small: it has been calculated that at the threshold for hearing the movement may be little more than 1% of the magnitude of a hydrogen atom. It is thus obvious that "mass" displacement of perilymph is not necessary to hearing. This makes more understandable the newer concept of the round window (see that section) as an adjuster for increased perilymphatic pressure, however generated, rather than as an integral part of the auditory mechanism.

Since all response of the basilar membrane

to sound, whatever the route of delivery to the ear, begins at the base of the cochlea and moves toward the cupula, it would seem that the traveling wave in the basilar membrane produced by stapedial movement is not provoked directly by the movement of the perilymph (which would first reach the basilar membrane at the helicotrema), but by pressure transmitted through the endolymph; since the tectorial membrane lies in the endolymph, it is necessarily subjected to the same changes in pressure that move the basilar membrane.

Mechanism of Stimulation

As reviewed by Davis ('57 and in Rasmussen and Windle), studies of electric potentials within the cochlea seem to indicate that it functions somewhat as a battery, releasing more or less electrical energy; sound waves increase the flow of electric current which then excites the nerve fibers ending around the hair cells. Because the cells supporting the hair cells rest on the basilar membrane, while the hairs have traditionally been described as attached to the tectorial membrane, the usual assumption has been that vibrations of the perilymph produced by sound are in turn translated into vibrations of the basilar membrane or the tectorial membrane and that the resulting mechanical distortion of the hairs, variously described as a push–pull or a shearing action that bends them, controls the release of the electric current.

Fig. 3-32. The basilar lamina or membrane. Note the steady increase in width from the base of the cochlea almost to the apex, and then its decrease. (Wever EG: Ann Otol Rhinol Laryngol 47:37, 1938)

The basilar membrane has long been regarded as the most likely vibrator and has been studied intensively, particularly by von Békésy. It may be recalled that although the bony cochlea gets smaller as it is traced toward the cupula, there is a disproportionate decrease in the width of the osseous spiral lamina, and the basilar membrane therefore increases in width between basal and apical turns (Fig. 3-32). Wever's measurements of 25 basilar membranes indicate an average minimal width of about 80 μm in the basal turn, an average maximal width of about 498 μm about ½ turn before the cupula is reached, and a decrease in width thereafter. Except for the final decrease, the meaning of which is not understood, this change in width is consonant with the concept that while high tones produce maximal movement of the basal or narrower portion of the basilar membrane, decreasing the pitch moves the place of maximal displacement of the membrane toward the cupula.

Various investigators have believed that the physical characteristics of the basilar membrane make it unlikely that it vibrates in response to sound and suggested that the tectorial membrane, which also increases in size from base to cupula, is the more likely mechanism for stimulating the hair cells. Von Békésy's studies of the basilar membrane indicate, however, that sound does set up vibration, in the form of traveling waves, in the basilar membrane, and his observations on the waves fit in well with current concepts of the localization of tones in the cochlea (see the following section). The most commonly accepted concept seems to be that it is the movement of the basilar membrane that distorts the hair cells and thus controls the stimulation of the nerve fibers. Davis (in Rasmussen and Windle) admitted the tectorial membrane as a vibrator also, remarking that since the tectorial membrane and the basilar membrane have different axes of rotation (because of the difference in their "hinges" of attachment) they slide past each other as they swing up and down, thus bending the hairs to and fro crosswise in the cochlea.

Dohlman (in Rasmussen and Windle) took sharp exception to the view that vibration of the basilar membrane with mechanical distortion of the hairs has anything to do with hearing. He found such a concept incompatible with the reception of sound in the saccule of fishes, in the cristae of the semicircular ducts after fenestration, and in the cochlea of birds (in which, he said, no shearing movement between basilar and tectorial membranes is possible) and also incompatible with the slight movement of the membrane (a fraction of the width of the hydrogen atom?) necessary to stimulate the hair cells. He therefore regarded the movements of the basilar membrane as merely allowing transmission of pressure between the two scalae and suggested that the fundamental mechanism is deformation of the tectorial membrane by the vibration of the endolymph.

Mygind ('65) likewise argued that the vibrations of the basilar membrane can have nothing to do with hearing; he agreed with Dohlman that it is deformation of the tectorial membrane that is responsible for stimulating the hair cells and suggested that this deformation results in bending of molecules of hyaluronic acid located between the apices of the cells of the spiral organ.

Obviously, if the inner hair cells have no attachment to the tectorial membrane, they must be stimulated in some other manner. Dallos and co-workers reported a different mechanism for these cells; from a study of cochlear microphonics in normal ears and in ears in which the outer hair cells of the basal turn had been destroyed, they concluded that although the response of the outer hair cells is proportional to the displacement of the basilar membrane, that of the inner cells is proportional to the velocity of the membrane—that is, the hairs here are moved by the viscous endolymph.

Localization in Cochlea

Concepts of localization of sound in the cochlea have been based largely on the role of the basilar membrane as the vibrator. Thus, while it has long been accepted that recognition of pitch depends in part upon the temporal sequence of stimulation of nerve fibers, it was held also that sounds of different frequency produced vibrations of different parts of the basilar membrane; this concept was later modified when it was realized that low tones produced widespread vibration of the basilar membrane, to the view that high tones are localized in the base of the cochlea, whereas there was probably no true localization for low tones. Von Békésy's observations of the behavior of the basilar membrane in response to sound, and his experiments with models, have allowed him ('56) to explain localization in a manner that fits most of the observations concerning it. Below a certain critical frequency, which varies with the animal species, the basilar membrane apparently vibrates as a whole, and pitch is discriminated purely by the temporal sequence of stimulation, but above this critical frequency traveling waves are produced in the membrane; the place of localization in the cochlea then depends upon the level at which these waves reach their maximum, which varies according to the frequency of the tone—toward the base of the cochlea for high tones, farther toward the cupula as the tone is lowered. Thus, high tones are confined to the base of the cochlea, since the traveling waves they set up do not extend beyond this; low tones, on the other hand, although their waves also begin at the base, extend upward toward the cupula and have their maximal effect there. "Place" is therefore more important for high tones than for low tones. (Mygind, '65, thought von Békésy's observations had nothing to do with hearing. While advocating deformation of the tectorial membrane as the fundamental stimulus, he suggested that the modiolus with its spiral bony lamina may be the vibrating and differentiating part of the cochlea.)

Several contributors to Rasmussen and Windle have offered corroboratory evidence that this is the type of localization, irrespective of whether the "traveling wave" of the basilar membrane is responsible for it or not. Davis reported that electrodes in the last turn of the cochlea detect a cochlear microphonic response to low tones only, while those in the basal turn detect a response to all tones, high or low. Similarly, Tasaki reported that the auditory fibers arising in the basal turn respond to tones of both high and low frequency, while those from the apical turn respond to low frequencies only. Finally, Schuknecht indicated that the louder the tone (therefore the greater the mechanical displacement of the vibratory mechanism), the greater the area of cochlea involved in hearing low tones. Cats with injury confined to the apical part of the cochlea, he reported, can still hear low tones, although there is a 40 decibel loss in acuity (such a lesion produced, as expected, no effect on high tones). He calculated that from the place of maximal stimulation in the apical turn for a given tone, each decibel increase in intensity spread the area of response to that tone 0.2 mm farther down the cochlea.

ANOMALIES

Polvogt and Crowe, in an investigation of the internal ears of persons who had been shown before death to have normal hearing, found among 17 cochleas one that had only two turns and another that had three full turns; they concluded that the number of turns of the cochlea is apparently of little importance provided the length of the cochlear duct is within normal limits.

Among these 17 apparently normally functioning cochleas, they also found several marked anomalies. In six there was a gap in the bony septum between the middle and apical turns, making a free communication between the scala vestibuli and the scala tympani and forming what is termed a scala communis cochleae; they said that there should be in this condition, theoretically, an impairment for low tones but no such impairment had been found. Two other ears showed arrested bony development at the apex of the modiolus, so that the nerve in the apical turn was supported purely by a fibrous band; although the lesion was unilateral in both, hearing had been equally acute on the two sides. In another patient there was a defect on one side in the bony wall of the middle turn of the cochlea, and some of the fibers of the facial nerve passed through a portion of the spiral ligament. Polvogt has even noted atrophy of the nerve in a portion of the basal coil of the cochlea, with no apparent effect upon the hearing; he stated that it is common to observe atrophy of the stria vascularis over short distances of any level of the cochlea, with no apparent effect upon hearing.

VESSELS AND NERVES

Arteries

The chief artery of the internal ear is the labyrinthine or internal auditory artery. This usually arises from the anterior inferior cerebellar artery (in Sunderland's 264 cases, from that in 83% and from the basilar in 17%; in Mazzonni's ['69] 100 cases, from the anterior inferior cerebellar in 80, from an accessory anterior inferior cerebellar in 17, and from the posterior inferior cerebellar in three). Several workers have noted that the artery or arteries (Mazzoni found more than one in 49 of his 100 cases, although not every one reached the internal ear) usually arise from an arterial loop which on its way to the cerebellum passes close to the porus or even for a variable distance into the internal acoustic meatus. Mazzoni found no origin other than from a loop or a loop collateral; in 60% it was closely related to the porus, and in 40% it extended well into the canal. The loop into the meatus has variable relations to the nerves there.

Although the labyrinthine artery supplies the dura and bone of the canal and its contained nerves, and a second vessel may do only this, its main branches run in the lining periosteum of the bony labyrinth and through the trabeculae between the membranous and bony labyrinth. There are apparently no anastomoses of any size in the ear between the vessels supplying the membranous labyrinth and those to the otic capsule itself; however, Mazzoni ('72) pointed out that the labyrinthine artery's branches to the bone around the internal meatus anastomose

Fig. 3-33. The arterial supply to the membranous labyrinth. (After Shambaugh GE: Am J Anat 32:189, 1923)

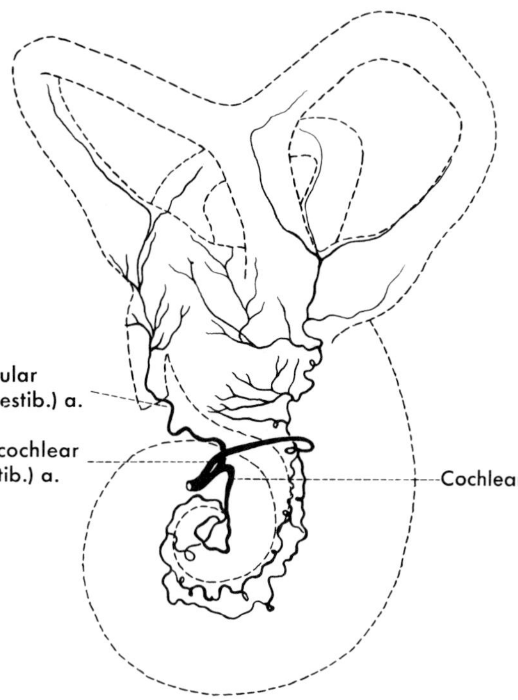

with other vessels supplying this bone and that even within the internal ear there are some anastomoses between branches of the labyrinthine artery to the bony septa between the cochlear coils and other vessels that supply the bony labyrinth.

The labyrinthine artery varies in its branches and their distribution. Most commonly, apparently, the first branch is to the utricle, perhaps a part of the saccule (Anson and Bast) and the anterior ends of the lateral and anterior semicircular ducts. This branch, variably called the anterior vestibular or the vestibular artery, may arise separately from the anterior inferior cerebellar artery (for instance, Blunt). Shambaugh ('23; Fig. 3-33) described and figured it as arising with the cochlear artery. The other two branches of the labyrinthine artery are the vestibulocochlear (also called posterior vestibular) and the cochlear artery (cochlear artery proper). The vestibulocochlear artery is distributed to most of the saccule, including the macula, to the posterior semicircular duct, posterior parts of the other ducts, part of the utricle (Anson and Bast), and part or all of the basal turn of the cochlea; the cochlear artery is then distributed to the remaining part of the cochlear duct. The common stem of these two arteries is usually called the common cochlear artery, but the terminology is confusing: Mazzoni ('69) quoted a study in which the labyrinthine artery was said to divide into anterior vestibular, common cochlear, and vestibulocochlear branches in two-thirds of cases and into anterior vestibular and vestibulocochlear in the remainder. There may be no stem common to the vestibulocochlear and cochlear arteries; Shambaugh described the vestibulocochlear as the first branch of the labyrinthine, and Hilger noted that the cochlear artery may be the first branch. Typically, the branch of the vestibulocochlear artery to the basal coil anastomoses with the cochlear artery, but it may fail to do so; if the cochlear artery is small, the vestibulocochlear may be the chief blood supply to all the cochlea.

Veins

Vestibular veins and a spiral vein of the modiolus give rise to the veins draining the labyrinth. Shambaugh described only two such

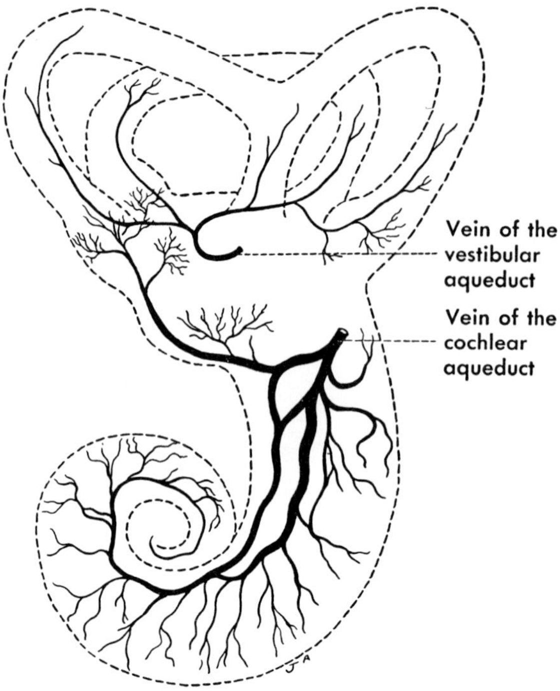

Fig. 3-34. The venous drainage of the membranous labyrinth. In this example, what is usually described as the labyrinthine vein (from the apical and middle turns of the cochlea) joins the vein of the cochlear aqueduct. (After Shambaugh GE: Am J Anat 32:189, 1923)

veins, one running by way of the cochlear canaliculus or aqueduct and the other by way of the vestibular aqueduct; he denied that a labyrinthine or internal auditory vein is usually present in the human ear. He stated that the vein of the cochlear aqueduct drains the entire cochlea and part of the vestibular apparatus, including the maculae of the sacculus and utriculus and parts of the anterior and lateral ducts, the vein of the vestibular aqueduct then draining only the remainder of the anterior and lateral ducts and the entire posterior one (Fig. 3-34). Anson and Bast described three veins: the labyrinthine or internal auditory vein, draining the apical and middle turns of the cochlea; the vein of the cochlear aqueduct, draining the basal turn of the cochlea, the saccule, and part of the utricle; and the vein of the vestibular aqueduct, draining the semicircular ducts and the major part of the utricle.

The labyrinthine vein is described as traversing the internal acoustic meatus and joining the inferior petrosal sinus; the vein of the cochlear aqueduct as traversing a separate canal parallel to the cochlear aqueduct and opening into the superior bulb of the internal jugular vein or into the inferior petrosal sinus; and the vein of the vestibular aqueduct as traversing that aqueduct (or paralleling it—Ogura and Clemis) and joining the sigmoid sinus (or sometimes as entering the inferior petrosal sinus).

Lymphatics

Lymphatics connected with the internal ear are not usually described.

Nerves

The nerve to the internal ear is the vestibulocochlear (statoacoustic, acoustic, auditory) or eighth cranial nerve, consisting functionally of vestibular and cochlear divisions (Fig. 3-35). In the internal meatus these two parts are closely associated, but at the lateral end of the meatus the trunk divides into three parts, two vestibular and one cochlear.

The fibers of the *vestibular nerve* arise from the vestibular ganglion (Scarpa's ganglion), which is situated at the lateral end of the internal acoustic meatus; centrally, the fibers become associated with the cochlear division of the nerve, from which they are separable grossly with difficulty or not at all; at the brain stem, however, they lie anterior (ventral) to the cochlear fibers. Peripherally, the vestibular nerve divides into two main divisions, a superior and an inferior, each bearing a part of the vestibular ganglion. The superior part of the ganglion gives nerves to the ampullae of the lateral and anterior canals and to the maculae of both the utriculus and the sacculus. The inferior part of the ganglion gives rise to a posterior ampullary nerve and a nerve to the sacculus; it also gives a communicating branch (called nerve of Oort; also vestibulocochlear nerve before the entire eighth nerve was so named) to the cochlear division of the eighth nerve. Hardy ('34) described the macula of the saccule as being innervated not only by both divisions of the vestibular nerve,

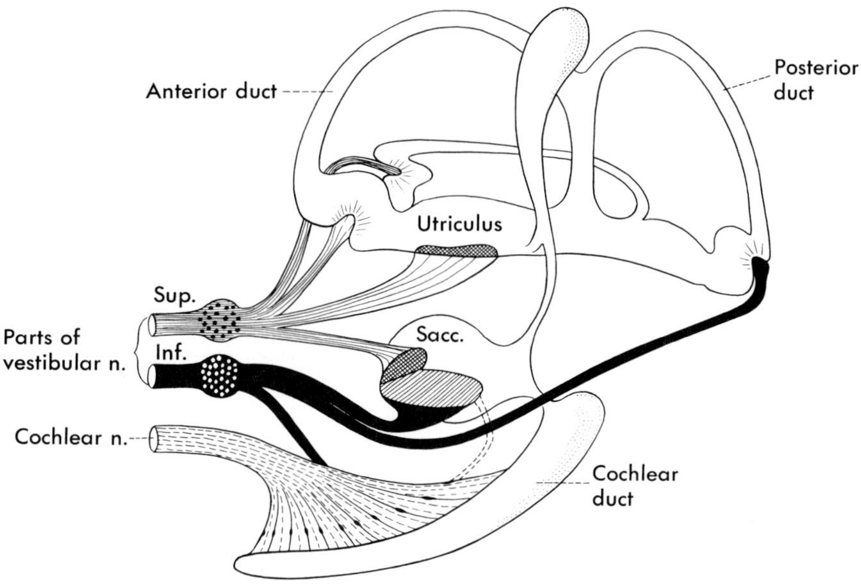

Fig. 3-35. Schema of the innervation of the labyrinth. The vestibulocochlear nerve of Oort, composed of olivocochlear fibers, is the branch between vestibular and cochlear nerves. The saccular nerve is represented by *broken lines.* (After Hardy M: Anat Rec 59:403, 1934, after de Burlet)

but also by the cochlear nerve through a branch (saccular nerve) passing from the base of the cochlea to the sacculus.

The cells of origin of the cochlear nerve form the *spiral ganglion,* situated in the coils of the modiolus at the base of the attachment of the spiral bony lamina; with the exception of the saccular nerve, the peripheral fibers pass through the lamina to perforate its free edge and end in the spiral organ of Corti.

Within the internal acoustic meatus the eighth nerve is usually connected by two filaments to the facial; it has been thought that through such communications the facial nerve may distribute fibers to the internal ear, although the communications more probably represent simply a short aberrant course of facial fibers with the eighth.

Rasmussen ('46, and in Rasmussen and Windle) described apparent efferent fibers arising from the superior olive, making their exit from the brain stem with the vestibular part of the nerve and joining the cochlear nerve by way of the vestibulocochlear communicating branch or nerve of Oort. Once tentatively identified as being autonomic fibers, it now seems clear that they have no postganglionic cell bodies associated with them and probably end in the spiral organ (of Corti); Kimura and Wersall cut some of them and found degenerating endings on the outer hair cells. Gacek (in Rasmussen and Windle) described similar fibers arising in vestibular nuclei and apparently distributed to the sacculus, utriculus, and all three ampullae. Sohmer has apparently shown that stimulation of the fibers to the cochlea depresses the cochlear potential; apparently the fibers from the vestibular nuclei act in a similar manner as a "feedback" to suppress the activity of the vestibular nerve, and Sala suggested they may also be excitatory, their effect varying with the functional state of the organ.

Speculation concerning *autonomic fibers* to the ear has centered largely around their possible role in the production of hydrops. Terayana and co-workers have apparently demonstrated postganglionic fibers from the superior cervical ganglion on some of the branches of the labyrinthine artery, but found none in the stria vascularis, where an autonomic supply might be expected to have the greatest effect on function. As already noted, no evidence of a parasympathetic supply to the ear has been found, and the observations of Clairmont and Jackson and of Suga and Snow indicate, at most, a very weak effect of the sympathetic supply to the ear. As one would anticipate, the labyrinthine artery thus seems to be similar to other cerebral vessels, a concept reinforced by Schneider's ('74b) report that CO_2 seems to be the most effective agent in increasing cochlear flow.

References

ALDRED P, HALLPIKE CS, LEDOUX A: Observations on the osmotic pressure of the endolymph. J. Physiol 98:446, 1940

ALFORD BR, COHN AM, IGARASHI M: Current status of surgical decompression and drainage procedures upon the endolymphatic system. Ann Otol Rhinol Laryngol 86:683, 1977

ALLAM AF: Pneumatization of the temporal bone. Ann Otol Rhinol Laryngol 78:49, 1969

ALTHAUS SR, HOUSE HP: Long term results of perilymph fistula repair. Laryngoscope 83:1502, 1973

ALTMANN F, FOWLER EP JR: Histological findings in Ménière's symptom complex. Ann Otol Rhinol Laryngol 52:52, 1943

ALTMANN F, WALTNER JG: New investigations on the physiology of the labyrinthine fluids. Laryngoscope 60:727, 1950

ANSON BJ: Stapedial, capsular and labyrinthine anatomy in relation to otologic surgery. Ann Otol Rhinol Laryngol 70:607, 1961

ANSON BJ, BAST TH: The surgical significance of stapedial and labyrinthine anatomy. Arch Otolaryngol 71:188, 1960

ANSON BJ, CAULDWELL EW, BAST TH: The fissula ante fenestram of the human otic capsule. I. Developmental and normal adult structure. Ann Otol Rhinol Laryngol 56:957, 1947

ANSON BJ, DONALDSON JA, WARPEHA RL, WINCH TR: The vestibular and cochlear aqueducts: Their variational anatomy in the adult human ear. Laryngoscope 75:1203, 1965

ANSON BJ, HARPER DG, HANSON JR: Vascular anatomy of the auditory ossicles and petrous part of the temporal bone in man. Ann Otol Rhinol Laryngol 71:622, 1962

ANSON BJ, NESSELROD JP: Endolymphatic and associated ducts in man. Arch Otolaryngol 24:127, 1936

ANSON BJ, WILSON JG: The form and structure of the endolymphatic and associated ducts in the child. Anat Rec 65:485, 1936

ANSON BJ, WARPEHA RL, RENSINK MJ: The gross and microscopic anatomy of the labyrinth. Ann Otol Rhinol Laryngol 77:583, 1968

ANSON BJ, WINCH TR: Vascular channels in the auditory ossicles in man. Ann Otol Rhinol Laryngol 83:142, 1974

ARENBERG K, BAYER RF: Therapeutic options in Ménière's disease. Arch Otolaryngol 103:589, 1977

ARENBERG IK, MAROWITZ WF, SHAMBAUGH GE JR: The role of the endolymphatic sac in the pathogenesis of endolymphatic hydrops. Acta Otolaryngol (Suppl 275):7, 1970

ARENBERG IK, SPECTOR GJ: Endolymphatic sac surgery for hearing conservation in Meniere disease. Arch Otolaryngol 103:268, 1977

ATKINSON M: Ménière's syndrome. The basic fault? Arch Otolaryngol 44:385, 1946

BALLANCE C, DUEL AB: The operative treatment of facial palsy by the introduction of nerve grafts into the Fallopian canal and by other intratemporal methods. Arch Otolaryngol 15:1, 1932

BARNETT SB, KOSSOFF C: Round window ultrasonic treatment of Meniere disease. Arch Otolaryngol 103:124, 1977

BASEK M: Anomalies of the facial nerve in normal temporal bones. Ann Otol Rhinol Laryngol 71:382, 1962

BAST TH: The utriculo-endolymphatic valve. Anat Rec 40:61, 1928

BAST TH: Function of the utriculo-endolymphatic valve: Two cases of ruptured saccules in children. Arch Otolaryngol 19:537, 1934

BAST TH: The utriculo-endolymphatic valve and duct and its relation to the endolymphatic and saccular ducts in man and guinea pig. Anat Rec 68:75, 1937

BAST TH: Development of the otic capsule. V. Residual cartilages and defective ossification and their relations to otosclerotic foci. Arch Otolaryngol 32:771, 1940

BAST TH: Development of the otic capsule. VI. Histological changes and variations in the growing bony capsule of the vestibule and cochlea. Ann Otol Rhinol Laryngol 51:343, 1942

BAST TH: Development of aquaeductus cochleae and its contained periotic duct and cochlear vein in human embryos. Ann Otol Rhinol Laryngol 55:278, 1946

BAST TH, ANSON BJ: The Temporal Bone and the Ear. Springfield, Ill, Charles C Thomas, 1949

BECKER SW, BRUNSCHWIG A: Sinus preauricularis (fistula preauricularis congenita). Am J Surg ns 24:174, 1934

VON BÉKÉSY G: Current status of theories of hearing. Science 123:779, 1956

BELEMER JJ: The vessels of the stria vascularis: With special reference to their function. Arch Otolaryngol 23:93, 1936

BELLUCI RJ: Trends and profiles in stapes surgery. Ann Otol Rhinol Laryngol 88:708, 1979

BLUNT MJ: The blood supply of the facial nerve. J Anat 88:520, 1954

BOS JH, JONGKEES LBW, PHILIPSZOON AJ: On the action of linear acceleration upon the otoliths. Acta Otolaryngol 56:477, 1963

BOSMAN DH: The distribution of the chorda tympani in the middle ear in man and two other primates. J Anat 127:443, 1978

BRYAN V, BUCY PC: Vestibular nerve section for Meniere's disease. A followup and comparative study. Arch Otolaryngol 97:115, 1973

BUNNELL S: Surgical repair of the facial nerve. Arch Otolaryngol 25:235, 1937

CANTEKIN EL, DOYLE WJ, REICHERT TJ, PHILLIPS DC, BLUESTONE CD: Dilation of the eustachian tube by electrical stimulation of the mandibular nerve. Ann Otol Rhinol Laryngol 88:40, 1979

CHOWARD C–H: Acousticofacial anastomoses in Meniere's disease. Arch Otolaryngol 101:296, 1975

CLAIRMONT AA, JACKSON RT: Sympathetic stimulation and otic blood flow. Ann Otol Rhinol Laryngol 83:84, 1974

CODY DTR, SIMONTON KM, HALLBERG OE: Automatic repetitive decompression of the saccule in endolymphatic hydrops (tack operation). Laryngoscope 77:1480, 1967

COHEN L, FOX SL: Atresia of the external auditory canal. Arch Otolaryngol 38:338, 1943

CONGDON ED, ROWHANAVONGSE S, VARAMISARA P: Human congenital auricular and juxta-auricular fossae, sinuses and scars (including the so-called aural and auricular fistulae) and the bearing of their anatomy upon the theories of their genesis. Am J Anat 51:439, 1932

COSTELLO MJ, SHEPARD JH: Supernumerary external ears. Arch Otolaryngol 29:695, 1939

CRARY WG, WEXLER M, BERLINER K: The abnormally patent Eustachian tube. Alterations in psychological status resulting from polytef paste injections. Arch Otolaryngol 105:21, 1979

CRIFO S, CRIFO C: Free amino acid content of perilymph. Arch Otolaryngol 93:65, 1971

CROWE SJ, HUGHSON W, WITTING EG: Function of the tensor tympani muscle: An experimental study. Arch Otolaryngol 14:575, 1931

CUSHING H: The sensory distribution of the fifth cranial nerve. Bull Johns Hopkins Hosp 15:213, 1904

DALLOS P, BILLONE MC, DURRANT JD, WANG C–Y, RAYNOR S: Cochlear inner and outer hair cells: Functional differences. Science 177:356, 1972

DAVENPORT CB, MILLES BL, FRINK LB: The genetic factor in otosclerosis. Part I. Problem, method of study and results; Part III. General. Arch Otolaryngol 17:135:503, 1933

DAVIS H: Biophysics and physiology of the inner ear. Physiol Rev 37:1, 1957

DOHLMAN G: Investigations in the function of the semicircular canals. Acta Otolaryngol (Suppl 51):211, 1944

DRUSS JG: Incisions of drum membrane in otitis media. Am J Surg ns 36:102, 1937

DRUSS JG, ALLEN B: Congenital fistula of the neck communicating with the middle ear. Arch Otolaryngol 31:437, 1940

ELIES W, PLESTER D: Basilar impression. A differential diagnosis of Meniere's disease. Arch Otolaryngol 106:232, 1980

FARRIOR JB: The sublabyrinthine exenteration of the petrous apex. Ann Otol Rhinol Laryngol 51:1007, 1942

FARRIOR JB: The radical mastoidectomy: Anatomical considerations in surgical technique. Surg Gynecol Obstet 89:328, 1949

FAVILL J: An explanation of the mechanism of induced rotary and vertical nystagmus. Arch Neurol Psychiatr 13:479, 1925

FISCH U, ESSEN E: Total intratemporal exposure of the facial nerve. Pathologic findings in Bell's palsy. Arch Otolaryngol 95:335, 1972

FLUUR E: Influences of semicircular ducts on extraocular muscles. Acta Otolaryngol (Suppl 149), 1959

FOLEY JO: The cytological processes involved in the formation of the scalae of the internal ear. Anat Rec 49:1, 1931

GOODHILL V: Stapes Surgery for Otosclerosis. New York City, Paul B Hoeber, 1961

GRAVES GO, EDWARDS LF: The Eustachian tube: A review of its descriptive, miscroscopic, topographic and clinical anatomy. Arch Otolaryngol 39:359, 1944

GREER JA, CODY DTR, LAMBERT EH, WEALAND LH: Experimental facial paralysis: Influence of decompression. Ann Otol Rhinol Laryngol 83:582, 1974

GREISEN O: Aberrant course of facial nerve. Arch Otolaryngol 101:327, 1975

GUILD SR: The circulation of the endolymph. Am J Anat 39:57, 1927

GUILD SR: Histologic otosclerosos. Ann Otol Rhinol Laryngol 53:246, 1944

GUILD SR: The glomus jugulare, a nonchromaffin paraganglion in man. Ann Otol Rhinol Laryngol 62:1045, 1953

GUSSEN R: Pathogenesis of Bell's palsy. Retrograde epineural edema and postedematous fibrous compression neuropathy of the facial nerve. Ann Otol Rhinol Laryngol 86:549, 1977

HAGENS EW: Anatomy and pathology of the petrous bone: Based on a study of fifty temporal bones. Arch Otolaryngol 19:556, 1934

HAMBURGER GA, WERSÄLL J: Vascular supply of the tympanic membrane and the ossicular chain. Acta Otolaryngol (Suppl 188):308, 1964

HANSON JR, ANSON BJ, STRICKLAND EM: Branchial sources of the auditory ossicles in man. Arch Otolaryngol 76:200, 1962

HARDY M: Observations on the innervation of the macula sacculi in man. Anat Rec 59:403, 1934

HARDY, M: The length of the organ of Corti in man. Am J Anat 62:291, 1938

HILGER JA: Vasomotor labyrinthine ischemia. Ann Otol Rhinol Laryngol 59:1102, 1950

HINOJOSA R, RODRIGUEZ-ECHANDIA EL: The fine structure of the stria vascularis of the cat inner ear. Am J Anat 118:631, 1966

HIRSCH IJ: The Measurement of Hearing, pp 242–248. New York City, McGraw-Hill, 1952

HOUGH JVD: Partial stapedectomy. JAMA 187:697, 1964

HOUSE HP: Polyethylene in middle ear surgery. Arch Otolaryngol 71:926, 1960

HOUSE WF: Oval window and round window surgery in extensive otosclerosis: A preliminary report. Laryngoscope 69:693, 1959

HOUSE WF: Subarachnoid shunt for drainage of hydrops. A report of 146 cases. Laryngoscope 75:1547, 1965

HOUSE WF, FRAYSSE B: Revision of the endolymphatic subarachnoid shunt in Meniere's disease. Arch Otolaryngol 105:599, 1979

HUGHSON W: A summary of round window graft operations performed for deafness. Ann Otol Rhinol Laryngol 49:384, 1940

HUNT JR: The sensory field of the facial nerve: A further contribution to the symptomatology of the geniculate ganglion. Brain 38:418, 1915

HYNDMAN OR: Physiology of the vestibular labyrinth. Arch Otolaryngol 29:759, 1939

JEPPSSON P-H: Studies on the structure and innervation of taste buds. An experimental and clinical investigation. Acta Otolaryngol (Suppl 259):5, 1969

JONGKEES LBW: On peripheral facial nerve paralysis. Arch Otolaryngol 95:317, 1972

KAMERES DB: Electromyographic correlation of tensor tympani and tensor veli palatini muscles in man. Laryngoscope 88:651, 1978

KELEMEN G, LAFUENTA AD, OLIVARES FP: The cochlear aqueduct: Structural considerations. Laryngoscope 89:639, 1979

KETTEL K: Abnormal course of the facial nerve in the Fallopian canal. Arch Otolaryngol 44:406, 1946

KETTEL K: Bell's palsy: Pathology and surgery; a report concerning fifty patients who were operated on after the method of Ballance and Duel. Arch Otolaryngol 46:427, 1947

KIMURA RS: Hairs of the cochlear sensory cells and their attachment to the tectorial membrane. Acta Otolaryngol 61:55, 1966

KIMURA RS, SCHUKNECHT HF: Effect of fistulas on endolymphatic hydrops. Ann Otol Rhinol Laryngol 84:271, 1975

KIMURA R, WERSÄLL J: Termination of the olivocochlear bundle in relation to the outer hair cells of the organ of Corti in guinea pig. Acta Otolaryngol 55:11, 1962

KOBRAK H: Influence of the middle ear on labyrinthine pressure. Arch Otolaryngol 21:547, 1935

KOBRAK HG: Round window membrane of the cochlea: Experiments demonstrating its physical responses. Arch Otolaryngol 49:36, 1949

KOBRAK HG: The Middle Ear. Chicago, University of Chicago Press, 1959

KULLMAN GL, DYCK PJ, CODY DTR: Anatomy of the mastoid portion of the facial nerve. Arch Otolaryngol 93:29, 1971

LAWRENCE M: Recent investigations of sound conduction: Part I. The normal ear. Ann Otol Rhinol Laryngol 59:1020, 1950

LEDOUX A: Les liquides labyrinthiques. Acta Otorhinolaryngol Belg 4:216, 1950

LEMPERT J: Endaural, antauricular surgical approach to the temporal bone: Principles involved in this new approach; summary report of 1,780 cases. Arch Otolaryngol 27:555, 1938

LEMPERT J: Bone-dust-free Lempert fenestra novovalis. A new evolutionary development of the surgical treatment of clinical otosclerosis. Arch Otolaryngol 47:280, 1948

LEMPERT J, WOLFF D: Otosclerosis: Theory of its origin and development. Arch Otolaryngol 50:115, 1949

LEMPERT J, WEVER EG, LAWRENCE M, MELTZER PE: Perilymph: Its relation to the improvement of hearing which follows fenestration of the vestibular labyrinth in clinical otosclerosis. Arch Otolaryngol 50:377, 1949

LIDEN G, NORDLUND B, HAWKINS JE JR: Significance of the stapedius for understanding speech. Acta Otolaryngol (Suppl 188):275, 1964

LIM DJ: Vestibular sensory organs: A scanning electron microscopic observation. Arch Otolaryngol 94:69, 1971

LIM DJ: Fine morphology of the tectorial membrane. Its relationship to the organ of Corti. Arch Otolaryngol 96:199, 1972

LINDEMAN H: The Fallopian canal: An anatomical study of its distal part. Acta Otolaryngol (Suppl 158):204, 1960

LINDEMAN H: Some histological examinations of the incus and stapes with especial regard to their vascularization. Acta Otolaryngol (Suppl 188):319, 1964

LINDSAY JR: Osteomyelitis of the petrous pyramid of the temporal bone. Ann Surg 122:1060, 1945

LINDSAY JR: Histologic observations on the healing of labyrinthine fistulas in monkeys. Arch Otolaryngol 43:37, 1946

LINDSAY JR: Effect of obliteration of the endolymphatic sac and duct in the monkey. Arch Otolaryngol 45:1, 1947

LINDSAY JR, KOBRAK H, PERLMAN HB: Relation of the stapedius reflex to hearing sensation in man. Arch Otolaryngol 23:671, 1936

LINDSAY JR, SCHUKNECHT HF, NEFF WD, KIMURA RS: Obliteration of the endolymphatic sac and the cochlear aqueduct. Ann Otol Rhinol Laryngol 61:697, 1952

LUPIN AJ: The relationship of the tensor tympani and

tensor veli palatini muscles. Ann Otol Rhinol Laryngol 78:792, 1969

MADDOX HE III: Endolymphatic sac surgery. Laryngoscope 87:1676, 1977

MAHONEY T, VERNON J, MERKLE M: Function of the acoustic reflex in discrimination of intense speech. Arch Otolaryngol 105:119, 1979

MAYBAUM JL, GOLDMAN JL: Primary jugular bulb thrombosis: A study of twenty cases. Arch Otolaryngol 17:70, 1933

MAZZONI A: Internal auditory canal: Arterial relations at the porus acusticus. Ann Otol Rhinol Laryngol 78:797, 1969

MAZZONI A: Internal auditory supply to the petrous bone. Ann Otol Rhinol Laryngol 81:13, 1972

MELTZER PE: The mastoid cells: Their arrangement in relation to the sigmoid portion of the transverse sinus. Arch Otolaryngol 19:326, 1934

MEURMAN Y: Zur Anatomie des Aquaeductus Cochleae nebst einigen Bermerkungen über dessen Physiologie. Acta Soc Med Fenn Doudecim sB 13:1, 1930

MEYER JG, CRABTREE JA: The facial nerve coursing inferior to the oval window. Arch Otolaryngol 102:744, 1976

MEYERS PW: Jugular vein compression and elevation of perilymphatic fluid pressure. Arch Otolaryngol 98:314, 1973

MISURYA VK: Functional anatomy of the tensor palati and levator palati muscles. Arch Otolaryngol 102:265, 1976

MYERSON MC, RUBIN H, GILBERT JG: Anatomic studies of the petrous portion of the temporal bone. Arch Otolaryngol 20:195, 1934

MYGIND SH: Further labyrinthine studies: I. Affections of the humoral system of the labyrinth; II. On the labyrinthine transformation of the acoustic vibrations to pitch-differentiated nervous impulse. Acta Otolaryngol (Suppl 68), 1948

MYGIND SH: Functional mechanism of the labyrinthine epithelium. Arch Otolaryngol 82:452, 579, 1965; 83:3, 1966

NAGER GT, NAGER M: The arteries of the human middle ear, with particular regard to the blood supply of the auditory ossicles. Ann Otol Rhinol Laryngol 62:923, 1953

NATHANSON SA, JACKSON RT: Vidian nerve and the Eustachian tube. Ann Otol Rhinol Laryngol 85:83, 1976

OGURA Y, CLEMIS JD: A study of the gross anatomy of the human vestibular aqueduct. Ann Otol Rhinol Laryngol 80:813, 1971

PALVA T, DAMMERT K: Human cochlear aqueduct. Acta Otolaryngol (Suppl 246):5, 1969

PARKINSON D: Intracranial pheochromocytomas (active glomus jugulare). Case report. J Neurosurg 31:94, 1969

PASTORE PN, ERICH JB: Congenital preauricular cysts and fistulas. Arch Otolaryngol 36:120, 1942

PATTEE GL: An operation to improve hearing in cases of congenital atresia of the external auditory meatus. Arch Otolaryngol 45:568, 1947

PERLMAN HB, LINDSAY JR: The utriculo-endolymphatic valve. Arch Otolaryngol 24:68, 1936

POHLMAN AG: The reactions in the ear to sound. Ann Otol Rhinol Laryngol 50:363, 1941

POLVOGT LM: Histologic variations in the middle and inner ears of patients with normal hearing. Arch Otolaryngol 23:48, 1936

POLVOGT LM, CROWE SJ: Anomalies of the cochlea in patients with normal hearing. Ann Otol Rhinol Laryngol 46:579, 1937

PROCTOR B: Anatomy of the Eustachian tube. Arch Otolaryngol 97:2, 1973

PULEC JL: Abnormally patent Eustachian tubes: Treatment with injection of polytetrafluoroethylene (Teflon) paste. Laryngoscope 77:1543, 1967

RAMBO JHT, WOLFF D, FREEMAN G: A research study of the effect of the autonomic nervous system on the internal ear. Ann Otol Rhinol Laryngol 62:1149, 1953

RASK-ANDERSEN H, STAHLE J, WILBRAND H: Human cochlear aqueduct and its accessory canals. Ann Otol Rhinol Laryngol 86 (Suppl 42), 1977

RASMUSSEN GL: The olivary peduncle and other fiber projections of the superior olivary complex. J Comp Neurol 84:141, 1946

RASMUSSEN GL, WINDLE WF: Neural Mechanisms of the Auditory and Vestibular Systems. Springfield, Ill, Charles C Thomas, 1960

RICHARDS L: Congenital atresia of the external auditory meatus. Ann Otol Rhinol Laryngol 42:692, 1933

RITTER FN, LAWRENCE M: A histological and experimental study of cochlear aqueduct patency in the adult human. Laryngoscope 75:1224, 1965

ROOD SR, DOYLE WJ: Morphology of the tensor veli palatini, tensor tympani, and dilator tubae muscles. Ann Otol Rhinol Laryngol 87:202, 1978

ROSEN S: Palpation of the stapes for fixation. Arch Otolaryngol 56:610, 1952

ROSEN S: Principles of stapes surgery. Arch Otolaryngol 71:182, 1960a

ROSEN S: A new phenomenon in the physiology of hearing in otosclerosis surgery. Arch Otolaryngol 71:64, 1960b

ROSS MA: Functional anatomy of the tensor palati: Its relevance in cleft palate surgery. Arch Otolaryngol 93:1, 1971

SADE J, LEVY E, CHACO J: Surgery and pathology of Bell's palsy. Arch Otolaryngol 82:594, 1965

SAITO H, RUBY RFR, SCHUKNECHT HF: Course of the sensory component of the nervus intermedius in the temporal bone. Ann Otol Rhinol Laryngol 79:960, 1971

SAITO R, IGARASHI M, ALFORD BR, GUILFORD FR: Anatomical measurements of the sinus tympani: A study of horizontal serial sections of the human temporal bone. Arch Otolaryngol 94:418, 1971

SALA O: The efferent vestibular system. Acta Otolaryngol (Suppl 197):5, 1964

SANDO I, EGAMU T, HARADA T: Course and contents of the paravestibular canaliculus. Ann Otol Rhinol Laryngol 89:147, 1980

SCHERRER FW: Calcification and ossification of the external ears. Ann Otol Rhinol Laryngol 41:867, 1932

SCHNEIDER E–A: A contribution to the physiology of the perilymph. Part I: The origin of perilymph. Ann Otol Rhinol Laryngol 83:76, 1974a

SCHNEIDER E–A: A contribution to the physiology of the perilymph. Part II: Cochlear and cerebral blood flow. Ann Otol Rhinol Laryngol 83:247, 1974b

SCHNEIDER E–A: A contribution to the physiology of the perilymph. Part IV: Effect of histamine on the cochlear microcirculation. Ann Otol Rhinol Laryngol 84:228, 1975

SCHREINER L: Experimentelle Untersuchung uber die Bildungsstätten und den Stoffaustausch der Perilymph. Arch Otolaryngol (Suppl 212):5, 1966

SCHUKNECHT HF: Pathology of Ménière's disease as it relates to the sac and tack procedures. Ann Otol Rhinol Laryngol 86:677, 1977

SCHUKNECHT HF, KIMURA RS: Functional and histological findings after obliteration of the periotic duct and endolymphatic sac in sound-conditioned cats. Laryngoscope 63:1170, 1953

SHAMBAUGH GE: On the structure and function of the epithelium in the sulcus spiralis externus. Arch Otology 37:538, 1908

SHAMBAUGH GE: Blood stream in the labyrinth of the ear of dog and of man. Am J Anat 32:189, 1923

SHEA DA, CHOLE RA, PAPARELLA MM: The endolymphatic sac: Anatomical considerations. Laryngoscope 89:88, 1979

SHEA JJ JR: Fenestration of the oval window. Ann Otol Rhinol Laryngol 67:933, 1958

SHEA JJ, EMMETT JR, MOORE RE: The surgical treatment of Ménière's disease. Laryngoscope 89:1244, 1979

SILVERSTEIN H, DAVIS DG, GRIFFIN WL JR: Cochlear aqueduct obstruction changes in perilymph biochemistry. Ann Otol Rhinol Laryngol 78:532, 1969

SIMPSON GI III, KONRAD HR, TAKAHASHI M, HOUSE J: Immediate postembolization excision of glomus jugulare tumors. Advantages of new combined techniques. Arch Otolaryngol 105:639, 1979

SMITH A: Capillary areas of the membranous labyrinth. Ann Otol Rhinol Laryngol 63:435, 1954

SMITH CA: Electron microscopy of the inner ear. Ann Otol Rhinol Laryngol 77:629, 1968

SMITH CA, LOWRY OH, WU M-L: The electrolytes of the labyrinthine fluids. Laryngoscope 64:141, 1954

SOHMER H: The effect of contralateral olivocochlear bundle stimulation on the cochlear potentials evoked by acoustic stimuli of various frequencies and intensities. Acta Otolaryngol 60:59, 1965

STEINHAUSEN W: Neuere Untersuchungen zur Anatomie und Physiologie der Cupula terminalis in den Bogenausgangsampullen des inneren Ohres. Z Laryngol Rhinol Otol 26:29, 1935

STELTER GP, BAST TH, ANSON BJ: The development and adult anatomy of the air-cells in the petrous part of the temporal bone. Q Bull Northwestern Med School 34:23, 1960

STREETER GL: The factors involved in the excavation of the cavities in the cartilaginous capsule of the ear in the human embryo. Am J Anat 22:1, 1917a

STREETER GL: The development of the scala tympani, scala vestibuli and perioticular cistern in the human embryo. Am J Anat 21:299, 1917b

SUGA F, SNOW JB: Adrenergic control of cochlear blood flow. Ann Otol Rhinol Laryngol 78:358, 1969a

SUGA F, SNOW JB: Cholinergic control of cochlear blood flow. Ann Otol Rhinol Laryngol 78:1081, 1969b

SUH KW, CODY DTR: Obliteration of vestibular and cochlear aqueducts in the guinea pig. Laryngoscope 84:1352, 1974

SULLIVAN JA, HODGES WE: Effect of differential mobility of the windows of the cochlea on the mechanism of hearing. Arch Otolaryngol 49:63, 1949

SUNDERLAND S: The arterial relations of the internal auditory meatus. Brain 68:23, 1945

SZENTAGOTHAL J: The elementary vestibuloocular reflex arc. J Neurophysiol 13:395, 1950

TABB HG, NORRIS CH, HAGAN WE: Round window ultrasonic irradiation for Meniere's disease with ENG monitoring. Laryngoscope 88:1460, 1978

TERAYANA Y, YAMAMOTO K, SAKAMOTO T: Electron microscopic observations on the postganglionic sympathetic fibers in the guinea pig cochlea. Ann Otol Rhinol Laryngol 77:1152, 1968

TONNDORF J, TABOR JR: Closure of the cochlear windows: its effect upon air- and bone-conduction. Ann Otol Rhinol Laryngol 71:5, 1962

TREMBLE GE: Pneumatization of the temporal bone. Arch Otolaryngol 19:172, 1934

TSCHIASSNY K: The site of the facial nerve lesion in cases of Ramsay Hunt's syndrome. Ann Otol Rhinol Laryngol 55:152, 1946

TURNER JS, SAUNDERS AZ, PER–LEE JH: A 10-year profile of Meniere's disease and endolymphatic shunt surgery. Laryngoscope 83:1816, 1973

ULRICH H: Die Funktion der Otolithen, geprüft durch direkte mechanische Beeinglussung des Utriculusotolithen am lebenden Hecht. Arch ges Physiol 235:545, 1935

VIDIĆ B: The origin and the course of the communicating branch of the facial nerve to the lesser petrosal nerve in man. Anat Rec 162:511, 1968

VILSTRUP TH, JENSEN CE: Three reports on the chemical composition of the fluids of the labyrinth. Ann Otol Rhinol Laryngol 63:151, 1954

VINDIC B, YOUNG PA: Gross and microscopic observations on the geniculo-tympanic nerve in man (abstr). Anat Rec 154:436, 1966

WALLS EW: The laxator tympani muscle. J Anat 80:210, 1946

WALTNER JG: Barrier membrane of the cochlear aqueduct: Histologic studies on the patency of the cochlear aqueduct. Arch Otolaryngol 47:656, 1948

WEVER EG: The width of the basilar membrane in man. Ann Otol Rhinol Laryngol 47:37, 1938

WEVER EG, BRAY CW: The tensor tympani muscle and its relation to sound conduction. Ann Otol Rhinol Laryngol 46:947, 1937

WEVER EG, LAWRENCE M: The functions of the round window. Ann Otol Rhinol Laryngol 57:579, 1948

WEVER EG, LAWRENCE M, SMITH KR: The middle ear in sound conduction. Arch Otolaryngol 48:19, 1948a

WEVER EG, LAWRENCE M, SMITH KR: The effects of negative air pressure in the middle ear. Ann Otol Rhinol Laryngol 57:418, 1948b

WIGAND ME, TRILLISCH K: Surgical anatomy of the sinus epitympani. Ann Otol Rhinol Laryngol 82:378, 1973

WILLIAMS HL: Diagnosis and treatment of endolymphatic hydrops. (Ménière's disease). J Iowa State Med Soc 37:1, 1947

WILLIAMS HL: Latent or dormant disease in the pneumatic cell tracts of the temporal bone. Trans Am Acad Ophthalmol Otolaryngol 70:545, 1966

WILSON JG: The nerves and nerve endings in the membrana tympani of man. Am J Anat 11:101, 1911

WILSON JG, ANSON BJ: Form and structure of an area of otitic sclerosis in the temporal bone of an adult. Arch Otolaryngol 18:291, 1933

WITTMAACK K: The pathogenesis of otosclerosis. Arch Otolaryngol 14:186, 1931

WLODKA J: Studies on aqueduct patency. Ann Otol Rhinol Laryngol 87:22, 1978

WOLFERMAN A: The present status of therapy of Bell's palsy: A critical evaluation. Ann Otol Rhinol Laryngol 83 (Suppl 12), 1974

WOLFF D: Significant anatomic features of the auditory mechanism with special reference to the late fetus. Ann Otol Rhinol Laryngol 42:1136, 1933

WOLFF D: The microscopic anatomy of the Eustachian tube. Ann Otol Rhinol Laryngol 43:483, 1934a

WOLFF D: Significant anatomic features of the auditory mechanism with special reference to the late fetus. Ann Otol Rhinol Laryngol 43:193, 1934b

WOLFF D: Otosclerosis: Hypothesis of its origin and progress. Trans Am Acad Ophthalmol (Suppl), p. 11, 1950

WOOD–JONES F, I–CHUAN W: The development of the external ear. J Anat 68:525, 1934

Chapter 4
THE NOSE AND PARANASAL SINUSES

The External Nose

The external nose has a roughly pyramidal shape and for this reason is frequently referred to as the nasal pyramid. The upper attachment of the nose at the forehead is defined as the root, and the free lower angle as the apex or tip; the rounded ridge connecting these two parts is the dorsum nasi, the upper part of which, supported by the nasal bones, is the bridge (Fig. 4-1). The lateral surfaces of the nose are referred to simply as the sides, while the blendings of these sides with the tissues of the face proper form the margins of the nose (nasofacial angles). The sides are expanded below to form the rounded alae, uniting with the upper lip at the nasolabial sulci. The alae and the septum bound the two external openings (nares, nostrils, or anterior or external nares) at the base of the nasal pyramid.

Both the dorsum and the tip of the nose vary in shape as seen in profile; thus the former may be straight ("Grecian nose") or present a projecting angle ("Roman nose") or a depression ("pug nose"); a prominent tip is characteristic of the "Roman" and "pug" noses; a downwardly turned tip converts the "Roman" into the "Jewish" or "aquiline" nose. These variations in the shape of the nose are familial and individual; the width of the nose at the alae, as compared with the length of the nose (given as the nasal index, the greatest width times 100, divided by the greatest length) varies racially, however—white persons have the lower nasal index (longer, higher noses), black persons the higher index (short, broad noses), and Orientals are intermediate in this regard.

The muscles about the nose are described in connection with the other facial muscles (see Chap. 6).

SKELETON OF NASAL PYRAMID

The skeleton of the external nose is partly bony and partly cartilaginous and membranous. The bony parts to which the cartilaginous and membranous ones are attached surround the *piriform aperture,* the anterior opening of the nose in the prepared skull. Inferiorly, the aperture is bounded by the alveolar processes of the two maxillae; where they come together in the midline they project upward to form the anterior nasal spine, to which the mobile septum is attached. Lateral to the spine, particularly in Caucasians, is a sharp crest at the junction of the anterior and nasal surfaces of the bone; this may project several millimeters above the floor of the nasal cavity and present difficulty in elevating mucosa from that floor.

Each lateral boundary of the piriform aperture is a thin edge of bone provided below by the body of the maxilla and above by its frontal process. The aperture is completed by the

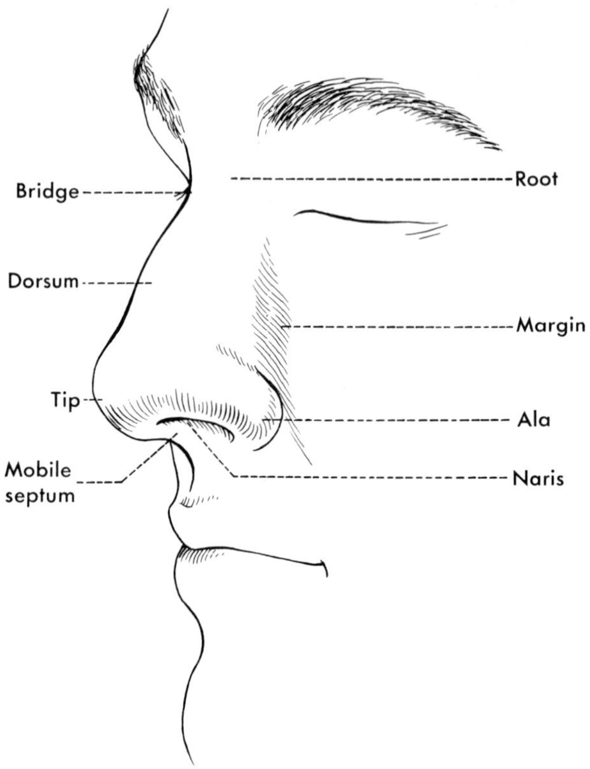

Fig. 4-1. External parts of the nose.

lower free edges of the two nasal bones. The boundaries of the piriform aperture therefore consist of only two bones on each side, the maxilla below and laterally and the nasal bone above.

Nasal Bones

The nasal bones are usually narrower and thicker above and wider and thinner below; they articulate firmly above with the nasal part of the frontal bone, which extends downward behind, deep to the nasal bones and the frontal processes of the maxillae to form a shelflike support to these, thus adding strength to the bridge of the nose. The lateral border of each nasal bone is beveled—in about the upper half of this border the bevel is on the inner surface of the bone, while in the lower part the bevel is on the outside. Thus, the lateral articulation is also particularly firm, since the upper part of this border rests upon the anterior surface of the frontal process of the maxilla, while the lower part is locked behind this process.

The medial articular surfaces of the nasal bones, where the two come together, are wider, for each projects downward and backward into the nasal cavity so that the two projections together form a crest, tapering from above downward; this crest contributes to the nasal septum and articulates with the spine of the nasal process of the frontal bone, the perpendicular plate of the ethmoid bone, and the septal cartilage of the nose. Even without the bracing action of the septum in the middle, the bridge of the nose possesses considerable structural strength because of its arched form and the strength of its articulations.

The lower edges of the nasal bones, where they help bound the piriform aperture, are beveled on their inner surfaces. This bevel accommodates the upper borders of the lateral nasal cartilages, which extend upward behind the nasal bones and are firmly bound to them by dense connective tissue. Care must be taken, in making the intercartilaginous incision of Jacques Joseph, not to separate the cartilages from the bones.

Exceptionally, the nasal bones may be fused in the midline, or they may even be completely absent, their place being taken by enlarged frontal processes of the maxilla. Minor variations in their shape and size are common. It should be remembered also, in examining roentgenograms of the face for possible fracture of the nasal bones, that the nasal bones may be multiple—that is, a bone may be naturally divided into two parts—and that accessory bones may occur at their lines of articulation.

Nasal Cartilages

The lower part of the nasal pyramid, largely cartilaginous, is usually described as being formed by five major cartilages and a variable number of smaller ones (Fig. 4-2). The larger cartilages are the two lateral (upper lateral or triangular) cartilages above, the greater or major alar (lower lateral or lobular) cartilages below, and the median unpaired septal (quadrangular) cartilage. In their upper portions, the lateral nasal cartilages are continuous with each other across the midline and also with the septal cartilage; therefore the lateral nasal and septal cartilages are frequently described together as a single cartilage, the nasoseptal cartilage.

The *lateral nasal cartilage* is somewhat triangular: its upper and posterior edge, the base of the triangle, is thin and articulates closely with the nasal and maxillary bones, extending upward deep to the former for several millimeters. Its lateral and inferior border, the hypotenuse of the triangle, likewise thin, is commonly curved laterally at a variable angle and overlapped by the upper border of the greater alar cartilage. Its medial border is thick, both above where it is continuous with the septal cartilage and below where it forms a free edge lateral to the midline of the nose. As in the case of the bridge of the nose, the part of the nasal pyramid formed by the lateral nasal cartilages possesses a certain intrinsic strength of its own, since the cartilages of the two sides are continuous with each other above and therefore actually form a single, highly curved piece of cartilage (Fig. 4-3); in addition, this arch is supported at its middle

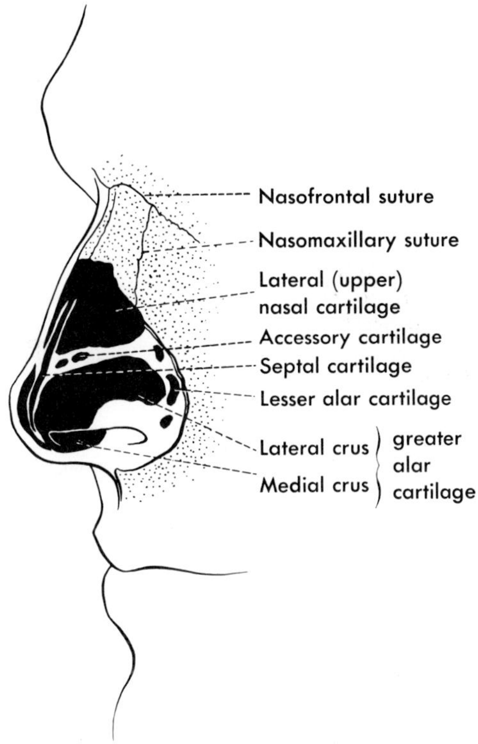

Fig. 4-2. Skeleton of the nose. More common clinical terminology is upper lateral nasal cartilage for the lateral nasal and lower lateral nasal, or lobular, for the greater alar cartilage. The greater alar cartilage has been drawn downward to show the accessory cartilages.

by its continuity with the septal cartilage. The *septal cartilage* is discussed with other elements of the nasal septum.

The *greater alar cartilages* are thin curved plates that bound and hold open the nasal apertures. They are thus responsible for the shape of the tip of the nose, and plastic surgery involving them has become relatively common. Some of the numerous variations of these cartilages have been discussed by Daley.

Each greater alar cartilage consists of a medial and a lateral crus. The lateral crus is the broader and extends from the tip into the ala of the nose; where it overlaps the lateral nasal cartilage it may be curved medially. Whereas Drumheller apparently regarded as constant an overlap between the lateral nasal and greater alar cartilages, with an outward curve

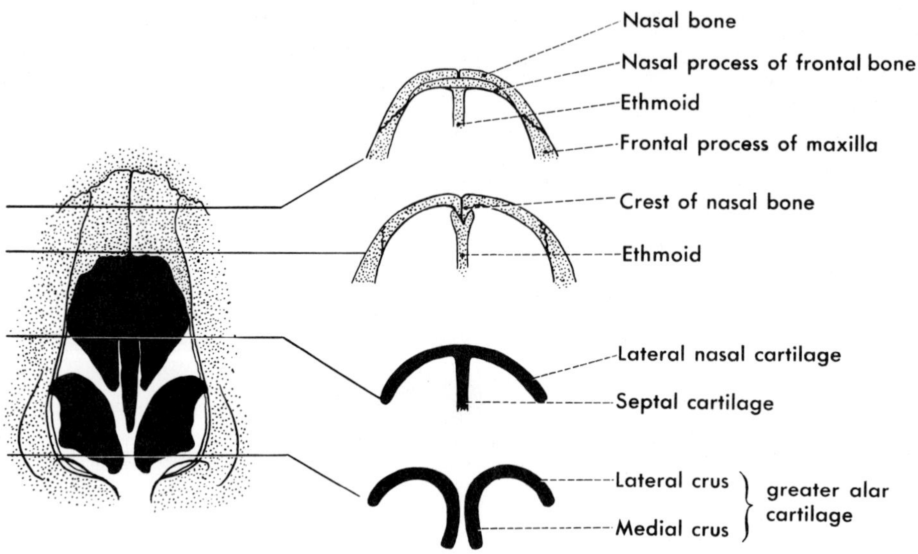

Fig. 4-3. Some structural features in the support of the nose.

of the former more or less interlocking with an inward curve of the latter, Dion and co-workers found these conditions in only 52% of 46 sides. In 17% there was no overlap; the remaining 31% had overlap but no interlocking, since 20% had a slight inward curve of the lower end of the lateral nasal cartilage and an outward curve of the greater alar cartilage, and in 11% neither cartilage curved.

Regardless of overlap and interlocking, the dense connective tissue that unites the greater alar cartilage to the lateral nasal cartilage and to the lower part of the septal cartilage is rather limited toward the midline, because the edges of these cartilages are usually close together here, but more laterally the fibrous membrane uniting lateral nasal and alar cartilages is broader and usually contains one or more small accessory (sesamoid) cartilages within it. The lateral crus varies in the extent to which it continues posterolaterally into the ala, but is in all persons united to the maxilla by a lateral continuation of the fibrous membrane; except when the lateral crus is unusually large and closely approaches the maxilla, this membrane forms the most posterolateral part of the ala and has embedded in it several minor (lesser) alar cartilages. Variations in the numbers and relative sizes of the cartilages of the nose are easily accounted for by their development—they form at first one continuous cartilaginous mass, in which separate elements are produced by disappearance of cartilage in certain areas and its replacement by connective tissue; it is this tissue that forms the fibrous membrane uniting the cartilages.

The medial crus of the greater alar cartilage, more slender than the lateral, is continuous with the lateral at the apex of the nose, the curve between the two forming the lateral portion of the apex; the medial crus runs downward and backward in the free edge of the nasal septum. The two medial crura are, at the very tip of the nose, separated from each other by a rather wide groove, which can easily be palpated. As they run into the septum, however, they become closely associated and together form the support of the mobile septum, which lies below or distal to the cartilage of the nasal septum. The medial crura are loosely united to each other and to the lower edge of the septal cartilage by connective tissue; the mobile septum can thus easily be pushed to one side, so that the firm lower edge of the septal cartilage can be felt. At their posterior free edges, the medial crura turn slightly outward again.

BLOOD SUPPLY

The arteries on the external surface of the nose vary, but typically include a lateral nasal from the angular (upper part of the facial, or external maxillary) artery and branches from the dorsal nasal artery, a branch of the ophthalmic (Fig. 4-4); this latter perforates the orbital septum above the medial palpebral ligament and runs downward on the side of the nose to anastomose, variably, with the nasal branch of the facial or the angular artery, giving off in its course a branch to the lacrimal sac. These vessels, which may vary inversely in size, are supplemented laterally by twigs from the infraorbital branch of the (internal) maxillary artery and inferiorly, over the nasal septum, by twigs from the arteries of the nasal septum (which largely, however, supply the interior of the nose and are a common source of nosebleed). At the junction between nasal bone and nasal cartilage, the small external nasal branch of the anterior ethmodial artery emerges to supply skin along the dorsum to the apex. The chief veins of the nose end in the angular and ophthalmic veins.

INNERVATION

The skin of the root, the bridge, and the upper part of the side of the nose is supplied by twigs from the supratrochlear and infratrochler branches of the ophthalmic nerve (Fig. 4-5). The skin on the side of approximately the lower half of the nose is supplied through the infraorbital branch of the maxillary nerve; the external nasal branch of the anterior ethmoidal nerve (from the nasociliary portion of the opthalmic) emerges between the nasal bone and the lateral nasal cartilage to supply skin over the dorsum of this part of thenose down to the tip.

The Nasal Cavity

The paired cavities of the nose, divided from each other by the nasal septum, are also known as the nasal fossae; in frontal section they are roughly triangular in shape, being narrow above and wider below. Their external openings are the nares (anterior or external nares), while their posterior or pharyngeal openings are the choanae (posterior or internal nares). The major *walls* of each nasal cavity consist of a medial wall, formed by the nasal septum, a lateral wall, and an inferior wall or floor. The medial and lateral walls are partly cartilaginous and membranous, partly bony, but the floor is entirely bony. The roof is formed by the cribriform plate, and is quite narrow—about 5 mm at its widest, or posterior, end.

The portion of the nasal cavity corresponding to the ala of the nose is the *vestibule;* it is lined with skin containing vibrissae, or hairs, and sweat and sebaceous glands. On the lateral nasal wall there is a ridge, the limen nasi (limen vestibuli), formed by the lower end of the lateral nasal cartilage; this separates the vestibule from the remainder of the nasal cavity and coincides with the area in which there is a gradual transition from the skin of the vestibule to the mucous membrane of the remainder of the cavity. There is no similar marking on the septum, but a few millimeters inside the naris there is a rather abrupt transition between skin and mucous membrane; because the mucosa is thicker and less easily torn at this mucocutaneous line, incisions for performing a submucous resection are usually started here.

Except in the vestibule, each nasal cavity is lined by *mucous membrane* that is tightly attached to the periosteum and perichondrium of its walls. The greater part of this mucous membrane is highly vascular, contains numerous mucous and serous glands, and is covered by pseudostratified columnar ciliated epithelium. This constitutes the respiratory mucosa (Schneiderian membrane). The mucosa over the medial surfaces of the middle and inferior conchae is particularly thick and contains in its lamina propria a plexus of large veins (plexus cavernosi concharum) which have been likened to erectile tissue because of the rapidity with which they can become engorged.

A limited upper part of the nasal cavity,

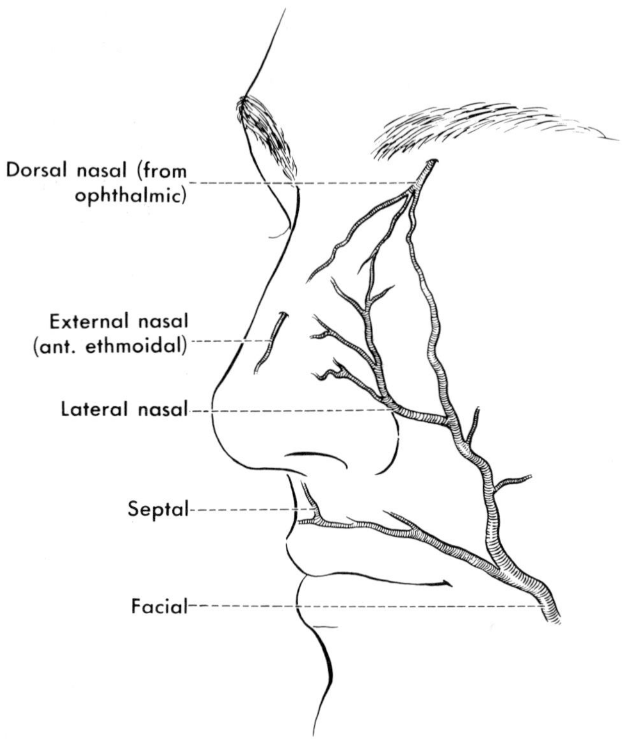

Fig. 4-4. Chief arteries about the external nose.

bounded largely by the superior nasal concha and the lateral wall above this, and by a corresponding part of the septum, is lined by the olfactory mucous membrane, which is less vascular (yellowish rather than pink), covered by a nonciliated epithelium, and contains the modified nerve-cell bodies that give rise to the olfactory nerve fibers; its glands are apparently serous. The mucosa of the paranasal sinuses is similar to respiratory mucosa, but less vascular, thinner, and more loosely attached to the bony walls.

The presence of the large inferior and middle conchae much increases the area of the respiratory mucosa, so that air drawn into the nose may be warmed and humidified before reaching the trachea. The marked vascularity of the respiratory mucosa, and its abundant glands, adapt it especially for this purpose, and in quiet respiration the air currents are primarily around the inferior and middle conchae, therefore against this mucosa. The thinner olfactory mucosa is adapted for the reception of odoriferous stimuli and is above

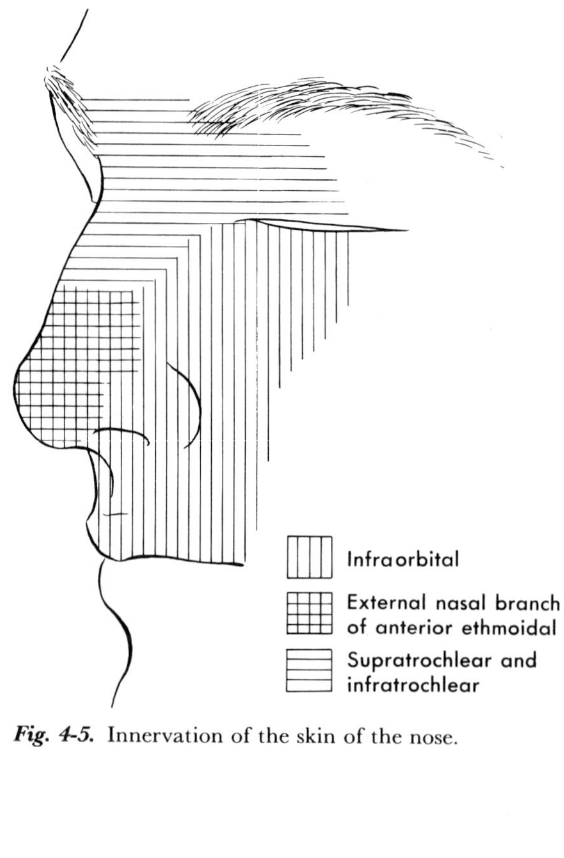

Fig. 4-5. Innervation of the skin of the nose.

the main stream of air flow—hence the necessity to sniff, and thus create eddies toward the roof of the nasal cavities, in attempting to recognize or appreciate odors.

Irritation of the nasal mucous membrane leads to reflex vasodilation of the vessels within it, and increased secretion by its glands, and may be followed by sneezing, also a reflex act. Since sneezing involves voluntary muscles, however, this (but not the turgescence and increased secretion of the mucous membrane) may often be voluntarily inhibited; it is also well known that pressure upon the mobile part of the nasal septum tends to inhibit sneezing.

Ciliary Activity

The ciliated epithelium of the respiratory mucosa and the paranasal sinuses as a whole has a beat so directed that mucus is propelled from smaller toward larger cavities and carried away from the margin of the olfactory epithelium; in general, therefore, the beat of the cilia is downward and backward toward the nasopharynx (Hilding, '32; Figs. 4-6 and 4-7).

Hilding divided the respiratory mucous membrane into anterior and posterior parts on the basis of its activity. He found that in the anterior third of the nose the cilia are relatively inactive, so that it requires an hour or more for droplets of ink placed here to enter the inferior and middle meatuses. However, in the posterior two thirds the highly active cilia may move the material into the nasopharynx in 3–10 minutes. Hilding noted that mucus is moved apparently not only by ciliary activity directly, but also by gravity and by the traction resulting either from ciliary movements or from the pharynx as in swallowing; the greatest rate of movement produced by these combined factors was in the inferior and middle meatuses.

Ciliary activity may be drastically affected by extremes of temperature and by various chemicals. Proetz found that temperatures of 40 ° C greatly retard and those of 43 ° or 44 ° completely inactivate ciliary movement; temperatures below 18 ° C were found also to decrease ciliary movement, and all movement ceased when the temperature was lowered to between 12 ° and 7 °; upon warming the mu-

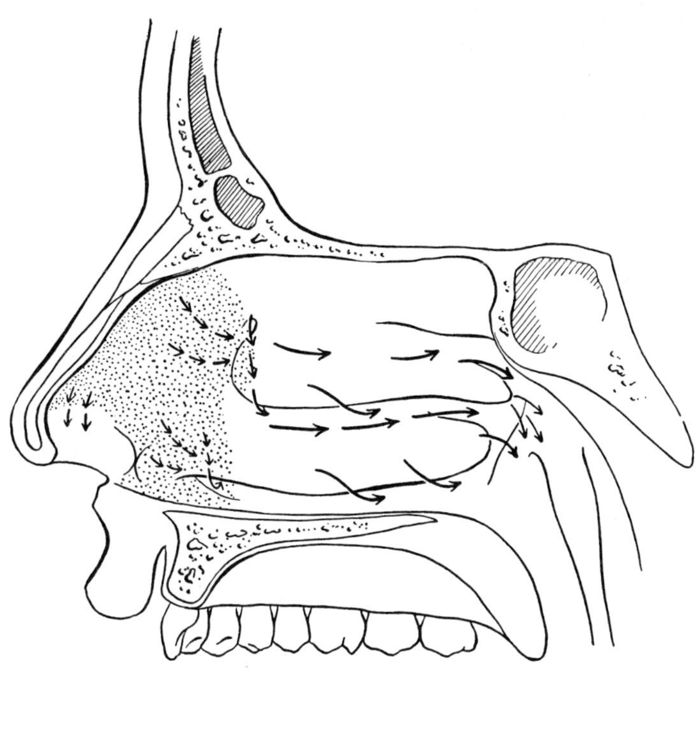

Fig. 4-6. Direction of flow of mucus (direction of ciliary beat) on the lateral wall of the nose. The area of less active ciliary movement is *stippled.* (After Hilding A: Arch Otolaryngol 15:92, 1932)

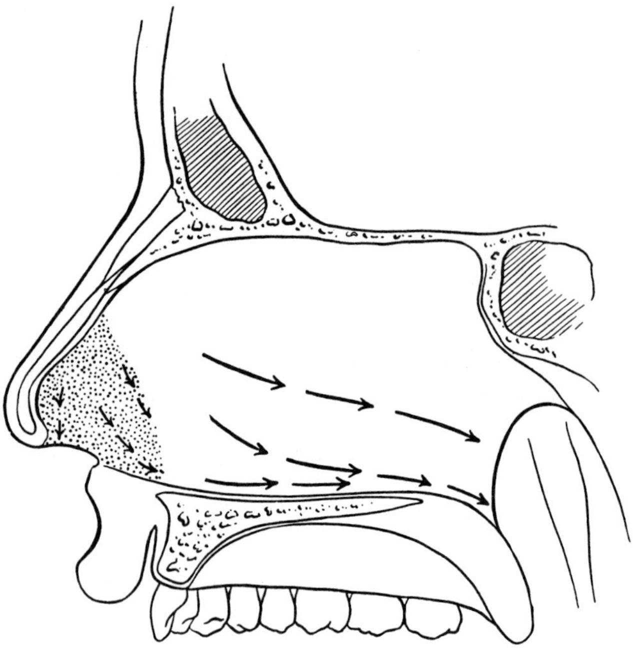

Fig. 4-7. Direction of flow of mucus (ciliary beat) on the nasal septum, the area of less active ciliary action being *stippled.* (After Hilding A: Arch Otolaryngol 15:92, 1932)

cous membrane again, however, the ciliary action was resumed and at 18 ° C was found to be much more rapid than at the beginning of the experiments, reaching as much as 10 or more beats per second. In contrast, the cilia exposed to excessive heat did not resume their function even after several hours at more nearly normal temperatures. This work, carried out both upon the rabbit and upon man, apparently indicates that while inspiration of cold air does not actually damage the cilia, nasal douches exceeding 40 ° C may produce severe damage. Drugs introduced into the nasal cavity for various purposes may also have a deleterious effect on ciliary action.

NASAL SEPTUM

The larger part of the nasal septum (Fig. 4-8) is formed by the perpendicular plate of the ethmoid bone posteriorly and the septal cartilage anteriorly; below, the vomer, the medial crura of the greater alar cartilages, the membrane connecting them to the septal cartilage, and the nasal crests of the maxillary and palatine bones complete the septum.

The *perpendicular lamina or plate of the ethmoid* forms the upper third or more of the nasal septum; it is continuous above with the lamina cribrosa or horizontal portion of the bone, thus dividing the nasal roof into right and left sides. It articulates anteriorly and above with the frontal and nasal bones, posteriorly with the crest of the sphenoid, posteroinferiorly with the vomer, and anteroinferiorly with the septal cartilage.

The extent of the perpendicular plate anteriorly and inferiorly, that is, the level of its junction with the septal cartilage, varies; its lowest point may be 1 cm or so above the lower free edge of the nasal bones, but it may extend to the edge of these bones or actually not articulate with them at all, but only with the nasal spine of the frontal bone (Schaeffer, '20). The septal cartilage varies inversely with the size of the perpendicular plate of the ethmoid.

The *vomer* completes the nasal septum posteriorly and inferiorly; it articulates above with the body of the sphenoid bone and with the perpendicular plate of the ethmoid and below with the nasal crests of the maxillary and palatine bones. Its anterior border is grooved for the reception of the septal cartilage.

The nasal spine of the frontal and the crests of the maxillary and palatine bones contribute little to the nasal septum; they serve as points of articulation for the other elements of the septum, and hence help to complete it about its edges.

Septal Cartilage

The septal (quadrangular) cartilage, continuous with the lateral nasal cartilages toward the bridge of the nose but separate from their diverging lower ends, forms the cartilaginous portion of the septum; it sends a thin posterior (sphenoidal) process backward between the vomer and ethmoid. Its articulations are peculiar anatomically, for cartilage does not usually articulate directly with bone, and important clinically because they allow movement. Posteriorly and above, the thin edge of the septal cartilage is received into a groove on the edge of the perpendicular plate of the ethmoid, and its inferior edge also fits into a similar groove on the vomer and the nasal crest (Fig. 4-9). There is some interchange of connective tissue fibers between perichondrium and periosteum, but most of the fibers of each pass around the edge of the cartilage or bone to become continuous with the perichondrium or periosteum of the opposite side. Thus, in elevating the mucoperiosteum for exposure of the nasal septum, both perichondrium and periosteum must be torn or cut at the borders of the cartilage. The looser connective tissue between the cartilage and bone, in which there may be fat (Aymard), allows for mobility of the cartilaginous septum; the joint permits the base of the cartilage to rotate as the cartilage is flexed, thus minimizing the danger of fracture of it from pressure on the dorsum of the nose and permitting marked deviation of the septum without dislocation.

The tongue-and-groove articulation of the septal cartilage with the vomer and maxilla tends to be distorted by some flaring of the lips of the groove, and this is particularly true of the nasal crest for a variable distance behind the anterior nasal spine. Here the lips of the groove tend to be almost horizontal, so that they form laterally projecting edges (called the "premaxillary wings") at the upper border of the nasal crest, and the septal cartilage also tends to expand laterally (Fig. 4-10). As a result, the groove may be almost

Fig. 4-8. Composition of the nasal septum.

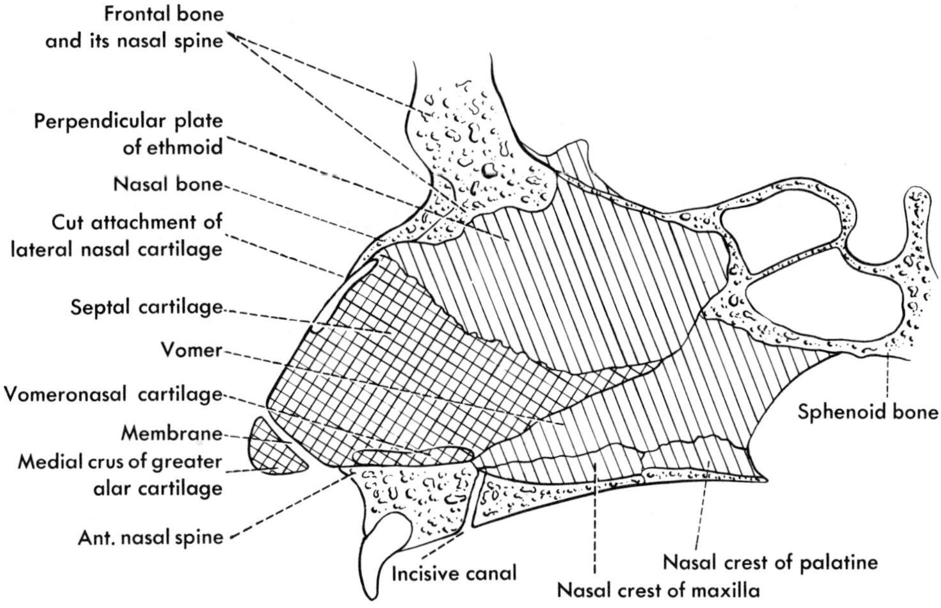

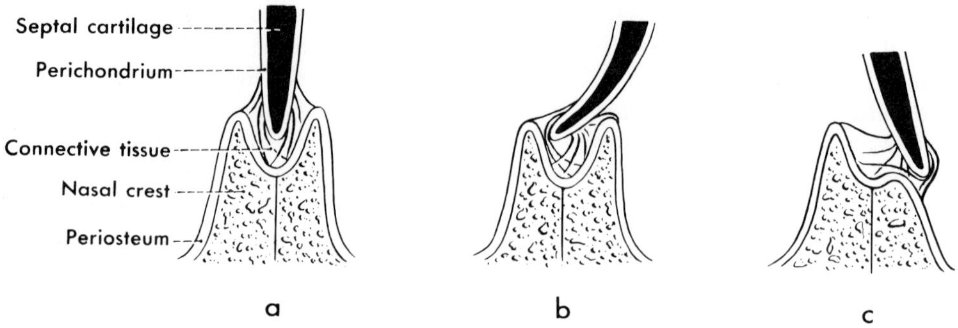

Fig. 4-9. Method of articulation of the septal cartilage so as to allow flexion of it as shown in (*b*); (*c*) indicates how dislocation is favored by defective development of the nasal crest. The width of the articulation is exaggerated.

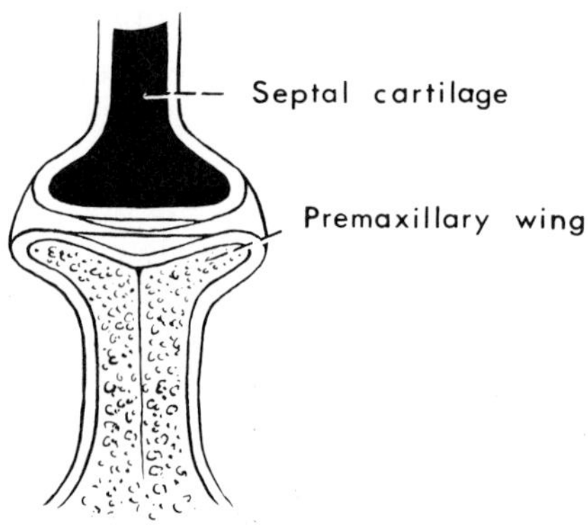

Fig. 4-10. The articulation of the septal cartilage in the anterior or premaxillary region.

obliterated, and the articulation between the broadened edges of the septal cartilage and nasal crest almost flat. Because of these lateral projections, elevation of the mucoperiosteum without tearing it is particularly difficult; Klaff and Cottle and co-workers have discussed the anatomy here and described surgical techniques that they have found useful.

Because the strength of the articulation between septal cartilage and bone depends on the articular surfaces rather than on continuity between perichondrium and periosteum, improper development or fracture of one lip of the bony groove (or a premaxillary wing) will allow dislocation of the cartilage.

Mobile Septum

The lowest part of the mobile septum (pars mobilis septi nasi) is frequently referred to as the columella or columna and is supported by the medial crura of the greater alar cartilages. As already noted, these are connected by membrane to the lower edge of the septal cartilage; this membrane with its covering skin forms the pars membranacea, or membranous septum, a part of the mobile septum. However, the terms mobile septum and membranous septum are often used as synonyms.

Submucous Resection

The entire septum, bony, cartilaginous, and membranous, is covered by a continuous layer of mucosa and skin; in submucous resections this can be incised at the mucocutaneous line by pushing the mobile septum to one side. The problem of maintaining continuity of the mucoperiosteum at the junctions of the various elements of the septum has already been mentioned.

Deviated septa usually involve the septal cartilage and perpendicular plate of the ethmoid, and sometimes the nasal crests of the maxilla and palatine, but the vomer is less often affected. In spite of the fact that the union of the two lateral nasal cartilages in a curve on the dorsum of the nose gives added strength to this region, the support to the dorsum of the nose here, and to a much greater extent below, where the lateral nasal cartilages have freely projecting apices, must depend upon the septal cartilage; a depressed area of the dorsum ("saddle nose") may follow too great removal of the septal cartilage, although it has been denied, and affirmed, that this is due entirely to the removal of support. The argument on this point seems to be the extent to which some support from the septal cartilage is necessary to support the soft tissues of the dorsum below the union of the lateral nasal cartilages and is therefore one concerning relative strength following an operation. Several writers have emphasized the role that scar tissue and its retraction may play in the formation of "saddle nose"; although the medial crura of the greater alar cartilages (columella), after submucous resection, adequately support the tip of the nose, the tip also may be involved by scar tissue and drawn inward and upward.

Vomeronasal Cartilage

On either side of the septal cartilage along its inferior border there may be a small bar of cartilage, the vomeronasal cartilage; this is associated with a small opening leading into the rudimentary vomeronasal organ (of Jacobson); in some animals this organ gives rise to an olfactory (vomeronasal) nerve and is highly specialized, but in man it is usually simply a short tubular sac, 2–6 mm long, lined by epithelium continuous with that of the nasal cavity proper.

Innervation and Blood Supply

The chief nerves of the nasal septum are from the anterior ethmoidal and maxillary nerves, the latter by way of branches of the pterygopalatine (sphenopalatine) ganglion; they are described in connection with the innervation of the nasal cavity as a whole. The arteries of the nasal septum include branches from the sphenopalatine artery, from both ethmoidal arteries, and from the facial.

LATERAL NASAL WALL

In contrast to the simple medial wall, the lateral nasal wall is complicated in its anatomy. It bounds most of the paranasal sinuses and receives the openings from these, and bears three or four nasal conchae or turbinates, delicate projecting scrolls of bone covered by mucous membrane (Fig. 4-11). These conchae are named, from below upward, the inferior, middle, and superior conchae; when a fourth one is present it lies above the others, is the smallest of all, and is the supreme concha.

The air spaces, or meatuses, beneath and lateral to the conchae are named according to the conchae to which they are related: thus

Fig. 4-11. Lateral nasal wall.

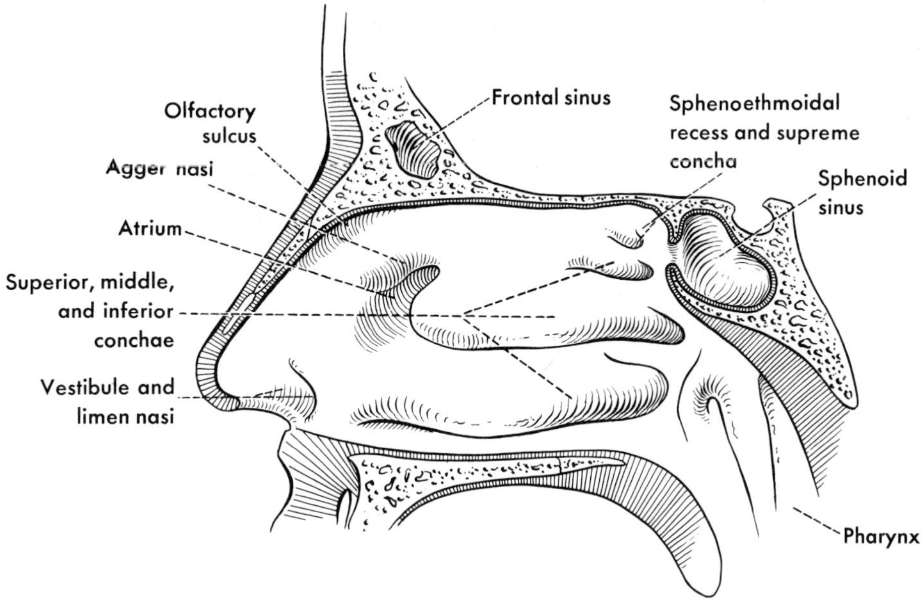

the inferior meatus lies below the root of attachment of the inferior concha, and lateral to its freely projecting portion, between this and the lateral wall of the nose; similarly, the middle meatus lies below and lateral to the middle concha and above the root of attachment of the inferior concha, and the superior lies below and lateral to the superior concha. The supreme meatus, when present, is usually a barely perceptible furrow below the supreme concha.

The *inferior and middle conchae,* the prominent projections of the lateral wall, both begin anteriorly at approximately the level of the vertical plane of the forehead and extend, one below the other, almost to the choana. Anterior to these two conchae the lateral wall of the nose above the limen nasi is more delicately marked. About halfway between the anterior end of the middle concha and the dorsum of the nose is a slight projection, the *agger nasi,* said to be the remnant of an additional concha found in lower animals, but important in man as marking the location of ethmoidal air cells frequently designated the agger nasi cells. The passageway above the agger nasi, the olfactory sulcus, leads to the uppermost part of the nasal cavity and aids in conducting air to this upper, olfactory, portion; below and posterior to the agger nasi, leading to the middle meatus proper, and above and anterior to the attached end of the inferior concha, is a shallow depression known as the atrium of the middle meatus.

The *superior concha,* about half the length of the other two, begins at about the middle of these and extends backward to become about coterminous with them. The part of the nasal cavity between the uppermost concha, superior or supreme as the case may be, and the anterior surface of the body of the sphenoid bone is the *sphenoethmoidal recess;* the opening of the sphenoid sinus is typically through the posterior wall (sphenoid concha, body or "face" of the sphenoid) of this recess.

The three nasal conchae converge somewhat toward each other posteriorly, and the remaining part of the nasal cavity behind their posterior ends is the nasopharyngeal meatus. It opens into the nasal part of the pharynx through the choana.

INFERIOR NASAL CONCHA AND INFERIOR MEATUS

The inferior nasal concha is an independent bone, covered with thick mucous membrane which contains a plexus cavernosus. It is so arched that the inferior meatus, lying below and lateral to it, is narrowed both anteriorly and posteriorly, but is both wider and higher at its middle.

The only structure of importance in the inferior meatus is the opening of the nasolacrimal duct (Fig. 4-12). This ostium is usually in the anterior portion of the lateral wall of the inferior meatus; it has been variously described as being placed more commonly quite high, just below the attached border of the lower concha, or more commonly somewhat below this level. Costen ('32) found it more commonly high and well forward, and pointed out that accessory openings into the maxillary sinus through the inferior meatus should therefore be made as far back as possible in order to avoid scarring or destruction of this ostium.

The shape of the opening varies considerably, from rounded to slitlike; it may open upon a papillalike formation or into a shallow fossa or deep groove (Fig. 4-13), and occasionally the opening may be duplicated (Schaeffer, '12). It has been stated that when the opening is high it tends also to be wide, while when it is located lower it is more apt to be slitlike. When the opening is slitlike the nasolacrimal duct runs obliquely through the mucous membrane and is usually then protected by a fold of mucous membrane, the plica lacrimalis or valve of Hasner. The lacrimal sac and the nasolacrimal duct are described in connection with the orbit.

MIDDLE NASAL CONCHA AND MIDDLE MEATUS

The middle nasal concha, also large, is a part of the ethmoid bone and overhangs the important and complicated middle nasal meatus (Fig. 4-14). The anterior end of the middle concha has its line of attachment running almost vertically upward, to join the remainder of the concha at an angle or genu; the

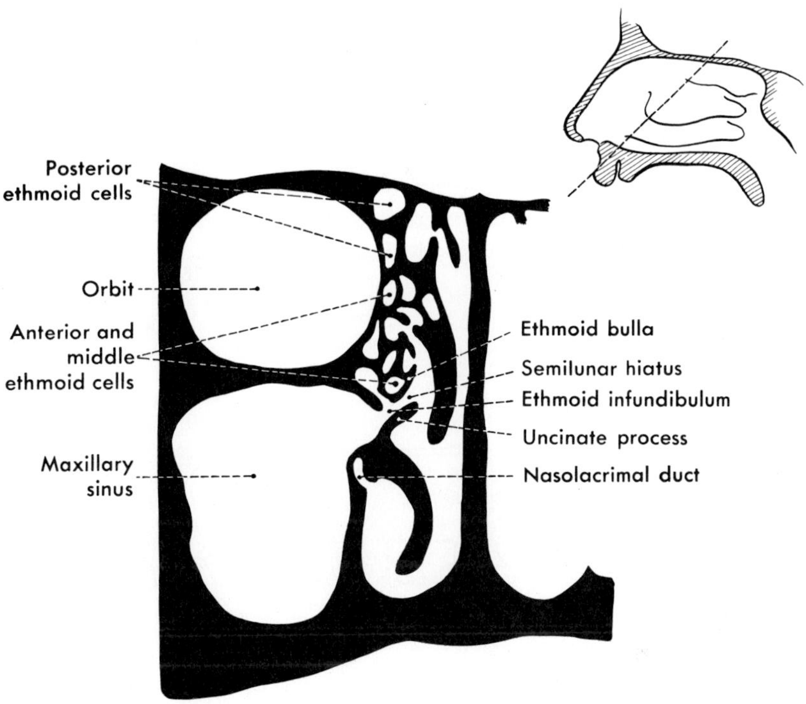

Fig. 4-12. A hypothetical section through the lateral nasal wall, *slanted* as indicated in the smaller diagram, indicating some of its relations.

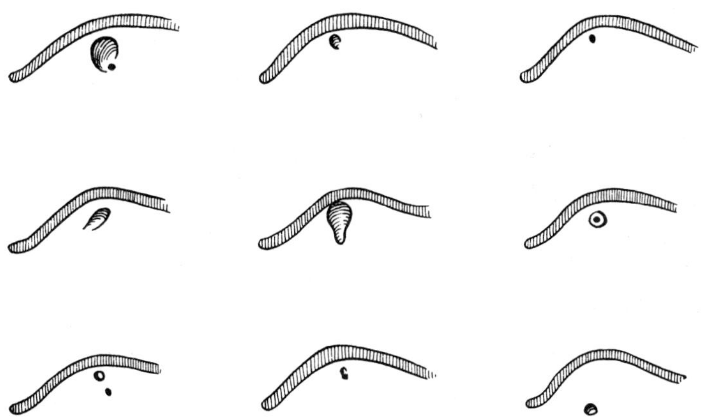

Fig. 4-13. Variations in the size, shape, and position of the opening of the nasolacrimal duct. The cut base of the inferior concha is indicated. (After Schaeffer JP: Am J Anat 13:183, 1912)

highest part of the middle meatus lies below the genu and is known as the *frontal recess.* The frontal recess may receive directly the opening from the frontal sinus and the openings of some of the anterior ethmoidal cells. From the frontal recess the middle meatus runs downward and backward, and this portion is then referred to as the descending ramus; it is marked by the ethmoidal bulla, the uncinate process, and the semilunar hiatus. The base of the middle concha may be invaded by one or more ethmoid cells, usually of the posterior ethmoid group.

Ethmoidal Bulla

The ethmoidal bulla, a rounded projection of the lateral wall of the middle meatus under cover of the middle concha, contains one or more ethmoid cells (known variously as middle, anterior, or bullar ethmoid cells); these

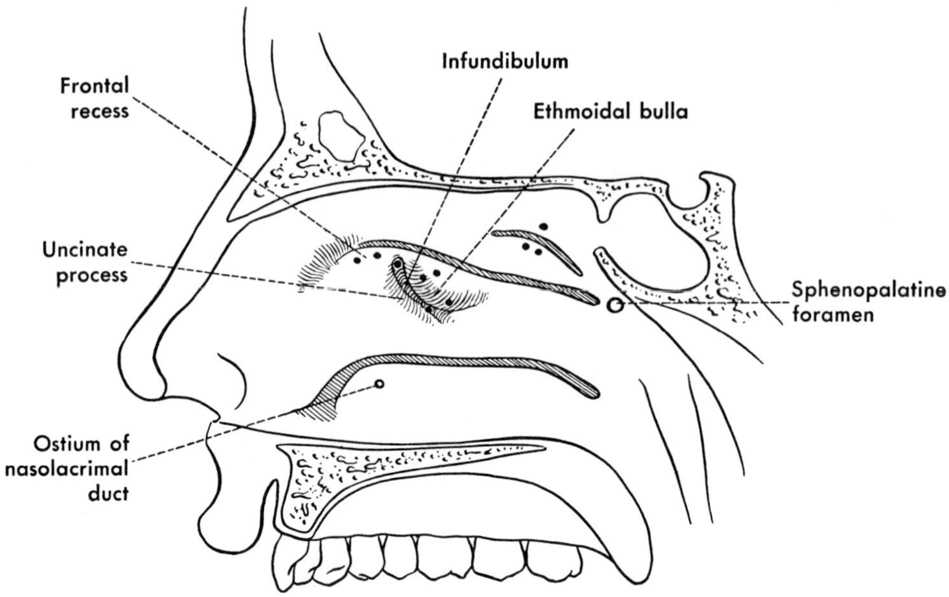

Fig. 4-14. Lateral nasal wall with the conchae removed. Variable locations of the ostia of the ethmoid cells are indicated by the *black dots*—anterior ethmoid cells may open into the frontal recess or the ethmoidal infundibulum; middle cells open on or above the bulla; and posterior cells open into the superior or supreme meatus.

open into the middle meatus upon or above the bulla. Above the bulla, between it and the attached base of the middle concha, is the suprabullar recess or furrow.

The ethmoidal bulla is bordered below by an aperture termed the *semilunar* (or *ethmoidal*) hiatus, which lies between the bulla above and a sharp ridge of bone, the uncinate process, below; the semilunar hiatus is the opening into a groove known as the ethmoidal infundibulum (uncinate or unciform groove).

Ethmoidal Infundibulum

From the semilunar hiatus the ethmoidal infundibulum extends downward, thus intervening between the lateral nasal wall and the ledge of bone forming the uncinate process. The ethmoidal infundibulum varies in depth (from 0.5 to 10 mm, averaging about 5mm) and therefore the upwardly projecting uncinate process varies in its height. Anterior ethmoid cells (infundibular cells) open into the ethmoidal infundibulum, and so also does the maxillary sinus; the accessibility of the ostium of this sinus therefore depends primarily upon the depth of the ethmoidal infundibulum and upon the closeness with which the uncinate process hugs the lateral wall—that is, also upon the width of the infundibulum.

Posteriorly, the ethmoidal infundibulum becomes continuous with the middle meatus; anteriorly, it either ends blindly at the ostia of anterior ethmoid cells or is prolonged upward to groove the frontal recess and is then directly continuous with the ostium of the frontal sinus. It is the funnel-shaped anterior end of the infundibulum that earns it its name, and sometimes the name is restricted to this part; the remainder of it is then referred to as the uncinate groove, or even as the semilunar hiatus, a term otherwise used to denote only the opening rather than the groove itself. The *medial wall* of the infundibulum is, of course, the bony uncinate process; the *lateral wall,* a part of the lateral nasal wall, is in places entirely membranous, and in perhaps as many as 40% of cases has accessory ostia leading into the maxillary sinus. The lateral nasal wall may also be largely membranous behind and below the uncinate process, between it and the inferior concha. This is the area that is removed in the transantral approach to the sphenoid sinus (Mosher, '03).

SUPERIOR NASAL CONCHA AND SUPERIOR MEATUS

The superior nasal concha is short, only about half the length of the middle, and is a much less prominent projection; its mucous membrane is thinner and less vascular than that of the other conchae and, unlike those, includes olfactory epithelium containing the sensory cells of the olfactory nerve. The superior meatus is a narrow channel between the superior concha and the posterior half of the middle concha; most posterior ethmoid cells open into this meatus.

SUPREME NASAL CONCHA

The supreme concha is said to be present unilaterally or bilaterally in about 60% of bodies, but even when present is usually a rather slight fold of the lateral nasal wall. The shallow groove below it is then spoken of as the supreme meatus and is said to receive the ostium of a posterior ethmoid cell in about 75% of the cases in which it is present.

NERVES

Olfactory Mucous Membrane

The olfactory mucous membrane, containing the cells of origin of the olfactory nerve fibers, is limited to approximately the upper third of the nasal mucous membrane, thus primarily to the region of the superior concha, a corresponding portion of the septum, and the roof of the nasal cavity; because of the greater length of the nasal cavity inferiorly, the olfactory mucous membrane actually occupies much less than a third of the entire epithelium (Fig. 4-15). Smith ('41, '42) has shown that the number of fibers in the olfactory nerve begins to decrease shortly after birth and continues to do so at the rate of about 1% a year; he found that 55% of adults had lost more than three-fifths of their complement of olfactory nerve fibers and that 13% had lost all of them; only 29 of 163 adult noses that he examined seemed to have a normal number of fibers, some 10,000 or more. He concluded that the marked reduction observed in adults is due in part to the aging factor, but much more to pathological changes in the nasal mucosa.

GENERAL SENSATION

The general sensory supply to the mucous membrane of the nose is derived from both the ophthalmic and maxillary branches of the fifth nerve and possibly also from the greater petrosal branch of the seventh; the fibers from the latter two branches reach the nose largely through the pterygopalatine (sphenopalatine) ganglion and its branches.

Nasociliary Nerve

The branch of the ophthalmic division of the fifth nerve concerned with the innervation of the nasal mucous membrane is the nasociliary (nasal). The origin, course, and branches of this nerve have already been described in connection with the orbit; briefly summarized, the nasociliary nerve arises from the ophthalmic nerve while this lies in the lateral wall of the cavernous sinus, enters and traverses the orbit, giving off branches to structures here, and divides into two terminal branches, the infratrochlear nerve (to skin at the medial angle of the eyelids) and the anterior ethmoidal nerve.

The anterior ethmoidal nerve leaves the orbit through the anterior ethmoidal foramen, typically in company with the anterior ethmoidal artery, and enters the cranial cavity to run forward along the lateral margin of the cribriform plate. It leaves the cranium through the ethmoidal slit at the side of the crista galli to enter the nasal cavity and supply, through lateral and medial internal nasal branches, a limited upper and anterior portion (Figs. 4-16 and 4-17), including the area anterior to the superior concha and above the middle one, the atrium, the anterior ends of the middle and inferior conchae, and a corresponding portion of the nasal septum. After

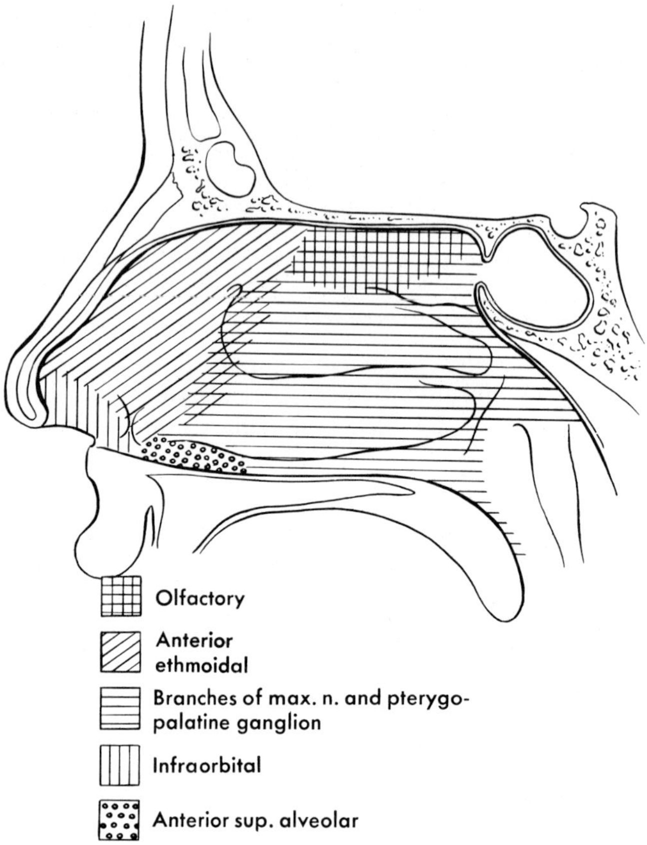

Fig. 4-15. General distribution of nerves to the lateral nasal wall.

giving off its internal nasal branches, the anterior ethmoidal nerve leaves the nasal cavity between the nasal bone and the lateral nasal cartilage to supply skin on the dorsum and tip of the nose.

Maxillary Nerve

The terminal or infraorbital branch of the maxillary nerve supplies a portion of the lining of the vestibule, through branches entering by way of the nares, and the anterior superior alveolar (dental) nerve gives off a nasal branch that pierces the lateral wall of the nose to supply an anterior part of the inferior meatus and a corresponding part of the floor of the nasal cavity. However, the chief supply from the maxillary nerve to the nasal cavity is through the *pterygopalatine* (formerly sphenopalatine) *ganglion,* one of the four major parasympathetic ganglia of the head. This is suspended from the maxillary nerve in the pterygopalatine (sphenomaxillary) fossa, which lies just lateral to a posterior part of the nasal cavity.

In the pterygopalatine fossa the maxillary nerve gives off pterygopalatine (sphenopalatine) nerves, usually two, that pass downward to the ganglion. Since the maxillary nerve fibers are sensory ones, however, they do not synapse in the ganglion. Instead, while some traverse it, most pass on its medial side; they then continue as the major component of the so-called branches of the ganglion, which contain also some postganglionic secretory and vasomotor fibers derived both from the pterygopalatine ganglion and from sympathetic fibers reaching it. The branches of the ganglion are enumerated later; it is the branches to the nasal cavity that are important here.

Most of the *nasal branches of the pterygopalatine ganglion* pass through the sphenopalatine foramen, located just behind the posterior end of the middle nasal concha, and are divisible into two sets, the lateral posterior superior (short sphenopalatine, superior nasal) and the medial posterior superior (septal) branches.

The lateral posterior superior nasal branches are a number of small twigs that are distributed to the mucous membrane over the larger posterior part of the superior and middle nasal conchae and help also to supply the posterior ethmoidal cells; the medial posterior superior nasal branches (Fig. 4-17) are distributed to the nasal septum, crossing the roof of the nasal cavity where it is formed by the anteroinferior face of the sphenoid bone. The largest of this group is the nasopalatine (long sphenopalatine) nerve; this runs downward and forward on the nasal septum, between the periosteum and the mucous membrane, to reach an incisive (anterior palatine) canal. Here it anastomoses with the nasal branch of the anterior superior alveolar nerve; the two nasopalatine nerves then pass through the median incisive canals, the left nerve usually anterior to the right, to emerge through the incisive foramina (foramina of Scarpa) that open into the incisive fossa on the lower surface of the bony palate. They are then distributed to the anterior portion of the hard palate behind the incisor teeth, where they communicate with the greater (anterior) palatine nerve.

The lower lateral part of the nasal cavity is also supplied by branches of the pterygopalatine ganglion, but less directly; as the greater palatine nerve passes downward in the greater palatine (pterygopalatine) canal within the lateral nasal wall it gives off small posterior inferior nasal branches which pass through the perpendicular plate of the palatine bone to supply the mucous membrane over the posterior part of the inferior nasal concha and the adjacent portions of the middle and inferior meatuses.

Nervus Terminalis

In addition to the branches already described, the nasal mucous membrane also receives the fibers of the nervus terminalis. According to

Fig. 4-16. Major nerve branches on the lateral nasal wall, the olfactory not being shown. The greater palatine canal and the position of the pterygopalatine (sphenopalatine) ganglion at its upper end are indicated by *broken lines.*

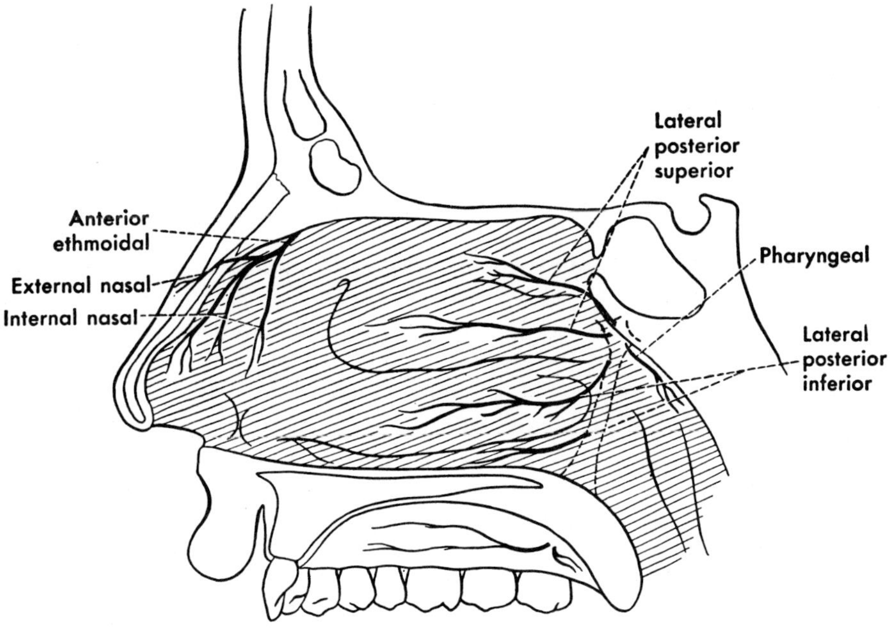

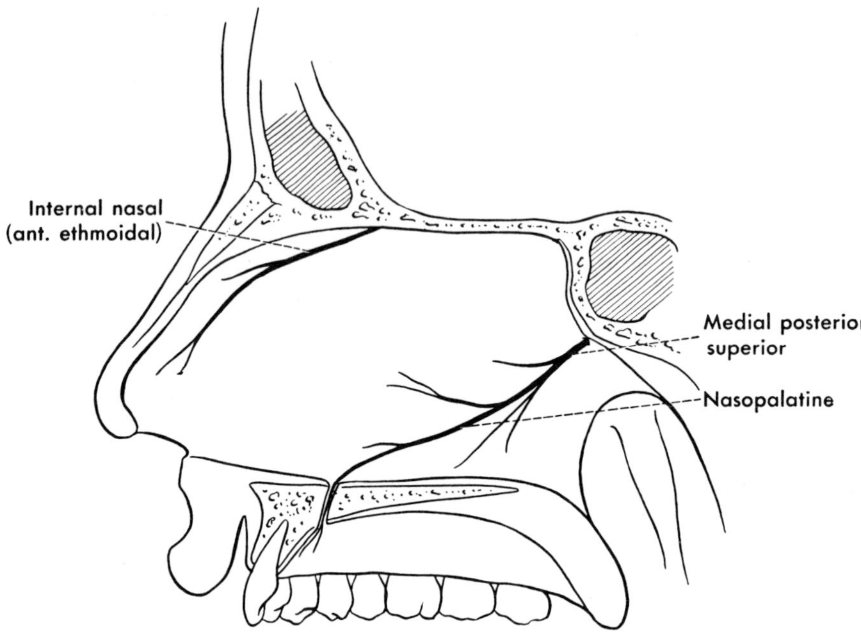

Fig. 4-17. Major nerve branches, other than the olfactory, on the nasal septum.

Brookover the ganglion cells of this nerve lie largely embedded in the dura between the crista galli and the olfactory bulb; its branches emerge into the nose through the cribriform plate to lie against the cartilaginous septum, deep to most of the vessels and other structures of the nasal mucosa. The plexus that they form in this location also contains nerve cells.

The function of the nervus terminalis is not known; it has been regarded as both sensory and autonomic. The ganglion cells of the nervus terminalis are said (Pearson) to be derived at least in part by migration of cells from the olfactory placode and perhaps also from migration of cells from the forebrain.

Greater Petrosal Nerve

Since the greater petrosal (greater superficial petrosal) nerve, the motor root to the pterygopalatine ganglion, contains in addition to its preganglionic fibers a goodly number of sensory ones, it is obvious that these fibers must be distributed through the branches of the ganglion. The greater petrosal nerve is derived from the seventh nerve, and the sensory fibers in it have their cells of origin in the geniculate ganglion of the facial. (Foley and colleagues found about 2000 sensory fibers in the facial nerve of the cat, of which from 27% to 40%, according to Foley and DuBois, go into the greater petrosal nerve.) The exact distribution of these sensory fibers of the facial nerve is not known, nor are their functions clearly understood, but evidence that some may reach the nasal cavity is quoted in the following section.

Sensation and Anesthesia of Nose

Cushing described the nasal mucous membrane as being completely anesthetic, even to such irritating stimuli as ammonia, following interruption of the fifth nerve. Lederer and Dinolt studied the noses of ten patients in whom the sensory root of the trigeminal nerve had been sectioned and reported that they found anesthesia of the affected side in nine patients and hypesthesia in one; they concluded that the first and second branches of the trigeminal nerve provide all the sensory innervation of the nose except perhaps that of a small area of the septum that receives branches of the nervus terminalis. Higbee, from an examination of 16 patients with section of the root of the fifth nerve, concluded likewise that the fifth nerve is the sole sensory supply to the nose.

These clinical reports are not, however, in complete agreement with the known anatomy; it seems possible that residual sensation in the more posterior part of the nose, to which sensory fibers of the facial nerve are apparently distributed, may have been overlooked—or that the greater petrosal nerve had been interrupted, as may easily occur, in the approach to the sensory root of the fifth nerve.

While some of the sensory fibers of the facial passing through the pterygopalatine ganglion may, as stated by Rowbotham, be gustatory fibers distributed by way of the palatine nerves to the soft palate, pain in the deeper nasal region and the sinuses has been described as a symptom of geniculate ganglion neuralgia (Hunt), and Fenton pointed out that reference of pain of apparent nasal origin to the auditory and mastoid regions can best be explained as due to the sensory distribution of the facial nerve to the nose. Christensen stated that not only do the lateral nasal and nasopalatine branches of the pterygopalatine ganglion contain sensory fibers from the seventh nerve as well as the fifth, but that these branches and the nasociliary nerve apparently also include some afferent components derived from the vagus and upper thoracic spinal nerves. Larsell and Burns indicated that myelinated fibers (therefore presumably afferent ones) leave the internal carotid plexus by way of the deep petrosal to reach the pterygopalatine ganglion, and suggested that they may have to do with the muscle pain sometimes present in the neck when the region of the pterygopalatine ganglion is diseased. The fifth nerve, therefore, probably cannot be regarded as the sole sensory supply to the nose; since, however, the other nerves concerned run with fifth nerve fibers as branches of the pterygopalatine ganglion, anesthetization of the ganglion affects all the nerve supply to the posterior part of the nose.

Because of the complexity of the nasal cavity, cocainization of the mucosa usually produces only partial anesthesia; Hill therefore advocated injection through the infraorbital foramen to infiltrate the maxillary nerve and the pterygopalatine ganglion and its branches (which have also been anesthetized through the greater palatine canal and through a lateral approach, anterior to the mandible or between its coronoid and condylar processes) and injection along the nasal wall of the orbit to block the nasociliary nerve at the anterior ethmoid foramen. These two injections, properly carried out should effectively interrupt all sensory fibers to the nasal cavity, regardless of their origin (and also fibers to most of the paranasal sinuses).

According to the investigations of McAuliffe and co-workers, the linings of the paranasal sinuses are relatively insensitive to pain, as is the mucous membrane of the nasal septum; the conchae are far more sensitive, and the ostia of the sinuses are the most sensitive of any of the nasal structures. The pain of so-called sinus headache apparently arises from inflammation and engorgement of the ostia and conchae, not from the sinuses themselves, and is referred to the forehead, top of the head, and eyes if it arises primarily from the higher part of the nasal cavity and to the region of the zygomatic arch, the temple, and the teeth and jaws if it arises from middle and inferior nasal structures. According to these workers' findings, pain referred to the back of the head and neck is not of nasal origin, but is secondary to sustained contraction of muscles of the scalp and neck.

AUTONOMIC INNERVATION

Some sympathetic fibers may enter the nose with the nasociliary nerve (Christensen), and the nervus terminalis may have an autonomic function, but the chief autonomic innervation of the nose, both sympathetic and parasympathetic, is through the pterygopalatine ganglion and its branches of distribution (Fig. 4-18).

The facial nerve supplies the parasympathetic fibers to the nose. The preganglionic fibers pass through the nervus intermedius portion of the facial nerve with the sensory

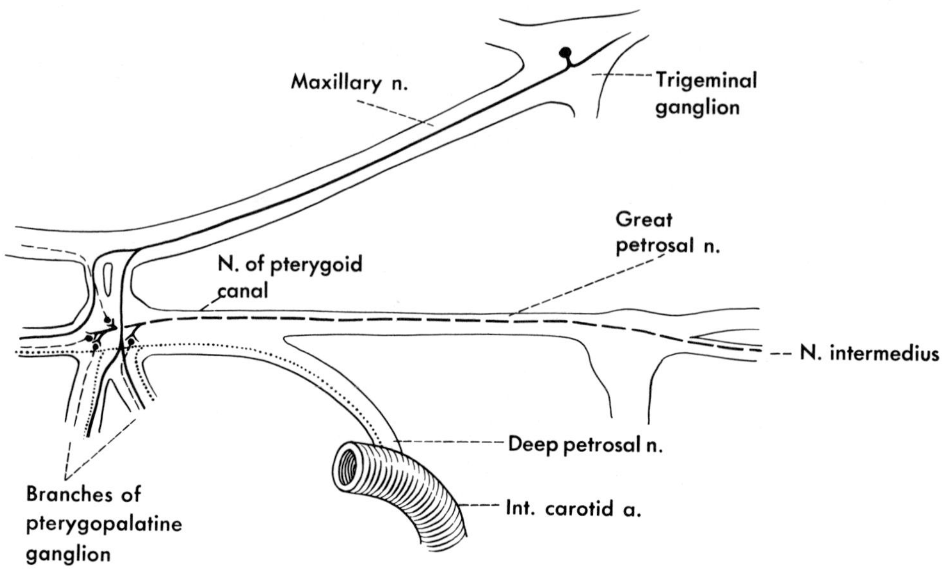

Fig. 4-18. Connections of the pterygopalatine ganglion. Preganglionic parasympathetic fibers to the ganglion are represented by *long broken lines;* postganglionic fibers originating in the ganglion, by *short broken lines;* and sympathetic postganglionic fibers traversing the ganglion, by *dotted lines.* Afferent fibers in the great and deep petrosal nerves are omitted, but those of the maxillary nerve that help form the branches of the ganglion are represented by *solid lines.*

fibers, leave the facial at its geniculum to form a part of the greater petrosal nerve, and end in the pterygopalatine ganglion. The postganglionic fibers, arising from the cells of the ganglion, are then distributed with the sympathetic and sensory fibers through the branches of the ganglion. The greater petrosal nerve contains both vasodilator and secretory fibers.

The sympathetic fibers to the nose undoubtedly arise, like the fibers to other parts of the head, from upper thoracic segments of the cord, and perhaps occasionally from the eighth cervical segment; these preganglionic fibers then ascend to reach the superior cervical ganglion in which they synapse. Postganglionic fibers from the superior cervical ganglion pass along the internal carotid artery as the chief component of the internal carotid plexus, and those destined for the nose leave this plexus as the deep petrosal nerve, which joins the greater petrosal to form the nerve of the pterygoid canal (Vidian nerve); the sympathetic fibers pass through the pterygopalatine ganglion without synapse to be distributed with its branches. These fibers are primarily vasoconstrictors, but according to Larsell and Fenton they have been said to include also some vasodilators. Rooker and Jackson concluded, on the basis of their experiments, that the nasal vessels are normally kept tonically contracted by sympathetic impulses.

Pterygopalatine Ganglion

The *pterygopalatine* (sphenopalatine, nasal, or Meckel's) *ganglion* lies deeply in the pterygopalatine (sphenomaxillary) fossa, suspended from the maxillary nerve by connections called the pterygopalatine nerves; it is immediately adjacent to the sphenopalatine foramen in the lateral nasal wall and thus just posterior to and perhaps slightly above the posterior end of the middle nasal concha. It is usually below the floor of the sphenoid sinus, but if this is small or narrow it may lie lateral to the sinus. Since its posterior superior nasal branches pass through the sphenopalatine foramen, these may be blocked by the application of a cocainized pledget of cotton to the lateral wall of the nose behind the middle concha.

Injection about the ganglion and the sensory fibers associated with it has been carried out through the foramen or the thin bone about it; by injecting deeply backward along the infraorbital canal; by the lateral, paramandibular, approach to the pterygopalatine fossa; and by injecting upward through the greater palatine (pterygopalatine) canal. Both the infraorbital and palatine approaches are potentially dangerous, for in the first the needle may penetrate the often very thin roof of the infraorbital canal and enter the orbit, while in the second if the needle is too long it may miss the bony landmark, the face of the sphenoid bone, and pass through the inferior orbital fissure into the orbit or even through this and the superior orbital fissure into the cranial cavity. The palatine approach, with a needle of such length that it cannot reach the orbit, is often now preferred.

Anatomically, the ganglion is easily accessible in the cadaver, after making a sagittal section of the head, through the removal of the thin bone around the sphenopalatine foramen, but surgically this approach is difficult because of the very limited intranasal space here and dangerous because of the pres ence of the terminal branches of the (internal) maxillary artery in the fossa. Averbukh and co-workers described a *palatine approach,* removing a posterior portion of the hard palate, dissecting the nasal mucous membrane loose from the lateral nasal wall, and following this bony wall up to the inferior surface of the body of the sphenoid; here they removed the upper and inner wall of the pterygoid canal to expose the nerve of the canal and then followed this to the ganglion. They stated they had carried out this procedure only on the cadaver, but their surgical friends had found it to be an appropriate approach.

Sewall advocated a *transantral approach,* pointing out that this offers access not only to the ganglion but to the maxillary nerve, the branches of the maxillary (internal maxillary) artery, and even to the sphenoid sinus and the orbit. In this approach he opened the maxillary sinus through the canine fossa (as for a Caldwell–Luc operation), removed the mucosa and the bone in the posterior wall of the sinus, and incised the periosteum to expose the contents of the fossa. This approach to the pterygopalatine fossa has been used by Golding–Wood for section of the nerve of the pterygoid canal (see following paragraphs) and by a number of workers for ligation of the sphenopalatine artery or its parent stem, the maxillary, in certain cases of nosebleed.

The *roots* of the pterygopalatine ganglion, motor, sympathetic, and sensory, have necessarily been discussed in part with the branches of distribution of the ganglion; the following is a summary of these roots and a description of their courses. The *motor or preganglionic root* of the ganglion is the greater petrosal nerve; after leaving the geniculate ganglion this nerve typically passes through a short canal (less than 2.5 mm long in more than 35% of 100 sides studied by Rhoton and co-workers, with no canal present in 30%, and a maximum of 6.0 mm). It then, if there is a canal, emerges at the hiatus (hiatus of the canal for the greater petrosal nerve, formerly called the hiatus of the facial canal, or Fallopian hiatus) into the middle cranial fossa; here it lies in a groove on the anterolateral surface of the petrous portion of the temporal bone and passes forward and medially deep to the trigeminal ganglion to leave the fossa by way of the foramen lacerum. At the foramen lacerum the greater petrosal nerve lies immediately lateral to the internal carotid artery and, in this location, is joined by the deep petrosal nerve consisting of fibers derived from the carotid plexus. The union of the greater and deep petrosal nerves forms the nerve of the pterygoid canal, or Vidian nerve, which then passes forward in the base of the pterygoid process and usually in the floor of the sphenoid sinus to enter the pterygopalatine fossa and end in the ganglion.

The *sympathetic root* of the pterygopalatine ganglion is the deep petrosal; this conducts postganglionic sympathetic fibers to the nerve of the pterygoid canal and thence to the branches of the ganglion. Obviously, motor and sympathetic roots of the ganglion are combined from about the foramen lacerum forward.

The *sensory "root"* of the pterygopalatine ganglion is actually multiple, as it consists not only of the usually named pterygopalatine nerves from the maxillary, but of the divers sensory fibers contained in the nerve of the pterygoid canal. As already noted, the majority of the latter fibers are of geniculate ganglion origin, but some are said to be derived from the tenth nerve (for example, Christensen), and nerve fibers of afferent type have also been seen in the deep petrosal nerve (Larsell and Fenton); it has been stated that the latter may represent sensory fibers from upper thoracic spinal nerves.

Surgical approaches to the nerve of the pterygoid canal have already been mentioned. Golding-Wood and others have used the transantral approach to section it in treating chronic vasomotor rhinitis. Earlier, Ziegelman suggested intracranial section of the greater petrosal, by the temporal approach, as a technique for controlling the marked swelling of the nasal mucous membrane sometimes observed in head injuries.

The *branches of distribution* of the pterygopalatine ganglion are to the sphenoid sinus and the posterior ethmoid cells, to the nose, to the hard and soft palates, to the choanae, and to the uppermost part of the pharynx. The nasal branches have already been described, but the other branches will be described briefly here.

The *orbital (ascending) branches* are two or three tiny twigs, which are described as entering the orbit through the inferior orbital (sphenomaxillary) fissure to pass upward in the periosteum on the inner wall of the orbit to the posterior ethmoidal foramen. In their course they supply the orbitalis muscle with sympathetic fibers and perhaps sometimes supply the lacrimal gland with parasympathetic (secretory) and sympathetic fibers (Chap. 2), but otherwise they leave the orbit by way of this foramen and are distributed to ethmoid cells and the sphenoid sinus.

In addition to the nasopalatine nerve, which supplies an anterior portion of the hard palate, three palatine nerves are usually described: the greater (anterior) palatine nerve and two lesser palatines. The *greater palatine nerve* leaves the inferior border of the pterygopalatine ganglion and passes downward in the pterygopalatine fossa, which it leaves by way of the greater palatine (pterygopalatine) canal. It emerges from the canal at the greater or posterior palatine foramen and divides into several branches that pass forward on the hard palate to supply its mucous membrane and the gums on the inner aspect of the upper jaw. It is accompanied in its course by the greater palatine artery; within the palatine canal it gives off the posterior nasal nerves to the posterior portions of the middle and inferior meatuses and the inferior concha. The greater palatine nerve is said to derive its sensory fibers both from the maxillary and the greater petrosal nerves; its autonomic fibers are distributed to the glands and blood vessels of the palate.

The two *lesser palatine nerves* almost parallel the greater; they traverse the same or two lesser palatine canals that open a little behind the greater foramen. Both are believed to contain sensory fibers derived from the facial nerve by way of the greater petrosal. One (the "middle") supplies a lateral portion of the soft palate and sends a few twigs to the tonsil; the other (the "posterior") supplies most of the soft palate and may perhaps also reach the tonsil.

The *pharynegeal branch* of the pterygopalatine ganglion is a small branch, which after passing backward and somewhat medially in company with a corresponding branch of the sphenopalatine artery is distributed to the upper posterior part of the choanae, to the upper portion of the pharynx, and to the lining of the sphenoid sinus. Its sensory fibers are said to be derived from the maxillary nerve; evidence has been quoted (see Chap. 3) that the pharyngeal branch reaches as far as the opening of the auditory tube in about 50% of cases, while in the remainder this opening is supplied by the pharyngeal branch of the glossopharyngeal nerve.

The distribution of the sensory fibers believed to derive from the tenth cranial and upper spinal nerves is said to be to the nose.

Reports that certain atypical facial pains involving various cranial nerves or their branches could be abolished by alcohol injection or removal of the pterygopalatine (sphenopalatine) ganglion led to these pains being reported as "sphenopalatine ganglion neuralgia." In spite of much discussion, the anatomy and physiology of these pains have never been established, but subsequent experience has indicated that they cannot be controlled by destruction or removal of the ganglion (Kerr). There is therefore no reason to perpetuate the never clearly explained concept of a connection between these pains and the ganglion.

VESSELS

The arterial supply to the nasal cavity is, like the nerve supply it parallels, derived from several sources and enters the nose in various locations. The maxillary (internal maxillary) artery, especially through its important sphenopalatine branch, supplies the larger posterior part of the nose; anterior and posterior ethmoid arteries, from the ophthalmic and therefore the internal carotid, supply an upper and anterior portion; branches of the facial (external maxillary), like the maxillary a branch of the external carotid, supply especially the vestibule of the nose.

Sphenopalatine Artery

The sphenopalatine artery passes with the posterior superior nasal nerves through the sphenopalatine foramen and, dividing into lateral and septal posterior nasal arteries, is distributed to the major portions of the conchae and a corresponding part of the nasal septum (Figs. 4-19 and 4-20). It is branches of this artery that are more commonly involved in postoperative bleeding, making it necessary to pack the area behind the middle concha or even occasionally to ligate the external carotid or the maxillary or sphenopalatine in the pterygopalatine fossa. Sewall apparently first described the transantral approach to the pterygopalatine fossa, an approach that has since been used to control epistaxis, to reduce bleeding in surgery of the posterior ethmoid cells or the sphenoid sinus, or to section the nerve of the pterygoid canal to alleviate chronic vasomotor rhinitis.

Fig. 4-19. Arteries of the lateral nasal wall.

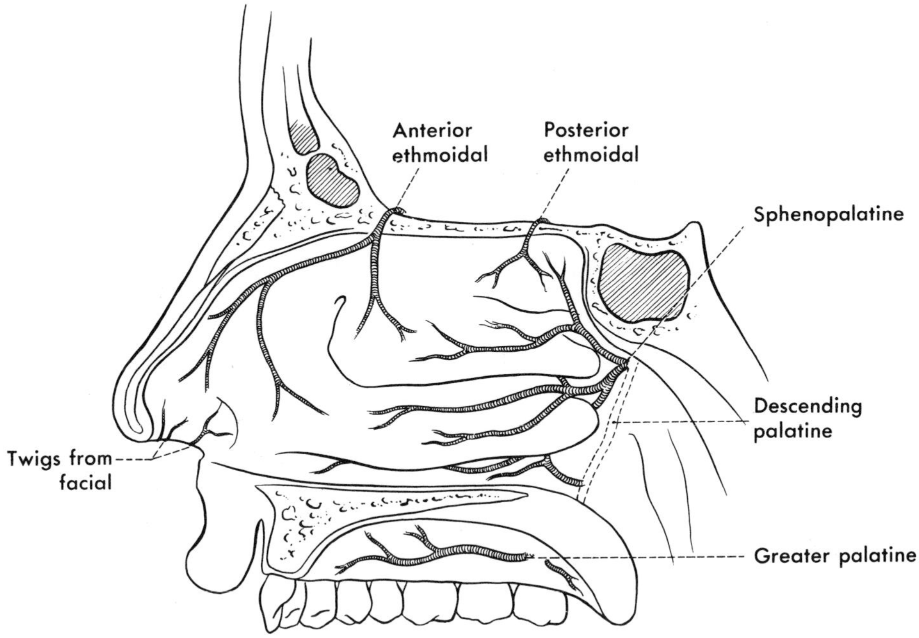

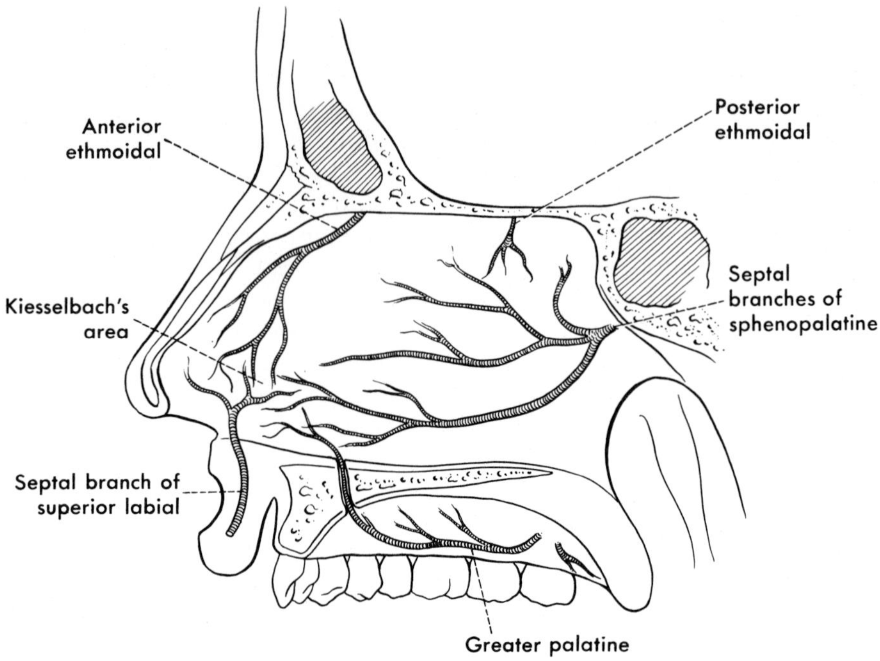

Fig. 4-20. Arteries of the nasal septum.

The larger lateral nasal branches of the sphenopalatine artery are those that run along the middle and inferior conchae. They are peculiarly placed in that for about the middle third of the lengths of the conchae, by which level the artery on the inferior concha has usually divided into two slightly divergent stems, they traverse canals in the conchae instead of lying in the mucosa as they do elsewhere. They are apparently surrounded in the canals by a venous plexus that drains the erectile tissue of the conchae, so that dilation of the arteries can cause congestion of the mucosa.

The septal part of the artery crosses the anterior face of the sphenoid bone before dividing into branches to the septum and is therefore often wounded in operations on the sphenoid sinus.

Descending Palatine

Another branch of the maxillary artery, the descending palatine, gives rise to greater and lesser palatine arteries that accompany the palatine nerves downward in the palatine canals and supply twigs to the lower posterior portion of the nasal cavity. The lesser palatine arteries supply the soft palate. The greater palatine, after emerging through the greater palatine foramen, turns forward and supplies the hard palate and the inner gums of the upper teeth; a terminal branch turns upward through the incisive fossa and the lateral incisive canal (of Stensen) opening into it, to supply a lower part of the nasal cavity (see Fig. 4-20).

Anterior and Posterior Ethmoidal Arteries

The anterior and posterior ethmoidal arteries enter the nasal cavity through its roof (see Fig. 4-20); they leave the ophthalmic artery within the orbit, course through the anterior and posterior ethmoidal canals, respectively, to obtain an intracranial course, and then turn downward through or anteromedial to the cribriform plate. The anterior ethmoidal artery is normally much the larger; it is distributed to about the anterior third of the lateral wall of the nose and to a similar portion of the septum and anastomoses with the sphenopalatine artery; a branch of the artery (external nasal branch) accompanies the external nasal branch of the anterior ethmoidal nerve, emerging between the nasal bone and the lateral nasal cartilage to run on the dorsum of the nose down to its tip. The posterior ethmoidal artery is limited largely in its distribution to the region of the superior concha and a corresponding portion of the septum.

Facial

The superior labial branch of the facial artery sends a branch into the vestibule, which is distributed largely to the septum. On the anterior portion of the septal cartilage this vessel anastomoses with septal branches of the sphenopalatine artery; the anastomosing branches between these vessels (probably joined variably by branches from the greater palatine and anterior ethmoidal arteries) are situated superficially in the mucosa, and the site at which they occur is known as *Kiesselbach's* (*or Little's*) *area.*

Epistaxis

According to Trible, it has been stated that 90%–99% of all bleeding from the nose is due to a local vascular defect and that 90% of this is from Kiesselbach's area. Since this area is easily accessible, such bleeding can frequently be stopped with simple packs or even by pressure upon the mobile septum. Other measures have included the use of silver nitrate and cutting through the posterior septal vessels about 1 cm posterior and above the bleeding point and applying packs.

Unfortunately, however, as Roy has emphasized, the majority of nasal hemorrhages in elderly people and in persons with arteriosclerosis arise from the neighborhood of the middle concha or from the "attic" of the nose—that is, from the sphenopalatine or ethmoidal vessels. If ordinary methods fail to control severe bleeding from these locations, surgical ligation is usually undertaken. Of 100 patients in a series reported by Spar and Hallberg requiring hospitalization for severe nosebleed, more than half bled from the anterior portion of the nasal septum, but nearly a third bled from a point high and posteriorly in the nose, usually from the sphenopalatine artery.

These workers found that this bleeding also was usually controllable by packs, but 11 patients required ligation of the external carotid artery. However, ligation of the external carotid cannot be expected always to control bleeding from the sphenopalatine artery, for there is a potentially abundant collateral circulation. Hirsch concluded that external carotid ligation produces too uncertain results and therefore advocated the transantral approach to the pterygopalatine fossa, through which the sphenopalatine artery can be ligated quite near the foramen. Durisch and Frable found ligation by the transantral approach preferable to prolonged packing, while Roberson and Reardon, and Wills and Russell, injected Gelfoam into the maxillary artery when other methods had failed. Rosnagle and co-workers found selective arteriography helpful in difficult cases, particularly after operation had failed to halt the bleeding.

Bleeding from high in the nose is not necessarily from the sphenopalatine artery, however; it may be from an ethmoidal artery (from the internal carotid). Surgical interruption of one or both of the ethmoid arteries is then a logical procedure if such a source can be established.

Weddell and co-workers reviewed the blood supply of the nose and the methods available for checking bleeding from this organ; they reported on seven patients in whom nosebleed was not controlled until the ethmoidal arteries of the affected side had been occluded. For this they used the orbital route, usually placing a clip upon the anterior ethmoidal artery before it enters its canal, and coagulating the usually smaller posterior ethmoidal artery as it approaches the medial orbital wall. Kirchner and co-workers have described the orbital relations and the variations of these vessels.

Veins

The veins of the nose begin in venous plexuses, which are so well developed on the inferior nasal concha, the inferior meatus, and the back part of the septum as to resemble erectile tissue; Batson found the nasal venous plexus to be so dense everywhere that the anatomy of the nasal wall could be recognized in corrosion specimens in which only casts of the veins remained. The definitive veins of the nose correspond approximately to the arteries. Thus, veins drain through the sphenopalatine foramen into the pterygoid plexus; others accompany the ethmoidal arteries to join the superior ophthalmic vein; and over the alar cartilages the nasal plexus is continuous with the subcutaneous plexus of the nose (Batson) and hence drains into the facial (anterior facial) vein or its tributaries.

Lymphatics

Lymphatics from the more anterior part of the nose communicate through the naris with lymphatics of the skin, but the larger lymphatics run posteriorly (Most). Some are said to pass to the tonsillar region and hence to deep cervical nodes directly, but the majority join the pharyngeal plexus and hence reach the retropharyngeal nodes. Robison ('44) described the lymphatics from the nose, upper pharynx, and middle ear as forming a plexus in front of the auditory (eustachian) tube before sending vessels to the retropharyngeal nodes; in consequence, he said, obstruction of the plexus or retropharyngeal nodes through infection originating from the nose and sinuses can lead to backing up of lymph into the middle ear cavity and the production of a serous otitis media.

The Paranasal Sinuses

The paranasal or accessory nasal sinuses, with the exception of the sphenoid sinus, begin their development as evaginations of the mucosa of the nasal cavity during the third and fourth fetal months. Although all the sinuses undergo their major expansion after birth, the maxillary sinuses and the ethmoid air cells are of appreciable size at birth, even though, as pointed out by Wasson, they are, of course, not filled with air. Aeration of the sinuses developed at birth takes place slowly, and Wasson stressed that great care must be exercised in attempting to diagnose their conditions from roentgenograms during the early weeks of life. The frontal sinus is a very small diverticulum at birth and undergoes almost all of its development thereafter; the sphenoid sinus at birth can hardly be said to exist as an entity, since it is represented merely by a posterosuperior part of the nasal cavity; it is not until about the third year of life that it actually invades the sphenoid bone.

Like the nasal cavity, the sinuses typically are lined by a pseudostratified columnar ciliated epithelium; as reported by Hilding ('44) the direction of flow induced by the ciliary beat in all of the sinuses seems to be a spiral centered at the ostium, material anywhere within the sinus tending therefore to be carried to the nose. McMurray has stressed also the effect of negative air pressure as an aid to ciliary drainage; he reported that inspiration produces a negative pressure in the maxillary sinus and that in the upright position with absence of the inspiratory air current, the ciliary action alone is not sufficient to drain it.

Latta and Schall pointed out that in acute irritation the basal cells of the epithelium lining the paranasal sinuses undergo hyperplasia and that this may continue until a true stratified columnar epithelium is present. Regeneration following degeneration and sloughing of an injured epithelium is said to be probably from the basal cells, which are not usually involved in such degeneration. Brownell reviewed the literature up to 1936 upon regeneration of the mucous membrane of the paranasal sinuses.

All the sinuses typically vary, among persons, in size and therefore in the extent to which they have invaded the neighboring bones; these, and the varying positions and sizes of their ostia, are of some importance to the rhinologist.

The mucosa of all the sinuses is thinner and less vascular than that of the nasal mucosa proper. Their arterial blood supply does not necessarily come entirely from the nasal arteries, but their chief venous drainage is through their openings into the nasal cavity, where their venous plexuses become continuous with the more dense plexuses of that cavity. Similarly, their relatively few lymphatics leave them through these same openings to join the lymphatics of the nasal cavity proper. Robison ('46) discussed the role of lymphatic drainage in allergic sinusitis and the advantages of increasing this drainage by pressure on the swollen mucous membrane.

Blanton and Biggs reviewed the literature relating to the possible function or functions of the paranasal sinuses and concluded that many of those suggested are speculative rather than factual, and that none has been universally accepted.

FRONTAL SINUS

The frontal sinus develops as one of several outgrowths from the region of the frontal recess, other similar outgrowths here representing anterior ethmoid cells. Two, three, or even more frontal sinuses on a side have been reported, although a source of confusion is that anterior ethmoid cells may invade the territory of the frontal sinus, and whether they should then be called frontal sinuses has been disputed. It makes no practical difference, particularly since the normal frontal sinus can be regarded as an anterior ethmoid cell that leaves the ethmoid labyrinth to expand instead in the frontal bone; the point is that several separate sinuses may occur in the frontal bone on one or both sides. They may lie one lateral to another, or one behind the other (Fig. 4-21). The frontal sinus likewise often contains septa, which partially subdivide its cavity, and may be so placed as to interfere seriously with its drainage. Aside from these, however, the sinus varies markedly in its development. It normally extends upward into the squama (vertical portion) of the frontal bone and also backward in the orbital part of this bone, between the inferior surface of the frontal lobe and the orbital contents (Fig. 4-22). Either the vertical or the horizontal portions of the sinus may be unusually developed and invade adjacent bones or may be poorly developed (Figs. 4-23 and 4-24); failure of the sinus to invade the vertical portion of the frontal bone has been mistaken for agenesis of the sinus, which is apparently quite rare (Schaeffer, '20). The two frontal sinuses are frequently asymmetrical in size, and the larger sinus may in such cases pass across the midline and even overlap the other. This makes it possible to open a normal sinus of the other side in attempting to reach a diseased one.

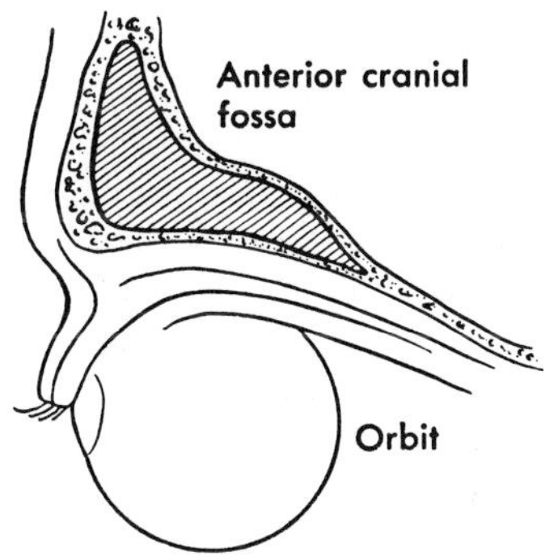

Fig. 4-22. Relations of the frontal sinus in sagittal section.

Fig. 4-21. Anteroposterior duplication of the frontal sinus, indicating the two types that may occur. In (*a*) are two separate sinuses; in (*b*) one is a diverticulum of the other.

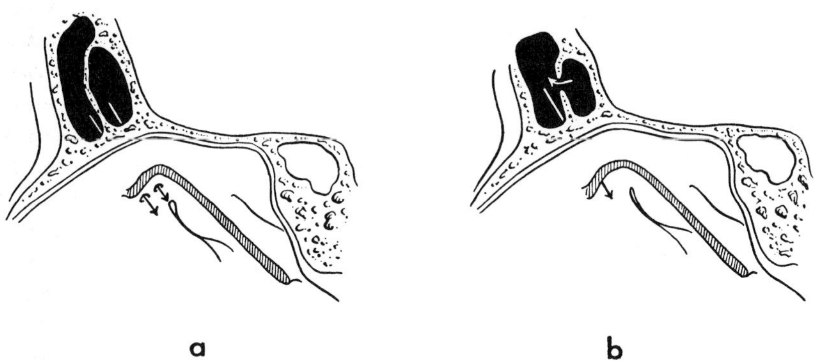

The important relations of the frontal sinus are to the anterior cranial fossa and the orbit. The bone separating the frontal sinus from the roof of the orbit and from the floor of the cranial cavity is often particularly thin; an operative perforation of it can easily occur. Wickliffe reported a case of abscess of the frontal lobe due to acute frontal sinusitis.

Frontonasal Connection

As noted by Schaeffer ('16), the frontal sinus is usually a derivative of the frontal recess directly or of one or more anterior ethmoid cells arising in frontal pits, but occasionally it arises from the frontal extremity of the ethmoidal infundibulum. In Kasper's investigation of 100 frontal sinuses he found that the frontonasal connection led to the frontal recess rather than to the ethmoidal infundibulum in 62%; in 4% the ethmoidal infundibulum and the frontal sinus were in direct anatomical continuity; in 34% more, however, he concluded that the sinus had arisen from an anterior ethmoid cell derived from the infundibulum, and thus in 38% the frontal sinus actually drained, directly or indirectly, into the infundibulum. Schaeffer ('16) stated that in approximately 50% of adults the ethmoidal infundibulum may act as a channel for carrying secretion or infection from the frontal sinus and some of the anterior ethmoid cells to the maxillary sinus, either because of continuity of the frontal sinus and the infundibulum or their intimate relation.

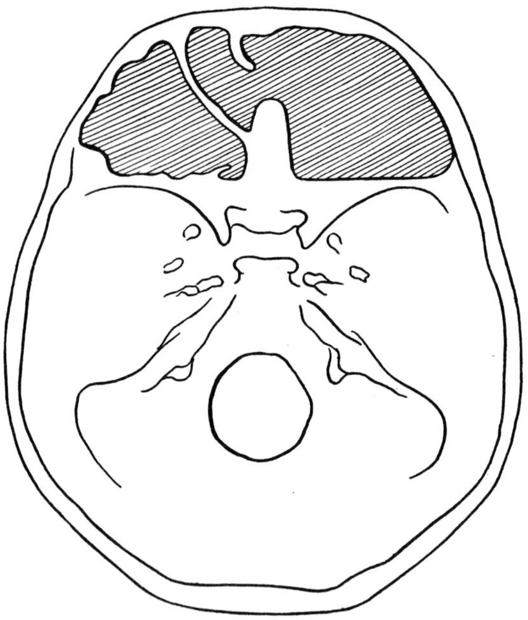

Fig. 4-23. Extreme supraorbital development of the frontal sinuses. (After Schaeffer JP: The Nose, Paranasal Sinuses, Nasolacrimal Passageways, and Olfactory Organ in Man. Philadelphia, Blakiston, 1920)

The connection of the frontal sinus with the nasal cavity may be by means of a narrow duct, or the sinus may open directly through an ostium (Fig. 4-25). Lillie and Simonton reported on a patient in whom at operation an anterior ethmoid cell was found to have invaded the frontal sinus so as to interfere with its ventilation and drainage, and Van Alyea ('41a, '44) stated that in approximately 50% of persons there are extra air cells (anterior ethmoid cells) impinging upon the frontal sinuses, about one third of them in the region of the frontal ostium. If the floor of the sinus is pushed up, the bulge is known as a frontal bulla; such a bulla may be due to encroach-

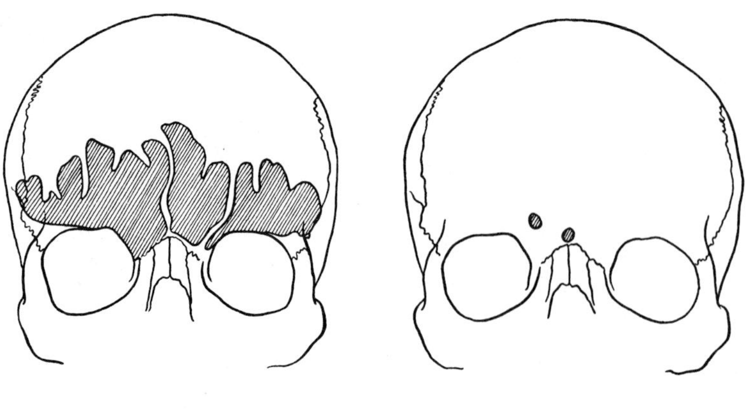

Fig. 4-24. Some extremes in the development of the frontal sinuses. On the left, the left sinus is double. (After Schaeffer JP: The Nose, Paranasal Sinuses, Nasolacrimal Passageways, and Olfactory Organ in Man. Philadelphia, Blakiston, 1920)

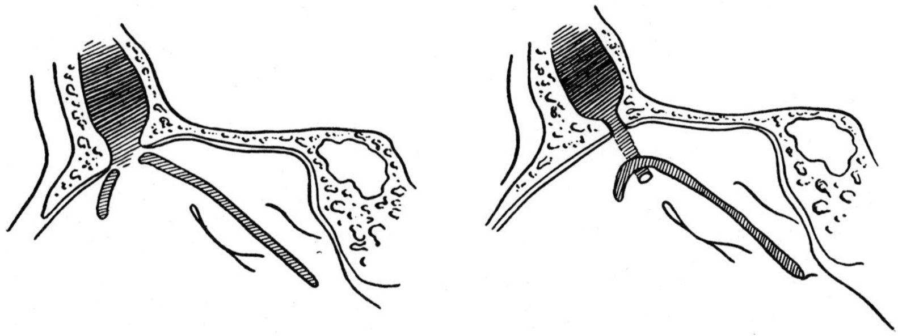

Fig. 4-25. The two chief types of drainage of the frontal sinus—by an ostium directly into the nose and by a (nasofrontal) duct.

ment by the other frontal sinus, but more frequently is formed by an ethmoid cell (Fig. 4-26).

Van Alyea concluded that in many instances such cells must influence drainage from the frontal sinus and that their presence is sometimes the sole reason for the maintenance of chronic infection in the sinus. According to Van Alyea, another anatomical factor interfering with drainage of the frontal sinus is a blocked middle meatus, usually caused by impingement of the middle concha against the lateral nasal wall.

From Van Alyea's ('46a) findings it appears that the frontal ostium should be relatively accessible for irrigation about 85% of the time, since in about 55% the frontal sinus drained directly into the frontal recess and in another 30% opened above but not into the infundibulum (Fig. 4-27). In 15%, however (his figures differ somewhat from Kasper's), he found the frontal sinus to be continuous with the ethmoidal infundibulum, and in about 1% it opened above the ethmoid bulla; in these instances the opening is apt to be inaccessible.

Blood Supply and Innervation

The chief blood and nerve supply to the frontal sinus reaches it through one or more small openings in the roof of the supraorbital foramen or notch. The supraorbital artery, as it passes from the orbit to the forehead, gives off a diploic branch to supply the diploë and the frontal sinus. The venous drainage, in addition to its connection with the veins of the nose, also parallels the arterial supply through a small vein that joins the connection between the supraorbital and superior ophthalmic veins. The connections of the venous drainage of the frontal sinus with other diploic veins and of the latter with the veins of the scalp and dura probably account for the spread of osteomyelitis in the frontal bone subsequent to disease of the frontal sinus (*e.g.*, Williams and Heilman). Similarly, the supraorbital branch of the ophthalmic nerve, as it passes across the supraorbital notch, gives off a branch to the diploë and the mucous membrane of the frontal sinus.

Fenton expressed the belief that referred pain from the ophthalmic division of the fifth nerve probably depends upon both the cutaneous and dural distribution of this nerve; frontal sinusitis has been said to give rise to supraorbital pain, which is consonant with the distribution of the ophthalmic nerve to the forehead and to the dura of the anterior cranial fossa. Reference of pain from the nose and the more posterior sinuses to the region of the ear and mastoid process, said to occur, has likewise been explained by the distribution of the fifth and seventh nerves to these regions.

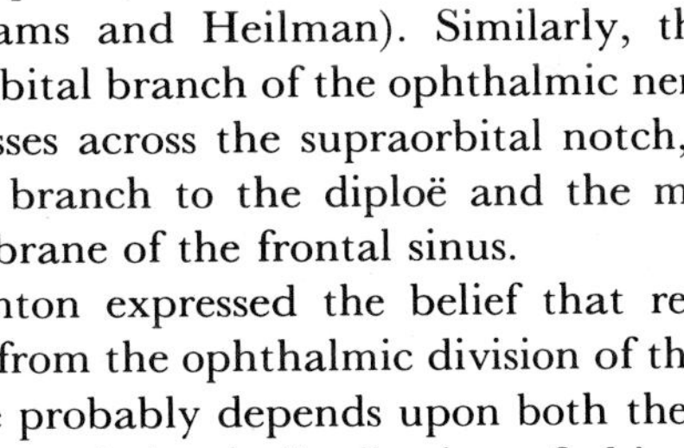

Fig. 4-26. A frontal bulla, in this case formed by an ethmoid cell, encroaching on the frontal sinus and its duct. (After Van Alyea OE: Arch Otolaryngol 34:11, 1941)

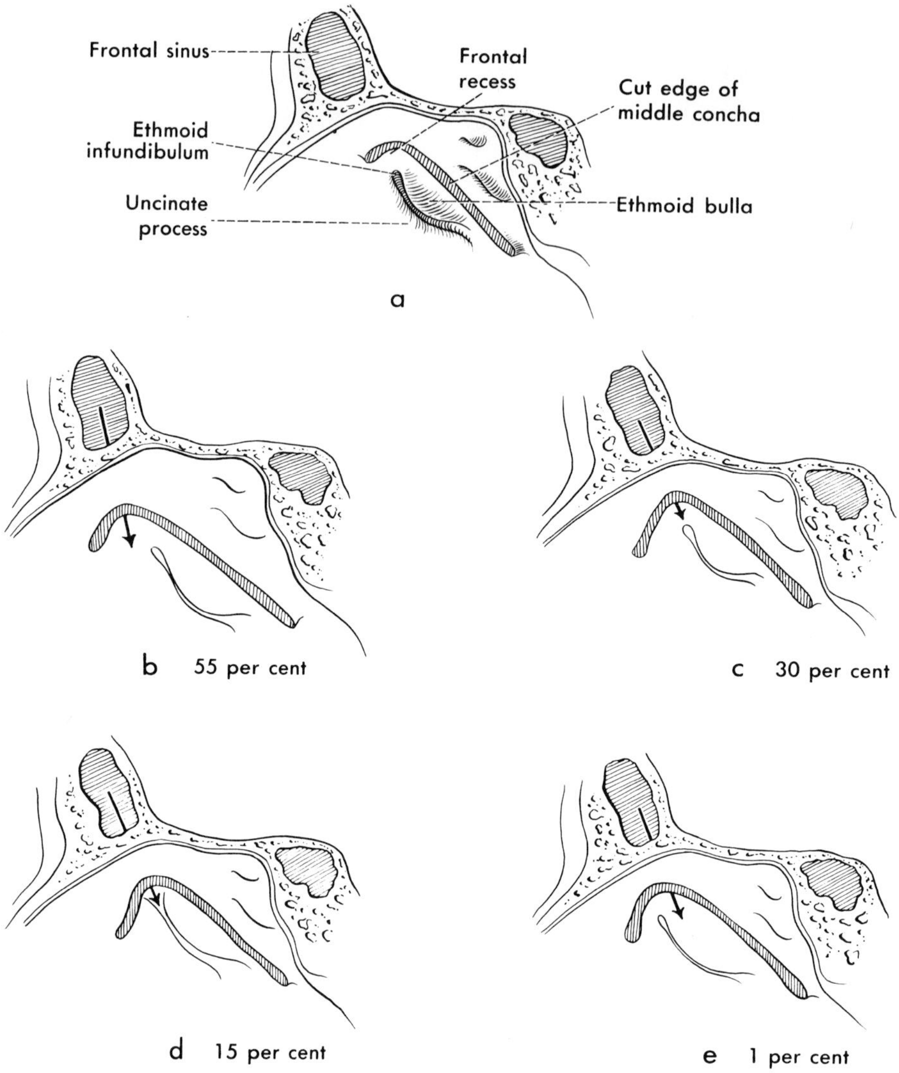

Fig. 4-27. Variations in the location of the opening of the frontal sinus. (*a*), The general anatomy depicted in these diagrams; in the remainder, which are adapted from Van Alyea (Ann Otol Rhinol Laryngol 55:267, 1946), the drainage of the sinus is indicated by the *arrow*. Approximate percentages of the various types, as given by Van Alyea, are noted. (*b*), Drainage into the frontal recess anterior to the infundibulum; (*c*), drainage above but not into the infundibulum; (*d*), drainage into the infundibulum; (*e*), drainage above the bulla.

ETHMOID SINUS

Each ethmoid sinus consists of a variable number of separate cavities, the ethmoid cells or ethmoid air cells (Figs. 4-28, 4-29, and 4-30) that honeycomb the ethmoid bone and thus lie between the upper part of the lateral nasal wall and the medial wall of the orbit. There may be as few as 3 or as many as 18 or so on each side; the greater the number, the smaller the cells. Mosher ('29) discussed various phases of the surgical anatomy of the ethmoid region; he noted that although the ethmoid cells typically together form a pyramid, with the wider base located posteriorly, the ethmoid labyrinth as a whole may form a relatively thin plate. He gave its average dimensions as 4–5 cm long (anteroposterior),

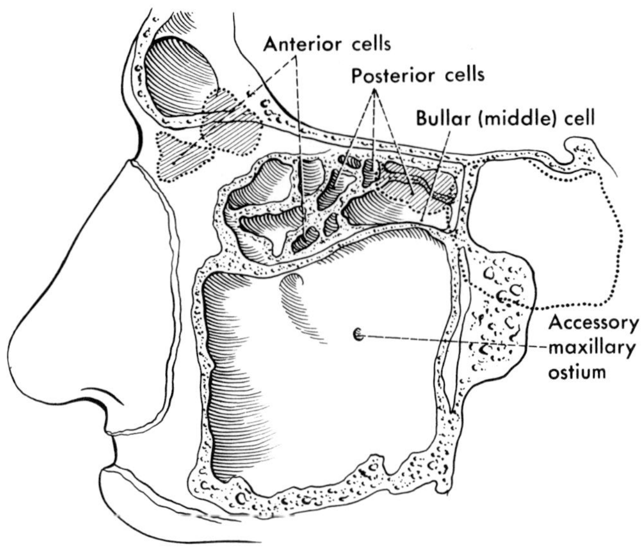

Fig. 4-28. Ethmoid cells from the lateral side. (After Van Alyea OE: Arch Otolaryngol 29:881, 1939)

2.5–3 cm high, and about 0.5 cm wide anteriorly and 1.5 cm posteriorly.

The ethmoid cells show great variety in their *openings,* since they may have arisen as diverticula above, upon, or below the bulla, or from any portion of the middle meatus, from the frontal recess, from the superior meatus, or from above and behind the superior nasal concha. On the basis of their openings, they are commonly divided into anterior ethmoid cells, which open into the middle meatus, and posterior ethmoid cells, which open into the superior or supreme meatus. In this classification the middle ethmoid cells of anatomists, which open on or above the ethmoid bulla, are a subgroup (bullar cells) of the anterior cells.

The ethmoid cells do not necessarily remain in the ethmoid bone. They may invade the squama (perpendicular part) of the frontal bone and thus encroach on a frontal sinus, or the orbital part of this bone, spreading variably in the roof of the orbit. Or they may invade the maxilla, encroaching on the maxillary sinus, or the sphenoid bone, where a large ethmoid cell may largely crowd out and replace the sphenoid sinus of its side.

Anterior Ethmoid Cells

The anterior ethmoid cells are classified into subgroups upon the basis of their location or that of their ostia; since, however, cells of a given origin frequently invade the territory usually occupied by cells of another origin, they are most easily classified by the location of their ostia. On this basis the anterior ethmoid cells are divided into frontal recess cells, opening into the frontal recess of the middle meatus; infundibular cells, opening into the ethmoidal infundibulum; and bullar (*middle*

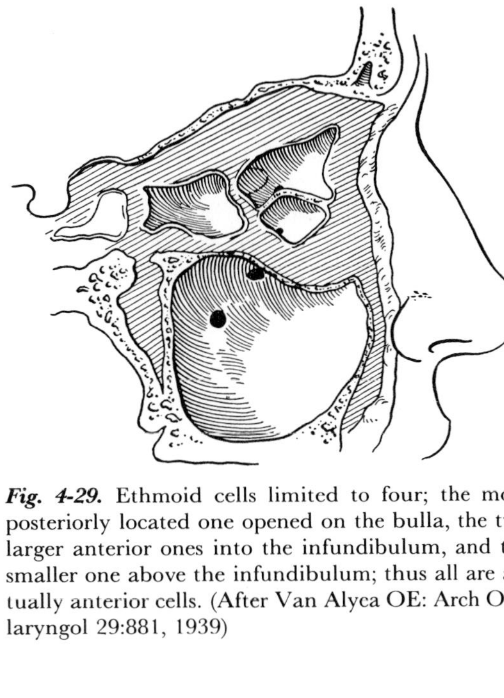

Fig. 4-29. Ethmoid cells limited to four; the more posteriorly located one opened on the bulla, the two larger anterior ones into the infundibulum, and the smaller one above the infundibulum; thus all are actually anterior cells. (After Van Alyea OE: Arch Otolaryngol 29:881, 1939)

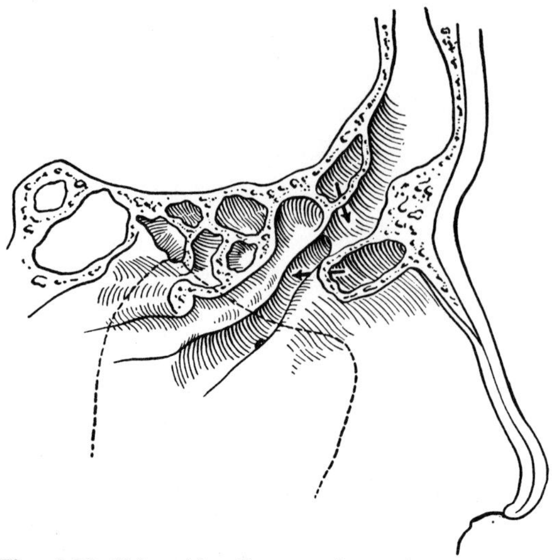

Fig. 4-30. Ethmoid cells seen from the nasal side. Only three anterior cells are shown here, opening respectively above the infundibulum, above the bulla and (an agger nasi cell) into the infundibulum. (After Van Alyea OE: Arch Otolaryngol 29:881, 1939)

ethmoid) cells, opening directly into the middle meatus either upon or above the ethmoidal bulla.

The three or four *frontal recess ethmoid cells* are related developmentally to the frontal sinus; one of these, especially, may encroach upon the frontal sinus, either upon its wall or at its ostium, although infundibular or bullar cells often also reach this region. Van Alyea ('41a) found 100 specimens in which ethmoid cells encroached upon the frontal sinus (frontal ethmoid or ethmofrontal cells) in 242 specimens examined; 56 of these were frontal recess cells, and 63 cells were of other origin (both types being present in 19 cases).

Van Alyea ('39) reported that in 100 specimens there was an average of three *infundibular cells,* the number varying from one to seven; the most constant cell he found was in the agger nasi, and encountered in 89 of 100 specimens, being represented in 12 of these by two cells. It is an agger nasi cell that is most frequently opened in the endonasal approach to the lacrimal sac. In 34 cases the cell of the agger nasi formed the upper blind termination of the infundibulum.

The number of *bullar* (*middle*) *cells* was also reported to average three, with a variation of from one to six; Van Alyea regarded them as the most constant of any of the ethmoid group.

As already stated, anterior ethmoid cells of one origin may encroach upon the territory of those of another origin; the most frequent example of this Van Alyea found was the extension of a "tip cell" (from the anterior end of the infundibulum) into the bulla (16%). Cells of either the anterior or the posterior group may encroach upon the maxillary sinus. Ethmoid cells may also extend into the middle or superior nasal conchae or into the uncinate process (Schaeffer, '10a); they open into the infundibulum, the middle meatus directly or into the superior meatus, and appear therefore to be sometimes anterior and sometimes posterior ethmoid cells.

Posterior Ethmoid Cells

There are usually one to seven posterior ethmoid cells; in his dissections Van Alyea found from none to six, but one or more were present in 96% of his specimens; in 67 of his 100 specimens there was a supreme concha present and in 38 of these one or more cells (postreme cells) opened into the supreme meatus.

The bullar and posterior ethmoid cells may encroach upon each other, the bullar cells spreading backward or the posterior cells forward so that these groups overlap; less frequently, infundibular and posterior ethmoid cells are in contact. The posterior cells may grow into the sphenoid bone and therefore become closely related to the sphenoid sinus. Mosher ('29) found such cells in about 25% of skulls and reported that they were especially likely to be overlooked in the transantral approach to the ethmoid labyrinth. Riggle reported three cases in which a posterior ethmoid cell had grown back into the body of the sphenoid bone and practically taken the place of the sphenoid sinus, the latter being a rudimentary cell rather than a large sinus. In 55 of 1600 specimens Dixon found posterior ethmoid cells that had invaded the superior part of the sphenoid bone to partially surround the optic canal; Van Alyea ('44) found a similar relation in five of 100 sides. In contrast to the tendency of ethmoid cells to invade the territory of the neighboring sinuses, these cells are relatively free from interference with their growth, except by each other, since they develop early.

Because the ethmoid cells as a whole are separated from the orbit only by the very thin orbital lamina (lamina papyracea) of the ethmoid bone, severe infections of them spread fairly readily to the orbit, and infections of the ethmoids have been said to be the most common cause of orbital cellulitis. Since one or more posterior ethmoid cells may be in close contact with the optic nerve as this traverses the optic canal, infections of these or of the sphenoid sinus, also closely related to the optic canal, have been held to be sometimes reponsible for retrobulbar neuritis.

Regardless of whether such a causal relationship exists, the anatomical relations are of importance to the surgeon. Goodyear ('48) expressed the belief that the sinuses should be carefully examined in all ophthalmic lesions but, in addition, pointed out that he had seen total blindness of an eye follow surgical opening of posterior ethmoid cells and of the sphenoid sinus; he described and illustrated a case in which the optic nerve passed through the middle of a posterior ethmoid cell, being completely surrounded by this and separated from the ethmoid mucosa by only a very thin cylinder of bone. In such cases as this, the danger of operative injury to the optic nerve during the process of opening and cleaning infected sinuses is obviously considerable, and in all cases it should be realized that the bony lamina about the optic canal may be very thin, even presenting dehiscences.

Blood Supply and Innervation

The arterial supply of the ethmoid cells is through the posterior lateral nasal branches of the sphenopalatine artery; in addition, the anterior ethmoid cells are supplied by twigs from the anterior ethmoidal artery (and in the case of some frontal ethmoid cells, probably also by the frontal branch of the supraorbital artery), while the posterior ethmoid cells are supplied by branches from the posterior ethmoidal artery. Thus, the ethmoid sinuses receive blood from both the external and internal carotid systems.

The venous plexuses of the ethmoid sinus join those of the nasal cavity proper.

The ethmoid cells possess only a few lymphatic capillaries. These pass at the ostia into the general nasal mucosa, whose lymphatics they join.

The *innervation* of the anterior and middle ethmoid cells is through the ophthalmic division of the trigeminal nerve, by way of the anterior ethmoidal branch of the nasociliary (nasal) nerve, and in the case of frontal ethmoid cells probably also by the twig given off by the supraorbital branch of the ophthalmic to the frontal sinus. The posterior ethmoid cells receive their innervation from the maxillary branch of the fifth (and from the facial?) by way of the lateral posterior superior nasal branches of the pterygopalatine (sphenopalatine) ganglion that supply the adjacent mucous membrane of the superior and middle nasal conchae; they are said to be innervated also by the orbital or ascending branches of the pterygopalatine ganglion. In addition, the posterior ethmoidal foramen may transmit a posterior ethmoidal branch of the nasociliary nerve, so that the posterior ethmoid cells may be innervated by both ophthalmic and maxillary divisions of the fifth; the posterior ethmoidal branch of the nasociliary is inconstant, however, apparently being less commonly present than absent.

SPHENOID SINUS

The sphenoid sinus regularly opens into the sphenoethmoidal recess, that is, above and behind the superior nasal concha (see Figs. 4-14 and 4-32). The ostium is usually in the posterior wall of the recess, but sometimes is in its lateral wall. Although the anlage of the sinus, represented by a posterior recess of the nasal cavity, is recognizable as early as the fourth month of fetal life, it is only after the third year that the sinus begins to grow and actually excavate the body of the sphenoid bone.

The degree of growth of the sinus into the sphenoid bone varies enormously (Fig. 4-31); the capacity of this sinus has been said to vary from 0.5–30 ml, with an approximate average of about 7.5 ml. The sphenoid sinus may be entirely limited to the body of the sphenoid

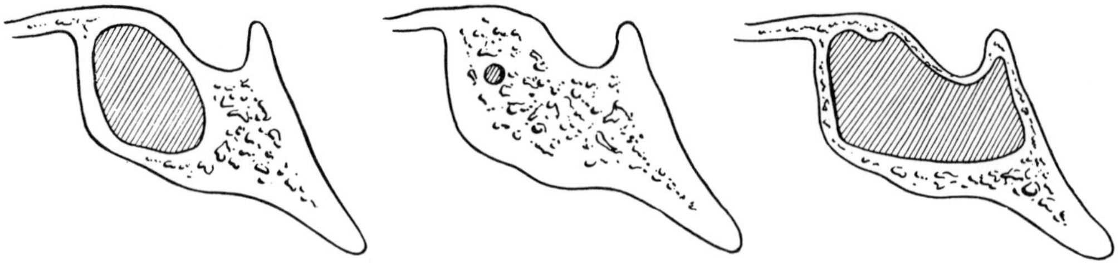

Fig. 4-31. Medium and extremes of development of the sphenoid sinus.

bone or may extend into the greater wing of the sphenoid, into the pterygoid process, and into the basilar portion of the occipital bone. The two sphenoid sinuses are rarely symmetrical in size and shape, and the septum between them is usually also asymmetrical in its position, especially posteriorly; sometimes the overlapping sinuses of the two sides lie above and below each other, rather than having the expected lateral relationship. The two sinuses rarely communicate with each other, although Dixon found five cases (in 1600 skulls) in which the septum was deficient. Judging from a case described by Costen ('30), infection of a sphenoid sinus that extends markedly across the midline may cause pain referred to the side to which it has extended.

Congdon stated that the sphenoid sinus is the most variable in form of any bilateral cavity or organ in the human body, and Van Alyea ('41b) found it to be more variable in its anteroposterior dimension than in any other. The 100 sinuses examined by Van Alyea varied from 4–44 mm in length, from 2.5–34 mm in width, and from 5–33 mm in height. Dixon, noting that the sinus is larger in black than in white persons (which is generally true of all the sinuses and their ostia), found the average height of the sinus to be 18–20 mm, the average width 15–17 mm, and the average length 19–22 mm, in 1600 skulls classified according to race and sex.

Relationships

The relationships of the sphenoid sinus are of particular importance and have been detailed especially by Schaeffer ('20), by Van Alyea, and by Dixon. Above, the sinus is related to the middle cranial fossa and the pituitary gland in the sella turcica (Fig. 4-32); laterally, it is related to the cavernous sinus and the internal carotid artery (Figs. 4-33 and 4-34); posteriorly, it is in relation to that portion of the posterior cranial fossa in which the pons lies; and inferiorly, it lies above the roof of the nasopharynx. The thickness of the walls of the sphenoid sinus depends largely upon the extent of the development of the sinus; according to Dixon they average about 0.5 mm, the inferior wall, or floor, being usually the thickest. The posterior wall, between the sinus and the posterior cranial fossa, may be of extremely heavy bone, however, as much as 1 cm thick.

When a sphenoid sinus is well developed it is usual to see some of the structures adjacent to its walls modeling these and thus appearing in more or less bas-relief within the sinus. The sphenoid sinus may extend into the anterior and posterior clinoid processes, so that the hypophysis projects markedly into the sinus and the wall of the sella turcica is particularly thin (Dixon). The internal carotid artery rather regularly indents the posterolateral wall (Fig. 4-35); Van Alyea found such a projection in 65 of 100 cases, and in 53 of these it was pronounced, while Dixon noted that actual gaps in the bony wall are not uncommon here.

Dixon found a very deep projection of the optic canal into the sinus in 7% of specimens, the wall of the optic canal being then very thin and sometimes presenting dehiscences; Van Alyea found a recognizable projection formed by the optic canal into the sphenoid in 40%. The maxillary division of the fifth nerve may also indent the sinus; Dixon noted this in 32 (of 1600), while Van Alyea found it far more frequently, in 40 of 100. Fujii and co-workers, investigating 25 heads with the

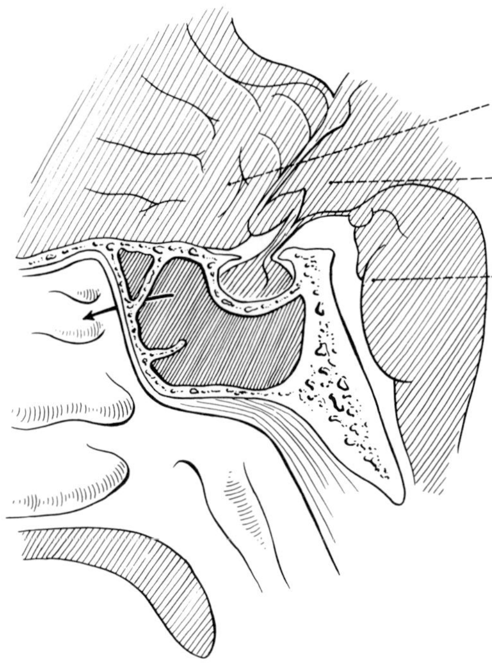

Fig. 4-32. Relations of the sphenoid sinus to the nose, brain, and hypophysis in a nearly sagittal section. In this diagram an anterior overlapping of the two sinuses is indicated, as is a septum partially subdividing the larger sinus.

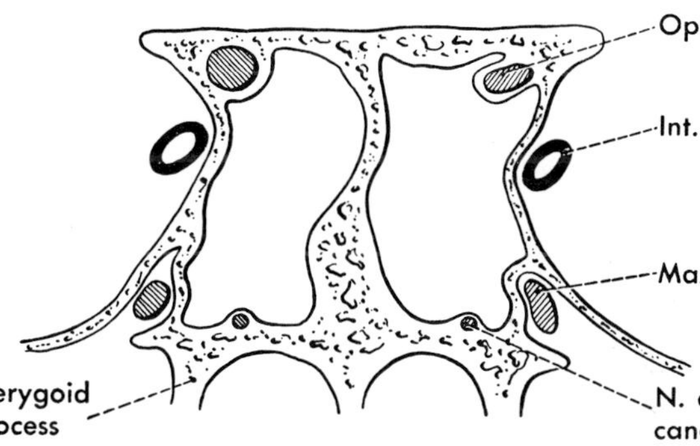

Fig. 4-33. Relations of the sphenoid sinuses in a schematic frontal section anterior to the hypophyscal fossa. The degree to which the structures related to the wall of the sinus project into this varies.

surgical microscope, recognized bulges made by the optic canals and carotid arteries in all but one side of one specimen and bulges made by the maxillary nerve in all but one side of two specimens. In more than half, the bony wall between the sinus and the nerve or vessel was thinner than 0.5 mm, and only dura and mucosa separated them in some instances. Dixon also noted that a ridge for the nerve of the pterygoid canal (Vidian nerve) was often seen in the floor of the sinus; Van Alyea found such a ridge to be pronounced in 36% and slight in 12%. In 18 instances Van Alyea found that the sphenoid sinus extended far enough anteriorly to produce a bulge in the wall of the maxillary sinus. Diverticula of the sinus may extend through bony dehiscences to lie directly against the dura (Schaeffer). These various relationships to important structures indicate the care that should be exercised in exploring the sinus.

As already noted, a posterior ethmoid cell may grow into the body of the sphenoid bone and lead to marked underdevelopment of the sphenoid sinus (Fig. 4-36); such posterior ethmoid cells may assume the relationship to the optic canal more frequently assumed by the sphenoid sinus. Vail pointed out that some-

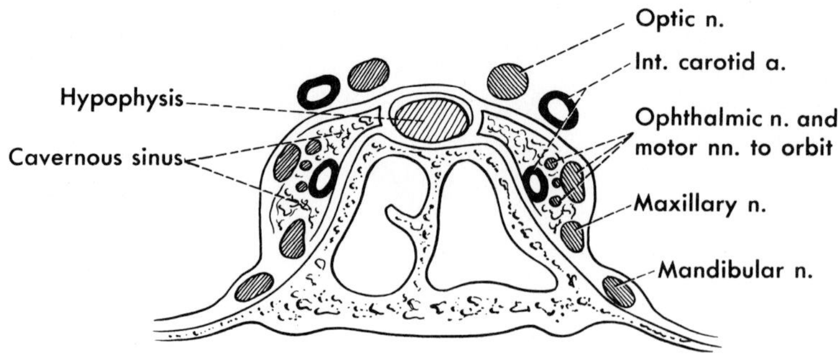

Fig. 4-34. Relation of the sphenoid sinuses in a schematic frontal section through the level of the hypophyseal fossa.

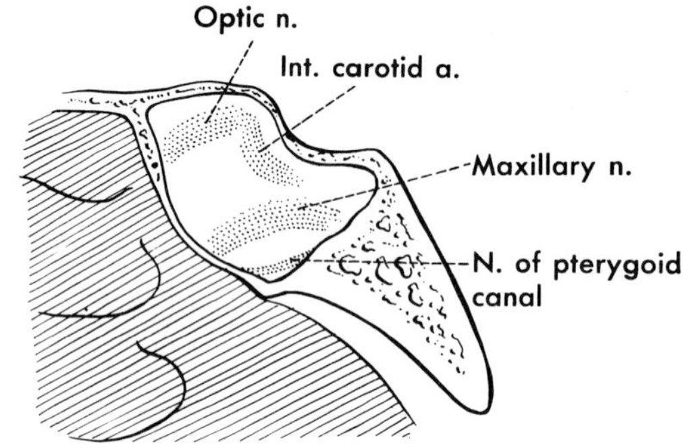

Fig. 4-35. Some markings that may be visible within the sphenoid sinus.

times the wall between the sphenoid sinus or a posterior ethmoid cell and the optic nerve is dense and nonporous, and sometimes it has large spongelike marrow spaces; he has visualized the sphenoid sinus by placing radiopaque oil within it and then, by putting the patient's head so that the upper outer angle of the sinus is dependent, has studied the relationship between the optic nerve and the sinus.

Drainage

Van Alyea stated that the sphenoid sinus has been said to be involved in 15% of all clinical cases of sinusitis and to be responsible for approximately 35% of all intracranial complications arising from the nose. He reported that examination of necropsy and cadaveric material has shown an incidence of low-grade chronic infection of these sinuses in approximately 13% to 14%. Poor drainage of any sinus obviously favors its infection, and Van Alyea emphasized the relation of anatomical variations to poor drainage of the sphenoid.

Normally, drainage from the sphenoid sinus in the erect posture is entirely through ciliary action, since the ostium is located typically well above (about 1.5 cm) the floor of the sinus. Gravity drainage obtainable by turning the face down may be interfered with by septa or rows of spurs projecting into it and partially subdividing it. Congdon found 122 bony septa or rows of spurs in 212 sinuses; Dixon found septa, partly bony and partly membranous, in 32% of his 1600 specimens, and Van Alyea found partial septa in 44 of his 100 cases. Obviously, whatever the actual incidence of such septa may be, it is high.

The ostium of the sinus in the fresh condition is rarely as large as the bony ostium seen on the dried skull, since it is normally blocked in part by a membranous septum, but among the variations interfering with drainage Van Alyea listed an ostium smaller than 2 mm in diameter, one particularly near the roof, and one encroached upon by a posterior ethmoid cell. The frequent presence of recesses and septa within the sinus, already mentioned, and the occurrence of a narrow sphenoethmoid recess through which heavy secretions pass only with difficulty, are also variations that may interfere with drainage. Pathologic blockage may result from thickening of the mucosa, hyperplasia, polyps, and the like.

Blood Supply and Innervation

The innervation and vascular supply of the sphenoid sinus are similar to those of the posterior ethmoid cells, that is, branches of the pterygopalatine ganglion and sphenopalatine artery, the posterior ethmoid artery, and sometimes a posterior ethmoid nerve. Its venous plexus and sparse lymphatics pass through its ostium to join those of the nasal cavity.

MAXILLARY SINUS

The maxillary sinus (antrum, of Highmore), usually the largest of the paranasal sinuses, is located in the body of the maxilla. Its antero-lateral wall is the anterior or facial surface of this bone, its posterior wall the infratemporal surface, and its medial wall the lateral wall of the nasal cavity. The roof of the sinus is also the floor of the orbit, and its floor is the alveolar and palatine processes of the maxilla (Fig. 4-37). The average dimensions of the maxillary sinus in the adult are said to be 33 mm high, 23 mm wide, and (in the anteroposterior direction) 34 mm long; it varies considerably in capacity, however, since it may extend into the various bony processes from the body of the maxilla or may have unusually thick walls, or there may be an inward bulging of the facial and nasal walls. The apex of the cavity usually extends into the zygomatic process of the maxilla and may extend even into the zygomatic bone itself.

In the same way, the relation of the maxillary sinus to the teeth is inconstant, depending upon the extent to which the sinus invades the alveolar process (Figs. 4-38 to 4-41). According to Schaeffer ('10b), the three most constant teeth in relation to it are the three

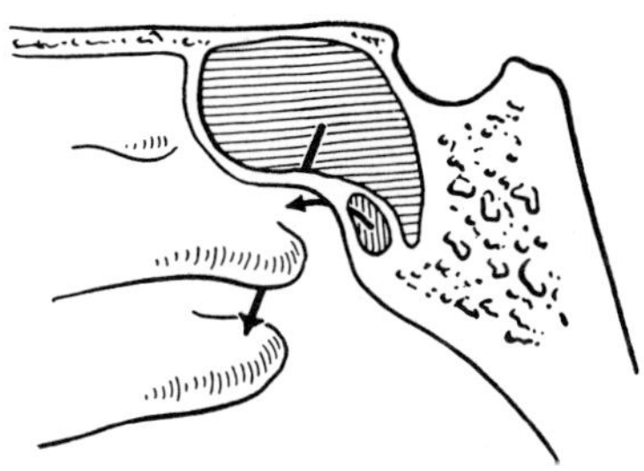

Fig. 4-36. A large posterior ethmoid cell largely surrounding and almost replacing a small sphenoid sinus. (After Riggle PP: Arch Otolaryngol 16:532, 1932)

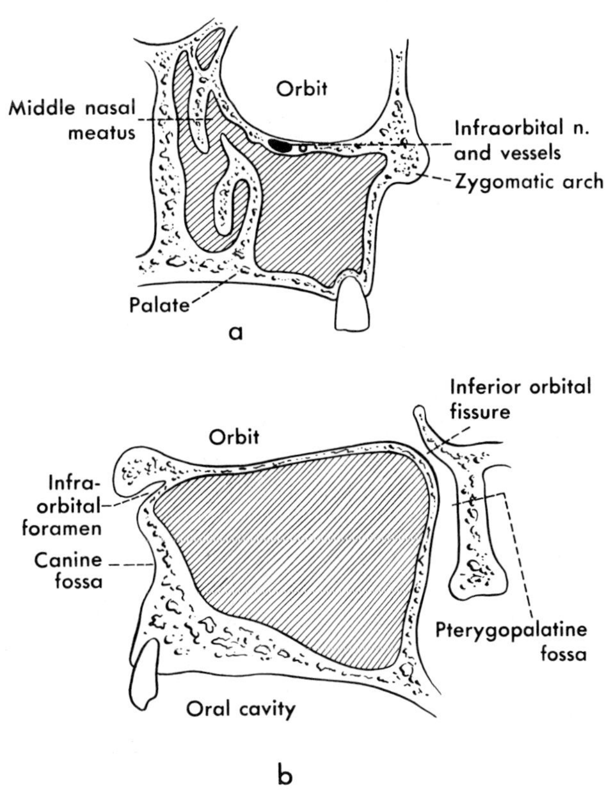

Fig. 4-37. Some relations of the maxillary sinus in (*a*) frontal section and (*b*) sagittal section.

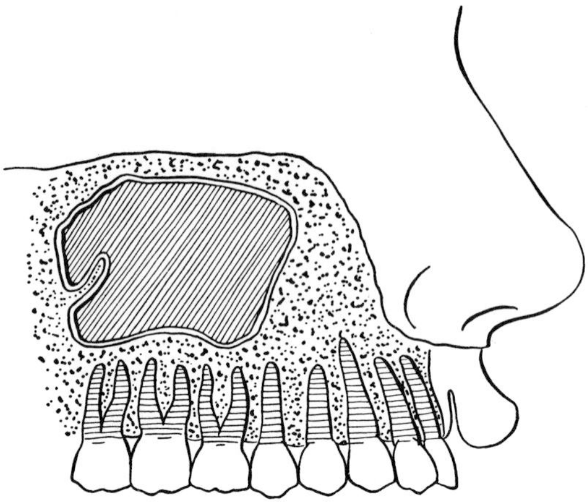

Fig. 4-38. Approximately average relations of the maxillary sinus and the roots of the teeth.

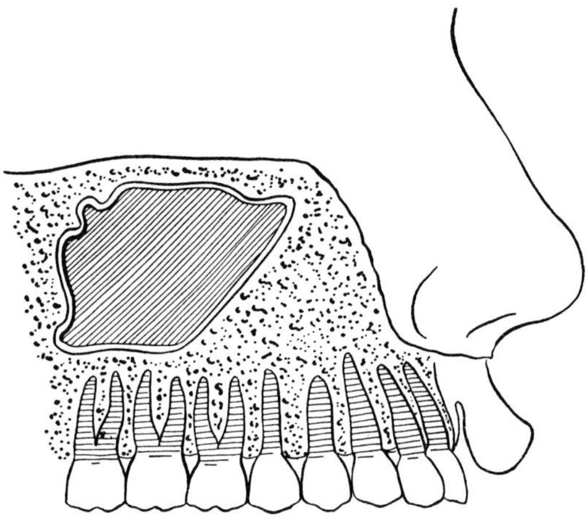

Fig. 4-39. Shortening of the ventrosuperior diagonal of the maxillary sinus, so that the only teeth in direct relation to it are the second and third molars. (After Schaeffer JP: Am J Anat 10:313, 1910)

molars (Fig. 4-38), although there may be fewer (Fig. 4-39); he found that only occasionally does the canine tooth have a direct relation to it, and in a slightly larger percentage the first and second premolars are similarly related. The floor of the sinus in relation to these teeth may be relatively smooth, or it may be modeled over the root of one or more teeth so that these project upward into the sinus, separated from it only by a very thin lamina of bone (Fig. 4-40) or even by no bone at all (Fig. 4-41).

Ridges and septa in the maxillary sinus are common; Schaeffer pointed out that larger ones may turn upward so that they prevent adequate drainage of the sinus from below, serving as small basins that catch the fluid. Occasionally the ridges may form complete septa, dividing a sinus into two portions, in which case one portion may empty by a small foramen into the other or by an accessory ostium into the nasal cavity; very rarely two maxillary sinuses develop in one bone. The

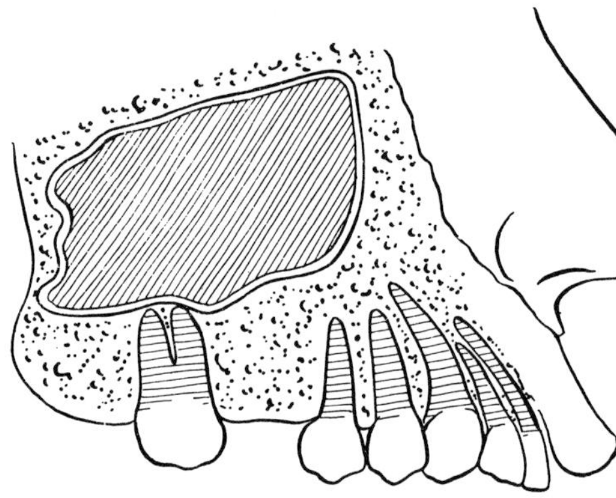

Fig. 4-40. Moderately thick alveolar process, with the roots of the second molar tooth very close to the mucous membrane of the maxillary sinus. (After Schaeffer JP: Am J Anat 10:313, 1910)

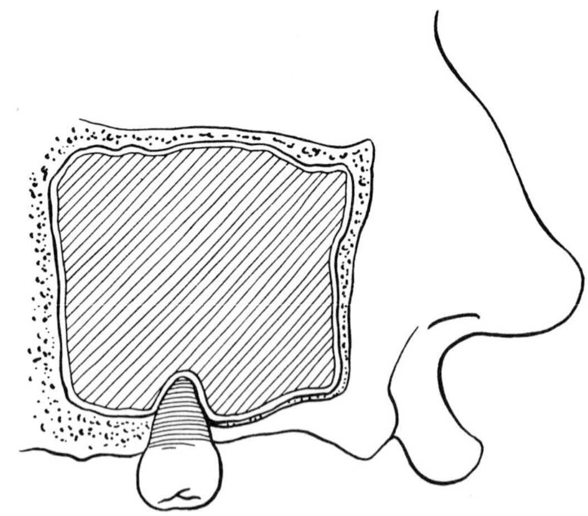

Fig. 4-41. Extreme development of the maxillary sinus, with marked thinning of the alveolar process; the root of the sole remaining tooth projects into the sinus, covered only by the mucous membrane of this. (After Schaeffer JP: Am J Anat 10:313, 1910)

infraorbital vessels and nerves, which run in the roof of the sinus, may be separated from the mucous membrane of the sinus by a bony wall of only papery thinness and are occasionally even in direct contact with it, as may be also the superior alveolar nerves, both anterior and posterior.

Berry investigated the roots of the teeth in 152 patients with maxillary sinusitis to see what percentage might have been caused by bad teeth. In only 11% did all the teeth seem healthy; 41% had a dead tooth in the floor of the sinus, and an origin of the infection from the tooth seemed possible; in 30% there was an abscessed tooth present here, and the origin seemed probable; in 18% the maxillary sinusitis was very definitely traced to an infection about the dental apex, and further sinus infection was obviated after the elimination of this primary infection.

The *innervation* and *blood supply* of the maxillary sinus is in part through small twigs derived from the posterior superior alveolar, infraorbital and anterior superior alveolar nerves and arteries. Its largest artery is a branch from the superior artery on the inferior concha, which enters the ostium of the sinus. Similarly, its chief venous and lymphatic drainage is through the ostium into the nasal cavity.

Ostium of Maxillary Sinus

The ostium of the maxillary sinus is on the highest part of the medial wall of the sinus and is therefore poorly placed from the point of view of free drainage; in addition, it opens not directly into the nasal cavity but into the narrow ethmoid infundibulum, inflammation of which can further interfere with drainage. Van Alyea ('46b) stated that the maxillary ostium in a healthy state is of ample size and provides adequate drainage for the sinus (with the aid of ciliary action) but that in any large group of specimens there may be found some with a complete absence of drainage because of adherence between the uncinate process and the ethmoidal bulla, with obliteration of the semilunar hiatus.

Simon insisted that the maxillary ostium is actually more frequently a canal, reporting that in 82.7% of 110 openings examined there were canals 3 mm or more in length, the average length being 5.55 mm. This surely indicates the thickness of the infundibular and sinus mucosa rather than the presence of a bony canal, since the ostium is usually located in that region of the lateral nasal wall which is partly very thin bone and partly membranous.

It is in this thin and partly membranous lateral nasal wall that accessory ostia may also be located; Myerson ('32a) found an incidence of accessory ostia of 30.7%; Schaeffer ('20), one of almost 41% in 295 sinuses. Van Alyea found accessory ostia in 23% of 163 specimens; Rosenberger observed them in only 4% of black and 17% of white persons, an incidence much lower than that reported by other workers.

The ostium of the maxillary sinus is commonly said to open into the posterior half of the ethmoidal infundibulum (that is, into the groove—"uncinate groove"—extending downward and backward from the infundibulum proper; under cover of the uncinate process); Van Alyea ('36), analyzing 163 specimens, found only nine ostia in the anterior third of the uncinate groove, 18 in the middle third, and 117 or 71.8% in the posterior third; the remaining 19 ostia opened at the extreme posterior tip of the groove and were visible, rather than hidden by the uncinate process (Fig. 4-42). Rosenberger agreed that the ostium opens into the lower or posterior third in about 70%.

Fig. 4-42. Variation in positions of the ostium of the maxillary sinus in relation to the anterior, middle and posterior thirds (*1, 2,* and *3*) of the ethmoidal infundibulum. The percentages given are those of Van Alyea (Arch Otolaryngol 24:553, 1936)

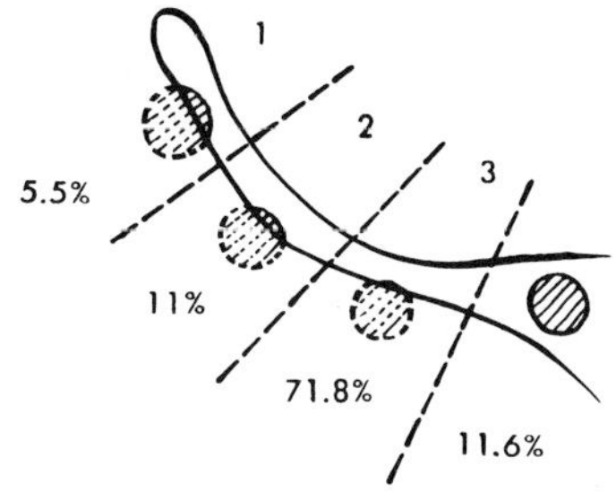

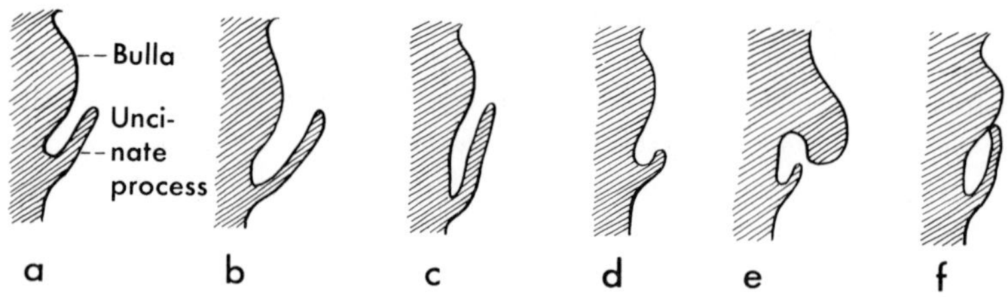

Fig. 4-43. Some variations in the semilunar hiatus and ethmoid infundibulum that would affect the accessibility of a maxillary ostium opening into it. In (*a*) the uncinate process is moderately tall and the hiatus moderately wide; in (*b*) there is an extreme development of the process but the hiatus is also especially wide; in (*c*) the hiatus is narrow; in (*d*) the uncinate process and therefore the infundibulum are rudimentary; in (*e*) the bulla overhangs the uncinate process; and in (*f*) the two are in contact.

The accessibility of the ostium varies both with its position and the shape of the infundibulum (Fig. 4-43). In 25% of Van Alyea's specimens the uncinate wall was high, and probing the ostium would apparently have been almost impossible in the living, while in eight cases it was noticeably low; in 18 the hiatus was especially narrow, and in 12 a low and overhanging bulla blocked the ostium.

Myerson found that the uncinate process varied in length from 14–22 mm and was usually about 5 mm high—that is, the ethmoidal infundibulum was therefore about 5 mm deep. In 114 lateral nasal walls he found the uncinate process to be absent 3 times, rudimentary in 10, moderately developed in 69, and pronounced, that is, more than 5 mm high, in 32; he noted that the ethmoidal bulla may bulge so as actually to touch the uncinate process and that the greatest width of the semilunar hiatus was 3 mm. The depth of the ethmoidal infundibulum was found in this study to vary from 0.5–10 mm.

Since the maxillary sinus is not only subject to infection, as are the other sinuses, from general nasal infection, but may also be infected from the teeth or from the frontal sinus or anterior ethmoid cells (Schaeffer, '16, found the ethmoidal infundibulum acting as a channel between these and the maxillary sinus in approximately 50% of adults), the accessibility of the maxillary ostium for catheterization and lavage becomes of considerable clinical importance. This obviously varies not only with the skill of the worker, but with the anatomy of the region—a high uncinate process and a deep and narrow ethmoidal infundibulum necessarily render access to the ostium difficult or impossible. Myerson ('32b) stated that in about 81% of clinical cases he was able to reach the normal ostium, but Rosenberger found that he could catheterize this in only about 46% of the skulls of white persons and 40% of the skulls of black persons; Van Alyea ('36) similarly concluded from his anatomical study of the location of the maxillary ostium that in nearly one half of the 163 specimens he examined it was so located that catheterization of the antrum would have been impossible or at least very difficult.

Approaches to Maxillary Sinus

Accessory, surgical, openings into the maxillary sinus may be either intranasal or facial. In making the former, it is necessary to penetrate the lateral nasal wall; this is commonly done through the inferior meatus, in which case the possibility of damage to the nasolacrimal ostium must be borne in mind, but has also been done above the attached edge of the inferior concha. Myerson ('32b) stated that there is no particular point in attempting to drain the maxillary sinus through a surgical opening in the inferior meatus, where one must traverse bone, when the sinus can be opened quite easily through the membranous portion of the lateral nasal wall; the former opening, however, has the advantage of better gravity drainage in the upright position, a factor that may be of considerable importance (see the last paragraph of this section).

In contrast, Goodyear ('49), stressing the importance of normal ciliary action about the natural ostium of the sinus, has strongly disapproved of interrupting this by antrostomies through the middle meatus and stated that a very large and permanent opening under the inferior concha should be made. Others have preferred the external approach to the maxillary sinus, finding that it usually gives a better chance of viewing the mucosa of the sinus, and the ostium, than does an opening through the wall of the inferior meatus. Simple punctures of the antral wall are of course most easily made through the thin wall of the middle meatus.

The extranasal approach to the maxillary sinus is usually through the canine fossa, the depression on the front of the maxilla extending from the infraorbital foramen (or the lower margin of the orbit) above to the alveolar process below; the anterior superior alveolar nerves and vessels run in the bony wall between the canine fossa and the maxillary sinus, and Mellinger, pointing out that their exact positions vary greatly, has found transmitted light to be often of value in locating these nerves and vessels as they cross the fossa.

Shapiro and co-workers, noting that operations upon the maxillary sinuses through the canine fossa are sometimes accompanied by anesthesia or paresthesia of the cheek, upper lip or teeth, studied the results following vertical incisions into the mucosa, as compared to horizontal ones, in regard to sparing of the infraorbital nerve. They stated that vertical incisions gave better exposure and less bleeding, but that there was apparently no appreciable difference insofar as sensation was concerned between the two incisions. They concluded that the course of the alveolar nerves, particularly the anterior, is such that damage to one or more of these is practically unavoidable in performing the usual radical type of operation on the sinus.

One should realize that artificial openings into the sinus are probably effective only through gravity drainage and that complete gravity drainage in the upright position frequently cannot be obtained by an intranasal approach, since the floor of the sinus may lie below the level of the nasal floor. Hilding ('41) concluded from experiments on rabbits that supplementary openings between the nasal cavity and the maxillary sinus may upset the normal physiology of the sinus, especially if they are made near the normal ostium, because of interference with the ciliary mechanisms; McMurray expressed the belief that even large normal accessory ostia may interfere with drainage from the maxillary sinus, by diminishing the normal negative pressure produced in the sinus by inspiration. King placed radiopaque oil in the maxillary sinus of a patient with an artificial opening and found that the oil usually did not flow out through the larger artificial opening but through the natural orifice, being drained therefore by ciliary action rather than by gravity. Hilding ('44) likewise stated that the direction of flow in the maxillary sinus due to ciliary activity is not altered by making a window into the inferior meatus. Nevertheless, the intranasal approach is generally regarded as providing satisfactory drainage in cases of chronic maxillary sinusitis.

Anomalies

Even though variations in the nasal cavities and sinuses are, as described under these structures, exceedingly common, major anomalies of the nose are relatively rare.

Congenital Atresia of Nose

Probably the most common of these is congenital atresia or obstruction of the nose. Blair classified patients with this anomaly into five general groups: (1) atresia at the (anterior) nares, (2) atresia at the choanae (posterior nares), (3) obstruction by some malformation within the passages, (4) total absence of the nasal passages, and (5) pharyngeal narrowing due to backward displacement of the maxilla. He pointed out that congenital obstructions of the nose vary greatly in their origin, but suggested that many are due to a partial reduplication of the nose.

Jervey described a patient with congenital *occlusion of both nares;* in this case there was a complete occlusion of the nasal passage about 6 mm back from each naris; the nose otherwise appeared normal. Jervey concluded that this condition must be the result of "faulty dislocations" of tissue before birth. He stated that this condition is apparently very rare.

Blair described atresia at the naris as characterized also by lack of the ala on that side and a turning downward of the inner canthus; this obviously differs from the case described by Jervey, but seems to correspond to what Blair and co-workers described as total absence of one lateral half of the external nose and of the naris. They reported several cases of such *hemiabsence,* accompanied by a marked downward inclination of the inner canthus and a flattening of the cheek; one patient was shown to have no maxillary sinus, and three had complete blockage of normal tear drainage. In the four patients on whom they operated there was also no evidence of a medial palpebral ligament.

Congenital *atresia of the choanae* is more common than that of the nares, but is nevertheless apparently rare; Wright and co-workers found 230 cases reported in the literature up to 1945, and Kazanjian found 19 cases among the records of more than 6200 patients at the hospital at which he worked. The condition may be unilateral or bilateral, and the obstruction may be a bony one or due to a membranous septum. Kearney, in reporting a case of unilateral atresia, stated that the obstruction is more frequently osseous than it is membranous, most commonly unilateral, and occurs more often in females and on the right side of the nose; Wright and co-workers gave a ratio of 90% bony to 10% membranous atresia. Colver noted both that unilateral atresia has been said to be more common and that bilateral atresia has been said to predominate 3:2. In four of Kazanjian's 19 cases the atresia was on the right side, in three it was on the left, and in three it was bilateral.

Colver pointed out that unilateral atresia, because it is less serious than bilateral, may remain unsuspected for years and suggested that bilateral atresia has probably been the unsuspected cause of many deaths within the first several postnatal days. It appears obvious that the true incidence of this condition is not known. Wright and co-workers listed three theories on the origin of choanal atresia: persistence of the upper part of the buccopharyngeal membrane, with fusion of the septum and the palatine processes to it; persistence of the nasobuccal membrane; and growth of the palatine processes at the choanae so that they fuse not only with the septum but also with the spenoid bone. They regarded none of these theories as proved. Stewart discussed several ways in which a persistent nasobuccal membrane might contribute to the deformity and stated that when the buccopharyngeal membrane persists the atresia is bilateral.

Membranous occlusions of the choanae may be relatively easily corrected, of course, or even spontaneous rupture may occur in early infancy. Bony atresia, however, offers more difficulty, although it must be corrected at least on one side if it is bilateral. According to Wright and co-workers, the occluding bone typically extends from the junction of hard and soft palates to the base of the sphenoid bone, and Colver found it reported as having been as much as 12 mm thick. Colver, and Wright and co-workers emphasized removal of the posterior portion of the nasal septum as fundamental to ensuring a free and permanent choanal opening, and Kazanjian also followed this procedure.

One operative approach has been along the nasal septum, following this up to the site of occlusion, but Wright and co-workers preferred an approach through the maxillary sinus, stating that this gave both a better exposure and more ready control of hemorrhage through ligation of the (internal) maxillary artery, if this should prove necessary. The transantral approach would, of course, not be feasible in the infant.

Cerebrospinal Rhinorrhea

Cerebrospinal rhinorrhea, although more commonly resulting from fractures of the skull, may also be due to congenital defects in the cribriform plate with the formation of a meningocele protruding into the nasal fossa; in this case it is known as spontaneous cerebrospinal rhinorrhea. Symptoms may not occur, or at least remain unnoticed, until adult life, and the condition must therefore be differentiated from allergic and vascular rhinorrhea.

Cerebrospinal rhinorrhea is discussed in more detail in Chapter 1; it need only be pointed out here that intranasal medication with silver nitrate does not attack the rhinorrhea at the logical point—the defect in the cribriform plate—and, since meningitis is a common consequence of spontaneous cerebrospinal rhinorrhea, an intracranial operation is usually indicated. The cerebrospinal rhinorrhea resulting from fractures of the skull usually disappears in some 6–8 weeks or less, as a result of spontaneous healing.

Other Anomalies

Anomalies of the external nose, besides hemiabsence, include cleft lip (harelip) and bifid nose. The former is discussed in Chapters 6 and 7. Bifid nose is characterized by separation at the mobile septum, that is, of the medial crura of the greater alar cartilages; more commonly the tip only is cleft, but the separation between the crura may be extreme, so that connective tissue and skin sink deeply between them.

References

AVERBUKH SS, BREVDA LS, LUBOTSKY DN, SEMENOVA OS: The palatine access to the ganglion sphenopalatinum and to the second branch of the trifacial nerve. Ann Surg 101:819, 1935

AYMARD JL: Some new points in the anatomy of the nasal septum, and their surgical significance. J Anat 51:293, 1917

BATSON OV: The venous networks of the nasal mucosa. Ann Otol Rhinol Laryngol 63:571, 1954

BERRY G: Further observations on dental caries as a contributing factor in maxillary sinusitis. Arch Otolaryngol 11:55, 1930

BLAIR VP: Congenital atresia or obstruction of the nasal air passages. Ann Otol Rhinol Laryngol 40:1021, 1931

BLAIR VP, BROWN JB, BYARS LT: Observations on sinus abnormalities in congenital total and hemi-absence of the nose. Ann Otol Rhinol Laryngol 46:592, 1937

BLANTON PL, BIGGS NL: Eighteen hundred years of controversy: The paranasal sinuses. Am J Anat 124:135, 1969

BROOKOVER C: The peripheral distribution of the nervus terminalis in an infant. J Comp Neurol 28:349, 1917

BROWNELL DH: Postoperative regeneration of the mucous membrane of the paranasal sinuses: A summary of the published investigations. Arch Otolaryngol 24:582, 1936

CHRISTENSEN K: The innervation of the nasal mucosa, with special reference to its afferent supply. Ann Otol Rhinol Laryngol 43:1066, 1934

COLVER BN: Congenital choanal atresia: Two cases of complete bilateral obstruction. Ann Otol Rhinol Laryngol 46:358, 1937

CONGDON ED: The distribution and mode of origin of septa and walls of the sphenoid sinus. Anat Rec 18:97, 1920

COSTEN JB: Relief of right-sided headache by resection of left sphenoid (anatomic variation of sphenoid sinuses): Report of a case. Ann Otol Rhinol Laryngol 39:1074, 1930

COSTEN JB: Anatomic phases involved in the surgery of the naso-antral wall and the floor of the mouth. Ann Otol Rhinol Laryngol 41:820, 1932

COTTLE MH, LORING RM, FISCHER GG, GAYNON IE: The "maxilla–premaxilla" approach to extensive nasal septum surgery. Arch Otolaryngol 68:301, 1958

CUSHING H: The sensory distribution of the fifth cranial nerve. Bull Johns Hopkins Hosp 15:213, 1904

DALEY J: Morphologic deformities of the lower lateral cartilages. Arch Otolaryngol 47:49, 1948

DION MC, JAFEK W, TOBIN CE: The anatomy of the nose. External support. Arch Otolaryngol 104:145, 1978

DIXON FW: A comparative study of the sphenoid sinus: A study of 1600 skulls. Ann Otol Rhinol Laryngol 46:687, 1937

DRUMHELLER GW: Topology of the lateral nasal cartilages: The anatomical relationship of the lateral nasal to the greater alar cartilage, lateral crus. Anat Rec 176:321, 1973

DURISCH LL JR, FRABLE MA: A surgical solution for posterior epistaxis. Surg Gynecol Obstet 133:669, 1971

FENTON RA: Pathways of referred pain from the nose. Am J Surg ns 42:194, 1938

FOLEY JO, DUBOIS FS: An experimental study of the facial nerve. J Comp Neurol 79:79, 1943

FOLEY JO, PEPPER HR, KESSLER WH: The ratio of nerve fibers to nerve cells in the geniculate ganglion. J Comp Neurol 85:141, 1946

FUJII K, CHAMBERS SM, RHOTON AL JR: Neurovascular

relations of the sphenoid sinus. A microsurgical study. J Neurosurg 50:31, 1979

GOODYEAR HM Ophthalmic conditions referable to diseases of the paranasal sinuses. Arch Otolaryngol 48:202, 1948

GOODYEAR HM Intranasal surgery of the maxillary sinus. Arch Otolaryngol 50:795, 1949

HIGBEE D: Functional and anatomic relation of sphenopalatine ganglion to the autonomic nervous system. Arch Otolaryngol 50:45, 1949

HILDING A: The physiology of drainage of nasal mucus. I. The flow of the mucus currents through the drainage system of the nasal mucosa and its relation to ciliary activity. Arch Otolaryngol 15:92, 1932

HILDING AC: Experimental sinus surgery: Effects of operative windows on normal sinuses. Ann Otol Rhinol Laryngol 50:379, 1941

HILDING AC: The physiology of drainage of nasal mucus. IV. Drainage of the accessory sinuses in man; rationale of irrigation of the infected maxillary sinuses. Ann Otol Rhinol Laryngol 53:35, 1944

HILL FT Local anestheisa for surgical treatment of the sinuses. Arch Otolaryngol 27:197, 1938

HIRSCH C: Ligation of the internal maxillary artery in patients with nasal hemorrhage. Arch Otolaryngol 24:589, 1936

HUNT JR: Geniculate neuralgia (neuralgia of the nervus facialis): A further contribution to the sensory system of the facial nerve and its neuralgic conditions. Arch Neurol Psychiatr 37:253, 1937

JERVEY JW JR: Congenital occlusion of both anterior nares. Ann Otol Rhinol Laryngol 53:182, 1944

KASPER KA: Nasofrontal connections: A study based on one hundred consecutive dissections. Arch Otolaryngol 23:322, 1936

KAZANJIAN VH: The treatment of congenital atresia of the choanae. Ann Otol Rhinol Laryngol 51:704, 1942

KEARNEY HL: Congenital bony atresia of the right posterior naris. Ann Otol Rhinol Laryngol 45:583, 1936

KERR FWL: Mechanisms, diagnosis, and management of some facial and cranial pain syndromes. Surg Clin North Am 43:951, 1963

KING E: A clinical study of the functioning of the maxillary sinus mucosa. Ann Otol Rhinol Laryngol 44:480, 1935

KIRCHNER JA, YANIGISAWA E, CRELIN ES JR: Surgical anatomy of the ethmoidal arteries. A laboratory study of 150 orbits. Arch Otolaryngol 74:382, 1961

KLAFF DD: The surgical anatomy of the antero-caudal portion of the nasal septum: A study of the area of the premaxilla. Laryngoscope 66:995, 1956

LARSELL O, BURNS EM: Some aspects of certain of the cranial nerves. Ann Otol Rhinol Laryngol 40:661, 1931

LARSELL O, FENTON RA: Sympathetic innervation of the nose: Research report. Arch Otolaryngol 24:687, 1936

LATTA JS, SCHALL RF: The histology of the epithelium of the paranasal sinuses under various conditions. Ann Otol Rhinol Laryngol 43:945, 1934

LEDERER FL, DINOLT R: Influence of avulsion of the trigeminal nerve on the human nose. Arch Otolaryngol 37:768, 1943

LILLIE HI, SIMONTON KM: Developmental extension of an anterior ethmoid cell within the frontal sinus: Report of a case. Arch Otolaryngol 32:32, 1940

MCAULIFFE GW, GOODELL H, WOLFF HG: Experimental studies on headache: Pain from the nasal and paranasal structures. Res Publ Assoc Nerv Ment Dis 23:185, 1943

MCMURRAY J: The intra-antral air pressure incident to the respiratory excursion and its effect on antral drainage. Arch Otolaryngol 14:581, 1931

MELLINGER WJ: The canine fossa. Arch Otolaryngol 31:930, 1940

MOSHER HP: The anatomy of the sphenoidal sinus and the method of approaching it from the antrum. Laryngoscope 13:177, 1903

MOSHER HP: Symposium on the ethmoid: The surgical anatomy of the ethmoidal labyrinth. Trans Am Acad Ophthalmol p. 376, 1929

MOST A: Ueber den Lymphgefässapparat von Nase und Rachen. Arch Anat Physiol p. 75, 1901

MYERSON MC: The natural orifice of the maxillary sinus. I. Anatomic studies. Arch Otolaryngol 15:80, 1932a

MYERSON MC: Natural orifice of the maxillary sinus. II. Clinical studies. Arch Otolaryngol 15:716, 1932b

PEARSON AA: The development of the nervus terminalis in man. J Comp Neurol 75:39, 1941

PROETZ AW: Effect of temperature on nasal cilia. Arch Otolaryngol 19:607, 1934

RHOTON AL JR, PULEC JL, HALL GM, BOYD AS JR: Absence of bone over the geniculate ganglion. J Neurosurg 28:48, 1968

ROBERSON GH, REARDON EJ: Angiography and embolization of the internal maxillary artery for posterior epistaxis. Arch Otolaryngol 105:333, 1979

ROBISON JM: Lymphangitis of the retropharyngeal lymphatic system. Arch Otolaryngol 40:385, 1944

ROBISON JM: Physiology and functional pathology of the lymphatic system applied to allergy of the nose and paranasal sinuses. J Allergy 17:53, 1946

ROOKER DW, JACKSON RT: The effect of certain drugs, cervical sympathetic stimulation and section on nasal patency. Ann Otol Rhinol Laryngol 78:403, 1969

ROSENBERGER HC: The clinical availability of the ostium maxillare: A clinical and cadaver study. Ann Otol Rhinol Laryngol 47:176, 1938

ROSNAGLE RS, ALLEN WE III, KIER L, ROTHMAN SLG: Use of selective arteriography in the treatment of epistaxis. Arch Otolaryngol 106:137, 1980

ROWBOTHAM GF: Observations on the effects of trigeminal denervation. Brain 62:364, 1939

ROY D: Some clinical observations on nasal hemorrhage. Ann Otol Rhinol Laryngol 42:1117, 1933

SCHAEFFER JP: On the genesis of air cells in the conchae nasales. Anat Rec 4:167, 1910a

SCHAEFFER JP: The sinus maxillaris and its relations in the embryo, child, and adult man. Am J Anat 10:313, 1910b

SCHAEFFER JP: Types of ostia nasolacrimalia in man and their genetic significance. Am J Anat 13:183, 1912

SCHAEFFER JP: The genesis, development, and adult anatomy of the nasofrontal region in man. Am J Anat 20:125, 1916

SCHAEFFER JP: The Nose, Paranasal Sinuses, Nasolacrimal Passageways, and Olfactory Organ in Man. Philadelphia, Blakiston, 1920

SEWALL EC: Surgical removal of the sphenopalatine ganglion: Report of three operations elaborating an original technic to expose the pterygopalatine fossa, command the internal maxillary artery and its terminals and the infraorbital nerve and its branches. Ann Otol Rhinol Laryngol 46:79, 1937

SHAPIRO SL, FABRICANT ND, STEPHAN RM: Sensory disturbances following radical operations on the antrums: With an evaluation of the vertical incision. Arch Otolaryngol 19:303, 1934

SIMON E: Anatomy of the opening of the maxillary sinus. Arch Otolaryngol 29:640, 1939

SMITH CG: Incidence of atrophy of the olfactory nerves in man. Arch Otolaryngol 34:533, 1941

SMITH CG: Age incidence of atrophy of olfactory nerves in man: A contribution to the study of the process of ageing. J Comp Neurol 77:589, 1942

SPAR AA, HALLBERG OE: Severe epistaxis and its management: Report of 11 cases in which the external carotid artery was ligated. Ann Otol Rhinol Laryngol 56:141, 1947

STEWART JP: Congenital atresia of the posterior nares. Arch Otolaryngol 13:570, 1931

TRIBLE GB: Epistaxis of antral origin: Report of cases. Arch Otolaryngol 10:633, 1929

VAIL HH: Retrobulbar optic neuritis originating in the nasal sinuses: A new method of demonstrating the relation between the sphenoid sinus and the optic nerve. Arch Otolaryngol 13:846, 1931

VAN ALYEA OE: The ostium maxillare: Anatomic study of its surgical accessibility. Arch Otolaryngol 24:553, 1936

VAN ALYEA OE: Ethmoid labyrinth: Anatomic study, with consideration of the clinical significance of its structural characteristics. Arch Otolaryngol 29:881, 1939

VAN ALYEA OE: Frontal cells: An anatomic study of these cells with consideration of their clinical significance. Arch Otolaryngol 34:11, 1941a

VAN ALYEA OE: Sphenoid sinus: Anatomic study, with consideration of the clinical significance of the structural characteristics of the sphenoid sinus. Arch Otolaryngol 34:225, 1941b

VAN ALYEA OE: Sphenoid sinus drainage. Ann Otol Rhinol Laryngol 53:493, 1944

VAN ALYEA OE: Frontal sinus drainage. Ann Otol Rhinol Laryngol 55:267, 1946a

VAN ALYEA OE: Maxillary sinus drainage. Ann Otol Rhinol Laryngol 55:754, 1946b

WASSON WW: Changes in the nasal accessory sinuses after birth. Arch Otolaryngol 17:197, 1933

WEDDELL G, MACBETH RG, SHARP HS, CALVERT CA: The surgical treatment of severe epistaxis in relation to the ethmoidal arteries. Br. J. Surg 33:387, 1946

WICKLIFFE TF: Abscess of the frontal lobe due to acute frontal sinusitis. Arch Otolaryngol 15:290, 1932

WILLIAMS HL, HEILMAN FR: Spreading osteomyelitis of the frontal bone secondary to disease of the frontal sinus: With a preliminary report as to bacteriology and specific treatment. Arch Otolaryngol 25:196, 1937

WILLS PI, RUSSELL RD: Percutaneous embolization to control intractable epistaxis. Laryngoscope 89:1385, 1979

WRIGHT WK, SHAMBAUGH GE JR, GREEN L: Congenital choanal atresia: A new surgical approach. Ann Otol Rhinol Laryngol 56:120, 1947

ZIEGELMAN EF: The surgery of the great superficial petrosal nerve: Its possible relation to some of the pathology of the nasal and paranasal mucous membranes. Ann Otol Rhinol Laryngol 43:1091, 1934

Chapter 5

FASCIA AND FASCIAL SPACES OF THE HEAD AND NECK

It seems probable that few subjects in gross anatomy have suffered so many diverse descriptions as the fascia and fascial spaces of the neck. As quoted by Grodinsky and Holyoke, "the cervical fasciae appear in a new form under the pen of each author who attempts to describe them," and, it may be added, the fascial spaces are apt also to appear under new names.

Despite this, however, a good deal of agreement exists on the fundamentals of the relationships that have been thought to be of clinical importance; moreover, although they are difficult to describe clearly, the major fascial layers and spaces are actually relatively simple. If the reader will recall that there is no generally accepted definition on how dense connective tissue must be before it can be regarded as forming a fascia and that fascial spaces are simply areas of relatively loose connective tissue, the reasons for many of the discrepancies in the various descriptions will be obvious.

Connective tissue naturally forms a padding between and around the various structures in the neck, and, as occurs also elsewhere in the body, the connective tissue tends to be somewhat more dense where it is arranged immediately about organs. When the fascia and spaces of the head and neck of normal bodies are investigated by dissection or injection, as they must be, the looser connective tissue intervening between organs is torn and pressed against these organs, thereby exaggerating any fascial layer already present. Thus, Parsons was unable to identify in histological sections the carotid sheath or most of the other fascial layers usually described in the neck unless the material had been partly dissected beforehand; he therefore regarded the usually described layers as being formed entirely artificially by pressing together loose connective tissue that really serves simply as a padding around all structures. Insofar as infections tend to localize in or spread through the looser connective tissue, however, these produce the same effects—that is, they exaggerate both the "spaces" and their fascial walls.

For practical purposes, it makes little difference to the surgeon operating in an uninfected neck whether, for instance, the connective tissue about the internal jugular vein, common carotid artery, and vagus nerve is described as forming a fascial sheath or not, if he realizes that these closely grouped structures are, of course, surrounded by connective tissue; at operation, he can easily dispose of this carotid sheath with one sweep of his scissors or, if he wishes to stay away from these structures, push them aside with their surrounding tissue. Of some practical importance, however, are the relations of the loose connective tissue areas, the fascial spaces, of the head and neck, both because of the structures that traverse or abut against them, and because they may become infected and harbor abscesses.

The interrelationships of these spaces have been regarded as especial importance in the spread of infection, and many surgeons have urged accurate knowledge of the anatomy of these spaces and prompt drainage of them when they become infected. On the other hand, other workers (*e.g.,* New and Erich) have doubted the importance of spread of infection through fascial spaces and have advocated conservative treatment, with initiation of drainage only after an abscess has pointed. According to this view, the anatomy of the fascial spaces is of little moment.

It is not within an anatomist's province to decide between these sharply divergent views as to pathologic processes and clinical treatments, and this account of the anatomy of these spaces and the surgical approaches to them is presented without prejudice in this regard. It should be noted that since the advent of antibiotics, the anatomy of these spaces is frequently less significant clinically.

As a description of the fascial spaces is not understandable without some discussion of the fascial layers, these also are described briefly. For a fuller description of the spaces and fascial layers, the detailed works of Coller and Yglesias ('35, '37) and of Grodinsky and Holyoke should be consulted.

Superficial Fascia

The subcutaneous tissue of the head and neck, usually referred to as the superficial fascia, resembles this tissue elsewhere in that it contains a variable amount of fat; its distinguishing characteristic, as compared with subcutaneous tissue in general, lies primarily in the fact that it encloses voluntary muscles in its deep portion. Thus, the platysma is embedded in this layer in the neck, while in the face the connective tissue is more dense and the muscles of expression lie within it. Except for the subcutaneous tissue of the eyelids, which is loose on both sides of the orbicularis oculi muscle and permits considerable accumulation of fluid, a corresponding layer in the scalp deep to the epicranial aponeurosis (sometimes known as the danger space of the scalp) and the area in the cheek occupied by the buccal fat pad, the subcutaneous tissue of the head and face is rather tightly attached about the muscles and to the underlying bone, and spaces can hardly be said to exist here. A space has been described in the neck between the superficial and deep fascial layers, but is of little practical importance.

Deep Fascia

Cervical fascia (deep fascia of the front of the neck) upon dissection forms several more or less distinct layers. If, however, it is borne in mind that fascias or connective tissue layers in general split to surround muscles and other soft tissues and then unite again on the other side, blending with other layers where possible, it should be apparent that any description of the fascia of the neck is largely an arbitrary one.

Some of the layers in the front of the neck, constituting the cervical fascia proper, are continued around the posterior aspect of the neck and thus are also nuchal fascia. The layers take their names, however, from their disposition anteriorly below the level of the hyoid bone. A simple concept is that fascia encircles the entire neck deep to the tela subcutanea (superficial fascia), while other layers, by their dispositions and fusions, form two unpaired and one paired compartment within this. The layer surrounding the entire neck is the *superficial* (anterior) *layer* (Fig. 5-1); another layer, the *prevertebral* (posterior, deep), encloses the vertebral column and its associated muscles, thus forming a vertebral compartment; condensations of connective tissue around the great vessels and vagus nerve on each side form the *carotid sheaths,* or neurovascular compartment; and the *pretracheal* (middle) *layer* of fascia passes across the front of the neck and, with the carotid sheaths laterally and the prevertebral layer posteriorly, forms the visceral compartment containing the trachea, esophagus, and associated structures. Other terms have also been applied to the superficial, pretracheal, and prevertebral layers or parts of them, and some descriptions of them seem needlessly complicated.

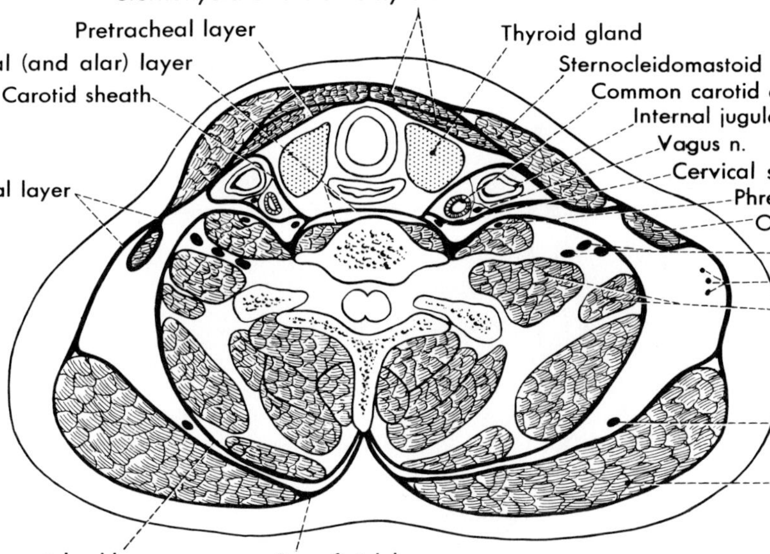

Fig. 5-1. Chief layers of the cervical fascia below the hyoid bone. Note the relations of the various nerves. The fascias and fascial spaces shown in this and the succeeding drawings on this subject are all, of course, both schematic and exaggerated.

The Fascia and Spaces Below the Hyoid Bone

FASCIA

Superficial Layer

The superficial, investing, or anterior layer of the cervical or deep fascia may be thought of as arising from the vertebral spinous processes and ligamentum nuchae and completely encircling the neck to attach again to these (Fig. 5-1). This fascia is usually described as enclosing the trapezius muscle, but the layer deep to the trapezius is thicker. At the anterior border of the trapezius the two layers unite to form a single layer that crosses the posterior triangle of the neck to the posterior border of the sternocleidomastoid muscle; in so doing it splits to surround that portion of the omohyoid muscle which extends across the posterior triangle and, uniting below this muscle, acts as a ligament that holds the omohyoid in a relatively constant relation to the clavicle. At the posterior border of the sternocleidomastoid muscle the superficial layer splits again to surround this muscle, and at the lateral borders of the infrahyoid or strap muscles sends thin laminae between the muscles before passing in front of them. Thus it crosses the midline and proceeds around the other side of the neck.

When the layer of fascia lying anterior to the strap muscles is traced upward, it may be seen to attach to the hyoid bone (Fig. 5-2). When it is traced downward between the two sternocleidomastoid muscles, it will be found to split into two layers just above the sternum; one attaches to the anterior and one to the posterior surface of the sternum, and the small space between these two layers is the *suprasternal space* (of Burns). It is essentially due to a failure of the fascial layers to unite between the closely adjacent sternal heads of the sternocleidomastoid muscle and therefore is bounded laterally by these heads; between them, the lower ends of the anterior jugular veins and their transverse connecting branch lie in this space, and there may sometimes be a lymph node here.

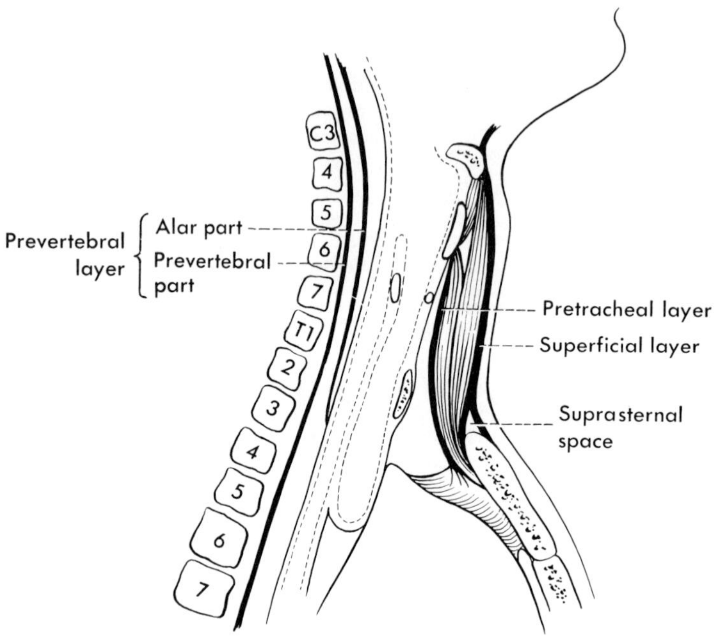

Fig. 5-2. The chief fascial layers of the neck below the hyoid bone, in longitudinal section. (After Grodinsky M, Holyoke EA: Am J Anat 63:367, 1938)

Pretracheal Layer

The pretracheal or middle layer of fascia is most commonly described as a layer that passes behind the intrahyoid or strap muscles, hence in front of the trachea and thyroid gland, and at the lateral edges of the muscles blends with the superficial layer (Figs. 5-3 and 5-4). Descriptions of it vary much, however, and it has also been described and illustrated as encircling the trachea, esophagus, and thyroid gland; sometimes "pretracheal fascia" has apparently been used synonymously with "visceral fascia," although this is well developed only on the pharynx and esophagus, thus behind the trachea. In contrast to the layer in front of the infrahyoid muscles, the connective tissue behind them is apt to be filmy and therefore not very obvious, except in the lowest part of the neck where it is thicker and may contain some fat; it is probably not so much the pretracheal layer as it is the muscles and all the fascia about them that should be regarded as forming the anterior wall of the visceral compartment; sometimes, in fact, the pretracheal layer is said to have two laminae, one the fascia between the infrahyoid muscles, and the other the lamina behind them.

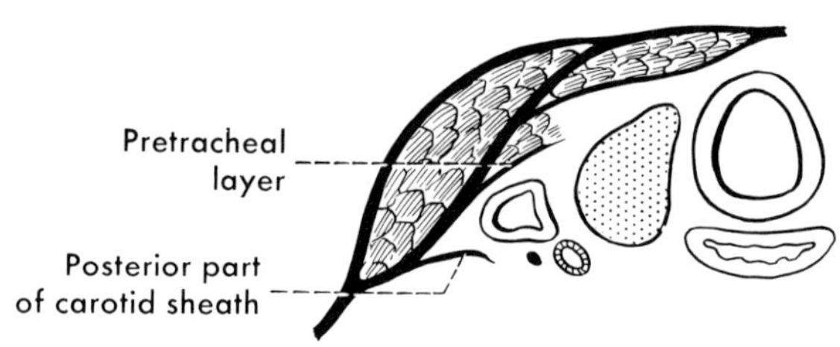

Fig. 5-3. Disposition of the anterior part of the superficial (anterior, investing) layer of fascia.

The fascia behind the infrahyoid muscles (*i.e.*, the pretracheal or middle layer) is, of course, continuous with the superficial layer of cervical fascia at the lateral borders of these muscles (Fig. 5-4). Traced downward, it follows these muscles behind the sternum to about their origin, where it fuses with the fibrous pericardium as this is prolonged out along the great vessels in the superior mediastinum. Above, this layer of fascia ends with the infrahyoid muscles; that is, it fuses to the thyroid cartilage and the hyoid bone. Theoretically, there are, of course, spaces among these muscles, and they have been grouped together and named (by a number, 2A) by Grodinsky and Holyoke; actually, however, there seems no more point in naming these than in naming a sternocleidomastoid space or a trapezius space—none is of clinical importance, and all consist merely of the potential planes of separation typically lying between adjacent muscles.

Prevertebral Layer

The prevertebral or posterior layer of fascia may, like the anterior layer, be thought of as beginning from cervical spinous processes and the ligamentum nuchae (see Fig. 5-1). It lies at first on the outer surface of those muscles of the back that extend into the neck (and is sometimes called the nuchal layer) and therefore immediately deep to the trapezius muscle and its surrounding superficial layer of fascia. Farther forward, it covers the scalene muscles before it passes in front of the vertebral column.

Deep to the posterior part of the trapezius, and in the upper part of the posterior triangle of the neck, the prevertebral layer is in contact with the superficial layer, and only a potential cleavage space exists between the layers. Even close to the middle of the posterior triangle there is little space between the two layers, so that the cutaneous branches of the cervical plexus that emerge at the posterior border of the sternocleidomastoid pierce the layers almost simultaneously. The accessory nerve as it crosses the posterior triangle, and a few lymph nodes along it, lie between the two layers, however. Lower down, the two layers are distinctly separated: the prevertebral layer follows the scalenes, while the superficial layer remains against the trapezius and attaches below to the clavicle.

This area of separation (Space 4A of Grodinsky and Holyoke) is occupied by loose connective tissue and fat, in which lie the subclavian vein and the lower end of the external jugular vein and through which run the transverse cervical and suprascapular vessels, the supraclavicular and suprascapular nerves, and the posterior belly of the omohyoid muscles. Except for these structures, the space is of little clinical importance, since it communicates with no other space in the neck. Grodinsky and Holyoke regarded it as extending to the posterior midline through the cleavage plane deep to the trapezius.

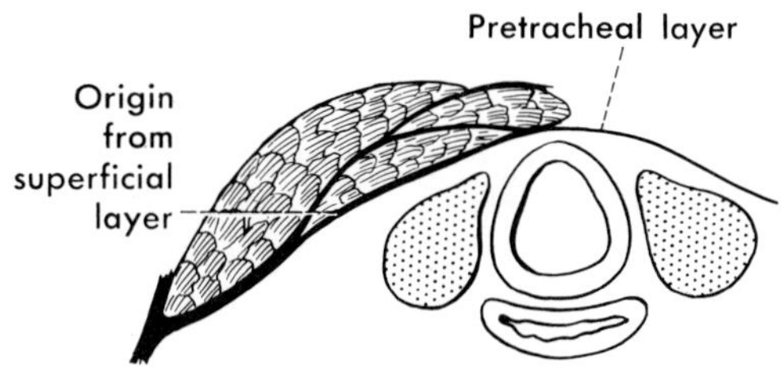

Fig. 5-4. The pretracheal or middle layer of fascia.

Deep to the sternocleidomastoid muscle, regardless of how closely they may have been associated, the superficial and prevertebral layers separate lateral to the internal jugular vein: the superficial layer continues forward anterolateral to the vein and the common carotid artery toward the infrahyoid muscles, and in so doing forms the anterolateral wall of the carotid sheath; the prevertebral layer, however, turns medially behind these vessels and the vagus nerve, from which it is separated by loose connective tissue and the posterior wall of the carotid sheath. That part of the prevertebral layer on the scalene muscles is usually called the scalene fascia.

Between this fascia and the anterior scalene muscle is the phrenic nerve, and deep to the fascia, between the anterior and middle scalenes, the brachial plexus takes form; some of the fascia is carried out around the plexus and the subclavian artery as they pass beyond the lateral edges of the scalenes, to form a sheath about them and blend with the axillary sheath.

Upon reaching the anterior tubercles of the transverse processes the prevertebral layer of fascia attaches to them, and at about this position has embedded in its anterior surface the cervical sympathetic trunk. From the transverse processes of one side the prevertebral layer crosses the midline to attach to the transverse processes of the other side, passing behind the esophagus and in front of the vertebral column. It is this part, between the transverse processes and across the front of the vertebral bodies, that gave the prevertebral layer its name (another name is retrovisceral fascia); Grodinsky and Holyoke have quite properly pointed out that there are really two layers here (Fig. 5-5), separated by

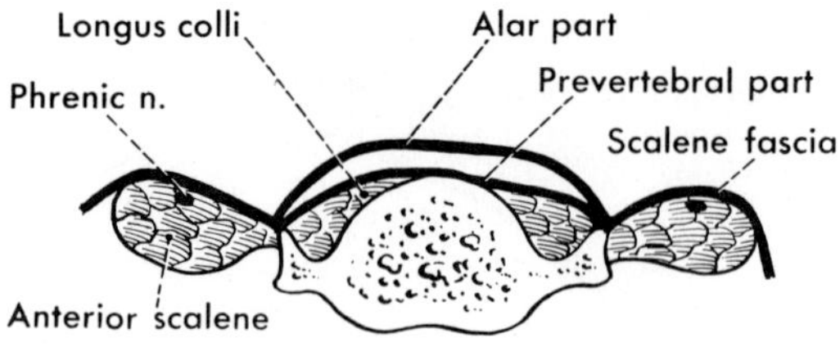

Fig. 5-5. Disposition of the anterior part of the prevertebral or posterior layer of the deep fascia.

extremely loose connective tissue; they termed the anterior lamina the alar fascia, reserving the term "prevertebral" for the posterior lamina only.

Carotid Sheath

The carotid sheath, surrounding the internal jugular vein, common carotid artery, and vagus nerve, is interposed between the superficial and pretracheal layers, on the one hand, and the prevertebral layer on the other. Its anterolateral wall is, in fact, the superficial layer of fascia deep to the sternocleidomastoid muscle and, to a lesser extent, the pretracheal layer where the infrahyoid muscles overlap the great vessels. Its posterior wall is formed by a lamina given off medially by the superficial layer as the latter reaches the level of the vessels (Fig. 5-6); this lamina passes behind the vessels and nerve to form the posterior wall of the sheath, and the medial wall of the sheath is them completed by fascia passing from the anterolateral to the posterior wall, between the contents of the sheath and the trachea and esophagus. The sheath is attached medially to the prevertebral layer of fascia. The sympathetic trunk lies behind it; within it are separate compartments for the vein, artery and nerve.

SPACES

Since the pretracheal layer of the cervical fascia, reinforced by the infrahyoid muscles, passes in front of the thyroid gland and the trachea, while the prevertebral layer passes between the vertebral column and the esophagus, these layers are the anterior and posterior walls of a compartment about the esophagus and trachea; laterally, the walls of this compartment are completed by the fascia on the deep surface of the sternocleidomastoid, where it is continuous with the pretracheal layer, and by the anteromedial wall of the carotid sheath as it extends from the fascia on the deep surface of the sternocleidomastoid to the prevertebral layer.

VISCERAL COMPARTMENT

The area of loose connective tissue surrounding the thyroid gland, trachea, and esophagus as a whole (Fig. 5-7) was long known as the visceral compartment. Around the upper parts of the trachea, esophagus, and thyroid gland the visceral compartment is usually described as completely surrounding these, but below the level at which the inferior thyroid artery enters the thyroid gland it is divided into two portions by denser connective tissue that attaches the esophageal wall laterally to the prevertebral layer of fascia (Fig. 5-8).

The anterior part of the compartment, more or less surrounding the trachea and lying against the anterior wall of the esophagus, has been known by a number of names, the more common ones being the previsceral space or the *pretracheal space.* The posterior part of the compartment, lying behind the esophagus and the lower part of the pharynx, has been variously termed the *retrovisceral,* retropharyngeal, retroesophageal, or postvisceral space. Grodinsky and Holyoke emphasized the continuity of the pretracheal and retrovisceral spaces (the visceral compartment) by

Fig. 5-6. Formation of the carotid sheath. (*1, 2,* and *3*), The superficial, pretracheal, and prevertebral layers of fascia, respectively.

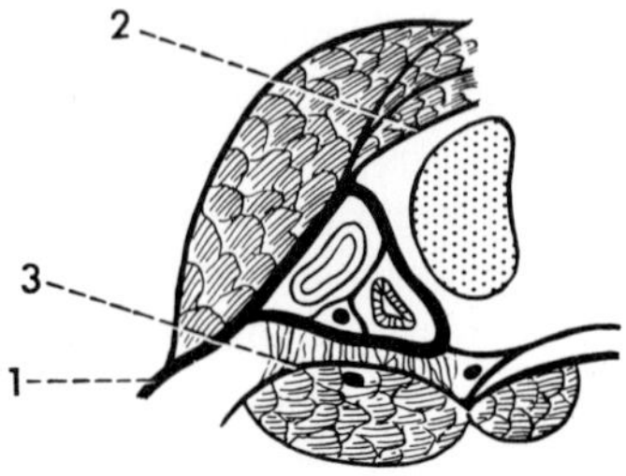

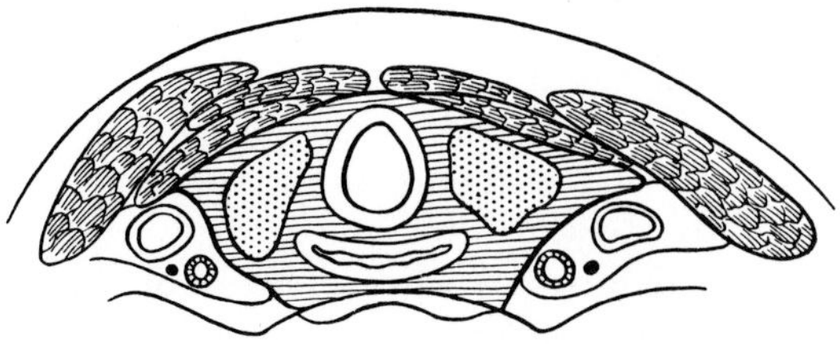

Fig. 5-7. The visceral compartment of the neck.

calling the two combined spaces part of their Space 3.

It is the visceral compartment that has usually been regarded as of particular importance in transmitting infections from the head and upper part of the neck to the mediastinum, since anatomically (as will be evidenced shortly) its posterior part is directly continuous above with the major spaces of the upper portion of the neck.

Pretracheal Space

The pretracheal portion of the visceral compartment is limited above by the attachments of the infrahyoid muscles and their fascia to the throid cartilage and to the hyoid bone; below, it continues into the anterior portion of the superior mediastinum. It extends to approximately the arch of the aorta, according to Grodinsky and Holyoke (*i.e.,* to about the level of the fourth thoracic vertebra), where the posterior surface of the sternum and the fibrous pericardium are united by denser connective tissue.

The pretracheal portion of the visceral compartment could be infected directly by anterior perforations of the esophagus or indirectly by spread from the retrovisceral portion, around the sides of the esophaghus and thyroid gland between the levels of the inferior thyroid artery and the oblique line of the thyroid cartilage. Both pretracheal and retrovisceral spaces descend into the superior mediastinum (Fig. 5-9). According to Pearse, however, infections descending from higher in the neck reach the mediastinum through the pretracheal space rather rarely, only some 8% of mediastinal suppurations of cranial or cervical origin having apparently traveled this route in the cases that he studied. This space has been opened by an incision anterior to the sternocleidomastoid muscle, carried medial to the carotid sheath (Hunt).

Retrovisceral Space

The posterior, retroesophageal, or retrovisceral portion of the visceral compartment in the neck is not limited above to the neck, since it extends upward behind the pharynx to the base of the skull (see Fig, 5-11); while this upper part is usually known as the retropharyngeal space, this terminology should not be allowed to obscure the fact that the retropharyngeal and retroesophageal spaces are just as continuous with each other as are their anterior walls, the pharynx and esophagus, and actually form a single retrovisceral space.

Inferiorly, the retrovisceral space usually extends, like the pretracheal, into the mediastinum. The lower level of extent varies, according to Grodinsky and Holyoke, from the sixth cervical to the fourth thoracic vertebral level, where the space is obliterated through fusion of the connective tissue on the posterior surface of the esophagus to the prevertebral layer of fascia. Coller and Yglesias ('37) gave the lower level of this space as being that of the bifurcation of the trachea, thus at about the fourth thoracic vertebral level. The latter workers also described a retroesophageal space below this level; apparently they believed it represents a lower, but separate, portion of the retrovisceral space, but it is probably the same as that described by Grodinsky and Holyoke as the "danger space," or Space 4.

Since the retrovisceral portion of the visceral compartment extends upward behind

Fig. 5-8. Subdivision of the visceral compartment into pretracheal (*horizontally lined*) and retrovisceral (*vertically lined*) portions by connective tissue attaching laterally to the esophagus, at a level below that shown in Fig. 5-7.

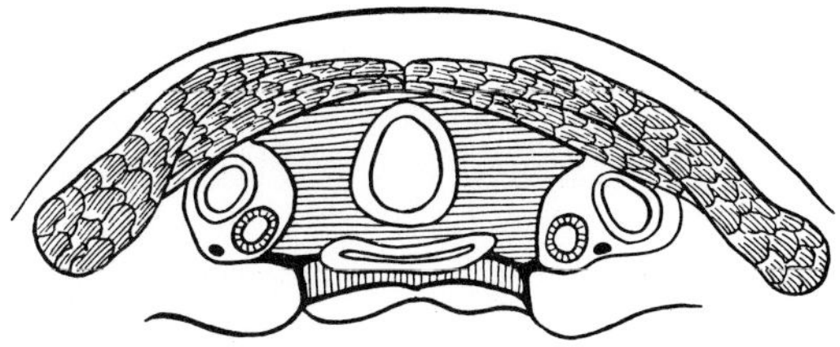

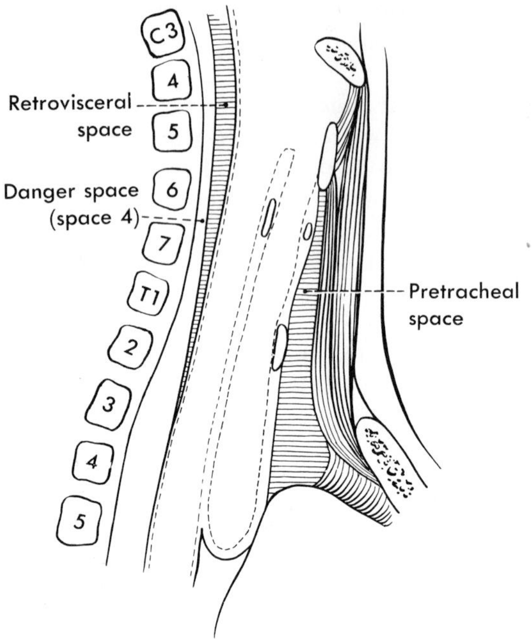

Fig. 5-9. The visceral compartment (horizontally lined) in a sagittal section of the neck. The danger space is also indicated. (After Grodinsky M, Holyoke, EA: Am J Anat 63:367, 1938)

the pharynx to the base of the skull, it has been described as an important pathway by which infections originating from various locations in the head and the upper portion of the neck reach the mediastinum; Pearse reported that this was the route followed in 71% of the patients he studied in whom infection had descended from the neck into the mediastinum. The retrovisceral space may also be infected directly from posterior perforations of the esophagus or by infections of the deep cervical nodes lying adjacent to it. The most convenient surgical approach to the retrovisceral space, according to Iglauer, is by an incision posterior to the sternocleidomastoid, carried medially behind the carotid sheath and its great vessels.

"Visceral Space"

The esophagus is enclosed in a connective tissue coat, or tunica adventitia, continuous above with the thin but slightly more obvious layer (buccopharyngeal fascia) on the pharynx, and delicate connective tissue also is adjacent to the surface of the thyroid gland and that of the trachea. This connective tissue is frequently referred to as a fascia, although it more nearly corresponds to the subperitoneal connective tissue of the serous coat of most of the digestive tract and can hardly be said to form a definite layer over the thyroid gland or the trachea.

If it must be designated as a fascia, it should probably be called the visceral fascia, as has been done by Grodinsky and Holyoke; that part on the thyroid gland is frequently referred to as the perithyroid sheath, thyroid fascia, or even, incorrectly, the thyroid capsule; and sometimes the entire layer in the neck, whether related to trachea, thyroid gland, or esophagus, is known as the pretracheal fascia.

Thus the term "pretracheal fascia" has been applied both to the connective tissue associated most closely with the posterior surface of the strap muscles (the so-called middle layer of the deep fascia) and to all the connective tissue surrounding and immediately adjacent to the viscera; correspondingly, the previsceral or pretracheal space has sometimes been described as lying behind the pretracheal fascia and sometimes as lying anterior or external to it.

Confusingly, also, in view of the widespread use of the term "visceral compartment" for the pretracheal and retrovisceral spaces, the potential space that may be imagined to exist between the visceral layer of fascia and the organs themselves is also referred to as the visceral space. Actually, such connective tissue as exists here to form a visceral fascia is firmly united to the structures it covers, and the visceral space in the latter sense does not really exist; infections lying deep to the fascia on the esophagus, for instance, do not tend to spread within this fascia up and down the esophagus, but perforate it to reach the visceral compartment.

Other Spaces

In the lower part of the neck there are two other compartments that may be of clinical importance (Fig. 5-10). One of these is the potential *cavity within the carotid sheath,* into which, according to Coller and Yglesias ('35), infections from the visceral space readily spread; indeed, they have grouped the two under the name "visceral-vascular space." These authors have conceived of the space of the carotid sheath as being, like the retrovisceral space, a pathway for the spread of infections from the upper to the lower part of the neck and into the mediastinum. Pearse apparently agreed with this, since he attributed 21% of mediastinal suppurations originating in the neck to spread along this pathway (as against 8% along the pretracheal space and 71% through the retrovisceral).

In contrast to this concept, Grodinsky and Holyoke found that injections within the carotid sheath were usually closely limited to the area of injection and never extended higher than the hyoid bone or lower than the root of the neck, since above and below these levels the sheath was too closely adherent to the vessels. They concluded that infections within it would tend to be localized within the cervical region and that it has no important relations to either the head or the thorax. They stated that infection within the carotid sheath usually arises from thrombosis of the internal jugular vein or from infection of those deep cervical lymph nodes that lie within the sheath.

The second space remaining to be described in the lower part of the neck, although not confined to that, lies between the two laminae of the prevertebral layer as it passes from the transverse processes on one side to those of the other, beteween the viscera and the vertebral column. Its presence has apparently been ignored in many studies of the fascial spaces of the neck, yet it is easily demonstrated in most dissections.

It will be recalled that after the prevertebral layer of fascia attaches to the transverse processes, it divides into two layers; the anterior layer, called alar fascia, forms the posterior boundary of the retrovisceral space, and the posterior layer, also called prevertebral fascia, consists of the connective tissue immediately against the anterior surfaces of the prevertebral muscles and the vertebral bodies. Between these two layers lies an area

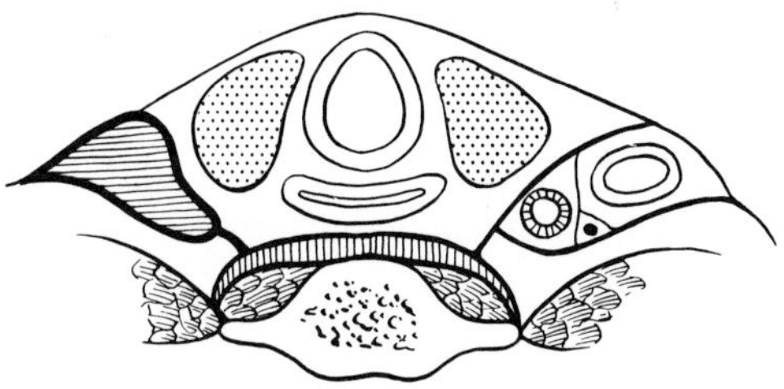

Fig. 5-10. The space of the carotid sheath (*horizontally lined*), and the "danger space" or space 4 of Grodinsky and Holyoke (*vertically lined*).

of delicate loose connective tissue, which Grodinsky and Holyoke called *Space 4* or the *danger space.*

It is possible that this has also been called the prevertebral space, but most descriptions of the latter space are not sufficiently clear to decide this; Grodinsky and Holyoke reserved the term "prevertebral space" for the potential cleavage plane existing between the "prevertebral" fascia and the vertebral bodies. Whatever the space between the alar and prevertebral layers is called, one should realize that it is not the same as that already designated as the retrovisceral space, and the confusion in terminology has been only more compounded by some authors' implications that the retrovisceral and prevertebral spaces are synonymous.

As noted by Grodinsky and Holyoke, the texture of the connective tissue within the danger space is such that it is almost an actual rather than a potential space. It has been thought to be of particular importance because it extends upward as far as the base of the skull and is said to extend downward as far as the diaphragm. Since the space is closed above, below, and laterally, it could be infected only through its walls: Grodinsky and Holyoke concluded that infections of this should arise most commonly from infections of the retrovisceral space, for in their injections behind the prevertebral fascia they found that there was little tendency to break forward into Space 4. It is quite possibly the lower portion of this space, below the bifurcation of the trachea, which Coller and Yglesias ('37) described as a resumption of the retrovisceral space. Infections within this space (Space 4) are not limited to the superior mediastinum, according to Grodinsky and Holyoke, but may extend throughout the length of the posterior mediastinum.

The Fascia and Spaces Above the Hyoid Bone

Above the level of the hyoid bone the fascial layers become perhaps slightly less complicated, since there is no pretracheal layer here, but the fascial spaces themselves become more complicated.

FASCIA

Superficial Layer of Fascia

The superficial or anterior layer of the cervical (deep) fascia extends from its attachment on the hyoid bone upward to the mandible (Fig. 5-11) and from the anterior border of the sternocleidomastoid upward to the mandible and zygomatic arch; in so doing, it passes across the superficial surface of the muscles of the floor of the mouth (the mylohyoid and the anterior belly of the digastric) and thus forms the floor of a potential space lying above the fascia and below the muscles. As it attaches to the mandible it is usually described as splitting into two layers, one attaching on the inner and one on the outer surface of the mandible.

More posteriorly, at the level of the submandibular (submaxillary) gland, the superficial layer splits to form a capsule about this gland; still more posteriorly, about the insertions of the masseter and medial pterygoid muscles, it splits to enclose these muscles and the intervening angle and ramus of the mandible, one portion, the masseteric fascia, following the external surface of the masseter to the zygomatic arch, and the other portion following the internal surface of the medial pterygoid to the pterygoid plate.

Finally, behind the angle of the jaw and anterior to the sternocleidomastoid, this layer of fascia passes upward toward the zygomatic arch as the parotid fascia; Grodinsky and Holyoke described it as splitting to form a capsule about the parotid gland before attaching to the zygomatic arch. The potential spaces formed by the splitting of the superficial layer of fascia are all closed spaces; those lying immediately deep to this fascia, however, communicate, or can be made to communicate, fairly freely with each other.

Fig. 5-11. Fascial layers and spaces above the hyoid bone, in a sagittal section. Note that the "danger space" and the retropharyngeal space show no interruption, but are continuous with the danger space and the retrovisceral space, respectively, below the level of the hyoid bone. (After Grodinsky M, Holyoke EA: Am J Anat 63:367, 1938)

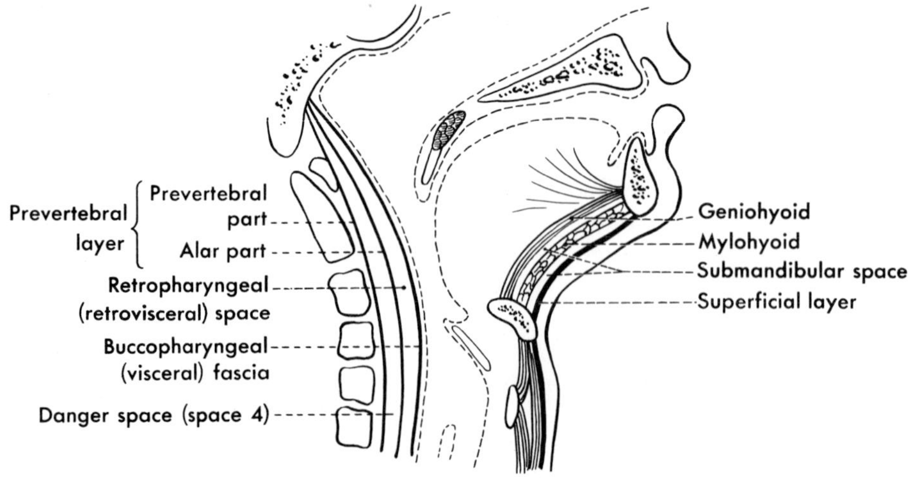

Prevertebral Layer of Fascia

The relations of the prevertebral or posterior layer of the cervical fascia are identical to those lower in the neck—that is, it encloses the vertebral muscles, is attached to the transverse processes of the vertebrae, and across the front of the vertebrae splits into alar and prevertebral portions enclosing the danger space or Space 4.

Buccopharyngeal Fascia

The only other layer of fascia to be considered here is that on the pharynx itself, called the buccopharyngeal fascia. It is continuous below with the visceral fascia on the esophagus, and above with fascia on the outer surface of the buccinator muscle. The thin buccopharyngeal fascia separates the muscular wall of the pharynx from certain potential spaces that largely surround it.

SPACES

The fascial spaces above the level of the hyoid have been especially investigated by Coller and Yglesias ('35), and these relationships have been also generally agreed upon by Grodinsky and Holyoke and by other workers who have had occasion to investigate some of them. The suprahyoid spaces may be divided into three categories: blind or intrafascial spaces formed by splitting of fascial layers; intercommunicating spaces surrounding the pharynx and lying between fascial laminae and the pharyngeal walls; and blind "spaces," potential only, within the pharyngeal wall deep to the buccopharyngeal fascia.

INTRAFASCIAL SPACES

The danger space has already been described as being formed by a splitting of the prevertebral layer of cervical fascia and as extending from the base of the skull into the thorax; although its upper portion is suprahyoid, the relationships of this portion do not vary from those already described and need not, therefore, be repeated here.

The other intrafascial spaces related to the upper part of the neck are all formed by a splitting of the superficial layer of cerfical fascia as it attaches to the skull: thus, as this layer reaches the lower border of a limited portion of the mandible, it is said to attach to the outer and inner surfaces of this; as it reaches the insertions of the masseter and medial pterygoid muscles on the outer and inner surfaces of the mandible, respectively, it splits to follow the free surfaces of these muscles; and as it reaches the submandibular and parotid glands it is usually described as surrounding these.

Since the potential spaces thus formed do not communicate with each other or with other spaces, infections within them can spread only through rupture of their walls. The only one of these four spaces containing any appreciable quantity of loose connective tissue is that associated with the muscles of mastication.

Space of the Body of the Mandible

The space of the body of the mandible, according to Coller and Yglesias, is formed by the attachment of the superficial layer of fascia to both the outer and inner surfaces of the body of the mandible; the attachment to the outer surface is at the lower border of the mandible, but the deeper leaflet is said to be easily elevated from the mandible up to the origin of the mylohyoid muscle (Fig. 5-12).

As an actual space in the sense of an area of loose connective tissue, this space does not exist; rather, it seems to be a potential cleavage plane between the fascia and the bone. This cleavage plane is apparently limited anteriorly by the attachment of the anterior belly of the digastric and posteriorly by the attachment of the medial pterygoid to the jaw; below, it is closed by the continuity of the fascial layers, and above, it is closed by the attachment of these layers to the mandible.

Apparently, an infection here may remain localized, may discharge into the mouth, or may spread to the masticator space. This "space" has been drained by an incision through the buccal gingival mucous membrane or by one externally along the inferior border of the mandible, the incision in either case being carried to the bone.

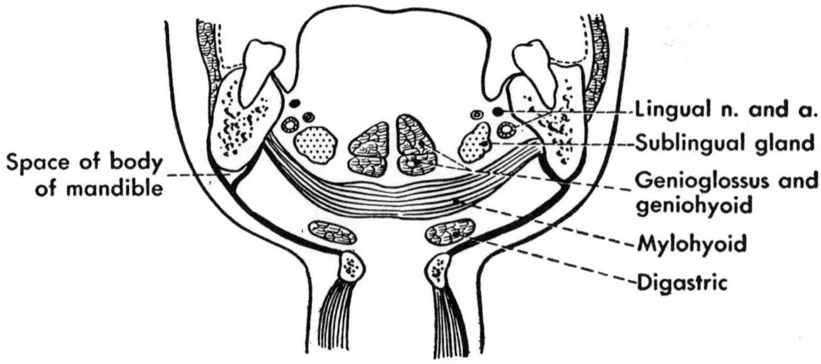

Fig. 5-12. Drawing of frontal section through the mandible and tongue, illustrating the attachment of the superficial layer of cervical fascia to the mandible, and some of the general anatomy here. Compare with Figs. 5-11 and 5-20.

Space of "Submaxillary" Gland

Terminology of the fascial spaces of the sublingual and submental regions is complicated and was, of course, suggested before there was general adoption of the more meaningful "submandibular" for "submaxillary." Changing "submaxillary" to "submandibular" in the terminology of the spaces can only add to the confusion; hence, the original terminology is used here.

The superficial layer of fascia splits to form a capsule around the submandibular or submaxillary gland and is usually described as enclosing a space, the submaxillary space (or, less confusingly, the space of the gland). This description is misleading: the submandibular gland does not lie loosely within a sheath from which it can be shelled out; rather, it and its associated lymph nodes are embedded in and fused with the fascial capsule. The submaxillary space, therefore, is really the substance of the gland and nodes (Fig. 5-13).

The outer layer of the capsule, the continuation upward of the main portion of the superficial layer of the cervical fascia, is strong, while the inner layer is thinner and is of course perforated by the duct of the gland; infections arising in the region of the gland generally break inward, therefore. The term "submaxillary space" has also been commonly used to describe the space between the superficial layer and the mylohyoid muscle.

Masticator Space

The masticator space is formed by the splitting of the superficial layer of cervical fascia to enclose the ramus of the mandible, the masseter, the medial pterygoid, and the lower portion of the temporal muscle (Fig. 5-14). Since these structures lie between the fascial layers on the outer surface of the masseter and the inner surface of the medial pterygoid, the loose connective tissue and fat about them forms the potential space. Its largest part is medial to the ramus of the mandible, between this and the medial pterygoid, and medial and anterior to the lower portion of the insertion of the temporal muscle.

The fascial walls of this space come together posteriorly, behind the ramus of the mandible. Anteriorly, a part of the masseteric fascia attaches to the mandible in front of the masseter muscle and to the insertion of the temporal muscle along the anterior border of the ramus, but another part passes in front of the ramus, across the outer surface of the buccal fat pad, to attach to the maxilla and the buccinator fascia below that, thus limiting the masticator space anteriorly (Gaughran). Superiorly it is limited deep to the temporal muscle by the origin of this muscle from the skull; superficially, by the muscle's origin from the temporal fascia. Deeply, anterior to the lateral pterygoid plate, it extends into the pterygopalatine fossa.

The masticator space thus lies largely among the muscles of mastication. It is traversed particularly by the mandibular nerve and the maxillary (internal maxillary) vessels, and largely filled by the buccal fat pad and its extensions posteriorly, upward, and medially. Kostrubala divided it into a number of parts, which he called buccal, infratemporal, and the like; and Gaughran pointed out that most

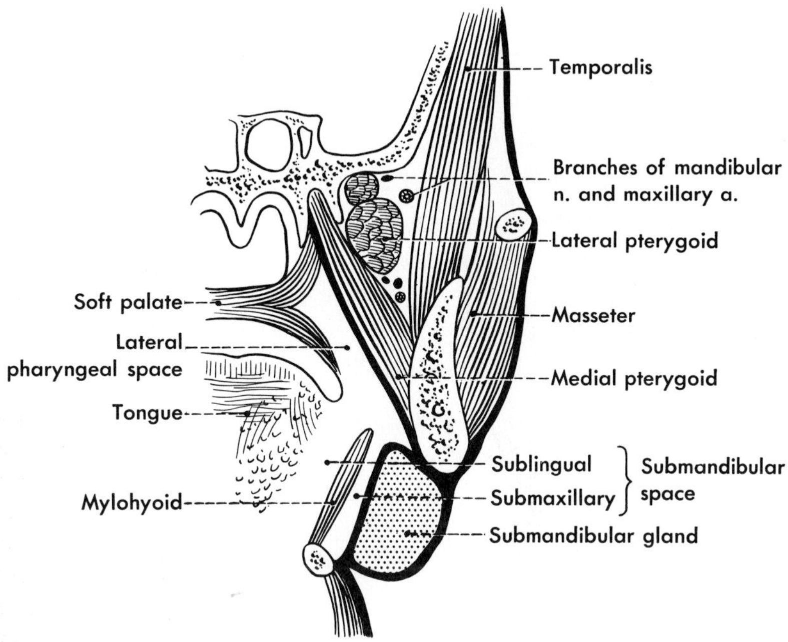

Fig. 5-13. Schema of half of a frontal section close to the angle of the jaw, to show the splitting of the superficial layer of fascia about the submandibular (submaxillary) gland and the muscles of mastication.

of the muscles in or bordering the space are enclosed by thin fascial layers that separate them from the fat pad and subdivide the masticator space into compartments that do not freely communicate with each other. These refinements do not alter the fundamental concept of the masticator space. However, as pointed out by Gaughran, its compartmentalization indicates that in opening the space, care should be taken to approach the particular compartment, whether that of the fat pad or of one of the muscles, that is harboring the infection.

Infections of the zygomatic or temporal bones may pass to the masticator space, and so may abscesses from the lower molar teeth (Coller and Yglesias); Hall and Morris reviewed 20 cases of infection of this space and found that they most commonly resulted from extraction of a lower molar tooth. Abscesses within this space may apparently point at the anterior aspect of the masseter muscle, either into the cheek or the mouth, or they may point posteriorly below the parotid gland.

Space of Parotid Gland

The parotid space (Fig. 5-15) encloses the parotid gland and its associated lymph nodes and the facial nerve and great vessels traversing it. Although Grodinsky and Holyoke found that injections into the parotid space broke out either superficially or deeply, and Coller and Yglesias regarded the tissue on the deep surface of the parotid gland as the stronger layer, both have pointed out that infections may readily pass deeply and therefore into the important lateral pharyngeal space lying deep to the parotid gland. As in the case of the submandibular gland, the parotid gland is attached to its surrounding fascia, and the parotid "space" is therefore not so much an anatomical as a clinical one; infections within it are infections of the gland or its nodes, not a cellulitis in loose connective tissue.

PERIPHARYNGEAL SPACES

These lie immediately posterior and lateral to the pharynx and extend forward into the sub-

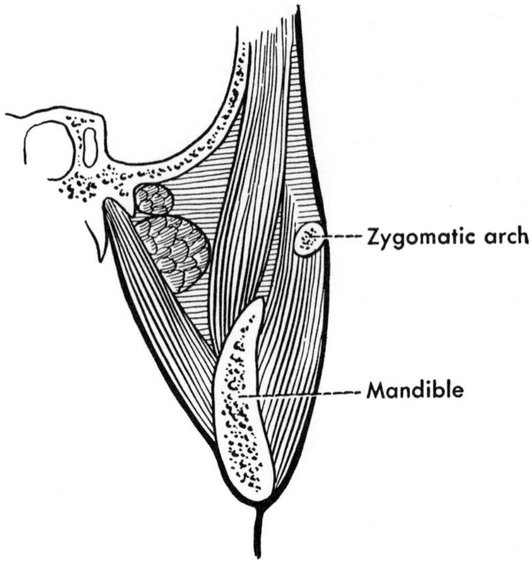

Fig. 5-14. The masticator space. Compare Figs. 5-13 and 5-18.

lingual region, so that together they actually form a ring about the pharynx (Fig. 5-16). They lie entirely deep to the superficial or anterior layer of fascia and communicate more or less freely with each other around the muscles and vessels traversing them.

Since they intervene between the intrafascial spaces and the mandible, on the one hand, and the pharynx, on the other, they are liable to infection from either of these sources by extension from them; moreover, these spaces are most intimately related to the lymph nodes receiving the drainage from the nose, throat, and jaw, so that abscesses within them may form from breakdown of nodes secondarily infected from their regions of drainage. It is one or more of these spaces that is more commonly infected in the neck; while, as already noted, clinical opinions vary sharply as to the extent to which spread of pus or of a cellulitis occurs along fascial planes, it has been emphasized by many workers that injections of inert materials into certain of these spaces spread readily into others, and that infections may spread rapidly from one space to another.

Retropharyngeal Space

The central or posterior space of this group, and the key to an understanding of such downward spread of infections of the head and neck within fascial spaces as occurs, is the retropharyngeal space. This is simply the upper portion of the retrovisceral space, already described in connection with the infrahyoid spaces; it, also, is logically termed the *retrovisceral space,* although the term "retropharyngeal" is more commonly used. (Like the lower part of the retrovisceral space, this upper part has also been called the prevertebral space.)

The retropharyngeal space is the area of loose connective tissue lying behind the pharynx and in front of the alar layer of the prevertebral fascia (Fig. 5-17). It extends upward between the pharynx and the vertebral column to the base of the skull; it extends downward behind the lower portion of the pharynx and the esophagus, as already described, to form the posterior portion of the visceral compartment of the neck, communicate with the pretracheal space, and end at about the level of the bifurcation of the trachea. The connections of this space are therefore such that injections within it may spread into either the anterior or the posterior portions of

Fig. 5-15. Part of a diagrammatic semifrontal section, slanted somewhat anteriorly from behind the ramus of the mandible, to show the relations of the superficial layer of fascia to the parotid gland.

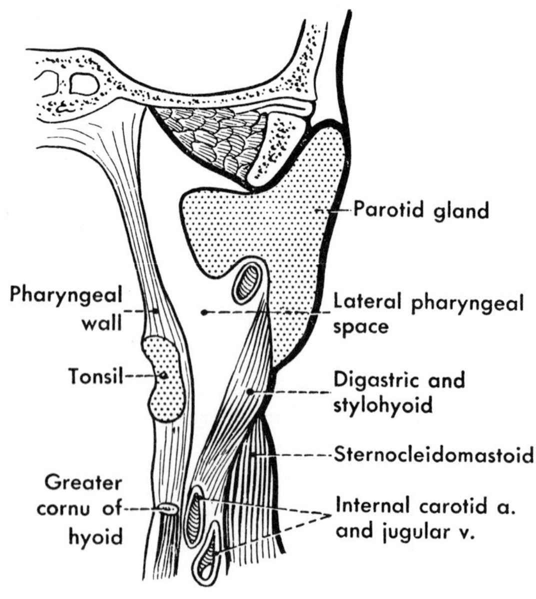

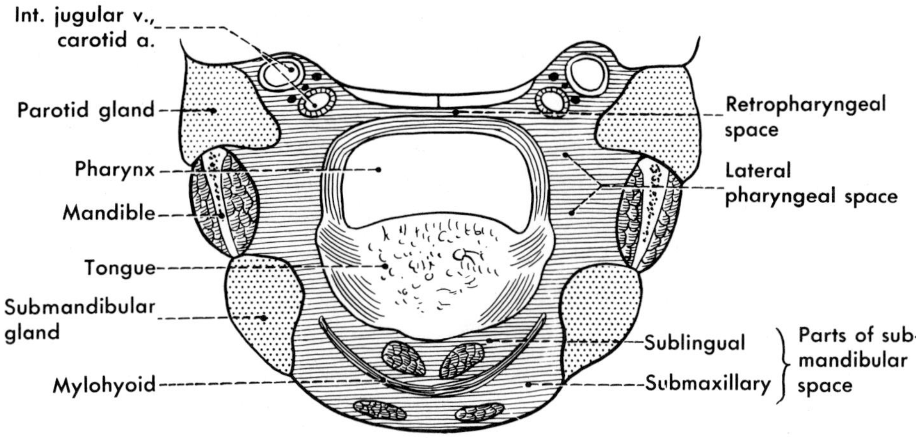

Fig. 5-16. The peripharyngeal spaces in an oblique section, slanting forward and downward. This illustrates the general relations of these spaces, and their continuity with each other. (After Grodinsky M, Holyoke EA: Am J Anat 63:367, 1938)

the superior mediastinum without breaking through its walls; Grodinsky and Holyoke found that their injections sometimes broke through the posterior wall of the retropharyngeal space, that is, through the alar fascia, and entered Space 4, the danger space, between the two lamellae of the prevertebral layer of fascia; this space, as already noted, extends from the base of the skull to the level of the diaphragm.

Because of these relations, the retropharyngeal space has been commonly regarded as a route through which infections of the mouth and throat reach the mediastinum. New and Erich stated that they had not found a single case of mediastinitis secondary to cervical infection in the 267 patients that they studied. Pearse, however, although he concluded that suppurative mediastinitis is due to spread from the neck in not more than one fifth of cases, reviewed 100 cases in which such spread was apparently the source of the mediastinitis and stated that in by far the majority of these (71%) the mediastinitis was from spread in the retrovisceral space.

Lateral Pharyngeal Space

The retropharyngeal space extends also laterally around the pharynx, and these lateral portions, although continuous with the retropharyngeal space, are usually spoken of as the

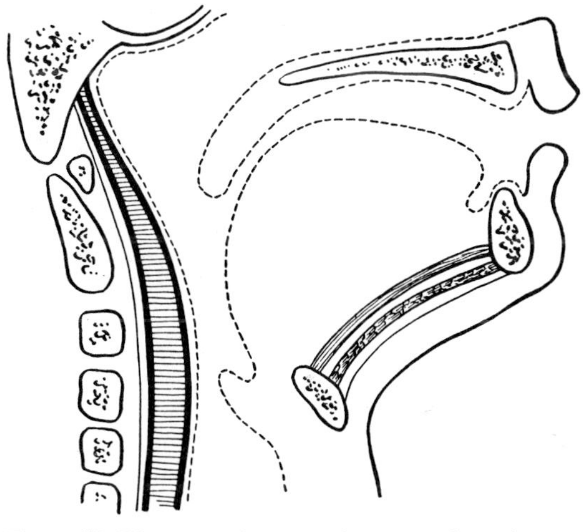

Fig. 5-17. The retropharyngeal or retrovisceral space in a sagittal section. Compare with Fig. 5-11.

lateral pharyngeal spaces (Fig. 5-18). The lateral pharyngeal space has been variously termed the parapharyngeal, peripharyngeal, pharyngomaxillary, pterygopharyngeal, pterygomandibular, and pharyngomasticatory space; it consists of loose connective tissue lateral to the pharynx and is bounded medially by the fascia of the pharynx itself, laterally by the pterygoid muscles and the sheath of the parotid gland.

Like the retropharyngeal portion of this visceral ring, the lateral pharyngeal space ex-

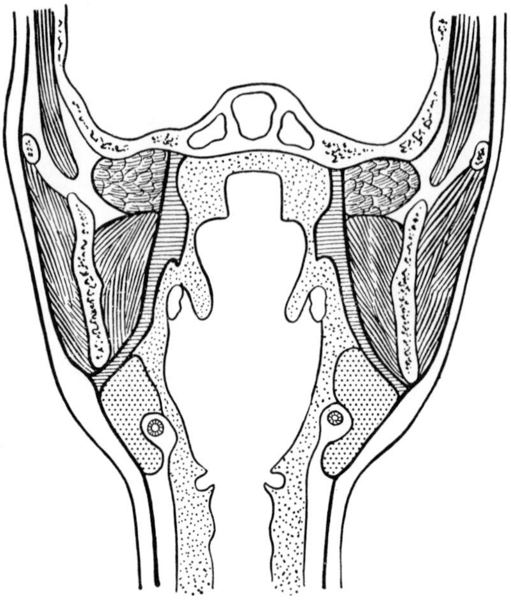

Fig. 5-18. The lateral pharyngeal space (*horizontally lined*) and its relations in frontal section. Lateral to it are the muscles of mastication and the masticator space, medially the pharyngeal wall including the tonsil, and below and medially the submandibular gland. (After Coller FA, Yglesias L: Surg Gynecol Obstet 60:277,1935)

tends upward to the base of the skull, but it does not extend inferiorly below the level of the hyoid bone, since it is limited here by the sheath of the submandibular gland and the attachments of this sheath to the sheaths of the stylohyoid muscle and the posterior belly of the digastric. This space is traversed by the styloglossus and stylopharyngeus muscles and is by some authors (*e.g.,* Hall) divided into a prestyloid and a poststyloid portion; both above and below these muscles it opens medially into the retropharyngeal space.

The posterior wall of the lateral pharyngeal space is formed by the carotid sheath, or at least by the connective tissue about the internal jugular and internal carotid vessels; many authors do not consider the carotid sheath to extend above the level of the hyoid bone. Anterosuperiorly the lateral pharyngeal space extends to the pterygomandibular raphe; anteriorly and inferiorly, above the submandibular gland, it communicates with the spaces associated with the floor of the mouth.

Because of its relationships, the lateral pharyngeal space is subject to infection from more sources than any other peripharyngeal space. Through its connection with the spaces about the tongue, it may receive and transmit to the retropharyngeal space infections originating here, as from the teeth; similarly, it is adjacent to the submandibular gland, and infections in this gland may spread into the space; likewise, both the masticator space and the parotid gland border the lateral pharyngeal space, and infections within either of these that perforate deeply instead of superficially will necessarily invade the lateral pharyngeal space. The tonsillar region of the pharynx is the medial wall of the lateral pharyngeal space, and infections originating about the tonsils may also involve this space.

The lateral pharyngeal space is thus subject to infection from several sources; among those workers who regard the fascial spaces as being of particular importance in the spread of infections, therefore, the lateral pharyngeal space is considered to be the route by which infections of diverse origins may be transmitted to the highway—the retropharyngeal space—that in turn leads to the mediastinum. Infections of the retropharyngeal, lateral pharyngeal, and sublingual spaces are most easily discussed together, therefore.

Submandibular Space

The anterior element of the peripharyngeal spaces is conveniently referred to as the submandibular space (Grodinsky and Holyoke). It actually consists of a group of spaces (Fig. 5-19) that, however, either communicate with each other or can be made to do so with relatively little pressure upon injection. The submandibular space as a whole is limited above by oral mucous membrane and the tongue, while its floor is the superficial layer of cervical fascia as it extends from the hyoid bone to the mandible. Its inferior extent is the attachment of the fascia to the hyoid.

Some of the subdivisions of the submandibular space are fairly obvious: the mylohyoid muscle, stretching across the floor of the mouth, divides the submandibular space into a portion above this muscle and a portion

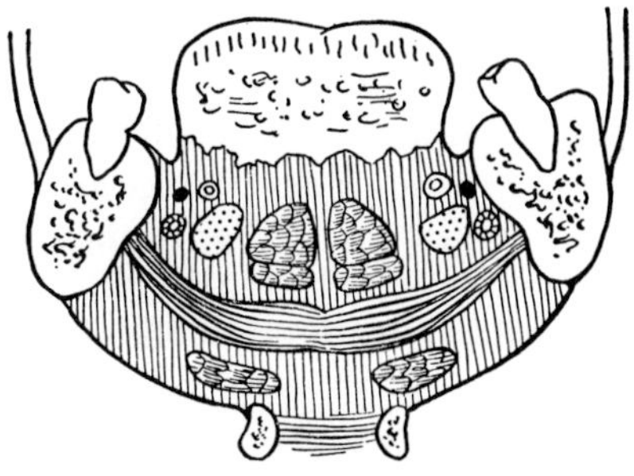

Fig. 5-19. The submandibular space in a frontal section through the floor of the mouth. The continuity between the spaces above and below the mylohyoid muscle, and their connection with the lateral pharyngeal spaces, may be seen in Figs. 5-13 and 5-16. For identification of other structures shown here, refer to Fig. 5-12.

below; about the submandibular gland, which lies partly above and partly below the posterior portion of the mylohyoid, these two subdivisions communicate with each other, however (Fig. 5-20).

The space below the mylohyoid muscle, often termed the *submaxillary space,* is sometimes said to be divided into subsidiary submental and submaxillary spaces by attachment of the superficial layer of fascia to the anterior belly of the digastric muscle; the submental space, corresponding to the triangle of the same name, lies medial to the anterior belly of the digastric, the submaxillary space lateral and posterior to it, but Grodinsky and Holyoke found that injections spread readily across the anterior belly of the digastric from one space to the other. These spaces consist only of an easy line of cleavage between the fascia and the muscles, unless they are abnormally distended.

The spaces above the mylohyoid muscle can be collectively termed the *sublingual space* and consist of the loose connective tissue lying between the muscles of the tongue and about the sublingual gland, the lingual and hypoglossal nerves, and a portion of the submandibular gland and its duct. The sublingual space is largely paired according to Williams, but the two sides communicate anteriorly. The sublingual space has been subdivided (*e.g.,* Coller and Yglesias, '35) into as many as three compartments, based upon their relationships to the genioglossus and geniohyoid muscles, but this subdivision seems needlessly complex.

INFECTIONS OF THE PERIPHARYNGEAL SPACES

Grodinsky ('39a) described infection of the peripharyngeal spaces as occurring more commonly in children less than 3 years of age, and it is frequently stated that such infections are relatively rare in adults. Beck ('42, '47) reported, however, that amng 100 cases he studied only 11 occurred in the age group 1 month to 2 years, 45 occurred in the age group 2 to 12 years inclusive, 12 in the age group 13 to 20 years inclusive, and 32 in the age group 21 to 81 years inclusive; in Boemer's series of 75 cases 26 were in adults; in Smith's series, there were more cases of retropharyngeal abscesses in adults than in children.

Grodinsky regarded the majority of the acute abscesses of the retropharyngeal and lateral pharyngeal spaces as resulting from infections of the nose, throat, and middle ear and stated that infection through lymphatic drainage to the nodes in these spaces is the most common method of involvement where trauma is not involved. Faier also stressed the relation of infection of the retropharyngeal and deep cervical nodes to involvement of these spaces, and Boemer reported such infec-

Fig. 5-20. Continuity of the sublingual and "submaxillary" parts of the submandibular space around the posterior border of the mylohyoid muscle, and adjacent to the submandibular gland, as it might be seen from the lateral aspect after removing the body of the mandible.

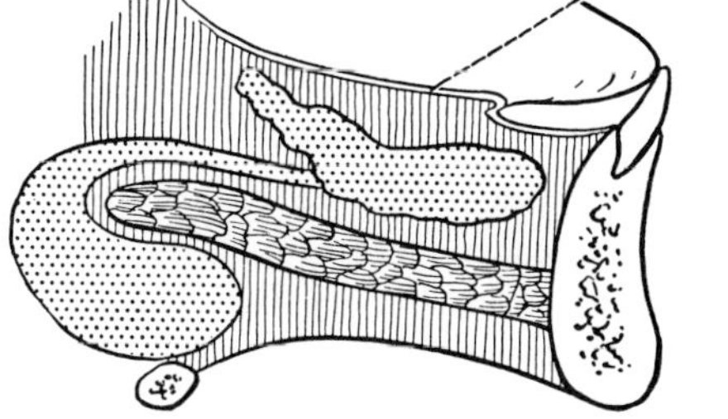

tion to be involved in all 49 of the children in his series. Beck found that about 50% of the infections involving fascial spaces in the neck arose from infections of the pharynx and tonsils; Lake also has stressed the relationship of infections of those spaces to infections of the tonsils. About 20% of the deep infections of the neck in Beck's series, and about one-third of the adult cases in Boemer's series, arose from infections of dental origin. Faier expressed the belief that infections of the ear are not a common cause of retropharyngeal abscess, but that infections of this origin, which may reach the space not only along lymphatics but also by following canals in the bone that transmit vessels and nerves, are frequently severe. Beck pointed out that infections within the petrous portion of the temporal bone may rupture directly into the lateral pharyngeal space, and infection at the tip of the mastoid process may follow the mastoid groove and extend along the styloid and digastric muscles to this space; in two of the adult cases from Boemer's series infection of the peripharyngeal spaces resulted from mastoid infections.

Beck described the lateral pharyngeal space as the one most frequently involved in deep infections of the neck, but not infrequently more than one space was involved simultaneously. In 12 cases of infection of the lateral pharyngeal space, for instance, the submaxillary region was also infected, and in eight cases of lateral pharyngeal infection the carotid sheath was also infected.

Lifschutz reported a case of fatal hemorrhage resulting from extension of retropharyngeal abscess to the deep vessels of the neck. While this is comparatively rare, he found that it has been reported as arising in the majority of cases from the internal carotid artery rather than from the jugular vein; he concluded that the vein is more often occluded by the infectious process than it is eroded to the point of hemorrhage. Havens emphasized that a sudden enlargement of a retropharyngeal mass may indicate erosion of a large vessel and that in such a case aspiration of the mass before its incision is wise; if forewarned by the presence of blood, ligation of the carotid may prevent fatal hemorrhage. Barlow expressed the belief that thrombosis of the jugular vein from a deep infection of the neck is probably not due to direct infection of the carotid sheath, but rather to the fact that infectious material follows tributaries of the internal jugular vein to reach the sheath.

Surgical Drainage

Opinions differ sharply as to the advisability and necessity of surgical drainage of fascial spaces. However, the lateral pharyngeal space and the retropharyngeal one may be approached through the mouth and opened by appropriate incisions through the pharyngeal wall; Iglauer concluded that this approach to the retropharyngeal space is usually preferable except for large abscesses, cellulitis, and infections due to caries of the vertebral column. Boemer stated that external drainage of the lateral pharyngeal space should be carried out in practically all cases. External drainage of the lateral pharyngeal space can be obtained by making an incision immediately behind the submandibular gland and passing upward along the medial or deep surface of the medial pterygoid muscle, into the lateral pharyngeal space; if desired, the stylohyoid and posterior belly of the digastric can be reflected to give entrance to this space. The retropharyngeal space can be drained externally by an incision behind the sternocleidomastoid muscle, carried medially behind the great vessels of the neck.

The sublingual space may be drained through the mucosa of the mouth or, as described by Williams, the entire submandibular space may be opened by an incision extending parallel to the medial border of the mandible and carried through the superficial layer of the cervical fascia and the mylohyoid and digastric muscles. If this opening is to be carried posteriorly, the facial artery and vein must be ligated and the submandibular gland may be removed. Tschiassny ('47) suggested a different approach: he used a transverse incision medial to the angle of the mandible to approach the submandibular gland; after cutting the fascia he ligated the submental branch of the facial artery, freed the submandibular gland and retracted it laterally; he then divided the mylohyoid muscle, sectioning its fibers at right angles, but did not extend the incision forward to involve the digastric or the geniogyoid muscles.

Ludwig's Angina

Ludwig's angina is the term given to the symptoms resulting from infection of the submandibular space and is characterized especially, therefore, by extreme hardness of the floor of the mouth. The infection here may eventually extend to the lateral pharyngeal space also and thence may enter the retropharyngeal space and even descend to the mediastinum. Beck estimated that in about 20% of the patients in his series deep infection of the neck originated from the submandibular space, and Boemer expressed the belief that in about a third of the adults in his series the infection had this origin. Trout ('40a) pointed out that the original criteria of Ludwig's angina are sufficiently indefinite to explain the various reports on the mortality, which has been said to be as high as 75% or as low as 5%; he found that the mortality reported at that time (before the use of antibiotics) ranged usually from about 30% to 75%.

As pointed out by Grodinsky ('39b), death from Ludwig's angina occurs as a result of suffocation due to edema of the mouth, tongue, and the glottis, from mediastinitis due to spread, or from septicemia or pneumonia. Grodinsky noted that infections starting from the nose and throat may spread from the retropharyngeal and lateral pharyngeal spaces into the submandibular and resemble Ludwig's angina in their later stages, but stated that such secondary extension is not characterized by the boardlike swelling of the neck centering about the floor of the mouth and by the ensuing elevation of the mucosa of the mouth and tongue and should not be classified as true Ludwig's angina.

Trout ('40a) stated that extraction of a lower molar tooth precedes Ludwig's angina in about 82% of cases. He also ('40b) agreed with most workers that the mylohyoid muscle tends to confine infections from these teeth to the deep tissues under the tongue and advocated cutting the superficial layer of fascia and the mylohyoid and digastric muscles to obtain adequate drainage; Taffel and Harvey also emphasized that this is the only way in which pressure can be relieved, and stated that the mylohyoid must be completely divided, as must the anterior belly of the digastric, regardless of whether pus or edema seems to be present or not. In contrast to this older concept, Erich's experience has been that premature incision of fascial spaces in cases of Ludwig's angina is harmful and that the appropriate spaces should be opened only after purulent material has formed. Houser stated that Ludwig's angina always starts with involvement of the sublingual space, that is, always lies at first above the mylohyoid muscle, and suggested therefore that the logical method of evacuating pus is through an intraoral incision; if this is done as a result of early diagnosis, he concluded, involvement of the neck may be avoided.

The accepted fact that in the majority of cases Ludwig's angina originates from infections about the lower molar teeth is usually explained as a consequence of the eccentric setting of these teeth, so that their roots are closer to the inner than the outer side of the jaw. Although the anterior teeth are also usually set eccentrically, they are nearer to the outer than the inner surface of the jaw, and therefore infections about their roots, if they erode the thinner area of bone, would tend to break outward rather than inward.

Tschiassny ('43) stated, however, that it is not necessarily true that the medial border of the alveolar socket of the molar teeth is thinner than is the outer border, and found some jaws in which the reverse was true; he concluded that the decisive factor is the depth to which the roots penetrate, for he regarded Ludwig's angina as regularly starting below the mylohyoid muscle. According to his investigations, the roots of the second and third molar teeth almost always reach downward at least as far as the level of the attachment of the mylohyoid muscle, and usually below it, while most of those of the first molar teeth, and usually all of those anterior to this, are located above this level. He found it proba-

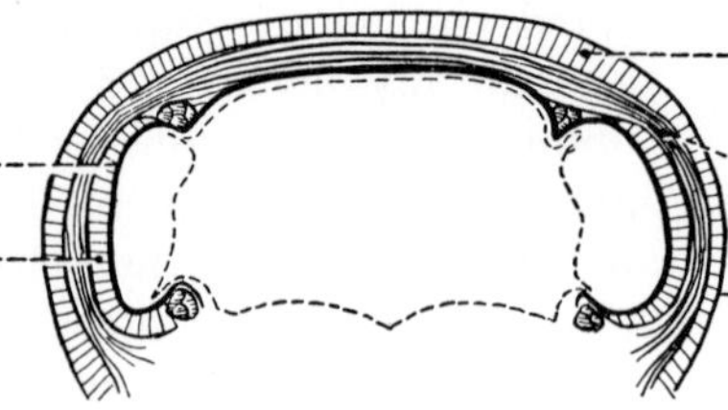

Fig. 5-21. Potential intrapharyngeal spaces, in a section at the level of the tonsil.

ble, therefore, that an infected second or third molar tooth might produce infection in the "submaxillary" space, which he would regard as true Ludwig's angina, while infections of the first molar or of a premolar would produce infections first in the sublingual space, which he apparently would not classify as Ludwig's angina. In a later ('47) article, however, he emphasized the importance of cutting the mylohyoid muscle for the relief of tension in Ludwig's angina; whether the infection starts above or below this muscle seems to be of little moment, therefore. Most workers have emphasized the involvement of the sublingual space.

INTRAPHARYNGEAL SPACES

Since the pharyngeal constrictors are covered by the buccopharyngeal fascia, a potential space theoretically exists between muscles and fascia. Infections here, like those within the visceral fascia lower in the neck, tend either to remain localized or to break through into the lateral pharyngeal and retropharyngeal spaces.

Paratonsillar Space

Of more importance as a space is the peritonsillar or paratonsillar space, the area of loose connective tissue lying in the tonsillar bed and uniting the capsule of the tonsil loosely to the underlying pharyngeal muscles (Fig. 5-21). Infections here (quinsy) produce bulging of the tissue about the tonsil and can most easily be drained by incisions through the mucosa under which they point; often the opening is made through the supratonsillar fossa. McCurdy concluded that immediate tonsillectomy is not only safe, but superior to drainage followed later by tonsillectomy.

According to Wood, injections into the paratonsillar spaces tend to spread longitudinally but not transversely; in his study they extended as high as the hard palate or the pharyngeal orifice of the auditory tube, and some extended downward as far as the piriform recess. Abscesses in the paratonsillar space may drain internally spontaneously; if they break laterally, they open into the lateral pharyngeal space.

References

BARLOW D: The surgical anatomy of the neck in relation to septic lesions: An investigation of the cervical connective tissue. J Anat 70:548, 1936

BECK AL: Deep neck infection. Ann Otol Rhinol Laryngol 51:592, 1942

BECK AL: Deep neck infection. Ann Otol Rhinol Laryngol 56:439: 1947

BOEMER LC: The great vessels in deep infection of the neck. Arch Otolaryngol 25:465, 1937

COLLER FA, YGLESIAS L: Infections of the lip and face. Surg Gynecol Obstet 60:277, 1935

COLLER FA, YGLESIAS L: The relation of the spread of infection to fascial planes in the neck and thorax. Surgery 1:323, 1937

ERICH JB: Personal communication to the author, 1966

FAIER SZ: Retropharyngeal abscess of otitic origin: Anatomy and pathogenesis with report of cases. Ann Otol Rhinol Laryngol 42:408, 1933

GAUGHRAN GRL: Fasciae of the masticator space. Anat Rec 129:383, 1957

GRODINSKY M: Retropharyngeal and lateral pharyngeal abscesses: An anatomic and clinical study. Ann Surg 110:177, 1939a

GRODINSKY M: Ludwig's angina: An anatomical and clinical study with review of the literature. Surgery 5:678, 1939b.

GRODINSKY M, HOLYOKE EA: The fasciae and fascial spaces of the head, neck and adjacent regions. Am J Anat 63:367, 1938

HALL C: The parapharyngeal space: An anatomical and clinical study. Ann Otol Rhinol Laryngol 43:793, 1934

HALL C, MORRIS F: Infections of the masticator space. Ann Otol Rhinol Laryngol 50:1123, 1941

HAVENS FZ: Severe hemorrhage secondary to retropharyngeal abscess. Am J Dis Child 58:1256, 1939

HOUSER KM: Ludwig's angina: Intra-oral incision in infections of the floor of the mouth. Arch Otolaryngol 16:317, 1932

HUNT WM: Periesophageal abscesses: The importance of early surgical interference. Ann Otol Rhinol Laryngol 48:128, 1939

IGLAUER S: Surgical approaches to deep suppuration in the neck and posterior mediastinum. Arch Otolaryngol 21:707, 1935

KOSTRUBALA JG: Potential anatomical spaces in the face. Am J Surg ns68:28, 1945

LAKE CF: Infection of the neck after tonsillectomy. Minn Med 30:851, 1947

LIFSCHUTZ J: Fatal hemorrhage resulting from the extension of a retropharyngeal abscess to the deep vessels of the neck. Arch Otolaryngol 14:149, 1931

MCCURDY JA JR: Peritonsillar abscess. A comparison of treatment by immediate tonsillectomy and interval tonsillectomy. Arch Otoloaryngol 103:414, 1977

NEW GB, ERICH JB: Deep infections of the neck: Collective review. Int Abstr Surg (Surg Gynecol Obstet) 68:555, 1939

PARSONS FG: On the carotid sheath and other fascial planes. J Anat Physiol 44:153, 1910

PEARSE HE JR: Mediastinitis following cervical suppuration. Ann Surg 108:588, 1938

SMITH JE: Retropharyngeal abscess with reference to abnormally large percentage of adult cases. Ann Otol Rhinol Laryngol 49:490, 1940

TAFFEL M, HARVEY SC: Ludwig's angina: An analysis of forty-five cases. Surgery 11:841, 1942

TROUT HH: Ludwig's angina. Arch Surg 41:532, 1940a

TROUT HH: Von Ludwig's angina. Surgery 8:1024, 1940b

TSCHIASSNY K: Ludwig's angina: An anatomic study of the role of the lower molar teeth in its pathogenesis. Arch Otolaryngol 38:485, 1943

TSCHIASSNY K: Ludwig's angina: A surgical approach based on anatomical and pathological criteria. Ann Otol Rhinol Laryngol 56:937, 1947

WILLIAMS AC: Ludwig's angina. Surg Gynecol Obstet 70:140, 1940

WOOD GB: The peritonsillar spaces: An anatomic study. Arch Otolaryng 20:837, 1934

Chapter 6
THE FACE

Bony Structure

In the broad sense, the bones of the face (Fig. 6-1) include the frontal and nasal bones, as well as the facial bones proper (maxilla, mandible, and zygomatic and palatine bones). Of these, the nasal bones are described with the skeleton of the nose; the palatine, which does not enter into the subcutaneous part of the facial skeleton, the mandible, and parts of the maxilla are considered in the following chapter.

The large facial part of the *frontal bone* is the squama, the vertical portion that forms the forehead and therefore the front of the calvaria. The pars orbitalis, in the roof of the orbit, is continuous with the squama at the supraorbital rim. Between the orbits the short pars nasalis extends downward to articulate medially with the two nasal bones and laterally with the frontal processes of the two maxillae (these lateral parts of the nasal process have also been called the maxillary processes). Lateral to each orbit a zygomatic process extends downward to articulate with the zygomatic bone.

The *zygomatic* (*malar*) *bone* sends its frontal process upward along the lateral margin of the orbit to articulate with the frontal bone and sends a temporal process backward to form the anterior end of the zygomatic arch. Its temporal surface, facing largely posteriorly, forms much of the anterior wall of the temporal fossa and bears the zygomaticotemporal foramen; its orbital surface faces the orbit; and its lateral (malar) surface forms the most prominent part of the cheek and presents the zygomaticofacial foramen. The zygomatic bone articulates with the maxilla medially on both its orbital and lateral surfaces and may house an upper posterior part of the maxillary sinus.

The body of the *maxilla,* containing the major part of the maxillary sinus, forms the lower border of the piriform aperture (the curved nasal notch), the two bones meeting in the midline to form the anterior nasal spine. From the lateral border of the piriform aperture the body extends posteriorly to form a lower part of the lateral nasal wall, and from the upper medial part of the body the frontal process extends upward along the side of the nose to articulate with the nasal, lacrimal, and frontal bones. The body also extends into the floor of the orbit, its orbital and anterior surfaces meeting to form the medial part of the inferior orbital rim. A variable distance below this rim the anterior surface of the body is pierced by the infraorbital foramen, and extending downward from this, lateral to the prominence of the canine tooth, is the concavity of the canine fossa. The anterior or facial surface of the maxilla curves laterally and backward to become continuous with the infratemporal or posterior surface. This surface presents several foramina for the poste-

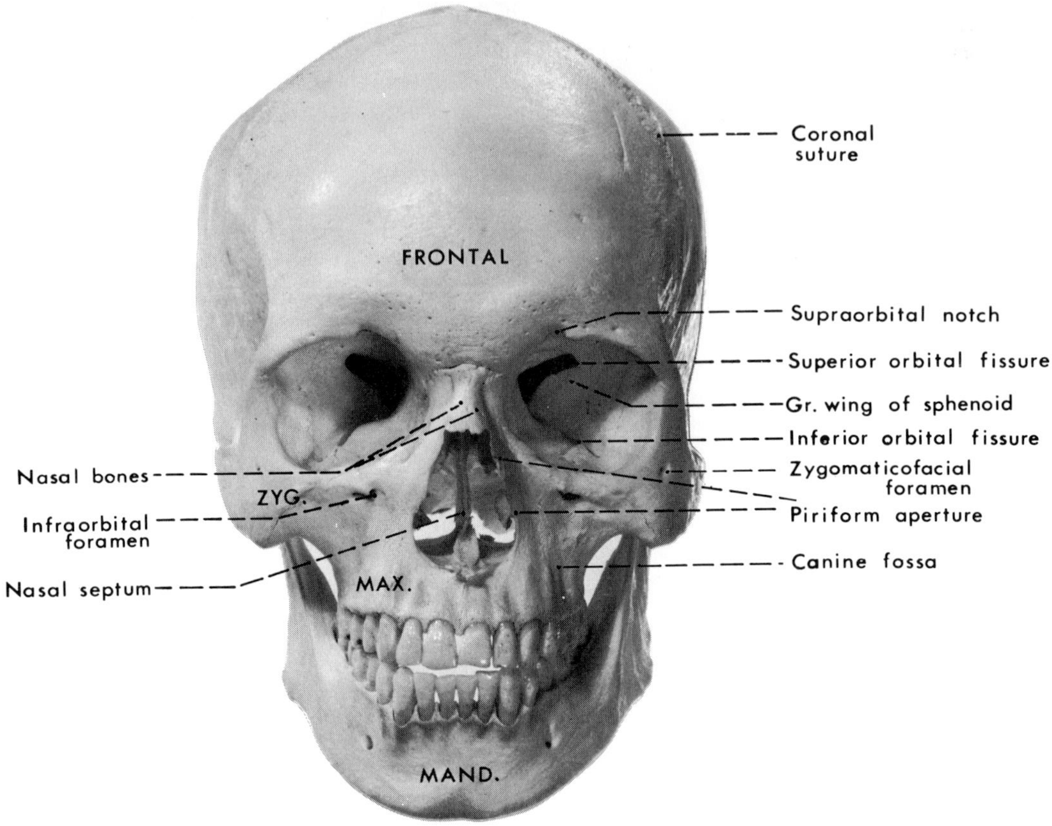

Fig. 6-1. Anterior view of the skull and facial bones, *MAX.* and *ZYG.*, the maxilla and zygomatic bone, respectively; *MAND.*, the mandible.

rior superior alveolar nerves and vessels, and below, above and behind the last molar tooth, it bears the swelling known as the maxillary tuber (tubercle). The alveolar process, bearing the teeth, is continuous with the body, and the palatine process, forming much of the bony palate, is continuous with the alveolar process and body; these parts are described in the following chapter on the jaws and palate.

Fractures of the frontal bone are, of course, considered skull fractures rather than facial ones; facial fractures involve primarily the maxilla or the zygomatic bone, with which fractures of the nasal bone are also sometimes grouped. From the brief account of the relations of these bones, it is obvious that facial fractures often involve the maxillary sinus and may extend into the orbit, the nose, and the palate. Fractures of the maxilla may involve any of the superior alveolar nerves, which traverse it, or the much larger infraorbital nerve with its extensive cutaneous distribution. McKenzie ('50), among others, has discussed fractures of the facial bones.

The Facial Muscles

The facial or mimetic muscles may be conveniently divided into five chief groups, concerned especially with the mouth, the nose, the orbit, the ear, and the scalp, respectively; in addition, the platysma muscle in the neck belongs to the facial group. The muscles are characterized by the fact that their chief action is upon skin, into which for the most part they insert. They all are innervated by the facial nerve and are therefore facial in innervation as well as position; besides these muscles the facial nerve supplies only three deeply lying ones: the stylohyoid, the posterior belly of the digastric (extending from the otic region to the hyoid bone), and the tiny stapedius muscle in the middle ear.

The muscles of the face vary considerably in their development and in the degree to which they interlace and have been subdivided and named in several different ways. In this chapter both the NA and common alternative names are used.

The subcutaneous connective tissue or "superficial fascia" of the face, as noted in the preceding chapter, closely invests the numerous facial muscles but is so bound down around them that, from the point of view of the spread of infections, fascial spaces in the face can generally be disregarded. The exception to this is in the lids, where loose connective tissue exists both superficial and deep to the palpebral portion of the orbicularis oculi muscle; the spaces here are discussed in the chapter on the orbit. Deep to and between the facial muscles, also enclosed in the connective tissue, are the nerves and blood vessels of the face. Except for the few deeper-lying facial muscles, the nerves to the muscles enter their deep surfaces and typically do so close to the posterior border of the muscle.

Platysma

The bulk of the platysma muscle lies in the superficial fascia of the neck, but its origin extends downward across the clavicle to the upper portion of the thorax, and its insertion similarly extends upward over the border of the jaw, superficial to the facial vessels, toward the lateral angle of the mouth; here it blends with some of the muscles connected with the lower lip, especially the depressor anguli oris (triangularis). Over the two platysma muscles just beneath the chin there may be superficially placed transverse muscle bundles to form a transversus menti muscle. The platysma is innervated by the cervical branch of the facial nerve, the lowest and also the least important branch of the facial.

ORAL MUSCLES

The muscles most directly concerned with movements of the lips and cheek include a lower group, an upper group, the orbicularis oris encircling the mouth, and, laterally, the buccinator (Fig. 6-2).

LOWER GROUP

The muscles of the lower group consist of the depressor anguli oris (triangularis), the de-

Fig. 6-2. Muscles about the mouth. On the left side of the face the superficial muscles are removed, and the lower part of the masseter is shown.

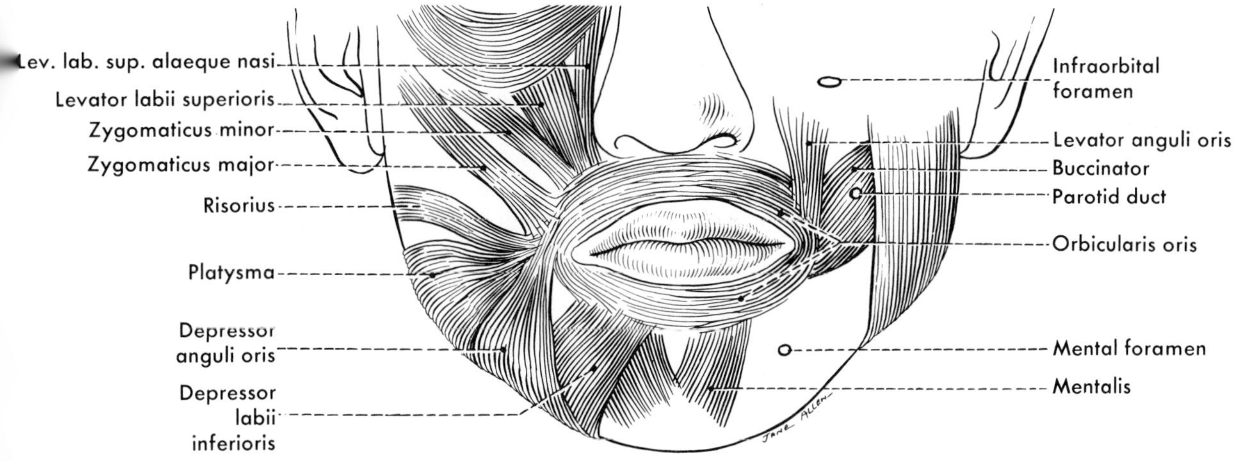

pressor labii inferioris (quadratus labii inferioris), and the mentalis.

Depressor Anguli

The depressor anguli oris or triangularis muscle, usually well developed, arises from the side of the body of the mandible below the canine, premolar, and first molar teeth; its fibers converge to the corner of the mouth, where they are partly inserted into skin and partly become continuous with the upper part of the orbicularis oris.

The posterior border of this muscle may be somewhat blended with the upper fibers of the platysma, which also converge toward the corner of the mouth. The transversus menti is sometimes regarded as an aberrant portion of the depressor.

Depressor Labii Inferioris

The depressor (quadratus) labii inferioris arises from the front of the mandible, in part under cover of anterior fibers of the depressor anguli; it passes upward in front of the mental foramen, the fiber bundles running somewhat medially so as to converge with the muscle of the other side, to pass among the fiber bundles of the lower portion of the orbicularis oris and insert into skin and mucosa of the lower lip.

Mentalis

The mentalis, the deeper-lying muscle of this group, arises from the mandible at the level of the root of the lower lateral incisor; in contrast to the preceding muscles it passes downward, to insert into the skin of the chin. It is a short, stout muscle placed deep to the depressor labii inferioris; the two muscles converge toward their insertions.

Innervation and Action

These three muscles of the lower group are innervated by the marginal mandibular branch of the facial and perhaps also from lower buccal branches. The actions of the two depressors are indicated by their names; the mentalis draws the skin of the lip upward and thus aids in forcing the lower lip against the gum and teeth and in protracting it. Damage to the branch of the facial nerve supplying these muscles can easily be inflicted during a block dissection of the neck for carcinomatous nodes, and, as is obvious from the actions of these muscles, such damage would upset the activity and symmetry of the mouth.

UPPER GROUP

The upper group of labial muscles consists of the risorius, the zygomaticus major (zygomaticus), the zygomaticus minor, the levator labii superioris, the levator labii superioris alaeque nasi (the zygomaticus minor and the two levators were once called the zygomatic, infraorbital, and angular heads, respectively, of the quadratus labii superioris), and the levator anguli oris (caninus).

Risorius

The risorius is a thin, usually poorly developed muscle that differs from most of the facial muscles in that it does not arise from bone, but in the subcutaneous tissue over the parotid gland. It is a bandlike or slightly triangular muscle lying at about the upper border of the facial portion of the platysma, partly overlapping this superficially and often separable from it with difficulty. From its origin the risorius runs forward and slightly downward, across the masseter muscle to the lateral corner of the mouth where it inserts into skin and mucosa. Like the facial portion of the platysma, this muscle crosses superficial to the facial vein and artery.

Zygomaticus Major and Minor

The *zygomaticus major* arises from the posterior portion of the lateral surface of the zygomatic bone below the orbicularis oculi, with which it may partially blend, and passes obliquely downward and forward to the corner of the mouth where it attaches to skin and mucosa. It is a superficially located, ribbonlike muscle, which passes superficial to the masseter and buccinator muscles and to the facial vein and artery.

The *zygomaticus minor* (zygomatic head of the quadratus labii superioris) arises from the zygomatic bone either below or deep to the orbicularis oculi, anterior to the origin of the zygomaticus major; it runs toward the mouth almost parallel to the latter muscle, lying anterior to and above it. It inserts into the upper lip just medial to the corner of the mouth.

Levators

The *levator labii superioris* (infraorbital head of the quadratus labii superioris) arises from the infraorbital margin of the maxilla under cover of the orbicularis oculi. It extends downward and somewhat medially to mingle with the orbicularis oris and insert into much of the skin of the lateral half of the upper lip.

The *levator labii superioris alaeque nasi* (angular head of the quadratus labii superioris) arises from the frontal process of the maxilla alongside the nose; part of it inserts into the skin and greater alar cartilage of the nose, but the major portion passes obliquely downward to the skin and musculature of the upper lip.

The third muscle of this group, the *levator anguli oris* (caninus), lies more deeply than do the others and is covered by them and the zygomaticus muscles. It arises from the canine fossa of the maxilla and runs downward to the corner of the mouth to attach to the skin, also sending fasciculi around the corner of the mouth to blend with the lower part of the orbicularis oris. The facial artery runs across the lower part of the muscle, and branches of the infraorbital nerve and vessels spread out in the connective tissue between it and the overlying muscles.

Innervation and Action

The risorius is innervated through a buccal branch of the facial nerve; the remaining, higher placed muscles are innervated through upper buccal or zygomatic branches. The risorius and the zygomaticus major and minor draw the corner of the mouth laterally and upward, but the risorius contributes the least to the upward movement, the zygomaticus minor the most. The levators are named from their actions. In lifting the nasal ala, the levator labii superioris alaeque nasi widens the naris.

ORBICULARIS ORIS

Under the term *orbicularis oris* are usually grouped a number of muscle bundles arranged in diverse directions; the major portion of the muscle, or the orbicularis oris proper, consists of fibers arranged in a sphincteric fashion about the mouth. To what extent true sphincteric fibers actually exist is not known; at least some of the fibers as they reach the corner of the mouth simply cross those of the other lip to insert into mucosa and skin, many of the deeper ones are prolonged into the lips from the laterally placed buccinator muscles, and some of the fibers of the upper lip seem to be a direct continuation of the depressor anguli oris (triangularis) muscle, while some of those of the lower lip seem to be correspondingly a continuation of the levator anguli oris (caninus). Lee, for instance, investigated the orbicularis muscle in a case of double harelip and found it well developed at the corner of the mouth; serial sections revealed, however, only a relatively few sphincteric muscle fibers here.

In addition to the sphincteric or pseudosphincteric fibers that encircle the mouth (some of them, in the midline of both upper and lower lips, inserting here also into the skin), the orbicularis oris contains the insertions of many of the muscles already described as reaching the upper and lower lips. For practical purposes, the latter muscles may be regarded as blending with the orbicularis, although most of them, as already noted, are usually described as inserting at least in part into skin and mucosa of the lip.

Finally, there are a few fibers in both upper and lower lips that are said to arise from the bony jaws at the levels of the canine teeth and pass laterally to insert into the corners of the mouth, and there are other deep fibers, said to be best developed in infants, which pass at right angles to most of the fibers, from the skin to the mucosa of the lip. Through the action of the whole and of the various parts, the orbicularis oris draws the lips together, draws inward the corners of the lips, and either purses them or draws them against the teeth.

It is innervated by the buccal portion of the facial nerve.

BUCCINATOR

The buccinator muscle (Fig. 6-3), forming the muscular substance of the cheek, arises from the posterior portion of the alveolar process of the maxilla, from the upper medial surface of the mandible at the junction of the body and ramus, just posteromedial to the last molar tooth, and from the pterygomandibular raphe (pterygomandibular ligament) which stretches from the medial pterygoid process to the inner surface of the mandible and represents the line of junction between the buccinator muscle and the superior constrictor of the pharynx. (Gaughran has denied that a pterygomandibular raphe, in the sense of a fibrous band, intervenes between the buccinator and superior constrictor muscles. Although noting certain changes in the character of the fascia and muscle along this line, he described the two muscles as directly continuous with each other.) The fibers pass forward to become continuous with fibers of the orbicularis oris in both lips and to pass through this to attach to the mucosa of the lips and, to a lesser extent, into the skin.

Fig. 6-3. The buccinator muscle. Note its continuity with fibers of the orbicularis oris and its union posteriorly, at the pterygomandibular raphe, with the superior constrictor.

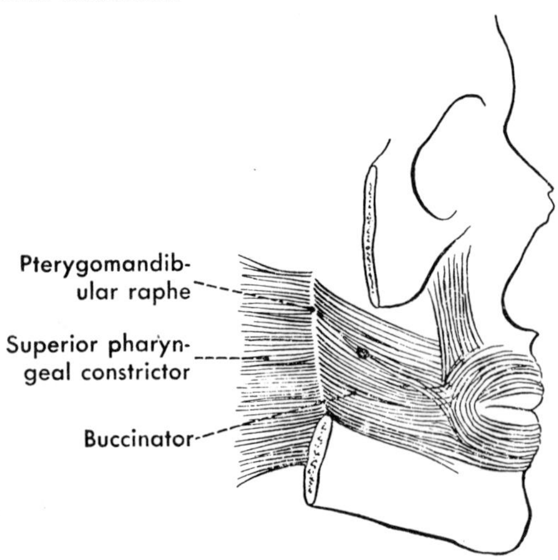

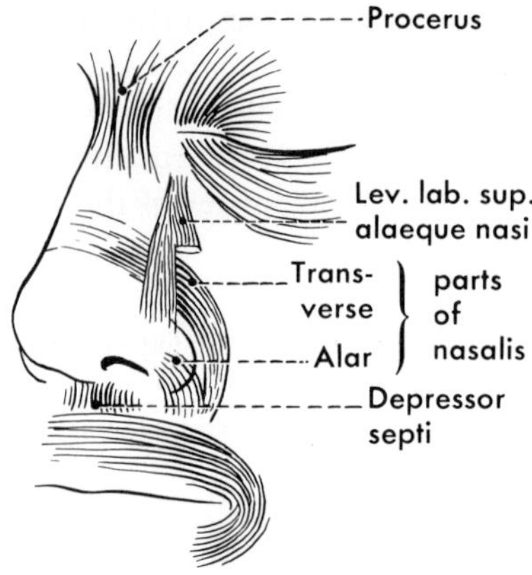

Fig. 6-4. The nasal group of facial muscles.

The muscle is crossed superficially by the facial artery and facial (anterior facial) vein. The buccal (buccinator) artery and the similarly named branch of the mandibular nerve emerge through the buccal fat pad, between the buccinator and the insertion of the temporal muscle. The parotid duct passes forward upon the muscle and pierces it to enter the mouth, and the transverse facial artery roughly parallels the duct, usually lying above it. The buccinator muscle is innervated through buccal branches of the facial nerve (not through the buccal or buccinator branch of the mandibular, which also lies on its outer surface and sends twigs through it to the buccal mucosa); it flattens the cheek against the teeth, thus minimizing the passage of food between teeth and cheek, and prevents the cheek from being distended unduly by air pressure, as when wind instruments are blown.

THE NASAL GROUP

The muscles connected with the external nares vary considerably in their development and have been variously described. Essentially, however, they consist of dilators and compressors of the nasal aperture (Fig. 6-4).

Nasalis

The best developed of these muscles is the nasalis, the chief or transverse part of which

Fig. 6-5. The facial muscles about the orbit.

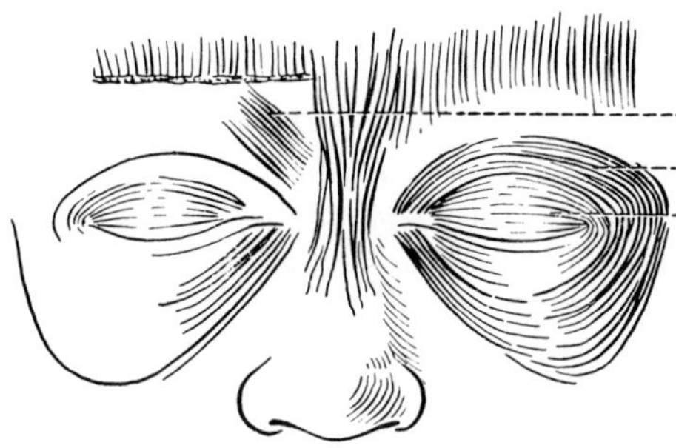

(compressor naris) is sometimes described as arising from the maxilla and inserting with the muscle of the opposite side into an aponeurosis across the dorsum of the nose and is sometimes said to arise from the aponeurosis and insert into skin in the nasolabial sulcus. The small alar part (dilator naris) arises from the maxilla over the lateral incisor and inserts into the greater alar cartilage. Its medial fibers tend to blend with the depressor septi, and it has been described as a part of that muscle. As indicated by their other names, the transverse and alar parts of the nasalis have been supposed to compress and dilate the naris, respectively. The transverse part has also been said to be a dilator, however.

Depressor Septi

The depressor septi arises from the incisive fossa of the maxilla, and its fibers extend upward to be inserted both into the septum and the posterior part of the ala of the nose. At its origin it lies deep to the orbicularis oris.

Dilator Muscles

Two small dilator muscles of the nose, an anterior and a posterior, are often described, although they are difficult to demonstrate because they consist of very delicate muscle bundles. The posterior dilator arises from lesser alar cartilages, and the anterior from the greater alar cartilage; both pass downward, the anterior in front of the posterior, close to the border of the nasal aperture to insert into skin near the margin.

In addition to the muscles just described, the *procerus* lies partly upon the dorsum of the nose; it is therefore sometimes described as a nasal muscle, but because its action is on the skin over the glabella, it is here described as one of the muscles about the orbit (although labeled in Fig. 6-4 rather than Fig. 6-5).

Innervation and Action

The muscles of the nose are supplied from the zygomatic and buccal branches of the facial nerve. The depressor septi narrows the naris and the levator labii superioris alaeque nasi dilates it; as already noted, opinions differ on the action of the nasalis on the naris.

THE ORBITAL GROUP

Orbicularis Oculi

The chief muscle about the orbit is the orbicularis oculi. This lies upon the bone surrounding the orbit and extends also into the lids; it is therefore divided into two major portions, an orbital and a palpebral one (Fig. 6-5).

The *orbital portion* arises from the more medial part of the medial palpebral ligament, from the nasal portion of the frontal bone and the frontal process of the maxilla; from this origin the fiber bundles spread out as they run superiorly and inferiorly about the orbit, so that they form a fairly broad ring extending beyond the orbital rim, which covers the origins of some of the other muscles in this neighborhood and may contribute to them. Apparently some of the fibers are truly circular.

The *palpebral portion* of the orbicularis oculi is described in detail in connection with the orbit; it need only be stated here that it arises largely from the medial palpebral ligament but also in part from the posterior lacrimal crest (these fibers at and close to their origin constituting a smaller third part, the pars lacrimalis) and that some fibers of the two lids interdigitate at the lateral palpebral raphe. The medial palpebral arteries, the larger arteries of the lids, lie deep to this portion at the medial corner of the eye.

The palpebral portion of the muscle is separated from the skin and from the deeper tissues by loose connective tissue, but the orbital portion is embedded in denser tissue; hence accumulations of fluid in the eyelids do not tend to spread much beyond the orbital rim.

Procerus

The procerus lies superficial to the nasal bone and arises from this and the upper part of the lateral (upper lateral) nasal cartilage; it attaches into the skin at the root of the nose.

Corrugator Supercilii

The corrugator supercilii is a small, deeply situated muscle arising from the frontal bone above the rim of the orbit close to the nasofrontal suture and extending laterally and upward to insert into the skin of the medial half of the eyebrow. It lies directly against the bone and is covered by the orbicularis oculi, the procerus, and the frontalis. The supraorbital vessels and nerves pass deep to it as they turn upward across the orbital rim into the forehead.

Innervation and Action

These muscles are supplied primarily by the zygomatic and temporal branches of the facial nerve; the orbicularis probably also receives fibers from an upper buccal branch. The orbicularis closes the eyelids, the palpebral portion alone being able to do this, but in forced closure the entire muscle contracts. Contraction of the orbital portion draws the skin of the forehead and the upper lid downward and the skin of the cheek and the lower lid upward, thus tending to shade the eyes. The procerus draws the skin of the forehead downward, producing transverse wrinkles at the root of the nose; the corrugator supercilii draws the eyebrows medially, producing vertical wrinkles in the forehead above the root of the nose.

According to Ehni and Woltman, it is in the orbicularis oculi that *hemifacial spasm* usually begins; it may slowly spread to adjacent muscles or to the entire side of the face. Women are said to be more often afflicted than men, in the proportion of about six to four, and children do not show it. Since the precipitating lesion was then thought to be probably in the facial nucleus or the proximal portion of the nerve, they advocated spinofacial anastomosis as probably the best remedy; later, Woltman and co-workers described two cases in which decompression of the nerve within the facial canal relieved the spasm. Pulec later apparently established edema as one cause, reporting that edema was present in six of eight patients with spasm and that decompression returned those with edema to normal but had no effect on the other two. Nosik and Weil performed partial section of branches of the nerve, weakening but not paralyzing the affected muscles, and Hori and co-workers coagulated the nerve percutaneously.

MUSCLES OF THE EAR AND SCALP

Extrinsic Muscles of Ear

There are three extrinsic muscles of the ear, that is, muscles arising from the head and inserting into the ear: an anterior, a superior, and a posterior (Fig. 6-6).

The *anterior auricular muscle* is a small triangular one that arises from the galea aponeurotica to extend downward and slightly backward and insert into the helix. The *superior auricular muscle,* lying just behind the anterior, is also triangular; it arises from the galea aponeurotica and descends to insert upon the auricle. The *posterior auricular muscle* arises from the base of the mastoid process and extends over the insertion of the sternocleidomastoid muscle to reach and insert upon the convexity of the concha.

The anterior auricular muscle is supplied

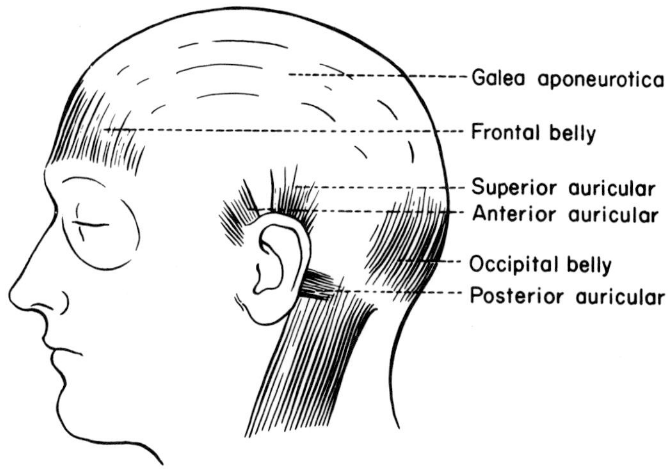

Fig. 6-6. Occipitofrontalis and auricular muscles.

by a temporal branch of the facial, the posterior by the posterior auricular branch, while the superior is supplied by both branches. The anterior muscle draws the ear forward and upward, the superior elevates it, and the posterior retracts and elevates, but the muscles are usually not subject to voluntary control in human beings. The muscles vary considerably in development, the most constant being the superior.

Muscles of Scalp

The *epicranius muscles,* paired muscles of the scalp, consist mostly of the occipitofrontalis muscles, with their frontal and occipital bellies. The frontal belly is a thin muscle covering with its mate the larger portion of the forehead. It has no attachments to bone, since it originates from the tendinous galea oponeurotica and inserts below into skin of the forehead, intermingled with the fibers of the procerus, corrugator supercilii, and orbicularis oculi.

The occipital belly, smaller than the frontal one, arises from the mastoid portion of the temporal bone and from the superior nuchal line on the occipital bone; unlike the frontal bellies, the occipital ones are usually separated by a considerable space, which is occupied by the galea aponeurotica. The fibers of the occipital belly, like those of the frontal, end in the galea. Since the latter is firmly attached to the skin of the scalp by dense connective tissue, but is loosely attached to the periosteum of the bone, contraction of the occipital and frontal bellies moves the scalp. Beneath the galea is the danger space of the scalp, an area of loose connective tissue in which fluid can accumulate and spread. Most of the nerves and vessels enter the scalp superficial to the galea; those leaving the orbit to run to the forehead lie at first, however, deep to the frontal belly, which they subsequently pierce to lie in the dense connective tissue between the skin and galea. The frontal part is supplied by a temporal branch of the facial nerve; the occipital, by the posterior auricular branch.

Above each ear, identifiable with difficulty and blending with the anterior and superior auricular muscles, are the small temporoparietal muscles, also part of the epicranius.

FACIAL PARALYSIS

Facial paralysis is a particularly disfiguring condition, since it leads not only to inactivity of one side of the face or a portion of this during the active movements of smiling, frowning, and the like, but also to an obliteration of the facial furrows that even in repose are so important in facial expression. Ultimately there is also distortion of the face from the pull of unopposed muscles. In cases of interruption of the facial nerve as a whole, the method of choice has been reanastomosis of the facial stem, since this may restore both some voluntary and some emotional response to the face. A procedure of secondary choice has been a spinofacial (*i.e.,* accessory–facial) or a hypoglossofacial anastomosis, which, if successful, restores some voluntary facial movements even though involuntary emotional responses are lacking. These procedures are discussed further in the section on the facial nerve. (The treatment of Bell's palsy has been discussed in the chapter on the ear.) In some cases, however, as discussed by Owens, they are not possible, and a plastic operation, with either fascial or muscle strips, must be performed if the condition is to be improved. It may also be necessary to remove excess sag in the skin.

Adams regarded muscle transplantation as much superior to the use of fascial strips or other nonactive means of suspension; he has transplanted anterior fibers of the masseter muscle into both lips and fibers of the temporal muscle into the eyelids and has used a part of the frontalis of one side to improve the forehead on the other. He stated that 2 to 3 months is required for the transplants to function properly, but that function improves as the patient reeducates the muscles. The improvement may continue for a year or more. Adour and co-workers also attached parts of the masseter to facial muscles with success, but obtained the best reinnervation (from the mandibular nerve) by stripping fascia from the masseter and suturing paralyzed muscles to its raw surface.

An associated movement of eyelid and mouth (jaw-winking phenomenon, inverted Marcus Gunn phenomenon, or Marin Amat syndrome) is sometimes seen after recovery from a facial palsy. Wartenberg ('48a) pointed out that this is an associated movement between the orbicularis oculi and the facial muscles around the mouth, and it is now generally accepted that it results from abnormal regeneration of the facial nerve. (Associated movements between the mandible and a ptotic eyelid, the original Marcus Gunn phenomenon, are usually congenital, and those between the orbicularis oculi and the mandible seem to be release phenomena following a supranuclear lesion—Wartenberg, '48b.)

Vessels

ARTERIES

FACIAL ARTERY

The facial or external maxillary artery, after its origin from the external carotid, has at first a cervical course, during which it is carried upward medial to the mandible and in fairly close contact with the pharynx; in this course it runs upward deep to the digastric (posterior belly) and stylohyoid muscles, then crosses above them to descend on the medial surface of the mandible, grooving or passing through the submandibular (submaxillary) gland as it rounds the lower border of the bone.

After its appearance on the external surface of the mandible, at about the anterior border of the masseter muscle, the artery runs upward and medially to the side of the nose and, as the angular artery, alongside the nose toward the medial angle of the eye (Fig. 6-7). In this course it lies successively deep to the platysma, risorius, and the zygomaticus major and minor, but crosses superficial to the buccinator and the levator anguli oris. Beyond the zygomaticus minor it passes either superficial or deep to the levator (infraorbital head of the quadratus) labii superioris, may be covered by the lower medial part of the orbicularis oculi, and runs up alongside the nose in the levator labii superioris alaeque nasi (angular head of the quadratus). It may be very tortuous. Above the mandibular margin it lies anterior to the facial vein, which takes a corresponding but straighter course on the face (and leaves the face by crossing the submandibular gland superficially, rather than by turning between the gland and the mandible). The chief branches of the artery on the face are the labial arteries, but it gives off twigs to the masseter muscle and unnamed branches that anastomose with the other arteries that reach the face.

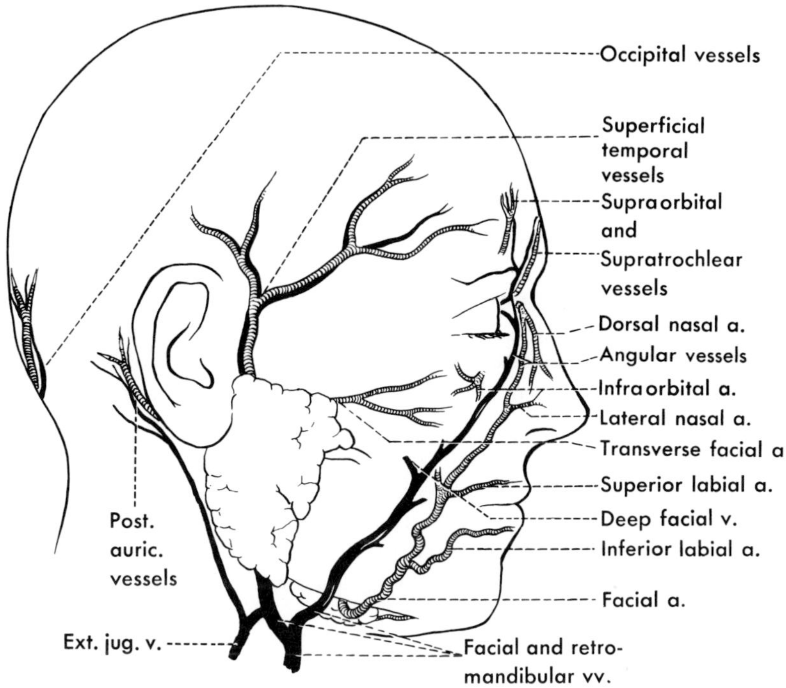

Fig. 6-7. Chief blood vessels of the face. These lie deep to the superficial facial muscles (see text), and only their general courses rather than their relations are indicated here.

Labial Arteries

The inferior and superior labial arteries run tortuously in the lips, in or behind the deeper fibers of the orbicularis oris and therefore fairly superficial in relation to the labial mucosa. Because of their tortuosity, they wind in and out in the deeper part of the muscle, sometimes lying in this and sometimes between the muscle and the mucosa. The tortuosity is more marked in the aged. In doing "lip shaves" for superficially located carcinoma of the lower lip, the inferior labial artery is therefore frequently uncovered just under the mucosa.

There may be two inferior labial arteries, in which case the lower arises from the facial below the level of the alveolar border of the mandible and courses farther from the free margin of the lip, and less close to the mucosa, while the upper has the course already described. The superior labial gives off a branch to the nasal septum, which is a common source of bleeding from the nose.

Angular Artery

The terminal portion of the facial artery, the angular, proceeds alongside the margin of the nose toward the medial angle of the eye. It usually gives rise to a lateral nasal artery that supplies the nose and may anastomose broadly on the dorsum of the nose with the dorsal nasal branch of the ophthalmic. Otherwise, the angular may give off several small branches to the nose.

In its course the facial artery has anastomoses with the buccal branch of the maxillary (internal maxillary), with the transverse facial artery (a branch of the superficial temporal), and with the infraorbital artery from the maxillary, while the angular portion frequently terminates by anastomosing broadly with the dorsal nasal branch of the ophthalmic artery at the medial side of the orbit. The anastomoses of the two facial arteries across the midline of the face through the labial vessels form an important part of the collateral circulation available after ligation of the external or common carotid artery on one side.

Abnormality of Facial Artery

The facial artery may be deficient and be replaced in part by other branches. Grönroos recorded a case in which it ended essentially in its submental branch, sending only a tiny continuation upward along the anterior border of the masseter muscle; in this case the buccal branch of the maxillary was very much enlarged and, after appearing on the cheek, turned upward in the usual position of the facial to supply labial, nasal, and angular branches. In occasional instances in the dissecting room the facial artery is found to end in its superior labial or lateral nasal branch, the upper part of its usual territory of distribution then being taken care of by various branches of the transverse facial, infraorbital, and dorsal nasal arteries.

OTHER ARTERIES

The other arteries to the face are those already mentioned as anastomosing with the facial artery. The *buccal* (buccinator) *branch of the maxillary artery* enters the cheek from a position deep to the ascending ramus of the mandible and the associated insertion of the temporal muscle to supply the buccinator muscle, adjacent mucous membrane and tissue lying external to the muscle; this is usually a small vessel.

The *transverse facial artery* arises from the base of the superficial temporal and, as its name implies, runs almost transversely across the face, lying upon the outer surface of the masseter muscle and below the zygomatic arch but above the parotid duct. This vessel gives off branches to the orbital region and anastomoses across the cheek with the facial.

The *infraorbital artery* emerges with the nerve of the same name from the infraorbital foramen, lying therefore at first deep to the levator (quadratus) labii superioris; it may have a large branch of communication with the facial artery, but in any case its branches tend to spread out over the infraorbital region, some of them also reaching the lower eyelid.

The *dorsal nasal artery,* one of the large terminal branches of the ophthalmic, penetrates the orbital septum in the upper eyelid and runs downward on the nose. A major branch may inosculate with the upper end of the angular, and another branch may similarly join the lateral nasal artery; if not, smaller anastomoses then typically unite the dorsal nasal to the angular and its branches.

In addition to these vessels to the major portion of the face, the supratrochlear (frontal) and supraorbital arteries, branches of the ophthalmic, turn upward around the superior orbital rim deep to the frontalis muscle, pierce it, and supply the skin on the forehead, while other branches of the ophthalmic deep to the orbicularis oculi form the palpebral arterial arches. Finally, the superficial temporal artery arises deep to or in the parotid gland and emerges between the condyle of the mandible and the external acoustic meatus to pass upward, superficially, immediately in front of the ear. In this position it is accompanied by the corresponding vein, a tributary of the retromandibular (posterior facial), and by the auriculotemporal branch of the mandibular nerve. This artery ascends in front of the ear over the root of the zygomatic process and divides into parietal and frontal branches that are distributed to the scalp and to the skin of the forehead; twigs from the frontal branch also reinforce the circulation to the skin of the eyelids.

Smaller branches, in addition to the transverse facial, include twigs to the parotid gland, anterior auricular branches to an anterior portion of the ear, a branch (middle temporal) to the temporal muscle, and a zygomaticoorbital artery that supplies the lateral part of the orbicularis oculi and sends twigs through the zygomatic bone and around the anterior border of the orbit to anastomose with the ophthalmic artery. Although the artery is subcutaneous in front of the ear, it is crossed superficially by the upper (temporal and zygomatic) branches of the facial nerve and by the anterior auricular muscle. The auriculotemporal nerve usually lies a little bit behind and deep to it; the superficial temporal vein frequently lies at first behind the artery, but the branches of the two vessels tend to intertwine on the forehead and temporal region.

VEINS

Facial Vein

The facial (anterior facial) vein is the chief vein of the face. It begins as the angular vein, in the angle between the nose and eye, where it communicates freely with the ophthalmic vein (see chapter on the orbit). Besides this communication, it also communicates with veins of the forehead, receives veins from the lids, the external nose, and the lips, and over the cheek communicates through the deep facial vein with the pterygoid plexus. Its tributaries thus correspond largely to the branches of the facial artery, which it approximately parallels, lying behind the artery and deep to most of the facial muscles. The facial vein runs across the surface of the submandibular gland (while the artery runs between the gland and the mandible) to end in the internal jugular vein. Before it does so it receives a part of the retromandibular (posterior facial) vein, which also contributes to the external jugular. The part of the facial vein below the junction of the retromandibular was once called the common facial vein.

Danger Area of Face

Since the veins of the head and face are without valves, the communications of the facial vein with the ophthalmic, the infraorbital, and deep facial veins have been of special clinical importance. The ophthalmic veins drain directly into the cavernous sinus (Fig. 6-8); the infraorbital and deep facial veins drain into the pterygoid plexus, which also, however, has connections with the cavernous sinus through the foramen ovale and, by way of the inferior orbital fissure, to the inferior ophthalmic vein and hence to the cavernous sinus.

Infections of the face, especially about the upper lip and the nose, may therefore carry considerable danger if unwisely handled. Areas of infection about the upper lip are al-

Fig. 6-8. Chief connections of the facial vein to the cavernous sinus.

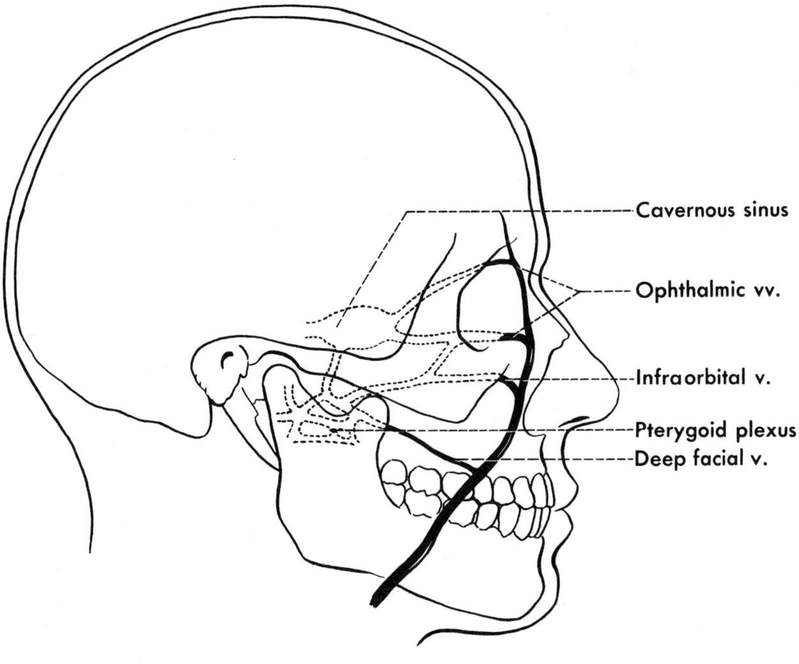

most impossible to immobilize; the actions of the muscles aid in rapid dissemination of septic material into adjacent venules with the production of facial vein thrombosis. Because of the connections of the facial vein, it has often been emphasized that infections above the area of the mouth should not be touched in an attempt to evacuate them (*i.e.*, Maes). Such attempts usually result in spread of infected blood or other septic material, without evacuating any large amount of pus and with added danger of transmitting infection to branches of the vein.

Treatment of such infections should be conservative; Maes stated that local surgical treatment should usually not be initiated, since the less that is done locally, the better. He regarded prophylactic ligation of veins, in order to prevent spread to the cavernous sinus, as of dubious value, although once the thrombosis has reached the cavernous sinus, surgical treatment is of little avail because of the prompt development of meningitis. Writing in the days before antibiotics, he found that it had been estimated that although only 7% of patients afflicted with cavernous sinus thrombosis recovered without surgical intervention, fewer than 7% recovered with such intervention. Later, Shaw concluded that the only effective treatment for cavernous sinus thrombosis is intensive and prolonged antibiotic therapy. Since 61% of the cases he reviewed originated from furuncles on the upper part of the face, and since only 23% of the patients who survived had thereafter no serious complications or residual abnormalities, he emphasized that antibiotics have not eliminated the danger from the danger area.

Smaller Veins

The smaller veins of the face correspond to the arteries. Thus, the transverse facial joins the superficial temporal; this empties into the retromandibular (posterior facial) vein which runs through the parotid gland; the deep facial vein, already mentioned as communicating with the pterygoid plexus of veins, parallels the buccal artery, and tributaries of the ophthalmic vein appear about the orbit.

LYMPHATICS

Nodes

The lymph nodes on the face are represented by a variable number of inconspicuous nodes. Anteriorly, approximately along the course of the facial artery (therefore once called facial or anterior facial nodes), are a few small nodes that are now divided into two groups (Fig. 6-9): buccal nodes, on the buccinator muscle, and mandibular nodes, on the mandible. There may be others higher (maxillary or infraorbital nodes). Posteriorly, many of the nodes are embedded in the outer surface of the parotid gland and are called *superficial parotid nodes.* Included with the superficial parotid nodes are one or two nodes in front of the tragus, formerly called the anterior auricular nodes.

Drainage from Face

The lymphatic vessels from the forehead empty into the superficial parotid nodes, as do those from the larger lateral part of both eyelids and from a part of the external nose; lymphatics from the medial corner of the eye, from the greater part of the external nose, and from most of the lower part of the face pass largely downward through or around the buccal and mandibular nodes to reach the submandibular (submaxillary) nodes; the lymphatics from the medial portion of the lower lip drain into the submental nodes. However, metastatic involvement of the submental nodes secondary to carcinoma of the lower lip is said to be relatively rare. The submental and submandibular nodes in turn drain into the superficial cervical and upper deep cervical nodes; the parotid nodes drain also into the upper deep cervical group.

McClure and Lam concluded, from a study of their own and others' cases, that if there is small early carcinoma of the lip with no palpable evidence of involvement of the cervical nodes, dissection of the nodes of the neck is unnecessary, although they also stressed the importance of repeated examination of the neck in the 2 years following operation. Figi stated that removal of squamous cell eipthe-liomas of the lip of grade 2 or 3 should be fol-

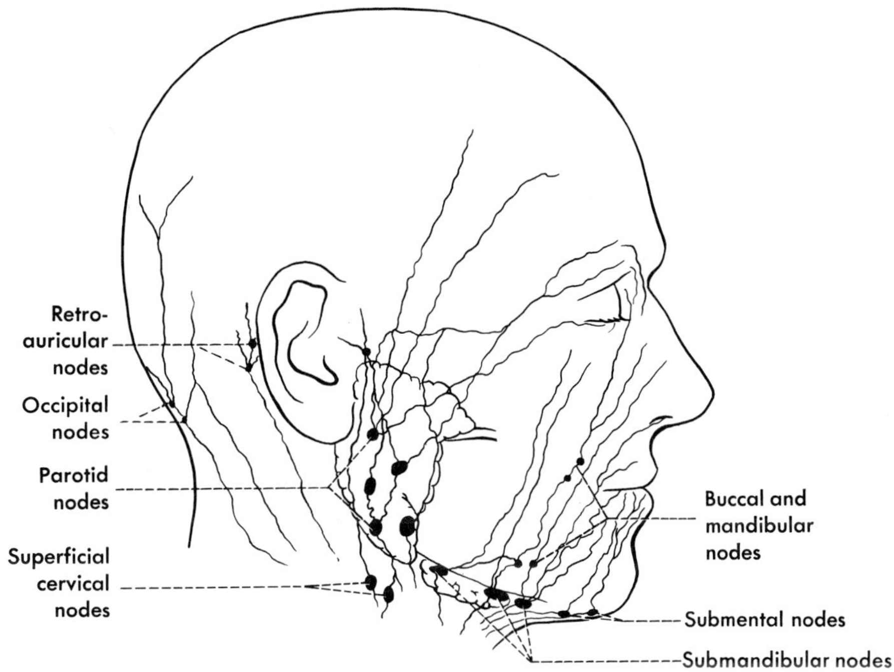

Fig. 6-9. General directions of flow of lymph from the face.

lowed by removal and examination of the submental and submandibular nodes and, if these are involved, by block dissection of the neck. This program was recommended regardless of the size of the primary lesion on the lip.

The Parotid Gland

The tough connective tissue about the parotid gland, united intimately to the surface of the gland, is formed by a splitting of the superficial layer of the deep fascia of the neck about the gland; as already noted, the deeper portion of the parotid gland lies in part adjacent to the lateral pharyngeal space, and this relationship presumably accounts for the sometimes severe consequences of parotitis. At its lower pole this gland is separated from the submandibular gland by a thickening of fascia termed the *stylomandibular ligament;* this represents a line along which the fascia superficial and deep to the parotid gland fuses before splitting again to pass around the submandibular gland.

The parotid gland, located anterior to and below the lower part of the ear (Fig. 6-10), extends subcutaneously backward over the anterior portion of the sternocleidomastoid muscle and forward over the masseter muscle, and also deeply behind the ramus of the mandible to lie between this and the external acoustic meatus and mastoid process. Superficial and deep parts of the gland have been described and regarded both as broadly continuous with each other and as being united only by an isthmus about which run the branches of the facial nerve; the relationship of the gland to the facial nerve constitutes the chief hazard of surgery of this gland and is discussed in detail in the following section on the facial nerve.

Superficial Relations

The outer surface of the parotid gland is practically subcutaneous and is related only to the skin and fascia and to a few inconstant superficial parotid nodes that may occur on

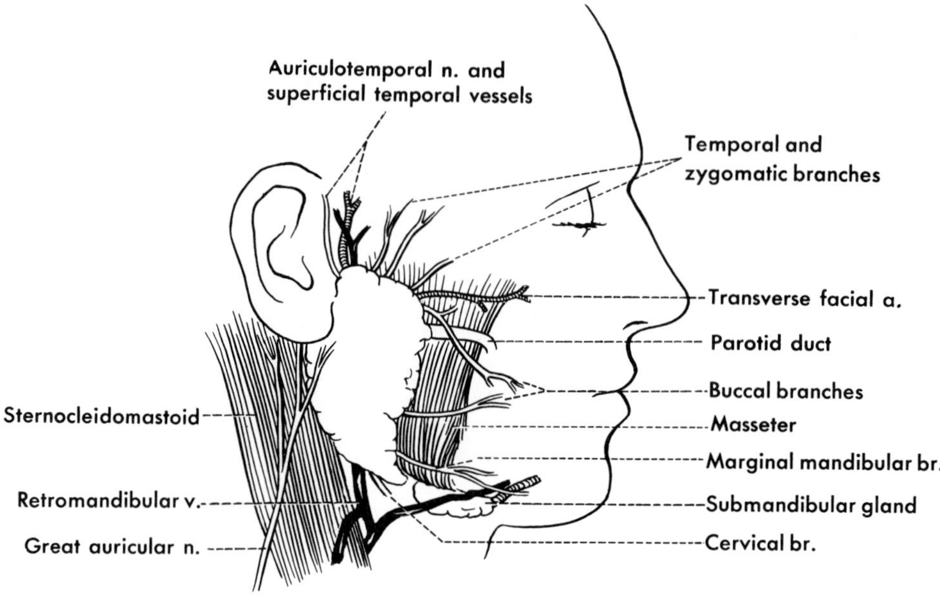

Fig. 6-10. Superficial relations of the parotid gland, and branches of the facial nerve on the face. The gland varies considerably in size and shape.

its surface; from the anterior edge of this superficial part the *parotid* (Stensen's) *duct* (Figs. 6-10, 6-11, and 6-16), often accompanied by accessory parotid tissue, passes forward across the masseter muscle, at a level about halfway between the zygomatic arch and the angle of the mouth. Beyond the anterior border of the masseter the duct turns medially to penetrate the buccinator muscle and open into the mouth through the small parotid papilla at about the level of junction of the root and crown of the upper second molar tooth. External fistulas of the parotid duct are troublesome: they tend to heal slowly because of the constant secretion of saliva. Surgical repair or drainage of an injured duct is necessary since it will not heal spontaneously. Glascock and Glascock, among others, have suggested ingenious methods of repair. A drain from the severed end of the duct into the mouth allows healing of it or the formation of a fistulous channel to the mouth, as the case may be. A more recent method successfully used by Mandour and co-workers to promote healing is tympanic neurectomy. In contrast to fistulas of the duct, fistulas from the parotid gland itself are rare; the lower part of the gland is, for instance, regularly amputated in neck dissections for carcinomatous lymph nodes, and no special precautions need be taken.

A short distance above the parotid duct, the transverse facial artery and vein appear at the anterior edge of the parotid gland and run forward usually above and roughly parallel to the duct, but sometimes descending to cross superficial to it. The parotid duct may also be crossed superficially by anastomosing branches of the facial nerve connecting zygomatic and buccal branches.

Also appearing at the anterior edge of the parotid gland are the major branches of the facial nerve which, having traversed the gland, radiate downward, forward, and upward across the masseter muscle and the zygomatic arch. The relations of the nerve within the gland are discussed in the following section. Obviously, vertical incisions over the anterior portion of the parotid gland carry with them great danger of injury to the parotid duct or some branch of the facial nerve, while horizontal incisions render either of these more unlikely.

At the upper border of the parotid gland

the temporal branches of the facial nerve, the superficial temporal artery and vein, and the auriculotemporal nerve, having traversed the parotid gland, cross the zygomatic arch subcutaneously. These branches of the facial nerve tend to lie anterior to the vessels and to cross them superficially; the superficial temporal vein lies slightly superficial to and behind the corresponding artery, while the auriculotemporal nerve lies deeper and usually slightly more posteriorly. All these structures, however, tend to be concentrated immediately in front of the ear.

The occurrence of redness and sweating on the temple in the area of distribution of the auriculotemporal nerve when eating (auriculotemporal or Frey's syndrome) is usual after parotidectomy. It is generally believed that this is caused by regeneration of the secretory fibers of the parotid gland, contained in the auriculotemporal nerve, to sweat glands in its area of distribution; thus the sweat glands respond to the nerve impulses that should reflexly provoke parotid secretion. This concept seems to have been proved by Ross's finding that a complete tympanic neurectomy will relieve the condition.

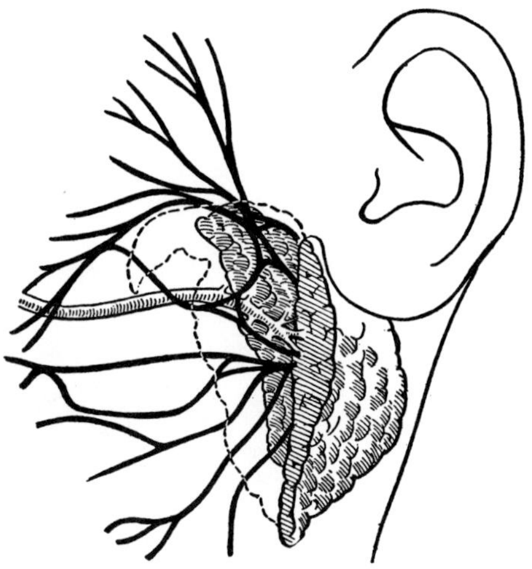

Fig. 6-11. Dissection of the parotid gland to show the duct and the facial nerve. This is supposed to show the separation of superficial and deep lobes of the gland by the facial nerve, but see text and Fig. 6-14. A part of the superficial "lobe," indicated by the *broken line,* has been cut away. (After McWhorter GL: Anat Rec 12:149, 1917)

Internal Relations

The important relations of the deep portion of the parotid gland (Figs. 6-12 and 6-15) are vascular ones. Adjacent to the medial portion of the gland as a whole, and typically embedded in it for some distance, are the external carotid artery and the retromandibular (posterior facial) vein; in this position the external carotid divides into the superficial temporal and (internal) maxillary arteries, and the retromandibular vein receives the corresponding venous tributaries. The deep auricular and transverse facial arteries also originate here, and the auriculotemporal nerve passes upward through or just deep to the upper portion of the gland.

Anteromedial to the deep portion of the gland are the internal jugular vein, the internal carotid artery, and the styloid process and its muscles; the posterior belly of the digastric is in contact with the gland posteriorly and medially.

Above and anteriorly, the deep portion of the gland is related to the ramus of the mandible and the posterior border of the medial pterygoid muscle; posteriorly it is related to the mastoid process and the external acoustic meatus. The pain produced by movements of the jaw in parotitis or mumps is due to a compression of the deep portion of the parotid gland by the ramus of the mandible, especially against the anterior wall of the external meatus.

Blood Supply and Innervation

The parotid gland receives its blood supply from the vessels traversing it, and its sensory and secretory nerve fibers are received from the auriculotemporal nerve, which also traverses it. The sensory fibers are true trigeminal nerve fibers, with cells of origin in the trigeminal ganglion, but the secretory fibers are derived originally from the glossopharyngeal nerve. These take a devious course to reach the parotid gland: the preganglionic parasympathetic fibers leave the ninth nerve by way of its tympanic branch to join the tympanic plexus on the promontory of the middle ear cavity, subsequently leaving the plexus

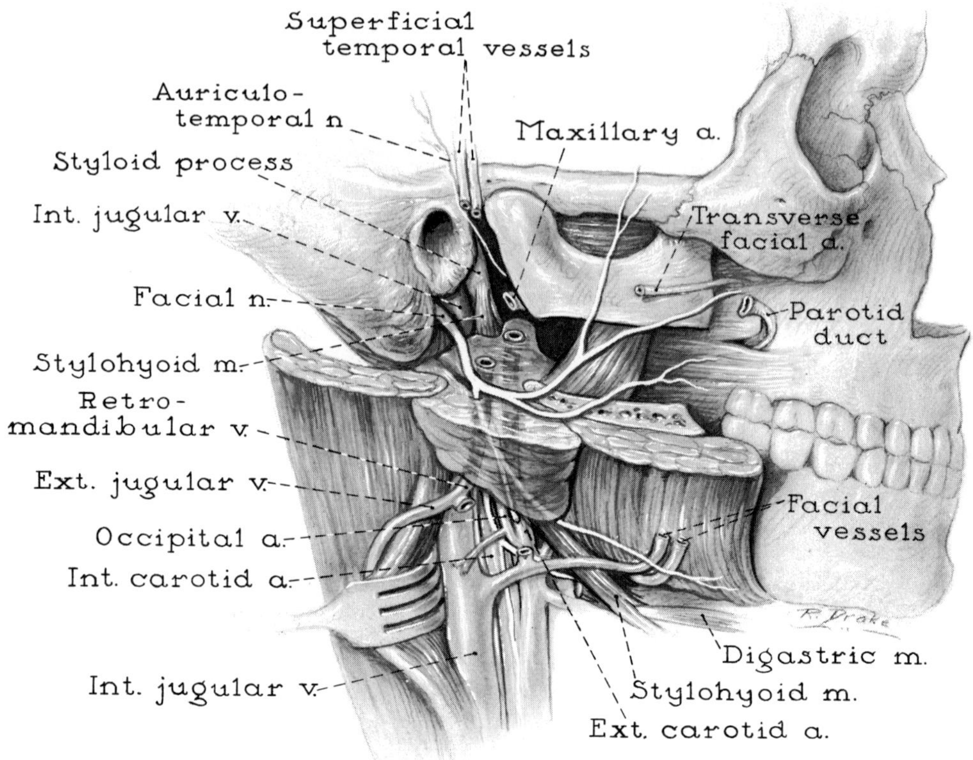

Fig. 6-12. Deep relations of the parotid gland, and relations of the facial nerve, retromandibular (posterior facial) vein, and external carotid artery in the gland. (Beahrs OH, Adson MA: Am J Surg 95:885, 1958)

through the lesser petrosal nerve; this nerve ends in the otic ganglion, where the preganglionic fibers make synapse (Fig. 6-13).

The otic ganglion lies just outside the foramen ovale on the medial surface of the mandibular nerve and gives off filaments to the roots by which the auriculotemporal nerve arises from the mandibular; these fibers then leave the auriculotemporal with sensory fibers to form the parotid branches of this nerve. Some writers have stated that preganglionic fibers from the facial nerve also reach the otic ganglion through connections from the facial to the tympanic plexus or between the greater and lesser petrosal nerves. The most common suggestion is that at least sometimes the preganglionic fibers from the ninth nerve reach the otic ganglion through its connection from the chorda tympani, having joined the facial nerve just outside the skull instead of traversing the tympanic and lesser petrosal branches of the ninth. Ross, however, stimulated both the tympanic nerve and the chorda tympani in eight normal persons and in five with Frey's syndrome and found fibers to the parotid gland in the tympanic branch only.

Dishell sectioned both the tympanic branch of the ninth nerve and the chorda tympani in a patient with intractable parotitis and reported prompt relief from pain followed by slow atrophy of the gland. Morgan combined ligation of the parotid duct and tympanic neurectomy in six patients; none had postoperative pain or swelling of the gland in the first 3 weeks, and one was free of symptoms until his death 2 years later. Of the others, followed for 3 years, two were completely free of symptoms, two had single attacks of pain and swelling at 3½ and 4 weeks, respectively, and one had an attack after 4 weeks and another after 6 months. Although Rosen and Konak obtained clinical improvement in their cases, only one of nine patients showed a marked decrease in parotid function, as indicated by scans before and after operation; they could demonstrate no permanent improvement from the operation. In contrast, Mandour and co-workers took biopsies at the time of operation and later; besides clinical relief in all their patients (some with fistulas that refused to heal, others with bilateral pain and swelling), they found definite degenerative and atrophic changes in the 6-month biopsy specimens.

Ross found that failure to obtain relief from the symptoms of Frey's syndrome through tympanic neurectomy was due to failure to section a low-arising branch to the plexus. Similarly, Porto and co-workers found such a branch, often covered by bone, in 10 of 20 dissections and suggested that failure of tympanic plexectomy to affect the parotid gland (their survey of the literature indicated success in only 79%) may lie in the presence of this hypotympanic branch of the tympanic nerve.

As is also the case with the other salivary glands, relatively little is known about the sympathetic supply to the parotid gland; sympathetic fibers from the superior cervical ganglion presumably reach it by way of the external carotid plexus and its branches (and Nordenfelt has apparently shown that in the cat, some of the fibers traverse the tympanic plexus, presumably therefore reaching the gland through the lesser petrosal nerve and its connections), but the functional significance of the sympathetic supply is not clearly understood. This question is discussed in relation to the salivary glands as a whole in Chapter 7.

Nerves

FACIAL NERVE

In connection with the facial nerve after it leaves the skull there are two questions to be considered: the relation of the nerve to the parotid gland, and the anatomical and functional distribution of the nerve on the face.

Relation to Parotid Gland

As the facial nerve emerges through the stylomastoid foramen, typically situated slightly posterolateral to the styloid process and hence between this and the mastoid process (but very superficially located in the infant), it almost immediately becomes related to the deeper and posterior aspect of the parotid gland. In the very occasional individuals in whom the nerve has an aberrant course and emerges from the anterior surface of the mastoid process, it still comes into immediate relation to this gland. In either case the nerve typically enters the substance of the parotid

Fig. 6-13. Schema of the parasympathetic innervation of the parotid gland. The preganglionic pathway is indicated by a *solid line;* the postganglionic one, by a *broken line.*

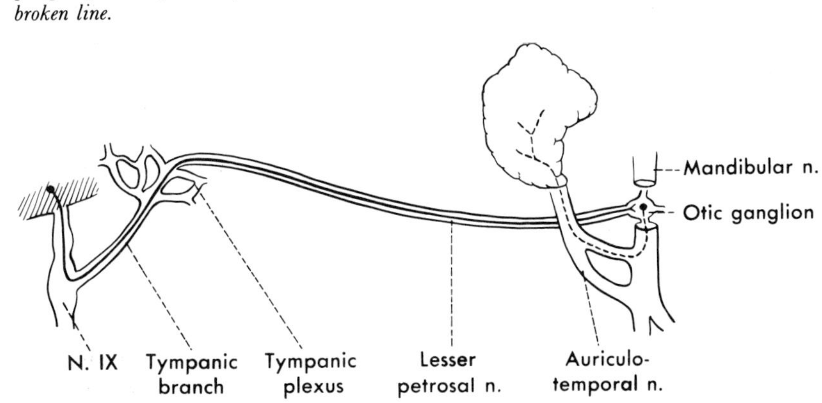

gland before dividing into its branches to the face. The first major division is often into an upper and a lower division ("temporofacial" and "cervicofacial" branches).

McWhorter described these two divisions of the facial nerve as passing between superficial and deep parts of the parotid gland, on either side of an isthmus that unites these two parts; according to his description the superficial and deep parts of the gland thus form lobes that are usually distinct and readily separable and have separate duct systems, except where they are united through the isthmus. According to this description, therefore, the facial nerve does not traverse the substance of the parotid gland, but the lobes of the gland are folded about the nerve. Hurford stated that of 11 specimens that were especially dissected for this purpose, there was evidence of a division into superficial and deep lobes in 10, and among the latter, seven showed a fairly clear isthmus and three showed a well-defined one. He added, however, that among these bilobed parotids there were only four in which the branches of the facial nerve remained strictly in the plane between the lobes and that careful dissection was required to demonstrate the isthmus in every case.

This writer has not seen a clear division along the plane of the branches of the facial nerve, although in the dissecting room it is certainly possible to produce such a plane about the main nerve branches by following the facial nerve through the gland. This is routinely done surgically, but does not of course prove that a natural subdivision of the gland exists. J. McKenzie adduced rather compelling evidence that this concept of deep and superficial lobes connected by an isthmus and otherwise separated by a plane in which lie the branches of the facial nerve is a simplification that does not actually exist (Fig. 6-14). He injected the ducts of the parotid system and upon subsequent dissection found that there was not only an "isthmus" between the upper and lower divisions of the facial nerve, but that there were multiple isthmi between and among the further branches of the facial nerve as it forms the parotid plexus within the gland; the isthmi that he described were not simply cases of contact between superficial and deep lobes, for he found that each isthmus had ducts connecting the deeper and the superficial parts. Rather than the parotid gland's being simply folded about the branches of the facial nerve, therefore, his results indicate that parotid tissue, ducts, and facial nerve are intimately interwoven; he compared the parotid gland to a climbing vine that had interwoven itself in the meshes of a trellis-work fence (the facial nerve). The pattern of the connecting isthmi between the more superficial and the deeper portions of the parotid gland was found to vary from one individual to another, just as there is also variation in the pattern of branching of the facial nerve. Superficial and deep lobes have not been accepted for the NA; the term *superficial and deep parts* refers only to the obvious fact that parotid tissue lies on both sides of the facial nerve. Gasser has adduced further evidence. He described the developing parotid gland as first growing deep to the temporofacial part of the nerve and superficial to the cervicofacial and then both parts; the superficial part then grows deeply between and around the various branches, thus eliminating a cleavage plane.

Any operation for removal of a portion of the parotid gland, unless it involves only a very superficial part, necessarily carries with it the danger of damage to the facial nerve, and complete removal of the gland without damage to the nerve is a delicate procedure. Hurford found that in five of the ten persons in whom he recognized a "deep lobe" (*i.e.*, one deep to the facial nerve), this was larger than the superficial lobe, in contrast to what is ordinarily stated; he expressed the belief that in such patients, removal of the part deep to the facial nerve without damaging that nerve would be difficult, and perhaps impossible when a tumor was present; he suggested that perhaps the superficial part is more likely to be the site of tumors than is the deeper part of the gland.

For total removal of the parotid gland, some surgeons have preferred to locate the main trunk of the nerve by following the marginal mandibular branch into the gland, thereafter removing first the superficial part of the gland and then the deep part. How-

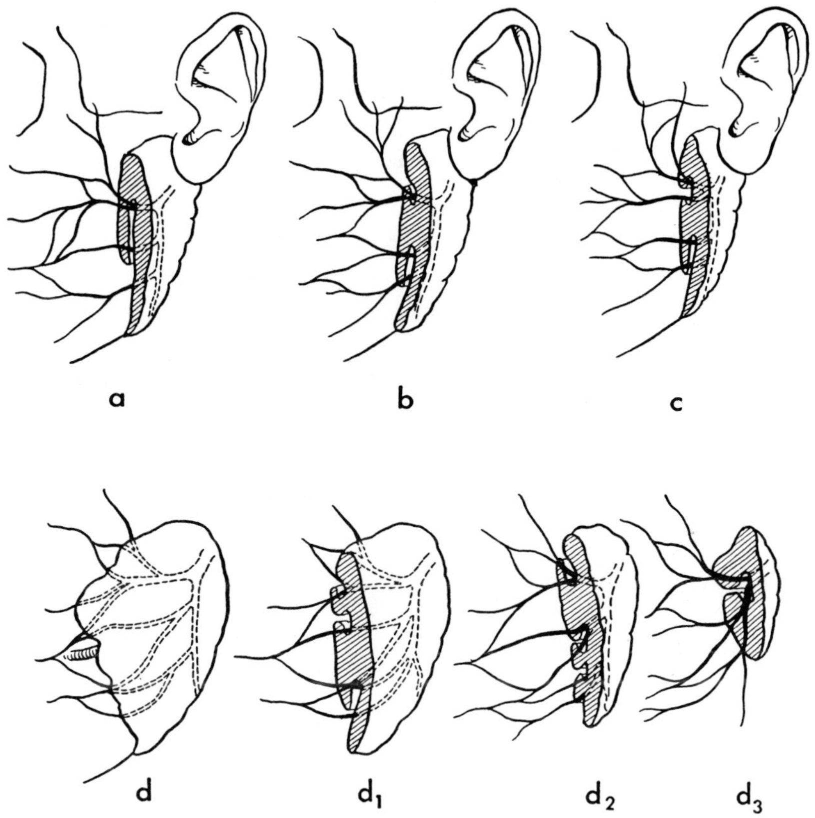

Fig. 6-14. Varying relations, each of which has been described as commonly prevailing, between the parotid gland and the facial nerve. In (*a*), (*b*), and (*c*) the nerve is shown passing between deep and superficial lobes of the gland, but the relationship of these lobes is variously depicted: (*a*), superficial and deep lobes united above, so that the gland is essentially folded over the nerve; (*b*), the two lobes united by an isthmus; (*c*), a combination of (*a*) and (*b*). In (*d*) and its *subnumbers* is shown the relationship according to McKenzie: gland and nerve intertwined, with no superficial and deep lobes, and the relationship varying according to the plane of section. (After McKenzie J: J Anat 82:183, 1948)

ever, Beahrs and Adson, and many surgeons, have preferred to identify the main stem of the nerve and trace it through the gland by careful dissection (Fig. 6-15). This demands accurate identification of the nerve, since damage to the main stem may obviously result in severe facial paralysis. Woods and Beahrs have described their technique in detail.

Surgical Repair of Facial Nerve

Sufficient damage to the main trunk of the facial nerve produces paralysis of the entire facial musculature on that side. Coleman ('44), in discussing this condition, pointed out that the paralysis is less obvious in the upper facial area than in the lower, because it is not exaggerated by the activity of the muscles on the normal side as is lower facial paralysis, especially about the mouth. In cases of injury to the facial stem as a whole, the nerve trunk is of sufficient size to allow end-to-end anastomosis, the use of a nerve graft, or a hypoglossofacial or accessory–facial (spinofacial) anastomosis. Unfortunately, operations upon the parotid gland are more likely to injure smaller branches of the facial nerve, whose surgical repair is difficult or impossible.

Duel and Tickle stated that surgical repair of the main stem of the facial nerve is worth-

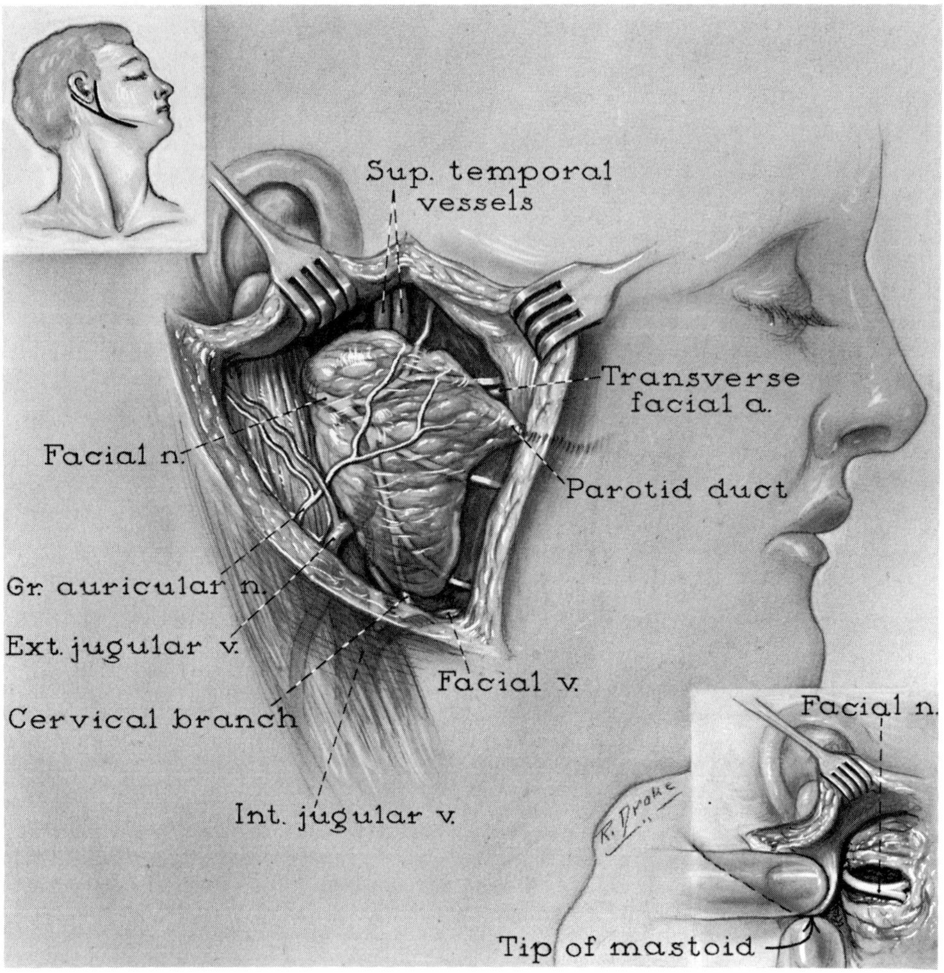

Fig. 6-15. Exposure of the parotid gland and the facial nerve. (Beahrs OH, Adson MA: Am J Surg 95:885, 1958)

while as long as the muscles will respond to galvanic current. They suggested an immediate attempt at repair following surgical injury, since the peripheral end retains its response to faradic stimulation for from 48 to 72 hours and can therefore be more easily located by this means. It is generally said that end-to-end anastomosis of the facial stem offers the best opportunity for successful repair; the anastomosis may sometimes be made possible by lifting the facial nerve from its bony canal in the mastoid and freeing it in its course into the parotid gland.

Duel and Tickle stated that one cannot expect perfect repair upon reanastomosis of the ends of the facial, since some neurons may always be expected to reach the wrong destination. Sadé emphasized that statement, noting that the optimum to be expected is excellent tone at rest, the ability to close the eye to a useful degree when the other is closed, and some ability to close the eye with mass movements of the face; he found a few patients who could smile faintly. Further, he reported about the same results from end-to-end repair and nerve grafting.

Howe and associates produced evidence that the *facial tic* sometimes occurring after injury to the nerve is due at least in part to bifurcation of some of the regenerated fibers in the nerve, so that one branch of a fiber goes to one muscle of the face and the other branch to another muscle; under these conditions, of course, there is no ability to relearn (because of the splitting of the axons), and the tic is therefore apparently permanent. Duel and Tickle have used previously degenerated femoral cutaneous or long thoracic nerves for nerve grafts.

In regard to anastomosis of the accessory or hypoglossal nerves with the facial in cases of facial paralysis, Coleman ('40) preferred the use of the hypoglossal. In his series he found that a hypoglossofacial anastomosis restored movement about the mouth in practically every case, even though function of the frontalis, far less important, was recovered in only one instance. In analyzing the results, he pointed out that function of muscle groups cannot be individualized after such an anastomosis, that emotional stimuli are registered only on the healthy side of the face, and that associated movements of the face may be produced by action of the tongue in individuals so treated. He stated that the ensuing hemiatrophy of the tongue caused no appreciable interference with either speech or swallowing.

Other workers prefer to use the proximal end of the (spinal) accessory nerve as a graft upon the distal end of the facial; this is subject to some of the same drawbacks found in hypoglossofacial union, but again the disability due to denervation of the sternocleidomastoid and trapezius muscles is apparently slight and usually not objected to by the patient, in comparison to the disfigurement of a facial paralysis. The phrenic nerve has also been used to provide motor fibers. Treatment by fascial strips or muscle transplants has already been mentioned.

Branches of Facial Nerve

As the facial nerve emerges from the stylomastoid foramen its first branch is the posterior auricular, which runs upward between the parotid gland and the anterior border of the sternocleidomastoid muscle and then in the notch between the external acoustic meatus and the mastoid process; this branch supplies the more posterior muscles of the facial group: the occipitalis, the auricularis posterior and part of the auricularis superior, and the intrinsic muscles of the auricle. It also contains afferent fibers, at least some of which are thought to be cutaneous sensory to skin over the mastoid process and the adjacent part of the ear. It communicates with the auricular branch of the vagus and with the great auricular and lesser occipital nerves from the cervical plexus.

The facial nerve then passes forward and downward lateral to the styloid process and its attached muscles, supplying branches to the posterior belly of the digastric and to the stylohyoid. After entering the parotid gland it commonly divides into upper and lower branches (sometimes called temporofacial and cervicofacial divisions); these may give off further branches and pass through the parotid gland without anastomosis, but often these divisions or their branches have multiple anastomoses and form a plexus (parotid plexus; pes anserinus) within the gland. A variable number of branches emerge from the gland to be distributed to the face (Figs. 6-10, 6-11, and 6-16 to 6-18).

Kempe explored the facial nerve by electrical stimulation from 10 mm above the stylomastoid foramen to its entrance into the parotid gland. He found only three distinct fascicles in the nerve; one was distributed exclusively to the orbicularis oris, another to the orbicularis oculi, and the third to all the other mimetic muscles.

Although it is customary and convenient to describe the branches of the facial nerve on the face as being divided into five sets—temporal, zygomatic (infraorbital), buccal, marginal mandibular (supramandibular), and cervical (inframandibular)—these terms indicate regions of distribution rather than individual branches. The temporal and zygomatic branches, from the upper (temporofacial) part of the nerve, are typically multiple; the marginal mandibular and cervical branches, from the lower (cervicofacial) part, are typically single, although the former may be double, particularly for part of its course. The buccal branches often arise either from the lower part of the nerve or

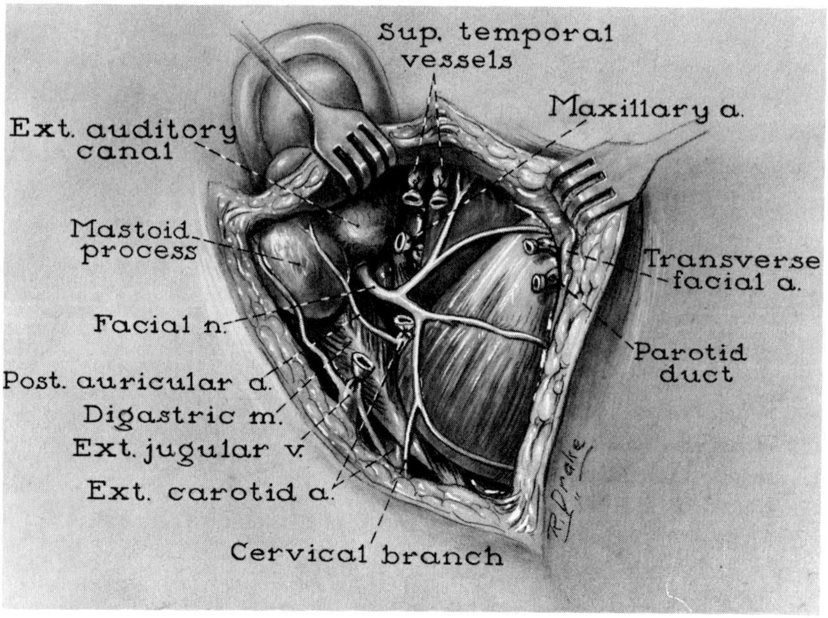

Fig. 6-16. Total removal of the parotid gland. (Beahrs OH, Adson MA: Am J Surg 95:885, 1958)

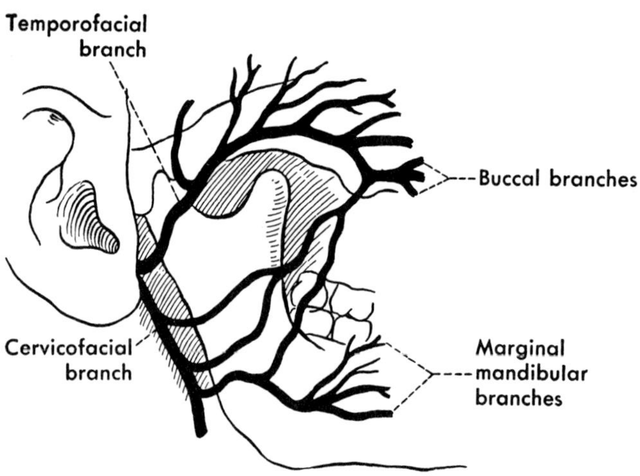

Fig. 6-17. Relationship of the facial nerve to the mandible. (After McCormack, LJ, Cauldwell EW, Anson BJ: Surg Gynecol Obstet 80:620, 1945, by permission of Surgery, Gynecology & Obstetrics)

from both upper and lower parts. However, anastomoses between upper and lower parts of the nerve within the parotid gland, and anastomoses between branches both here and more peripherally on the face, produce extremely variable patterns of distribution. McCormack and co-workers attempted to classify the facial nerves from 100 dissections according to the method of branching and the major anastomoses found, and divided them into eight types (Fig. 6-18); undoubtedly, the various patterns could also be allocated to more types, if desired, or simply divided into those that show no major anastomoses upon the face and those that do.

In general, the temporal branches supply the anterior auricular and a portion of the superior auricular muscle and the muscles upon the forehead, including the major portion of the orbicularis oculi; the zygomatic branches also supply the orbicularis oculi, the muscles about the nasal aperture, and most of the elevators of the upper lip; the buccal branches supply especially the musculature about both lips, the marginal mandibular the musculature of the lower lip, and the cervical branch supplies the platysma.

It is paralysis of the buccal and marginal mandibular branches that produces the most facial deformity, since they are concerned with the highly mobile lips; both the cervical and the marginal mandibular branches of the facial lie within the field of operation in radical dissections of the neck for metastatically involved lymph nodes, but it is the marginal mandibular branch that must be watched for and spared if the patient is not to suffer from a facial deformity following the operation.

The main portion of this branch runs, as its name implies, approximately along the lower margin of the mandible—sometimes above, sometimes below, this margin. The surgeon should take pains to identify and preserve this nerve if possible. It is frequently spared in accidental cuts upon the face in this region, since when it lies below the mandible it is protected by this bone. Both Marino and Pollack, among others, have discussed this branch. The former recommended resection of a portion of the nerve on the other side when one nerve had been injured, saying that the facial asymmetry produced by paralysis is evident only when the patient talks and that bilateral section abolishes the asymmetry without interfering with speech.

Sensory Fibers

Before emerging from the stylomastoid foramen, the facial nerve gives off its greater petrosal, stapedial, and chorda tympani branches. Presumably all the parasympathetic fibers of the facial nerve leave through the petrosal and chorda tympani branches, and so do the great majority of the sensory fibers—according to Foley and DuBois, from 27% to 40% of the afferent fibers in the facial

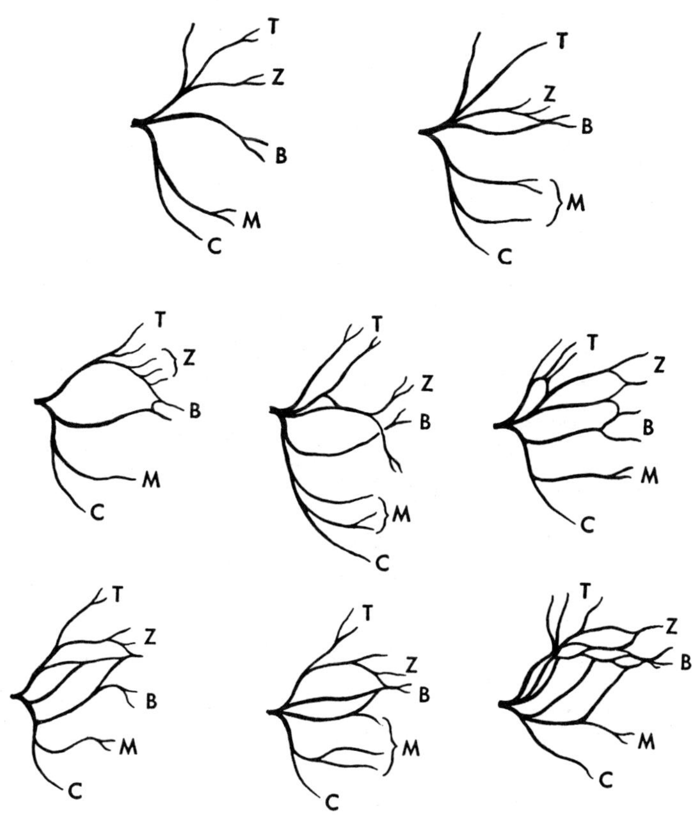

Fig. 6-18. Eight major types of branching and anastomosis of the facial nerve. (*T*), (*Z*), (*B*), (*M*), and (*C*) identify respectively the temporal, zygomatic, buccal, marginal mandibular, and cervical branches of the nerve. (After McCormack LJ, Cauldwell EW, Anson BJ: Surg Gynecol Obstet 80:620, 1945, by permission of Surgery, Gynecology & Obstetrics)

nerve of the cat leave it by way of the greater petrosal, from 45% to 62% leave by way of the chorda tympani, while a minority (11% to 15%) emerge with the facial through the stylomastoid foramen. Many of the latter are said to be distributed by way of the posterior auricular branch of the facial and, with the twigs of the facial that probably reach the external acoustic meatus and the middle ear cavity, presumably account for the otalgia that may be a symptom of geniculate ganglion neuralgia (Hunt); others, however, course toward the mimetic muscles and have variously been regarded as afferent to these muscles and as subserving deep pain from the face.

Maloney and Kennedy concluded that the facial nerve transmits pressure pain from the skin, muscles, and bones of the face. In contrast, Carmichael and Woollard could elicit no pressure pain after destruction of the trigeminal ganglion. Hunt described deep-seated pain from the face (prosopalgia) as being as characteristic of geniculate ganglion neuralgia as is pain in the ear. Although prosopalgia may be caused largely by conduction of pain over the greater petrosal nerve, some observations suggest that pain fibers run also in the superficial branches of the facial nerve. Surgeons commonly recognize that stimulation of the facial nerve in the region of the parotid gland may elicit pain, and Bruesch described the myelinated afferent fibers coursing in the peripheral branches of the facial nerve of the cat (he did not study unmyelinated fibers) as being of the size usually considered to be concerned with pain and temperature rather than muscle afferents and as ending primarily along blood vessels. Pain conducted by the seventh nerve has undoubtedly been confused with trigeminal neuralgia, and Lyman, noting that the term "atypical trigeminal neuralgia" has been used to describe a number of different conditions, reported that some of them were apparently cases of "sphenopalatine ganglion neuralgia." Although atypical facial pains are not yet understood, the concept of sphenopalatine (pterygopalatine) ganglion neuralgia has been generally abandoned. Resection of the ganglion, no longer regarded as a useful treatment for atypical facial pain, interrupts sensory trigeminal fibers from the nose as well as the sensory fibers in the greater petrosal nerve, and even if it were successful in relieving deep pain, it would prove nothing of the participation of the facial nerve in this pain.

Evidence that the afferent fibers in the superficial branches of the facial nerve may supply facial muscles is also contradictory. Cushing, in the first definitive study of the distribution of the trigeminal nerve, reported that movements of the face could not be appreciated after destruction of the ganglion or sensory root of the trigeminal nerve. Later, Carmichael and Woollard, and others, have maintained that movements and displacements of muscles are recognized after interruption of the trigeminal pathway. The relatively recent discovery (Kadanoff) of muscle spindles in the facial muscles has not clarified the afferent role of the facial nerve. For one thing, facial and trigeminal fibers mix together, and although Bowden and Mahran reported suggestive evidence that in the rabbit the trigeminal nerve supplies muscle spindles in facial muscles, they did not account for the sensory fibers in the facial nerve. Furthermore, what role, if any, afferent endings in muscle play in proprioception, or consciousness of movement, is not known.

Whatever the facts concerning the afferent innervation of the facial muscles, it seems clear that cutaneous impulses over the fifth nerve contribute considerably to appreciation of movement in the face, and the numbness following destruction of the fifth nerve upsets this appreciation. This is witnessed by the common experience of the effects of local anesthesia administered by the dentist and may sometimes explain also the occasional occurrence of a temporary facial paralysis following section of the fifth nerve—although the latter is usually blamed upon indirect damage to the facial nerve by way of its greater petrosal branch.

In summary, the anatomical evidence indicates that the facial nerve carries afferent fibers to the face and that they are of the type usually associated with pain; the clinical evidence regarding the role of the facial nerve in pain and proprioception from the face is somewhat contradictory, but seems to be especially in favor of the afferent fibers in the superficial branches of the nerve being largely proprioceptive. The question cannot be regarded as settled, and bizarre observations on this subject are sometimes made. For instance, Jaeger stated that he had sectioned the roots of the fifth and ninth cranial nerves, and the upper five cervical sensory roots, and found not only persistence of pain from the ear but preservation of touch over the entire side of the face and forehead on the side of operation.

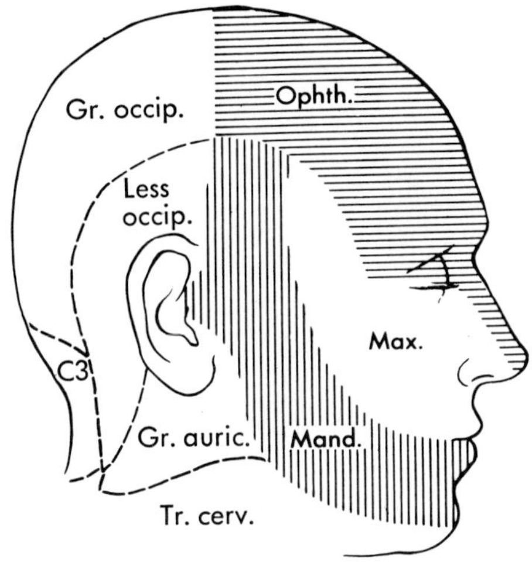

Fig. 6-19. Approximate cutaneous distribution of the three divisions of the fifth nerve and of cervical nerves to the face and scalp.

Lower and Upper Neuron Paralysis

In attempting to localize the lesion in cases of facial paralysis, it is well to remember that paralysis of the facial muscles on the whole of one side of the face can result only from lesions of the facial nucleus or nerve; supranuclear (upper motor neuron) paralysis of the facial nerve spares the orbicularis oculi and the muscles of the forehead, since the portion of the facial nucleus supplying these muscles is controlled by fibers from both cerebral cortices, and only in the instance of a bilateral supranuclear lesion, therefore, is voluntary control of these muscles abolished. Monrad–Krohn also pointed out that central lesions may abolish voluntary control of the facial muscles, yet leave them active in emotional responses, such as smiling at a joke.

THE TRIGEMINAL NERVE AND THE SENSORY SUPPLY TO THE FACE

The sensory nerve supply to the face is largely through the trigeminal or fifth cranial nerve. Branches from the cervical plexus supply a certain amount of skin along the lower border of the mandible and over its angle, and the seventh, ninth, and tenth nerves, with the cervical plexus, supply the greater portion of the external ear; the scalp behind the ear is supplied by cervical nerves, but otherwise the skin of the entire face and of the scalp anterior to the level of the ear is supplied through the branches of the trigeminal nerve (Fig. 6-19). Since the three primary branches of the fifth nerve separate before leaving the skull and have widely different courses thereafter, sensation over the face must be tested in the distribution of each major branch; this is commonly done by testing sensation on the forehead, the prominence of the cheek, and the chin.

Ophthalmic Branches

Each division of the trigeminal nerve sends several cutaneous branches to the face (Fig. 6-20). The largest branch of the ophthalmic nerve is the *supraorbital,* which makes its exit from the orbit through the supraorbital foramen or notch, situated on the upper margin of the orbit about a finger's breadth from the lateral side of the nose and often palpable. This nerve supplies most of the skin over the forehead and the scalp back to about the level of a line connecting the two external acoustic meatuses. A smaller branch of the ophthalmic, the *supratrochlear,* rounds the superior border of the orbit medial to the supraorbital nerve and supplies a limited medial region of skin over the forehead.

These branches, with the *infratrochlear* branch of the ophthalmic at the medial angle

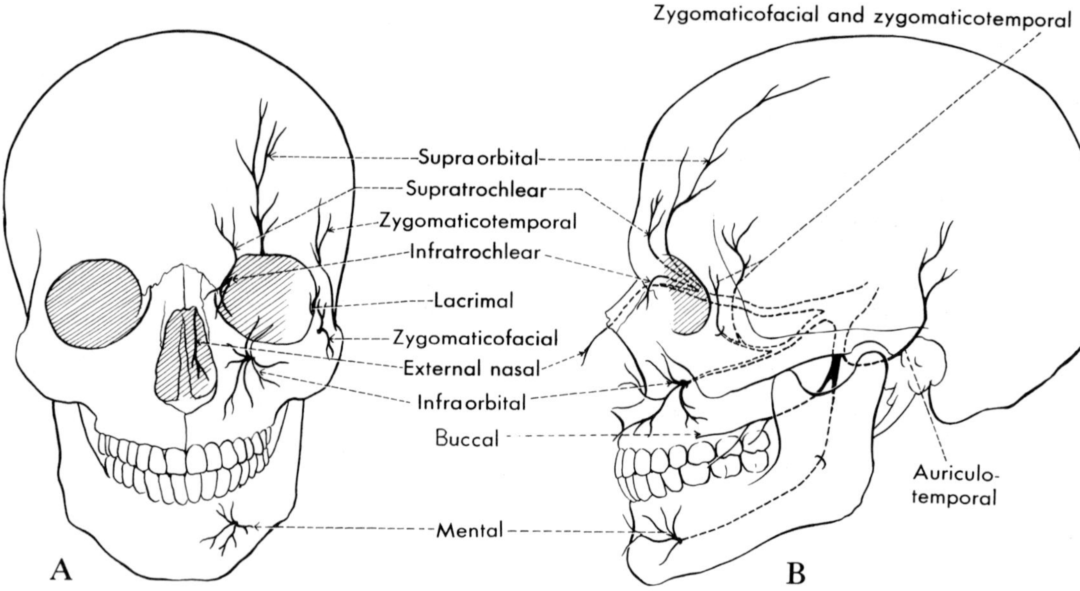

Fig. 6-20. (*A*), branches of the trigeminal nerve to the skin of the face. (*B*), The general courses of these nerves.

of the eye and the small *lacrimal* branch of the ophthalmic at the lateral angle, supply also the skin of the upper lid, the medial and lateral angles of the lids, and the bridge and sides of the root of the nose.

Another branch of the ophthalmic, the *anterior ethmoidal* branch from the nasociliary, supplies skin on the lower half of the dorsum of the nose down to the tip through its external nasal branch. The ophthalmic nerve thus supplies the forehead, the upper lid, and most of the nose, overlapping on the side of the nose with the distribution of the maxillary nerve.

Maxillary Branches

The maxillary branch of the trigeminal nerve is represented on the face by three branches: the infraorbital, the zygomaticofacial, and the zygomaticotemporal. As a whole, this branch of the trigeminal supplies the lower eyelid, the prominence of the cheek and the side of the nose, the upper lip, and a relatively small area extending upward on the temple just behind the eye.

The *infraorbital nerve,* by far the largest cutaneous branch of the maxillary, makes its exit through the infraorbital foramen in the maxilla, in company with the correspondingly named artery; this foramen may lie upon, but usually lies slightly lateral to a vertical line drawn downward from the supraorbital foramen. The nerve emerges under cover of the orbicularis oculi and the levator (quadratus) labii superioris and ramifies in the connective tissue and fat lying between the latter muscle and the levator anguli oris (caninus), anastomosing here with zygomatic branches of the facial nerve. Its branches are distributed upward to the skin of the lower lid, medially to the side but not the dorsum of the lower part of the nose, laterally to most of the prominence of the cheek, and downward to the upper lip.

The *zygomatic nerve* leaves the maxillary in the pterygopalatine fossa or the floor of the orbit, and its zygomaticofacial and zygomaticotemporal branches then diverge from each other to appear separately upon the face. The *zygomaticofacial nerve* makes its exit through a small foramen on the lateral aspect of the zygomatic bone, where it supplies a limited area of skin extending upward toward the lateral corner of the eyelids. The *zygomaticotemporal nerve,* after leaving the zygomaticofacial, passes more laterally to enter the anterior part of the temporal fossa, emerge around the anterior border of the temporal muscle, and supply skin over the anterior part of the temporal region.

Mandibular Branches

There are also three cutaneous branches of the mandibular nerve; through these branches the nerve supplies skin of the lower lip and of the chin, skin over the lateral aspect of the mandible and the fleshy part of the cheek, and an area extending upward in front of the ear and including much of the temporal region. According to Cushing, the latter area of distribution includes only a small part of the ear, primarily the tragus and the crus of the helix, but extends also into the anterior and upper parts of the external acoustic canal and to the external aspect of the tympanic membrane.

The *mental* branch of the mandibular nerve emerges through the mental foramen of the mandible, where it lies at first deep to the depressor (quadratus) labii inferioris; it immediately breaks up into a large number of twigs that are distributed to the chin and to the lower lip, extending laterally to about the level of the angle of the mouth. This nerve is a continuation of the inferior alveolar, which supplies the teeth of the lower jaw; the mental foramen lies laterally on the mandible, approximately on a vertical line drawn downward from the lateral border of the supraorbital notch.

The *buccal* (buccinator, long buccal) branch of the mandibular nerve passes deep to or through the lower part of the temporal muscle and through the buccal fat pad to appear on the mobile part of the cheek between the upper and lower jaws and supply the skin here; this nerve also supplies the mucosa on the inner surface of the cheek.

The *auriculotemporal nerve* courses medial to and then behind the temporomandibular articulation, to which it gives a branch, traverses the upper part of the parotid gland and supplies branches to this, and emerges to cross the zygomatic arch and supply the most anterior portion of the auricle and most of the temporal region. This branch also supplies portions of the external acoustic meatus and tympanic membrane; because of this distribution, pain about the mandible and tongue may be referred to the ear. Before it emerges from the parotid gland, the auriculotemporal has a branch of communication with the facial nerve.

The distribution of the mandibular nerve is the most common area of involvement in trigeminal neuralgia; the "trigger zone" is usually on the chin, the lip, or the tongue.

Other Suggested Pathways from Face

Helson reported that after total or subtotal section of the sensory root of the trigeminal nerve there was, over a period of years, a gradual return of sensation to the denervated area; he stated that sensitivity to deep pressure increased 100-fold, that there was pronounced improvement in the ability to localize a given point, and that extremely hot stimuli were usually reported as producing stinging or pricking sensations. He concluded that much of this functional return might be due to ingrowth of fibers from the cervical plexus or to sensory fibers in the facial nerve; there seemed to be also the possibility that afferent fibers in the cervical sympathetic trunk might be implicated.

The possible participation of the seventh nerve in sensation from the face has already been discussed in connection with this nerve; similarly, it has been suggested that sensory fibers may accompany sympathetic fibers to the face and serve also, with the trigeminal, for conduction of sensations from the face. Thus Helson found that although appreciation of light touch, two-point discrimination, and pain were completely abolished over the area of distribution of the fifth nerve by section of its sensory root, sensitivity to extreme heat was regularly present except in those patients who had had both a section of the fifth root and a thoracic sympathectomy; in the latter patients, heat was not at all appreciated, and the skin of the face could be seared without provoking a reaction from the patient.

Kuntz found experimental evidence of sensory fibers derived from the upper thoracic (but not upper cervical) nerves in the carotid plexuses and suggested they might be concerned with certain atypical facial neuralgias that are not relieved by section of the trigeminal nerve, but Cleveland, and Davis and Pollock, could not confirm the presence of sensory fibers associated with the upper part of the sympathetic trunk. Browder reported a case that seemed to indicate probable conduction of pain through fibers coursing with the sympathetic system to the head.

SYMPATHETIC DISTRIBUTION TO THE FACE

The sympathetic nerve supply to the face is usually assumed to be by way of the external carotid plexus, from the superior cervical ganglion. According to Hyndman and Wolkin, all the fibers for sweat glands of the face arise within the first five thoracic segments of the cord, and many are apparently concentrated in the second thoracic nerve.

Gardner pointed out that the external carotid plexus is not continuous with the common carotid plexus and therefore cannot be interrupted by stripping the adventitia from the common carotid artery. He described the external carotid plexus as originating in a fairly constant manner from at least two branches. One arises from the upper pole of the superior cervical ganglion and runs directly to the external carotid artery, usually on its medial side; the other descends posterior and medial to the external carotid as far as the superior thyroid artery, around which it passes to turn upward along the anterior surface of the external carotid. In addition, fibers from the upper part of the ganglion also enter the intercarotid plexus, which can contribute to the external carotid plexus.

In addition to the sympathetic fibers reaching the face along the vessels, it is possible that sympathetic fibers also reach the face by coursing with the several divisions of the fifth nerve, especially since these are so intimately related near the petrous apex to the internal carotid artery and its nerve plexus. Simonton and Gay concluded that there is considerable evidence that only the portions of the face supplied by the maxillary and mandibular nerves receive their sympathetic fibers through the external carotid plexus and that the portion supplied by the ophthalmic nerve receives its sympathetic fibers from the internal carotid plexus. As additional evidence for this view, they reported a case in which retraction of the internal carotid artery at operation was necessary, and following this retraction there was not only a temporary miosis but also a temporary anhidrosis of the frontal region on the side of operation; they interpreted this as indicating that the ophthalmic division of the face receives its sweat fibers intracranially from the internal carotid plexus.

Anomalies

Cleft Lip

Davis listed 31 types of congenital deformities of the face in a series of 1000 studied; in this series, however, the chief defect in all except 63 was cleft lip or cleft palate. Although cleft lips are commonly associated with clefts of the alveolar process and with cleft palates, they may occur separately; Ritchie estimated that some 60% to 70% of facial clefts involve also clefts of the alveolar process, and in Davis' series there were only 129 cases of cleft lip unassociated with cleavage of the alveolar process or palate, as opposed to 808 so associated.

By far the most common type of cleft lip is the unilateral one (Fig. 6-21); bilateral cleft lip (harelip) is distinctly less common, and median cleft lip is rare and apparently always associated with absence of the anterior central ("premaxillary") portion of the alveolar process (Davis). The occurrence of these types is discussed further in the next chapter, in connection with cleft palate.

The problem in cleft lip is that of restoring muscle continuity and function, as well as

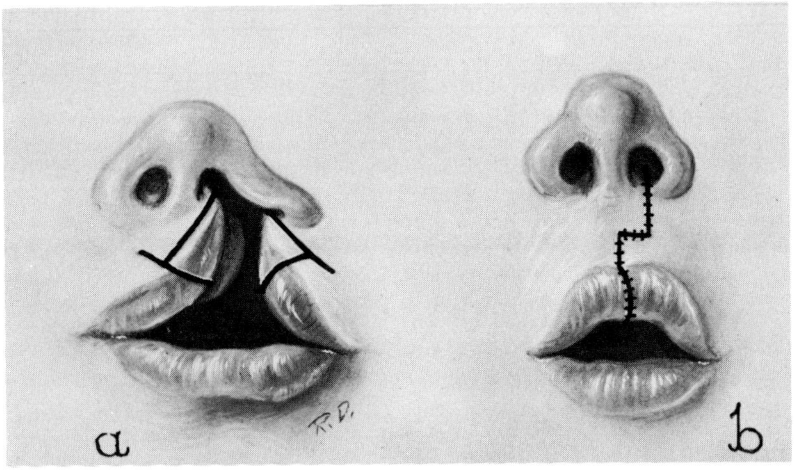

Fig. 6-21. The Le Mesurier procedure for the repair of cleft lip. (Litzow TJ: Journal Lancet 78:366, 1958)

cosmetic correction of the deformity. Peyton and Ritchie stated that there is no tendency for the deformity of cleft lip to increase with age of the child and that growth of the face seems to proceed at a nearly normal rate. Discussions of the technique of repair of cleft lip include those of Wang and of Millard. As noted in connection with repair of cleft palate (see Chap. 7), the cleft of the lip is usually corrected earlier than the palatine cleft when both are present, but a few workers repair the lip last.

Labial Pits

A relatively rare deformity, constituting only 1.4% of the patients in Davis' series, but nevertheless the most common other than facial clefts, is the occurrence of pits in the lower lip; as in a case described by Mason and coauthors, these may take the form of slits that produce a partial doubling of the lower lip. They are apparently due to improper development at the mucocutaneous junction.

Other Abnormalities

Other abnormalities associated with the face include abnormalities of the external ear, ranging from complete absence to macrotia or, most common of all, simple protrusion; preauricular appendages, sinuses and fistulas, palpebral coloboma; and malformations of the nose. Most of these have already been discussed.

References

ADAMS WM: Multiple muscle substitution for correction of facial paralysis. Am J Surg ns74:654, 1947

ADOUR KD, KLEIN JC, BELL DN: Trigeminal neurotization of paralyzed facial musculature. Modification of the Lexer–Rosenthal surgical procedure. Arch Otolaryngol 105:13, 1979

BEAHRS OH, ADSON MA: The surgical anatomy and technic of parotidectomy. Am J Surg 95:885, 1958

BOWDEN REM, MAHRAN ZY: Experimental and histological studies of the extrapetrous portion of the facial nerve and its communications with the trigeminal nerve in the rabbit. J Anat 94:375, 1960

BROWDER J: Do sympathetic nerves transmit painful impulses? Report of a case. Am J Surg ns18:100, 1932

BRUESCH SR: The distribution of myelinated afferent fibers in the branches of the cat's facial nerve. J Comp Neurol 81:169, 1944

CARMICHAEL EA, WOOLLARD HH: Some observations on the fifth and seventh cranial nerves. Brain 56:109, 1933

CLEVELAND DA: Afferent fibers in the cervical sympathetic trunk, superior cervical ganglion, and internal carotid nerve. J Comp Neurol 54:35, 1932

COLEMAN CC: Results of facio-hypoglossal anastomosis in the treatment of facial paralysis. Ann Surg 111:958, 1940

COLEMAN CC: Surgical lesions of the facial nerve: With comments on its anatomy. Ann Surg 119:641, 1944

CUSHING H: The sensory distribution of the fifth cranial nerve. Bull Johns Hopkins Hosp 15:213, 1904

DAVIS L, POLLOCK LJ: The role of the sympathetic nervous system in the production of pain in the head. Arch Neurol Psychiat 27:282, 1932

DAVIS WB: Congenital deformities of the face: Types found in a series of one thousand cases. Surg Gynecol Obstet 61:201, 1935

DISHELL WD: Tympanic neurectomy in chronic parotitis. Arch Otolaryngol 94:471, 1971

DUEL AB, TICKLE TG: The surgical repair of facial nerve paralysis: A clinical presentation. Ann Otol Rhinol Laryngol 45:3, 1936

EHNI G, WOLTMAN HW: Hemifacial spasm: Review of one hundred and six cases. Arch Neurol Psychiatr 53:205, 1945

FIGI FA: Malignant lesions of the face. Collect Papers Mayo Clin Mayo Found 38:410, 1946

FOLEY JO, DUBOIS FS: An experimental study of the facial nerve. J Comp Neurol 79:79, 1943

GARDNER E: Surgical anatomy of the external carotid plexus. Arch Surg 46:238, 1943

GASSER RF: The early development of the parotid gland around the facial nerve and its branches in man. Anat Rec 167:63, 1970

GAUGHRAN GRL: The pterygomandibular raphe—Anatomical artifact (abst). Anat Rec 184:410, 1976

GLASCOCK H, GLASCOCK H JR: Repair of traumatic fistulas of Stenson's duct. Surg Gynecol Obstet 65:355, 1937

GRÖNROOS H: Eine seltene Anordnung der Arteria maxillaris externa bei einem Erwachsenen. Anat Anz 20:9, 1902

HELSON H: The part played by the sympathetic system as an afferent mechanism in the region of the trigeminus. Brain 55:114, 1932

HORI T, FUKUSHIMA T, TERESO H, TABAKURA K, SANO K: Percutaneous radiofrequency facial nerve coagulation in the management of facial spasm. J Neurosurg 54:655, 1981

HOWE HA, TOWER SS, DUEL AB: Facial tic in relation to injury of the facial nerve: An experimental study. Arch Neurol Psychiatr 38:1190, 1937

HUNT JR: Geniculate neuralgia (neuralgia of the nervus facialis): A further contribution to the sensory system of the facial nerve and its neuralgic conditions. Arch Neurol Psychiatr 37:253, 1937

HURFORD FR: The surgical anatomy of the parotid gland. Br J Surg 34:186, 1946

HYNDMAN OR, WOLKIN J: Sweat mechanism in man: Study of distribution of sweat fibers from the sympathetic ganglia, spinal roots, spinal cord and common carotid artery. Arch Neurol Psychiatr 45:446, 1941

JAEGER R: Neuralgic syndromes of the head: Anatomy and physiology of sensory nerves to the head. Arch Otolaryngol 44:424, 1946

KADANOFF D: Die sensiblen Nervendigungen in der mimischen Muskulatur des Menschen. Z Mikrosk Anat Forsch 62:1, 1956

KEMPE LG: Topical organization of the distal portion of the facial nerve. J Neurosurg 52:671, 1980

KUNTZ A: Nerve fibers of spinal and vagus origin associated with the cephalic sympathetic nerves. Ann Otol Rhinol Laryngol 43:50, 1934

LEE FC: Orbicularis oris muscle in double harelip. Arch Surg 53:407, 1946

LYMAN HW: The sphenoid sinus and the sphenopalatine ganglion as factors in so-called atypical trigeminal neuralgia. Ann Otol Rhinol Laryngol 44:653, 1935

MAES U: Infections of the "dangerous area" of the face. Surgery 2:789, 1937

MALONEY WJ, KENNEDY RF: The sense of pressure in the face, eye, and tongue. Brain 34:1, 1911

MANDOUR MA, HELMI AM, EL-SHEIKH MM, EL-GAREM F, EL-GHAZZAWI E: Effect of tympanic neurectomy on human parotid salivary gland. Histopathologic, histochemical and chemical study. Arch Otolaryngol 103:338, 1977

MARINO H: Paralysis of the muscles of the chin. Surgical treatment. Surg Gynecol Obstet 96:433, 1953

MASON ML, ANSON BJ, BEATON LE: The surgical and anatomical aspects of a case of double lower lip. Surg Gynecol Obstet 70:12, 1940

MCCLURE RD, LAM CR: Should the neck nodes be dissected in patients with carcinoma of the lip? Ann Surg 125:658, 1947

MCCORMACK LJ, CAULDWELL EW, ANSON BJ: The surgical anatomy of the facial nerve with special reference to the parotid gland. Surg Gynecol Obstet 80:620, 1945

MCKENZIE J: The parotid gland in relation to the facial nerve. J Anat 82:183, 1948

MCKENZIE WR: Fracture of the facial bones. Arch Otolaryngol 52:257, 1950

MCWORTER GL: The relations of the superficial and deep lobes of the parotid gland to the ducts and to the facial nerve. Anat Rec 12:149, 1917

MILLARD DR JR: Refinements in rotation-advancement cleft lip technique. Plast Reconstr Surg 33:26, 1964

MONRAD–KROHN GH: On the dissociation of voluntary and emotional innervation in facial paresis of central origin. Brain 47:22, 1924

MORGAN WR: Parotid duct ligation and tympanic neurectomy in chronic recurrent parotiditis. Arch Otolaryngol 98:179, 1973

NORDENFELT I: On the sympathetic innervation of the parotid gland of the cat. Q J Exp Physiol 50:62, 1965

NOSIK WA, WEIL AA: Selective partial neurectomy in hemifacial spasm and the electrophysiologic selection of patients. J Neurosurg 13:596, 1956

OWENS N: The surgical treatment of facial paralysis. Plast Reconstr Surg 7:61, 1951

PEYTON WT, RITCHIE HP: Quantitative studies on

congenital clefts of the lip. Arch Surg 33:1046, 1936

POLLACK RS: Surgical significance of the marginal mandibular branch of the facial nerve. AMA Arch Surg 68:81, 1954

PORTO AF JR, WICKER J, PROUD GO: An anatomic study of the hypotympanic branch of Jacobson's nerve. Laryngoscope 88:56, 1978

PULEC JL: Idiopathic hemifacial spasm. Pathogenesis and surgical treatment. Ann Otol Rhinol Laryngol 81:664, 1972

ROSEN G, KONAK S: Scanning evaluation of parotid gland after tympanic neurectomy. Ann Otol Rhinol Laryngol 84:203, 1975

ROSS JAT: The function of the tympanic plexus as related to Frey's syndrome. Laryngoscope 80:1816, 1970

SADÉ J: Facial nerve reconstruction and its prognosis. Ann Otol Rhinol Laryngol 84:695, 1975

SHAW RE: Cavernous sinus thrombophlebitis: A review. Br J Surg 40:40, 1952

SIMONTON KM, GAY JR: Unilateral frontal anhidrosis and miosis occurring after petrous apicectomy (Ramadier technic): Report of a case. Arch Neurol Psychiatr 60:86, 1948

WANG MKH: A follow-up study of unilateral cleft lips repaired by the modified Le Mesuriere's and Tennison's technique. Plast Reconstr Surg 30:329, 1962

WARTENBERG R: Winking-jaw phenomenon. Arch Neurol Psychiatr 59:734, 1948a

WARTENBERG R: "Inverted Marcus Gunn phenomenon" (so-called Marin Amat syndrome). Arch Neurol Psychiatr 60:584, 1948b

WINCKLER G: Contribution à l'étude de la région orbitaire externe chez l'homme. Arch Anat Histol Embryol 24:167, 1937

WOLTMAN HW, WILLIAMS HL, LAMBERT EH: An attempt to relieve hemifacial spasm by neurolysis of the facial nerves: A report of two cases of hemifacial spasm with reflections on the nature of the spasm, the contracture and mass movement. Proc Staff Meet Mayo Clin 26:236, 1951

WOODS JE, BEAHRS OH: A technique for the rapid performance of parotidectomy with minimal risk. Surg Gynecol Obstet 142:87, 1976

Chapter 7

THE JAWS, PALATE, AND TONGUE

The Upper Jaw

The maxilla, the chief bony component of the upper jaw, has already been described in part with the face (see also Fig. 6-1). Its *body* contains the maxillary sinus and transmits the superior alveolar nerves and vessels. Certain general features of its anatomy, as well as its alveolar process, are necessarily of interest in connection with the upper jaw and teeth. On the anterior (facial) surface of the body a variable distance below the inferior orbital rim is the infraorbital foramen, and extending downward from this is the concavity of the canine fossa. The infratemporal or posterior surface, facing the infratemporal fossa, presents several alveolar foramina for the entrance of posterior superior alveolar nerves and vessels, and below, just above and behind the third molar tooth, is the maxillary tuber, a swelling that gives rise to a small part of the origin of the medial pterygoid muscle. Medially and below, the pyramidal process of the perpendicular plate of the palatine bone unites the two pterygoid plates of the sphenoid to each other and to the maxilla, but above there is a gap, the pterygomaxillary (sphenomaxillary, sphenopalatine) fissure between the pterygoid processes and maxilla (Fig. 7-1), that leads into the pterygopalatine fossa. Through this fissure the maxillary (internal maxillary) artery enters the pterygopalatine fossa to divide into its terminal branches. The foramen rotundum opens into the posterior wall of the pterygopalatine fossa, and therefore the maxillary nerve enters this fossa immediately after leaving the middle cranial fossa. The pterygoid (Vidian) canal also ends here, transmitting the nerve of the same name to the pterygopalatine (sphenopalatine) ganglion, which is here suspended from the maxillary nerve.

Branches of the maxillary artery and nerve, the latter also including branches from the ganglion, leave the pterygopalatine fossa in three principal directions: medially, to enter the nasal cavity through the sphenopalatine foramen; anteriorly, below the inferior orbital fissure, to run forward in the floor of the orbit and the roof of the maxillary sinus (zygomatic and infraorbital nerves and infraorbital vessels); and downward, both along the infratemporal surface of the body of the maxilla as posterior superior alveolar nerves and vessels and through the palatine canals to reach the palate as greater and lesser palatine nerves and vessels. (Of minor importance are the twigs from the ganglion and from the sphenopalatine artery that run posteriorly through the palatovaginal and vomerovaginal canals to the roof of the pharynx and the orbital twig or twigs from the ganglion that ascend through the inferior orbital fissure.)

The *alveolar process* of the maxilla bears the upper teeth and is continuous above with the body of the maxilla and medially with the

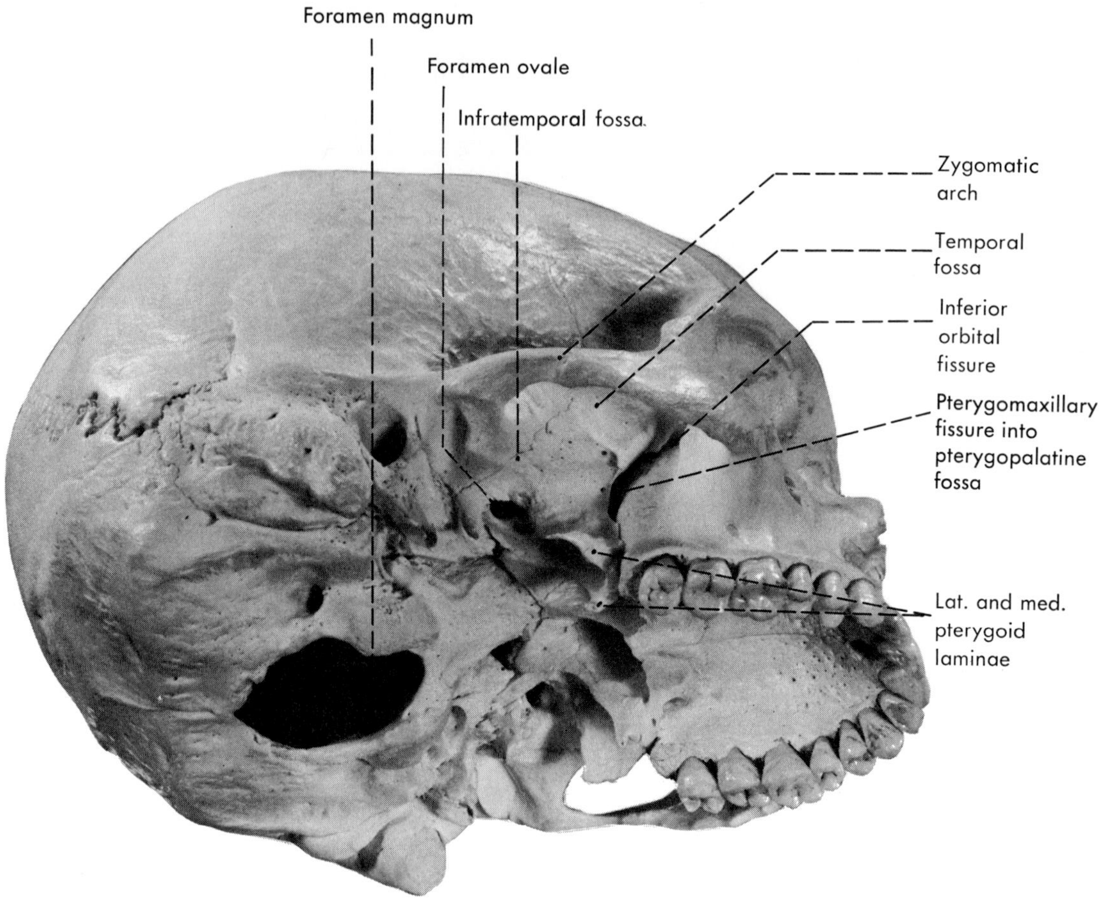

Fig. 7-1. Oblique view of the external base of the skull.

palatine process of that bone. Posteriorly, at the level of the second and third molar teeth, the alveolar process articulates medially with the horizontal (palatine) process of the palatine bone. The maxillary sinus (antrum) in the body of the maxilla typically extends into the alveolar process, most commonly coming into relationship with the apices of the three molars and the second premolar, but with much variation both in the number of the teeth to which it is closely related and the thinness of the bone between the mucosa of the sinus and the roots of the teeth (see Chap. 4, and Figs. 4-38 to 4-41). Broken roots of the molars are sometimes forced upward into the maxillary sinus in attempts to extract these teeth, and abscesses of these teeth may be a cause of maxillary sinusitis.

An adult's fully developed superior dental arch, beginning at the anterior midline, consists on each side of medial and lateral incisors, a canine, first and second premolars (bicuspids), and first, second, and third molars. The deciduous teeth consist of incisors, canine, and two molars that lie in the positions of the permanent premolars. The permanent third molar (dens serotinus, or wisdom tooth) is often defective in its direction of growth and may fail to erupt; it is also more likely to be congenitally absent than are other teeth. Sometimes there are supernumerary teeth behind the third molar.

The suture between the premaxilla and the maxilla in the fetus and infant extends just lateral to the lateral incisor tooth, and therefore the part of the alveolar arch bearing the incisor teeth is often referred to as the premaxilla. The alveolar arch as a whole presents sockets for the teeth, the dental alveoli, separated by interalveolar septa and partially subdivided, for those teeth with more than one root, by interradicular septa. The particularly hard bone lining an alveolus is called the lamina dura. The roots of the incisor and canine teeth are situated so far forward in the alveolar arch that the bone over them protrudes to form juga alveolaria; the canine jugum, bounding the canine fossa medially, is usually particularly prominent.

A tooth is largely formed from dentine, which is modified bone, but the crown, the normally exposed part of the tooth, is covered by much harder enamel (see Fig. 7-24). At the neck (collum) of a tooth the gingiva is attached, the enamel ceases, and the roots of the teeth extend into the alveolar process; when dentine is exposed through resorption of peridontal tissue, the entire exposed part of the tooth, whether covered by enamel or not, is known as the clinical crown, while the part deep to the gingival attachment is the clinical root. The incisors and canine have a single root each, the premolars one each that may have a bifid apex, and the molars three (those of the lower jaw typically two). Each root is united by cementum to a periodontal ligament or membrane (NA, periodontium) that represents the periosteum lining the alveolus. The apex of each root presents a foramen that leads through a root canal into a larger space (cavum coronale); canal and cavum coronale together constitute the *cavum dentis* and are occupied by the pulp of the tooth.

The crown of each tooth has an occlusal (masticatory) surface, a vestibular or facial (outer) surface, a lingual (inner) surface, and, except for the third molar, two surfaces of contact, mesial (medial or anterior) and distal (lateral or posterior), with other teeth. The occlusal "surfaces" of the incisors are more or less sharp incisal edges, and the canines end in blunted points. The upper canines or incisors sometimes show a marked thickening, the cingulum, on the lingual surfaces close to the gingiva. The occlusal surfaces of the premolars and molars are broad and have cusps (tubercles) that add to their grinding power but sometimes interfere with good occlusion between upper and lower teeth and may have to be reduced in height. The premolars have two cusps (hence "bicuspids"), the upper molars three or four, and the lower molars four or five. A ridge on a more posterior tooth running from the cusp toward the center is called a crista triangularis, and if two such ridges meet they form a crista transversalis.

The alveolar arch of the maxilla is typically somewhat larger than that of the mandible, and the premolars and molars of the upper jaw typically slant slightly outward toward their occlusal surfaces. The external walls of the alveoli, especially of the more anterior ones, are thin, but the medial walls are strengthened by the attachment of the palatine process. The third molar lacks this medial strengthening and is also close to the infratemporal (posterior) surface of the maxilla; its walls may therefore be particularly thin, and attempts to remove it may result in a maxillary fracture.

NERVES AND VESSELS

Except for some participation by the mandibular nerve in the innervation of the buccal gingival mucosa, the nerve supply of the upper jaw is entirely through the *maxillary nerve,* and, correspondingly, the chief blood supply is from the *maxillary artery.* Unlike the lower jaw, in which a single nerve branch (the inferior alveolar) supplies all the teeth, the branches to the upper teeth (superior alveolar nerves) are multiple and have divergent courses through the maxillary bone. Thus anesthesia of all the teeth on one side (except for the incisors, which may receive some innervation from the opposite side) can be obtained only by injection about the maxillary nerve as a whole. This can be done by delivering the anesthetic into the pterygopalatine (pterygomaxillary, sphenomaxillary, sphenopalatine) fossa. Routes that have been used include external or lateral ones, the needle being passed through the skin and either anterior to the coronoid process of the mandible or through the mandibular notch; an intraoral route, through the buccal mucosa deep to the mandible and the insertion of the temporalis muscle; the palatine route, up the greater palatine canal; and the infraorbital route, through the infraorbital foramen and canal. A successful injection of this type produces also anesthesia of most of the nasal cavity and palate.

Production of anesthesia of specific groups of teeth depends upon delivering the anesthetic in the immediate neighborhood of the various nerves before they enter the small bony canals in which they run in the maxilla and in injecting deep to the mucosa or periosteum of the alveolar process itself; this local infiltration not only produces anesthesia of the gingival mucosa, but also allows the anesthetic to reach the tooth sockets through the numerous small foramina for nerves and vessels that exist in the alveolar process. Submucous injections diffuse more widely, which is sometimes an advantage; subperiosteal injections should be made slowly, in order to minimize the pain produced by the elevation of the periosteum from the bone.

The nerve endings in and about the teeth have been described by Lewinsky and Stewart ('36a, '37), among others, and more recently by Bernick (*e.g.,* '57); although nerve fibers have been described as penetrating the dentine deeply, Bernick found only occasional fibers looping into the predentine or dentine for a short distance and then apparently returning to end among the odontoblasts. Of physiological importance is the generally agreed upon fact that many of the nerve fibers around the teeth are not concerned with pain but are large sensory fibers probably serving to control the strength of the bite.

Fig. 7-2. Innervation of the upper teeth. The posterior superior alveolar nerves may be seen leaving the maxillary nerve in the pterygopalatine fossa and coursing downward before entering the maxilla.

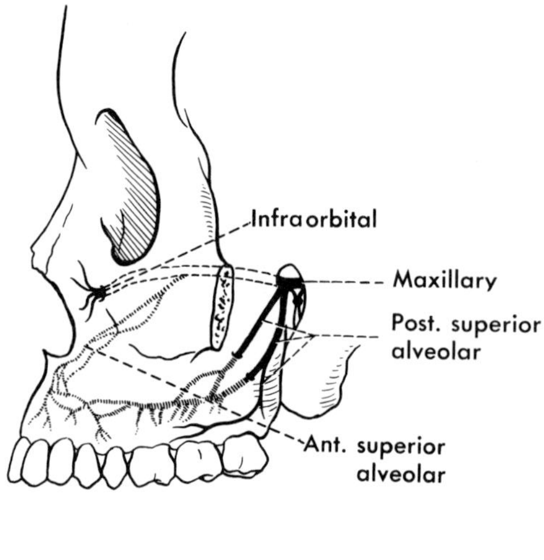

Innervation of Upper Teeth

The innervation of the upper teeth is probably not accurately known, since the nerves are small and difficult to trace in their bony canals, show some gross variability, and, moreover, enter into a plexus through which it is impossible to trace fibers of a given branch to a specific tooth. The superior alveolar (dental) nerves (Fig. 7-2) are usually described as consisting of three sets: a posterior, a middle and an anterior one.

Jones found three *posterior superior alveolar nerves* as a rule. These are given off from the infraorbital nerve just before it enters the infraorbital canal and therefore descend on the posterior wall of the maxilla, where this forms the anterior wall of the pterygopalatine fossa; one apparently usually does not enter the bone, but runs laterally to supply a posterior part of the gingiva and an adjacent part of the cheek, but the other two are true dental branches and enter canals on the posterior and lateral walls of the maxilla. They may be anesthetized before they enter their canals by passing a needle through the mucosa behind the last molar tooth and upward along the rounded infratemporal surface of the maxilla. As the jaw is opened widely, the tendon of insertion of the temporal muscle is brought in close contact with the maxilla, however, so that it may interfere with the passage of the needle here; this interference may be minimized by having the patient close his mouth somewhat, and even more successfully by combining this with deviation of the jaw toward the side to be injected. The posterior superior alveolar nerves that enter the maxilla course forward above the level of the roots of the teeth, the larger branch running, according to Jones, as far forward as the canine tooth. According to other descriptions, the posterior superior alveolar nerves supply only the molar teeth. The nerves also give off twigs to the mucosa of the maxillary sinus.

Jones regarded the *middle superior alveolar nerve* as rare, but Fitzgerald failed to find one in only 18% of dissections. It varied in origin and course, for although it arose from the infraorbital trunk in 54%, it sometimes arose from the maxillary in the pterygopalatine fossa and sometimes arose with the anterior superior alveolar nerve; depending on its origin it traversed the posterior, lateral, or anterior wall of the maxillary sinus. Fitzgerald traced the posterior alveolar nerves as far as the premolar teeth and described the middle alveolar nerve as joining the plexus over the premolars and being joined in turn at the level of the canine root by the anterior superior alveolar; thus there is overlap between the distribution of the middle and the other superior alveolar nerves. Roberts and Jorgensen described the posterior superior alveolar as extending no farther forward (in a child) than the second primary molar and said that the middle alveolar supplied the buccal root of this molar as well as teeth anterior to it.

Most texts describe two or three *anterior superior alveolar nerves,* although Jones stated that there is typically a single stem of origin. The nerve leaves the infraorbital canal in its middle or last portion and diverges laterally from the infraorbital nerve in the floor of the orbit; it then turns medially and downward on the anterior wall of the antrum to run alongside and then below the naris, reaching almost to the nasal spine. From this point descending branches spread downward in the anterior antral wall and form a superior alveloar plexus that is distributed to the roots of the canine and incisor teeth. The canine tooth has been said to be supplied, variably, from any of the three superior alveolar nerves or a combination of any two of them; presumably some of this reported variability arises from the absence of the middle alveolar, or its origin with or close to the posterior or anterior nerves, when it might therefore be regarded as a branch of one of these. The two plexuses apparently communicate across the midline, and the medial, and occasionally the lateral, incisors may receive a bilateral innervation.

As the anterior superior alveolar nerves run in the wall of the canine fossa, they (and the infraorbital nerve) are subject to damage in external antrotomies. Before they reach the teeth, the anterior superior alveolar nerves give off a nasal branch to a limited portion of the nasal cavity (inferior concha, inferior meatus) and twigs to the nasolacrimal duct.

Because the anterior superior alveolar nerves have no part of their course on the outer surface of the maxilla, they are difficult to inject; anesthesia of the incisors is therefore usually produced by infiltration rather than by nerve block. For a nerve block, the injection is typically into the infraorbital foramen, which lies usually on a vertical line passed through the supraorbital foramen or notch (this line also passing usually through the mental foramen on the lower jaw), but may lie slightly lateral to this line. The injection can be delivered directly into the infraorbital canal through the cheek or may be made intraorally into the region of the foramen.

In making the injection through the external surface of the cheek, it is well to remember that the canal does not usually approach the foramen in a straight posteroanterior line, but usually slants somewhat inward; thus the canal can be entered more readily if the syringe is directed slightly to the opposite side, so that the needle is directed posterolaterally. Also, care must be taken not to mistake the concavity on the front of the maxilla, the canine fossa, for the infraorbital foramen itself; the infraorbital foramen lies at the upper border of the fossa, and material delivered into the depths of the fossa, into or deep to the levator anguli oris (caninus) muscle, may not affect the dental nerves appreciably. If injection is made intraorally, the region of the infraorbital foramen may be reached by passing the needle upward through the mucosa above the root of the first bicuspid, until it strikes the orbital rim overhanging the canine fossa; in this approach, however, the needle should not be introduced for any distance into the infraorbital canal, since it would then necessarily impinge upon the roof of the canal, which may be exceedingly thin just behind

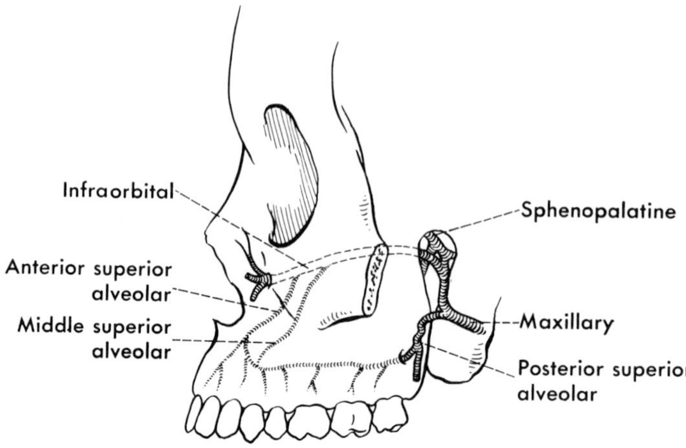

Fig. 7-3. Arteries to the upper teeth.

the orbital rim and allow passage of the needle into the orbit.

Blood Supply to Upper Teeth

The chief arteries to the teeth (Fig. 7-3) correspond approximately to the chief nerves already described. The *posterior superior alveolar (dental) artery* arises from the (internal) maxillary artery and descends on the posterior aspect of the maxilla, where its branches enter canals and run forward with the nerves above the upper teeth, supplying the molar and premolar teeth; like the nerves, these vessels give off branches to the gingiva, the cheek, and the maxillary sinus. The *anterior superior alveolar or dental artery* is a branch of the infraorbital artery (in turn a branch of the maxillary) and runs downward to supply the more anterior teeth and to anastomose with the posterior superior alveolar. A *middle superior alveolar artery* is also often described, leaving the infraorbital artery to course in the lateral wall of the maxillary sinus to the region of the canine tooth and anastomose with anterior and posterior arteries. Bleeding into the nose following extraction of a tooth, apparently resulting from rupture of one of these vessels in the floor of the antrum, may be quite troublesome (Trible).

Periodontal Membrane

As a branch of an alveolar nerve approaches the apex of a tooth, it divides into branches for the periodontal membrane or ligament; these in turn give off branches that enter the apical canal or canals and unite to form the nerve to the dental pulp. They then run distally toward the gingiva, supplying the periodontal membrane as they do so and overlapping with the gingival nerves (Fig. 7-4). These periodontal nerves are in turn joined by a second group (Lewinsky and Stewart, '36b; Bernick) also derived from the alveolar nerves but traversing the interalveolar septa rather than running distally from the apex. According to Lewinsky and Stewart ('37) these interalveolar nerves divide as they reach the periodontal membrane, so this membrane would appear to have a double innervation throughout.

Gingival Nerves and Vessels

Because of the overlap between gingival nerves and those of the periodontal membrane, local injection is necessary if complete anesthesia of a tooth and its adnexa is desired. As indicated for the nerves in Figure 7-5, the buccal gingival mucosa about the more posterior teeth is supplied by branches from the posterior superior alveolar nerves and vessels, but also receives twigs from the buccal (buccinator, long buccal) branch of the mandibular and the buccal artery (from the maxillary); nerve and artery run together deep to the temporal muscle and emerge on the cheek over the buccinator muscle, but give off branches that supply both the outer and inner surfaces of the cheek and the adjacent gingival mucosa. The labial gingival mucosa of the more anterior teeth is supplied through the descending or labial branches of the infraorbital nerve and vessels, which of course also supply the skin and mucous membrane of the upper lip.

The lingual (palatine) gingival mucosa of the upper teeth is supplied largely from the greater or anterior palatine nerves, which emerge from the greater palatine foramina and run forward on the hard palate to supply the major portion of this and the adjacent gum (Fig. 7-6). A small anteromedial portion of the gingiva and palate behind the medial incisors is supplied by the nasopalatine nerves, which emerge through the midline incisive fossa.

The buccal nerve may be blocked by anesthetic delivered through the mucosa on the inner surface of the temporal muscle or into the buccal fat pad from the external surface; the twigs from the labial branches of the infraorbital nerve may be reached either through injection at the foramen or deep to the mucosa of the anterior teeth. The nasopalatine and greater or anterior palatine nerves can be most easily blocked at the foramina through which they emerge onto the palate; thus the nasopalatine nerves, both emerging through the incisive fossa (the left usually anterior to the right) may be blocked by injection immediately behind and between the central incisors, while the greater palatine nerve may be blocked by injection at the greater palatine foramen, which lies close to the posterior border of the second molar tooth. Insertion of a needle into this foramen may cause injury to the greater palatine artery and give rise to troublesome hemorrhage; such insertion is not necessary to block the nerve, but is often done in order to inject around the maxillary nerve in the pterygopalatine fossa.

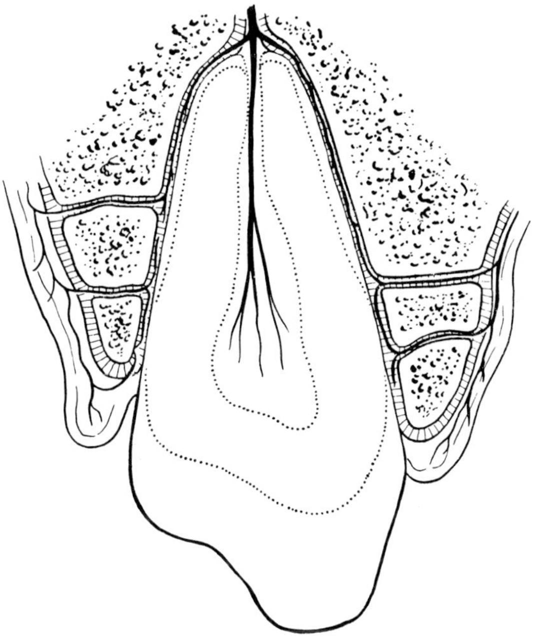

Fig. 7-4. The innervation of a single tooth. The nerve gives off branches to the periodontal membrane or ligament before it enters the pulp, and the nerves of the membrane and those of the gingiva communicate.

The Palate

The palate intervenes between the nasal and oral cavities; its functional efficiency, necessary to proper swallowing and to proper articulation of the sounds used in speech, depends upon the palate's ability to close off completely the nasal part of the pharynx and hence the nasal cavity from the mouth and oral pharynx. The musculature of the soft palate normally brings about apposition between the soft palate and the posterior pharyngeal wall; Wardill ('28) emphasized that this closure must be complete during certain stages of phonation and stated that a gap between the soft palate and the posterior pharyngeal wall of even 1 mm, so small that it cannot be detected upon examination, may be responsible for defective speech. Smith and McCabe found polytetrafluoroethylene (Teflon) injections into the nasopharynx to be effective in abolishing voice problems in 60% of patients; in another 18% the voice was less nasal after the injection.

Osseous Palate and Related Bony Structures

Understanding the palate and its musculature demands a knowledge of both the bony palate and the adjacent external base of the skull (Fig. 7-7).

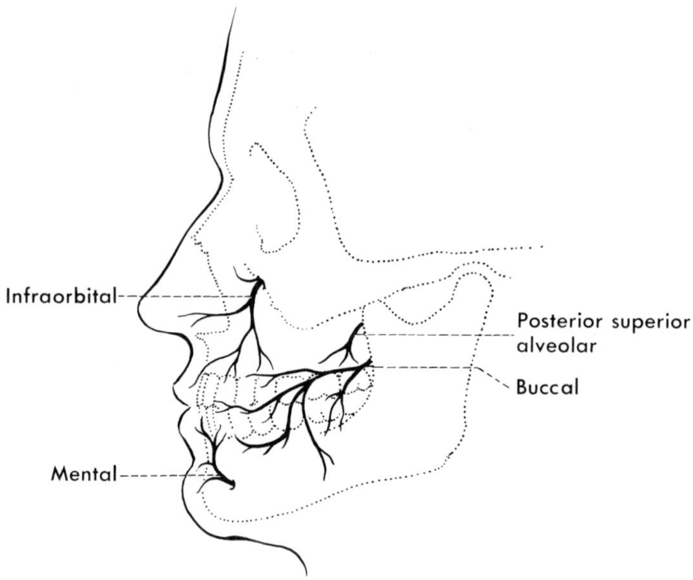

Fig. 7-5. Innervation of the buccal gingival mucosa and that of the adjacent cheek and lip. For the lower jaw it is entirely through the mandibular, but for the upper jaw only the posterior and anterior parts are innervated through the maxillary nerve.

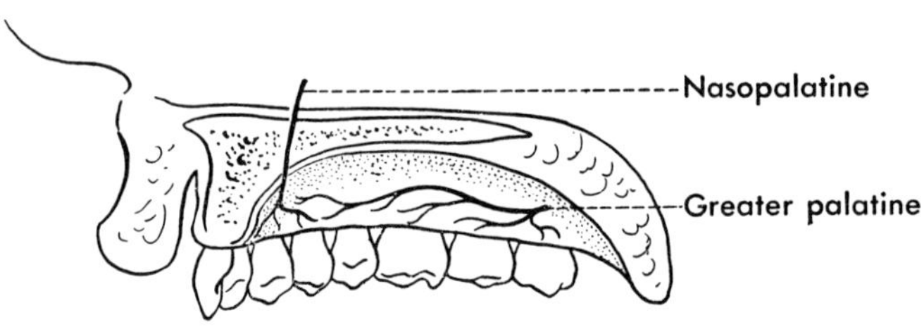

Fig. 7-6. Nerves to the inner or lingual gingival mucosa of the upper jaw.

The bony palate is formed by the palatine processes of the two maxillae and the horizontal (or palatine) laminae of the palatine bones. The maxillary part is continuous anteriorly and laterally with the alveolar process of the maxilla; laterally there is no demarcation between alveolar and palatine processes, but anteriorly in the midline the incisive fossa is usually thought of as dividing the two. This fossa, typically presenting the openings of four canals (two each, one for an artery and one for a nerve, derived from the incisive canal leading from each side of the nasal cavity), marks the junction in the midline between the maxilla proper and what was once the premaxilla. From the incisive fossa the suture that once existed between maxilla and premaxilla extended through the alveolar process just lateral to the lateral incisor tooth (Shepherd and McCarthy); facial clefts that extend no farther than the incisive fossa, although perhaps involving a very small premaxillary palatine process, are therefore usually described as alveolar or premaxillary rather than palatine clefts.

The bony palate presents two sutures: a longitudinal, between the bones of the two sides, and a transverse, between the maxillary and palatine parts. Bony mounds or ridges (torus palatinus) sometimes develop along the longitudinal suture, but are not limited to it. There may be similar formations laterally along the junction of palate and alveolar process or on either outer or inner surfaces of the alveolar processes of the maxilla or mandible. Their cause is not known.

The transverse palatine suture lies approximately at the level of the interspace between the first and second molar teeth; as it reaches the alveolar process, this part of the palatomaxillary suture turns backward along the medial side of the alveolar process, and at or behind the level of the second molar it widens to form the greater palatine foramen. More posteriorly, in the palatine bone itself, are lesser palatine foramina, typically two on each side, through which emerge the lesser palatine nerves and vessels. A little in front of its posterior margin the horizontal plate of the palatine bone may present a transverse elevation, the palatine crest, indicating the most anterior attachment of the tensor veli palatini to the bony palate. In the posterior midline the two palatine bones come together in a posterior nasal spine.

From the horizontal lamina of the palatine bone the perpendicular lamina extends upward as a posterior part of the lateral nasal wall, articulating with the maxilla and forming with that bone the greater palatine canal. (This lamina also helps to bound the sphenopalatine foramen and sends a small orbital

Fig. 7-7. A view of the external base of the skull, looking slightly forward. *MAX.* and *PAL.*, the palatine process of the maxilla and the horizontal lamina of the palatine bone, respectively; *V,* the vomer.

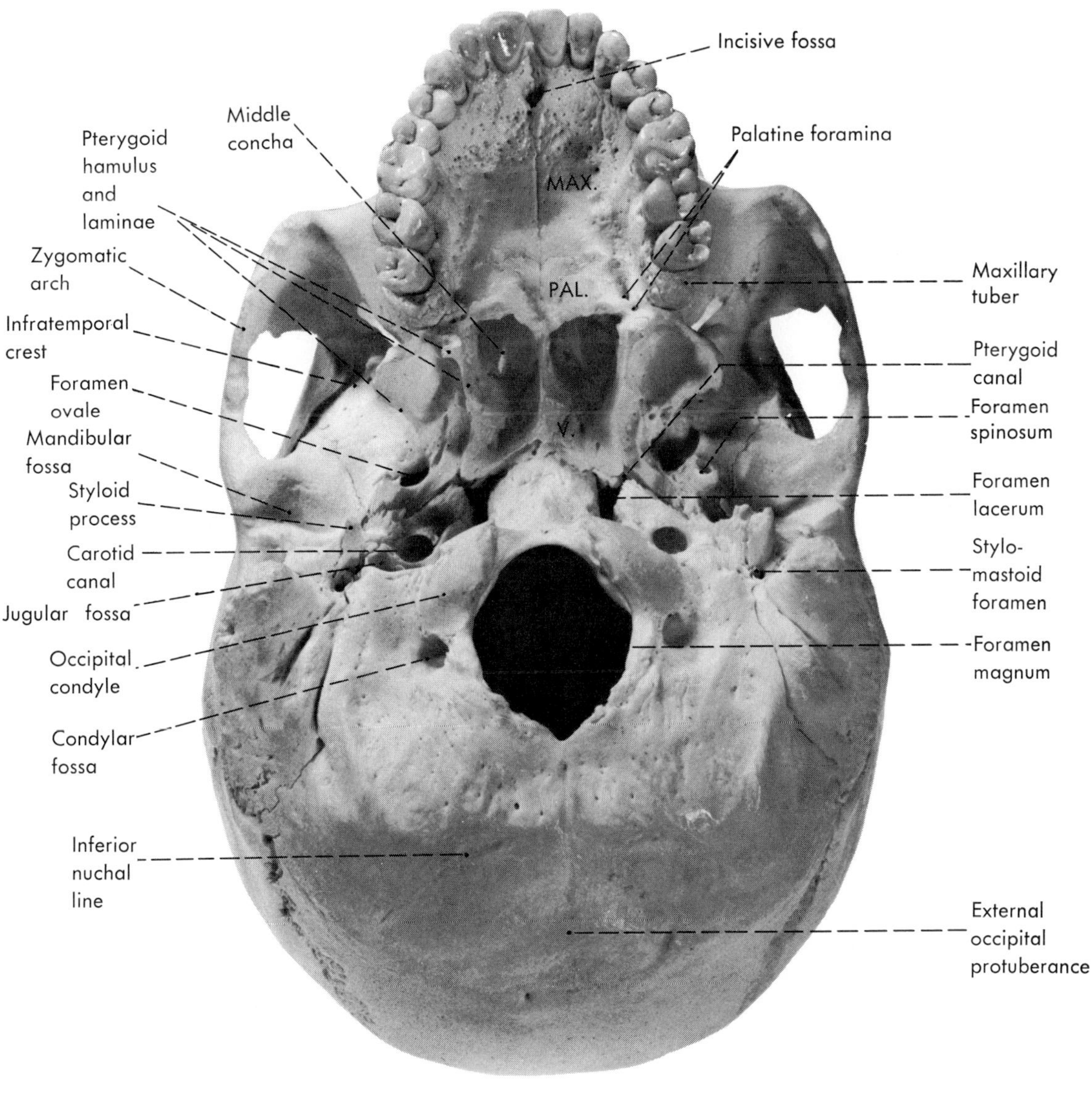

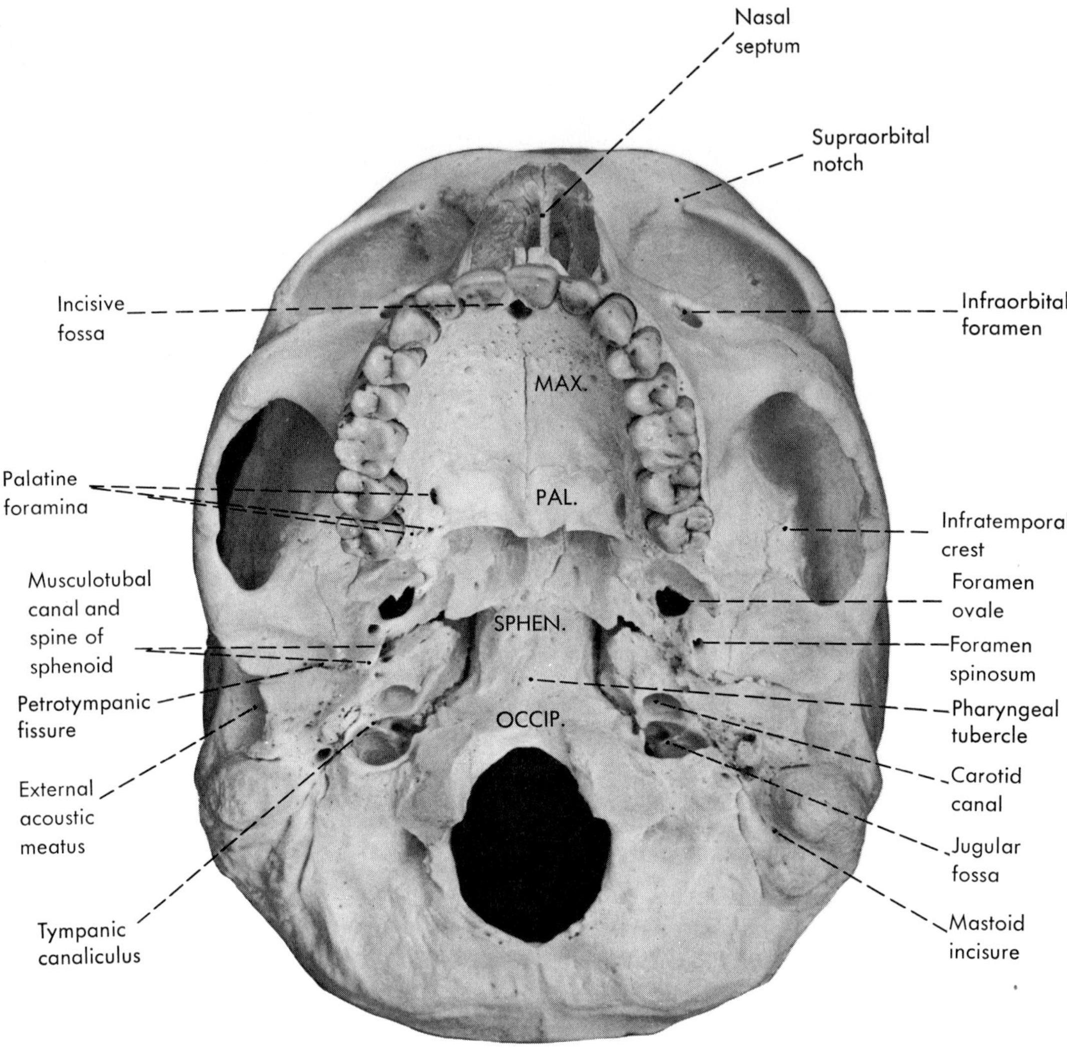

Fig. 7-8. Another view of the external base of the skull, this time looking slightly backward. *MAX.* and *PAL.*, as in the preceding figure; *SPHEN.* and *OCCIP.*, the bodies of the sphenoid and occipital bones. Certain features of the skull are identifiable in both figures; others are clearer in one than the other.

process into the floor of the orbit.) The most posterior part of the horizontal lamina is the pyramidal process, which articulates with the infratemporal or posterior surface of the maxilla and with both pterygoid plates (laminae) filling in an anterior triangular gap between these laminae.

Several features of the external *base of the skull* (Figs. 7-7 and 7-8) are of interest in connection with the palatine and jaw musculature. The medial and lateral laminae or plates of the pterygoid process project downward from the body and greater wing of the sphenoid bone; the concavity between them, with its anteroinferior wall completed by the pyramidal process of the palatine, is the pterygoid fossa. The lateral pterygoid plate, larger than the medial, gives attachment on its outer surface to the inferior head of the lateral pterygoid muscle, and to a part of the origin of the medial pterygoid muscle, and on its inner surface to most of the origin of the medial pterygoid. Above, the posterior border of the medial pterygoid plate presents two diverging edges that enclose between them the scaphoid fossa, which affords origin to anterior fibers of the tensor veli palatini. Below, the medial pterygoid plate ends in the slender hamulus, projecting downward and curving posteriorly and laterally; the pterygoid sulcus around its base lodges the anterosuperior part of the tendon of the tensor veli palatini, which covers most of the length of the hamulus as it turns around that to enter the palate.

Lateral to the lateral pterygoid plate is the infratemporal fossa. The greater wing of the sphenoid forms the roof of this fossa and gives rise to the upper head of the lateral pterygoid muscle; a more or less sharp ridge, the infratemporal crest, separates the infratemporal fossa from the large temporal fossa (giving rise to the temporal muscle) on the side of the skull. Anteriorly, the infratemporal fossa is separated from the maxilla by the inferior orbital fissure, continuous posteromedially with the pterygomaxillary (pterygopalatine) fissure. Posteriorly, the infratemporal fossa is bounded by the articular tubercle of the temporal bone, and behind the tubercle is the mandibular (glenoid) fossa. The back end of the zygomatic arch helps to form the lateral parts of the tubercle and fossa. Behind the fossa is the tympanosquamous fissure; medially this fissure divides into two (not always recognizable), an anterior petrosquamous and a posterior petrotympanic; the chorda tympani emerges through the latter.

Behind the lateral pterygoid plate at the medial border of the infratemporal fossa is the foramen ovale, the exit of the mandibular nerve, and posterolateral to that is the smaller foramen spinosum for the middle meningeal artery. Both foramina open into the middle cranial fossa. Posterior to the foramen spinosum and medial to the mandibular fossa is the sphenoid spine, varying considerably in height; a part of the sphenomandibular ligament and some fibers of the tensor veli palatini attach to the sphenoid spine. Medial to the sphenoid spine is the external opening of the musculotubal canal (for the tensor tympani muscle and the auditory tube); the cartilaginous portion of the auditory (eustachian) tube runs forward, medially, and downward from this opening.

Posteromedial to the foramen ovale, foramen spinosum, and sphenoid spine is the inferior surface of the petrous part of the temporal bone, which gives rise to most of the fibers of the levator veli palatini. The lower surface of this part presents the carotid canal for the internal carotid artery; behind it, between it and the occipital bone, is the jugular fossa leading upward to the jugular foramen (for the internal jugular vein and cranial nerves 9 to 11); the jagged aperture at the anteromedial end of the petrous temporal is the foramen lacerum.

Hard Palate

The hard palate, forming approximately three fourths of the entire length of the palate, is simply the bony palate covered by mucosa firmly adherent to the periosteum and containing numerous mucous palatine glands. On its oral surface the palatine mucosa may present a faint midline ridge indicating the fusion of its two sides. This ends anteriorly in an incisive papilla, particularly marked in infants, which overlies the incisive fossa. Anteriorly on each side, again more prominent in infants, are a variable number of slight ridges, the transverse palatine plicae.

Soft Palate

The soft palate (palatinum molle) is a fibromuscular shelf that projects backward and downward from the posterior edge of the hard palate (Fig. 7-9). During quiet respiration it is almost vertical, but during speech its anterior two thirds is raised to or above the level of the hard palate, and only its more posterior part then curves downward. The *velum palatini* is sometimes defined as the posterior portion of the soft palate, but in practice the term is used as a synonym for soft palate. The nipplelike uvula projects downward from its free edge in the midline, and laterally the palatine arches run downward in front of and behind the palatine tonsil, helping to bound the faucial isthmus. On the oral (inferior and anterior) aspect of the soft palate are numerous palatine glands continuous with those on the

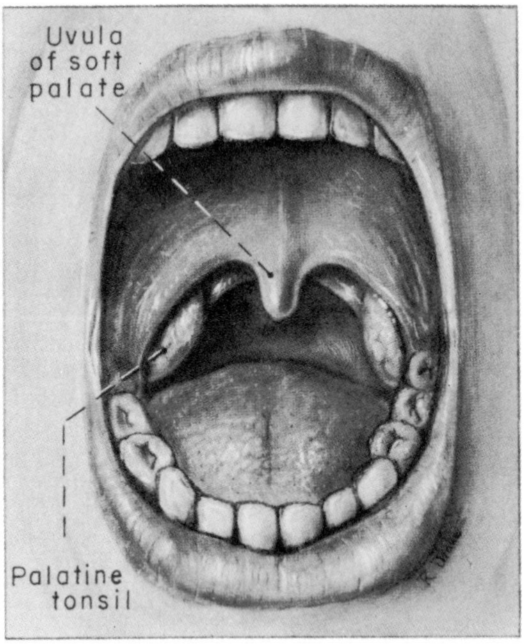

Fig. 7-9. The oral cavity, soft palate, and palatine tonsils. The palatoglossal arch partly hides the tonsil; the palatopharyngeal arch, visible above, is hidden by the tonsil below.

hard palate, and the bulk of the uvula is largely made up of glands; because of this, and its especially abundant lymphatics, the uvula readily becomes edematous.

The central layer of the soft palate is fibrous tissue, usually called the palatine aponeurosis, which attaches to the hard palate anteriorly and receives or gives rise to the palatine muscles. Muscles attach upon both surfaces of the aponeurosis, but the more anterior portion of the soft palate is largely membranous, consisting of the aponeurosis and the tendon of the tensor veli palatini muscle; the bulk of the muscle of the palate occupies a middle segment, and Oldfield stated that the posterior segment is not sufficiently muscular to move actively, but moves passively as a result of movement of the middle segment.

The uvula is sometimes surgically shortened, especially in the course of tonsillectomy; Richardson and Pullen, however, expressed the belief that it should not be traumatized needlessly, since they regarded it as important in moistening and removing mucus from the posterior pharyngeal wall and as a valuable aid in children in preventing middle ear disease. Froeschels pointed out normal variations in the shape of the uvula; he stated that the uvula shrinks after tonsillectomy, and this may obviate the clipping that sometimes seems indicated. The function of the uvula in helping to seal off the nasopharynx is illustrated by Wodak's case, in which the uvula was turned forward and adhered to the oral surface of the soft palate, so that the nasopharynx remained constantly open and the patient's speech was scarcely intelligible.

CLEFT PALATE

Developmentally, the palatine processes of the maxillae and of the palatine bones have arisen in shelves of tissue that have grown medially to meet the elements of the opposite side in the midline. The part of the alveolar process that bears the four incisor teeth, however, is formed by a median downgrowth (the medial nasal process) that also forms the lower part of the nasal septum and the midpart of the upper lip; this normally fuses with the maxillary alveolar processes and the remainder of the hard palate (Fig. 7-10). The line of fusion between the medial and lateral processes apparently occurs at about the levels at which the tooth germs for the lateral incisors will appear later; an unfused medial nasal process is often spoken of as the premaxilla (the part of the bone bearing the incisor teeth), and in failure of fusion the lateral incisor may be defective or may lie on the medial side of the cleft.

Because of the method of development of the palate, it occurs that cleft of the palate proper is always a defect in the midline, due to the failure of fusion of the palatine processes from the two sides; on the other hand, the alveolar process, like the upper lip, may show either a unilateral cleft or a bilateral cleft due to improper fusion of the medial nasal process with the lateral nasal and palatine processes or, rarely, a large midline defect due to failure of development of the medial nasal process. The exact method of the formation of cleft palate, particularly in regard to whether the various processes never meet or separate after meeting, is debatable, as indicated in several of the papers in the symposium edited by Pruzansky.

Incidence and Types

Maxillofacial clefts (cleft lip or "hare-lip" and cleft palate) are by far the most common facial anomalies; Smith and Johnson quoted other authors as having found that white children show cleft palate about once in 1000 births and black children once in 1800 births; Ivy concluded that cleft lip or palate occurs only about one fifth as frequently in nonwhites as in whites. Harelip and cleft palate are apparently also more common in the male than in the female. Peyton found in a series of 2352 patients with congenital malformation of the upper lip and palate that 64.3% were males.

Since cleft lip and cleft palate occur so frequently together, most of the figures do not break down these associated defects into their component parts. Actually, the lack of fusion may involve the entire hard and soft palates, the alveolar process, and the upper lip, thus combining cleft lip with cleft palate; it may involve the entire palate alone; or it may be limited to either the anterior or posterior part of the palate. According to Ritchie ('41) the alveolar process is involved in about 60% to 70% of all facial clefts. The alveolar process and the lip are apparently more commonly cleft on the left side than on the right, and a midline defect is fortunately rare. In bilateral cleft of the alveolar process the superior alveolar nerves, from the maxillary, cannot of course grow across the gap to reach the central incisors, and the latter are in these cases, according to Goyder, innervated by the "nasal" nerve (an old term for nasociliary, but here probably the nasopalatine). King, however, reported a case in which the incisor teeth received no innervation at all, although the nasopalatine reached the premaxilla and the superior alveolar nerves may have, since they joined the infraorbital nerves, and these sent branches above the cleft to reach the premaxilla.

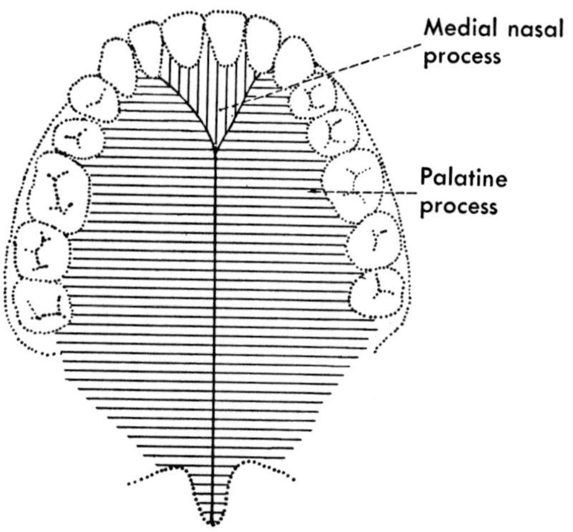

Fig. 7-10. Formation of the palate from two lateral and an anterior median process, thus determining the various types of cleft palate that may occur. Unless the medial nasal process fails to contribute to the upper jaw, clefts of the alveolar process must obviously be lateral ones, while clefts of the palate proper are necessarily midline.

Because of the sequence of closure of the palate, from before backward, the soft palate is usually cleft if the hard palate is; Goyder reported that tripartite palate (*i.e.,* bilateral alveolar clefts and complete cleft palate) occurred in 22.8% of his cases, bipartite palate in 31.9%, cleft of the soft and part of the hard palate in 28.7%, of the soft palate alone in 15.7%, and all other varieties in 0.9%. Davis, in 804 cases of cleft palate, found unilateral clefts in the lip and alveolar process in 74, unilateral clefts of the entire palate in 342, bilateral cleft palate in 139, partial posterior clefts of the palate in 240, and submucous clefts of the hard palate in 9. It is apparent from his table that clefts of the lip and alveolar process do not always correspond for he listed cases of unilateral cleft lip and bilateral cleft palate and cases of bilateral cleft lip and unilateral cleft palate.

Repair

For proper speech, as already noted, the palate must come in contact with the pharyngeal wall, and with a normal palate this contact is between the middle or muscular third of the palate and the sloping roof of the nasopharynx above the arch of the atlas (Calnan). Because cleft palates are abnormally short, repair that leads to normal speech must therefore not only repair the cleft so as to separate oral and nasal cavities, but also enable the palate to come in contact with the pharynx and completely separate these cavities. It is generally admitted that the most difficult problem is to establish satisfactory speech.

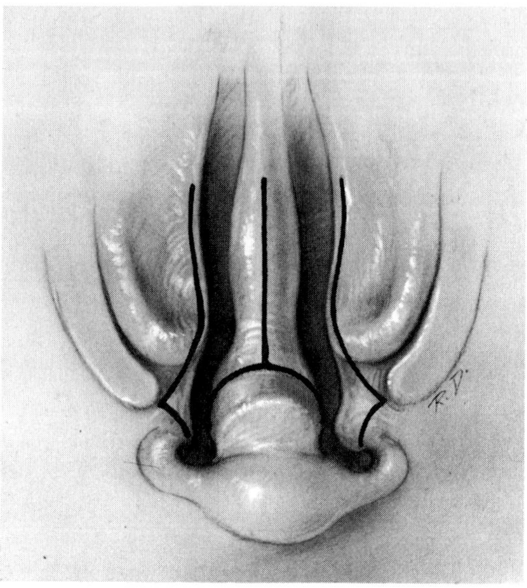

Fig. 7-11. Incisions for the Veau type of repair of a cleft of the alveolar process and anterior part of the palate. (Havens FZ: Arch Otolaryngol 48:9, 1948)

A Veau type of operation can be used to repair bilateral cleft lip and alveolar process (Figs. 7-11 and 7-12), but because of the too-short length of a cleft palate a "push-back" operation of the Wardill type has commonly been used to repair cleft of the palate proper (Fig. 7-13). This involves mobilizing the mucoperiosteum of the hard palate and moving it and the soft palate both to the midline and backward. Care is taken to leave the latter with as much mobility as possible. Most surgeons fracture the hamulus in order to free the tensor veli palatini (which passes around the hamulus into the front of the soft palate), and Ruding emphasized that the tensor should be completely removed from its attachment to the bony palate. He also advocated removal of the attachments of the levator veli palatini muscles, which in cleft palate parallel the cleft rather than run transversely to meet their fellow of the opposite side, and suturing the two together from before back in

Fig. 7-12. Suture of the flaps of mucous membrane. In this variation of the original Veau procedure, the raw lower surface is left uncovered. (Havens FZ: Arch Otolaryngol 48:9, 1948)

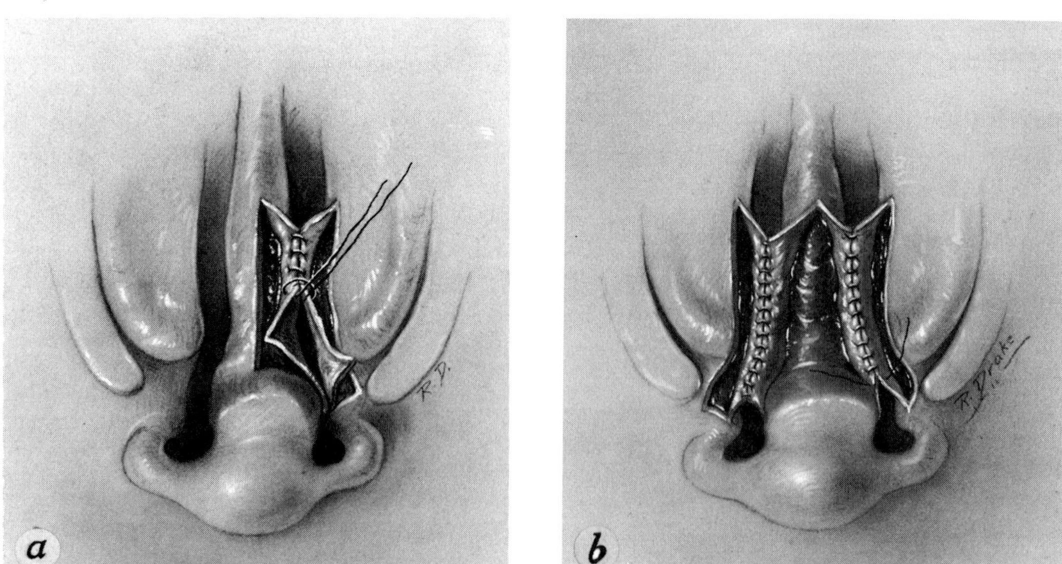

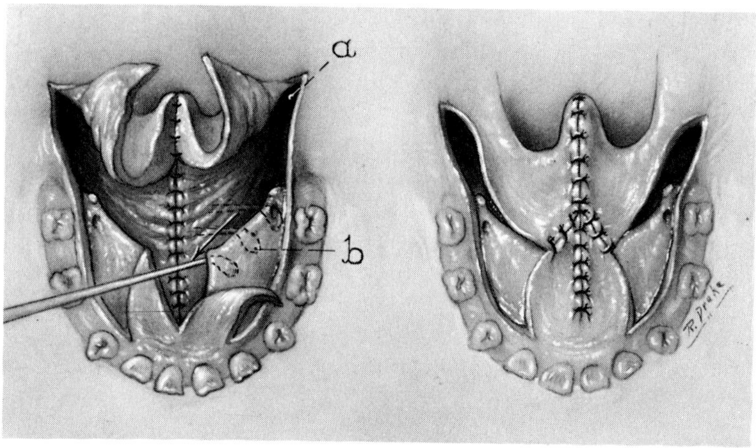

Fig. 7-13. *A*, the Wardill procedure for closure of a posterior defect of the palate. (*a*) calls attention to the posterior extension of the dissection, well behind the hamulus; (*b*), to the freeing of the soft tissue from the posterior and medial margins of the bony palate. (*B*), completed procedure. (Havens FZ: Arch Otolaryngol 48:9, 1948)

order to bring them into as transverse a position as possible, and swinging the greater palatine artery backward rather than sacrificing it. Many writers have emphasized the care with which the soft palate should be handled in order to minimize scarring of it and especially avoid damage to the levator veli palatini. Ritchie ('37), for instance, expressed the belief that mobility of the palate is probably more important than length in closure of the nasopharynx, and pointed out that apparently good anatomical results are not necessarily accompanied by good functional ones. Because the palate normally comes in contact with the adenoid tissue of the pharynx, removing that tissue decreases the chance of contact and should be avoided, unless absolutely necessary, in children with cleft palate repair.

Wardill ('37) advocated suturing the posterior pharyngeal wall in order to bring it forward for easier contact with a repaired cleft palate, and pharyngoplasties to accomplish the same purpose are now often used (Williams and Woolhouse; Moll and co-workers), sometimes at the time of repair of the cleft, sometimes as a secondary procedure when speech has proved unsatisfactory. Williams and Woolhouse described the single posterior pharyngeal flap, of the Rosenthal type, as giving best results in their hands. Moll and co-workers used pharyngoplasty not only in conjunction with cleft palate repair, but also as a primary procedure with congenitally short palate and other abnormalities; they found the results equally good regardless of the cause of improper closure, but less successful if carried out after age 15.

The age at which cleft palate should be repaired has been the subject of much disagreement, based primarily on growth studies of the palate in normal children, children with unrepaired cleft palate, and children with repaired cleft palate. Graber, for instance, suggested that surgical correction of cleft palate in early childhood may limit the growth potential of the maxillary denture, and Waldron in 1950 stated that the opinion of many surgeons and orthodontists experienced in this field was that surgical repair of the palate should be postponed until the end of the fifth year. This trend toward delayed closure has apparently largely disappeared, at least in the United States and Canada; Lewin found in a survey in 1964 that most complete clefts were repaired between the ages of 1 and 2 years and that few surgeons believed that repair of the hard palate interferes appreciably with maxillary growth; according to Subtelny it has been shown that repair of the soft palate is not incompatible with its proper growth. The tendency seems to be to emphasize repair of the soft palate before the age at which the child should learn to talk. Indeed, Crikelair and co-workers repaired the soft palate during early infancy and fitted the patient with an adjustable prosthesis through which they gradually brought the edges of the hard palate together. They did no surgical repair on the hard palate and repaired the cleft lip only after the hard palate was properly adjusted. Most surgeons repair unilateral cleft lip, preceding the repair of the palate, within the first 6 months after birth (Lewin).

Another technique said to give excellent results in repairing both bilateral and unilateral clefts is bone grafting of the alveolus and repair of the lip at an early age, about 5 to 6 months, followed by slow orthodontic adjustment (*e.g.*, Stellmach). The hard palate proper and the soft palate are then repaired after the alveolar arch has been brought into satisfactory alignment.

MUSCULATURE

The skeletal core of the soft palate is the palatine aponeurosis, a layer of connective tissue firmly attached anteriorly to the posterior margin of the hard palate and continuous laterally with the pharyngobasilar fascia (aponeurotic or submucous layer of the pharynx), although fading out posteriorly toward the free palatal margin. This aponeurosis is also continuous laterally with the tendon of the tensor veli palatini muscle, while upon its upper surface are arranged most of the other muscles associated with the soft palate. In a push-back operation upon the soft palate, the aponeurosis must be detached from the hard palate to allow the musculofibrous mass of the soft palate to be moved backward. This is done by mobilizing the periosteum of the hard palate, with which the aponeurosis is continuous; Havens found it useful to extend the dissection onto the medial surface of the medial pterygoid plate, after fracturing the hamulus.

The most superficial muscle fibers on the pharyngeal surface of the soft palate are mostly ones of the *palatopharyngeus* (pharyngopalatinus) muscle (Fig. 7-14) but include some sphincteric fibers (palatopharyngeal sphincter, see Fig. 8-11) associated with the superior pharyngeal constrictor and sometimes regarded as part of this, sometimes as part of the palatopharyngeus; other and more numerous fibers of the palatopharyngeus arise from the anterior or oral aspect of the palate. The anterior and posterior layers of this muscle enclose between them the musculus uvulae and the levator veli palatini and come together laterally to form a muscle bundle that runs downward into the pharynx (Fig. 7-15) to form the palatopharyngeal arch, also called the posterior pillar of the tonsil or fauces. This muscle is considered further in connection with the pharyngeal wall, to which it more properly belongs.

The *musculus uvulae* is situated immediately deep to the posterior (upper) layer of the palatopharyngeus and consists of longitudinally directed fibers that run downward into the uvula.

It is the fibers of the *levator veli palatini* that contribute the main muscular bulk of the soft palate; this muscle arises from the inferior surface of the petrous portion of the temporal bone in front of the carotid canal (and, according to Rohan and Turner, from a fascial sheet suspended here) and by a few fibers from the lower and medial part of the cartilage of the auditory (eustachian) tube, and runs downward, forward and medially, to curve into the soft palate below the ostium of the tube (where its bulge forms the torus levatorius); on entering the palate it lies between the musculus uvulae and the anterior layer of the palatopharyngeus. Most of the fibers from the two muscles meet in the midline, the more anterior ones attaching to the palatine apo-

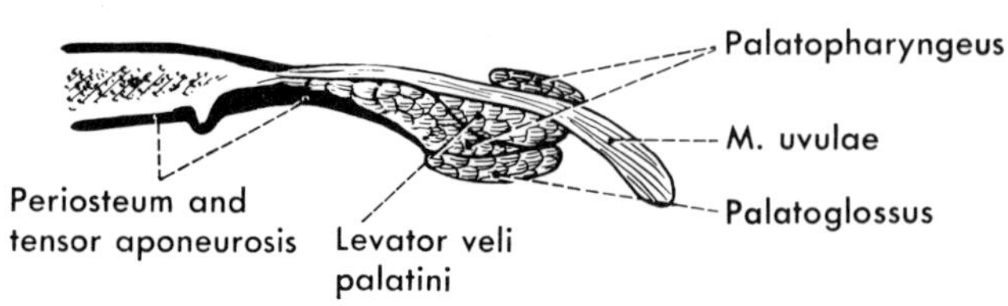

Fig. 7-14. The aponeuroticomuscular structure of the soft palate, in longitudinal section.

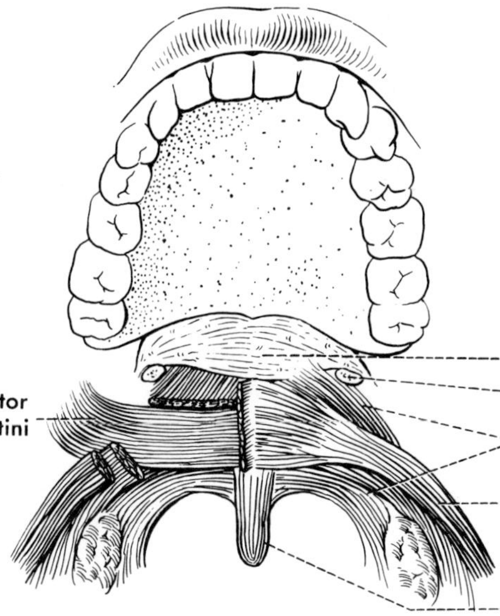

Fig. 7-15. Muscles of the soft palate from below. On the right side of the palate muscles are cut away to show the levator veli palatini.

neurosis. It is primarily this muscle that forms what Oldfield described as the "middle segment" of the soft palate.

The *tensor veli palatini* muscle, by its tendinous insertion into the anterior part of the palatal aponeurosis, contributes to the anterior, membranous, portion of the soft palate. This muscle lies slightly anterior and lateral to the levator veli palatini; about 75% of it originates from the lateral wall of the cartilaginous portion of the auditory tube (Ross), while the remainder has a bony origin that extends from the sphenoid spine to the scaphoid fossa (at the base of the medial pterygoid lamina). From this relatively broad origin it tapers as it passes downward between the medial pterygoid muscle and the medial pterygoid plate; its tendon, sufficiently broad to occupy most of the length of the hamulus, passes lateral to the hamulus and then turns at almost a right angle in front of that (being here provided with a bursa) to attach to the posterior edge of the bony palate, the inferior surface as far forward as the palatine crest, and its fellow tendon of the other side. The tendons of these muscles, therefore, form the main element in the aponeurotic anterior portion of the soft palate.

In repair of cleft soft palate, it is the tensor veli palatini that resists movement both toward the midline and backward; unlike the levators, which simply curve into the soft palate and can therefore be fairly readily approximated, the tensors of the two sides are held apart by the two hamuli around which they run (Fig. 7-16). Many surgeons have emphasized that these muscles may make closure of the palate exceedingly difficult or may subsequently, by their contraction, pull out the sutures; the tendons of the muscles may of course be cut, but a better method is to break the hamuli. When this is done, the tensors of the palate then take a course into the palate essentially similar to that taken by the levators; they thus both lose part of their tensing powers on the palate and possibly contribute to the elevation of this. The same result can be obtained by lifting the tendon off the hamulus instead of breaking the latter. However, Ruding regarded the tensor as being of no value as a levator after repair of cleft palate; he said that in cleft palate the only nonfibrotic part of the muscle is that which arises from the auditory tube (functioning to open the tube) and that the tendon is then small and rounded and sunk into the groove at the

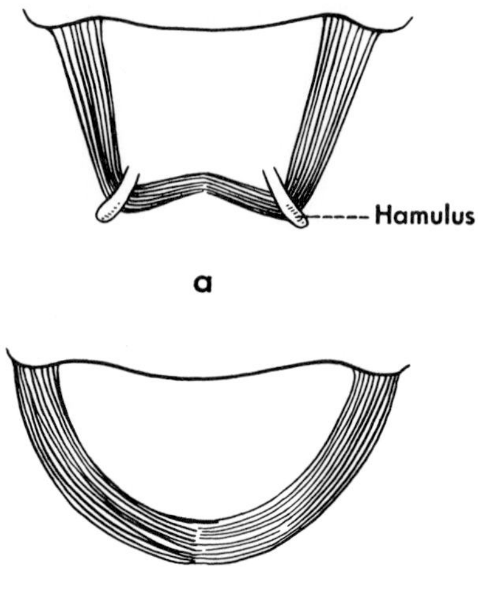

Fig. 7-16. The difference in the pull of the tensor veli palatini (*a*) and the levator (*b*). Note that by breaking the hamuli or removing the tendons from around them, the tensors can be released so that they have a course similar to that of the levators.

base of the hamulus instead of spread over it. Further, Ross reported that use of the tensor in repair of the palate in dogs resulted in the auditory tube's remaining partially closed; he advised, therefore, that the insertion of the muscle lateral to the greater palatine artery be preserved in order to allow the tensor to continue its function of helping to open the tube.

The most superficial muscle fibers on the oral surface of the soft palate form the small *palatoglossus* muscle. This curves downward to the lateral margin of the tongue anterolateral to the tonsil, raising the mucous membrane to form the palatoglossal arch (anterior pillar of the tonsil).

Actions

Acting upon the soft palate, the palatopharyngeus and the palatoglossus muscles draw this downward and hence narrow the faucial isthmus. The musculus uvulae draws the uvula upward and forward. The levator has the action implied by its name, and by raising and retracting the soft palate brings it in contact with the posterior wall of the pharynx; the tensor veli palatini tenses the soft palate and also somewhat depresses it. The tensor is usually regarded as being the primary opener of the auditory tube, and the levator also is sometimes said to have this function.

Because of the soft palate's action in closing the pharynx in swallowing, paralysis of it (Fig. 7-17) is particularly disabling; it allows food, and especially liquids, to pass upward into the nasal cavity rather than descend to the esophagus. Misurya reported success in restoring function to the paralyzed palate through transferring the tensors so that they became elevators of the soft palate.

NERVES AND VESSELS

Motor Nerves

The tensor veli palatini muscle receives, on its lateral surface where it is adjacent to the mandibular nerve, a branch from this nerve

Fig. 7-17. Paralysis of the soft palate and pharynx of the right side.

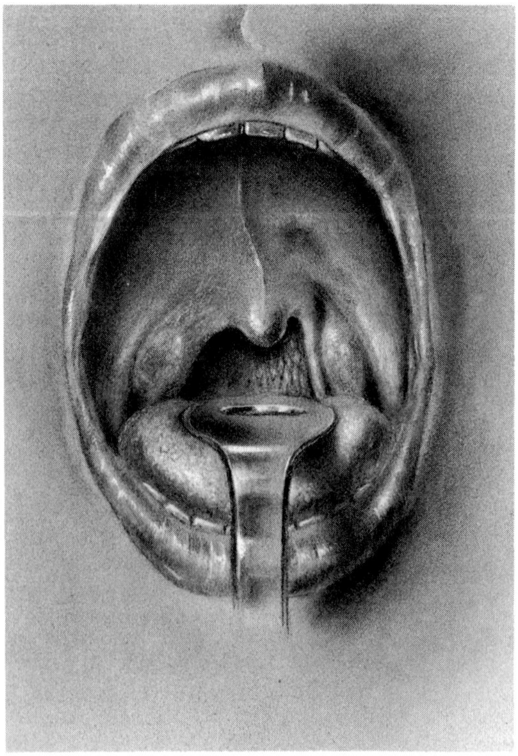

by way of the otic ganglion; with the exception of this muscle, the remaining muscles of the soft palate are innervated by an ascending branch, not readily demonstrable, from the pharyngeal plexus (Sprague, and others). The motor fibers in the pharyngeal plexus are derived from the vagus nerve, probably reaching it in large part through the so-called cranial or bulbar rootlets of the accessory nerve. The once widely held concept that the palatine muscles are innervated by the facial nerve (through the greater petrosal and lesser palatine nerves) is no longer tenable in view of the work of Rich and others and many clinical observations, nor have voluntary motor fibers ever been demonstrated in the greater petrosal nerve.

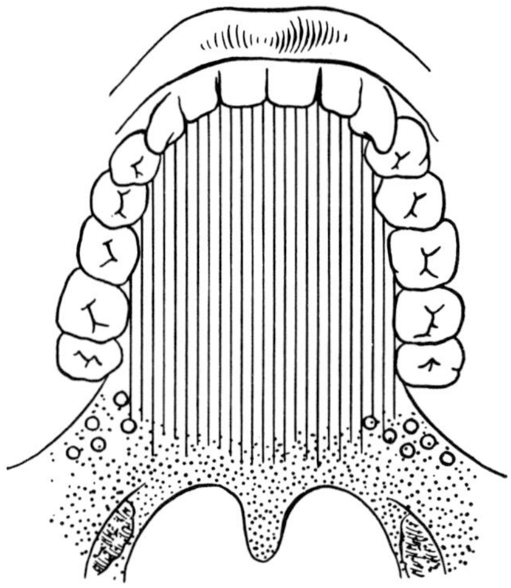

Fig. 7-18. Probable distribution of the fifth nerve (*lines*), seventh nerve (*circles*), and ninth nerve (*stipple*) on the palate.

Sensory Nerves

The sensory nerves of the palate are usually described as branches of the pterygopalatine (sphenopalatine) ganglion and therefore might consist of (1) fibers from the facial nerve (through the greater petrosal) that traverse the ganglion, (2) fibers of the maxillary nerve, or (3) fibers of posterior cranial or upper spinal nerves, which are said to reach the pterygopalatine ganglion by way of the nerve of the pterygoid canal (Vidian nerve). Actually, the details of the sensory innervation of the palate have apparently not been satisfactorily determined and may indeed vary somewhat among individuals, but the major innervation is from the trigeminal nerve (Fig. 7-18), and the palate may be a "trigger zone" in trigeminal neuralgia.

The greater palatine nerves are often described as consisting of both maxillary and facial nerve fibers, and the other palatine nerves as largely, at least, composed of facial nerve fibers. However, Cushing ('04) described anesthesia of the entire oral surface of the palate, including the uvula, and of at least the major part of the upper and posterior surface, as resulting from destruction of the fifth nerve only; Dandy described anesthesia of the uvula, of a narrow strip on the anterior surface of the soft palate, and of most of the posterior surface of the soft palate following section of the ninth nerve only; and Hunt described eruptions on the soft palate, just in front of the palatoglossal arch, associated with herpes of the geniculate ganglion. The sensory fibers are undoubtedly accompanied by secretory and vasomotor fibers from the pterygopalatine ganglion.

The pterygopalatine ganglion and its branches have been discussed in some detail in connection with the nose. There are three palatine branches of the ganglion (Fig. 7-19). Of these, the *greater* (anterior) *palatine nerve* is by far the largest; it passes downward through the greater palatine (pterygopalatine) canal in the lateral nasal wall, in company with the chief blood supply of the palate, the greater palatine artery. While in this canal the nerve supplies twigs to the lower posterior part of the nasal cavity. The nerve and its accompanying vessels emerge on the hard palate through the greater palatine foramen (posterior palatine foramen) at approximately the level of the third upper molar tooth (Fig. 7-20). The nerve then breaks up into branches that are distributed to the hard palate and to the palatine gingivae. A small area immediately behind the incisor teeth is supplied by terminal branches of the nasopalatine nerve, also a branch from the pterygopalatine ganglion.

The *lesser palatine nerves,* sometimes divided

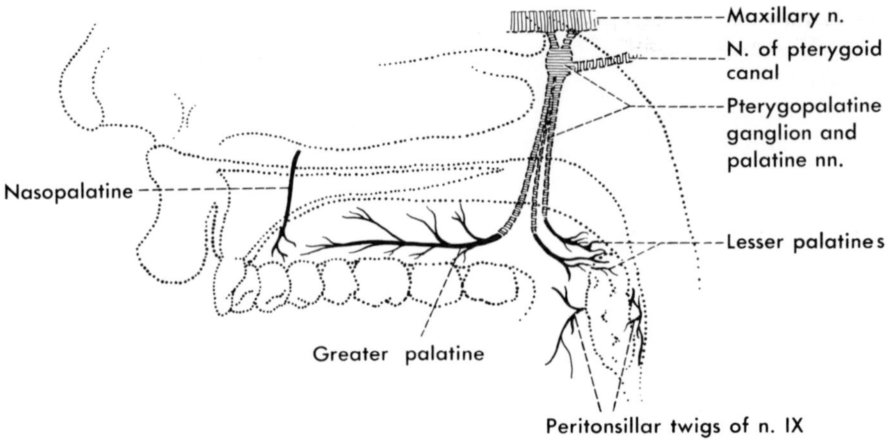

Fig. 7-19. Nerves of the palate from the medial aspect. Except for ascending twigs from the tonsillar plexus, all of them are so-called branches of the pterygopalatine ganglion.

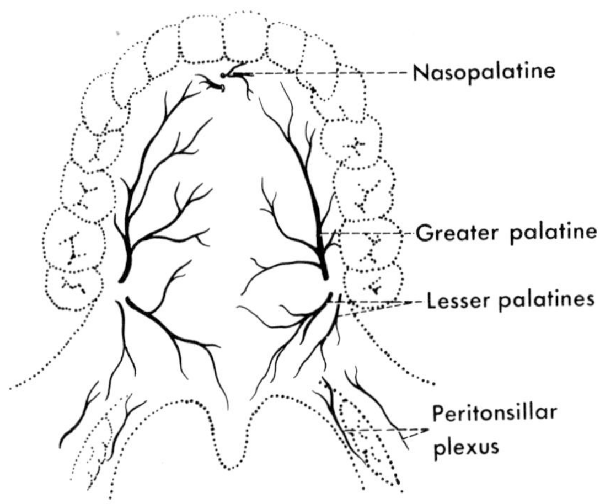

Fig. 7-20. Nerves of the hard and soft palates.

into posterior or small and middle or external, pass downward parallel to the greater and emerge through lesser palatine foramina situated almost at the posterior edge of the hard palate, behind the greater palatine foramen; they turn backward and downward to be distributed to the soft palate and to the adjacent area about the upper pole of the tonsil. They are often described as consisting largely of facial nerve fibers, both for general sensation and for the taste buds on the soft palate. The ninth nerve fibers that, according to Dandy, are distributed to the palate could reach this by way of connections between the tympanic plexus or lesser petrosal nerve and the greater, through the tonsillar branch of the ninth, or perhaps by other routes.

Arteries

The palate has an abundant arterial supply, its chief artery being the descending palatine, from the maxillary (Fig. 7-21). The main branch of the descending palatine, the *greater* (anterior) *palatine,* emerges on the hard palate with the greater palatine nerve and runs forward on the lateral aspect of the hard palate at the junction of this and the alveolar process. The descending palatine artery also gives off *lesser palatine* arteries, which are distributed to the soft palate. The greater palatine artery may be a source of serious hemorrhage in repair of cleft palate or may, as it emerges onto the hard palate, prevent proper mobilization of the soft tissues; probably the majority of surgeons purposely sacrifice it, if expedient, at the time of operation, but Rud-ing said it should be carried posteriorly with the mucoperiosteal flap, and some surgeons have advocated its preliminary ligation with sufficient lapse of time to allow the development of collateral circulation before repair is undertaken. This collateral circulation is provided both anteriorly and posteriorly, but especially from the latter position into the soft palate. Anteriorly, the greater palatine artery passes upward through an incisive (anterior palatine) canal to anastomose in the nasal cavity with septal branches of the sphenopalatine artery; posteriorly, the lesser palatine arteries anastomose on the soft palate with the palatine branch of the ascending pharyngeal artery (from the external carotid), the ascending palatine branch of the facial, and twigs from the tonsillar branch of the dorsalis linguae.

To recapitulate, the palate receives its blood supply from the descending palatine (a branch of the maxillary artery) and from the ascending pharyngeal, facial, and lingual arteries. The branches of the descending palatine enter the palate from above, as does the ascending palatine, which passes above the upper border of the superior pharyngeal constrictor; the remainder enter it from below and laterally. The anastomoses of these vessels in the nasal cavity and on the soft palate apparently suffice to supply the palate after ligation of the greater palatine artery.

Veins and Lymphatics

The veins of the palate drain both into the pterygoid plexus and, in the case of the soft palate, into the pharyngeal plexus; the latter in turn communicates with the pterygoid plexus and drains also into the internal jugular vein. A particularly large vein, the external palatine, leaves the soft palate and runs downward and forward in the bed of the tonsil before it penetrates the superior pharyngeal constrictor to end, variably, in the facial, a pharyngeal vein, or other vein in the neighborhood.

The lymphatics from the hard palate leave this at the level of the last molar tooth to pass downward and outward and end in the upper deep cervical nodes; those of the soft palate, which are especially numerous in the uvula, also pass largely into the upper nodes of the deep cervical chain, but those from the upper or posterior surface of the soft palate run also in part with pharyngeal lymphatics to end in the retropharyngeal (lateral pharyngeal) nodes.

The Mandibular Region

Many of the superficial relationships of the mandible have already been discussed in connection with the face; here are discussed primarily the mandible itself and its musculature and nerves and the structures lying deeper to it, in the infratemporal fossa.

Although the mandible is not developed primarily from the cartilage of the mandibular arch (Meckel's cartilage), since it is largely a derm bone, it does develop from the mesenchyme around Meckel's cartilage and therefore from the same source as the cartilage.

Fig. 7-21. Schema of the arteries of the palate. The ascending palatine, after supplying the tonsil, actually goes much higher and then loops downward to the soft palate.

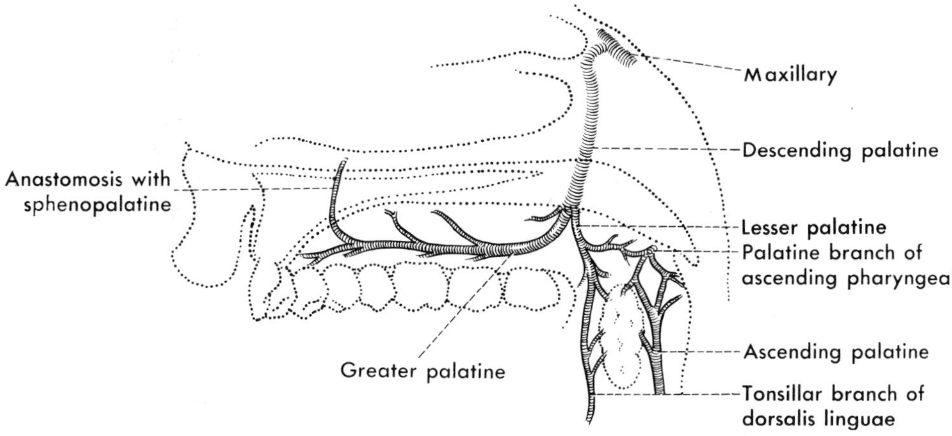

The musculature of the mandible is likewise developed from this arch; hence the muscles most intimately connected with the lower jaw, including the mylohyoid and the anterior belly of the digastric in the suprahyoid region, are innervated by the mandibular branch of the trigeminal nerve. Although Meckel's cartilage disappears from within the mandible, a fibrous remains of it stretches from the region of the mandibular foramen to that of the tympanosquamous fissure as the sphenomandibular ligament, and the dorsal end of Meckel's cartilage projects into the middle ear cavity to give rise to most of the malleus and incus (hence the innervation of the tensor tympani by the mandibular nerve).

Mandible

The mandible consists of the tooth-bearing body and the ramus that extends upward from the angle (Fig. 7-22). The ramus, including the angle, is covered externally by the masseter muscle (which in turn is crossed by the facial nerve and parotid duct), and internally, between it and the medial pterygoid muscle, are the inferior alveolar and lingual nerves. Overlapping the posterior border of the ramus is the parotid gland, and within this, paralleling this border, is the upper portion of the external carotid artery; the superficial temporal branch of this artery emerges from the parotid gland behind the temporomandibular joint, while its maxillary (internal maxillary) branch runs transversely deep to the ramus. Just in front of the insertion of the masseter muscle the facial artery and its accompanying vein cross the lower border of the body of the mandible, which, except for the thin facial muscles, is subcutaneous. Inferiorly and medially, the angle and posterior part of the body of the mandible are related to the submandibular (submaxillary) gland, and medially the anterior part of the mandible is adjacent to the sublingual gland and the nerves, vessels, and muscles of the tongue.

On the outer surface of the angle (the tip of which is the gonion) there may be a masseteric tuberosity, and on its inner surface a pterygoid one (raised by the insertion of the medial pterygoid). Above, the ramus ends posteriorly in a condylar process and anteriorly in a coronoid one, separated by the mandibular notch. The condylar process consists of neck and head (condyle); anteriorly on the neck is the pterygoid fovea for the lateral pterygoid muscle. The thinner and sharper coronoid process receives the insertion of the temporal muscle. On the inner surface of the

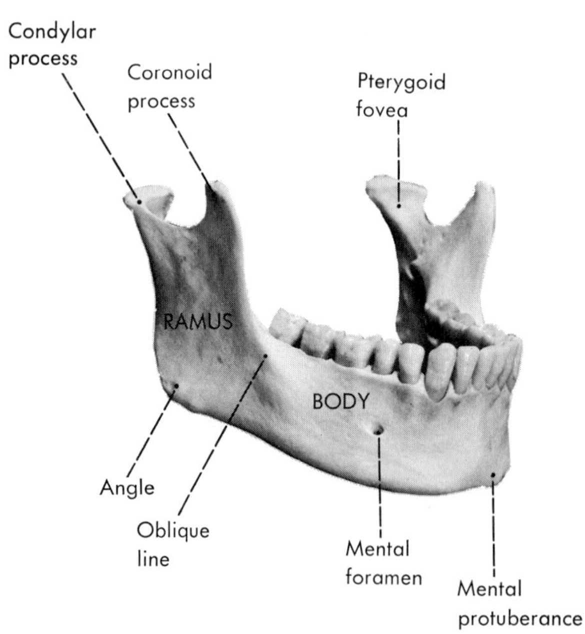

Fig. 7-22. Anterolateral view of the mandible.

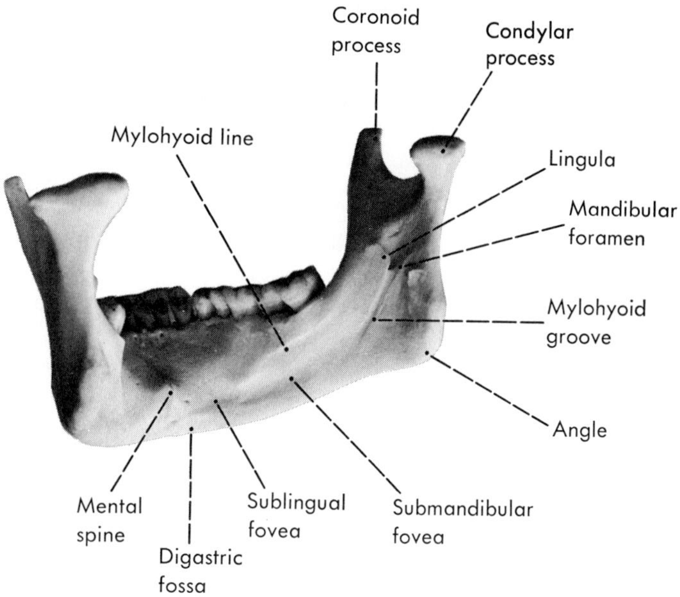

Fig. 7-23. Inner surface of the mandible.

ramus, about its middle, is the mandibular foramen through which the inferior alveolar nerves and vessels enter the mandibular canal, and overhanging it above and anteriorly is a sharp ledge of bone, the lingula (Fig. 7-23). The mylohyoid groove, for the mylohyoid nerve and vessels, leads obliquely downward and forward from the mandibular foramen; this is sometimes covered, according to Ossenberg, by a thin layer of bone (the mylohyoid bridge), more commonly in American Indians than in other groups. This bone is said to be another remnant of Meckel's cartilage.

Externally on the body of the mandible the oblique line runs downward and forward from the anterior border of the ramus. The mental foramen, for the exit of the mental nerve and vessels from the mandibular canal, lies at about the level of the interspace between the two premolars and, in an adult with full dentition, about midway in the height of the body. Anteriorly in the midline the rounded base of the body swells out to form the mental protuberance, surmounted by the smaller mental tubercle.

The inner surface of the body bears several markings indicating its soft tissue relations. On each side of the midline, or often fused together in the midline, is a mental spine (genial tubercle) that may be divided into upper and lower parts for the attachments of the genioglossus and geniohyoid muscles, respectively; on each side of this is a sublingual fovea for the sublingual gland, and below these on the base are the digastric fossae for the anterior bellies of the digastric. Behind the sublingual fovea the mylohyoid line, the attachment of the mylohyoid muscle, runs upward and backward, and below this line is the submandibular (submaxillary) fovea for the gland of the same name.

The body of the mandible is supplied by the inferior alveolar artery, supplemented anteriorly by vessels in the geniohyoid, genioglossus, and anterior belly of the digastric (Castelli). The lower part of the ramus, down to the angle, is supplied both by the inferior alveolar and by vessels in the muscles attaching to it; the condylar process is supplied by vessels from the joint capsule and the lateral pterygoid muscle, while the coronoid process receives its blood supply from vessels in the temporal muscle.

The *teeth* of the mandible correspond in structure (Fig. 7-24), name, and number to those of the upper jaw, and the same names (dental alveoli, interalveolar septa, and the

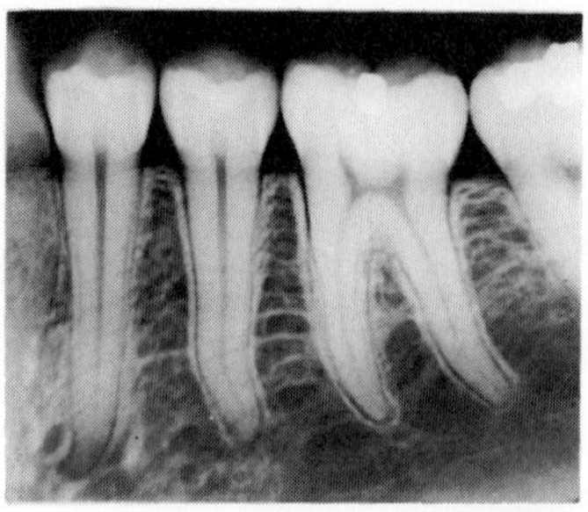

Fig. 7-24. Roentgenogram of the two premolar teeth and the first molar tooth of the lower jaw. The root canals within the teeth show clearly, as does the "lamina dura," the thin layer of dense bone that lines the alveolus. The *dark line* between the latter and the tooth is the periodontium (periodontal membrane or ligament). (Stafne EC: Oral Roentgenographic Diagnosis, 3rd ed. Philadelphia, WB Saunders, 1969)

like) apply to the alveolar part of the mandible. The more anterior teeth, like all the teeth of the upper jaw, are set closer to the labial than the lingual surface; the more posterior teeth are usually set closer to the lingual surface, but since they slant inward the apices of their roots are closer to the buccal surface of the bone than are their necks.

Although the mandible obviously varies considerably in form and proportions, anomalies of it are apparently rare. Congenital absence of the ramus has been reported (Kazanjian); Hrdlička described a number of cases in which the condylar head was more or less doubled.

Fractures of the mandible are relatively common, and the most frequent of these lie between the premolar (bicuspid) teeth and the ramus, while fractures of the ramus and the condyle occur less often (Moorehead). In these cases the long (anterior) fragment is always displaced downward (by the pull of the suprahyoid muscles), while the short or posterior fragment is displaced upward by the muscles that close the jaw.

Various aspects of the surgery of the mandible, including its plastic surgery, have been discussed by Byars. He also pointed out that the mandibular periosteum possesses a very considerable osteogenic capacity, so that relatively large defects may be healed if the periosteum can be left in place. Defects may, however, require a bone graft in order to restore the necessary strength to the mandible, and removal of a large amount of bone often makes the wearing of a denture impossible. Millard reported a satisfactory cosmetic and functional replacement of one entire side of the body using skin of the forehead as the bed for the bone graft.

Condylectomy produces changes in the mandible and in other parts of the skull. They have been attributed in part to loss of a growth site in the mandible, but according to Sarnat and Muchnik they are due entirely to loss of the normal physiology of the temporomandibular joint, for identical changes occur in the adult monkey after condylectomy.

Temporomandibular Joint

The parotid gland not only overlaps the mandible and the masseter muscle externally, but also extends deeply behind the mandible between its ramus and the external acoustic meatus, the mastoid process, and the posterior belly of the digastric muscle; wide opening of the jaw as a strictly hinge movement is therefore impeded by the lack of space behind the ramus of the mandible, and the temporomandibular articulation is so formed that it allows for protraction of the jaw concomitant with opening.

The temporomandibular joint (Fig. 7-25) is peculiar, for a synovial one, in that the articular surfaces are fibrocartilage or fibrous tissue instead of hyaline cartilage. According to Moffett and co-workers, the joint regularly shows evidence of remodeling, resorption of articular tissue and bone in one area and addition in another, which is accentuated by the change in bite as teeth are lost.

The *articular capsule* is attached to the temporal bone around the edges of the mandibular (glenoid) fossa and the articular tubercle in front of that fossa and to the neck of the mandible. It is thin and loose between the temporal bone and the articular disk (meniscus) that lies within the joint, but between the disk and the mandible it is reinforced medially and laterally by strong tight fibers (Sicher)—an arrangement that provides for simultaneous sliding movements of the disk

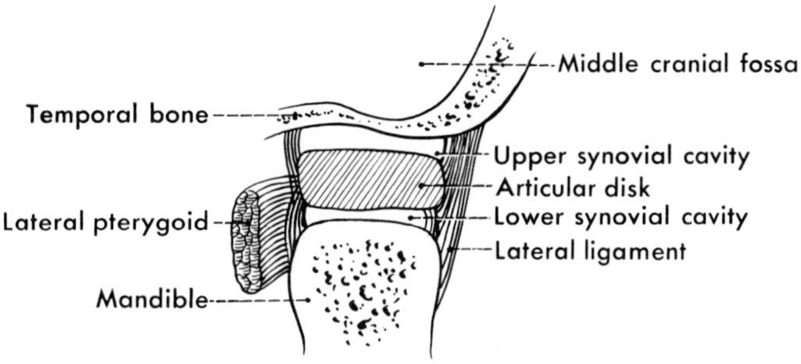

Fig. 7-25. Schema of the temporomandibular joint in a frontal section.

and the mandible, but hinge movement only between the two. Laterally, the capsule is reinforced by the *lateral* (temporomandibular) *ligament,* attached above to the articular tubercle and the adjacent lower margin of the zygomatic arch and running downward and backward to the lateral and posterior aspect of the neck of the mandible.

The other two ligaments of the mandible, the stylomandibular and the sphenomandibular, are not particularly closely related to the capsule of the joint, but they do influence mandibular movement. The *stylomandibular ligament,* a part of the cervical fascia between the parotid and submandibular glands, extends between the styloid process and the lower part of the posterior border of the ramus of the mandible; the *sphenomandibular ligament* (representing fibrous remains of a part of the cartilage—Meckel's—of the mandibular arch) runs downward and forward from the sphenoid spine and the tympanosquamous fissure to the lingula and the adjacent inner surface of the ramus. The maxillary artery and its middle meningeal and inferior alveolar branches pass between the ligament and the mandible, as do the auriculotemporal and inferior alveolar nerves. According to Lord, both the stylohyoid and sphenomandibular ligaments act as pivots through which, by protracting the jaw, the lateral pterygoid muscle also actively opens the mouth.

The *articular disk* (meniscus) has been described both as fibrocartilage and as pure fibrous connective tissue. Many surgeons have stated that they were sometimes unable to find cartilage cells in disks that had been removed; Moffett and co-workers described the disk as being, like the articular surfaces of the mandible and temporal bone, a mixture of fibrous connective tissue and fibrocartilage, the cartilage cells occurring only where there was particular pressure.

The disk apparently varies in shape in different jaws (Fig. 7-26), for the temporomandibular joint seems to be particularly plastic in its development (Angel). In general, however, it is thick where it intervenes between the upper surface of the head of the mandible and the mandibular fossa; it is thinnest where it is in contact with the posterior slope of the articular tubercle and may be perforated here. More important are its attachments; anteriorly, medially, and laterally it blends with the articular capsule, but posteriorly it is united to this capsule by thin connective tissue that is covered by synovial membrane and contains nerves and vessels (Sicher; Moffett and co-workers).

Subluxation of the jaw is not rare and occurs most frequently in young women (Doherty and Doherty). It may be caused by a relaxed ligament or a loose disk; constant clicking or snapping may occur when the jaw is moved. Complete dislocation involves a forward slipping of the condyle and disk to the front of the articular tubercle, where it is then held by spasm of the muscles. Silver and Simon ('61) regarded removal of the disk and section of the upper part of the lateral pterygoid muscle as the most satisfactory treatment for recurrent dislocation; Morris described plication of the joint capsule or limiting forward movement by a strip of temporal fascia as being often satisfactory for the repair of snapping jaw.

Tears are not of the meniscus itself, but of

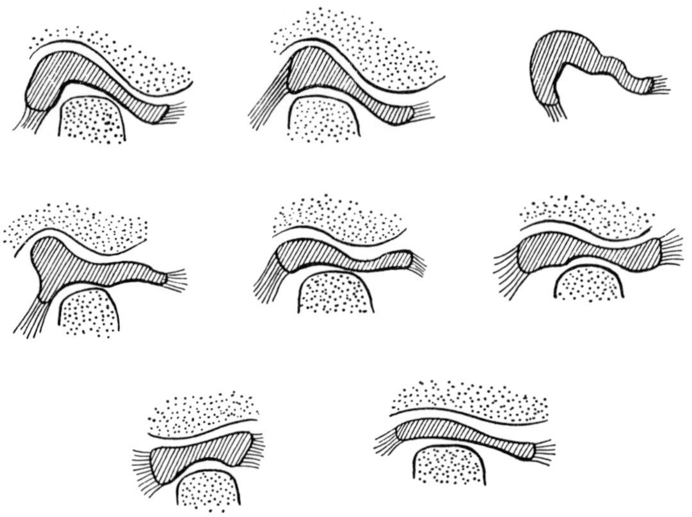

Fig. 7-26. Variations in the shape of the articular disk (meniscus) of the temporomandibular joint. (After Lubosch W: Anat Anz 29:417, 1906)

the loose connective tissue that attaches the disk posteriorly to the capsule (Wakely, and others); this may be a cause of clicking or snapping jaw but may also lead to forward displacement of the disk, so that it blocks the forward movement of the mandible necessary to opening the mouth and thus locks the jaw. Removal of the disk in such cases has been reported to produce an excellent functioning joint (Wakely; Silver and associates, '56); later, Silver and Simon ('63) reported that in only a few patients was condylectomy necessary to relieve persistent pain after meniscotomy, and they found no postoperative arthritic changes in patients followed for as long as 14 years. Georgiade used autogenous dermal grafts to replace disks that had to be removed. Sprinz concluded, from experiments on rabbits, that with condylectomy a meniscus that shows any damage at all should be removed, but a completely undamaged one should be left in place. However, Silver and co-workers ('73) stated that condylectomy with removal of the disk is the only cure for an arthritic temporomandibular joint. In operations upon the joint, the relationships of the facial nerve and superficial temporal vessels are particularly important.

The pressure produced by the closers of the jaw is normally exerted almost entirely on the molar teeth, and the temporomandibular joint receives little pressure; the thinness of the bone of the mandibular fossa, the fact that this may even be lined by synovial membrane containing nerves and vessels (Moffett and co-workers), and the sparseness of fibrocartilaginous tissue in the joint all indicate that it is not designed to bear pressure. Thus *malocclusion,* or any factor that leads to trismus (spastic contraction of the muscles), may so upset the normal relations in the joint that pressure is exerted on the sensitive synovial membrane. (Sicher noted that this can lead to degenerative changes in the membrane and, as a consequence of loss of synovial fluid, degenerative changes in the joint as a whole.) The pain produced by trismus is localized only in part in the joint, usually also in the tongue, the side of the head, and other places, and may be associated with disturbances of hearing and balance (Costen's syndrome). Costen explained the distribution of pain as being perhaps caused by pressure on the auriculotemporal and chorda tympani nerves, but pain in the tongue has not been reported from stimulation of the chorda tympani. Sicher regarded the pain as originating from spastic muscles, saying that pain caused by spasm of the temporal, masseter, pterygoid, and digastric and geniohyoid muscles is localized by the patient in the ear and temporal region, the cheek, the throat, and the tongue, respectively. Seaver explained the disturbance in functions of the ear as resulting from the spread of the spasticity to muscles of the palate and ear innervated by the mandibular nerve.

The *nerve supply* to the temporomandibular joint is largely from the auriculotemporal nerve, which runs below and behind it, but the masseteric nerve is said to send a twig to it (to the articular capsule rather than into the joint?). According to Tanasesco the *lymphatics* from the joint empty into three groups of nodes, namely, the superficial parotid (preauricular), deep parotid (intraparotid), and the subdigastric (part of the deep cervical group), but most of the lymph from the joint is said to reach the deep cervical group.

MUSCULATURE

The musculature most intimately concerned with the mandible and its movements is a group of four muscles, namely, the masseter, the temporal, and the two pterygoids, lateral and medial (external and internal). In addition, some of the suprahyoid muscles are attached to the mandible and assist in opening it, although not commonly classed as mandibular muscles.

As has already been described, the superficial layer of the cervical fascia splits at the lower margin of the mandible to follow the external surface of the masseter and the internal surface of the medial pterygoid muscle, so that the masseter, the pterygoids, and a portion of the temporalis are enclosed within a fascial compartment, usually referred to as the masticator space. The buccal fat pad, limited superficially by fascia passing across it from the masseter to the buccinator muscle and the maxilla, extends backward between the masseter and the buccinator, and then between the ramus of the mandible and the insertion of the temporal muscle laterally and the medial pterygoid medially, to become continuous with the looser connective tissue and fat between the muscles in the masticator space. Through this looser connective tissue run the branches of the mandibular nerve and of the maxillary artery, and in it also lies the pterygoid plexus of veins.

Masseter

The masseter muscle rather completely covers the ramus of the mandible, arising by a superficial part from the lower border of about the anterior two thirds of the zygomatic arch and by a deep part from the inner surface of the arch and the lower surface of the posterior third; some deep fibers may arise from the heavy fascia over the temporal muscle or may be intermingled with that muscle. The superficial part of the muscle slants backward and downward, largely covering the deep part (Fig. 7-27), and the muscle as a whole inserts

Fig. 7-27. The (*A*), masseter and (*B*), temporal muscles.

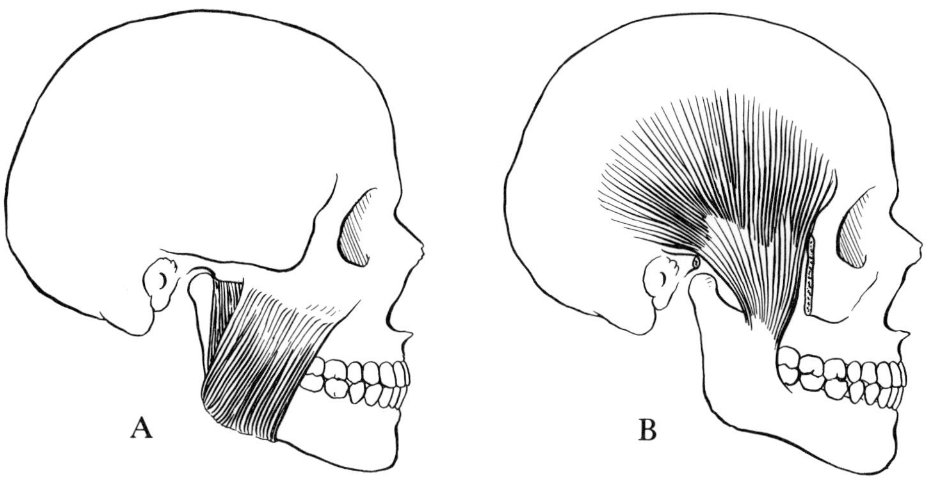

into most of the lateral surface of the ramus except the condylar process, from the lower part of the coronoid process to the base of the angle. Its *nerve supply,* the masseteric nerve from the mandibular, enters the deep surface of the muscle by passing through the mandibular notch in company with the masseteric artery from the maxillary and a corresponding vein. In sectioning the ramus of the mandible, in order to reset it and overcome undue protraction or retraction of the mandible, the masseter muscle may be reflected upward from its insertion as far as the mandibular notch, but not above this point without injury to its nerve supply. In any operations in this area, it should be remembered that the masseter is crossed by both the parotid duct and by branches of the facial nerve.

The masseter is sometimes so enlarged, unilaterally or bilaterally, that the enlargement has been mistaken for a tumor; Gurney, reporting cases of hypertrophy of the muscle, concluded it was due to a nervous habit of grating the teeth or clenching the jaw during emotional stress.

Temporal

The temporal muscle has an extensive origin from the temporal fossa on the side of the skull and from the covering temporal fascia; it inserts into the upper and anterior borders of the coronoid process laterally, but most of its insertion is on the medial side of the ramus, including the entire medial side of the coronoid process and an anterior part of the ramus extending approximately to the level of the occlusal surfaces of the lower molar teeth. The lower part of this insertion is largely tendinous and is the part that is forced against the maxillary tuber when the mouth is opened wide, making difficult the injection of the posterior superior alveolar nerves.

The *nerves* to the temporal muscle are typically two, an anterior and a posterior deep temporal, but there may be only one or there may be three. These nerves pass above the upper head of origin of the lateral pterygoid muscle, between this and the bone forming the margin of the infratemporal fossa; or the anterior nerve, or both, emerge between the two heads of the lateral pterygoid muscle. They run upward between temporal muscle and bone before entering the muscle.

The temporal nerves are usually accompanied by corresponding deep temporal arteries, but these again vary in number; the arteries, from the maxillary (internal maxillary), pass across the outer surface of the upper head of the lateral pterygoid to join the nerves on the deep surface of the temporal muscle; they supply both this and the adjacent bone. In addition, the smaller middle temporal artery arising from the superficial temporal close to the level of the zygomatic arch pierces the temporal fascia and runs between muscle and bone, supplying both and anastomosing with the deep temporals.

Lateral Pterygoid

In contrast to the two preceding muscles, the lateral (external) pterygoid muscle lies almost in the horizontal plane. It arises by two heads, a superior and an inferior (Fig. 7-28). The superior head, the smaller, arises from the infratemporal surface of the greater wing of the sphenoid (roof of the infratemporal fossa) and inclines somewhat downward as it proceeds posteriorly to its insertion; the inferior head arises from much of the lateral surface of the lateral pterygoid plate, and therefore many of its fibers run upward. Both heads are largely hidden by the dense pterygoid plexus of veins, which passes between them and around the lower border to lie also deep to the muscle. They are usually innervated by branches from the buccal nerve as this passes between them.

The two heads blend at their insertion, but the superior head inserts primarily into the articular capsule and disk, the inferior into the pterygoid fovea on the neck. According to Harpman and Woollard a portion of the tendon of the lateral pterygoid muscle in the human embryo passes through the temporomandibular joint to continue to the malleus, and later gives rise to the disk; hence the insertion of the muscle into the disk is readily understandable.

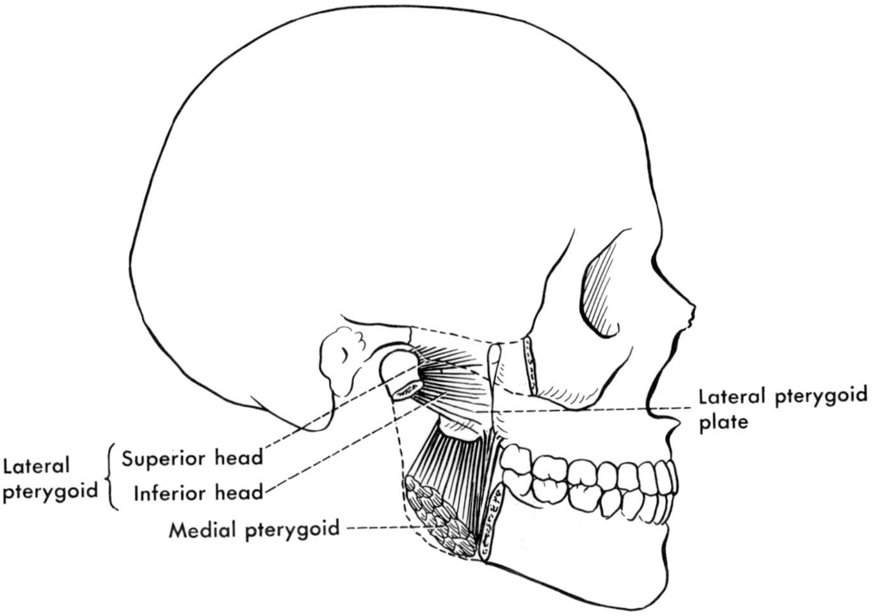

Fig. 7-28. Lateral and medial pterygoid muscles after removal of part of the mandible.

Relations of Nerves and Vessels

The mandibular nerve emerges from the foramen ovale just deep to the lateral pterygoid muscle, and its branches are therefore intimately related to this muscle: the masseteric and often one or both deep temporal nerves pass above the superior head; the auriculotemporal nerve passes backward deep to the superior head and medial to the temporomandibular articulation; the buccal nerve emerges from between the two heads of the muscle to cross the lower head and run toward the cheek; the lingual and inferior alveolar nerves pass deep to both heads to emerge beneath the lower border of the inferior head and run across the external surface of the medial pterygoid muscle (Fig. 7-29).

The main stem of the maxillary artery, after it has passed forward medial to the mandible, also comes in intimate contact with the lateral pterygoid muscle; it may run superficial to both heads of this muscle, or may pass deep to the inferior head and emerge through the cleft between the two heads, but in any case it then runs forward approximately in the groove between the heads to reach the posterolateral aspect of the maxilla and the pterygopalatine fossa. As already mentioned, the lateral pterygoid muscle is almost surrounded by the pterygoid plexus.

Medial Pterygoid

The medial (internal) pterygoid muscle, the deepest of the four major paired muscles of the mandible, arises mostly from the internal surface of the lateral plate or lamina of the pterygoid process, with some origin from the pyramidal process of the palatine bone (which unites the two laminae) and from the maxillary tuber. The small bundle of fibers arising from the pyramidal process and from the tuber partly covers lower anterior fibers of the lateral pterygoid. The muscle runs downward, laterally, and posteriorly, to insert into the medial surface of the ramus between the mylohyoid sulcus and the posterior and inferior borders of the angle; it thus forms a sling with the masseter on the outside of the jaw. This muscle receives its nerve supply from the mandibular nerve by a short branch lying deep to the inferior alveolar and lingual

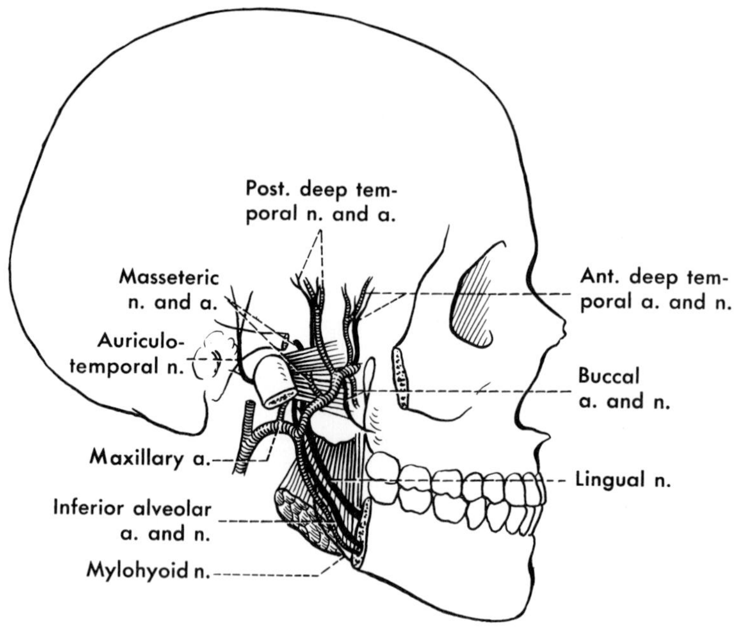

Fig. 7-29. Some neural and arterial relations of the pterygoid muscles. The middle meningeal artery arises just above the leader on the maxillary artery.

branches and usually traversing or running against the otic ganglion.

The lateral surface of the medial pterygoid muscle forms the deep boundary of the looser tissue of the masticator space and is crossed by the inferior alveolar and lingual nerves and the inferior alveolar artery. The lingual nerve is joined at about the upper border of the muscle by the chorda tympani, and the lingual and inferior alveolar nerves then descend almost parallel with each other across the muscle, the lingual nerve being the more anterior. The medial surface of the medial pterygoid muscle with its covering fascia lies adjacent to the important lateral pharyngeal space and commonly serves as a guide for the finger in opening this space.

Innervations and Actions

The nerves to these four muscles all arise close together, from the mandibular nerve soon after it has emerged from the foramen ovale; the masseteric and posterior deep temporal nerves may arise by a common stem, and the anterior deep temporal may arise from the buccal; or the pattern may vary, the deep temporals arising by a common stem between the origin of the masseteric and the buccal. The branches to the two heads of the lateral pterygoid usually arise in common with the buccal nerve; the branch to the medial pterygoid usually proceeds, as noted, through the otic ganglion.

The masseter, temporal, and medial pterygoid muscles all act to close the jaw forcefully. The medial pterygoid and masseter muscles both abduct (deviate) the jaw to the opposite side, and the temporal abducts it to the same side (Vitti and Basmajian), so that proper synchronization of muscles gives rise to the grinding movement of chewing. The pull of the lateral pterygoid is obviously forward instead of up; therefore it is not a closer but a protractor of the mandible. It can be assisted in this action by the masseter and the medial pterygoid. The inferior head of the lateral pterygoid pulls the mandible forward and slightly downward, thus helping it to slide forward onto the articular tubercle; the superior head is usually credited with pulling the articular disk forward, but Sicher said that this is unnecessary because the disk normally moves passively with the mandible and that the function of the superior head, with its slight upward pull, is to hold the disk in contact with the articular tubercle. McNamara found (in electromyographic recordings from monkeys) that the two heads do not act synchronously. The superior head, he found, acts only when the mouth is closed or being closed, while the inferior head acts, as has long been stated, to open the mouth. In contrast, Lehr and Owens studied the muscle electromyographically in 14 human subjects and reported that the two heads function together in all movements. Retraction is brought about largely by the posterior fibers of the temporal.

It has already been noted that protraction is necessary to the wide opening of the mouth, to which the lateral pterygoids must therefore be regarded as contributing. According to Lord, protraction so tightens the stylomandibular ligament and perhaps the sphenomandibular ligament (other workers have emphasized the latter especially) that the mandible rotates around these ligaments during protraction, and therefore the mouth opens wider. The digastric muscles help in opening the mouth, particularly in wide opening.

Other Muscles of Mandibular Region

Other muscles associated topographically with the mandible include the digastric, the mylohyoid, and muscles arising from the styloid process as well as certain muscles of the tongue (see Figs. 7-34, 7-36, and 7-38). These muscles belong to the suprahyoid, lingual, and pharyngeal groups and will therefore not be discussed here other than to point out that the posterior belly of the digastric and the stylohyoid muscle pass down in close relationship to the posterior border of the ramus of the mandible, but inclining somewhat more forward than it so that they pass particularly close to its angle before attaching to the hyoid bone.

Although the external jugular vein and its chief tributary, the retromandibular (posterior facial) vein, lie superficially behind and below the mandible after the retromandibular emerges from the parotid gland, the external carotid artery lies more deeply and is separated from the mandible by the posterior belly of the digastric and by the stylohyoid muscle. The other muscles originating from the styloid process run deep to the internal surface of the medial pterygoid muscle, thus through the connective tissue of the lateral pharyngeal space, which they partially subdivide.

NERVES

Most of the more posterior cranial nerves, and the great vessels extending upward from the neck into the head, are necessarily more or less related to the posterior and deeper aspects of the mandible; they are, however, more intimately concerned with the face, neck, tongue, and pharynx and are therefore discussed in connection with these parts. It is the mandibular branch of the trigeminal nerve that is most closely related to the mandible, both topographically and because it is the motor supply to the muscles of the mandible and the sensory supply to the lower teeth. Similarly, the maxillary artery is the chief supply of the lower jaw, just as it is that of the upper jaw and of the nose.

MANDIBULAR NERVE

The mandibular nerve, as it leaves the skull through the foramen ovale, consists of two rootlets, a large laterally lying sensory one that contains fibers derived from the trigeminal ganglion, and a medial motor (and proprioceptive) root that is the direct continuation of the motor root (portio minor) of the nerve and is much smaller. As these two parts emerge through the foramen ovale in intimate contact they unite to form the mandibular nerve, a very short trunk that lies deep (medial) to the upper portion of the lateral pterygoid muscle and lateral to the tensor veli palatini muscle. In this position the otic ganglion lies on the medial side of the nerve, between it and the tensor, and here also the nerve breaks up into a number of branches. Most of the muscular branches of the nerve leave it at its anterior border, while its larger sensory branches are from the posterior part of the nerve; it is frequently described, therefore, as forming two divisions: an anterior,

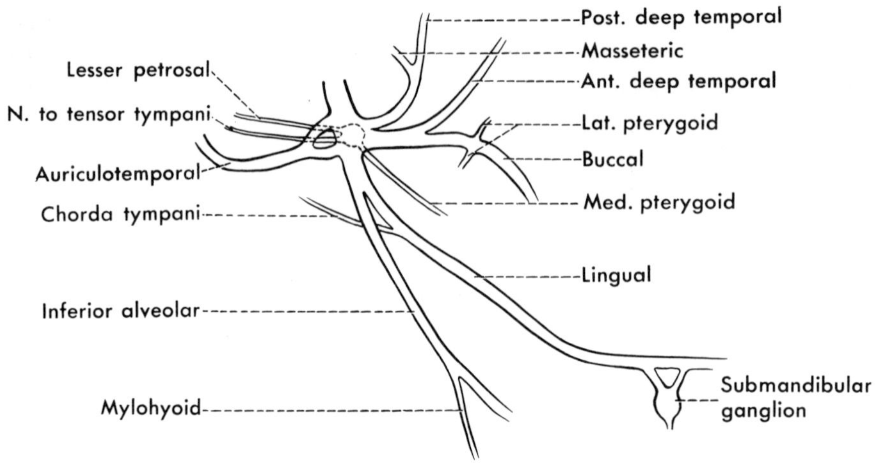

Fig. 7-30. The mandibular nerve in the region deep to the ramus of the mandible. In this lateral view the otic ganglion, situated on the medial side of the mandibular nerve, is *outlined.* The branch of the mandibular to the tensor veli palatini, which passes medially through the otic ganglion, is not shown.

primarily motor one, and a posterior, mostly sensory one. This division is not, however, very obvious.

The first branches given off anteriorly are usually the *masseteric* and the *anterior* and *posterior deep temporal nerves* (Fig. 7-30); the first and last mentioned pass above the upper border of the superior head of the lateral pterygoid muscle, between this and the bone, to be distributed to the muscles indicated by their names. The anterior deep temporal nerve may also have this course, but often emerges with the buccal nerve from between the two heads of the lateral pterygoid.

The *buccal* (buccinator) *branch* of the mandibular nerve also arises more or less in common with the preceding branches and passes forward and laterally to emerge between the two heads of the lateral pterygoid, turn downward across the lateral surface of the lower head, and come in contact with the medial surface of the insertion of the temporal muscle. Running forward and downward across this it penetrates the buccal fat pad and thus lies on the buccinator muscle in the cheek; some of its fibers are distributed to the skin here, while others penetrate the muscle to be distributed to the mucosa on the inner surface of the cheek and to the adjacent gingivae.

As it passes between the two heads of the lateral pterygoid muscle, the buccal nerve typically gives off a muscular branch to each, but thereafter is purely sensory. Thus, although it anastomoses with buccal branches of the facial nerve on the cheek, it is the facial nerve fibers that supply the buccinator muscle, not the buccal mandibular ones.

As already mentioned, the *otic ganglion,* a parasympathetic ganglion connected primarily with the parotid gland, is located on the medial side of the mandibular nerve just outside the foramen ovale. Two of the smaller muscular branches of the mandibular nerve typically pass through or in close association with the otic ganglion and are described as branches of this, and a third may be so related; actually, they consist of voluntary motor (and proprioceptive) fibers that have joined the mandibular by way of the motor root.

Of these three branches, the *nerves to the medial pterygoid, tensor tympani,* and *tensor veli palatini* muscles, respectively, the largest, not listed as a branch of the ganglion, is the nerve to the medial pterygoid muscle; this runs forward and downward into the upper edge of the muscle. A tiny branch is given off posteriorly from the otic ganglion and runs parallel to the lesser petrosal nerve to supply the tensor tympani muscle. Finally, a twig proceeds medially from the otic ganglion to enter the tensor veli palatini muscle, the only muscle of the soft palate innervated by the fifth nerve.

Obviously, these branches have nothing to do functionally with the otic ganglion, but are simply associated with it; the function of the otic ganglion, as of all autonomic ganglia, depends upon preganglionic and postganglionic autonomic fibers. The otic ganglion receives preganglionic fibers not from the mandibular nerve, but from the ninth nerve by way of the lesser petrosal (lesser superficial petrosal) nerve (which may contain a few fibers from the vagus, said by Vidíc to first join the facial nerve and then leave it to join the lesser petrosal.) Most of its postganglionic fibers are given off into the auriculotemporal nerve, to reach the parotid gland. There is also a twig of communication with the sympathetic plexus on the middle meningeal artery (often called the sympathetic root of the ganglion, but officially called the communicating ramus with the meningeal branch of the mandibular); similar tiny communicating rami with the nerve of the pterygoid canal and with the chorda tympani have been described. The importance of these communications is not known; through the last two, the otic ganglion could receive preganglionic fibers from the facial nerve or send postganglionic fibers to the submandibular and sublingual glands and to the nose and palate; sometimes, apparently, taste fibers from the tongue take this pathway to the geniculate ganglion, traversing the greater petrosal instead of the tympanic part of the chorda tympani (see Fig. 7-41).

The three remaining branches of the mandibular nerve are largely sensory. From its posterior border the *auriculotemporal nerve* arises, typically by two stems between which runs the middle meningeal artery; these stems unite posterior to the artery, and the single auriculotemporal nerve thus formed then continues to pass posteriorly, medial to the temporomandibular joint, and behind the joint turns upward to pass between this and the external acoustic meatus. This nerve supplies no muscles; its course has already been described.

The two largest branches of the mandibular nerve are the *inferior alveolar* (dental) and the *lingual nerves*. These typically arise by a common stem that continues the downward direction of the mandibular nerve, but they soon separate and cross the outer surface of the medial pterygoid muscle, lying at first between this and the lateral pterygoid and then between the medial pterygoid and the ramus of the mandible. On the surface of the medial pterygoid they diverge slightly, the lingual nerve passing somewhat more anteriorly while the inferior alveolar runs more directly downward. Because of their importance, these two nerves are described in some detail.

Innervation of the Teeth

Before it enters the mandibular foramen on the inner surface of the mandible the *inferior alveolar nerve* gives off its mylohyoid branch, which runs downward in a groove on the inner surface of the bone and passes below the mylohyoid muscle to supply this and the adjacent anterior belly of the digastric. The inferior alveolar nerve enters the mandible through the mandibular foramen and is distributed to the teeth of the lower jaw.

Hensel emphasized that the inferior alveolar nerve lies in contact with the mandible before it enters the mandibular foramen. This foramen lies approximately 1.5 to 2 cm below the mandibular notch, but according to Hensel the nerve is in immediate contact with the mandible for about 4.5 mm ($\frac{3}{16}$ in) above the spine of the lingula and for a similar distance below this before it actually enters the mandible. He found the distance from the bottom of the mandibular notch to the lingula varying from 10.5 mm to 15 mm ($\frac{7}{16}$ to $\frac{10}{16}$ in) and pointed out that if the distance through which the nerve is in contact with the mandible above the lingula is subtracted from this, there is relatively little leeway in which a cut can be made across the ramus without involving either the mandibular notch or the inferior alveolar nerve. In cases in which the ramus of the mandible is to be sectioned in order to correct abnormal protraction or retraction, therefore, he advocated making the cut 7.5 mm (5⁄16 inch) below the mandibular notch.

According to Starkie and Stewart the *distribution* of the inferior alveolar nerve *to the teeth* is as follows: soon after it enters the foramen, the nerve gives off an alveolar branch that runs parallel to the main stem, to supply the teeth posterior to the incisors; at or before the level of the mental foramen other fibers leave the nerve to form an intricate plexus external to the roots of the incisor teeth, while the remainder of the nerve leaves the mandible through the foramen. Starkie and Stewart found that the canine might be supplied either from the main alveolar branch or from the incisor plexus. In one third of the cases they investigated they found no overlap between the incisor plexuses of the two sides, but in the others there was some; they felt that their anatomical observations agreed with clinical ones indicating that when such overlap between the nerves of the two sides is present, it usually does not extend beyond the central incisor and never beyond the lateral one on the opposite side.

The *mental nerve* is a large branch of the inferior alveolar that leaves the interior of the mandible to supply the skin of the chin and lower lip, the mucosa of the lip, and the adjacent gum. The mental foramen may usually be found on the vertical line drawn downward from the supraorbital notch. In regard to its relation to the teeth, it lies in about 50% of cases at the level of the root of the second premolar (bicuspid) tooth, between the two premolars in about 20% to 25%, and posterior to the second premolar in about 24%; in approximately 1% to 2% each it lies as far forward as the first premolar or as far back as the first molar (Tebo and Telford; Fig. 7-31). Except for the distribution of the mental nerve, the inferior alveolar nerve does not participate very much in supplying the gingival mucosa; the buccal gingivae behind the area of the mental nerve are supplied primarily by twigs of the buccal nerve, the lingual gingivae primarily by branches of the lingual nerve. Gabriel pointed out that the mental foramen is higher than the mandibular canal posterior to it, and therefore resorption of the alveolar process in the edentulous mandible will expose the nerve at the mental foramen before that in the canal.

Fig. 7-31. Variations in the relation of the mental foramen to the roots of the teeth. Percentages according to Tebo HG, Telford IR: Anat Rec 106:254, 1950.

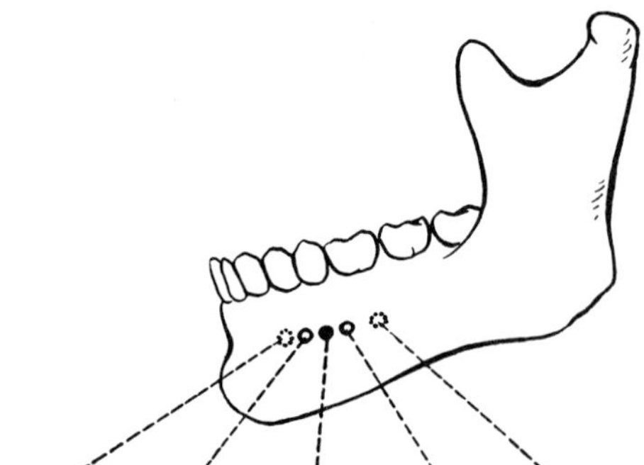

Much of the bone about the alveoli of the more posterior teeth is dense, and good *anesthesia* here is best obtained by injection about the inferior alveolar nerve before it enters the mandible. This is usually done from the mouth by inserting the needle behind the oblique ridge that can usually be felt running upward from behind the last molar tooth on the inner border of the mandible; after the needle strikes the ramus of the mandible it can be guided upward along the bone to about the midpoint of the ramus, which corresponds approximately to the mandibular foramen. The gingival mucosa, with its different innervation, requires local infiltration.

Lingual Nerve

After leaving the inferior alveolar nerve, the lingual nerve receives the chorda tympani from the facial; this branch contains both sensory fibers from the geniculate ganglion and preganglionic parasympathetic fibers. Beyond this point the lingual nerve therefore contains fibers for general sensation that are true fifth nerve fibers, fibers for taste derived from the facial nerve, and autonomic fibers destined for the submandibular (submaxillary) ganglion.

The nerve proceeds downward across the

outer surface of the medial pterygoid muscle, anterior to the inferior alveolar, with which it may communicate; it then leaves the masticator space to pass downward lateral to the styloglossus muscle, as this curves forward into the tongue, and then itself curves forward into the loose connective tissue above the mylohyoid muscle and lateral to the musculature of the tongue. Its course and connections here are described in connection with the tongue. In brief, most of the parasympathetic fibers leave the lingual nerve at the level of the submandibular gland to end in the similarly named ganglion, suspended from the nerve as it courses just above the gland; the fibers from the ganglion are distributed to the submandibular and sublingual glands, while the sensory fibers in the lingual nerve are distributed to the anterior two thirds of the tongue.

VESSELS

MAXILLARY ARTERY

The main stem of the external carotid artery ascends approximately parallel to the posterior border of the ramus of the mandible, after having passed deep to the posterior belly of the digastric and the associated stylohyoid muscle; under cover of the angle of the mandible it gives off the facial (external maxillary) artery that curves forward and downward over the submandibular gland and around the lower margin of the mandible.

Above this level it gives off the posterior auricular artery (if this does not arise from the occipital), and at approximately half to two thirds of the distance between the lower border of the angle and the temporomandibular joint it divides deep to or within the parotid gland into its two terminal branches, the superficial temporal and the maxillary (Fig. 7-32). The further course of the superficial temporal through the parotid gland and to the scalp has already been described.

The maxillary artery (internal maxillary when the facial is called external maxillary) leaves the external carotid at about a right angle, passing almost horizontally between the ramus of the mandible and the sphenomandibular ligament. In front of the ramus the artery passes upward across the lower part of the lateral pterygoid, but may cross either superficial or deep to the lower head of this muscle; in either position, it is embedded in the pterygoid plexus of veins.

Among 200 arteries, Lauber found only 16 passing lateral to the lower head of the lateral pterygoid rather than medial to it, but in the United States, at least, this course is seen more commonly than the deeper one. Lasker and co-workers reported a difference according to race, finding the artery deep to the muscle in 46% of 147 sides in white people, and in 31% of 61 sides in Negroes (Fig. 7-33).

Fig. 7-32. Schema of the maxillary artery and its branches. The latter vary somewhat, especially in the order of their origins. Some of their relations are shown in Fig. 7-29.

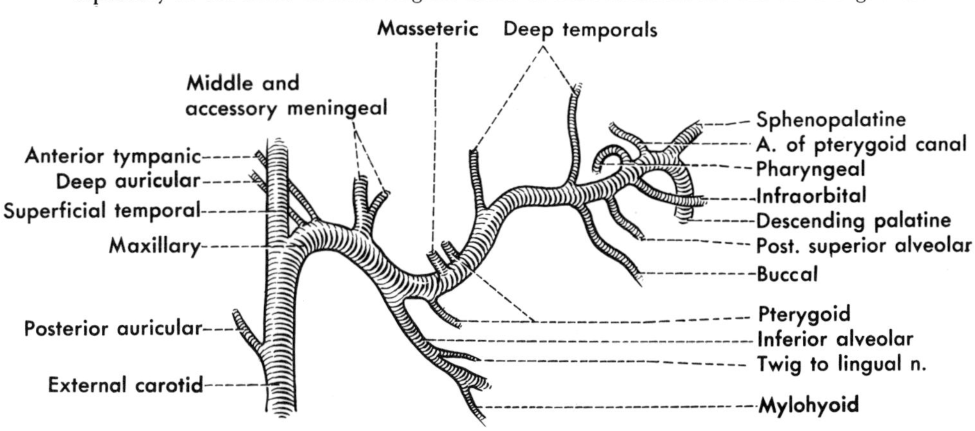

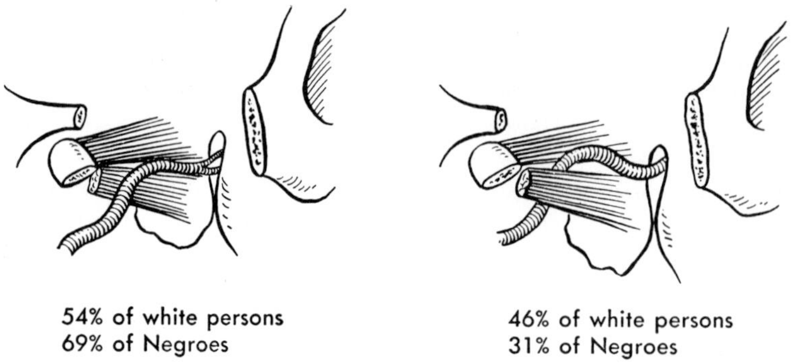

Fig. 7-33. The two relations of the maxillary artery to the lateral pterygoid muscle. Percentages according to Lasker GW, Opdyke DL, Miller H: Anat Rec 109:119, 1951.

If the artery passes medial to the lower head it emerges between the two heads, and its course is thereafter the same, lying upward and forward lateral to the groove between the two heads, or across the lateral surface of the upper head, of the lateral pterygoid.

First Part

While it lies medial to the mandible (the so-called first part of the artery) the maxillary typically gives off four branches: small *deep auricular* and *anterior tympanic arteries,* which pass upward to aid in the supply of the external acoustic meatus and the middle ear cavity; the *middle meningeal artery,* which runs upward and slightly forward to pass between the two heads of origin of the auriculotemporal nerve and enter the skull by way of the foramen spinosum, situated just posterolateral to the foramen ovale; and the *inferior alveolar* (*dental*) *artery,* which passes downward and forward, gives off a mylohyoid branch that accompanies the nerve of the same name, and then enters the mandible with the inferior alveolar nerve to be distributed to the teeth and the surrounding bone and to send a small branch through the mental foramen to anastomose with other vessels on the face. Details of its distribution in the mandible have been described by Castelli.

In this position also, the maxillary may give off an *accessory middle meningeal artery,* which may be distributed only to adjacent muscles and bone (Baumel and Beard), but when best developed runs through the foramen ovale with the mandibular nerve to supply that nerve and the trigeminal ganglion and adjacent dura. Frequently the accessory middle meningeal is a branch from the middle meningeal artery itself.

Second Part

While the artery is related to the lateral pterygoid muscle (the so-called second part of the artery) it gives off its muscular branches—*masseteric,* two *deep temporals,* and *buccal,* which accompany the nerves of the same names, and also *branches to both pterygoid muscles.* If the main stem of the artery lies deep to the inferior head of the lateral pterygoid, some of these branches necessarily pass between the two heads of the muscle to reach their destinations; others leave the artery after it has emerged between the two heads. In either case, therefore, the main branches of this part of the artery run on the lateral surface of the lateral pterygoid muscle.

Third Part

The terminal (third or pterygopalatine) portion of the artery is that which lies against the posterolateral aspect of the maxilla and passes deeply between the two heads of the lateral pterygoid muscle to enter the pterygopalatine fossa and lie lateral to the pterygopalatine (sphenopalatine) ganglion.

This portion gives off the *posterior superior alveolar artery,* described in connection with the upper jaw; the *infraorbital artery,* accompanying the similarly named nerve through the infraorbital groove and canal and giving off orbital and anterior superior alveolar branches before it emerges on the face to anastomose with vessels there; the *descending palatine artery,* described in connection with the palate; the *artery of the pterygoid canal,* passing backward with the nerve of the pterygoid canal (Vidian nerve) to be distributed to a small upper portion of the pharynx and to a part of the auditory tube; sometimes a tiny *pharyngeal branch* to the upper part of the pharynx; and, finally, the important *sphenopalatine artery,* distributed to the nose and described in connection with this.

The most common source of hemorrhage from the maxillary artery is its sphenopalatine branch. This has been controlled directly by exposure and ligation of the third part of the artery through a transantral approach or indirectly by ligation of the external carotid. Except in operations on the parotid gland or the ramus of the mandible, the maxillary artery is practically inaccessible from a lateral approach.

Anomalies

Aside from the variation in its position in relation to the inferior head of the lateral pterygoid muscle, variations or anomalies of the maxillary artery are apparently very rare. The maxillary has been seen to arise by a common stem with the facial (*e.g.,* Delitzin), in which case it ascended internal to and parallel with the long axis of the ramus of the mandible, instead of crossing this at right angles as usual. Usually the maxillary lies lateral to the major branches of the mandibular nerve in its course, but among 200 subjects investigated by Lauber there were 42 in which he regarded this artery as following an abnormal course; in some of these the artery passed medial to all the larger mandibular branches, in others it passed between the lingual and inferior alveolar nerves, and in a few it passed through a bifurcation of the inferior alveolar nerve.

VEINS

The *pterygoid plexus* is a dense venous plexus lying on both sides of the lateral pterygoid muscle and largely surrounding the maxillary artery. It receives tributary veins (meningeal, alveolar, sphenopalatine, and the like) corresponding to the branches of the artery, and its chief drainage is by way of the maxillary vein. However, it is also connected to the facial vein by the deep facial vein, which runs with the buccal artery and nerve, with the pharyngeal plexus of veins, and with the cavernous sinus by a plexus in the foramen ovale, by connections with the inferior ophthalmic vein through the inferior orbital fissure, and by a plexus in the foramen rotundum and the sphenoid emissary vein (of Vesalius) when these are present. It sometimes transmits infections to the cavernous sinus, particularly from the "danger area" of the face.

The *maxillary* (internal maxillary) *vein,* often double, originates from the posterior border of the pterygoid plexus and runs posteriorly with the first part of the maxillary artery between the sphenomandibular ligament and the mandible. It joins the superficial temporal vein to form the retromandibular vein, which then descends in the substance of the parotid gland, superficial and parallel to the external carotid; in the gland, it passes deep to the facial nerve and then divides below to contribute both to the external jugular and the facial vein.

The Suprahyoid and Lingual Regions

The nerve–muscle relationships in the suprahyoid region are complicated but are largely understandable upon the basis of the probable embryology of this region. Representing as it does the junction of head and neck, it is a region of overlap between the voluntary muscles derived from various branchial arches and innervated by the cranial nerves concerned with these arches and voluntary muscle developed from occipital and upper cervical myotomes.

Thus we have the mandibular arch represented not only by the muscles of mastication but also by the anterior belly of the digastric and by the mylohyoid muscle; these are innervated by the mandibular nerve, the nerve of the first branchial arch. The musculature derived from the second branchial arch is largely the facial (mimetic) muscles, but is represented in the suprahyoid region by the posterior belly of the digastric muscle and by the stylohyoid muscle, both innervated by the facial nerve. Musculature of the third branchial arch, innervated by the glossopharyngeal nerve, has almost disappeared and is represented solely by the stylopharyngeus muscle; although it is in the suprahyoid region, it is described in this book with the muscles of the pharynx. The musculature of the remaining branchial arches, supplied by the tenth nerve (and by the eleventh, developmentally a part of the tenth), is represented by the muscles of the palate, pharynx, and larynx (tensor veli palatini and stylopharyngeus excepted) and by the sternocleidomastoid and trapezius muscles.

Finally, both the extrinsic and intrinsic muscles of the tongue are believed to be developed largely from occipital myotomes in series with the cervical ones from which the infrahyoid (strap) muscles of the neck develop; hence they are innervated largely by the hypoglossal nerve, not a branchiomeric nerve but a cranial one in direct series with the upper spinal nerves. The hypoglossal nerve represents essentially several upper cervical nerves that have lost their sensory roots, fused together and, through fusion of bony vertebral elements to the skull, attained a cranial exit.

The geniohyoid muscle represents the suprahyoid continuation of the strap muscles of the neck, is apparently derived from upper cervical myotomes, and is innervated by fibers from upper cervical nerves that travel with the hypoglossal.

The skin of the suprahyoid region is innervated by upper branches from the transverse cervical (transversus colli, cutaneous colli) nerve, derived from the second and third cervical nerves. Deep to the skin and the platysma muscle the cervical branch of the facial nerve descends to supply the platysma muscle, incidentally anastomosing with sensory branches of the transverse cervical nerve; also under cover of the platysma muscle, and running approximately along the lower margin of the mandible (but sometimes above this and sometimes definitely in the suprahyoid region), is the marginal mandibular branch of the facial, supplying musculature connected with the lower lip. It is this branch that is especially endangered in the removal of the fascia and glands of the suprahyoid region during a radical dissection of the neck for carcinoma.

LYMPH NODES

Deep to the platysma the structures of the suprahyoid region are covered by the superficial (anterior, investing) layer of the deep fascia of the neck, which in the narrow region between the ramus of the mandible and the sternocleidomastoid muscle is continued upward to ensheath the parotid gland and more anteriorly is reflected from the hyoid bone to the mandible, being attached to both; it is this part that encloses also the submandibular (submaxillary) gland. Embedded in the deep surface of this fascia are the superficial parotid lymph nodes and the submandibular and submental ones (see Fig. 6-9).

Since the fascia is not cleanly dissectible from the parotid and submandibular glands, it is impossible to be sure of removing all lymph nodes by dissection of this fascia. Therefore it is customary, in block dissections of the neck, to remove the lower tip of the parotid gland with its associated fascia and to remove the submandibular gland in its entirety with its associated fascia; in the latter procedure, the facial vein and artery, which cross the submandibular gland, must be sacrificed. Removal of the submental nodes is assured by removing the anterior belly of the digastric muscle and all fascia on the lower surface of the mylohyoid.

Submental Lymph Nodes

The submental lymph nodes are a few small ones that are, however, of considerable importance because they receive the lymphatics from the medial part of the lower lip, the chin, and the more anterior part of the tongue. According to Stahr there are typically three submental nodes on each side, including both those which lie against the mylohyoid muscle between the two anterior bellies of the digastric and those (usually described now as anterior nodes of the submandibular group) which lie posterolateral to the anterior belly, upon the mylohyoid muscle as it forms the floor of the submandibular or digastric triangle. It is in order to remove these nodes that the external surface of the mylohyoid muscle must be cleanly dissected.

Submandibular Lymph Nodes

The submandibular lymph nodes are larger and more numerous than the submental; they form a chain of from three to eight nodes situated largely along the upper border of the submandibular gland, between this and the mandible, but also extending lateral or superficial to the gland, and often in front of and behind it.

According to Stahr, some of these nodes are constant and some are much less constant; the largest and most constant is a node close to the point at which the facial artery crosses the mandible. These nodes receive the efferent vessels from the buccal and mandibular (facial) nodes, along the facial vein and artery, and therefore either directly or indirectly drain a large part of the face, receiving vessels from the nose, cheek, upper lip, lateral part of the lower lip, lateral parts of the anterior portion of the tongue, and much of the gingival mucosa. Their efferent connections are in part to the superficial cervical nodes (along the external jugular vein), but are largely into the upper deep cervical nodes along the internal jugular vein.

POSTERIOR PART OF SUPRAHYOID REGION

The narrow posterior part of the suprahyoid region is a direct continuation upward of the neck and therefore contains the internal jugular vein, the internal and most of the external carotid artery, and the more posterior cranial nerves (*i.e.,* the seventh and the ninth to twelfth). The vascular and nerve relations here are described in connection with the neck, but this region also contains the posterior belly of the digastric muscle and the origins of the muscles arising from the styloid process. The sternocleidomastoid muscle and the ramus of the mandible largely overlap it, and much of the remaining interval is occupied superficially by the parotid gland; deep to this are the posterior belly of the digastric and the stylohyoid muscle (Fig. 7-34).

The digastric and stylohyoid muscles originate a short distance apart, from the mastoid notch and from the styloid process, respectively, but as they run downward toward the angle of the mandible and the hyoid bone they converge; the important point here concerning them is that they cross lateral or superficial to the great ascending arteries and the internal jugular vein (see Fig. 7-36) and also to the descending tenth, eleventh, and

Fig. 7-34. Some relations in the suprahyoid region, with the sternocleidomastoid reflected backward.

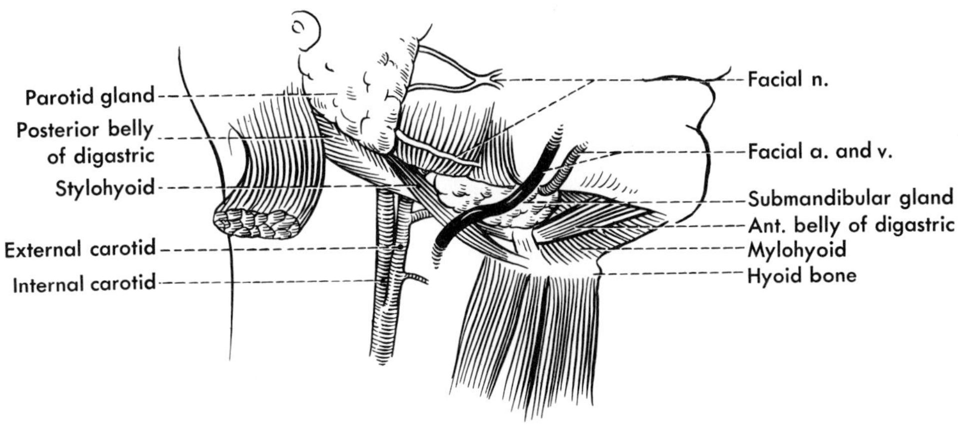

twelfth cranial nerves; at the level of the angle of the mandible one can cut down to these muscles without fear of damaging vascular or neural structures. Pisk did report two cases in which the stem giving rise to the maxillary and superficial temporal vessels passed superficial to the stylohyoid and the posterior belly of the digastric, but this is certainly exceedingly rare and should be readily appreciable before operation, since the artery in this case lies subcutaneously and should be easily palpable.

Much higher, under cover of the parotid gland and the posterior edge of the ramus of the mandible, these muscles are crossed superficially by the posterior auricular artery; also, the facial nerve appears in front of the digastric to run lateral to the styloid process and its attached muscles, while the external carotid artery passes between the stylohyoid muscle and the styloglossus to enter the deep surface of the parotid gland.

HYOID BONE

The hyoid bone, to which are attached both infrahyoid and suprahyoid muscles and the superficial layer of the deep fascia in the front of the neck, effectively separates the anteriorly located suprahyoid and infrahyoid fascial compartments (see Fig. 5-11). Because it is held in place primarily by the muscles that attach to it, it is freely movable and thus takes part in movements of swallowing and phonation. The thyrohyoid membrane stretches between it and the larynx. The sternohyoid, omohyoid, and thyrohyoid muscles attach to it below, the middle constrictor of the pharynx attaches laterally, and the stylohyoid muscle, the fascial sling of the digastric tendon, and the mylohyoid, geniohyoid, and genioglossus muscles all attach to it above.

The greater cornu can usually be palpated by lateral displacement of the bone toward the side to be palpated, but may lie under cover of the sternocleidomastoid muscle. The lesser cornu, at the level of junction of the greater cornu and the body, typically projects upward for only a few millimeters. The stylohyoid ligament attaches to the lesser cornu and represents an unossified part of the skeleton of the second pharyngeal arch, upper portions of which are represented by the styloid process and the stapes of the middle ear, lower portions by the lesser cornu and part of the body of the hyoid. Variations in the ossification of this intermediate segment lead to much variation in the appearance of the styloid process (Fig. 7-35); occasionally the entire ligament is replaced by bone. Chi–Min described bilateral replacement of the ligament by an anomalous muscle.

SUPRAHYOID MUSCULATURE

The suprahyoid musculature discussed here includes the digastric and stylohyoid muscles, the mylohyoid, and the geniohyoid; the remaining muscles in this region are muscles of the tongue.

Digastric Muscle

The digastric muscle (Figs. 7-34 and 7-36) consists of two bellies. The posterior belly arises from the mastoid (digastric) notch on the medial side of the mastoid process and passes downward and forward toward the hyoid bone, where the muscle fibers give place to a tendon; this tendon perforates the lower fibers of the stylohyoid muscle, is held downward against the hyoid by an aponeurotic attachment or fascial sling, and is continuous anteriorly with the anterior belly of the digastric. From its attachment to the hyoid bone the tendon between the two parts of the digastric curves forward and upward to be replaced by the fleshy fibers of the anterior belly, which are attached to the digastric fossa on the lower inner surface of the mandible. Sometimes there are aberrant fibers uniting the two anterior bellies of the digastric (du Bois–Reymond) or accessory muscles, derived either from them or the mylohyoid (Waern).

At its origin the posterior belly of the digastric lies deep to the sternocleidomastoid muscle; the facial nerve emerges from the stylomastoid foramen to pass anterior to and to supply this belly before curving forward lateral to the styloid process and the muscles at-

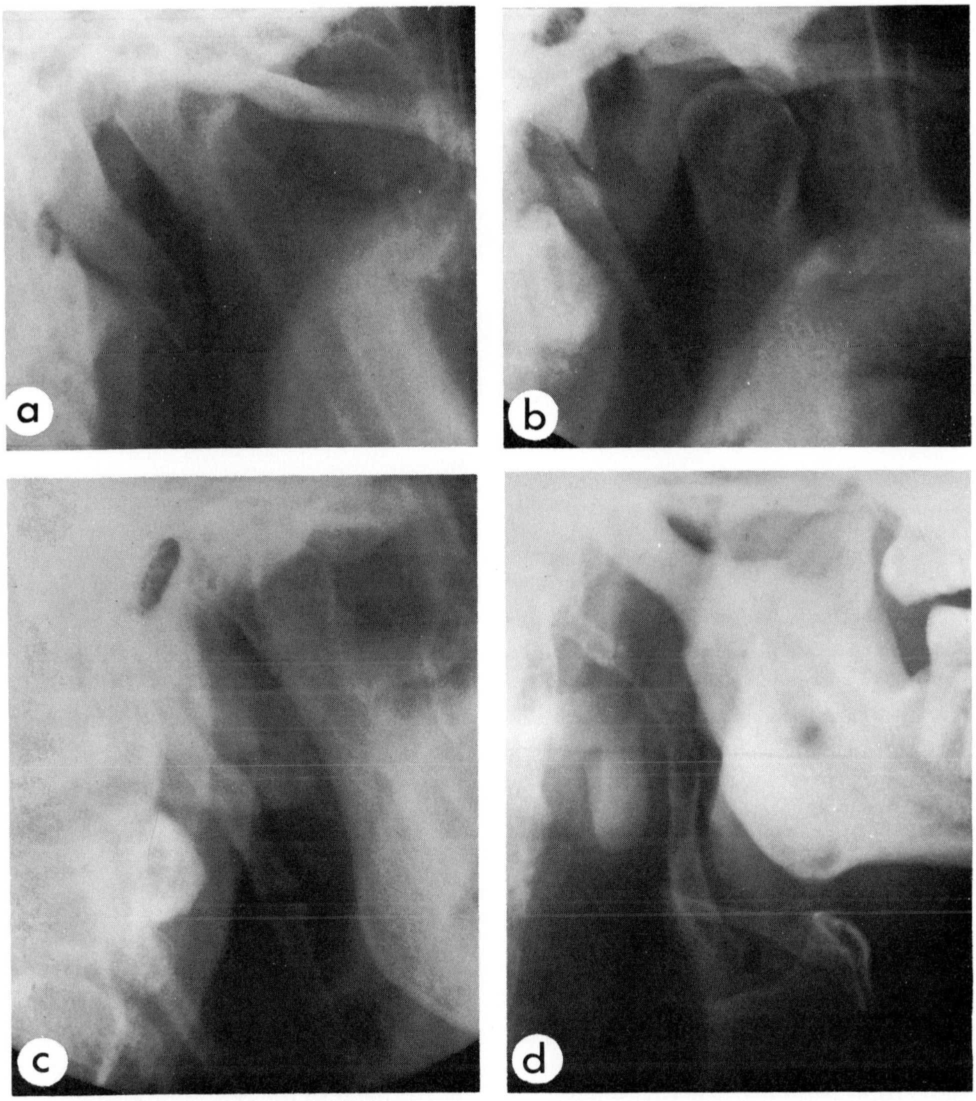

Fig. 7-35. Some variations in the styloid process: (*a*), short and heavy; (*b*), tapering and slender; (*c*), a jointlike junction between two parts; (*d*), extending to the lesser cornu of the hyoid bone. (Stafne EC, Hollinshead WH: Oral Surg 15:1195, 1962, by permission of the CV Mosby Company)

tached to it. Deep to the upper part of the digastric are the upper fibers of origin of the levator scapulae, and the occipital artery runs between these two muscles. The posterior border of the muscle is overlapped above by the insertions of the splenius capitis and semispinalis capitis muscles. Anterolaterally, the muscle is in relation to the deep portion of the parotid gland. The accessory nerve, internal jugular vein, hypoglossal nerve and external carotid artery all lie deep to the posterior belly of the digastric behind the angle of the jaw; associated with the internal jugular vein in this position are the upper deep cervical lymph nodes. The internal carotid artery and the vagus nerve also lie here, but more deeply.

The intermediary tendon of the digastric is associated superficially with the submandibular gland; the facial artery, after extending upward deep to the muscle, descends to

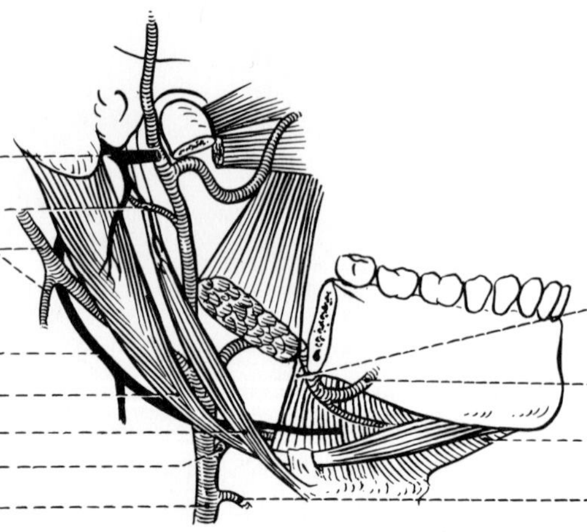

Fig. 7-36. Deeper relations in the suprahyoid region, after removal of the parotid and submandibular glands and much of the mandible. The stylohyoid and the posterior belly of the digastric are shown more widely separated than they actually are.

emerge along the upper border of the submandibular gland between this tendon and the lower margin of the mandible. The anterior belly of the digastric lies upon the lower surface of the mylohyoid muscle and is covered by the superficial layer of deep fascia as this is reflected from the hyoid bone to the chin. It is innervated by a branch from the mylohyoid nerve, in turn a branch of the inferior alveolar nerve (from the mandibular division of the trigeminal).

The anterior belly of the digastric divides the region between the hyoid bone and the mandible into two triangles: the submandibular (submaxillary, digastric) triangle posteriorly and the submental or suprahyoid triangle anteriorly. The former lies between the posterior and anterior bellies of the digastric on the one hand, as they converge to attach to the hyoid bone, and the margin of the mandible on the other; the latter triangle lies between the anterior belly of the digastric, the hyoid bone, and the midline of the submental region.

Submandibular Triangle

The submandibular (submaxillary) triangle is largely occupied by the major portion of the submandibular gland, with its associated lymph nodes; the facial vein crosses the gland superficially, and the facial artery emerges between the gland and the base of the mandible. Anteriorly, the floor of the triangle is the mylohyoid muscle, while posteriorly the hyoglossus appears here; a process of the submandibular gland leaves the triangle with the submandibular duct by passing forward above the posterior border of the mylohyoid muscle, while in the posterior part of the triangle, against the hyoglossus muscle, the hypoglossal nerve with its accompanying vein runs forward to the tongue. Close to the angle formed by the two bellies of the digastric the lingual artery is accessible before it disappears deep to the posterior border of the hyoglossus muscle. The surfaces of these muscles in the triangle, and the hypoglossal nerve, must be carefully cleaned of their connective tissue in complete block dissections of the neck.

Submental Triangle

The submental triangle contains nothing other than the submental lymph nodes (already mentioned), the mylohyoid nerve and vessels, and the submental branch of the facial artery and a corresponding vein (the latter joining the facial vein and also communicating with the anterior jugular vein). The floor of this triangle is the mylohyoid muscle; the relatively small amount of loose connective tissue lying between the covering fascia and the mylohyoid muscle, sometimes referred to as the submental space, is a part of the several anteriorly situated fascial spaces grouped together as the submandibular space.

Stylohyoid, Mylohyoid, and Geniohyoid Muscles

The *stylohyoid muscle* arises from the posterolateral aspect of the styloid process and descends anterior to the posterior belly of the digastric to split into two parts, between which the tendon of the digastric runs, and then attach to the lateral portion of the body or the adjacent part of the greater cornu of the hyoid bone. It is related laterally to the parotid and submandibular glands and medially to the great vessels of the neck. The muscle is innervated by a twig from the facial nerve, often arising with the nerve to the posterior belly of the digastric.

The *mylohyoid muscle* extends almost transversely from the inner surface of the mandible toward the midline. The most posterior fibers insert upon the body of the hyoid bone; the remainder insert with those of the muscle of the other side into a median raphe that extends from the hyoid bone almost to the mandible. The two muscles together form a diaphragm below the geniohyoid muscle and the muscles of the tongue. Superficially, or below, the mylohyoid is in contact with the anterior belly of the digastric and with a major part of the submandibular gland; the submental artery runs along its external surface, as does also the mylohyoid branch of the mandibular nerve (and its accompanying artery), which innervates both this muscle and the anterior belly of the digastric. Between the mylohyoid muscle and the overlying superficial layer of cervical fascia is the potential space sometimes involved in Ludwig's angina. Above, the muscle is in contact with the geniohyoid muscle and with the muscles of the tongue and the sublingual spaces about these; the submandibular gland rounds the posterior border of the muscle to lie also in part in the connective tissue above it, where are also the submandibular duct, the sublingual gland, the lingual and hypoglossal nerves, and the vessels of the tongue. The mylohyoid muscle may have a gap between some of its fibers, through which a part of the sublingual gland, or fat, may bulge into the submental region. Gaughran found such herniations, usually close to the mandible, in more than a third of 324 sides examined.

The *geniohyoid muscle* (see Fig. 7-39) lies on the upper surface of the mylohyoid, arising from the lower part of the mental spine and the adjacent internal surface of the mandible and inserting into the body of the hyoid bone just lateral to the midline; above it is the genioglossus muscle, and lateral to it is the sublingual gland, and submandibular duct and part of the gland, and the hypoglossal nerve. The lingual nerve and lingual artery also lie lateral to, but largely above the level of, the muscle. The geniohyoid is supplied by a twig from the hypoglossal nerve, which is usually stated to contain fibers derived from the first cervical nerve rather than true hypoglossal fibers.

Actions

All four muscles of the suprahyoid group elevate the hyoid bone, and therefore the floor of the mouth and the base of the tongue, and thus take part especially in swallowing. The digastric particularly, and presumably the mylohyoid and the geniohyoid, also aid in opening the mouth. According to Munro the posterior belly of the digastric probably acts alone in elevation of the hyoid bone, while both bellies act upon the mandible.

TONGUE

The posterior or attached portion of the tongue, through which the nerves and vessels enter, is the root. Its free surface faces posteriorly, where it is the anterior wall of the oral pharynx. Mucosa completely surrounds the remainder of the tongue, the body and apex.

Over the root the lingual mucosa becomes smoothly continuous with the mucosa of the floor of the mouth except in the anterior mid-

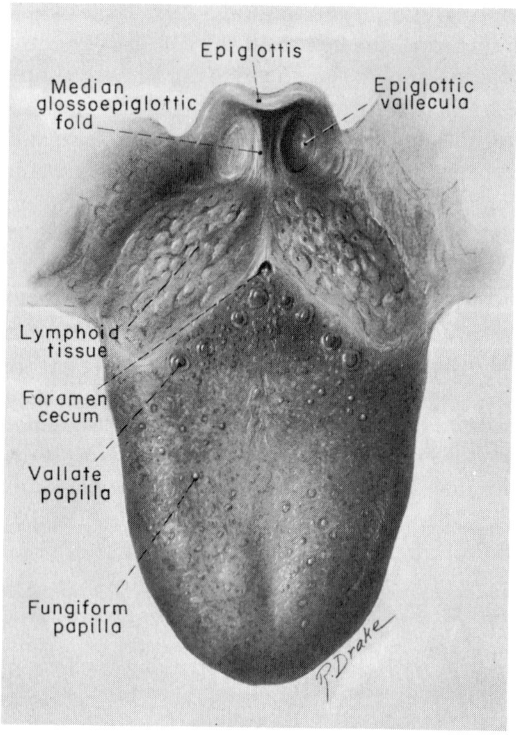

Fig. 7-37. The dorsum of the tongue. (Rankin FW, Crisp NW: West J Surg Obstet Gynecol 40:105, 1932)

line where a fold, the frenulum, is present. The mucosa of the dorsum, or the upper surface of the body, is shaggy, presenting several types of papillae; of these, the most important are the large vallate (circumvallate) papillae, bearing numerous taste buds and arranged in a V at about the junction of the anterior two thirds and posterior third of the tongue (Fig. 7-37). (The fungiform papillae are also provided with taste buds, but the more numerous conical and filiform papillae apparently have none. The foliate papillae, largely on the margins of the tongue, are heavily studded with taste buds.) Behind the vallate papillae on each side is a groove, the sulcus terminalis, that approximately separates the root from the dorsum. The two sulci terminate in the midline at the foramen cecum, a pit indicating the point of outgrowth of the thyroid gland. Most of the pharyngeal surface of the tongue is occupied by lymphoid tissue, constituting the lingual tonsil.

The tongue bears both mucous and serous glands. Mucous glands open onto the posterior surface of the tongue and into the crypts of the lingual tonsil. Serous glands, often called posterior lingual or von Ebner's glands, open into the mucous glands and also into the trenches of the vallate papillae. Anteriorly on the lower surface of the tongue, covered by a thin layer of longitudinal muscle, is the seromucous (largely mucous) anterior lingual gland (Fig. 7-38), opening onto the lower surface of the tongue by some five or six ducts. Its cystic enlargement is called a ranula.

Immediately behind the tongue is the epiglottis, connected to the tongue by a median and two lateral glossoepiglottic folds; the depressions between the folds are the epiglottic valleculae. More laterally, running downward in front of the palatine tonsil, is the palatoglossal arch (see Fig. 7-9).

Extrinsic Muscles

The extrinsic muscles of the tongue are the genioglossus, the hyoglossus, and the styloglossus; the palatoglossus (glossopalatinus) is sometimes described as a muscle of the tongue and sometimes as one of the soft palate, but receives its innervation from the vagus, as do other muscles of the soft palate; in this book it has been described as a palatine muscle. Of these muscles the two larger, the genioglossus and the hyoglossus, extend upward into the tongue from the sublingual region, while the styloglossus and palatoglossus enter the tongue from above and laterally.

The *genioglossus* arises immediately above the geniohyoid from the upper part of the mental spine and the adjacent internal surface of the mandible and forms a broad fanlike sheet in the sagittal plane, the fibers radiating from their origin so that the more anterior ones reach the anterior portion of the tongue, others extend back to the posterior portion, and the lowest insert into the hyoid bone (Fig. 7-39).

The *hyoglossus muscle* arises from a lateral part of the body of the hyoid bone and from its entire greater cornu, with sometimes a slip (chondroglossus) from the lesser cornu of the hyoid; its fibers pass upward and slightly

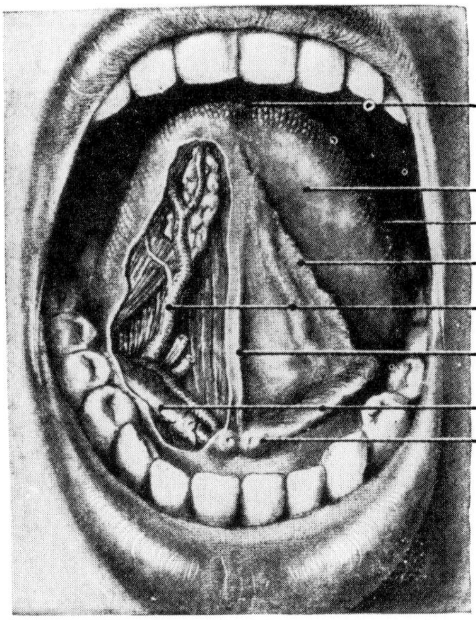

Fig. 7-38. Lower surface of the tongue, and the floor of the mouth. The lingual vein here is the vena comitans nervi hypoglossi or ranine vein. (After Spalteholz, from Anson B (ed): Morris' Human Anatomy, 12th ed, p 1238. New York, McGraw-Hill, 1966)

forward, lateral to the posterior part of the genioglossus, to interlace with the styloglossus muscle and the intrinsic muscles of the tongue.

The hypoglossal nerve and the lingual vein pass across the lateral (superficial) surface of the hyoglossus muscle in their course toward the tongue, but the lingual artery passes deep to it, between it and the genioglossus. Also lateral to the hyoglossus, between it and the inner surface of the mandible, are, from below upward, the submandibular duct and the associated deep part of the gland, the lingual nerve, and the sublingual gland; the latter, and the nerves and vessels to the tongue, also extend forward beyond the anterior border of the hyoglossus to lie on the lateral surface of the genioglossus.

The *styloglossus muscle* arises from the lower end of the styloid process and usually also from the upper part of the stylohyoid ligament and, broadening as it runs downward and forward, passes deep to the medial pterygoid muscle and to the lingual nerve to reach the tongue; here many of its fibers continue superficially on the lateral aspect of the tongue as a longitudinal bundle, while others interdigitate with fibers of the hyoglossus and pass more deeply into the tongue. Like the hyoglossus, the lower part of this muscle is related laterally to a portion of the sublingual gland.

The small *palatoglossus* (*glossopalatinus*) *muscle,* not truly a muscle of the tongue, runs downward from the soft palate to the posterolateral side of the dorsum of the tongue, forming as it does so the palatoglossal arch. It receives its innervation from the pharyngeal plexus.

Intrinsic Muscles

The intrinsic muscles of the tongue are complicated bundles of interlacing fibers, among which are connective tissue septa; the strongest is the midline *septum linguae,* which lies between the muscles, nerves, and vessels of the two sides, rather effectively separating the deeper structures of the tongue except close to its tip, where transverse fibers cross from one side to the other (Strong), and the deep lingual artery anastomoses with its fellow of the opposite side. Superior and inferior longitudinal muscles are described. The first lies just deep to the mucous membrane of the entire length of the dorsum of the tongue and runs forward and laterally to attach into the mucous membrane and intermingle with other fibers; the second lies on the lower part of the tongue, between the genioglossus and the hyoglossus as they enter the tongue, and

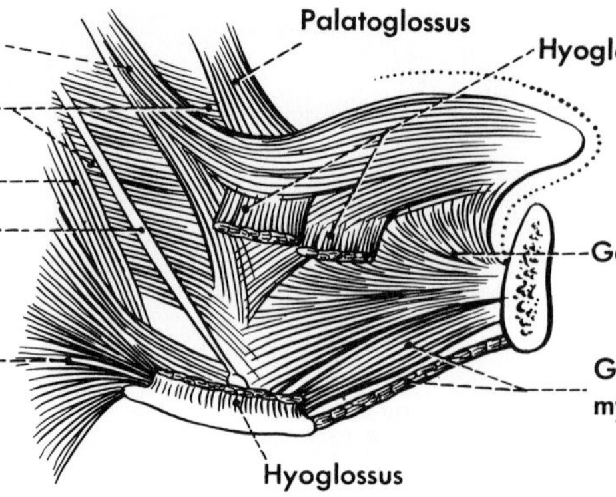

Fig. 7-39. Musculature of the tongue, with the hyoglossus largely removed.

blends anteriorly with longitudinal fibers contributed by the styloglossus. The remaining intrinsic fibers, contributing much to the bulk of the tongue, are called transverse and vertical, although both are oblique (Strong). The transverse fibers run obliquely from the septum to the side and the lateral part of the upper surface of the tongue, and the vertical fibers run from the upper to the lateral surface.

Actions and Innervation

The intrinsic muscles are particularly involved in producing changes in shape of the tongue involved in speaking and swallowing. The extrinsic muscles also help to modify the shape of the tongue, but especially assist in pulling the tongue forward, backward, upward, or downward, while both sets of muscles may act with others of the suprahyoid group, for instance, the mylohyoid, in such functions as swallowing.

Both intrinsic and extrinsic muscles of the tongue are supplied by branches of the hypoglossal nerve that are believed to contain true hypoglossal fibers. It is well known that paralysis of the hypoglossal nerve causes a deviation of the tongue to the paralyzed side when this organ is protruded, and this is usually accepted as being due to the actions of both the intrinsic muscles and the genioglossus muscle of the sound side. The upper fibers of the genioglossus can also retract the protruded tongue, at the same time depressing its tip.

The styloglossus is the chief retractor of the tongue, raising its margins at the same time; the hyoglossus aids in retraction and depresses the tongue, especially its margins. The palatoglossus is primarily a narrower of the faucial isthmus.

NERVES AND VESSELS

The carotid and internal jugular vessels and the nerves associated topographically with them are discussed in connection with the neck; the present section deals with the more strictly suprahyoid nerves and vessels, primarily therefore those of the tongue.

INNERVATION OF THE MUSCLES OF THE TONGUE

The mylohyoid nerve (from the mandibular), its associated artery (from the maxillary), and the submental branch of the facial artery have already been described as extending superficial to the mylohyoid muscle; the important nerves and vessels above the mylohyoid muscle are those connected with the tongue. As already stated in the discussion of the musculature of the tongue, the motor innervation to this is entirely by way of the hypoglossal nerve.

The afferent innervation to the musculature of the tongue has been said to be through fibers of the first and second cervical nerves that accompany the hypoglossal (Corbin and

Harrison) and through the lingual nerve (Barron). Porter recorded impulses consequent to deformation of the tongue in the cat from the lingual nerve, but not from the hypoglossal.

To what extent movements of the tongue in man depend upon afferent endings in the muscle—whether lingual or cervical–hypoglossal—and to what extent on touch from the mucous membrane, is not clear. Weddell and associates stated that after anesthetization of the lingual nerves in human beings there is no ataxia or tendency to bite the tongue, while Rowbotham stated that bilateral section of the fifth nerve markedly interferes with movements of the tongue at first, but there is much subsequent improvement. Carleton reported that anesthesia of the mucous membrane of the human tongue gave rise to complete loss of sense of position of the tongue in four of eight instances in which this was tried, partial loss in three, and no loss in one.

TASTE AND GENERAL SENSATION ON THE TONGUE

The afferent supply to the mucous membrane of the tongue is almost entirely through the lingual and glossopharyngeal nerves, although the superior laryngeal branch of the vagus sends twigs upward to reach the epiglottis and a small area on the posterior part of the tongue. The fibers of the lingual nerve are distributed to approximately the anterior two thirds of the tongue, that is, the area in front of the vallate (circumvallate) papillae, while those of the glossopharyngeal are distributed to the posterior third of the tongue, including the vallate papillae (Fig. 7-40).

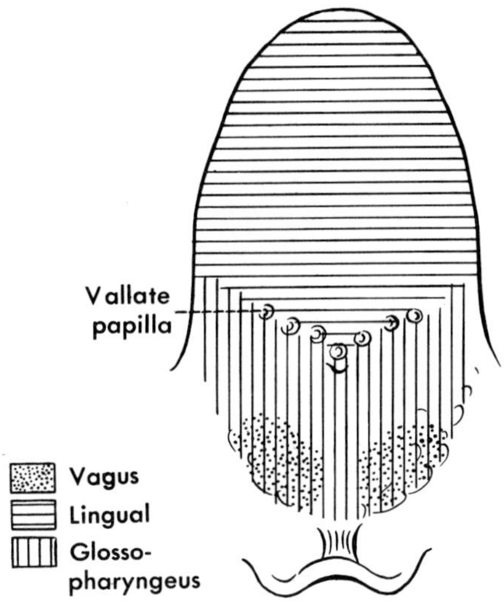

Fig. 7-40. Sensory innervation of the tongue.

Both nerves contain fibers both of general sensation (touch, pain, temperature) and of taste; the taste fibers in the lingual nerve are derived from the geniculate ganglion of the facial and join the lingual by way of the chorda tympani, while the taste fibers of the glossopharyngeal enter the brain stem with it and have their cells of origin situated in the inferior (petrosal) ganglion of the ninth. The fibers of general sensation in the lingual nerve are derived from cells in the trigeminal ganglion and are therefore true trigeminal fibers, while the fibers of general sensation in the ninth enter the brain stem with the ninth nerve.

It is the distribution of the ninth nerve to the posterior part of the tongue and to the adjacent pharyngeal wall that accounts for the distribution of pain here in *glossopharyngeal neuralgia;* this pain is often initiated by such actions as swallowing, eating, or talking and usually arises from the tonsillar-pharyngeal-lingual region, with or without extension to the ear, but may occur first in the ear and extend to the tongue and throat.

Intracranial section of the ninth nerve for glossopharyngeal neuralgia produces both an anesthesia and a loss of taste on the posterior third of the tongue (Peet, and others). Perhaps there may sometimes be an overlap between the regions of distribution of the ninth and lingual nerves, for Peet reported that in one case of section of the ninth there was a slight diminution of taste on the anterior two thirds of the tongue, and stated that sometimes also the fifth nerve overlapped (for general sensation) onto the territory of the ninth. Weddell and co-workers reported a little overlap for pain and touch close to the lower gum. Glossopharyngeal neuralgia has, however, apparently been mistaken for trigeminal neuralgia.

The taste fibers in the ninth nerve, distributed especially to the numerous taste buds in the vallate papillae, are far more important than are those in the seventh, but nevertheless the loss of taste due to section of one glossopharyngeal nerve is apparently not noticed by the patient; likewise, it is well known that loss of taste through section of the chorda tympani in operations on the stapes in the middle ear cavity is often not complained of by the patient. However, patients may complain of a total loss of taste following tonsillectomy, even though it is certain that in these cases taste to the anterior two thirds of the tongue has not been disturbed and that it is only the glossopharyngeal nerve whose function has been temporarily interrupted by edema about the nerve following the tonsillectomy. Arey and co-authors stated that the taste buds of the vallate papillae undergo considerable degeneration during the later years of life and may even almost entirely disappear after the seventieth year. Lalonde and Eglitis found almost as many taste buds on the soft palate, pharynx, epiglottis, and larynx as in the vallate papillae in a newborn infant.

Hayes and Elliott reported only a little overlap between the distribution of the ninth and chorda tympani nerves in the kitten and a little overlap between the two sides near the tip of the tongue, but according to Weddell and co-workers the lingual nerve of human beings does not cross the midline insofar as taste fibers are concerned.

Course of Taste Fibers

Although much of the peripheral course and distribution of the taste fibers is generally agreed upon, the route by which they enter the brain stem was long a matter of dispute. As long ago as 1903 Cushing took note of the fact that taste fibers from the tongue had been stated to enter the brain with the fifth nerve and examined taste sensation in a number of patients in whom the trigeminal ganglion had been removed. Although he found that removal of the ganglion sometimes seemed to abolish or to lessen the acuity of taste over the anterior portion of the tongue temporarily, it never did so permanently or completely. Rowbotham also agreed that the fifth nerve is not a part of the taste path into the brain stem, but stated that section of it may upset taste, apparently because of the removal of the normal accompanying sensation of touch.

Lewis and Dandy investigated patients in whom the fifth, seventh, or ninth nerves had been divided subtentorially or in whom both fifth and seventh nerves had been divided at the same time. In their cases they found that only section of the glossopharyngeal abolished taste on the posterior third of the tongue, while intracranial section of the seventh, similarly, immediately and permanently abolished taste over the anterior two thirds of the tongue on the operated side; they concluded, therefore, that the taste pathways are those described in modern texts, namely, that the fibers associated peripherally with the ninth nerve enter the brain stem with this, while taste fibers associated peripherally with the lingual and chorda tympani nerves enter the brain stem with the facial nerve.

Schwartz and Weddell agreed that section of the sensory root of the fifth nerve does not affect taste, but reported a case of loss of taste in one patient following excision of the greater (superficial) petrosal nerve and two cases of preservation of taste over the anterior two thirds of the tongue following section of the chorda tympani. Since they also found, in the majority of individuals, loss of taste after injury to the chorda tympani or to the facial nerve in the canal, they concluded that the chorda is the usual pathway for taste from the anterior two thirds of the tongue, but that in some individuals the greater petrosal nerve forms a part of this peripheral pathway (Fig. 7-41). Trigeminal fibers do join the tractus solitarius, which is concerned with visceral sensation, including taste; however, the evidence seems to be clear that a taste pathway entering the brain stem through the trigeminal, if it exists, is of no real importance.

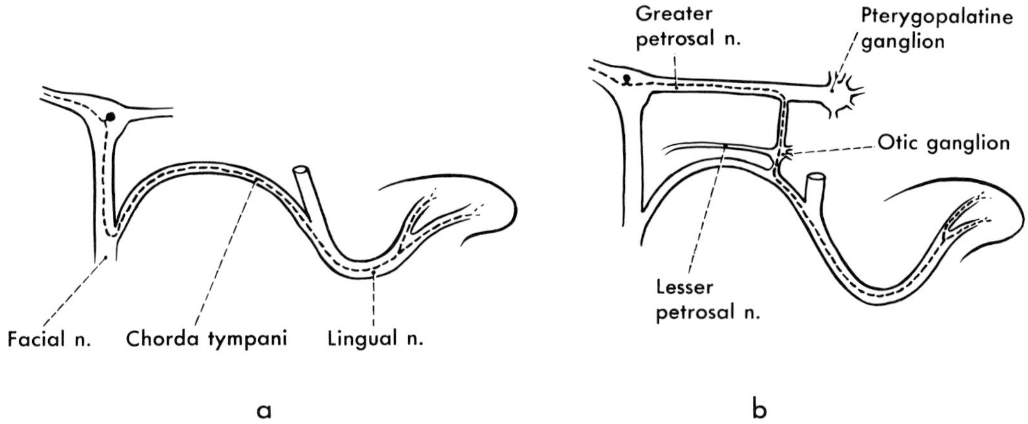

Fig. 7-41. (*a*), The usually accepted pathway for taste from the anterior two thirds of the tongue; (*b*), the pathway that apparently exists in some individuals, according to Schwartz and Weddell. In both figures the taste fibers are shown entering the central nervous system by way of the facial nerve; in (*b*), however, they are shown taking an unusual course to the geniculate ganglion of this nerve through minute connections between the chorda tympani, the otic ganglion, and the greater petrosal nerve.

COURSES AND RELATIONS OF THE NERVES TO THE TONGUE

The lingual and hypoglossal nerves and the lingual vessels enter the tongue from behind and, as they run forward, curve up into it. The glossopharyngeal nerve also enters from behind, but lies deeper and is limited in its distribution to the posterior part of the tongue.

Lingual Nerve

The lingual nerve receives the chorda tympani of the facial at about the level of the posterior border of the medial pterygoid muscle; the chorda tympani emerges from the petrotympanic fissure and passes downward and forward in a short course deep to the lateral pterygoid before joining the lingual. The lingual nerve then passes across the lateral surface of the medial pterygoid muscle, between this and the ramus of the mandible, to lie on the lateral surface of the styloglossus muscle as the latter curves forward into the tongue; in this position it bears the submandibular (submaxillary) ganglion and then curves forward lateral to the hyoglossus and the other muscles of the tongue (Fig. 7-42). It then lies somewhat between the submandibular duct below and the sublingual gland above, and also lateral to the sublingual gland. It sends off a number of small branches to this gland and across it to the gingival mucosa and the floor of the mouth and then passes below the gland, turning around the submandibular duct and the lower border of the sublingual gland to come in close connection with the hypoglossal nerve and the sublingual artery as these lie medial to the sublingual gland. The lingual nerve then runs upward between the sublingual gland and the genioglossus muscle into the tongue.

Hypoglossal Nerve

The hypoglossal nerve typically emerges between the internal carotid artery and the internal jugular vein to descend in the lateral groove between them, to a point not far above the carotid bifurcation. It then gives off the superior root of the ansa cervicalis (the descendens hypoglossi of the ansa hypoglossi) and curves forward across the lateral surfaces of both internal and external carotid vessels. In this course it usually passes downward until just lateral to the origin of the occipital artery and then turns forward, being held at the turn by the sternocleidomastoid branch of

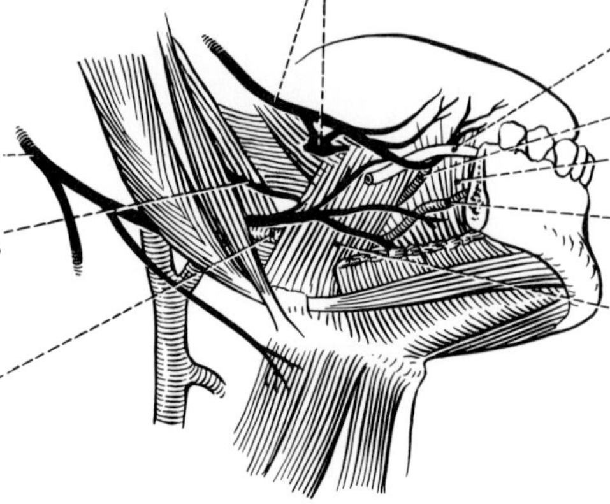

Fig. 7-42. Chief neural and vascular relations of the sublingual region, with the mandible and the submandibular and sublingual glands removed. Note that the hypoglossal nerve, submandibular duct, and lingual nerve all lie on the lateral surface of the hyoglossus muscle, but that the lingual artery passes deep to this muscle.

the occipital artery so that it hooks around these vessels before it runs forward above the hyoid bone. Before giving off the root of the ansa it receives fibers from the first or second cervical nerve, many of which go into the ansa.

The nerve passes forward deep to the stylohyoid muscle and to the posterior belly of the digastric and enters the posterior part of the submandibular or digastric triangle, where it comes to lie upon the lateral surface of the hyoglossus muscle, under cover of the submandibular gland. Here it is accompanied by the largest tributary of the lingual vein, the vena comitans nervi hypoglossi. It passes forward upon the hyoglossus and disappears above the posterior border of the mylohyoid muscle; here, beneath the tongue, it curves forward and upward on the lateral surface of the genioglossus muscle, anastomoses with the lingual nerve, and enters the musculature of the tongue.

After giving off the root of the ansa cervicalis (ansa hypoglossi) at about the level of its forward turn, the hypoglossal nerve gives twigs to the thyrohyoid and geniohyoid muscles, which are usually believed to contain only cervical nerve fibers, and branches to the extrinsic muscles of the tongue before ending in the intrinsic muscles.

Glossopharyngeal Nerve

The glossopharyngeal nerve, after giving off its carotid and pharyngeal branches and its lone motor (stylopharyngeal) branch, passes lateral to the stylopharyngeus muscle, and under cover of the styloglossus and hyoglossus muscles gives off branches to the tonsil and to the posterior third of the tongue. In this position it lies close to the lower pole of the tonsil.

Finally, the superior laryngeal branch of the *vagus* is said to send a few ascending twigs to the base of the tongue, presumably from the branches that also supply the epiglottis.

LINGUAL ARTERY AND VEIN

The lingual artery (Fig. 7-43) is typically the second branch arising anteriorly from the external carotid artery and may arise either below or under cover of the posterior belly of the digastric and the angle of the jaw. From this origin it runs forward or curves somewhat upward, gives off its tonsillar branch, and runs forward above the hyoid bone medial to the hypoglossal nerve. It passes deep to the posterior part of the digastric tendon and leaves the submandibular triangle by disappearing deep to the posterior border of the hyoglossus muscle.

Since ligation of its stem of origin from the external carotid may leave a good collateral circulation (because in about a fifth of cases this stem is a common one for both the lingual and facial arteries), the lingual artery is usually ligated above the hyoid bone if interruption of the entire vessel is desired. In seeking the lingual artery here, after pushing the submandibular gland upward, the hypoglossal nerve is identified as it runs across the digastric triangle and if necessary is pushed somewhat upward to form a triangle between it and the angle of union of the two bellies of the digastric muscle. The triangle thus made is Lesser's triangle, and slightly more deeply within the posterior part of this is the lingual artery; it is visible for a short distance before it passes deep to the hyoglossus muscle.

At the posterior border of this muscle, the lingual artery gives off a small suprahyoid branch and then runs forward deep to the muscle, lying immediately above and parallel to the hyoid bone, and with the middle constrictor of the pharynx, the stylopharyngeus muscle, and the stylohyoid ligament medial to it. In this position it is typically not accompanied by a vein, except posteriorly by the dorsal lingual vein, since the main lingual vein runs with the hypoglossal nerve lateral to the hyoglossus muscle.

Soon after it passes under cover of the hyoglossus muscle the lingual artery gives off one or, more commonly, two *dorsal lingual* branches to the posterior portion of the tongue, and at about the anterior edge of this muscle divides into its terminal branches, the sublingual and the deep lingual or ranine.

The *sublingual artery* is the smaller branch; continuing the course of the chief vessel it passes along the side of the genioglossus muscle between this and the sublingual gland and supplies both the gland and the muscles of the tongue. Beneath the mucous membrane of the anterior part of the floor of the mouth it anastomoses with its fellow of the opposite

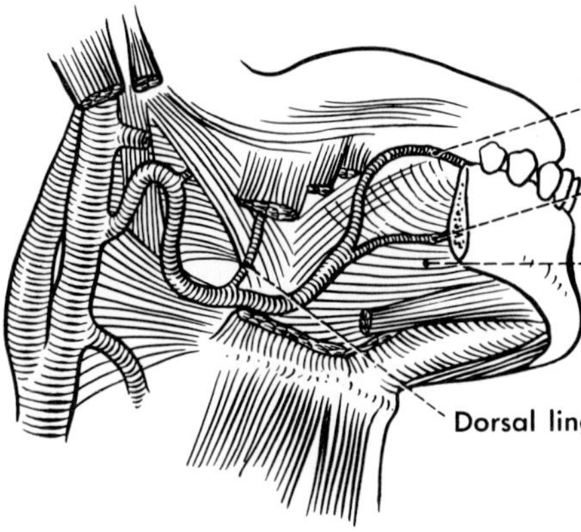

Fig. 7-43. Course and main branches of the lingual artery. For other relations see Fig. 7-42.

side, and this anastomotic branch gives off the artery to the frenulum; in cutting the frenulum for tongue-tie, this artery may have to be ligated.

From its origin the *deep lingual artery* turns upward and then once again forward, so that it runs not below the mucous membrane of the floor of the mouth but on the inferior surface of the tongue; its accompanying veins are here conspicuous when the mouth is open and the tongue lifted so that its undersurface can be examined. Its branches pass upward into the tongue with the terminal branches of the lingual and hypoglossal nerves. In operations involving excision of only a portion of the tongue, the deep lingual artery can be identified through an incision of the mucous membrane in the lateral part of the floor of the mouth, and ligated as it emerges from the cover of the hyoglossus muscle to lie on the lateral surface of the genioglossus. The anastomoses between the arteries of the two sides are largely submucous and are minute, so that unilateral operations upon the tongue can be carried out with practically no bleeding after ligation of the lingual artery on that side; bleeding is more free at the tip of the tongue, where the largest anastomoses between the two deep lingual arteries occur.

Lingual Vein

Small veins from the sides and dorsum of the tongue unite to form paired veins, sometimes at their origin termed *deep lingual veins,* that accompany the deep lingual and then the lingual artery, receive dorsal lingual veins, and unite to enter into the internal jugular or facial vein. A vein that runs separately from the lingual artery begins close to the tip of the tongue; it has been called the lingual vein and the deep lingual vein, but since it accompanies the hypoglossal nerve across the outer surface of the hyoglossus muscle, its better name is the vena comitans nervi hypoglossi. Because it usually forms a prominent swelling on the lower surface of the tongue, it is also known as the ranine vein. This vein receives the sublingual vein and ends either by uniting with the other lingual vein or independently in the facial (common facial).

LYMPHATIC DRAINAGE OF THE TONGUE

Knowledge of the lymphatic drainage of the tongue is of particular importance in metastasis of carcinoma of the tongue; although all the lymphatics of the tongue eventually drain into the deep cervical nodes, they take divergent courses from different portions of the tongue and may or may not end in regional nodes before reaching the deep cervical group (Fig. 7-44).

Fig. 7-44. Lymphatic drainage of the tongue. (After Jamieson JK, Dobson JF: Br J Surg 8:80, 1920)

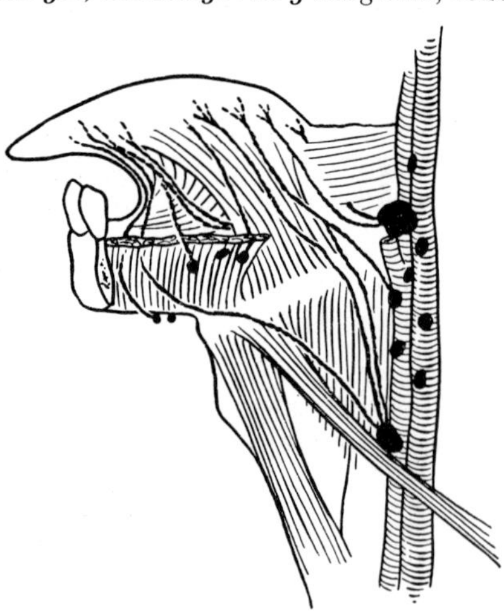

Lymphatic vessels from the tongue behind the vallate papillae run downward toward the hyoid bone and curve backward to penetrate the pharyngeal wall and enter nodes of the upper part of the deep cervical chain, especially a particularly large node (jugulodigastric) situated on the lateral aspect of the internal jugular vein at about the point at which this vessel is crossed by the lower border of the posterior belly of the digastric muscle.

The lymphatics from the anterior two thirds of the tongue are divided into two sets (Jamieson and Dobson): marginal and central vessels. The *marginal vessels* drain the outer thirds of the upper surface of the tongue, the margin itself, and the sides of the lower surface. The more anterior of these, originating behind the tip of the tongue, pass downward and penetrate the mylohyoid muscle to end in the submandibular (submaxillary) lymph nodes; the posterior ones course downward behind the edge of the mylohyoid muscle to reach the deep cervical lymph nodes directly, largely ending in the prominent node (juguloomohyoid node) situated at about the point at which the omohyoid muscle crosses the internal jugular vein.

Of the *central vessels,* those from the tip of the tongue pass downward through the mylohyoid muscle to end either in the submental nodes or course downward across the hyoid bone and end directly in nodes of the deep cervical chain, especially again in the juguloomohyoid node. The central vessels from the remainder of the upper surface of the tongue run down largely between the two genioglossus muscles; some of the more anterior ones drain, as do the marginal lymphatics, into the submandibular nodes, while others bypass these nodes and go directly to either the upper or the lower deep cervical nodes. An important fact stressed by Jamieson and Dobson is that the central vessels from one side of the tongue go to both deep cervical chains (Fig. 7-45); thus while a carcinoma originating on the lateral margin of the tongue may be expected to metastasize first to nodes on the side of origin, carcinoma from the central portion of the root, body, or the tip of the tongue is apt to involve the nodes along both internal jugular veins.

It should be apparent from the above description that metastatic carcinoma from the tongue may involve first the submental nodes, or (more frequently) the submandibular nodes, or any of the nodes of either the upper or lower parts of the deep cervical chain; thus failure to find enlargement of the submental or submandibular nodes upon palpation, or even failure to discover involvement of them at biopsy, does not indicate that carcinoma has not metastasized; it may instead have affected nodes of the deep cervical chain directly. It is because of the widespread locations of the nodes receiving the lymphatics of the tongue that block dissections of the neck are made in operations for lingual carcinoma. In these dissections the sternocleidomastoid muscle, the internal jugular vein, the digastric and stylohyoid muscles and the submandibular gland, and all the connective tissue associated with these structures are removed, usually *en bloc,* while the surfaces of the deeper muscles are cleaned to the bare muscle and the arteries and important motor nerves are carefully cleaned and spared—really a careful anatomical dissection of these parts.

It is worth noting, incidentally, that none of the lymphatics of the tongue reach the superficial nodes associated with the facial artery or the parotid gland (Jamieson and Dobson). In carcinoma of the tongue, therefore, in contrast to that of some portions of the face, excision of the lower pole of the parotid gland offers no advantage except to render more accessible upper members of the deep cervical chain of nodes.

Jamieson and Dobson stated that unilateral block dissection of the neck should be efficacious for early carcinoma of the lateral border of the tongue, but because of the bilateral lymphatic drainage bilateral dissection should be performed if the tip of the tongue, the frenulum, or the dorsum is involved. Block or "radical" dissection of the neck is discussed in Chapter 9.

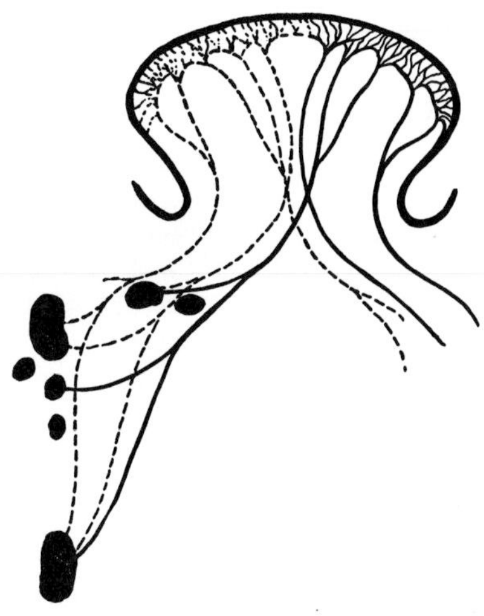

Fig. 7-45. Lymphatic drainage of the tongue, illustrated in a cross section. Note the bilateral drainage of the central vessels. (After Jamieson JK, Dobson JF: Br J Surg 8:80, 1920)

SUBMANDIBULAR AND SUBLINGUAL GLANDS

SUBMANDIBULAR GLAND

The submandibular (submaxillary) gland occupies most of the similarly named triangle and expands beyond this over the superficial surfaces of the anterior and posterior bellies of the digastric. Its posterior border is close to the lower part of the parotid gland at the angle of the jaw, where it is separated from this gland by the stylomandibular ligament. The submandibular gland is crossed superficially by the facial vein and sometimes by the marginal mandibular branch of the facial nerve. The large submandibular lymph nodes lie along the superficial upper border of the gland, between it and the mandible; other nodes may lie anterior and posterior to the gland.

In removing the gland the facial vein is sacrificed, but the marginal mandibular branch of the facial nerve is preserved if possible in order to avoid distortion of the mouth. The facial artery passes across the upper surface of the gland, usually grooving it deeply before rounding the lower border of

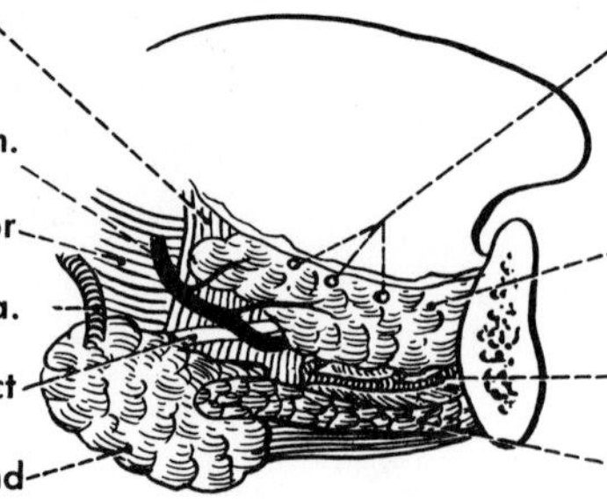

Fig. 7-46. Submandibular and sublingual glands from the side, after removal of the mandible.

the mandible, and must also be sacrificed in removal of the gland; it is therefore ligated both in the neck and at the inferior border of the mandible as it emerges between this and the gland.

The anterior portion of the submandibular gland lies directly against the mylohyoid muscle (Fig. 7-46), and the mylohyoid nerve is then the only important structure related to the deep surface of this portion of the gland; more posteriorly, however, the hypoglossal nerve lies between the gland and the hyoglossus muscle, and this nerve should therefore be sought and spared in removal of the gland. Higher up, medial to the mandible and above the level of the submandibular gland (unless this extends upward unusually deeply), runs the lingual nerve in its course toward the tongue.

In removing the gland from its extension medial to the mandible, care should be taken that the lingual nerve is not pulled downward into the field of operation and injured here; this is possible because the submandibular (submaxillary) ganglion with its roots and branches forms a connection between the gland and the lingual nerve. Total removal of the gland involves also dissection about the posterior border of the mylohyoid muscle, since a deep portion of the gland rather regularly follows the submandibular duct in this position, so that the gland is typically folded around the posterior edge of the muscle.

Submandibular Duct

The submandibular duct (Wharton's duct) runs forward above the mylohyoid muscle between the inner surface of the mandible and the lateral surfaces of the hyoglossus and genioglossus muscles. The duct lies lateral to the hypoglossal nerve and at first below the lingual nerve; as the lingual nerve descends, it crosses lateral to the duct. Both the duct and the lingual nerve then pass around the lower border of the sublingual gland to lie medial to this, the lingual nerve at the same time passing below the duct and then crossing it medially (thus having looped almost completely around it) to turn upward on the genioglossus muscle, while the duct continues forward.

In the terminal portion of its course, therefore, the submandibular duct is in immediate contact with the sublingual gland, lying at first below and then medial to it; at the anterior end of this gland it may receive a so-called major sublingual duct (Bartholin's duct) before it opens into the mouth. In the terminal part of its course the duct lies immediately below the mucosa of the floor of the mouth, which it raises slightly to form the sublingual caruncle or papilla just lateral to the frenulum. The submandibular duct is inelastic, and obstruction of it may cause intense pain.

SUBLINGUAL GLAND

The sublingual gland lies in the floor of the mouth below the mucosa (where it forms a sublingual fold) and above the mylohyoid muscle; laterally it is related to the mandible, and medially it lies against the styloglossus and hyoglossus muscles posteriorly and the genioglossus anteriorly; the glands of the two sides almost meet in the midline in front of the anterior fibers of the genioglossus. The lingual nerve may descend lateral to the posterior end of the gland and, at any rate, runs along its lower border and sends small branches across its lateral surface toward the mucosa of the floor of the mouth; farther forward the lingual nerve and the submandibular duct curve around the lower border of the gland to reach its medial surface. Between the anterior part of the gland and the genioglossus muscle the lingual and hypoglossal nerves both curve upward into the tongue, the sublingual artery runs forward to supply the gland and adjacent muscles, and the submandibular duct continues along the upper medial border of the sublingual gland.

Sublingual Ducts

Approximately a dozen or more small sublingual ducts (ducts of Rivinus) pass from the upper border of the gland to the sublingual fold to empty directly into the buccal cavity; sometimes some of the more anterior sublingual ducts unite together to form a larger duct, the "major" sublingual duct or duct of Bartholin, which joins the submandibular duct to empty with this at the sublingual caruncle. The major sublingual duct has been variously described as of usual occurrence, as occurring on only about 50% of sides, or as being rare in human beings; in any event, the variations in both its presence and size indicate that it is not always truly "major."

CALCULI AND CYSTS

Calculi in the submandibular and sublingual glands or their ducts are apparently not particularly common, but according to Hamlin those of the submandibular gland and its duct comprise about 75% of all salivary calculi. He stated that they occur about three times as commonly in males as in females; they can be demonstrated easily by roentgen ray. The operative approach to the submandibular duct and the sublingual gland and its ducts is through the mucosa of the floor of the mouth on the side of the tongue; since the duct and gland are in intimate contact with both the lingual and hypoglossal nerves, the anatomical relations here are of considerable importance. Cystic tumors of the floor of the mouth, usually called ranulas (as are cystic formations on the lower surface of the tongue—the term refers to their likeness to a frog's belly), are usually due to obstruction of a duct of the sublingual gland. They must be treated by removal of the gland, best done through the use of a Mallis electrocoagulator (Fischer).

SMALLER GLANDS

In addition to the three great pairs of salivary glands, there are of course numerous smaller glands opening into the mouth. These include palatine, labial and buccal (Fig. 7-47), the so-called lesser sublingual (or alveololingual) glands in the lateral part of the floor of the mouth, glands (of Weber) between the submandibular duct and the lateral surface of the tongue, and lingual glands, including the anterior lingual glands of Blandin–Nuhn, already described and shown in Figure 7-38.

INNERVATION

Both the submandibular and sublingual glands are innervated by fibers from the submandibular (submaxillary) ganglion, which accompany sensory fibers of the lingual nerve. The preganglionic fibers for these glands are derived at least primarily from the facial nerve; like the fibers for taste to the anterior two thirds of the tongue, they leave the facial nerve through the chorda tympani and join the lingual nerve to course thereafter with it: at the level of the submandibular ganglion they leave the lingual nerve to end in this ganglion (Fig. 7-48).

Some authors have regarded these glands as also receiving an innervation through the ninth nerve. Routes by which ninth nerve fibers could reach the submandibular ganglion (or seventh nerve fibers reach the otic ganglion) have been described as connections between the ganglion or greater petrosal branch of the seventh and the tympanic plexus or lesser petrosal nerve (ninth), or between the two petrosal nerves, and the communication between the chorda tympani and

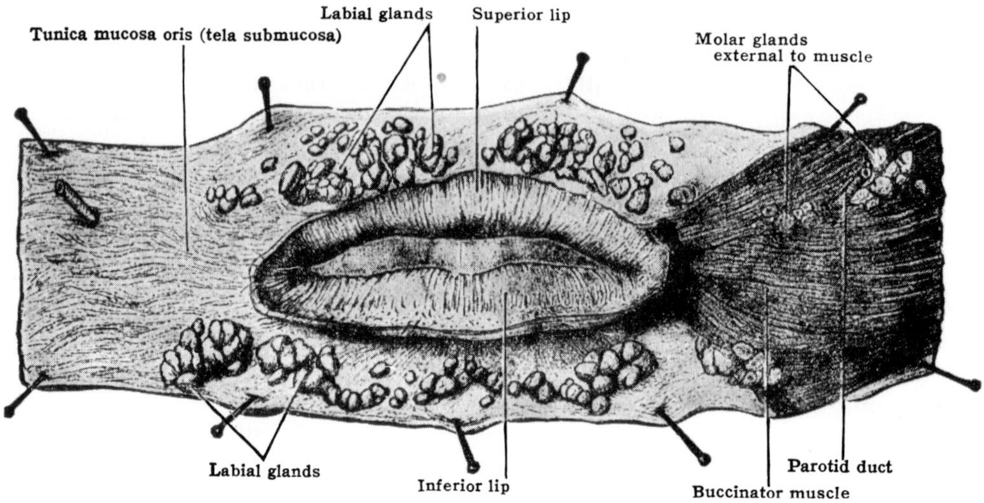

Fig. 7-47. Labial and buccal glands. The molar glands are buccal members of the outer surface of the buccinator muscle. (After Toldt, from Anson B, ed: Morris' Human Anatomy, 12th ed, p 1234)

the otic ganglion. However, many observers (*e.g.*, Dandy; Peet) have been unable to detect any effect on the submandibular and sublingual glands from section of the ninth nerve. Apparent denervation of these glands by section of the chorda tympani is generally acknowledged to produce only a diminution rather than a cessation of secretion, and additional tympanic plexectomy does not change this, so the glands can apparently function without an innervation. This is in contrast to the parotid gland, the denervation of which, as discussed in Chapter 6, is reported to produce cessation of function followed by atrophy.

Sympathetic fibers from the superior cervical ganglion are also said to reach these glands, either by way of other nerves or along the vessels. However, the exact function of this apparent double innervation, parasympathetic and sympathetic, is not clearly understood; stimulation of parasympathetic fibers is said to produce a watery secretion from these glands, while stimulation of sympathetic fibers is said to produce a viscid one. Kuntz and Richins stated that in cats, dogs, and monkeys the parasympathetic fibers far exceed the sympathetic in number. In contrast to some workers who have stated that sympathetic fibers are distributed primarily to mucous cells and parasympathetic ones to serous cells, Kuntz and Richins concluded that both serous and mucous cells are innervated predominantly or exclusively by parasympathetic fibers and that the sympathetic fibers supply mainly or exclusively blood vessels. Snell concluded that all elements in the gland, including blood vessels, are innervated by parasympathetic fibers, and he reported that sympathectomy produced no clear change even in the perivascular plexuses. Rawlinson concluded that variations in blood flow are of considerable importance in the secretory activity of the glands, and it has been stated that the control of salivary secretion is almost entirely brought about by vasodilatation or vasoconstriction in the glands.

ANOMALIES

Major anomalies of the sublingual and lingual regions are rare. Although the lingual artery varies in its origin, not uncommonly arising with the facial artery and sometimes arising with the superior thyroid, it usually shows no major variation in its course. According to Kren a few cases with marked anomalies of the artery have been described; he himself described a case in which one lingual artery was missing and was replaced in the tongue by a branch from the other lingual artery that crossed above the two genioglossus muscles. Other rare variations that he found described in the literature include bilateral absence with replacement by branches from the facial artery, a lingual artery coursing on the inferior surface of the mylohyoid muscle and penetrating this muscle to reach the tongue, and a lingual artery passing across the outer surface of the hyoglossus muscle, in company with the hypoglossal nerve, instead of running deep to this muscle.

Either the submandibular or the sublingual gland may very rarely be missing, and the sublingual gland varies considerably in size. Bilateral absence of the submandibular gland has been reported, as well as a gland lying entirely above the mylohyoid muscle (Reid). When the mylohyoid muscle is deficient, as sometimes occurs, the sublingual gland may bulge through this opening and therefore appear superficial to the muscle.

As evidenced by its ducts, the sublingual gland has developed by a number of outgrowths from the oral mucosa and apparently also sometimes from the region of the anterior end of the submandibular duct; the variations in the latter connection have already been mentioned. The large, single, submandibular duct is apparently quite constant (except, of course, in agenesis of the gland) but very rarely a marked anomaly of this duct may occur. Thus Waller described a case (in an embryo) in which a submandibular duct opened into the pharynx near the faucial isthmus, and Rose described a case in which the chief duct seemed to bifurcate, one part emptying as usual at the sublingual papilla while the second limb, of equal size, appeared to empty in the floor of the mouth opposite the second molar tooth; the anterior limb in this case was blocked by a calculus, and Rose was not sure that the posterior limb might not have been a fistulous tract.

According to Bell and Millar, *macroglossia* is usually defined as an enlargement of the tongue due to one of two conditions: either a lymphangiomatous enlargement or a true hypertrophy of the lingual muscles. They stated that lymphangioma is by far the most common cause of macroglossia; congenital hypertrophy of the musculature of the tongue is said to be usually associated with cretinism, mongolism, and idiocy. Janse described a case associated with cleft palate and micrognathia; in this case, the enlarged tongue produced difficulty in breathing. Handley, and Bell and

Fig. 7-48. The usually accepted course for the secretory parasympathetic fibers to the submandibular and sublingual glands. The preganglionic pathway is indicated by *broken lines;* the postganglionic one, originating in the submandibular ganglion, by *dotted lines.*

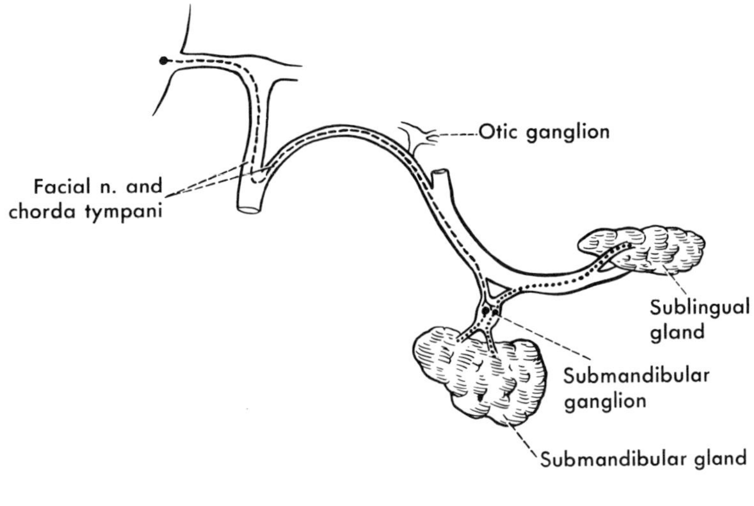

Millar, have discussed surgical techniques for dealing with macroglossia.

Fistulas of the body of the tongue are much rarer than thyroglossal cysts and fistulas, according to Montgomery. He reported a case and outlined the area in which such fistulas might be expected to be present if, as seems reasonable, they result from improper fusion of the swellings (paired, from the first branchial arches, and the unpaired tuberculum impar) normally forming the body of the tongue.

THYROGLOSSAL DUCTS; LINGUAL THYROIDS

Since the thyroid gland develops as a median outgrowth from the floor of the pharynx in the region that also gives rise to the tongue, evidence of this origin is typically found in the posterior part of the midline of the body of the tongue in the form of a pit, the *foramen cecum,* that marks the cranial end of the usually evanescent thyroglossal duct. In a large series of children from a few weeks old up to the age of 10 years, Marshall found that the foramen cecum was almost absent in 38% of these, was a slight pit up to about 3 mm (⅛ inch) long in 46%, and was about 6 mm (¼ inch) long in 15%. Persistence of a more caudal portion of the thyroglossal duct may also occur, however, and such portions may give rise to sinus tracts, cysts, or fistulas along the course of growth and migration of the thyroid gland to its definitive position in the neck.

If a long thyroglossal duct has retained its connection at the foramen cecum, it forms a *sinus tract* that extends among the muscles of the tongue; commonly it ends blindly somewhere in the neck, but it may through infection attain an external fistulous opening in the neck.

As emphasized by a number of workers, treatment of thyroglossal sinuses and fistulas depends upon complete removal of the tract, which is closely associated with the hyoid bone. (Sharp disagreement exists on the exact relation: Pemberton and Stalker stated that it is variable, the duct sometimes passing in front of, sometimes behind, and sometimes through the bone—Fig. 7-49. Ellis and van Norstrand claimed that it always passes in front, although often looping up also behind the bone. Sadé and Rosen published a photograph of a histologic section showing thyroid tissue associated with a duct in the middle of the bone.) Because of the close relationship, and because the portion above the hyoid bone is often small and friable, it is customary to remove also the central segment of the body of the hyoid and the tissue above this in the midline between the muscles of the tongue (Fig. 7-50).

In their series of 293 cases, Pemberton and Stalker reported that 66.2% of these sinuses occurred in males and 33.8% in females. They, Marshall and Becker, and others have reported a very low rate of recurrence (about 1%) when the hyoid bone and the midline tissue up to the foramen cecum were removed. Sadé and Rosen found, among 20 such resected specimens, no tract above the hyoid bone in six, but in only three of the remaining 14 was there a single duct, while in 11 there were three or more ducts here. Their epithelial linings varied, as did those of the cysts, but most showed mucous cells or numerous mucous glands. In six of the 20, thyroid tissue was associated with a duct.

Isolated segments of the thyroglossal duct, which have lost their connection to the foramen secum, may persist along the path of growth of the thyroglossal duct and give rise to *cysts.* Most cysts derived from the thyroglossal duct lie below the level of the hyoid bone, but Hubert found five located above the hyoid among 43 cases, and Marshall and Becker reported 61 of 310 occurred above the hyoid. Cysts of the thyroglossal duct above the hyoid bone may be situated in the base of the tongue (where, however, they are said to be even rarer than lingual thyroids), in the floor of the mouth, or immediately above the hyoid bone (Bailey).

Lingual Thyroid

The most striking anomaly resulting from the development of the thyroid gland from the

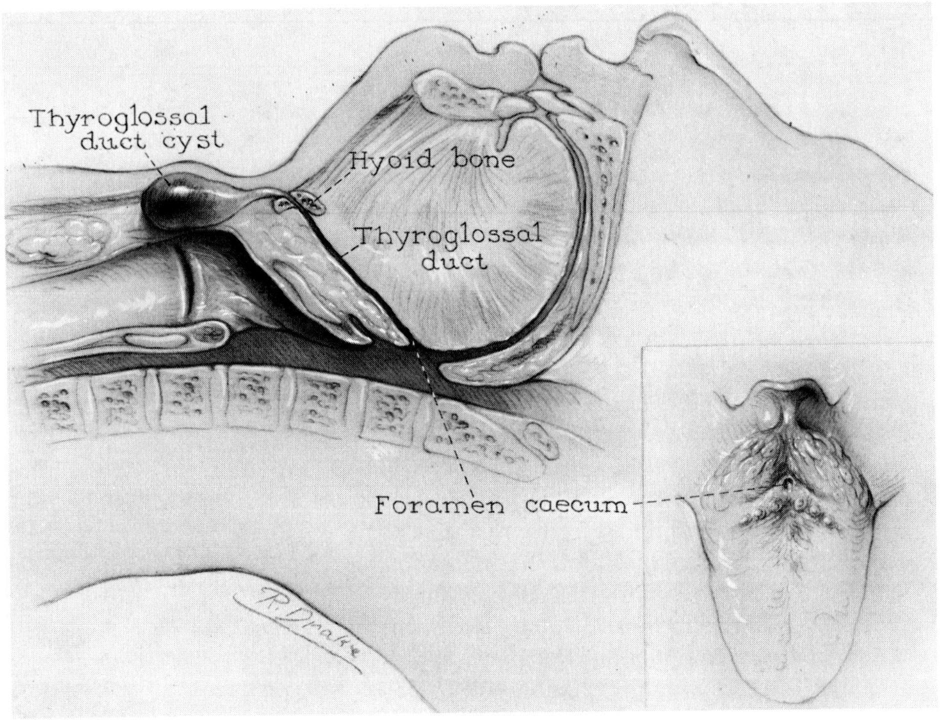

Fig. 7-49. A sagittal section showing the position of the thyroglossal duct and cyst; the inset shows the opening at the foramen cecum. (Pemberton J de J, Stalker LK: Ann Surg 111:950, 1940)

lingual portion of the pharynx is the retention of thyroid tissue here to form a lingual thyroid gland (Fig. 7-51). It may vary in size from a few nodules to a large mass occupying much of the posterior dorsum of the tongue. In recent years radioactive iodine has sometimes been used both to confirm the diagnosis of lingual thyroid without biopsy (which may give rise to troublesome hemorrhage) and to investigate the occurrence of thyroid tissue in the neck.

According to Goetsch the ectopic thyroid tissue was on the dorsum of the tongue in 142 of 144 cases of lingual thyroid reviewed from the literature, and buried in the body of the tongue in only 2; since the tissue is usually located beneath the mucosa at the back of the tongue, the usual symptoms are those of benign tumors of the tongue, that is, interference with swallowing, perhaps some dysphonia, and even respiratory symptoms.

Fig. 7-50. Dissection of the thyroglossal duct, removing a segment of the hyoid bone; the inset shows the incision. (Pemberton J de J, Stalker LK: Ann Surg 111:950, 1940)

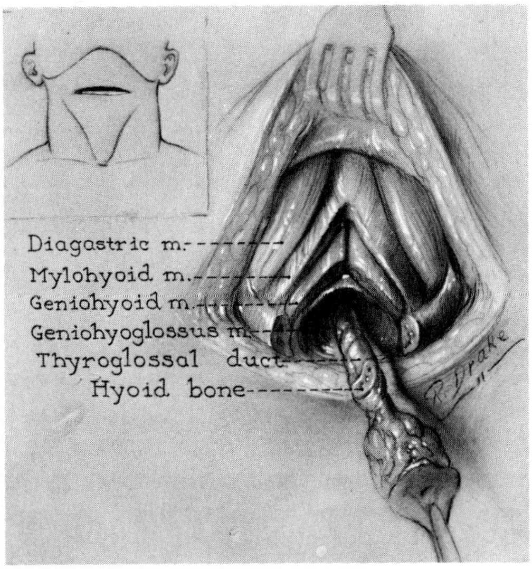

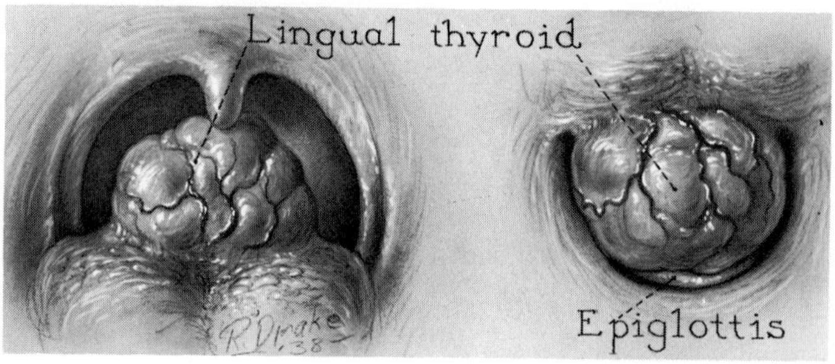

Fig. 7-51. Two views of a lingual thyroid.

Lingual thyroid is apparently found more commonly in females than in males, about 75% of the cases having been reported as occurring in females according to Ray (approximately the same sex ratio as seen in goiter), while Perlman found it had been estimated to occur perhaps eight times more frequently in females than in males. Lingual thyroid glands have frequently not been noted until the physiological enlargement of the gland occurs at puberty and sometimes have gone unnoticed until pregnancy; it may be, therefore, that the reported sex ratio is based not upon their actual occurrence, but rather upon the fact that they enlarge less frequently in the male and are therefore often not recognized.

Removal of a lingual thyroid has usually been through the oral route and incomplete, with control of postoperative myxedema, if necessary, by thyroid replacement therapy. The condition has also been treated by radioactive iodine, but Ward and co-workers concluded that the liability of a lingual thyroid to pathologic change is not sufficiently known, and they recommended complete excision through the lateral pharyngeal wall. Skolnik and co-workers, noting that the lingual thyroid is the only functioning thyroid tissue in some 70% to 80% of patients, reported three cases in which they moved the gland to a submandibular or suprahyoid position.

References

ANGEL JL: Factors in temporomandibular joint form. Am J Anat 83:223, 1948

AREY LB, TREMAINE MJ, MONZINGO FL: The numerical and topographical relations of taste buds to human circumvallate papillae throughout the life span. Anat Rec 64:9, 1935

BAILEY H: Thyroglossal cysts and fistulae. Br J Surg 12:579, 1925

BARRON DH: A note on the course of the proprioceptor fibres from the tongue. Anat Rec 66:11, 1936

BAUMEL JJ, BEARD DY: The accessory meningeal artery of man. J Anat 95:386, 1961

BELL HG, MILLAR RG: Congenital macroglossia: Report of two cases. Surgery 24:125, 1948

BERNICK S: Innervation of teeth and periodontium after enzymatic removal of collagenous elements. Oral Surg 10:323, 1957

DU BOIS-REYMOND R: Beschreibung einer Anzahl Muskelvarietäten an einem Individuum. Anat Anz 9:451, 1894

BYARS LT: Preservation and restoration of mandibular function and contour. Ann Surg 127:863, 1948

CALNAN JS: Movements of the soft palate. Br J Plast Surg 5:286, 1953

CARLETON A: Observations on the problem of the proprioceptive innervation of the tongue. J Anat 72:502, 1938

CASTELLI W: Vascular architecture of the human adult mandible. J Dent Res 42:786, 1963

CHI-MIN W: An anomalous muscle in place of the stylohyoid ligament of a Chinese. Anat Rec 107:375, 1950

CORBIN KB, HARRISON F: The sensory innervation of the spinal accessory and tongue musculature in the Rhesus monkey. Brain 62:191, 1939

COSTEN JB: Glossodynia: Reflex irritation from the mandibular joint as the principal etiologic factor; study of ten cases. Arch Otolaryngol 22:554, 1935

CRIKELAIR GF, BOM AF, LUBAN J, MOSS M: Early orthodontic movement of cleft maxillary segments prior to cleft lip repair. Plast Reconstr Surg 30:426, 1962

CUSHING H: The taste fibers and their independence of the N. trigeminus: Deductions from thirteen cases of Gasserian ganglion extirpation. Bull Johns Hopkins Hosp 14:71, 1903

CUSHING H: The sensory distribution of the fifth cranial nerve. Bull Johns Hopkins Hosp 15:213, 1904

DANDY WE: Glossopharyngeal neuralgia (tic douloureux): Its diagnosis and treatment. Arch Surg 15:198, 1927

DAVIS WB: Congenital deformities of the face: Types found in a series of one thousand cases. Surg Gynecol Obstet 61:201, 1935

DELITZIN S: Arteria maxillaris communis. Arch Anat Physiol p 268, 1890

DOHERTY JL, DOHERTY JA: Dislocation of the mandible. Am J Surg 38:480, 1937

ELLIS PDM, VAN NORSTRAND AWP: The applied anatomy of thyroglossal tract remnants. Laryngoscope 87:765, 1977

FISCHER ND: Personal communication to the author, 1981

FITZGERALD MJT: The occurrence of a middle superior alveolar nerve in man. J Anat 90:520, 1956

FROESCHELS E: Uvula and tonsils. Arch Otolaryngol 50:216, 1949

GABRIEL AC: Some anatomical features of the mandible. J Anat 92:580, 1958

GAUGHRAN GRL: Mylohyoid boutonnière and sublingual bouton. J Anat 97:565, 1963

GEORGIADE NG: The surgical correction of temporomandibular joint dysfunction by means of autogenous dermal grafts. Plast Reconstr Surg 30:68, 1962

GOETSCH E: Lingual goiter: Report of three cases. Ann Surg 127:291, 1948

GOYDER FW: The anatomy and treatment of cleft palate. Br J Surg 1:259, 1913

GRABER TM: Craniofacial morphology in cleft palate and cleft lip deformities. Surg Gynecol Obstet 88:359, 1949

GURNEY CE: Chronic bilateral benign hypertrophy of the masseter muscles. Am J Surg ns73:137, 1947

HAMLIN FE: Calculus of the submaxillary gland and duct. Arch Otolaryngol 10:177, 1929

HANDLEY S: The technique of Butlin's operation of marginal resection of the tongue. Br J Surg 1:42, 1913

HARPMAN JA, WOOLLARD HH: The tendon of the lateral pterygoid muscle. J Anat 73:112, 1938

HAVENS FZ: Cleft lip and palate: Some technical procedures used in its treatment, with especial reference to closure of complete cleft of the anterior half of the palate. Arch Otolaryngol 48:9, 1948

HAYES ER, ELLIOTT R: Distribution of the taste buds on the tongue of the kitten, with particular reference to those innervated by the chorda tympani branch of the facial nerve. J Comp Neurol 76:227, 1942

HENSEL GC: The surgical correction of mandibular protraction, retraction, and fractures of the ascending rami. Surgery 2:92, 1937

HRDLIČKA A: Lower jaw: Double condyles. Am J Phys Anthropol 28:75, 1941

HUBERT L: Thyroglossal cysts and sinuses: Analysis of 43 cases. Arch Otolaryngol 45:105, 1947

HUNT JR: The sensory field of the facial nerve: A further contribution to the symptomatology of the geniculate ganglion. Brain 38:418, 1915

IVY RH: The influence of race on the incidence of certain congenital anomalies, notably cleft lip-cleft palate. Plast Reconstr Surg 30:581, 1962

JAMIESON JK, DOBSON JF: The lymphatics of the tongue: With particular reference to the removal of lymphatic glands in cancer of the tongue. Br J Surg 8:80, 1920

JANSE HM: Macroglossia with an associated micrognathia causing respiratory difficulty. Ann Otol Rhinol Laryngol 39:99, 1930

JONES FW: The anterior superior alveolar nerve and vessels. J Anat 73:583, 1939

KAZANJIAN VH: Congenital absence of the ramus of the mandible. J Bone Joint Surg ns21:761, 1939

KING TS: The anatomy of hare-lip in man. J Anat 88:1, 1954

KREN O: Über eine seltene Anomalie der Arteria lingualis nebst einer kurzen Zusammenfassung ihrer häufigsten topographischen Varietäten. Anat Anz 73:107, 1931

KUNTZ A, RICHINS CA: Components and distribution of the nerves of the parotid and submandibular glands. J Comp Neurol 85:21, 1946

LALONDE ER, EGLITIS JA: Number and distribution of taste buds on the epiglottis, pharynx, larynx, soft palate and uvula in a human newborn. Anat Rec 140:91, 1961

LASKER GW, OPDYKE DL, MILLER H: The position of the internal maxillary artery and its questionable relation to the cephalic index. Anat Rec 109:119, 1951

LAUBER H: Ueber einige Varietäten im Verlaufe der Arteria maxillaris interna. Anat Anz 19:444, 1901

LEHR RP JR, OWENS SE JR: An electromyographic study of the human lateral pterygoid muscle. Anat Rec 196:441, 1980

LEWIN ML: Management of cleft lip and palate in the United States and Canada: Report of a survey. Plast Reconstr Surg 33:383, 1964

LEWINSKY W, STEWART D: The innervation of the dentine. J Anat 70:349, 1936a

LEWINSKY W, STEWART D: The innervation of the periodontal membrane. J Anat 71:98, 1936b

LEWINSKY W, STEWART D: The innervation of the periodontal membrane of the cat, with some observations on the function of the end-organs found in that structure. J Anat 71:232, 1937

LEWIS D, DANDY WE: The course of the nerve fibers transmitting sensation of taste. Arch Surg 21:249, 1930

LORD FP: Observations on the temporomandibular articulation. Anat Rec 7:355, 1913

MARSHALL CF: Variations in the form of the thyroid gland in man. J Anat Physiol 29:234, 1895

MARSHALL SF, BECKER WF: Thyroglossal cysts and sinuses. Ann Surg 129:642, 1949

MCNAMARA JA JR: The independent functions of the

two heads of the lateral pterygoid muscle. Am J Anat 138:197, 1973

MILLARD RR JR: A new approach to immediate mandibular repair. Ann Surg 160:306, 1964

MISURYA VK: Functional veloplasty in bilateral palatal palsy. Acta Otolaryngol 102:436, 1976

MOFFETT BC JR, JOHNSON LC, MCCABE JB, ASKEW HC: Articular remodeling in the adult human temporomandibular joint. Am J Anat 115:119, 1964

MOLL KL, HUFFMAN WC, LIERLE DM, SMITH JK: Factors related to the success of pharyngeal flap procedures. Plast Reconstr Surg 32:581, 1963

MONTGOMERY ML: Congenital fistulae of the body of the tongue. Ann Surg 100:68, 1934

MOOREHEAD FB: Fractures of the jaws and their management. Am J Surg ns38:474, 1937

MORRIS JH: Chronic recurring temporomaxillary subluxation: Surgical consideration of "snapping jaw" with report of a successful operative result. Surg Gynecol Obstet 50:483, 1930

MUNRO RR: Activity of the digastric muscle in swallowing and chewing. J Dent Res 53:530, 1974

OLDFIELD MC: Cleft palate and the mechanism of speech (Arris and Gale lecture). Br J Surg 29:197, 1941

OSSENBERG NS: The mylohyoid bridge: An anomalous derivative of Meckel's cartilage. J Dent Res 53:77, 1974

PEET MM: Glossopharyngeal neuralgia. Ann Surg 101:256, 1935

PEMBERTON J DEJ, STALKER LK: Cysts, sinuses and fistulae of the thyroglossal duct. Ann Surg 111:950, 1940

PERLMAN HB: Lingual thyroid gland. Arch Otolaryngol 19:594, 1934

PEYTON WT: The dimensions and growth of the palate in the normal infant and in the infant with gross maldevelopment of the upper lip and palate: A quantitative study. Arch Surg 22:704, 1931

PISK E: Über eine seltene Varietät im Verlaufe der Arteria carotis externa beim Menschen und beim Hunde. Anat Anz 45:373, 1914

PORTER R: Lingual mechanoreceptors activated by muscle twitch. J Physiol 183:101, 1966

PRUZANSKY S (ed): Congenital Anomalies of the Face and Associated Structures. Proceedings of an International Symposium. Springfield, Charles C Thomas, 1961

RAWLINSON HE: The changes in the cells of the striated ducts of the cat's submaxillary gland after autonomic stimulation and nerve section. Anat Rec 63:295, 1935

RAY BS: Lingual thyroid. Arch Surg 37:316, 1938

REID D: Notes on the salivary glands. Anat Rec 32:295, 1926

RICH AR: The innervation of the tensor veli palatini and levator veli palatini muscles. Bull Johns Hopkins Hosp 31:305, 1920

RICHARDSON GS, PULLEN EM: The uvula: Its structure and function and its importance. Arch Otolaryngol 47:379, 1948

RITCHIE HP: Congenital clefts of the face and jaws: A report of the operations used and a discussion of results. Surg Gynecol Obstet 73:654, 1941

RITCHIE WP: Cleft palate: A correlation of anatomic and functional results following operation. Arch Surg 35:548, 1937

ROBERTS WH, JORGENSEN NB: A note on the distribution of the superior alveolar nerves in relation to the primary teeth. Anat Rec 141:81, 1961

ROHAN RF, TURNER L: The levator palati muscle. J Anat 90:153, 1956

ROSE B-H: Bifurcation of the submaxillary duct. Am J Surg ns17:257, 1932

ROSS MA: Functional anatomy of the tensor palati: Its relevance in cleft palate surgery. Arch Otolaryngol 93:1, 1971

ROWBOTHAM GF: Observations on the effects of trigeminal denervation. Brain 62:364, 1939

RUDING R: Cleft palate: Anatomic and surgical considerations. Plast Reconstr Surg 33:132, 1964

SADÉ J, ROSEN G: Thyroglossal cysts and tracts: A histological and histochemical study. Ann Otol Rhinol Laryngol 77:138, 1968

SARNAT BG, MUCHNIK H: Facial changes after mandibular condylectomy in the adult monkey. J Anat 108:323, 1971

SCHWARTZ HG, WEDDELL G: Observations on the pathways transmitting the sensation of taste. Brain 61:99, 1938

SEAVER EP JR: Temporomandibular joint malocclusion and the inner ear: A neuromuscular explanation. Ann Otol Rhinol Laryngol 46:140, 1937

SHEPHERD WM, MCCARTHY MD: Observations on the appearance and ossification of the premaxilla and maxilla in the human embryo. Anat Rec 121:13, 1955

SICHER H: Structural and functional basis for disorders of the temporomandibular articulation. J Oral Surg 13:275, 1955

SILVER CM, SIMON SD: Operative treatment for recurrent dislocation of the temporomandibular joint. J Bone Joint Surg 43-A:211, 1961

SILVER CM, SIMON SD: Meniscus injuries of the temporomandibular joint: Further experiences. J Bone Joint Surg 45-A:113, 1963

SILVER CM, SIMON SD, LITCHMAN HM: The surgical treatment of the arthritic temporomandibular joint. Surg Gynecol Obstet 136:251, 1973

SILVER CM, SIMON SD, SAVASTANO AA: Meniscus injuries of the temporomandibular joint. J Bone Joint Surg 38-A:541, 1956

SKOLNIK EM, YEE KF, GOLDEN TA: Transposition of the lingual thyroid. Laryngoscope 86:785, 1976

SMITH AE, JOHNSON JB: Surgery of the cleft palate. Am J Surg ns45:93, 1939

SMITH JK, MCCABE BF: Teflon injection in the nasopharynx to improve velopharyngeal closure. Ann Otol Rhinol Laryngol 86:559, 1977

SNELL RS: The histochemical appearances of cholinesterase in the parasympathetic nerves supplying the submandibular and sublingual salivary glands of the rat. J Anat 92:534, 1958

SPRAGUE JM: The innervation of the pharynx in the Rhesus monkey, and the formation of the pharyngeal plexus in primates. Anat Rec 90:197, 1944

SPRINZ R: The role of the meniscus in the healing process, following excision of the articular surfaces of the mandibular joint in rabbits. J Anat 97:345, 1963

STAHR H: Die Zahl und Lage der submaxillaren und submentalen Lymphdrüsen vom topographischen und allgemein-anatomischen Standpunkte. Arch Anat Physiol p 444, 1898

STARKIE C, STEWART D: The intramandibular course of the inferior dental nerve. J Anat 65:319, 1931

STELLMACH RK: Bone grafting of the alveolus followed by orthopedic alignment of the alveolar arch in infants with complete cleft lip and cleft palate. Oral Surg 16:897, 1963

STRONG LM: Muscle fibers of the tongue functional in consonant production. Anat Rec 126:61, 1956

SUBTELNY JD: A review of cleft palate growth studies reported in the past 10 years. Plast Reconstr Surg 30:56, 1962

TANASESCO JG: Lymphatiques de l'articulation temporo-maxillaire. Anat Anz 41:460, 1912

TEBO HG, TELFORD IR: An analysis of the relative positions of the mental foramina. (abstr) Anat Rec 106:254, 1950

TRIBLE GB: Epistaxis of antral origin: Report of cases. Arch Otolaryngol 10:633, 1929

VIDIĆ B: The origin and the course of the communicating branch of the facial nerve to the lesser petrosal nerve in man. Anat Rec 162:511, 1968

VITTI M, BASMAJIAN JV: Integrated actions of masticatory muscles: Simultaneous EMG from 8 intramuscular electrodes. Anat Rec 187:173, 1977

WAERN A: Contribution à la connaissance des muscles interposés entre les ventres antérieurs des digastriques chez l'homme et présentation d'un projet de dénomination pour ceux-ci. Arch Anat Histol Embryol 20:325, 1935

WAKELEY CPG: The mandibular joint. Ann R Col Surg Eng 2:111, 1948

WALDRON CW: Management of unilateral clefts of the palate. Plast Reconstr Surg 5:322, 1950

WALLER WH: Ectopic submaxillary ostium near the isthmus of the fauces. Anat Rec 58:111, 1934

WARD GE, CANTRELL JR, ALLAN WB: The surgical treatment of lingual thyroid. Ann Surg 139:536, 1954

WARDILL WEM: Cleft palate. Br J Surg 16:127, 1928

WARDILL WEM: The technique of operation for cleft palate. Br J Surg 25:117, 1937

WEDDELL G, HARPMAN JA, LAMBLEY DG, YOUNG L: The innervation of the musculature of the tongue. J Anat 74:255, 1940

WILLIAMS HB, WOOLHOUSE FM: Comparison of speech improvement in cases of cleft palate after two methods of pharyngoplasty. Plast Reconstr Surg 30:36, 1962

WODAK E: Congenital anteflexion of the uvula accompanied by a swelling of the soft palate: Its correction by plastic methods. Ann Otol Rhinol Laryngol 53:581, 1944

Chapter 8

THE PHARYNX AND LARYNX

The Pharynx

The wall of the pharynx consists of mucosa and voluntary muscle, with a more or less well-defined connective tissue layer, the pharyngobasilar fascia (pharyngeal aponeurosis), interposed between the two and a fascial layer (buccopharyngeal fascia) on the external surface of the muscles. The pharyngobasilar fascia is well developed above, where it substitutes for a muscular wall in the pharyngeal recess and may also be distinguished at the level of the tonsil, for which it forms a fibrous bed; lower down, it becomes less and less distinguishable as a layer and finally disappears. It is attached above to the pharyngeal tubercle and the occipital bone lateral to that, and then across the petrous portion of the temporal bone just in front of the external opening of the carotid canal. Close to the sphenoid spine, its attachment turns anteriorly and medially, along the cartilaginous part of the auditory tube (therefore between temporal and sphenoid bones) and then across the tube to reach the sphenoid bone at the level of the medial pterygoid lamina, where it attaches to the entire length of this lamina. In this course it separates the levator veli palatini, which arises below and medial to the auditory tube, from the tensor which arises on the lateral side. Below these upper attachments it is said to be attached to the pterygomandibular raphe, the posterior end of the mylohyoid line on the mandible, the hyoid bone, and the thyroid cartilage.

The mucosa of the pharynx varies in structure. That of the nasal part is, for the most part, ciliated and resembles the mucosa of the nose; in addition, it contains the lymphoid nodules that constitute the pharyngeal tonsil or adenoids. In the remainder of the pharynx the epithelium is stratified squamous (like that of the mouth, with which it is continuous), and the mucosa tends to be tightly attached to the pharyngobasilar fascia.

The muscular wall of the pharynx with its thin covering of buccopharyngeal or visceral fascia is separated from the prevertebral fascia by an area of loose connective tissue that constitutes the retropharyngeal space; the lateral continuations of this connective tissue on both sides between the pharynx and the medial pterygoid muscles form the lateral pharyngeal spaces. These spaces are discussed in Chapter 5. Posterolateral to the pharynx lie the ninth through the twelfth cranial nerves, the internal jugular and the carotid vessels and their branches, and the cervical sympathetic trunk.

NASAL PART

The nasal part of the pharynx (naso-pharynx) is continuous anteriorly, through the choanae, with the nasal cavities; its floor is the

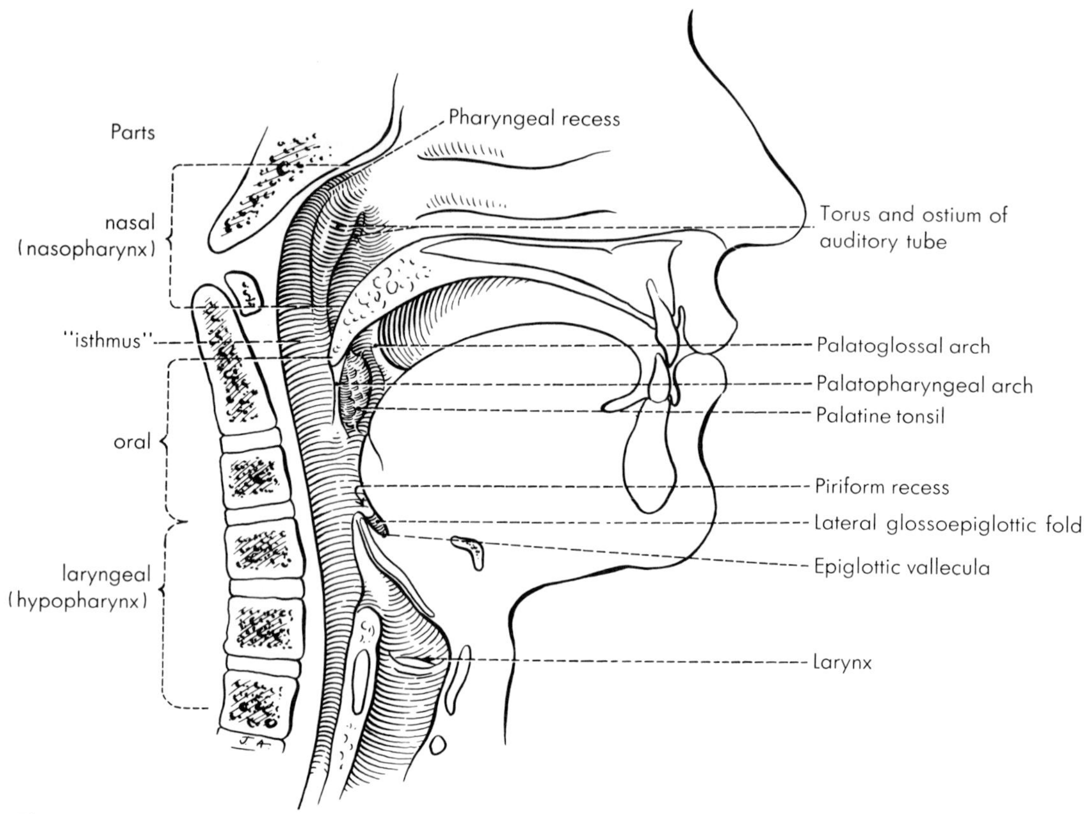

Fig. 8-1. General schema of the pharynx and larynx in sagittal section. Subdivisions of the pharynx are indicated on the left.

upper surface of the soft palate, while its fornix or roof, the mucosa of which is at first attached closely to the base of the skull, slopes downward and backward to become continuous with its posterior wall (Fig. 8-1). At the level of the palate the nasal part of the pharynx narrows and then continues as the oral part, the narrowing often being called the pharyngeal isthmus; the "isthmus" is bounded anteriorly by the soft palate, laterally by the palatopharyngeal arches, and posterolaterally and posteriorly by the pharyngeal wall. Above the isthmus the posterior part of the nasal pharynx is particularly wide, the lateral extension on each side constituting the pharyngeal recess (fossa of Rosenmüller, sinus of Morgagni).

In front of and below the pharyngeal recess, only slightly above the level of the floor of the nasal fossa, is the opening of the auditory (eustachian or pharyngotympanic) tube. The anterior lip of the pharyngeal ostium of the tube is not well marked, but the posterior lip is elevated by the projecting cartilage of the tube to form the torus tubarius.

Pharyngeal Tonsil

The walls of the pharyngeal recess and the roof of the nasal pharynx in contact with the sphenoid and occipital bones contain in children and sometimes in adults a large amount of lymphoid tissue, which constitutes the pharyngeal (naso-pharyngeal) tonsil. This tissue extends to a variable extent into the posterior wall and downward toward the orifice of the auditory tube; the wall of the tube itself is also sometimes described as containing lymphoid tissue. Enlargements of the pharyngeal tonsil (adenoids) may interfere with nasal breathing and, through obstruction of the auditory tube, lead also to deafness because of the subsequent absorption of air from the tympanic cavity. Unlike the palatine and lingual tonsils, the pharyngeal tonsils have no epithelial crypts extending into them, although glands open through them (Arey).

Stenosis of the nasal part of the pharynx (not to be confused with congenital choanal atresia) can be produced by scarring and contracture of the posterior pharyngeal wall in the region of the adenoids or below and may extend across toward the soft palate to occlude the pharynx. Although rare, a majority of cases apparently have resulted from operations on the tonsils and adenoids (Kazanjian and Holmes).

Pharyngeal Bursa and Hypophysis

In the posterior midline of the pharynx, surrounded by the pharyngeal tonsil, there may be a saclike depression, the pharyngeal bursa (Fig. 8-2); according to Dorrance it lies just above the uppermost fibers of the superior constrictor muscles and usually extends upward and backward toward the occipital bone; although it has apparently been mistaken for a persistent remnant of the craniopharyngeal canal, it lies below this. Dorrance found the pharyngeal bursa to be more common in the embryo than in the adult. It is said to arise at the point of adhesion between the anterior end of the notochord and the pharyngeal endoderm (Snook). It may be the site of cyst formation (Tornwaldt's cyst), which may become infected.

Farther upward and forward, deep to the mucosa or in the periosteum close to the vomerosphenoid articulation, and therefore corresponding in position to the diverticulum (Rathke's pouch) from which the anterior lobe of the hypophysis is developed, there apparently regularly occurs a pharyngeal hypophysis. Melchionna and Moore found this in 51 of 54 persons in whom it was searched for, but unless it becomes cystic or gives rise to a tumor it is usually not grossly visible.

PARS ORALIS

The pars oralis or oral pharynx is continuous anteriorly, through the faucial or oropharyngeal isthmus, with the oral cavity; the boundaries of the isthmus are the posterior border of the soft palate above, the palatine arches and intervening tissue laterally, and the dorsum of the tongue inferiorly. Below the fauces the anterior wall of the pharynx is the posterior or pharyngeal part of the dorsum of the tongue. Behind this the epiglottis projects upward and is connected to the tongue by three low folds: the paired lateral glossoepiglottic (pharyngoepiglottic) folds and a median glossoepiglottic fold. The shallow depression between each lateral glossoepiglottic fold is the

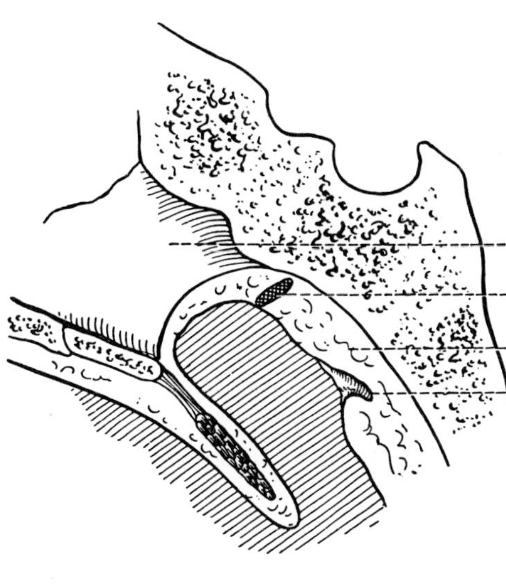

Fig. 8-2. Pharyngeal bursa and pharyngeal hypophysis as seen in a sagittal section of the head of an infant. (After Dorrance GM: Arch Otolaryngol 13:187, 1931)

epiglottic vallecula (see Fig. 7-37); foreign bodies sometimes lodge here.

Lingual Tonsil

On the posterior part of the dorsum of the tongue lie irregular nodules of lymphoid tissue (see Fig. 7-37), into which dip small and shallow tubules or crypts of the surface epithelium. These nodules collectively form the lingual tonsil and, if enlarged, may be partially removed at tonsillectomy; since, unlike the palatine tonsil, they have no well-developed deep surface or bed, it is impossible to remove them completely.

Fauces and Palatine Arches

The lateral component of the fauces (the term includes both the isthmus and its walls) is largely occupied by the palatine tonsil, the relationships of which are discussed in more detail in a following section. It fills the space between the palatine arches, also called the tonsillar pillars or the pillars of the fauces. These are the rather inconspicuous (unless the tongue is pulled forward) palatoglossal arch (glossopalatine arch, anterior pillar of the fauces) between the soft palate and the tongue, and the more prominent palatopharyngeal arch (pharyngopalatine arch, posterior pillar of the fauces) extending from the soft palate downward into the posterolateral wall of the pharynx. These arches are formed by the palatoglossus and palatopharyngeus (glossopalatine and pharyngopalatine) muscles, respectively.

Waldeyer's Tonsillar Ring

The lingual tonsils anteriorly, the palatine tonsils laterally, and the pharyngeal tonsils (adenoids) posterosuperiorly together form a ring of lymphoid or adenoid tissue about the upper end of the pharynx, known as Waldeyer's tonsillar ring (Fig. 8-3). Superficially located lymphoid nodules also extend upward along the palatopharyngeal arch and behind the soft palate to the pharyngeal recess to form a component of Waldeyer's ring; the tissue here is known as the lateral pharyngeal band or lateral band and has been the site of an hemangioma (Lillie and Heck). Lym-

Fig. 8-3. Waldeyer's tonsillar ring.

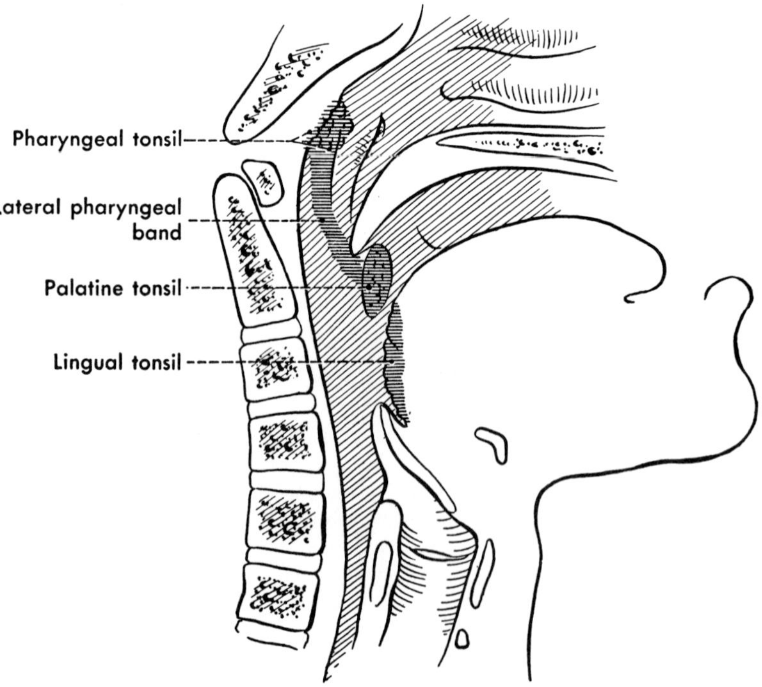

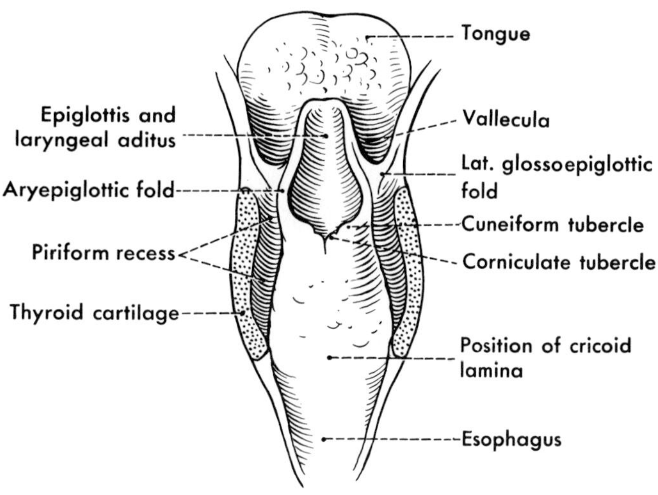

Fig. 8-4. Lower part of the pharynx from behind, with its posterior wall removed.

phomas may apparently originate in any component of the ring, or they may occupy the entire ring (Devine).

Posterior Wall

The posterior wall of the nasal and oral parts of the pharynx is intimately related above and posteriorly to the anteroinferior surface of the base of the occipital bone and posteriorly to approximately the first three cervical vertebrae. If the neck is hyperextended, the potential retropharyngeal space is obliterated and the wall of the pharynx forced against the vertebral bodies—hence, in adenoidectomy, the neck should be in flexion in order to prevent injury to them. Protrusion of a portion of one of these bones against the posterior pharyngeal wall has apparently sometimes been mistaken for a tumor.

LARYNGEAL PART

The laryngeal part of the pharynx, or hypopharynx, extends from just above the level of the hyoid bone superiorly to the lower border of the cricoid cartilage inferiorly, narrowing rapidly to become continuous with the esophagus. Its posterior wall shows no markings, and neither do its lateral walls, which attach to the hyoid bone and the thyroid cartilage. Its anterior wall is formed laterally by mucosa on the medial surface of the thyroid cartilage, and centrally or medially by the larynx and its appendages (Fig. 8-4): above is the epiglottis and the aditus of the larynx. Below the aditus the anterior wall of the pharynx is also the posterior wall of the larynx, the pharyngeal wall here being the posterior surfaces of the arytenoid cartilages and their connecting fold and the lamina of the cricoid cartilage and its attached muscles. These are covered posteriorly only by the mucosa of the pharynx; it is this mucosa at the junction of the pharynx and esophagus that is reflected backward in the approach to the arytenoid cartilage in operations of the King type.

Piriform Recesses

Lateral to the epiglottis are the lateral glossoepiglottic (pharyngoepiglottic) folds, which form the anterolateral boundary between oral and laryngeal parts of the pharynx. Below these folds the laryngeal pharynx or hypopharynx extends forward around the sides of the larynx, between this and the thyroid cartilage; these bilateral expansions are the piriform recesses, sinuses, or fossae (Figs. 8-4 and 8-5). Foreign bodies not infrequently lodge here, and both the piriform recesses and the epiglottic valleculae should therefore be investigated before such bodies are sought in the larynx, trachea, or esophagus.

According to Schugt, markings within the piriform recess are not constant. The internal

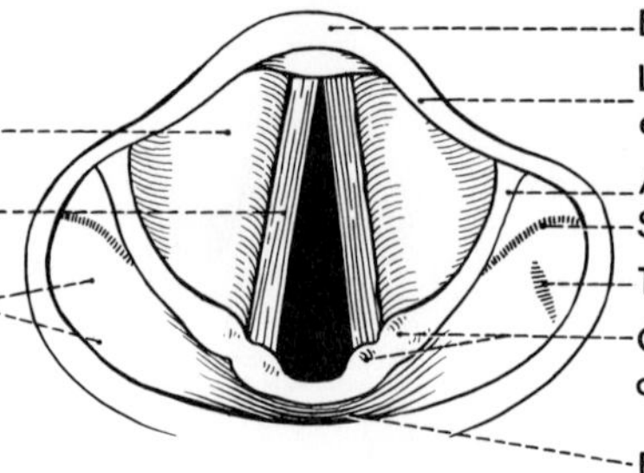

Fig. 8-5. Piriform recesses, pharynx, and larynx from above. Within the right recess is indicated the positions in which the internal branch of the superior laryngeal nerve and the superior border of the thyroid cartilage may form folds protruding into it; the position of the nerve alone is indicated in the left recess. Below and between the piriform recesses the pharynx narrows toward its junction with the esophagus (After Schugt HP: Arch Otolaryngol 31:626, 1940)

branch of the superior laryngeal nerve usually produces a slight fold (plica nervi laryngei) in the anterior part of the floor of the recess, as it runs here after penetrating the thyrohyoid membrane; regardless of the fold, the nerve can be blocked here by applying local anesthetic to the mucosa of the recess. Another variable but common marking is the upper border of the thyroid cartilage, which may be visible as a yellowish white line. Schugt found anomalies rare, but one of the most common was the projection of the superior cornu of the thyroid cartilage into the recess. Also running in the wall and across the floor of the piriform recess are the superior laryngeal artery and vein and many of the lymphatics from the larynx. Abscesses of the larynx may point in the piriform recess, and edema of the pharynx and larynx may so elevate the mucosa as to obliterate it.

PALATINE TONSIL

The palatine tonsil represents the largest accumulation of lymphoid tissue in Waldeyer's ring and, in contrast to the lingual and pharyngeal tonsils, constitutes a compact body having on its deep surface a definite though thin capsule. Tonsillar crypts, blind tubules from the epithelium on the surface of the tonsil, extend deeply into this tissue. According to Minear and co-workers the tonsillar crypts vary considerably in shape, but the majority are roughly platelike; since they originate as solid epithelial growths and the lumen usually develops first at the growing end, cystic crypts may be formed or the mouths of the crypts may be so narrowed that it is difficult for them to empty. Minear and co-workers stated that degeneration of the tonsil with increasing age usually begins in the lower half.

Capsule

The so-called capsule of the tonsil is usually regarded as a specialized portion of the pharyngobasilar fascia that covers the deep surface of the tonsil and extends into it to form septa that conduct the nerves and vessels. The tonsil is not, therefore, easily separable from its capsule, but the capsule in turn is united largely by loose connective tissue to the muscles of the pharynx; thus the tonsil can be partly displaced from its normal position, and its removal is most easily accomplished by separating the capsule from the muscle through this loose connective tissue, the peritonsillar or paratonsillar space. After repeated attacks of quinsy or peritonsillar infection, the tissue becomes dense and unites the tonsillar capsule firmly to the muscle. Painter apparently regarded the tonsillar capsule and the pharyngobasilar fascia as separate layers of connective tissue, for he described separating them in the course of a tonsilectomy; he emphasized that the capsule must be removed to obtain a clean dissection of tonsillar tissue.

Tonsillar Relations

The free surface of the tonsil is covered with mucosa from which are derived the tonsillar crypts, and in connection with this surface there are somewhat variable *folds* and *fossae* (Fig. 8-6). The triangular fold is a layer of mucosa and connective tissue, the latter said to be continuous with the capsule of the tonsil, that lies just behind the palatoglossal arch; this fold may adhere to the medial surface of the tonsil or may be separated from it by an anterior tonsillar fossa. Rarely there may be a similar fold posteriorly, and in about 40% of persons there is a semilunar fold and a supratonsillar fossa at the upper pole of the tonsil.

Since the tonsil lies in the depression (tonsillar fossa) between the palatoglossal and the palatopharyngeal arches (see Fig. 7-9), the unenlarged tonsil is largely covered by the former arch when viewed from the oral aspect; the enlarged tonsil varies much in its relations in different individuals, sometimes much of it being exposed whereas in other cases much of it is buried, depending upon the direction in which it has expanded. More of the buried tonsil can be displayed by pushing on the palatoglossal arch, and in children especially a buried tonsil may virtually "pop out" when the mucosa behind which it has expanded is cut. The upper portion of the tonsil was said by Hett and Butterfield to be especially well developed during the early years of life, so that approximately the upper third of the tonsil extends up under cover of the soft palate; they stated that by middle age much of this higher tissue has undergone atrophy.

On its deep surface the tonsil is held in place, somewhat below its middle, by fibers of the palatopharyngeus muscle that attach into it (Todd and Fowler) and that have apparently been termed the triangular ligament or the retrotonsillar or transverse plica; according to Todd and Fowler, it is the presence of these fibers that has especially given trouble in snare enucleation of the tonsil and has been responsible for leaving tonsillar tags.

The *muscular relations* of the tonsillar bed (Fig. 8-7) are variously described and illustrated and probably vary somewhat among individuals. The superior constrictor of the pharynx is frequently described and illustrated as forming the larger part of the tonsillar bed, with perhaps a variable posterior portion of this bed formed by more anterior fibers of the palatopharyngeus muscle; Todd and Fowler stated, however, that the tonsillar bed is formed entirely by the palatopharyngeus muscle and that although the superior constrictor is related to the upper part of the tonsil, it ends well above its lower part. This seems actually of no particular importance; in the region of the tonsil it is difficult to distinguish between the fibers of the superior constrictor and those of the palatopharyngeus,

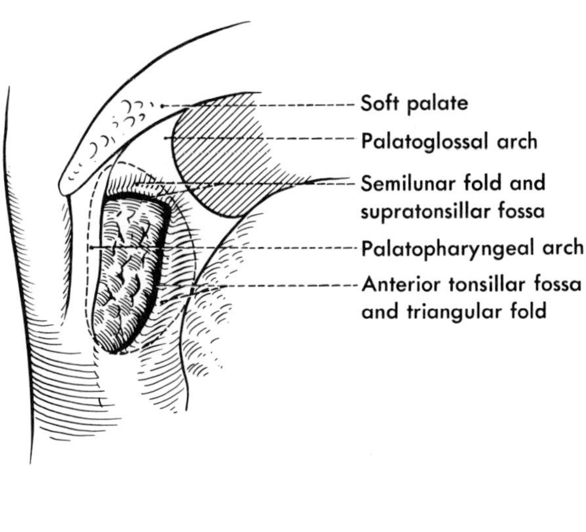

Fig. 8-6. Folds and fossae about the tonsil. The size of the tonsil is indicated by the *broken lines;* its free surface is shown here partly covered by well-developed semilunar and triangular folds.

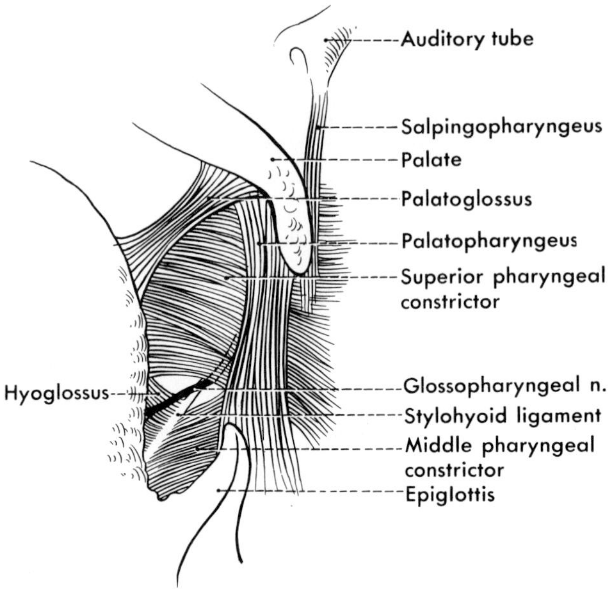

Fig. 8-7. Bed of the tonsil. It is here shown as being formed in great part by the superior constrictor. The course of the glossopharyngeal nerve lateral to the tonsillar bed and the appearance of this nerve and the stylohyoid ligament beneath the lower border of the superior constrictor are indicated; since the stylohyoid ligament is a continuation of the styloid process, a long bony process is necessarily closely related to the tonsil.

and both muscles probably vary considerably in their development.

In any case, the muscular wall is thin, and immediately against it on the outer wall of the pharynx is the glossopharyngeal nerve. Thus this nerve can be directly injured if the tonsillar bed is pierced, and not uncommonly it is temporarily affected by edema following tonsillectomy, producing a transitory loss of taste over the important posterior third of the tongue.

Also related to the lateral wall of the pharynx in the tonsillar region are the stylopharyngeus and styloglossus muscles, the stylohyoid ligament and sometimes the styloid process, the ascending pharyngeal artery, and the ascending palatine branch of the facial artery. Usually there is a considerable space (the lateral pharyngeal space) between the posterolateral wall of the pharynx at the tonsillar level and the external and internal carotid vessels and their main branches; sometimes, however, the facial artery or even the lingual may extend higher than usual and, in making the turn downward, bulge into close relation to the lower pole of the tonsil. The facial artery is apparently the more common important *vascular relation* of the tonsil, but various other arteries have also been implicated as a cause of pulsation of the pharynx.

According to MacCready, a visible pulsation of a large artery on the posterolateral wall of the pharynx is not uncommon, but as a rule the vessel itself is not identified—it has been thought to be an ascending pharyngeal, an ascending palatine, an aberrant vertebral, or a tortuous internal carotid. The pulsation has been mistaken for that of an aneurysm, since it may increase with age. MacCready reported a case in which he excised a portion of a pulsating vessel through an incision in the posterior pharyngeal wall and found that there was no aneurysm here, but simply a tortuosity of the involved vessel; he did not identify the vessel other than to find that it was not the internal carotid.

Cairney found tortuous internal carotids in 12 out of 36 adults examined and pointed out that usually such arteries lie nearer the pharynx than normal and sometimes in actual contact with it; he expressed the belief that pulsation of the posterior or lateral pharyngeal wall was probably due to such vessels. Since he found similar tortuosities in 5 of 20 fetuses, they obviously are not due entirely to changes with age.

Styloid Process

Of the other relations of the pharyngeal wall at the tonsillar level, the only one of any importance seems to be the possible relation to the styloid process. Eagle ('48, '49) pointed out that the normal styloid process does not reach the tonsillar fossa, but in about 4% of individuals the process is elongated. In such cases, he reported, the tip may be so deviated as to impinge upon either the internal or the external carotid artery, but more frequently it impinges upon the tonsillar region to produce pharyngeal pain, perhaps with extension to the ear.

Eagle found these symptoms primarily in persons who had had tonsillectomies, but Fritz reported that only 11 of 43 patients with pain due to an elongated styloid process had had tonsillar operations. These workers and Härmä, who found the condition to be more common than generally thought, agreed that most elongated styloid processes do not cause symptoms and that diagnosis can easily be made by palpation over the tonsillar fossa.

Babbitt reported a case of spontaneous fracture of the styloid process, the symptomatology of which closely resembled that of simple protrusion of an elongated process against the tonsillar area. He stated that the normal styloid process is about 30 mm long and that among 2000 skulls examined by another worker only 11 measured more than 38 mm (1½ inches) and 1 more than 76 mm (3 inches). Routine roentgenograms of this region show many variations in the breadth and length of the styloid process and in the development of bony elements that partially or even completely replace the stylohyoid ligament (see Fig. 7-35). Balasubramian regarded fractures as the most common cause of pain associated with long styloid processes.

Arteries and Veins of Tonsil

The arteries of the tonsil (Fig. 8-8) enter it especially toward its lower pole, with branches also at its upper pole, according to most descriptions. There are typically three arteries at the lower pole: the tonsillar branch of the dorsal lingual artery anteriorly, that of the ascending palatine artery (a branch of the facial) posteriorly, and the tonsillar branch of the facial artery between these (entering the tonsillar bed lower down). Toward the upper pole of the tonsil a twig from the ascending pharyngeal artery enters the posterior aspect of the tonsil, and the lesser palatine artery descends to the upper pole. The tonsillar branch of the facial is the largest.

The veins of the tonsil are usually described as forming a plexus about the capsule; the chief venous drainage is probably downward by way of the tonsillar branch of the lingual vein into this latter vein, with other vessels connecting to the pharyngeal plexus. A vein from the soft palate (the external palatine or paratonsillar vein) runs downward across the upper portion of the tonsillar bed before piercing the superior constrictor to join the pharyngeal plexus or some other neighboring vessel; this vein may be the chief source of bleeding from tonsillectomy.

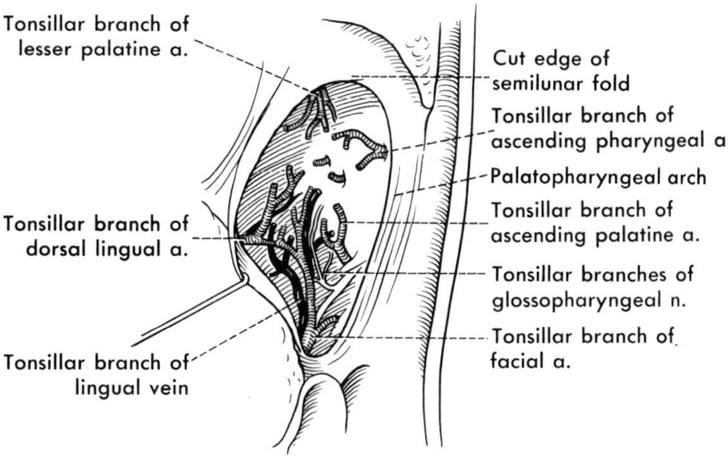

Fig. 8-8. Blood vessels and nerves of the tonsil, as usually identified. (After Truex RC, Kellner CE: Detailed Atlas of the Head and Neck. New York, Oxford University Press, 1948)

Nerve Supply

The nerve supply of the tonsillar region is through the tonsillar branches of the glossopharyngeal nerve about the lower pole of the tonsil and through descending branches from the lesser palatine nerves; the sensory fibers in the latter are usually assumed to be seventh nerve fibers coursing through the pterygopalatine (sphenopalatine) ganglion.

The nerves thus form a circle about the tonsil, and adequate local anesthesia of this region necessitates infiltration beneath the mucosa all about the tonsil. The possibility of damage to the main trunk of the ninth nerve, which lies immediately outside the tonsillar fossa, should be borne in mind; the nerve is sometimes seen crossing the fossa after removal of the tonsil.

Lymphatics

The lymphatic vessels leaving the tonsil are said to communicate with the submandibular (submaxillary) and superficial cervical lymph nodes, but the majority apparently pass to upper deep cervical nodes and especially to one located just behind the angle of the mandible; it is this node that has been called the tonsillar lymph node, because it is so frequently enlarged in tonsillar infections.

MUSCULATURE OF THE PHARYNX

The musculature of the pharynx (Figs. 8-7, 8-9, and 8-10) consists of three overlapping constrictors, superior, middle, and inferior, and three more longitudinally directed muscles. Each constrictor inserts with the corresponding muscle of the opposite side in the pharyngeal raphe of the posterior midline, while the longitudinal muscles, the palatopharyngeus, salpingopharyngeus, and stylopharyngeus, fan out on the inner surface. With the exception of the stylopharyngeus, which is innervated by the ninth cranial nerve, the muscles of the pharynx, like most of those of the palate, are innervated by the tenth nerve through the pharyngeal plexus.

Superior Constrictor

The superior constrictor of the pharynx, quadrilateral in shape, arises from the hamulus of the medial pterygoid process, from the

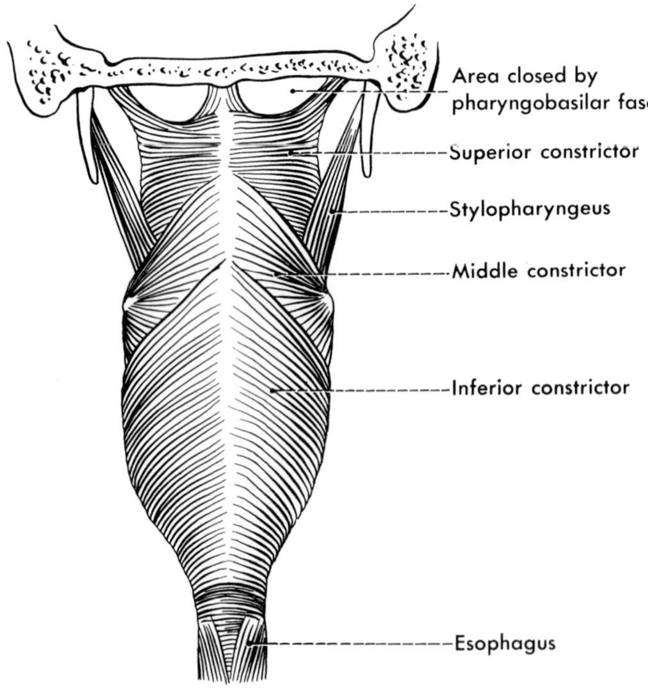

Fig. 8-9. Pharyngeal constrictors and stylopharyngeus from behind. The area closed by the pharyngobasilar fascia is the posterolateral wall of the pharyngeal recess. For other views of the pharyngeal muscles see Figs. 8-7 and 8-10.

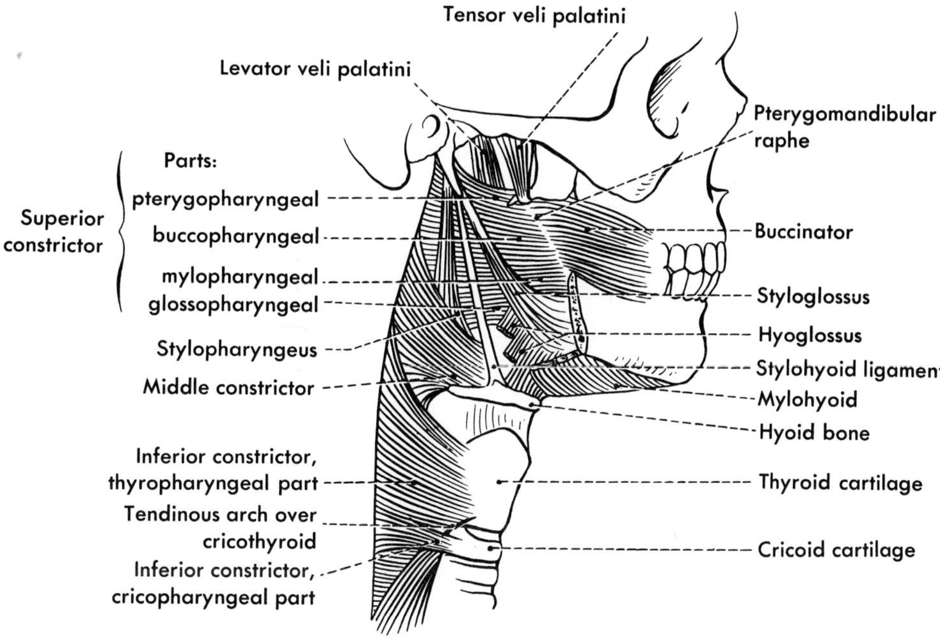

Fig. 8-10. Pharyngeal constrictors and related muscles from the side. In order to show the origin of the middle constrictor, that part of the hyoglossus arising from the greater cornu of the hyoid bone is here depicted as if cleanly cut away from this. Several parts of the constrictors are indicated. Note that the gap above the superior constrictor, from which the pharyngobasilar fascia has been removed, is nevertheless here largely hidden by the levator and the tensor of the palate.

so-called pterygomandibular raphe (where it meets the buccinator muscle or, according to Gaughran, is continuous with that muscle rather than the two being united through a raphe) and a posterior part of the mylohyoid line on the mandible, and from the side of the root of the tongue. Pterygopharyngeal, buccopharyngeal, mylopharyngeal, and glossopharyngeal parts are thus described. The pterygopharyngeal part is sometimes described as arising from the medial lamina of the pterygoid process, but Oldfield and others have denied this attachment. Of more importance is the fact that some inner fibers at the upper border of the superior constrictor attach to the palatine aponeurosis (Fig. 8-11); since they arise with palatopharyngeal fibers, they have been described as a part of this muscle. Whillis regarded these fibers as important in the formation of the ridge of Passavant and suggested that they be called the palatopharyngeal sphincter.

The muscle fibers of the superior constrictors of the two sides insert largely into each other in the posterior midline, thus forming the upper part of the pharyngeal raphe; the upper fibers arch somewhat upward, however, to attach also to the pharyngeal tubercle on the occipital bone; the lower fibers arch somewhat downward and are overlapped by the middle constrictor. As the upper fibers run from the level of the hamulus to the pharyngeal tubercle, there is left above their concave upper border an area deficient in muscle; here the posterior wall of the pharyngeal recess (Rosenmüller's fossa, sinus of Morgagni) is formed only by the mucosa of the pharynx and a relatively thick part of the pharyngobasilar fascia to which is fused the thinner buccopharyngeal or external fascia. Farther forward, the levator veli palatini, lying between the pharyngobasilar fascia and the mucosa, and the tensor veli palatini, lying outside the fascia, largely fill the gap above the constrictor (see Fig. 8-10). The auditory tube enters the pharynx above the attach-

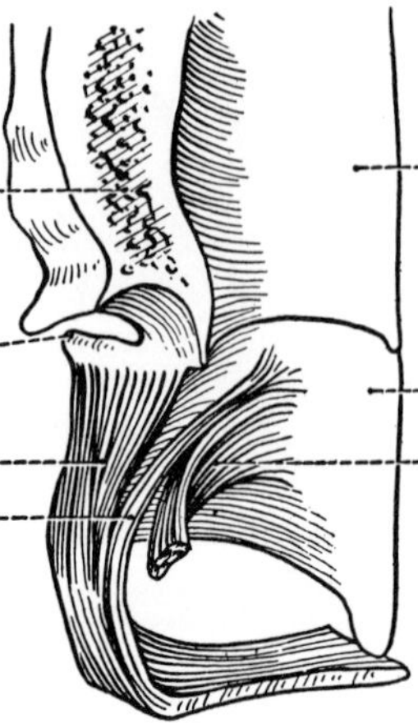

Fig. 8-11. Palatopharyngeal sphincter in a view, from above, of a horizontal section just above the floor of the nasal cavity. (After Whillis J: J Anat 65:92, 1930)

ment of the pharyngobasilar fascia, between the levator and the tensor.

Middle Constrictor

The middle constrictor of the pharynx is a fan-shaped muscle that arises from the greater and lesser cornua of the hyoid bone and the lower end of the stylohyoid ligament and is described as consisting of cerato-pharyngeal (from the greater cornu) and chondropharyngeal parts, of little importance. Between this muscle at its origin and the lower border of the superior constrictor there is laterally a gap, but as the upper fibers curve markedly upward and the lower ones curve similarly downward, the middle constrictor comes to overlap posteriorly the lower portion of the superior constrictor and is in turn overlapped below by the inferior constrictor.

At the posterior border of the lateral gap the stylopharyngeus muscle insinuates itself between the superior and middle constrictors to lie between the latter muscle and the mucous membrane of the pharynx; farther forward, fibers of the palatopharyngeus muscle lie internal to the upper two constrictors and bridge some of the space between the lower fibers of the superior constrictor and the upper fibers of the middle one. The diverging muscle fibers of the middle constrictor insert with those of the other side into the pharyngeal raphe.

Along or slightly above the level of the upper border of the middle constrictor the glossopharyngeal nerve curves forward into the tongue, and lateral to the muscle at its origin lie the lingual artery and the hyoglossus muscle.

Inferior Constrictor

The inferior constrictor of the pharynx arises on the lateral surface of the thyroid cartilage from the oblique line that also gives attachment to the thyrohyoid and sternothyroid muscles, from a tendinous line extendinig from the thyroid cartilage to the cricoid cartilage and arching across the cricothyroideus muscle, and from the posterior part of the arch of the cricoid cartilage. Thus it is divisible into thyropharyngeal and cricopharyngeal parts. The upper fibers arch markedly upward, to overlap a large part of the middle constrictor, while the middle fibers ascend less obliquely and the lower fibers are approximately horizontal or even arched slightly downward. Some of the latter blend with the circular fibers of the esophagus.

The cricopharyngeal part of the inferior constrictor, frequently called the *cricopharyngeus muscle,* lies at the junction of pharynx and esophagus. When the region of the "cardiac sphincter," that is, the gastroesophageal junction, has been abolished by resection, it is apparently the cricopharyngeus that confines regurgitation to the esophagus (Puppel). Fyke and Code have shown that the muscle, with perhaps some adjacent esophageal fibers, is tonically contracted in the resting condition, maintaining a zone of increased pressure that disappears only when swallowing is initiated.

Palatopharyngeus

The palatopharyngeus (pharyngopalatinus) muscle (see Figs. 8-7 and 7-15) arises chiefly from the anteroinferior surface of the palate as far forward as the posterior edge of the bony palate and is somewhat between the levator veli palatini above and the palatoglossus and tensor veli palatini below; as the fibers run downward, they are joined by others arising from the upper surface of the posterior part of the palate, and many of them become aggregated into a conspicuous band that forms the palatopharyngeal arch (posterior palatine arch, posterior pillar of the fauces); the fibers gradually fan out to form a thin and scattered layer attaching to the pharyngobasilar fascia, some few also attaching to the posterior border of the thyroid cartilage. As indicated by its alternative name, pharyngopalatinus, the origin and insertion of the muscle are sometimes regarded as the reverse of those given here.

Salpingopharyngeus

The salpingopharyngeus (see Fig. 8-7) is a rather small muscle slip that arises from the pharyngeal end of the cartilage of the auditory tube and inserts into the wall of the pharynx; it lies immediately adjacent to the mucous membrane and may form a slight fold in the mucosa, the salpingopharyngeal fold, extending downward from the orifice of the auditory tube. It is frequently spoken of as being of considerable importance in swallowing movements, but its morphology does not support this; it is ordinarily poorly developed, and de Paula Assis stated that it was absent in 30% of 50 subjects dissected. In only six was it a strong muscle, and in the remainder it was extremely thin. Since it lies adjacent to the posterior fibers of the palatopharyngeus muscle, de Paula Assis considered it actually part of this muscle.

Stylopharyngeus

The final muscle to be considered with the pharynx is the stylopharyngeus. This arises from the medial side of the base of the styloid process and passes downward and medially between the external and internal carotids to lie against the posterolateral surface of the superior constrictor of the pharynx and enter the pharynx between this muscle and the middle constrictor; here it spreads out between the latter muscle and the pharyngobasilar fascia.

MOVEMENTS OF THE PHARYNX AND THE SOFT PALATE

During Speech

Coordinated movements of the palate and pharynx normally occur during both speech and swallowing, and are therefore most easily discussed together. As pointed out in the discussion of the soft palate, apposition of the soft palate to the posterior wall of the pharynx is necessary for proper speech; this involves movements of both the pharynx and palate.

According to Calnan ('53), closure of the nasopharynx during speech is rapid and involves contraction of the palatopharyngeus (and salpingopharyngeus) muscles, which can be seen to raise the salpingopharyngeal folds; contraction of the levator veli palatini, which brings the middle, muscular, segment of the soft palate upward and backward into contact with the posterior pharyngeal wall and decreases the width of the nasopharynx by raising salpingopalatine folds (exaggerations of the torus levatorius) that curve into the soft palate from below the auditory tubes; and contraction of the superior constrictor, which narrows the pharynx from side to side. (Fritzell reported that although the palatopharyngeus usually showed some activity during speech, it was not consistently active during nasopharyngeal closure; and he found that the palatoglossus actively lowered the palate for nasal speech.) Older accounts describe the soft palate as coming in contact with a fold of the posterolateral wall of the pharynx, Passavant's fold or ridge, but Calnan ('53, '58) denied that this ridge is usually related to normal speech. He described the soft palate as coming in contact with the posterior naso-

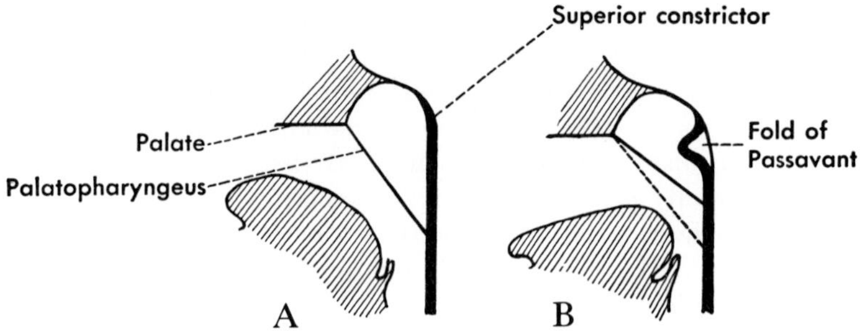

Fig. 8-12. (*A*), Formation of the fold of Passavant. (*B*), The upward movement of the pharyngeal wall with contraction of the palatopharyngeus, the relaxed position of the muscle being indicated by the *broken line* and its contracted one by the *solid line.* (After Browne D: Br J Surg 20:7, 1932)

pharyngeal wall above the level of the ridge and said that the ridge is seen infrequently in a speaker who does not have cleft palate; further, he found that it developed too slowly and fatigued too easily to be useful in the rapid repetitive pharyngeal closure necessary to normal speech.

In Swallowing

Muscular activity during swallowing, as analyzed by cinefluorographic and electromyographic techniques (Ramsey and co-workers; Basmajian and Dutta) follows in general the pattern that would be expected upon introspection. As the bolus is moved from the oral cavity into the relaxed pharynx, primarily by pressure of the tongue successively against the hard and then the soft palate (Ramsey and co-workers), the palate is tensed and elevated (the tensor, inactive in phonation, becoming active during swallowing) and then drawn downward toward the tongue by the action of the palatopharyngeus (and palatoglossus?) muscles, thus stripping the bolus toward the epiglottic valleculae. The soft palate elevates again, carrying the palatopharyngeal arches (posterior pillars) back, and the pharynx is elevated and develops a fold at the level of the ridge of Passavant. As the soft palate meets the fold, the entrance into the nasopharynx is sealed, and palate and fold move downward to carry the bolus with them. As determined electromyographically, the three pharyngeal constrictors contract in the expected order from above downward, thus stripping remains of the bolus into the lower relaxed portion of the pharynx and hence into the esophagus. The cricopharyngeal part of the inferior constrictor, which is tonically contracted in the resting condition, relaxes early in the act of swallowing (Fyke and Code).

The ridge of Passavant, described also by Calnan as developing primarily in an effort to seal off completely the nasal portion of the pharynx, has been said to develop either from passive folding of the pharyngeal wall brought about by elevation of the pharynx through the palatopharyngeus muscles (Fig. 8-12) or as an active fold produced by the palatopharyngeal constrictor and perhaps the upper fibers of the superior pharyngeal constrictor. The statement of Ramsey and co-workers that a fold in this position moves downward during swallowing and is usually succeeded by a similar fold lower down would seem to indicate that it is produced in part by localized activity of constrictor fibers.

As the larynx rises and moves forward in swallowing, the epiglottis folds backward and downward, and as it does so the epiglottic valleculae are emptied of any material remaining in them (Ramsey and co-workers). This protective movement of the epiglottis over the larynx has been both affirmed and denied by previous workers; its mechanism, whether active or passive, is not understood. A part of the epiglottis's function is simply to divert the bolus around the larynx through the piriform recesses; the bolus is usually unequally divided, with the larger part passing through one piriform recess and a smaller part passing through the other. Ramsey's group noted that the epiglottis does not fit tightly enough over the larynx to prevent the entrance of liquid into that and that with training one can learn to swallow liquids (by gravity?) with no movement of the epiglottis. According to Ramsey and co-workers, it is the vestibular folds, and sometimes the vocal folds, that halt the passage of liquid into the larynx; the liquid is then stripped out as the pharynx is emptied. Similarly, patients who have had the epiglottis removed because of carcinoma have no difficulty in swallowing.

Naffziger and associates discussed the great difficulty in swallowing after paralysis of the last four cranial nerves. Two important causes of this, in their opinion, are a lateral bulging of the paralyzed side of the pharynx and the inability of the pharynx to move upward because of the paralyzed stylopharyngeus muscle. However, most surgeons reporting upon the effect of glossopharyngeal nerve section for neuralgia have reported that paralysis of the stylopharyngeus muscle alone produces no defective movement of the pharynx; the palatopharyngeus thus apparently suffices when it is not affected also. Naffziger and co-workers devised a reparative operation that they tried upon both the monkeys and human beings, as follows: To strengthen the weakened pharyngeal wall, they cut the carotid sheath laterally but left it attached to the prevertebral fascia and brought it forward around the pharynx to fasten to the anterior midline of the larynx and the cricoid cartilage; as a substitute for the paralyzed levators of the pharynx, they attached the posterior belly of the digastric to the thyroid cartilage. The patient so treated is said to have swallowed very well after a little training.

PHARYNGEAL DIVERTICULA

Sinuses and fistulas of the pharynx connected with the development of the branchial pouches are discussed in connection with the neck in Chapter 9; these most commonly open into the region of the tonsillar fossa. The pharyngeal diverticulum referred to here is that encountered in the lower portion of the pharynx (Fig. 8-13) close to its junction with the esophagus and variously termed pharyngeal, hypopharyngeal, esophageal, pharyngoesophageal, and Zenker's diverticulum. Like the majority of diverticula found elsewhere in the digestive tract, these are commonly believed to be pulsion diverticula through an area of congenital weakness in the muscular coat, although causes other than congenital anatomical weakness have been suggested.

It is generally agreed that hypopharyngeal diverticula arise most commonly in the region of the "cricopharyngeus muscle" (*i.e.*, the cricopharyngeal part of the inferior constrictor), although opinions differ considerably on the exact site of origin of these diverticula. The several views advanced are that the diverticula arise in the midline below the cricopharyngeus, between two sets of fibers composing it, or above it; that they arise laterally below the cricopharyngeus or above it; and that they arise also in higher locations. It seems not impossible that all the various accounts of this are correct and that there are several potential sites of herniation, although some of these are so close together that it hardly seems worth arguing which might be the site of a diverticulum.

Just below the cricopharyngeal fibers is a triangular area, known as Laimer's triangular area (Fig. 8-14), or the Laimer–Haeckermann area, in which the posterior esophageal wall is thin because the longitudinal fibers of the esophagus are here swinging laterally and forward around the esophagus to reach an attachment on the cricoid cartilage. This area is therefore bounded laterally by the diverging bundles of longitudinal fibers and above by cricopharyngeal fibers; only the circular musculature of the esophagus forms the wall here, and it is conceivably a potentially weak area. However, Shallow stated that there were no recorded cases from this region.

Killian described the "cricopharyngeus muscle" as being composed of two parts: a superficial and upper oblique portion—frequently referred to as the constrictor por-

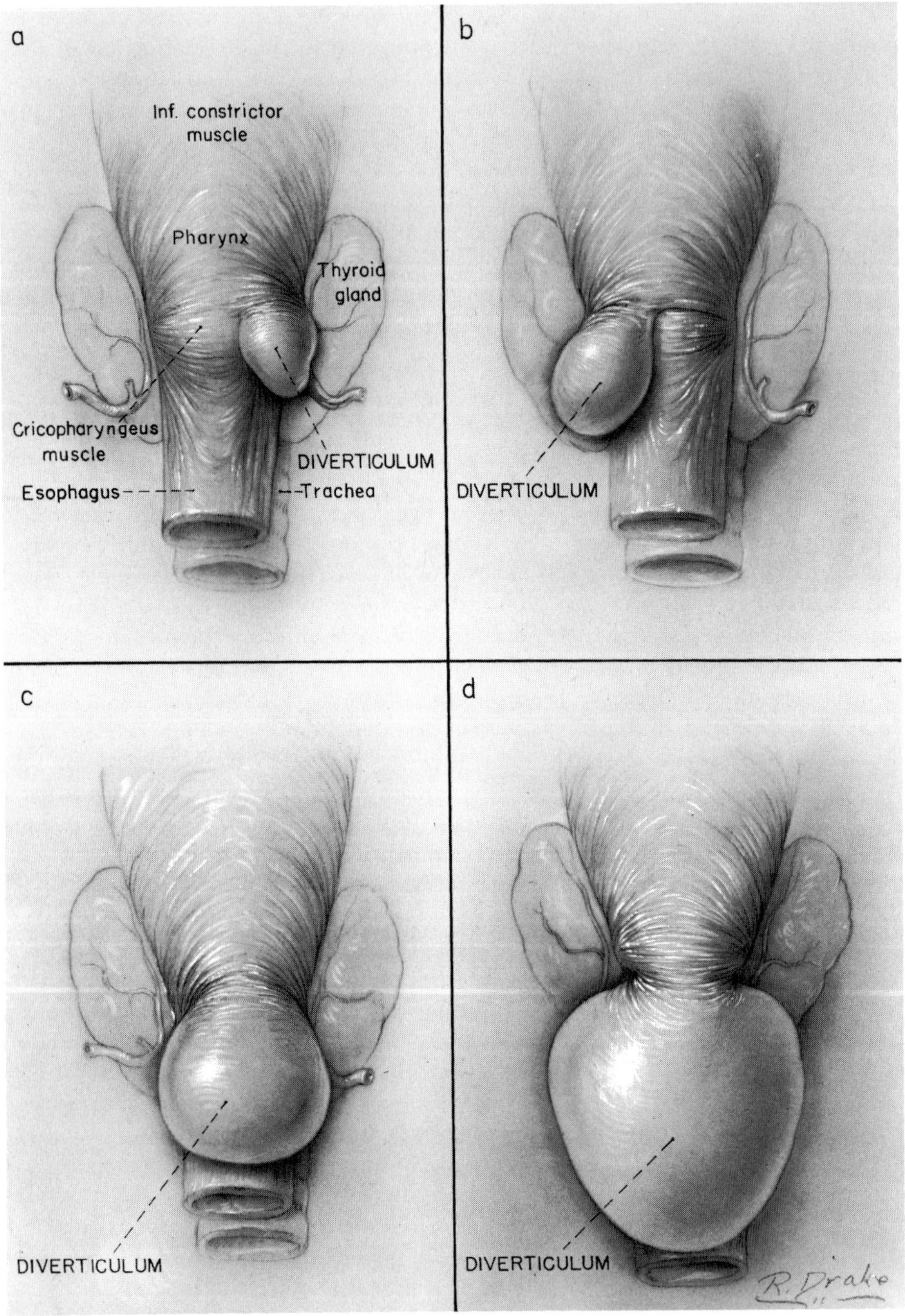

Fig. 8-13. Pharyngoesophageal diverticula. (Payne WS, Clagett OT: Current Problems in Surgery. Chicago, Year Book Medical Publishers, 1965. Used by permission of Year Book Medical Publishers)

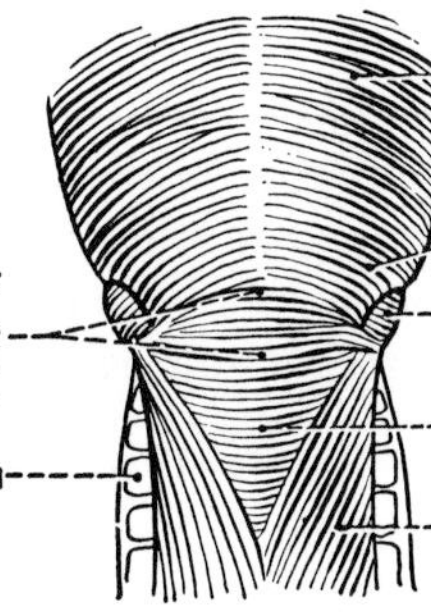

Fig. 8-14. Posterior wall of the lower part of the pharynx. The cricopharyngeal part of the inferior constrictor is somewhat variably described and may be difficult to delineate exactly, since its upper fibers blend with the lower fibers of the rest of the inferior constrictor, while its lower ones blend with the circular fibers of the esophagus.

tion—and a lower and deeper transverse or sphincteric portion. He regarded all hypopharyngeal diverticula as occurring between these two components. Raven agreed with this concept and stated (without citing evidence) that the upper or constrictor fibers of the cricopharyngeus receive their vagal innervation through the pharyngeal plexus and the lower or sphincteric ones through the recurrent laryngeal, implying that asynergia between these two parts might be a precipitating cause for herniation. Sunderland and Swaney regularly found a branch from the inferior laryngeal nerve to the inferior pharyngeal constrictor, but did not describe its distribution; Levitt and co-workers found no recurrent nerve innervation for the cricopharyngeus of the dog. Regardless of the details of their innervation, it seems obvious that if tonically contracted cricopharyngeal fibers fail to relax properly during swallowing, unusual pressure will be built up above them.

Other workers (*e.g.,* Moersch and Judd; King, '47) regarded the site of herniation as being above all the cricopharyngeal fibers, between these and the thyropharyngeal part of the inferior constrictor. It is not clear, however, that this apparent difference in the location in which the diverticula are said to develop is not primarily one of definition of the cricopharyngeal part of the muscle; after all, it is simply a lower portion of the inferior constrictor, and its upper oblique fibers so blend with the lower thyropharyngeal fibers that they are difficult to differentiate. It appears possible that the more obvious transverse or sphincteric fibers have often been regarded as constituting the "cricopharyngeus muscle."

If this is true, then there is actually no difference in opinion between those who state that the diverticulum occurs through the pars cricopharyngeus and those who claim that it occurs between this and the pars thyropharyngeus. Babcock and Jackson stated that herniation might occur either through or above the pars cricopharyngeus. In any event, however, it is difficult to believe that this is a matter of any real clinical importance, despite the emphasis that has been placed upon it.

Although the midline diverticulum is generally granted to be most common, several writers have also described other areas of herniation in this region.

Between the lowest cricopharyngeal fibers (which attach laterally to the cricoid cartilage) and the upper circular fibers of the esophagus (which circle anteriorly) is the gap that transmits the inferior laryngeal nerve, a branch (inferior laryngeal) of the inferior thyroid artery, and some lymphatics, and that is otherwise occupied by a small pad of fat. This is sometimes known as the Killian–Jamieson area (Fig. 8-15). Moersch and Judd described diverticula from this area, and Shallow described it as a common (35%) site of origin of pharyngeal diverticula. The distinction between diverticula here and most posteriorly may be of some importance because of the close relation between the neck of the sac and the inferior laryngeal nerve, although by the time diverticula about the pars cricopharyngeus attain the diameter of the esophagus, as they are said to do commonly, the relation between the nerve and the neck of the sac must necessarily be rather close.

Most of the diverticula arising in the hypopharynx, regardless of the exact site of their development, protrude to the left; in so doing, they come to lie in the visceral compartment or space, between the prevertebral or alar fascia and the pretracheal fascia with the associated infrahyoid muscles.

Shallow apparently believed that not even the so-called midline diverticula are strictly this, for although he recognized the majority as arising between the cricopharyngeus (apparently the transverse portion) and the rest of the inferior constrictor (Fig. 8-16), he regarded the diverticulum as actually originating along a branch of the inferior thyroid artery that enters in this location and therefore the weak places here as being bilateral rather than midline. He stated that of 79 cases of pharyngeal diverticula in various positions, the majority were in this location on the left side, and 4 were in a similar position on the right. Among these cases, the next most common area involved was said to be the gap between the cricopharyngeus and the upper circular fibers of the esophagus (Killian–Jamieson area), the diverticula arising here in slightly more than 35% of the patients; in these, hoarseness preceding or following the operation was not unusual. He stated also, however, that occasional diverticula arise higher, through the thyropharyngeal part of the inferior constrictor, where another branch of the inferior thyroid artery penetrates the pharyngeal wall; he regarded this as the least common locus.

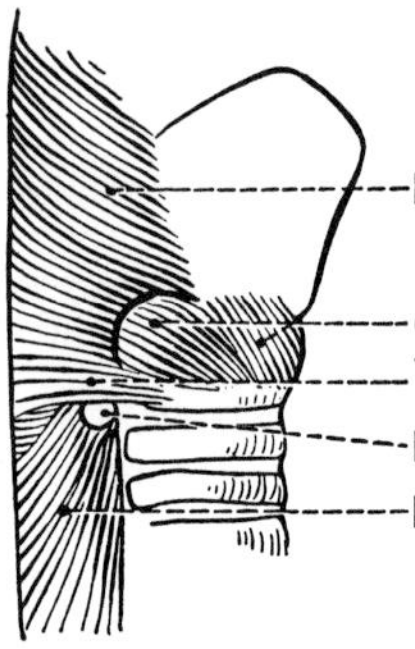

Fig. 8-15. The Killian-Jamieson area, the gap beneath the inferior constrictor for the entrance of the inferior laryngeal nerve and vessels into the larynx.

Morley pointed out that pharyngeal diverticula occur as a rule in late middle age, and King ('47) stated that he had never heard of one occurring before age 15. They are apparently more common in men than women, according to King about three times so, and King doubted that they are actually due to areas of congenital weakness. He said that the most common locus represents a point where the lower border of the cricoid cartilage rests against the cervical vertebral column, and he expressed the belief that pressure from the

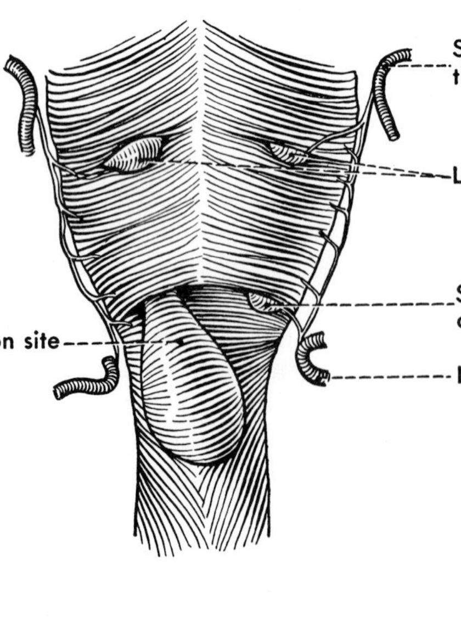

Fig. 8-16. Common sites of hypopharyngeal diverticula according to Shallow. (After Shallow TA: Surg Gynecol Obstet 62:624, 1936, by permission of Surgery, Gynecology & Obstetrics)

mature male larynx produces the weakness here.

Furstenberg ('47) also doubted that all cases of these diverticula are due to congenital weakness; he regarded esophageal obstruction as a possible cause and described a patient in whom he observed the development of such a diverticulum following obstruction of the upper part of the esophagus. This would agree with the theory advanced by Killian that spasm, or lack of relaxation, of the sphincteric fibers of the cricopharyngeus may be a cause of diverticula, a concept now widely accepted.

In many patients, symptoms of pharyngeal diverticula have been present for from 5 to 10 years before they present themselves for operation. These symptoms may include some degree of dysphagia associated with regurgitation, and there may be complaints of disagreeable noises on swallowing; regurgitation may be accompanied by violent coughing and choking (Morley). Enlargement of the pouch gradually obstructs the passage of food through the esophagus, so that even though in early cases the weight may be well maintained, in late or neglected cases there may be considerable emaciation.

The standard surgical treatment has been removal of the sac, but in recent years Holinger and Johnston, and others, have reported good results from an endoscopic approach in which they sectioned by cautery the cricopharyngeus fibers below the mouth of the diverticulum. Calcaterra and co-workers sectioned the cricopharyngeus, adjacent upper fibers of the esophagus, and lower fibers of the thyropharyngeal part of the inferior constrictor through a low cervical approach in cases of dysphagia due to a malfunctioning cricopharyngeus even when there was no sac.

There seems to be, however, no unanimity as to the cause or best method of dealing with the diverticulum. Macbeth stated that there is no cricopharyngeal spasm, but rather a weak thyropharyngeus, and disapproved of cutting or using diathermy on the cricopharyngeus. Cross, in contrast, reported that the diverticulum is caused by a high resting pressure and poor relaxation on the part of the cricopharyngeus; he regarded excision of the sac as probably the best treatment, but stated that myotomy is sometimes necessary. Holinger and Schild used a variety of treatments, but regarded diverticulectomy as giving best results; for older, severely debilitated patients, however, they preferred diverticulotomy. Giffin and co-workers preferred diverticulectomy also, but stated that the cricopharyngeus must be cut. Zuckerbraun and Bahna regarded cricopharyngeal myotomy as the only treatment necessary.

Diverticulectomy was once routinely done in two stages, the first stage including merely the mobilization of the diverticulum and its implantation in the superior portion of the wound, while the second included the actual excision of the sac and the closure of its opening. The one-stage operation (Fig. 8-17), in which the sac is mobilized and removed at one operation, is now more favored, although Giffin and co-workers preferred the two-stage operation for aged patients with huge sacs. Belsey regarded removal of the pouch as unnecessary, saying that even large ones can be treated by suturing upside down to the anterior longitudinal ligament.

NERVES

Pharyngeal Plexus

The motor nerve supply to the pharynx, and the sensory supply to most of it, is through the pharyngeal plexus, which is formed by the pharyngeal branches of the ninth and tenth cranial nerves (Fig. 8-18). The glossopharyngeal nerve at first descends between the internal carotid artery and the internal jugular vein and then passes forward lateral to the artery to lie behind the styloid process and the muscles attached to this and run downward and forward medial to, then behind, and then lateral to, the stylopharyngeus muscle. The pharyngeal branch of the glossopharyngeal, sometimes multiple instead of single, arises from the upper part of the nerve between the jugular foramen and the point at which the nerve passes across the posterior surface of the stylopharyngeus muscle.

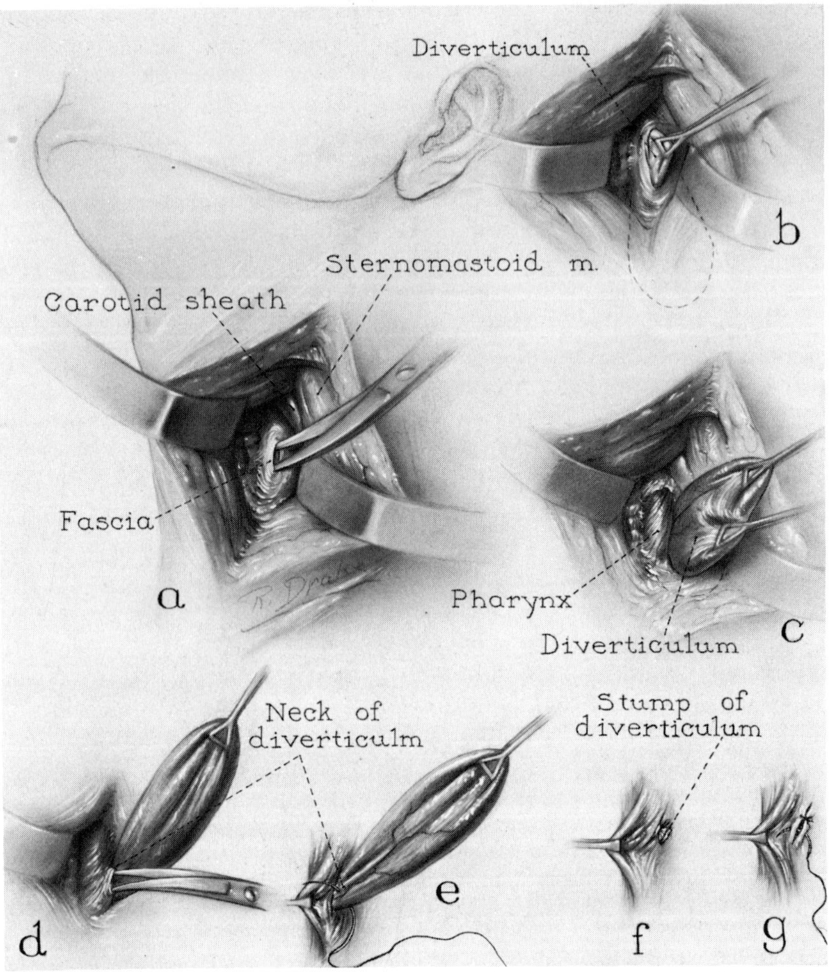

Fig. 8-17. One-stage removal of a pharyngoesophageal diverticulum. (Harrington SW: Surg Gynecol Obstet 69:364, 1939, by permission of Surgery, Gynecology & Obstetrics)

This branch of the nerve extends downward, medially, and forward to join the pharyngeal branch of the vagus on the lateral surface of the superior constrictor muscle and form with it the pharyngeal plexus. The pharyngeal branch or branches of the vagus similarly leave this nerve high and run forward lateral to the internal carotid artery, usually slightly below the level of the pharyngeal branch of the ninth nerve, to pass between internal and external carotids to the posterior or posterolateral surface of the pharynx. The pharyngeal branches of the ninth and tenth nerves may join as they cross the internal carotid, or more peripherally upon the pharynx.

The glossopharyngeal nerve contributes only sensory fibers to the pharyngeal plexus, while the vagus, which has already received the internal ramus of the accessory, contributes all the voluntary motor fibers. Fibers from the superior cervical ganglion also join the plexus.

The *motor fibers* in the pharyngeal plexus leave the brain stem through the lower vagal and the cranial accessory roots (the latter joining the vagus through the internal ramus of the accessory nerve) and are distributed to all the muscles of the pharynx and soft palate, with the exception of the tensor veli palatini (supplied by the mandibular branch of the fifth), the stylopharyngeus (supplied by the ninth nerve as it passes this muscle), and perhaps the lower part of the inferior constrictor, which may be supplied by the external branch of the superior laryngeal nerve (vagus) and the recurrent laryngeal branch of the vagus.

Some of the *sensory fibers* of the glossopharyngeal nerve join the pharyngeal plexus to reach the pharynx, and others form the tonsillar branches, given off as this nerve passes medial to the hyoglossus muscle. The glossopharyngeal nerve has traditionally been described as supplying sensory fibers to almost the entire pharynx (Fig. 8-19) from the level of the opening of the auditory tube downward. Above the opening of the tube a limited upper part of the nasopharynx is supplied by maxillary nerve fibers that reach it through the pharyngeal branch of the pterygopalatine (sphenopalatine) ganglion, and the superior laryngeal branch of the vagus may supply either more or less of the lower part of the pharynx than indicated in Figure 8-19. Further, there is evidence (see the following paragraph) that the pharyngeal branch of the vagus, once thought to be motor only, contains sensory fibers.

The distribution of the glossopharyngeal nerve to the pharynx and the posterior part of the tongue accounts for the initiation of pain in *glossopharyngeal neuralgia* (Chap. 1) by chewing and swallowing movements and its alleviation by section of the ninth nerve (usually done intracranially in order to interrupt also the glossopharyngeal fibers to the ear). Although high section of the nerve denervates the stylopharyngeus muscle, one of the elevators of the pharynx, it is generally agreed that no difficulty in swallowing or abnormality of pharyngeal movement can be detected thereafter; this agrees with the concept that the motor fibers in the pharyngeal plexus are vagal ones. There is, however, no longer agreement that, with the exceptions noted above, the glossopharyngeal nerve is the sole sensory supply to the pharynx. Although glossopharyngeal section has usually been said to produce anesthesia and analgesia of

Fig. 8-18. Schema of the nerves and arteries of the pharynx. The arterial twigs from tonsillar and palatine vessels are not shown.

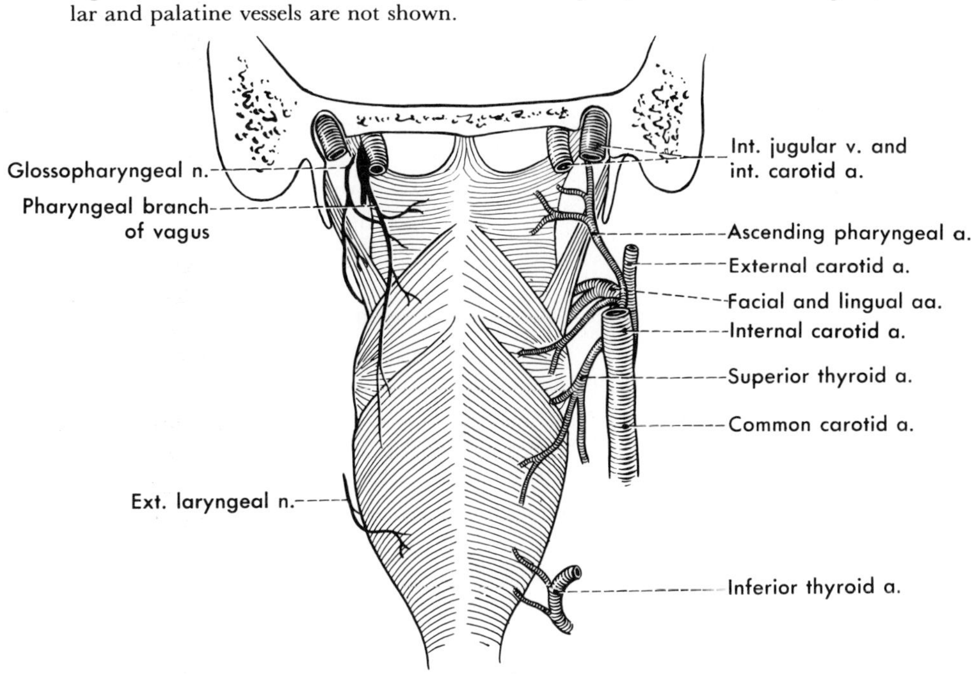

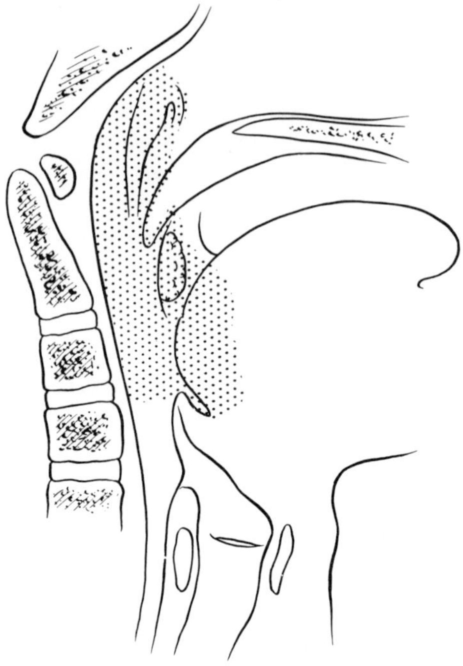

Fig. 8-19. Approximate sensory distribution of the ninth nerve to the pharynx.

most of the pharynx, and abolition of the gag reflex, it has also been stated that the primary result is analgesia only and that loss of touch and of the gag reflex depends upon injury or purposeful section of the upper (largely sensory) vagal rootlets, which are also sometimes involved in "glossopharyngeal" neuralgia. Thus the vagus presumably contributes many sensory fibers, as well as the motor ones, to the pharyngeal plexus.

VESSELS

Arteries

The blood supply of the pharynx (see Fig. 8-18) is largely from the ascending pharyngeal and superior thyroid arteries. The ascending pharyngeal, the smallest branch of the external carotid, passes upward between the external and internal carotids and inclines medially to run along the posterolateral aspect of the upper portion of the pharynx, which it supplies before it enters the jugular foramen. The superior thyroid artery as it descends gives off at about the level of the superior horn of the thyroid cartilage a branch to the pharynx, and the lowermost part of the pharynx is supplied by branches from the inferior thyroid artery. Finally, the pharyngeal fornix is supplied by twigs from the artery of the pterygoid canal (Vidian artery, from the maxillary) and by twigs from the sphenopalatine branch of the maxillary artery; in addition, the pharyngeal wall in the neighborhood of the tonsil is presumably also supplied by the tonsillar arteries already described.

Veins

The veins of the pharynx form a plexus on the posterior wall of this organ, this plexus communicating above with the pterygoid plexus of veins and below with the superior thyroid and lingual veins or directly with the facial or internal jugular. Batson described a second pharyngeal plexus lying close to the mucous membrane of the pharynx; this is said to be particularly dense, composed of veins that are themselves some 1 to 3 mm in diameter and arranged in two parts, one on the anterior and the other on the posterior wall of the pharynx. The anterior plexus is said to communicate with the posterior one around the sides of the pharynx, to communicate with the superficial veins of the tongue, and to drain both into the superior laryngeal and the esophageal veins; Batson stated that the plexus on the posterior wall has a well-marked upper border but connects below with the veins of the esophagus and also with the plexus on the outside of the constrictors. The plexus was described as particularly dense and ringlike around the entrance to the larynx, and Batson suggested that it might be concerned with pressure adjustments accompanying swallowing, since it seemed to be much larger than is necessary for the tissue here.

Lymphatics

The pharyngeal mucosa is rich in lymphatics, especially about the pharyngeal tonsil. According to Most, the efferent lymphatics of the pharynx leave this in three areas: those from the roof and most of the posterior wall of the pharynx pass through the pharyngeal wall to the lateral pharyngeal node (a constant lymph node lying just medial to the internal carotid artery close to where this enters the skull, and against the lateral wall of the pharynx); the lymphatics from the tonsillar region drain laterally directly into the deep cervical nodes; the lymphatics from the laryngeal part of the pharynx group together in the piriform recess, pierce the thyrohyoid membrane, and unite with other lymphatics to go to the deep cervical nodes.

The lymphatic plexus on the posterior pharyngeal wall receives most of the lymphatics from the nasal cavity, but near the ostium of the auditory tube some of these lymphatics diverge to flow downward to the tonsillar region. Although it is customary to refer to the drainage of the pharynx as being into retropharyngeal nodes, the paired "lateral nodes" are usually the only ones seen in routine dissections. Robison described the retropharyngeal nodes as being divisible into medial and lateral ones, both of which undergo atrophy, typically leaving only one lateral node on each side; Most stated that although the lateral nodes are constant, truly retropharyngeal ones are small and inconstant.

The Larynx

SKELETAL STRUCTURE

Thyroid Cartilage

Since the larynx, as a portion of the air passageway, must remain open except when it is reflexly or voluntarily closed, it is surrounded by cartilages that prevent its collapse and also give attachment to its muscles. The large and prominent thyroid cartilage forms most of the anterior and lateral walls of the larynx (Figs. 8-20 to 8-26). The upper border of this cartilage is attached above to the hyoid bone by the thyrohyoid membrane, with its thickened middle and edges (the latter often containing small triticeal cartilages) that are called the median and lateral thyrohyoid ligaments; its inferior horns articulate with the sides of the cricoid cartilage. As its name implies, the thyroid cartilage is shield-shaped and open behind. Since it forms a major portion of the wall of the larynx, approaches to the larynx are often made through it. Jackson and Jackson ('35a) stated that laryngofissure, or cutting through the cartilage, and removing the inner perichondrium and everything inward of that in an area of carcinoma, is effective in a high percentage of carefully chosen cases; this approach, of course, avoids the necessity of removing the entire larynx. Some anatomical relations of particular importance in total laryngectomy for carcinoma, in which the hyoid bone is also removed, are shown in Figures 8-20 to 8-23.

Cricoid Cartilage

The lower portion of the laryngeal wall, and the entrance to the trachea, is supported by the cricoid cartilage which, unlike the thyroid cartilage, forms a complete ring. It is the only supporting element of the larynx and trachea that extends completely around the air passage; because injury to it may be followed by perichondritis and eventual stenosis of the larynx, tracheostomies are usually made below it. The cricoid cartilage is shaped essentially like a signet ring, its anterior and lateral portions (arch) being a relatively narrow band, while posteriorly it expands to form a lamina some 2 to 3 cm in vertical length. Laterally and inferiorly, at the junction of arch and lamina, is the cricothyroid articulation; this may be a small rounded depression that receives the inferior horn of the thyroid cartilage, but according to Maue and Dickson only about 20% of the facets are well defined, and some articulations are through soft tissue only.

The cricoid and thyroid cartilages are also united to each other, anterior to this articulation, by the cricothyroid ligament (cricothyroid membrane, cricovocal membrane). In the anterior midline a part of this ligament can be seen attaching to the lower border of the thyroid cartilage (Fig. 8-24), and this part alone has been called the cricothyroid or the middle cricothyroid ligament. Most of the cricothyroid ligament, however, projects up-

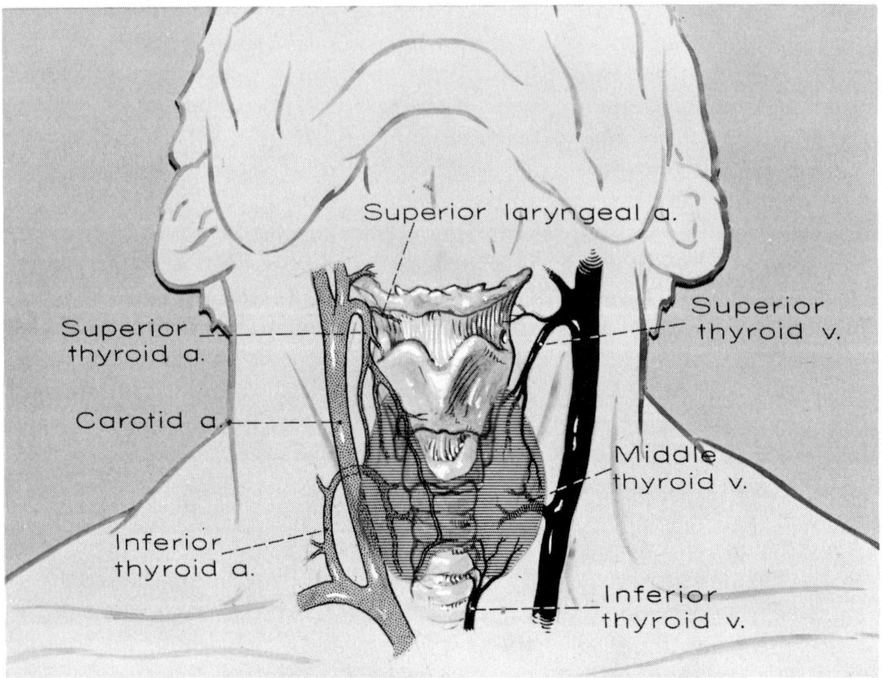

Fig. 8-20. Important blood vessels related to laryngectomy. (Schwartz AW, Hollinshead WH, Devine KD: Surg Clin North Am 43:1063, 1963)

Fig. 8-21. Suprahyoid dissection for total laryngectomy. (Schwartz AW, Hollinshead WH, Devine KD: Surg Clin North Am 43:1063, 1963)

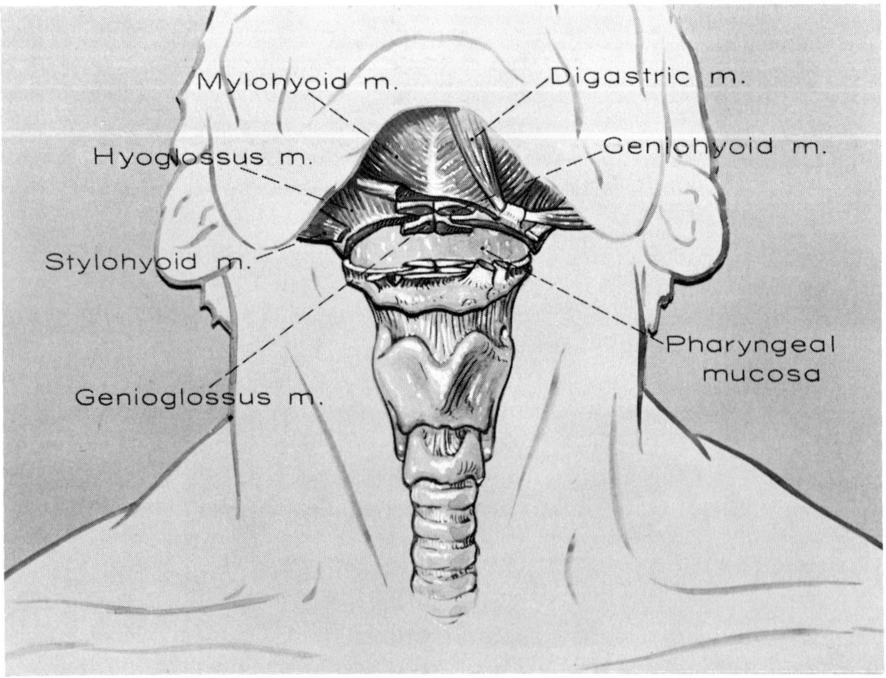

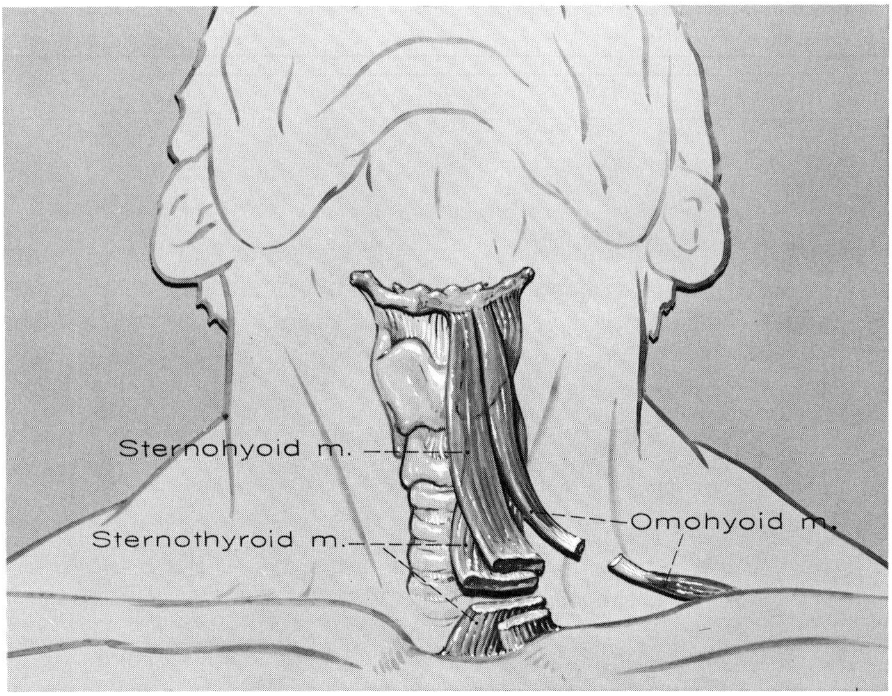

Fig. 8-22. Infrahyoid dissection for total laryngectomy. (Schwartz AW, Hollinshead WH, Devine KD: Surg Clin North Am 43:1063, 1963)

Fig. 8-23. Ligation and division of thyroid vessels, and division of the thyroid gland, in preparation for removal of the thyroid lobe on the side of the lesion. (Schwartz AW, Hollinshead WH, Devine KD: Surg Clin North Am 43:1063, 1963)

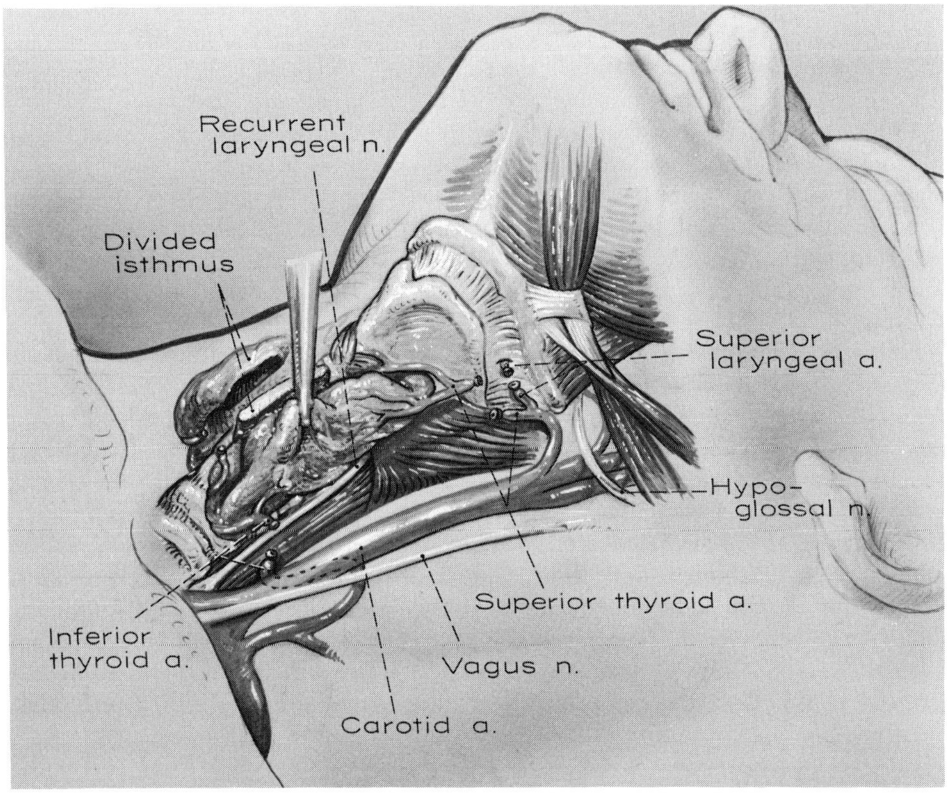

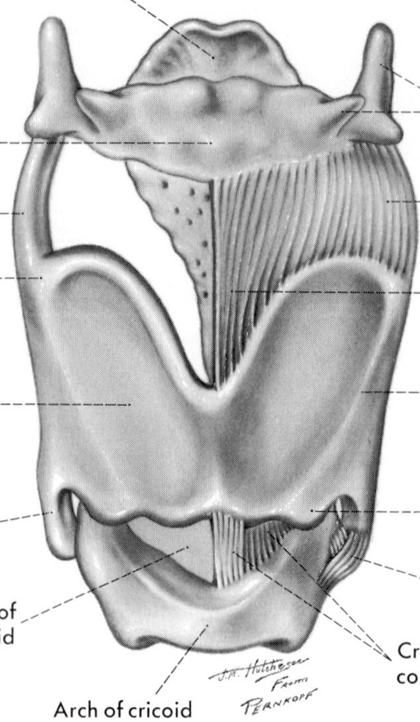

Fig. 8-24. Cartilages and ligaments of the larynx from the front.

ward deep to the thyroid cartilage to form the conus elasticus.

The posterior portion of the cricothyroid ligament is overlapped externally by the cricothyroid muscle, which also connects these two cartilages, but anterolaterally it is subcutaneous, and emergency tracheostomies have been made through it. Because of the cricothyroid articulation the cricoid and thyroid cartilages cannot be actually moved away from each other, despite the elasticity of the cricothyroid ligament. However, the rotatory action possible at the cricothyroid articulation allows the cricoid and thyroid cartilages to be tilted upon each other in such a fashion as to increase or decrease the distance between the thyroid cartilage and the upper border of the cricoid cartilage; Mitchinson and Yoffey ('48) expressed the belief that there may be also some slight sliding at this joint.

The lower border of the cricoid arch is united to the first tracheal cartilage by the cricotracheal ligament, similar in structure to the ligaments (anular ligaments) between the tracheal cartilages.

Arytenoid Cartilages

On the upper surface of the lamina of the cricoid cartilage, and therefore located posteriorly, are the paired arytenoid cartilages (Figs. 8-25 and 8-26), particularly important because it is to these that the posterior ends of the vocal ligaments are attached. The base of the arytenoid cartilage is roughly triangular, and the cartilage as a whole is a somewhat misshapened triangular pyramid. In addition to its base, therefore, it presents three surfaces: a posterior, an anterolateral, and a medial. The posterior surface is somewhat concave and receives the attachment of the arytenoid muscles; the anterolateral surface gives attachment to the vestibular ligament and to the thyroarytenoid and vocalis muscles and lodges a group of mucous glands; and the medial or laryngeal surface is covered by laryngeal mucous membrane. The base of the arytenoid cartilage projects laterally to form a

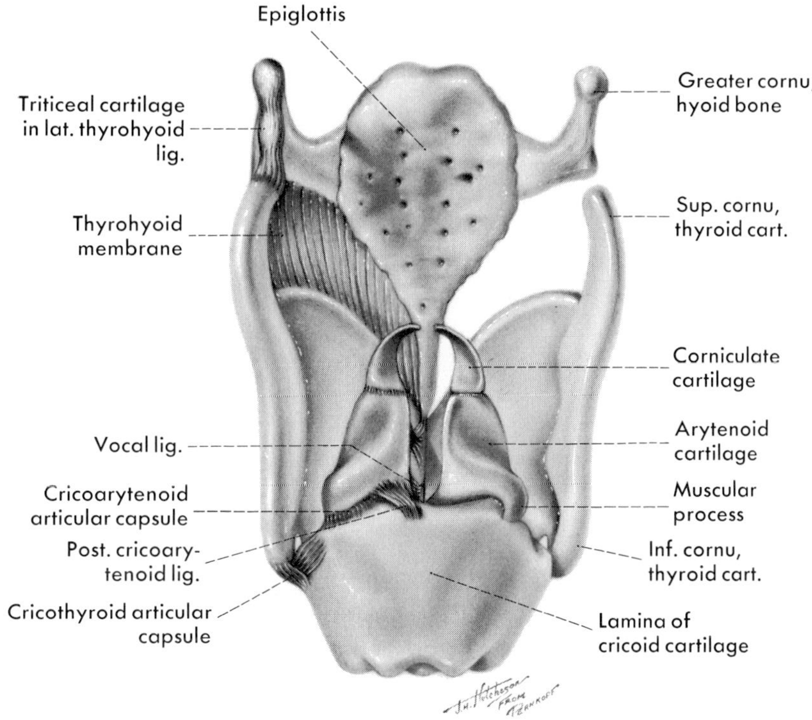

Fig. 8-25. Cartilages and ligaments of the larynx from behind. The quadrangular membrane and the position of the cuneiform cartilage are shown in Fig. 8-27.

Fig. 8-26. Cartilages and ligaments of the larynx in sagittal section.

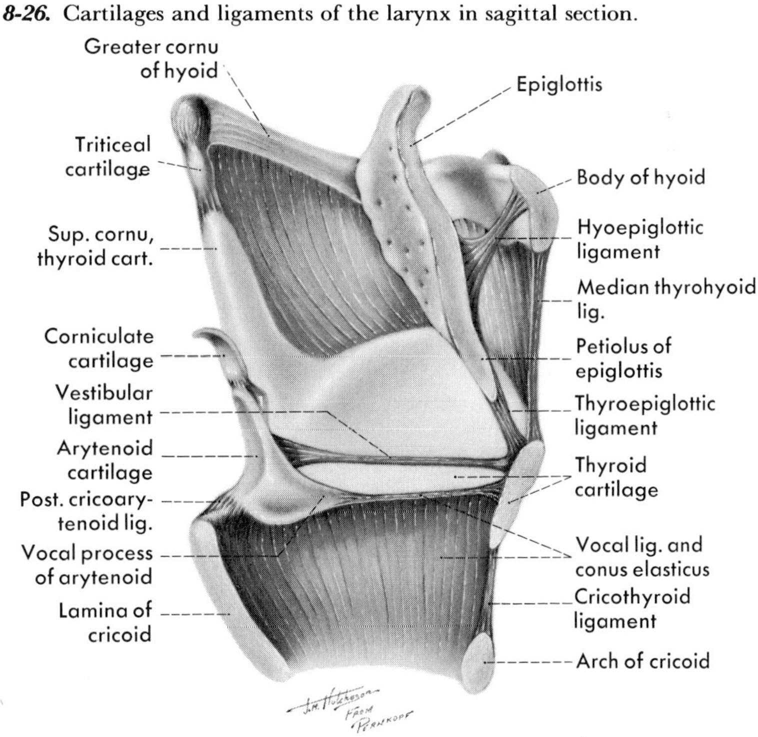

short, broad muscular process to which both cricoarytenoid muscles attach, and anteriorly to form the longer and more slender vocal process to which the vocal ligament is attached. On the lower surface of the base is the articular surface for the cricoarytenoid joint.

The cricoarytenoid joint, between the base of the arytenoid cartilage and the lateral part of the lamina of the cricoid cartilage, varies considerably in shape (Maue and Dickson), and it also varies in its placement on the cricoid cartilage both among specimens and between sides (Sellars and Keen). The greatest movement of this joint is a downward and lateral or upward and medial sliding of the arytenoid cartilage on the cricoid cartilage. According to Sellars and Keen, the sliding movement is accompanied by a rocking one and a twisting one. Thus in adduction the arytenoid cartilages slide upward and medially toward each other; an inward rocking of the cartilages approximates the vocal processes further and depresses them, while the cartilages also twist to bring the processes into complete adduction. From adduction, the arytenoids slide downward and laterally, thus apart, as they rock backward and finally twist to move the vocal processes outward. The rocking and twisting of the arytenoids produce what is essentially a rotation, with the muscular processes being moved farther apart and the vocal processes therefore moved closer together in adduction and in opposite directions in abduction.

Corniculate Cartilages

The apices of the arytenoid cartilages are bent somewhat posteriorly and medially and are in turn surmounted by the small corniculate cartilages (of Santorini), generally regarded as rudimentary and nonfunctional in human beings. According to Negus these cartilages, which necessarily move forward with the arytenoid cartilages during closure of the larynx, aid in opening the esophagus in lower animals; Fishman reported an individual in whom esophageal fibers attached to them.

Cuneiform and Epiglottic Cartilages

In the aryepiglottic folds are usually, but variably, a pair of small isolated cartilages, the cuneiform cartilages (of Wrisberg). The curved, very thin epiglottic cartilage, forming the anterior wall of the laryngeal aditus, is anchored anteriorly to the posterior surface of the body of the hyoid bone by a hyoepiglottic ligament, and its pointed inferior stalk (petiolus) is attached to the inner surface of the thyroid cartilage by a thyroepiglottic ligament. The freely projecting upper portion of the epiglottic cartilage is covered on all sides with mucous membrane, but the lateral edges of the lower portion have the aryepiglottic folds attached to them. The potential space between the anterior surface of the epiglottis and the inner surfaces of the thyrohyoid membrane and thyroid cartilage has been described as the preepiglottic space; carcinoma of the larynx may invade this space, which is therefore sometimes routinely removed by resection of a portion or all of the hyoid bone and total removal of the epiglottis, when there is a carcinoma of the larynx. Dayal and co-workers found that dye injected into the preepiglottic space does not spread from one side to the other, but they could find no definite septum, only a condensation of fibroelastic tissue, between the two sides. In none of ten specimens examined for lymph nodes was one present.

ELASTIC TISSUE

Quadrangular Membrane

The elastic tissue of the larynx deserves special description, since it forms what may be called the intrinsic ligaments of the larynx and especially the vocal ligament. Lewis has described this tissue in considerable detail. Essentially, it consists of two fundamental parts. The upper part of this elastic skeleton of the larynx, the quadrangular membrane (Fig. 8-27), extends from the sides of the epiglottic cartilage to the corniculate and arytenoid cartilages; with the mucous membrane covering it, the quadrangular membrane forms the aryepiglottic fold, thus the wall between the piriform recess and the larynx.

The quadrangular membrane is usually not well pronounced except in its upper part, being a thin sheet interposed between two layers of mucosa with which are associated delicate bundles of muscle. It is described as forming inferiorly the vestibular ligaments that help produce the vestibular or false vocal folds; Lewis described, from microscopic studies, a thin elastic membrane continuous above with the quadrangular membrane and below with the conus elasticus and thus forming a third or intermediate segment of elastic tissue enclosing the laryngeal ventricle.

Conus Elasticus and Vocal Ligaments

The conus elasticus is a much more strongly developed layer of elastic tissue than the quadrangular membrane. It arises from the upper border of the arch of the cricoid cartilage and sweeps medially and upward (Figs. 8-26 and 8-28). In the anterior midline, deep to the superficial part that is called the cricothyroid ligament, the two sides of the conus elasticus come together and are attached to the inner surface of the thyroid cartilage; posteriorly, the conus attaches to the movable arytenoid cartilage and its vocal process on each side; and between the thyroid cartilage and the tips of the vocal processes the conus forms two thickened projecting edges, the vocal ligaments. The vocal ligaments with their covering epithelium form the edges of the vocal folds and are commonly referred to as the vocal cords; the point at which the cords come together anteriorly is known as the commissure or anterior commissure of the larynx.

The vocal ligaments consist therefore of thickened bands of elastic tissue attached anteriorly to the inner aspect of the thyroid cartilage at the midline, and posteriorly to the apices of the vocal processes of the arytenoid cartilages (Figs. 8-26 and 8-29); the anterior ends of the vocal ligaments are thus fixed, but the posterior ends move laterally or medially with the vocal processes of the arytenoid cartilages. The cords are parallel with each other only when they are completely adducted and in any other position form two sides of a variable triangle, with the apex at the anterior midline of the larynx and the base an imaginary line drawn between the tips of the vocal processes.

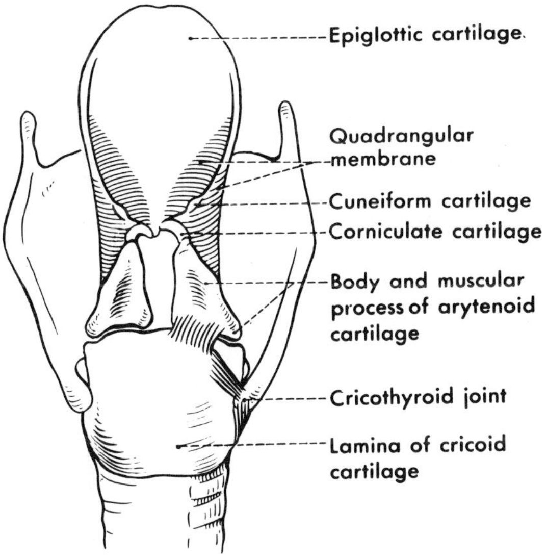

Fig. 8-27. The quadrangular membrane and the larynx from behind.

GENERAL FORM AND RELATIONS

EXTERNAL RELATIONS

Anterolaterally, the thyroid and cricoid cartilages are in relation externally to the upper poles of the thyroid gland (Fig. 8-20 and 8-23), and the isthmus of this gland lies across

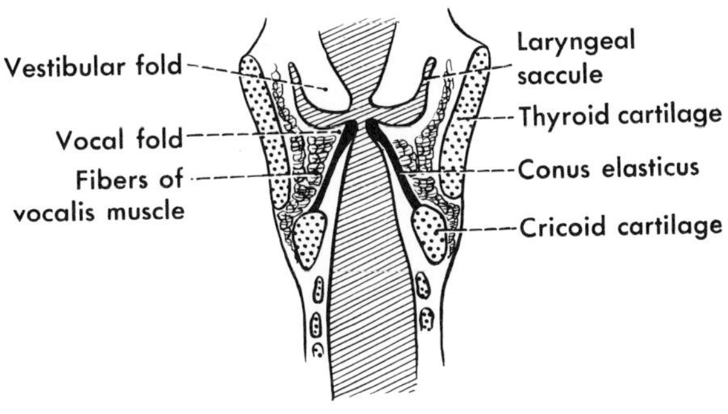

Fig. 8-28. The larynx and the conus elasticus in a frontal section.

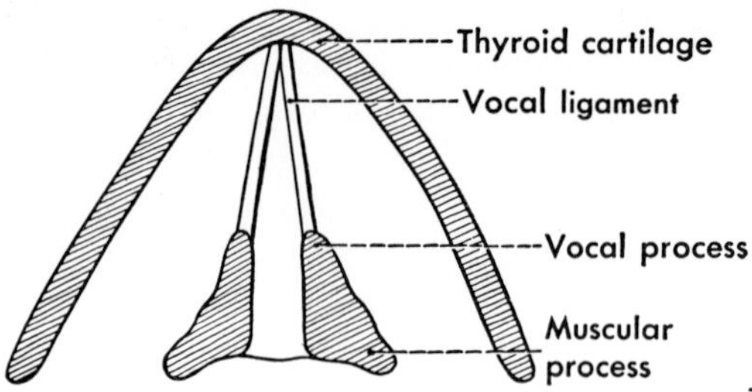

Fig. 8-29. Attachments of the vocal cords, seen from above.

or just below the cricoid; the thyroid gland is attached, through its adherent zone or suspensory ligament, either to the lateral surface of the cricoid cartilage or to the tracheal rings immediately below this. The cricoid cartilage is covered also by the infrahyoid or strap muscles of the neck, which largely cover the thyroid cartilage; otherwise the thyroid cartilage is practically subcutaneous. Posterolaterally, the larynx is related to the carotid sheath and its contents.

The recurrent laryngeal nerve, ascending approximately in the groove between the trachea and esophagus, passes just behind the cricothyroid articulation to enter the larynx, accompanied by a branch of the inferior thyroid artery. The palpable tip of the inferior horn of the thyroid cartilage thus serves as a guide to the point of entrance of this important nerve into the larynx. Descending medial to both the external and internal carotid arteries, the superior laryngeal nerve divides into two branches, internal and external; the internal branch curves forward across the lateral surface of the thickened posterior border (the lateral thyrohyoid ligament) of the thyrohyoid membrane and lies between the thyrohyoid muscle and the lateral surface of the membrane, with the accompanying superior laryngeal artery and vein (from the superior thyroid vessels).

The external laryngeal branch of the superior laryngeal nerve descends upon the lateral surface of the inferior pharyngeal constrictor as this muscle sweeps forward to attach to the oblique line on the thyroid cartilage, and here frequently gives off a twig to the lower portion of the constrictor; in the upper part of its course it is immediately medial to the superior thyroid artery and can easily be injured here in clamping the artery. The external laryngeal branch then swings forward across the outer surface of the lower part of the inferior constrictor, between this and the upper pole of the thyroid gland, to supply the cricothyroid muscle.

INTERNAL ANATOMY

The laryngeal aditus (Fig. 8-30), or entrance into the larynx, is formed anteriorly by the epiglottis, laterally by the aryepiglottic folds, and posteriorly by the tips of the corniculate cartilages and the upper border of the arytenoideus muscle—these, of course, covered by mucosa. Between the anterior surface of the freely projecting portion of the epiglottis and the base of the tongue are the median and lateral glossoepiglottic folds, enclosing the shallow epiglottic valleculae. The aryepiglottic folds are formed by the upper free edges of the quadrangular ligament and its covering mucosa and extend from the lower part of the sides of the epiglottic cartilage to the arytenoid and corniculate cartilages; lateral to them, and thus between them and the inner surface of the thyroid cartilage, are the piriform recesses of the pharynx. As the aryepiglottic folds swing posteriorly and medially toward the corniculate cartilages they typically present small swellings, the cuneiform or cuneate tubercles, caused by the presence of the cuneiform cartilages. Posteriorly, on either side of the midline, the corniculate tubercles formed by the projecting corniculate cartilages may be recognized. The depression between these is the interarytenoid incisure (notch).

The posterior surfaces of the corniculate and arytenoid cartilages and the lamina of the cricoid cartilage, covered by mucosa, form the anterior wall of the laryngeal pharynx. The potential gap between the arytenoid cartilages is largely occupied by the arytenoideus muscle; between the mucosa of the pharynx and the lamina of the cricoid cartilage the longitudinal musculature of the esophagus, swinging forward in two bands, one on either side, forms a single (cricoesophageal) tendon that receives on its sides the attachment of the circular muscle at this level and inserts upon the midline crest of the lamina; the larger part of the lateral portions of the posterior surface of the lamina is occupied by the cricoarytenoideus posterior muscles.

The larynx extends from the level of the aditus to the lower border of the cricoid cartilage, where it is continuous with the trachea (Fig. 8-31); it is divided into three parts, vestibule, ventricle, and infraglottic or subglottic cavity (cavum infraglotticum), by the vestibular and vocal folds. The upper portion of the larynx, the vestibule, extends from the aditus to the vestibular folds. It is bounded by the epiglottis, the aryepiglottic folds, the arytenoid cartilages above the level of the vestibular folds, and the arytenoideus muscles.

Vestibular Folds

The vestibular folds (ventricular folds or bands, false vocal cords, superior vocal cords; Fig. 8-32) are two thick folds of mucous membrane with a thin center of connective tissue (the vestibular ligament); they are attached at their anterior ends to the thyroid cartilage just below the attachment of the epiglottis and at their posterior ends to the bodies of the arytenoid cartilages. They thus move with the arytenoid cartilages and can be brought together to close the slit between them (the rima vestibuli) and help seal the air passage.

From above, the false vocal cords are simply inward bulges of the mucosa, which hide much of the true vocal folds; their lower free edges are, however, undermined by a recess, the ventricle of the larynx (ventricle of Morgagni, laryngeal sinus), which extends between the vestibular and vocal folds. At the anterior end of the ventricle, under cover of the vestibular folds, there is a diverticulum known as the laryngeal saccule or appendix of the laryngeal ventricle (see Fig. 8-28); this extends upward between the vestibular fold and the inner surface of the thyroid cartilage and contains a number of mucous glands whose

Fig. 8-30. The larynx, from above.

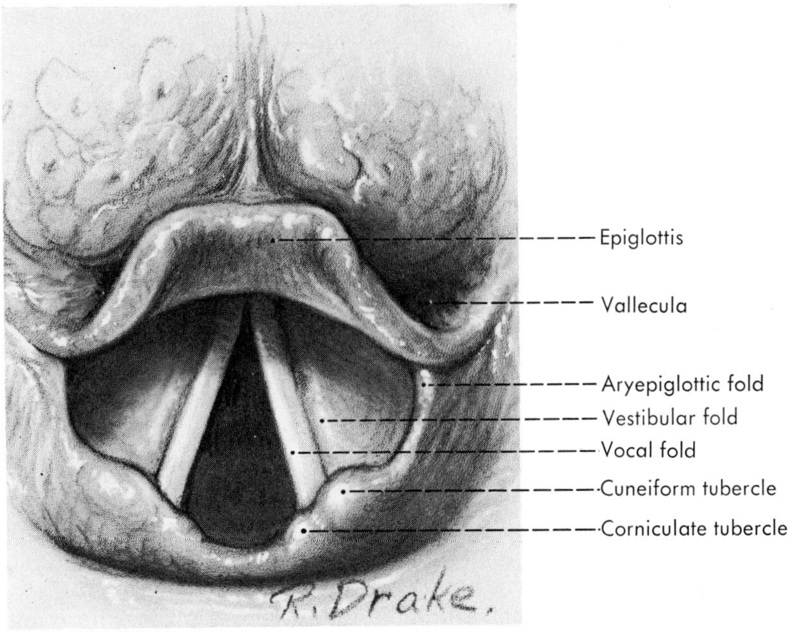

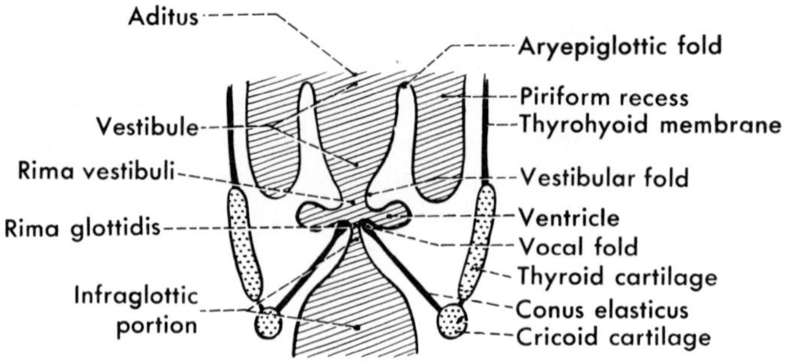

Fig. 8-31. Cavity of the larynx and its subdivisions in a frontal section. See also Fig. 8-26.

secretion is believed to lubricate the vocal folds. In addition to the mucous membrane, the wall of the ventricle and of the saccule consists of the thin thyroarytenoideus muscle, the contraction of which probably aids in expressing the secretion from the saccule, and a delicate elastic lamina.

The size of the laryngeal saccule varies so markedly that it is difficult to establish a norm, but it usually does not extend above the upper border of the thyroid cartilage; when it is markedly elongated it does so extend and may penetrate the thyrohyoid membrane to lie deep to the infrahyoid muscles.

Laryngocele

An enlargement of the laryngeal saccule is usually referred to as diverticulum of the larynx or as laryngocele. The incidence of laryngocele is unknown, and is indeed impossible to determine unless only true cystic formations of this are counted or an arbitrary limit is placed upon the size of a so-called normal saccule; Campbell noted, for instance, that one worker reported 85 cases of laryngocele, while another of wide experience had seen only two.

Large bilateral laryngeal saccules that extend markedly beyond the thyrohyoid membrane (a normal development in some animals—*e.g.,* the howling monkey) are apparently sometimes congenital and do not necessarily present symptoms; they are occasionally found in the dissecting room and are usually regarded as representing an atavism. The view that all laryngoceles are congenital seems to have been largely abandoned, however, in favor of the view that the majority develop as a result of dilation of a saccule by increased air pressure. Yarington and Frazer, for instance, noting that most patients with laryngocele are over 50 years of age, implicated loss of tone by the supporting tissues and increased air pressure caused by such activities as singing and glass blowing.

The clinical symptoms described by Taylor in a case of unilateral diverticulum included recurring hoarseness and swelling on the side of the neck, to some extent therefore resembling those of carcinoma of the larynx. The swelling associated with an enlarged ventricular saccule may be entirely internal and present itself within the larynx above the false vocal cord, bulging the aryepiglottic fold (internal laryngocele); it may be outside the thyrohyoid membrane and therefore present itself in the neck (external laryngocele); or there may be both an internal and an external laryngocele. Internal laryngoceles, which can occlude the air passage if they become large enough, are being reported in increasing numbers, according to Yarington and Frazer; they advocated an external approach through the thyrohyoid membrane in order to ensure complete removal of the sac.

Glottis and Vocal Folds

Below the ventricle of the larynx are the true focal folds. The vocal folds and the slit (rima glottidis) between them constitute the glottis. The edges of approximately the anterior three fifths of the vocal folds are formed by the vocal cords (the vocal ligaments and their covering epithelium), and the space between them is the intermembranous portion of the rima (rima glottidis vocalis); those of about the posterior two fifths of the vocal folds, on the other hand, are formed by the vocal processes of the arytenoid cartilages, and the space between them is therefore the intercartilaginous (or so-called respiratory) part of the rima.

The glottis is normally the narrowest portion of the larynx, but its shape and size vary greatly with movements of the arytenoid cartilages and the vocal folds. In quiet respiration the intermembranous part of the rima glottidis is triangular (Fig. 8-33), the two vocal folds extending from the separated ends of the vocal processes to their midline attachment on the inner surface of the thyroid cartilage; under these conditions, the vocal processes are parallel with each other, and therefore the intercartilaginous portion of the rima is approximately rectangular.

The intermembranous portion of the rima may be closed to bring the tips of the vocal processes together first, but more commonly the arytenoid cartilages are adducted so as to close both intermembranous and intercartilaginous portions. In extreme abduction of the vocal cords in forced inspiration the vocal processes may turn outward so as to convert the rima into a somewhat diamond-shaped opening, the widest parts being the ends of the vocal processes to which the ligaments are attached. Finally, in forcible closure of the air pathway, as in fixing the thorax by holding one's breath, the arytenoid cartilages are both adducted and drawn forward; thus the cartilage and the vestibular fold of one side meet those of the other in the midline, obliterating the lower part of the vestibule. This action occurs also in swallowing, so that the vestibular folds tend to prevent further passage of any material entering the vestibule.

Subglottic (Infraglottic) Cavity

The subglottic or infraglottic cavity (cavum infraglotticum), sometimes included as a part of the glottis, is the third part of the laryngeal cavity, extending from the glottis to the lower border of the cricoid cartilage. It is bounded laterally by the conus elasticus and therefore expands laterally as it is traced downward; just below the glottis it is slitlike if the cords are adducted, but at the level of the cricoid cartilage it has become circular.

Mucosa of Larynx

The epithelium of the mucosa of the larynx is largely of the respiratory type (columnar ciliated) such as is found in the trachea, but a portion of it is stratified squamous epithelium like the lining of the pharynx. The squamous

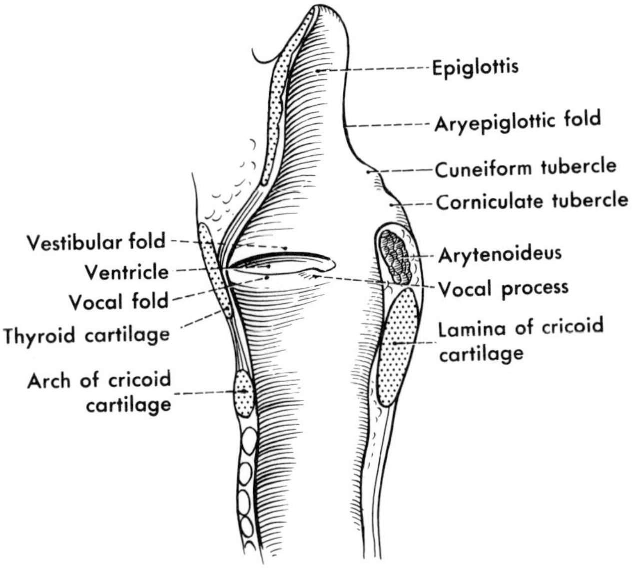

Fig. 8-32. Interior of the larynx.

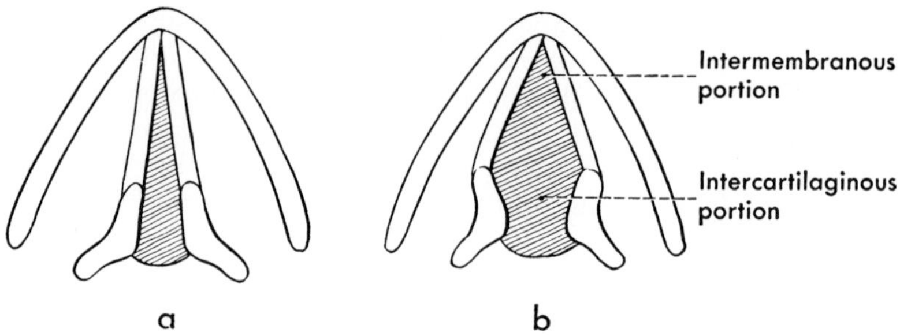

Fig. 8-33. Rima glottidis; (*a*), quiet respiration, and (*b*), forced respiration. Note that in (*b*) the arytenoid cartilages as a whole are abducted, and their vocal processes are directed somewhat laterally.

epithelium is largely limited to the upper half of the posterior surface of the epiglottis, the upper part of the aryepiglottic folds, and the vocal folds themselves, although scattered patches of such epithelium may be found elsewhere above the glottis; below the glottis the epithelium is entirely of the respiratory type. Mucous-secreting glands are numerous, especially in front of the arytenoid cartilages and in the ventricular saccules, but are absent from the edges of the vocal folds.

Miller investigated the spread to the larynx of fluid injected into the submucosa of the pharynx. He reported that fluid so injected spread through the pharyngeal wall to the lateral glossoepiglottic (pharyngoepiglottic) folds and was there temporarily halted as it accumulated in the submucosa of the piriform recess, including that of the aryepiglottic fold; it spread over the posterior or pharyngeal aspect of the cricoid lamina (it did not cross the midline here and extended only a little below the cartilage). The fluid also extended into the epiglottic vallecula and crossed the midline to cover the entire anterior surface of the epiglottis. The fluid reached the larynx by spreading downward over the edge of the aryepiglottic fold and passed downward deep to the mucosa of the vestibule, obliterating the ventricle but halting abruptly at the vocal cords. The close attachment of the mucosa to the vocal ligaments prevented the spread of any appreciable amount of fluid into or below them. Clinically, this attachment explains the independent occurrence of supraglottic and subglottic edema; in acute laryngotracheal bronchitis, for instance, the laryngeal edema is sharply limited above by the cords.

Pressman and co-workers agreed that the subglottic submucous compartment is a separate one, but divided the submucous space above the cord into a number of compartments that they described as communicating neither with the adjacent ones above and below nor across the midline. Welsh and coworkers, however, said that dye injected submucously above the cords did spread to the other side (mostly posteriorly because of the tight attachment of the mucosa to the epiglottis) and that dye injected into the cord, although never crossing at the commissure of the cords, can spread to the other side through the subglottic tissue.

MUSCULATURE

EXTRINSIC MUSCLES

The muscles acting upon the larynx fall into two groups: the extrinsic muscles, which tend to move the larynx as a whole, and the intrinsic muscles, which are confined to the larynx and modify the size of the aditus and of the rima glottidis and the length and position of the vocal cords. The extrinsic muscles include a number of diverse groups, such as the infrahyoid or strap muscles (omohyoid, sternohyoid, sternothyroid, and thyrohyoid), which, with the exception of the thyrohyoid, are depressors of the larynx; the stylohyoid, digastric, and mylohyoid, which with the thyrohyoid elevate the larynx; the stylopharyngeus and palatopharyngeus muscles, which tend to elevate the pharynx and therefore the larynx and are otherwise important in modifying the air passage above the larynx; and the middle and inferior constrictors of the pharynx, which arise from the hyoid bone and the larynx.

The exact parts that these various extrinsic muscles play in the physiology of the larynx, that is, in respiration and phonation, are certainly not clearly understood, but certain points appear fairly clear. In the first place, the larynx as a whole may participate in respiratory movements (Mitchinson and Yoffey, '47). The movements here are, however, not constant and probably not particularly important. In 23 individuals investigated by Mitchinson and Yoffey the larynx rose in five during inspiration, but descended in four, and there was no appreciable vertical displacement in the remaining 14 as determined by roentgenogram; the maximal movement was about 6 mm (¼ inch), the hyoid bone and larynx moving together. In 18 out of the 23 individuals there was an appreciable forward movement of the hyoid bone, the tongue, and the epiglottis in deep inspiration, this movement obviously tending to enlarge the aditus.

INTRINSIC MUSCLES

The intrinsic muscles of the larynx are the cricothyroid, posterior and lateral cricoarytenoids, arytenoid (interarytenoid), thyroarytenoid, and vocalis muscles, plus various portions of these that have received special names (Fig. 8-34). Of these, the cricothyroid appears on the lateral and outer aspect of the larynx, while the others are covered either by the thyroid cartilage or by mucous membrane; corresponding to this difference in position is one of innervation, for the cricothyroid is innervated by the external branch of the superior laryngeal nerve, while the other muscles are innervated by the inferior laryngeal nerve, the continuation into the larynx of the recurrent laryngeal nerve. All the muscles except the arytenoid are paired.

Cricothyroid

The cricothyroid muscle (see Fig. 8-15) arises from the external surface of the arch of the cricoid cartilage and is partially divisible into two portions: an anterior and superior portion (pars recta) that runs upward and laterally toward the inferior border of the thyroid cartilage, and a posterior and inferior (oblique part) that runs somewhat more obliquely backward. The more anterior portion inserts on the lower border of the thyroid cartilage, and some of the more posterior fibers are similarly inserted, but others extend backward to attach to the inferior horn and the inner surface of the thyroid cartilage. The muscle tilts the cricoid lamina backward through the cricothyroid joint, thus lengthening, tensing, and adducting the vocal cords.

Fig. 8-34. Internal muscles of the larynx: (*a*), from behind after removal of the pharynx and esophagus; (*b*), from the side, after partial removal of the thyroid cartilage.

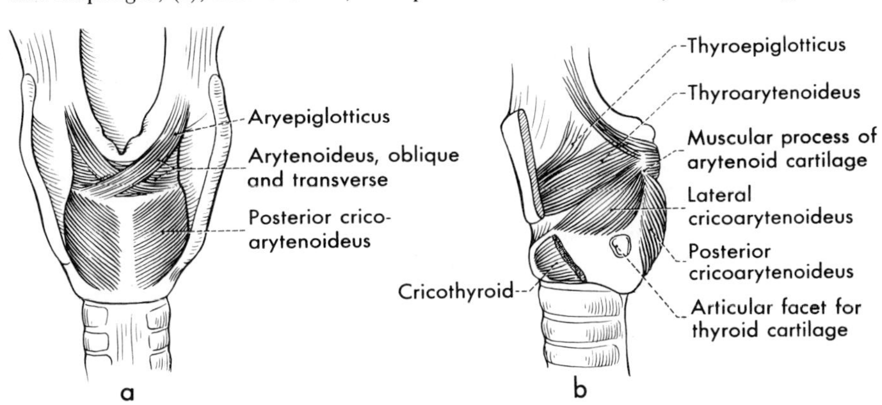

Posterior Cricoarytenoid

The posterior cricoarytenoid muscle arises from the posterior surface of the lamina of the cricoid cartilage, largely covering a lateral half of this and being in turn covered by the mucosa of the anterior wall of the pharynx; between this muscle and its mate of the other side is the tendinous attachment (cricoesophageal tendon) of the esophagus to the midline crest. From its origin on the lamina the fibers of the posterior cricoarytenoid pass obliquely laterad and upward to insert into the muscular process of the arytenoid cartilage; contraction of the muscle therefore pulls the muscular process posteriorly, medially, and downward, sliding the arytenoid cartilage laterally and also tilting it to abduct the vocal cord.

Sometimes associated with the lower border of this muscle is an aberrant muscular slip, the ceratocricoid muscle, stretching between the lamina of the cricoid cartilage and the inferior cornu of the thyroid cartilage; Hetherington found this in 7.6% of sides in black persons and 12.9% of sides in white persons. Presumably it is an antagonist of the cricothyroid muscle.

Lateral Cricoarytenoid

The lateral cricoarytenoid muscle arises from the upper border and outer surface of the cricoid arch and extends upward and backward to insert upon the anterior surface of the muscular process of the arytenoid cartilage; by its contraction, therefore, it pulls the muscular process forward and laterally, thereby sliding the arytenoid cartilage upward on the cricoid cartilage; it is therefore the antagonist of the posterior cricoarytenoid muscle, adducting the vocal cords.

Arytenoidei

The transverse arytenoideus (interarytenoideus) muscle consists of fibers stretching between the posterior surfaces of the two arytenoid cartilages and is therefore unpaired. It approximates the two arytenoid cartilages and thus assists in closing the posterior (intercartilaginous) portion of the glottis. On the posterior surface of the relatively heavy transverse arytenoideus there can usually be demonstrated two thin bands (oblique arytenoidei) that cross the transverse fibers obliquely, passing from the muscular process of the arytenoid upward and posteriorly to cross each other in the midline and run toward the tip of the opposite arytenoid cartilage. Some of the fibers insert here, while others fan out upon the quadrangular membrane, those running in the aryepiglottic folds being termed the aryepiglottic muscle, while other parts, of no practical significance and hardly demonstrable, have been given such names as thyroarytenoideus obliquus and arymembranosus. The arytenoid muscle as a whole is covered posteriorly by the mucosa of the laryngeal part of the pharynx; anteriorly, between the arytenoid cartilages, it is in contact with laryngeal mucosa.

Thyroarytenoideus

Above the lateral cricoarytenoid muscle is a thin, frequently poorly developed and variable muscle that lies between the thyroid lamina and the elastic layer of the larynx largely above the conus elasticus. It arises on the inner surface of the thyroid cartilage where the laminae come together and extends backward to insert upon the lateral surface of the arytenoid cartilage. It draws the arytenoid cartilage forward and medially, being in this respect somewhat of an antagonist to the cricothyroid and posterior cricoarytenoid muscles. It also adducts the vestibular (false vocal) folds and, with the transverse arytenoideus, is the sphincter mechanism at the level of these folds.

This muscle, sometimes called the thyroarytenoideus externus, has also been grouped, as a part of a single thyroarytenoideus muscle, with the constant and better developed fibers (vocalis muscle, or thyroarytenoideus internus) of similar attachment that lie deeper and in close contact with the conus elasticus. Some few fibers of the thyroarytenoideus muscle can usually be traced upward upon the quadrangular membrane to or toward the lateral border of the epiglottis, and constitute the thyroepiglottic muscle.

Vocalis

The vocalis muscle (also called thyroarytenoideus internus) is a deeper, thicker, and better developed thyroarytenoideus muscle; in contrast to the thyroarytenoideus proper (thyroarytenoideus externus), from which, however, it is scarcely separable in a dissection, it lies against and is intimately adherent to part of the conus elasticus. Like the thyroarytenoideus, its fibers arise from the inner surface of the thyroid cartilage at the angle formed by the junction of the laminae. Although once described as inserting into the vocal ligament, it seems now accepted (Floyd and co-workers; Sonesson) that most of the fibers of the vocalis parallel the ligament and insert into the vocal process and the base of the arytenoid lateral to the process, with no attachment at all to the ligament; inferiorly, however, some fibers apparently do attach to the outer surface of the conus elasticus below the vocal ligament. Fibers blending laterally with the upper border of the muscle and extending upward around the ventricle, sometimes regarded as an extension of the vocalis but usually as a part of the thyroarytenoideus (externus), have been called the *ventricular muscle.*

By its action the vocalis muscle could draw forward the arytenoid cartilage, thus shortening the vocal fold (cord); normally, however, this action is opposed by the cricothyroideus, and the vocalis muscle thus adducts and tenses the vocal fold. Through the fibers that insert into the conus elasticus, it can apparently draw the vocal folds upward and forward, bringing the edges rather than the broader inferomedial surfaces of the folds into contact with each other and making easier the escape of air between the apposed vocal folds.

MOVEMENTS OF THE VOCAL CORDS

The precise role of the intrinsic laryngeal muscles in the production of various tones is not understood, although it is generally believed that these muscles, or at least the cricothyroid and vocalis, are responsible for the finer modulations in tones. It is certainly the intrinsic muscles that are primarily responsible for the variations in the size of the rima glottidis and therefore for the opening or closing of the entance into the trachea. In the following account the actions of the muscles are given as they relate to movements of the vocal folds and guarding the air passage, but without prejudice as to the specific role in phonation played by variations in length and tenseness of the cords.

The *cricothyroid muscle* tilts the lamina of the cricoid cartilage backward and has also been said to tilt the thyroid cartilage forward; in so doing, it lengthens the distance between the angle of the thyroid cartilage, to which one end of each vocal cord is attached, and the arytenoid cartilage, to which the other end of each vocal cord is attached and which necessarily moves largely with the cricoid cartilage. This muscle by its contraction therefore lengthens the vocal cords and, in the process of so doing, tenses them. Of great clinical importance is the fact that lengthening the vocal cords straightens and therefore adducts them, in accordance with the principle that a straight line is the shortest distance between two points. With all muscles of the larynx intact this does not necessarily happen, since it may be opposed by other muscles, but if the cricothyroid alone is acting adduction must occur; this is of extreme importance in the development of respiratory symptoms following bilateral paralysis of the inferior laryngeal nerve.

According to Suzuki and co-workers, artificial dyspnea produces a marked increase in the activity of the cricothyroid muscle, particularly during inspiration (when its action is said to allow the posterior cricoarytenoid to decrease laryngeal resistance more than the latter muscle can do when it acts alone), and acute bilateral nerve paralysis produces strong activity that can be abolished by tracheostomy or the administration of oxygen. Its activity is also controlled by vagal fibers from the lungs (Fukuda and Kirchner), since if the recurrent nerves are cut the cords (in cats) assume a paramedian position, but if the vagus nerve is then cut or blocked in the tho-

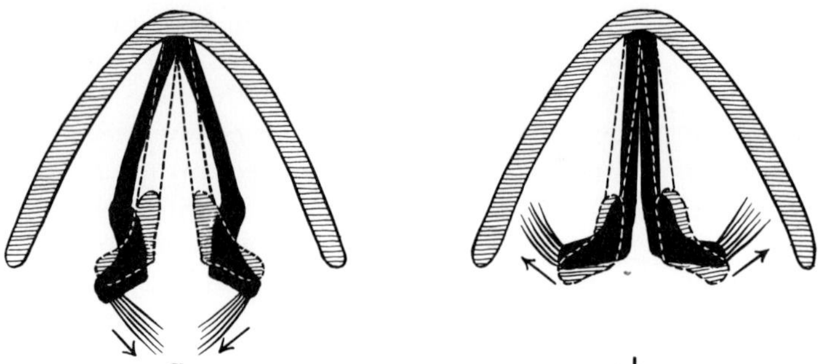

a b

Fig. 8-35. Actions of the cricoarytenoid muscles in opening and closing the rima glottidis. In (*a*) the contraction of the posterior cricoarytenoid is shown to move the arytenoid cartilages apart and direct the vocal process lateralward, therefore to abduct the vocal cords; in (*b*) the lateral cricoarytenoid is shown to have an opposing action, adduction of the vocal cords.

rax the cord moves laterally to a cadaveric position.

The *posterior cricoarytenoid muscle* is the most important one of the larynx, since it is the only abductor of the vocal cords. It pulls the arytenoid cartilage downward so that it slides laterally, rocks backward, and twists, further separating the vocal processes (Fig. 8-35*a*); the vocal processes necessarily carry with them the posterior ends of the vocal cords, and therefore both the intercartilaginous and the intermembranous portions of the rima are widened. Suzuki and Kirchner ('69b) reported that this muscle is driven rhythmically by the respiratory center, and therefore its activity, like that of the diaphragm, is affected by changes in oxygen, carbon dioxide, and the like.

The *lateral cricoarytenoid muscle* likewise acts upon the arytenoid cartilage, but since it pulls forward, it slides the cartilage upward and medially, rocks it forward, and twists it, thus bringing the vocal cords together in the midline or adducting them (Fig. 8-35*b*). Although it is the most direct antagonist of the cricoarytenoideus posterior, it is only one of several muscles that adduct the cords.

The *arytenoideus muscles,* especially the heavy transverse one, obviously bring the two arytenoid cartilages closer together in the midline and, in so doing, therefore close the posterior or intercartilaginous portion of the rima.

The *vocalis* supplies the sphincteric action at the level of the vocal folds, just as the thyroarytenoid does at the level of the vestibular folds, being helped in this action by the transverse arytenoideus and by other muscles that steady the arytenoid cartilage. With the arytenoid cartilages fixed, the vocalis not only adducts the vocal fold but also produces variations in its tenseness, thus varying the pitch of sounds produced by air escaping through the glottis. Also, in a manner not fully explained, the vocalis can hold posterior segments of the folds in firm adduction while allowing air to escape only between anterior segments, thus further modifying pitch.

The *thyroarytenoideus muscles* bring the vestibular folds together to seal off the larynx at this level and, if the arytenoid cartilages are not fixed, these muscles draw the cartilages forward, thus shortening the vocal folds. The various smaller slips of muscle, such as the thyroepiglotticus and aryepiglotticus, presumably assist in narrowing the aditus of the larynx. Pressman ('41) discussed the three levels of sphincteric action (aditus, vestibular or false vocal folds, and vocal folds) that protect the trachea and lungs and quite properly pointed out that although muscular action at the aditus is relatively weak, the two sets of folds have valvelike forms that add greatly to the effectiveness of the larynx in resisting pressure. Thus the vestibular folds, because of their downwardly directed concavity, are especially adapted to resist pressure outward, as in holding the breath, while the vocal folds when adducted resist especially downward pressure.

Production of Tones and Changes in Pitch

The extrinsic muscles, primarily the infrahyoid muscles and the digastric, stylohyoid, and thyrohyoid above the larynx, probably participate in the production of tones and changes in pitch in several ways (*e.g.*, Sokolowsky). They afford fixation to the thyroid cartilage; they raise the larynx for an average distance of about one vertebra in producing high tones (Mitchinson and Yoffey, '48) and lower it for low tones, so that the supraglottic space is shortened or lengthened for high and low tones, respectively (Fig. 8-36), in a fashion reminiscent of the use of organ pipes; and in raising the larynx they move the thyroid cartilage to a more vertical position, thus elongating the vocal cords, while in lowering it they shorten the cords (Griesman). Apparently one of the factors in producing the higher pitched, grating voice in psychophonasthenia is elevation of the larynx because of spasmodic contraction of the extralaryngeal muscles.

The intrinsic laryngeal muscles produce their effects by adducting the cords, varying their length and tenseness, and varying the aperture through which air escapes. Finally, pitch is also related to the pressure under which air escapes through the glottis. Indeed, one theory has been that the larynx functions essentially like a reed instrument, pitch depending upon the air pressure and the size of the aperture through which the air escapes. According to this theory, changes in length and tenseness of the vocal cords would be purely adventitious and of no importance in the production of various tones. This concept is no longer accepted, and Buchtal reported that increasing the air pressure is much less effective than contraction of the vocalis muscle in raising pitch.

Another theory has been that the vocal cords act essentially like piano strings, vibrating as a whole and with the pitch of the voice determined by the length of the cord and the tension it is subjected to by lengthening. It is now well documented, however, that the cord vibrates as a whole only for the lower tones (the difference in length between men's and women's cords apparently accounts for the difference in pitch of their voices) and that,

Fig. 8-36. Semischematic line drawings indicating the changes in position of the larynx between (*a*), humming a low tone, and (*b*), humming a high tone. Note the upward and forward movement of the pharynx and larynx and the lengthening of the vocal cords in (*b*). (After the reproductions of the roentgenograms of Mitchinson and Yoffey: J Anat 82:88, 1948)

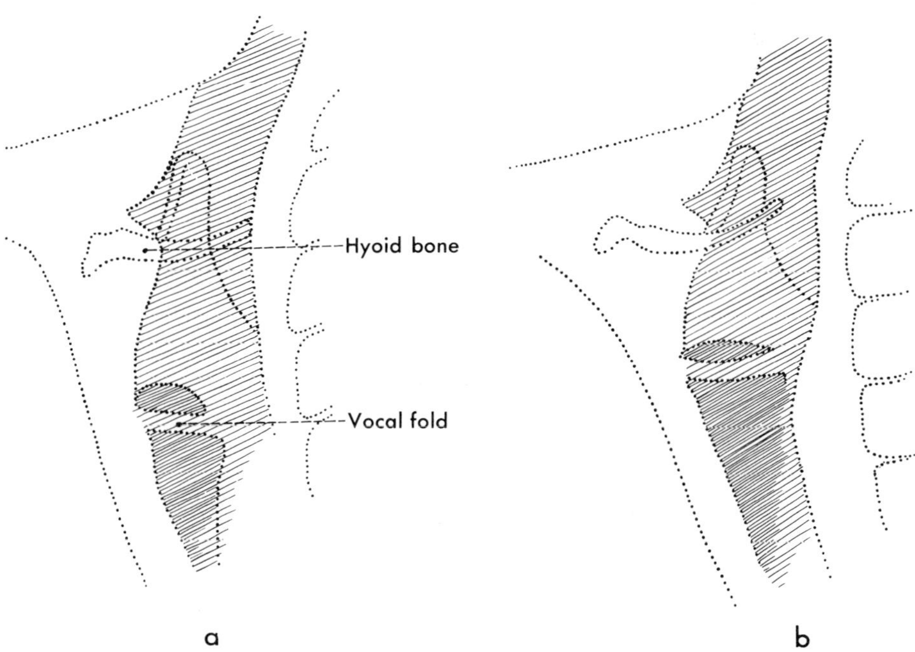

instead of lengthening for low tones as this theory would anticipate, the cords lengthen in producing high tones. Mitchinson and Yoffey found an average increase of 5 mm in length of the vocal cords in going from low to high tones, and practiced singers showed greater lengthening, to a maximum of 9 mm, the cord more than doubling in length in one case. Regarding increased length of the vocal ligament causing sufficient tension of it to produce the rise in pitch attained by the human voice, Strong compared the elasticity of the ligament to that of a rubber band rather than of a vibratory metal string; thus stretching the ligament could not produce a rise in pitch comparable to that produced by increased tension on a piano or violin string. The tension of the vocal fold as a whole is apparently an active process, brought about by contraction of the vocalis muscle against the pull of the cricothyroid muscle (Buchthal; Fink); it is therefore in part independent of the length and can be varied while any given length is maintained.

Although simple stretching of the vocal ligament cannot account for a rise in pitch, length and tenseness of the vocal folds as a whole do affect pitch. Buchthal quoted evidence that when air under constant pressure is passed through the larynx of a dog, stimulation of the cricothyroid muscles, which adduct, tense, and lengthen the folds, can produce a rise in pitch of more than an octave. In normal phonation, however, at least one other factor must be added, and that is the length and mass of the vibratory segment. Pressman's studies in particular (Pressman, '42; Pressman and Kelemen) have shown that after the folds, vibrating as a whole, have reached their maximal tenseness, further increases in pitch are produced by firm adduction of the vocal folds beginning posteriorly, so that only increasingly short anterior segments of the folds vibrate. Thus, as indicated also by the clinically observed lowering in pitch produced by polypoid thickening of the cords, and the lower pitch associated with use of the thicker vestibular folds for phonation, the rate of vibration probably depends in part upon the mass of the vibrating segments.

There are thus a number of variable factors that aid in determining pitch, and presumably one may substitute in part for another; Pressman, for instance, stated that the laryngeal picture may vary considerably even for the same note sung by individuals with similar voices. The exact mechanism by which the intrinsic laryngeal muscles produce and vary these factors is probably not completely understood, but the general outlines seem to be clear. Phonation is preceded by adduction of the cords, which involves an approximately simultaneous increase in activity of most or all of the intrinsic laryngeal muscles (Buchthal). The important muscles in this movement, however, would be the lateral cricoarytenoids, the transverse arytenoid, and the vocalis muscles; the posterior cricoarytenoids presumably help only to steady the arytenoid cartilage; the cricothyroids prevent unwanted shortening of the vocal cords by the vocalis or lengthen the vocal cords against the pull of the vocalis, thereby also helping to adduct the cords. Finally, the tone produced depends primarily on the activity of the cricothyroid and vocalis muscles.

The role of the cricothyroid seems fairly clear. By resisting forward tilting of the lamina of the cricoid, it allows the vocalis to build up more tension, and by lengthening the cord it not only adds to that tension but also thins the cord so that a given length of it has less mass (Fig. 8-37); thus it can contribute to raising pitch in two ways. It is, however, generally agreed that the vocalis muscle is primarily responsible for variations in pitch. Part of its action is apparently through the tenseness that the fibers paralleling the vocal ligament impart to the fold as a whole, and Buchthal reported that activity in the muscle may increase as much as 600% with increase in pitch.

Relatively recent reports that the vocalis muscle produces the vibrations of the cords by contracting repetitively at a rate that is governed by the impulses per second sent over the inferior laryngeal nerve have not been confirmed (*e.g.*, Floyd and co-workers; Buchthal); moreover, for higher tones this would demand a rate of stimulation that pro-

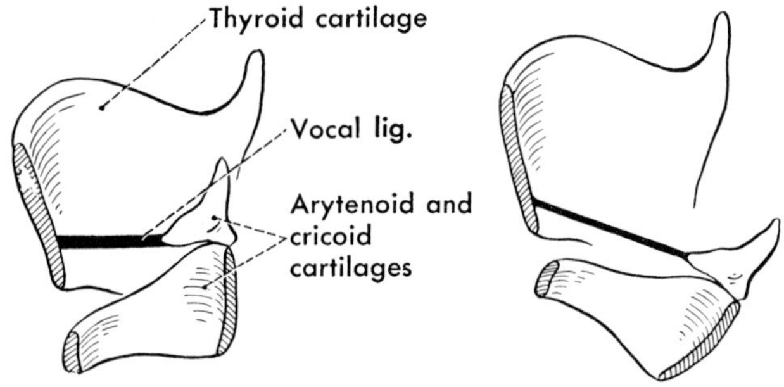

Fig. 8-37. Schema indicating the production of a longer and thinner vocal cord by tilting the cricoid cartilage.

duces tetany rather than repetitive contraction of voluntary muscle. The concept that the cords are vibrated by the air current, and not vice versa, seems to be well established, as does a relation between tenseness of the cords and pitch. As yet incompletely explained, however, is the manner in which the vocalis governs the length of vibrating segments of the cords while helping to hold posterior segments in adduction (Fig. 8-38).

On the basis of descriptions of insertion of the vocalis into the vocal ligament, now generally admitted to be eroneous, Strong described the vocalis as tensing part of the ligament and relaxing another part of the ligament so that it could vibrate, and Pressman suggested that the vocalis abducted a variable anterior segment of the cords so that air could pass between and vibrate them. A newer concept, consonant with the insertion of part of the vocalis into the conus elasticus below the vocal ligament, is that the vocalis draws a part of the conus upward (Floyd and co-workers). This would result in the thin edges of the vocal ligaments being in contact, rather than the entire surfaces, and could have two results: easier passage of air betweeen these thin edges, and a faster rate of vibration that involves only the edge rather than the entire mass of the vocal fold.

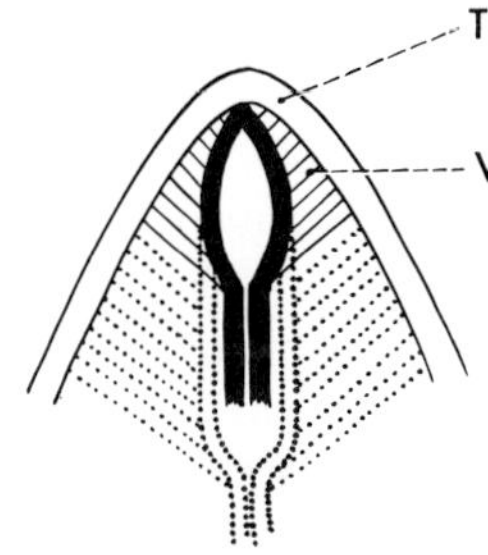

Fig. 8-38. How pitch is varied after the vocal folds have reached their maximal tension. For the highest tones the folds are adducted throughout most of their length, but certain fibers of the vocalis (*solid lines*) allow air to escape between small anterior segments of the folds; for lower tones, but still above the highest that can be obtained through vibration of the folds as a whole, more fibers of the vocalis allow a greater length of the vocal folds to separate (*dotted lines*) and therefore to vibrate.

Vicarious Voice

Despite the fact that the vocal cords are normally the chief vibrating mechanism for the production of voice, it should be noted that other mechanisms can be used to produce vibrations and can therefore substitute for the normal one. Jackson and Jackson ('35b) stated that phonation by the vestibular folds many occur if the voice is forced during acute laryngitis and that the same thing occurs after one or both true vocal cords have been destroyed by disease. They described the voice resulting from vestibular fold phonation as being usually somewhat deeper in pitch and rougher in quality than the normal voice, but stated that it can be especially trained until phonation is smooth and not far from normal in pitch. They also noted, however, that after surgical removal of the vocal cord phonation with the vestibular folds is usually not so good, because of scars or injury to them. Morrison, Kallen, and others emphasized the fact, now routinely considered as a part of postoperative care, that many laryngectomized persons can be taught to produce a voice by allowing air to escape past a narrowed portion of the digestive tract—the cardiac end of the stomach, the upper end of the esophagus, or a portion of the pharynx.

VESSELS AND NERVES

Arteries and Veins

The arteries of the larynx consist of two pairs, the superior and inferior laryngeal arteries (Fig. 8-39). The superior laryngeal artery arises from the superior thyroid artery as this runs downward toward the upper pole of the thyroid gland; it runs approximately horizontally across the posterior portion of the thyrohyoid membrane in company with the internal branch of the superior laryngeal nerve and penetrates the membrane below the nerve to run downward in the submucosa of the lateral wall and floor of the piriform recess and supply mucosa and musculature of the larynx. The inferior laryngeal artery, a branch of the inferior thyroid (from the subclavian artery via the thyrocervical trunk), passes with the inferior laryngeal nerve to the back of the cricothyroid articulation and enters the larynx through a gap here (the Killian–Jamieson area) deep to the lower border of the inferior constrictor muscle (see Fig. 8-15); within the larynx it branches and supplies mucosa and musculature, and anastomoses with the superior laryngeal artery.

Fig. 8-39. Nerves and arteries of the larynx, lateral view.

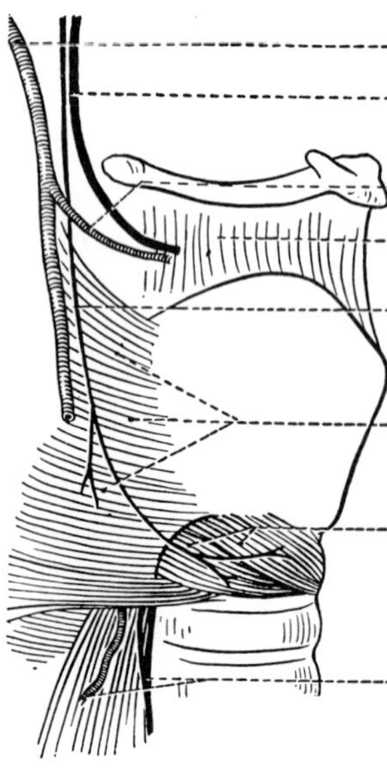

At the level of the cricothyroid membrane the superior thyroid artery also gives off its cricothyroid (cricoid) branch, which runs transversely across the membrane, somewhat closer to the thyroid than to the cricoid cartilage; sometimes, though apparently rarely, this may send a small branch through the cricothyroid membrane to anastomose with the laryngeal arteries. Emergency tracheostomies done through a transverse incision close to the upper border of the cricoid cartilage avoid the cricothyroid branch; this approach should be used only in a true emergency, however, since the incision must be carried through the conus elasticus and is so close to the vocal cords that it predisposes to laryngeal stenosis.

Superior and inferior laryngeal veins parallel the arteries; they join the superior and inferior thyroid veins, respectively.

Lymphatics

The lymphatics of the larynx are numerous except over the vocal folds, where the mucosa is thin and tightly attached to the vocal ligament; the lymphatics are largely divided by the vocal folds into a superior and an inferior group. The efferent channels from the superior group pass through the floor of the piriform recess with the superior laryngeal artery and then upward with it and join the nodes of the upper portion of the deep cervical chain; some are said to reach prelaryngeal nodes (Hendrick). Some efferents from the inferior group are said by Hendrick to reach pretracheal and paratracheal nodes (the latter along the recurrent nerves); others pass downward with the inferior artery and recurrent nerve to join lower deep cervical nodes, some of them going as far as the supraclavicular nodes.

Nerves

The innervation of the larynx is likewise by two sets of nerves, superior and inferior laryngeal, but these two nerves are of different functional importance—the superior laryngeal nerve is largely sensory and secretory and supplies only one muscle of the larynx, the cricothyroid; the inferior laryngeal nerve, the continuation upward of the recurrent laryngeal branch of the vagus, also contains afferent and secretory fibers, but is largely voluntary motor and supplies all the other intrinsic muscles of the larynx. Each nerve usually has associated with it a small glomus or paraganglion, apparently similar to the carotid body (Lawson and Zak); the superior glomus receives its blood from the superior laryngeal artery, and the inferior one receives its blood from the inferior laryngeal. Their locations correspond, according to Lawson and Zak, to the positions of the occasional glomus tumors occurring in the larynx.

The *superior laryngeal nerve* leaves the vagal trunk at the lower end of the inferior (nodose) ganglion and passes downward medial to both the internal and external carotid arteries. As it curves downward and forward toward the larynx it divides into a small external and a large internal branch. The external branch of the superior laryngeal nerve (external laryngeal nerve) continues downward and forward on the lateral surface of the inferior constrictor, to which it may give a branch, and ends in the cricothyroid muscle. This branch usually lies very close to the superior thyroid artery. The internal branch (internal laryngeal nerve) curves forward upon the thyrohyoid membrane in company with the superior laryngeal artery, penetrates this membrane, and breaks up into a number of branches that are distributed to a small portion of the posterior aspect of the root of the tongue, the epiglottic valleculae and epiglottis, the piriform recess, the vestibule, vestibular folds, and ventricle, plus the posterior laryngeal and anterior pharyngeal walls at the level of the cricoid cartilage.

One branch of the internal laryngeal nerve normally runs downward on the posterior surface of the larynx to supply lower pharyngeal mucosa and anastomose with an ascending branch from the inferior laryngeal nerve; branches from this penetrate the transverse arytenoid muscle to reach the mucosa of the posterior laryngeal wall, and several authors (*e.g.*, Berlin and Lahey; Nordland) maintained upon the basis of dissections that the superior laryngeal nerve sometimes or always supplies the transverse arytenoid. Obviously, the fact that branches from the nerve may penetrate the muscle does not necessarily indicate that these are motor branches, and the problem in a dissection is further complicated by the fact that branches from the inferior laryngeal nerve usually anastomose with the superior (Fig. 8-40) and that the branch of the inferior laryngeal to the arytenoid muscle may run through or deep to the posterior cricoarytenoid muscle; in view of these facts, and the difficulty of accurate dissection of the smaller branches of the nerves in the larynx, gross anatomical observations upon this question are of relatively little value. Further, the clinical and physiological fact (the adduction of the vocal cords that usually follows bilateral paralysis of the inferior laryngeal nerves) that these observations purport to explain can be explained completely by other uncontroverted facts.

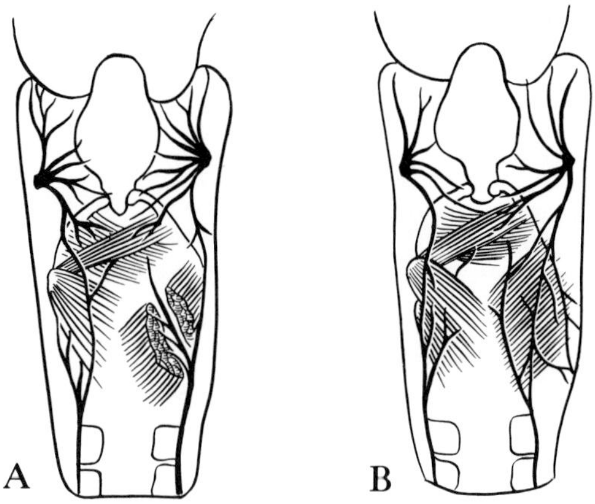

Fig. 8-40. Simple (*A*) and complicated (*B*) anastomoses between the superior and inferior laryngeal nerves. (After Lemere F: Anat Rec 54:389, 1932)

The vagus nerve, and usually the superior laryngeal directly, receive communications from the superior cervical ganglion, and both the superior and inferior laryngeal nerves probably contain both sympathetic and parasympathetic fibers. Johnson showed on the cat that stimulation of either of these nerves causes secretion of tracheal mucus, and expressed the belief that such stimulation is re-

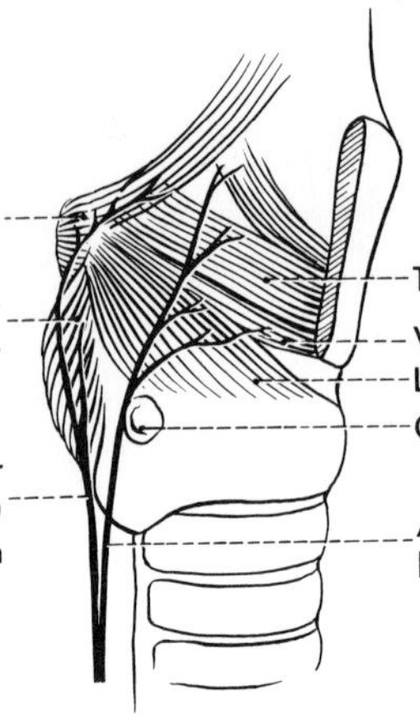

Fig. 8-41. One type of distribution of the inferior laryngeal nerve.

sponsible for the annoying accumulations of mucus sometimes occurring in the course of human thyroidectomies. He stated that mucous secretion was much reduced simply by taking care to prevent injury to the superior laryngeal nerve. In the dog (Lemere, '32b) there are ganglion cells, presumably parasympathetic, along the course of the internal laryngeal nerve, and Ramaswamy and Kulasekaran reported that a ganglion usually occurs on this nerve in human beings, apparently distributing many of its fibers to the saccule (which is particularly rich in glands).

The *recurrent laryngeal nerve* runs upward more or less in the groove between the trachea and esophagus, giving off branches to these structures, and is closely related to the inferior thyroid artery and the posterior aspect of the thyroid gland; these relationships are considered in connection with the thyroid gland, and it only need be mentioned here that injury to both recurrent nerves is a serious consequence of thyroid surgery.

The inferior laryngeal nerve, the upper end of the recurrent, enters the larynx with the corresponding artery just behind the cricothyroid articulation, dividing into at least two branches, typically an anterior and a posterior one (Fig. 8-41), either before or after it enters the larynx; sometimes the branches are multiple. The anterior branch, frequently referred to as the adductor branch, is usually described as passing upward and somewhat forward on the lateral surfaces of the lateral cricoarytenoideus and thyroarytenoideus muscles; it supplies these muscles, the vocalis, and the aryepiglotticus. The posterior branch, the so-called abductor branch, would then supply the cricoarytenoideus posterior (the abductor of the vocal cord) and the arytenoidei, transverse and oblique, which are adductors. However, Sunderland and Swaney found not only much variation in the branching of the nerve, but variation in the distribution even when there were two distinct branches; the anterior branch, for instance, sometimes innervated all the muscles, and the posterior branch contained only sensory fibers, and sometimes the more posterior muscles were innervated by both branches.

Despite the occasional description of the superior laryngeal nerve as innervating the transverse arytenoid muscle, no experimental work indicating this has ever been brought forward; instead, work such as that of Lemere ('32a, '33) on dogs, DuBois and Foley on cats, and the observations of Clerf and Suehs on people all indicate the accuracy of the description just given. Although Vogel still maintained that the human being differs from the dog in having a double innervation of the transverse arytenoideus, from both superior and inferior laryngeal nerves, the best evidence seems to indicate that all the voluntary motor fibers of the superior laryngeal nerve go into the external branch; that the cricothyroid is therefore the only laryngeal muscle supplied by the superior laryngeal nerve, which is otherwise sensory and secretory to the larynx; and that the inferior laryngeal nerve, while containing sensory fibers, is largely motor and supplies all the muscles of the larynx except the cricothyroid. Apparently (Mayet) the inferior laryngeal sometimes also helps to supply the cricothyroid.

Brocklehurst and Edgeworth reported that, in the monkey, muscle-afferent (proprioceptive) fibers constitute about 30% of the total in the external branch of the superior laryngeal nerve, while similar fibers constitute only about 3% of the total in the inferior laryngeal nerve. Nevertheless, Keene reported finding muscle spindles in all the intrinsic laryngeal muscles of human beings. It should be obvious upon reflection, however, that the voluntary control of tone depends not upon these muscle afferents but upon hearing; hence, for instance, the monotonous tone employed by the extremely deaf, and the lamentable efforts of the "tone deaf" to follow a tune.

The usual concept that the internal branch of the superior laryngeal nerve supplies all the afferent fibers to the supraglottic portion of the larynx, and the recurrent nerve all those to the infraglottic portion, is not entirely true, at least in the cat, according to Suzuki and Kirchner ('68, '69a). They found that an anterior part of the infraglottic mucosa is supplied by the external branch of the superior laryngeal.

PARALYSIS OF THE LARYNGEAL MUSCLES

Injuries to Superior Laryngeal Nerves

The external branch of the superior laryngeal nerve, lying as it usually does under cover of the superior thyroid artery, is easily subject to damage when a clamp is put upon this vessel during the course of a thyroidectomy; it has undoubtedly often been so injured, but because injury to it, even bilateral, does not have dramatic or even dangerous aftereffects comparable to those of damage to the inferior laryngeal nerve, damage to it has probably frequently been overlooked. Paralysis of the superior laryngeal nerve or its external branch causes, of course, a paralysis of the cricothyroid muscle; since this muscle is primarily responsible for gross changes in the length of the vocal cords, bilateral damage to the nerves, although seldom seen, can be expected to make it difficult or impossible to sing high tones and, since all the load of tensing the cords is then thrown on the less strong internal muscles, should cause easy tiring of the voice with some hoarseness and weakness. The quality of the voice is sometimes described as "flabby."

Lemere ('33), on the dog, and Clerf and Suehs, on human beings found that section of both superior laryngeal nerves or their external branches produced little or no effect on the appearance of the resting glottis; Lemere stated, however, that since the tension of the vocal cords was lost, these wrinkled on attempted phonation and caused hoarseness; Clerf and Suehs reported that the immediate result of section of the superior laryngeal nerves was usually a loss of voice, even though the cords met in the midline.

Injuries Involving Recurrent Nerves

Injuries to the recurrent nerve, or injuries to both recurrent and superior laryngeal nerves, produce several different and understandably confusing clinical pictures. Lemere's ('33) very careful experimental work on the dog, in which he cut various nerves and combinations of nerves, has served to clear up much of the confusion surrounding this subject; Clerf and Suehs, in analyzing 250 cases of injury to the nerves of the larynx, were apparently able to transfer Lemere's findings directly to human beings. This is not surprising in view of the close similarity which apparently exists between the larynx of the dog and of the human being.

One of the difficulties in assessing damage to the movements of the vocal cords is that the cords may be in abduction, adduction, or anywhere between, including the midway or cadaveric position. The cadaveric position implies, however, that no muscles whatever are acting upon the vocal cord and should therefore be expected to be found only in paralysis of both the inferior and superior laryngeal nerves. Obviously, the difference between a true cadaveric position and a more medial one may not be very marked and may pass unnoticed unless the examiner is experienced. Since in any event it is difficult during laryngoscopy to estimate or record the exact relation of a cord to the midline, many experienced workers have preferred to classify "fixed" (paralyzed) cords into only two main groups: paralyzed straight cords and paralyzed bowed cords. The true cadaveric position of the cord is the extreme example of this latter.

As pointed out by Lemere, the musculature of the larynx serves three main functions: protection in swallowing, the maintenance of a patent airway, and vocalization. The first function depends on all the intrinsic muscles of the larynx except the cricothyroid and the posterior cricoarytenoid; respiration, involving as it does the opening of the air passage, depends entirely upon the activity of the posterior cricoarytenoid; vocalization normally involves all the laryngeal muscles except the posterior cricoarytenoid, and probably that muscle also to an extent sufficient to prevent anterior displacement of the arytenoid cartilage as the vocal fold is lengthened and tensed.

Although Lemere emphasized that identical nerve lesions do not always give exactly the same picture in the glottis, his various combinations of lesions apparently explain most of the various clinical pictures that have been reported. In brief, paralysis of one recurrent nerve will cause some dyspnea upon exertion if the cord is fixed in the adducted position, but the voice will be good because the healthy cord can meet it in the midline; conversely, there will be hoarseness but no dyspnea if the paralyzed cord is not adducted. (Schramm and colleagues have used Gelfoam injections to produce a temporry improvement in adduction; Montgomery reviewed the history of injections into the larynx to produce adduction and described his technique of producing permanent adduction by the injection of Teflon; and Tucker ['77] reported that he had achieved some active adduction through a nerve-muscle implant into the thyroarytenoideus in patients in whom he judged Teflon injections would prove inadequate. Later, May and co-workers reported using the Tucker technique successfully in treating various impairments of laryngeal function, by selecting the appropriate muscle for reinnervation.)

In bilateral paralysis of the recurrent nerves in dogs, according to Lemere, both cords are adducted to about midway between the cadaveric position and the completely adducted position, and there is therefore both dyspnea and hoarseness. In people, depending upon the degree of adduction of the cords, there may be immediate and profound dyspnea, or the dyspnea may be slight and the voice correspondingly poor. In the latter case, however, improvement in the voice is regularly accompanied by increasing dyspnea, for both are due to gradual adduction of the vocal cords. Also, the least protection in swallowing would be expected to occur in bilateral paralysis, since all the muscles that normally close the larynx are inactive; while difficulty in swallowing is typical of recent paralyses, patients somehow accommodate to this lack of sphincteric action and after a time have little or no trouble in swallowing. The dyspnea of bilateral recurrent nerve paralysis can, according to Lemere, be somewhat alleviated by anesthetizing or sectioning the superior laryngeal nerves, since the cords then recede to a position of complete rest in the wider cadaveric position; however, he noted that under the latter circumstances dyspnea becomes marked upon exertion and that the incoming air during deep inspiration tends to appose the cords, because their upward curve below the ventricles makes them act as valves to incoming air. The results of Lemere's experiments are shown diagrammatically in Figure 8-42.

The clinical application of these observations is obvious. They agree with the well-known fact that paralysis of one recurrent nerve may be largely symptomless, since the normal vocal cord can be abducted to widen the airway, and adducted for phonation. They explain especially, without recourse to an anomalous innervation of the arytenoid muscle, the gradual adduction of the vocal cords that may follow injury to the recurrent nerves—as the cricothyroid muscles contract unopposed, the cords gradually become straightened and adducted. If this occurs the voice improves, thus signifying a gradual occlusion of the airway.

The variability of the state of the vocal cords following paralysis is well known and depends in part upon the time lapse between the onset of the paralysis and the examination; this is only one factor, however, since paralyzed cords may be either straight or bowed, may or may not gradually become adducted, or may be adducted immediately after the onset of paralysis. These differences presumably depend upon variations in the extent of the paralysis, but are difficult to analyze. They will vary with the degree of injury to the nerve, with whether the entire inferior laryngeal or only a major or a subsidiary branch is involved, and with the state of the unparalyzed muscles. Certainly not all factors are understood. There is, for example, no ready explanation for the fact that paralysis of the left recurrent nerve caused by a lesion in the mediastinum is typically not followed by adduction of the vocal cord, while paralysis caused by injury to the nerve near the larynx always carries with it the danger of gradual adduction (Havens).

Havens has also stated that a vocal cord that moves normally immediately following a thyroidectomy may within a few days to a week become completely paralyzed. This late postoperative paralysis is presumably due to edema affecting the nerves and rarely is permanent; if bilateral, however, it may be accompanied by such severe dyspnea that a tracheostomy is necessary.

Clerf and Suehs stated that about 90% of paralyses of the vocal cord are due to injury to the peripheral nerves and that central lesions account for only 10%. Furstenberg ('37) pointed out that the laryngeal muscles, like other muscles commonly acting bilaterally, receive their innervation from nuclei (nucleus ambiguus) that are in turn bilaterally innervated from both cerebral cortices; thus a supranuclear (corticobulbar) lesion will not produce paralysis of the laryngeal muscles unless such a lesion is bilateral. Laszlo and Fiertz discussed methods for assessing, by electrical

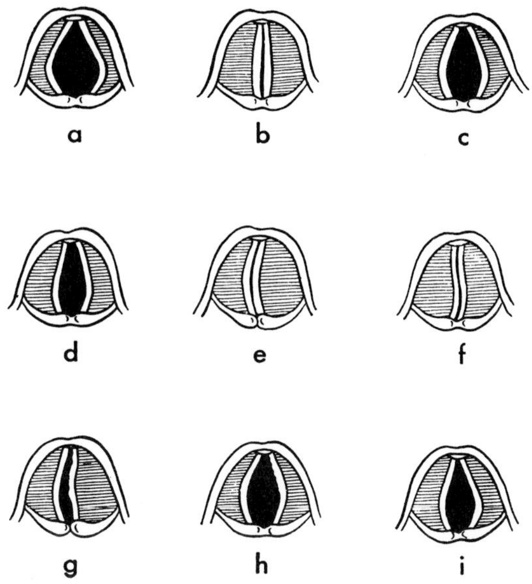

Fig. 8-42. Effects upon the vocal cords of the dog of sectioning various laryngeal nerves. (*a*) and (*b*), Complete abduction and complete adduction, respectively, of normal vocal cords; (*c*), the cadaveric position, also characteristic of complete interruption of all four laryngeal nerves; (*d*), slight adduction from the cadaveric position, which may become more marked after bilateral paralysis of the recurrent nerves; (*e*), attempted phonation with a unilateral paralysis of the recurrent nerve, the intact vocal cord contacting the paralyzed one; (*f*), attempted phonation with bilateral paralysis of the external branch of the superior laryngeal nerve, with both cords therefore improperly tensed; (*g*), attempted phonation with paralysis of the left recurrent nerve and the external branch of the right superior laryngeal nerve; the right vocal cord has moved toward the left one, but is lax; (*h*), paralysis of the left recurrent nerve and of the external rami of both superior laryngeal nerves, in quiet respiration; the left cord is fixed in the cadaveric position but the right cord, although relaxed, can move to the midline; and (*i*), bilateral paralysis of the recurrent nerves and paralysis of the external ramus of the right superior laryngeal nerve; neither cord can be moved, but the left one is partially adducted through the unopposed action of the cricothyroideus, while the right one is fixed in the cadaveric position. (After Lemere F: Arch Otolaryngol 18:413, 1933)

stimulation, the extent of damage to the recurrent nerve.

It has long been known that partial damage to the inferior laryngeal nerve may result in more damage to the abductor muscle (the posterior cricoarytenoid) than it does to the other muscles, and this fact, that the abductor fibers of the recurrent laryngeal nerve seem to be affected sooner and more profoundly than are the adductor fibers, has been termed Semon's law. In experiments on dogs, Lemere ('34) showed that after 4 months of age the adductors are stronger than the abductor; he reported also that the nerve fibers to the abductor are larger and have a lower threshold of stimulation and that they are less resistant to damage than are the adductor nerve fibers of the recurrent nerve. The latter is in line with the usually observed fact that larger fibers are more easily injured than smaller ones. When it is also recalled that the abductor is a single muscle, while the adductors are many, the increased susceptibility of the abductor muscle as stated in Semon's law seems readily explainable. Murtagh and Campbell concluded from physiological studies, however, that the fibers to the abductor (in goats and cats) are smaller than those to the adductors and probably also very few in number.

Bach and co-workers reported degenerative changes in the laryngeal muscles in senile individuals and concluded that these were sufficient to explain the peculiarities of the senile voice; they regarded the degeneration as due largely to poor vascular nutrition. In the five cases they investigated, they found the abductor muscles to be more affected than the adductors. Whether the two facts are related is not apparent, but Lemere ('34) found in the dog that the abductor contained a considerably larger percentage of red muscle than did the other laryngeal muscles, which contained more admixture of red and white muscle fibers or white fibers alone.

Surgical Improvement of the Airway

The dyspnea resullting from bilateral injury to the recurrent nerves may be so severe as to interfere seriously with the activity of the individual or indeed threaten death, and for these cases several operations have been devised. Although, as noted by Lemere ('34), some improvement in the airway can be produced by interruption of the superior laryngeal nerves, this would result in complete denervation of the larynx and abolition of the cough reflex and of protection to the larynx during swallowing; it probably should not be attempted surgically.

The surgical methods that have been used to relieve the dyspnea resulting from bilateral paralysis of the vocal cords include a permanent tracheostomy, cordectomy or cordotomy by an intralaryngeal punch operation (portions of the arytenoid cartilage also being sometimes thus removed), laryngofissure with removal of one cord (*e.g.,* Lore), and lateral displacements of the vocal cords. Permanent tracheostomy is of course to be avoided if possible, and laryngofissure with cordotomy or cordectomy is likely to be followed by the production of scar tissue that not only interferes with phonation but also produces again a stenosis of the larynx.

The method of choice has usually been one that produces a permanent lateral displacement of the posterior end of one of the vocal cords; although such operations are difficult technically, they will, if properly done, relieve the dyspnea permanently and yet allow good phonation. However, the results achieved by Tucker ('78) through transplanting a nerve–muscle pedicle into the posterior cricoarytenoid—return of abduction in 6 to 12 weeks, with no effect on the voice, and long-term success in 40 of 45 patients (88.8%)—offer an alternative to a permanent lateral displacement. In his cases the cord abducts during inspiration because the omohyoid muscle, the donor of the transplant, normally contracts during inspiration. Applebaum and co-workers used this technique on six patients, all of whom recovered vocal cord function; May and co-workers also reported success with this method.

Among the methods for producing permanent abduction of the vocal cords, the older operation and probably the most difficult, although offering at the same time good results without particular danger of scar tissue formation, is the King ('39) operation or a modification thereof; this consists of incising the inferior constrictor of the pharynx along the posterior border of the thyroid cartilage and behind the cricothyroid ligament, working upward and medially between the mucosa of the anterior surface of the pharynx and the lamina of the cricoid cartilage until the arytenoid cartilage comes into view, and detaching the arytenoid cartilage from its muscular and ligamentous attachments so that it can be moved laterally. In the original King operation the arytenoid cartilage was sutured to the omohyoid muscle, on the theory that contraction of this muscle during inspiration would produce the desired abduction of the cords, while between the inspiratory movements the cords would be adducted and therefore the voice would be better preserved. The use of the omohyoid muscle proved to have no apparent advantages, and therefore the arytenoid cartilage has been more commonly attached to the thyroid cartilage instead of to the omohyoid muscle. This procedure has been described by Seed and by Shirer, among others. Seed emphasized that the suture attaching the arytenoid cartilage to the thyroid should be through the vocal process, so as to produce good abduction of the cord, rather than through the stronger and more obvious muscular process. He also pointed out that before the suture is finally tied the vocal cord should be visualized by laryngoscopy to be sure of optimal placement of the cord. When this operation is successfully carried out, good and permanent opening of the air passage is secured, and the voice is said not to be appreciably affected.

Another extralaryngeal approach to the arytenoid cartilage is by way of a window made in the thyroid cartilage (Kelly; Wright). Kelly apparently preferred this to the approach through the inferior constrictor, pointing out that scar tissue left after thyroidectomy may complicate the latter approach. He removed the arytenoid cartilage and fixed the cord by a suture attaching it to the cartilage about the window in the thyroid cartilage. He noted that the suture must not be placed too far forward on the cord if a satisfactory voice is to be obtained. Kelly and Wright both stated that this operation is simpler and more easily performed than is the King operation.

Although the King operation and its variants have usually given excellent results, their technical difficulties are considerable. For those surgeons trained in suspension laryngoscopy, the intralaryngeal approach of Thornell has apparently completely replaced the King operation, because of its relative simplicity and high percentage of excellent results. The Thornell operation involves direct visualization of the larynx through the mouth, the approach being made through this route. An incision is made over the arytenoid cartilage and into the aryepiglottic fold, the arytenoid is removed by submucous resection, and the posterior part of the vocal cord is displaced laterally into the area from which the cartilage was removed; after it has been held in this position by an obturator for 3 to 7 days, it becomes permanently fixed here by scar tissue. With increased experience, the obturator has been omitted, as unnecessary, by most surgeons. The operation can be performed only while using a suspension apparatus, and Thornell warned that surgeons without this equipment and training in its use should continue to use the King operation or one of its modifications. Kirchner used a modification of the Thornell procedure; instead of removing the arytenoid cartilage, he removed by cautery a wedge of muscle and drew the cord laterally by sutures passed into the larynx through the skin. The sutures were then anchored to the skin so that the tension on them could be adjusted easily.

ANOMALIES

Congenital anomalies of the larynx are apparently rare; in addition to the presence of an enlarged laryngeal saccule, discussed earlier in this chapter, congenital stenosis of the larynx has been reported occasionally, as has also the presence of congenital cysts. Rare cases of doubling of the vocal cords have been reported (*e.g.*, Frank and Malev). Congenital stenosis of the larynx, usually referred to as a congenital diaphragm or web of the larynx (Fig. 8-43), has been reported by, among others, Clerf, and Poe and Seager. Apparently a congenital diaphragm typically unites the anterior portions of the vocal cords, especially

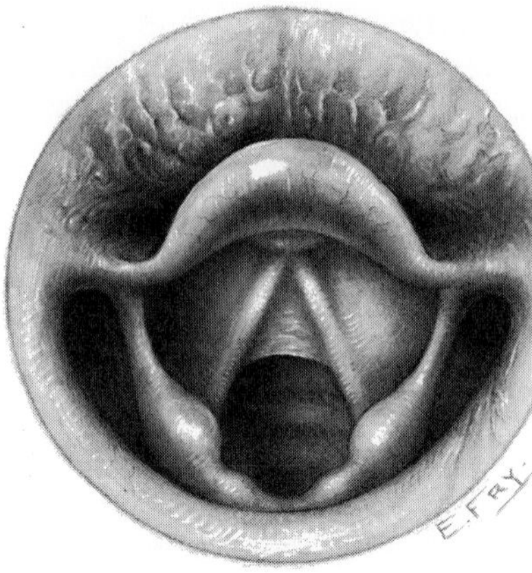

Fig. 8-43. A congenital diaphragm of the larynx.

just behind the anterior commissure where the cords come together to attach to the thyroid cartilage. In accordance with this, the diaphragm is apt to be thickest anteriorly and to thin out markedly posteriorly. It has been regarded as a partial persistence of the fusion normally existing between the primitive vocal cord of the embryo. Diaphragms or webs of the larynx may also be acquired, due to adhesions following injury.

Cysts of the larynx are relatively rare, according to Myerson, and the large majority are located on either the epiglottis or the vocal cords, although they may occur also in the ventricle, on the arytenoid cartilage, the aryepiglottic fold, or the vestibular fold. The usual symptoms are said to be hoarseness or dyspnea; Myerson found that at least some of them are congenital and noted that their management in young infants is difficult. According to Kleinfeld, congenital laryngeal cysts are rare and he found only about 15 cases reported in the literature up to that time. He described them as usually thin-walled and containing a milky fluid and commonly arising from a broad base in the region of the laryngeal aperture, often from the aryepiglottic fold.

Tucker ('37) described a case of laryngoptosis resulting from congenital downward displacement of both the larynx and the hyoid bone. He stated that no other reported cases of laryngoptosis had apparently been due to downward displacement of the hyoid. In the case he reported, the hyoid bone lay just above the jugular (suprasternal) notch and the larynx lay approximately level with it, with the cricoid being situated behind the sternum. The larynx failed to rise normally in swallowing. At operation a tendinous mass replacing the sternohyoid and sternothyroid muscles on one side was found and sectioned, and the hyoid bone and larynx immediately ascended to an almost normal position.

References

APPLEBAUM EL, ALLEN GW, SISSON GA: Human laryngeal reinnervation: The Northwestern experience. Laryngoscope 89:1784, 1979

AREY LB: On the development, morphology and interpretation of a system of cryptanalogues in the pharyngeal tonsil. Am J Anat 80:203, 1947

BABBITT JA: Fracture of the styloid process and its tonsil fossa complications, with report of a case. Ann Otol Rhinol Laryngol 42:789, 1933

BABCOCK WW, JACKSON C: The single stage operation for pulsion diverticulum of the oesophagus. Surg Gynecol Obstet 53:667, 1931

BACH AC, LEDERER FL, DINOLT R: Senile changes in the laryngeal musculature. Arch Otolaryngol 34:47, 1941

BALASUBRAMIAN S: The ossification of the stylohyoid ligament and its relation to facial pain. Br Dent J 116:108, 1964

BASMAJIAN JV, DUTTA CR: Electromyography of the pharyngeal constrictors and levator palati in man. Anat Rec 139:561, 1961

BATSON OV: Veins of the pharynx. Arch Otolaryngol 36:212, 1942

BELSEY R: Functional disease of the esophagus. J Thorac Cardiovasc Surg 52:164, 1966

BERLIN DD, LAHEY FH: Dissections of the recurrent and superior laryngeal nerves: The relation of the recurrent to the inferior thyroid artery and the relation of the superior to abductor paralysis. Surg Gynecol Obstet 49:102, 1929

BROCKLEHURST RJ, EDGEWORTH FH: The fibre components of the laryngeal nerves of Macaca mulatta. J Anat 74:386, 1940

BUCHTHAL F: Electromyography of intrinsic laryngeal muscles. Q J Exp Physiol 44:137, 1959

CAIRNEY J: Tortuosity of the cervical segment of the internal carotid artery. J Anat 59:87, 1924

CALCATERRA TC, KADELL BM, WARD PH: Dysphagia secondary to cricopharyngeal muscle dysfunction. Surgical management. Arch Otolaryngol 101:726, 1975

CALNAN JS: Movements of the soft palate. Br J Plast Surg 5:286, 1953

CALNAN JS: Modern views on Passavant's ridge. Br J Plast Surg 10:89, 1958

CAMPBELL MD: Laryngocele. Arch Otolaryngol 44:219, 1946

CLERF LH: Congenital stenosis of the larynx: Report of three cases. Ann Otol Rhinol Laryngol 40:770, 1931

CLERF LH, SUEHS OW: Paralysis of the larynx. Ann Otol Rhinol Laryngol 50:762, 1941

CROSS FS: Esophageal diverticula: Related neuromuscular problems. Ann Otol Rhinol Laryngol 77:914, 1968

DAYAL VS, BAHRI H, STONE PC: Preepiglottic space. An anatomic study. Arch Otolaryngol 95:130, 1972

DEVINE KD: Personal communication to the author, 1967

DORRANCE GM: The so-called bursa pharyngea in man: Its origin, relationship with the adjoining nasopharyngeal structures and pathology. Arch Otolaryngol 13:187, 1931

DUBOIS FS, FOLEY JO: Experimental studies on the vagus and spinal accessory nerves in the cat. Anat Rec 64:285, 1936

EAGLE WW: Elongated styloid process: Further observations and a new syndrome. Arch Otolaryngol 47:630, 1948

EAGLE WW: Symptomatic elongated styloid process: Report of two cases of styloid process-carotid artery syndrome with operation. Arch Otolaryngol 49:490, 1949

FINK BR: Tensor mechanism of the vocal folds. Ann Otol Rhinol Laryngol 71:591, 1962

FISHMAN LZ: Laryngeal and esophageal atavism in man as indicated by the probable phylogenesis of the hypopharyngeal receptacle, concerned in the act of deglutition. Ann Otol Rhinol Laryngol 44:139, 1935

FLOYD WF, NEGUS VE, NEIL E: Observations on the mechanism of phonation. Acta Otolaryngol 48:16, 1957

FRANK DI, MALEV M: Double vocal cord. Arch Otolaryngol 29:713, 1939

FRITZ M: Elongated styloid process: A cause of obscure throat symptoms. Arch Otolaryngol 31:911, 1940

FRITZELL B: The velopharyngeal muscles in speech. An electromyographic and cinéradiographic study. Acta Otolaryngol (Suppl 250):5, 1969

FUKUDA H, KIRCHNER JA: Changes in the respiratory activity of the cricothyroid muscle with intrathoracic interruption of the vagus nerve. Ann Otol Rhinol Laryngol 81:532, 1972

FURSTENBERG AC: An anatomical and clinical study of central lesions producing paralysis of the larynx. Ann Otol Rhinol Laryngol 46:39, 1937

FURSTENBERG AC: Diverticulum of the esophagus. Ann Otol Rhinol Laryngol 56:305, 1947

FYKE FE JR, CODE CF: Resting and deglutition pressures in the pharyngoesophageal region. Gastroenterology 29:24, 1955

GAUGHRAN GRL: The pterygomandibular raphé. Anatomical artifact (abstr). Anat Rec 184:410, 1976

GIFFIN CS, SNYDERMAN SC, THOMAS JR: Treatment of pharyngoesophageal diverticula. Ann Otol Rhinol Laryngol 81:799, 1972

GRIESMAN BL: Mechanism of phonation demonstrated by planigraphy of the larynx. Arch Otolaryngol 38:17, 1943

HÄRMÄ R: Styalgia. Clinical experience of 52 cases. Acta Otolaryngol (Suppl 224):149, 1967

HAVENS FZ: Personal communication to the author, 1951

HENDRICK JW: Surgical treatment of larynx and hypopharynx cancer. Arch Otolaryngol 79:269, 1964

HETHERINGTON J: The kerato-cricoid muscle in the American white and Negro. Am J Phys Anthropol 19:203, 1934

HETT GS, BUTTERFIELD HG: The anatomy of the palatine tonsils. J Anat Physiol 44:35, 1909

HOLINGER PH, JOHNSTON KC: Endoscopic surgery of Zenker's diverticula. Experience with the Dohlman technique. Ann Otol Rhinol Laryngol 70:1117, 1961

HOLINGER PH, SCHILD JA: The Zenker (hypopharyngeal) diverticulum. Ann Otol Rhinol Laryngol 78:679, 1969

JACKSON C, JACKSON CL: Malignant disease of the larynx: Its treatment by laryngofissure and laryngectomy. Am J Surg ns30:3, 1935a

JACKSON C, JACKSON CL: Dysphonia plicae ventricularis: Phonation with the ventricular bands. Arch Otolaryngol 21:157, 1935b

JOHNSON J: Effect of superior laryngeal nerves on tracheal mucus: An experimental study on the relationship. Ann Surg 101:494, 1935

KALLEN LA: Vicarious vocal mechanisms: The anatomy, physiology and development of speech in laryngectomized persons. Arch Otolaryngol 20:460, 1934

KAZANJIAN VH, HOLMES EM: Stenosis of the nasopharynx and its correction. Arch Otolaryngol 44:261, 1946

KEENE MFL: Muscle spindles in human laryngeal muscles. J Anat 95:25, 1962

KELLY JD: Some problems in the surgical treatment of bilateral abductor paralysis of the larynx. Ann Otol Rhinol Laryngol 53:461, 1944

KILLIAN G: Ueber den Mund der Speiseröhre. Z Ohrenh 55:1, 1908

KING BT: A new and function-restoring operation for bilateral abductor cord paralysis: Preliminary report. Trans Am Laryngol Assoc 61:264, 1939

KING BT: New concepts of the etiology and treatment of diverticula of the esophagus. Surg Gynecol Obstet 85:93, 1947

KIRCHNER FR: Endoscopic lateralization of the vocal cord in abduction paralysis of the larynx. Laryngoscope 89:1779, 1979

KLEINFELD L: Laryngeal cysts in the newborn. Arch Otolaryngol 19:590, 1934

LASZLO AF, FIERTZ CO: Prognosis of the recurrent nerve paralysis following thyroidectomy. Arch Otolaryngol 42:372, 1945

LAWSON W, ZAK FG: The glomus bodies ("paraganglia") of the human larynx. Laryngoscope 84:98, 1974

LEMERE F: Innervation of the larynx. I. Innervation of laryngeal muscles. Am J Anat 51:417, 1932a

LEMERE F: Innervation of the larynx. II. Ramus anastomoticus and ganglion cells of the superior laryngeal nerve. Anat Rec 54:389, 1932b

LEMERE F: Innervation of the larynx. III. Experimental paralysis of the laryngeal nerves. Arch Otolaryngol 18:413, 1933

LEMERE F: Innervation of the larynx. IV. An analysis of Semon's law. Ann Otol Rhinol Laryngol 43:525, 1934

LEVITT MN, DEDO HH, OGURA JH: The cricopharyngeus muscle, an electromyographic study in the dog. Laryngoscope 75:122, 1965

LEWIS DD: The elastic tissue of the human larynx. Am J Anat 4:175, 1905

LILLIE HI, HECK WE: Hemangioma of the lateral pharyngeal band: Report of a case. Ann Otol Rhinol Laryngol 57:519, 1948

LORE JM: A suggested operative procedure for the relief of stenosis in double abductor paralysis: An anatomic study. Ann Otol Rhinol Laryngol 45:679, 1936

MACBETH R: Pharyngeal diverticulum. Ann Otol Rhinol Laryngol 80:662, 1971

MACCREADY PB: So-called aneurysm of the ascending pharyngeal artery: Report of one case which developed under observation. Ann Otol Rhinol Laryngol 44:513, 1935

MAUE WM, DICKSON DR: Cartilages and ligaments of the adult human larynx. Arch Otolaryngol 94:432, 1971

MAY M, LAVOROTO AS, BLEYAERT AL: Rehabilitation of the crippled larynx: Application of the Tucker technique for muscle–nerve reinnervation. Laryngoscope 90:1, 1980

MAYET A: Zur Innervation des M. cricothyroideus. Anat Anz 103:340, 1956

MELCHIONNA RH, MOORE RA: The pharyngeal pituitary gland. Am J Pathol 14:763, 1938

MILLER MV: Edema of the larynx: A study of the loose areolar tissues of the larynx. Arch Otolaryngol 31:256, 1940

MINEAR WL, AREY LB, MILTON JT: Prenatal and postnatal development and form of crypts of human palatine tonsil. Arch Otolaryngol 25:487, 1937

MITCHINSON AG, YOFFEY JM: Respiratory displacement of larynx, hyoid bone and tongue. J Anat 81:118, 1947

MITCHINSON AGH, YOFFEY JM: Changes in the vocal folds in humming low and high notes: A radiographic study. J Anat 82:88, 1948

MOERSCH HJ, JUDD ES: Diagnosis and treatment of pharyngo-oesophageal diverticulum. Surg Gynecol Obstet 58:781, 1934

MONTGOMERY WW: Laryngeal paralysis. Teflon injection. Ann Otol Rhinol Laryngol 88:647, 1979

MORLEY J: Pharyngeal diverticula. Br J Surg 33:101, 1945

MORRISON WW: The production of voice and speech following total laryngectomy: Exercise and practice for the production of the pseudovoice. Arch Otolaryngol 14:413, 1931

MOST A: Ueber den Lymphgefässapparat von Nase und Rachen. Arch Anat Physiol p 75, 1901

MURTAGH JA, CAMPBELL CJ: The respiratory function of the larynx. II. The elementary physiology of the recurrent laryngeal nerve. Ann Otol Rhinol Laryngol 57:465, 1948

MYERSON MC: Cysts of the larynx. Arch Otolaryngol 18:281, 1933

NAFFZIGER HC, DAVIS C, BELL HG: Paralysis of deglutition—surgical correction. Ann Surg 128:732, 1948

NEGUS VE: Function of the cartilages of Santorini. J Anat 63:430, 1929

NORDLAND M: The larynx as related to surgery of the thyroid based on an anatomical study. Surg Gynecol Obstet 51:449, 1930

OLDFIELD MC: Cleft palate and the mechanism of speech. Br J Surg 29:197, 1941

PAINTER AM: Important points in tonsillectomy. Ann Otol Rhinol Laryngol 39:815, 1930

DE PAULA ASSIS JE: Observações anatômicas sôbre o musculus salpingopharyngeus. Rev brasil Otorino-laringol 15:169, 1947

POE DL, SEAGER PS: Congenital laryngeal web: Its eradication. Arch Otolaryngol 47:46, 1948

PRESSMAN JJ: Sphincter action of the larynx. Arch Otolaryngol 33:351, 1941

PRESSMAN JJ: Physiology of the vocal cords in phonation and respiration. Arch Otolaryngol 35:355, 1942

PRESSMAN JJ, DOWDY A, LIBBY R, FIELDS M: Further studies upon the submucosal compartments and lymphatics of the larynx by the injection of dyes and radioisotopes. Ann Otol Rhinol Laryngol 65:963, 1956

PRESSMAN JJ, KELEMAN G: Physiology of the larynx. Physiol Rev 35:506, 1955

PUPPEL ID: The role of esophageal motility in the surgical treatment of mega-esophagus. J Thorac Surg 19:371, 1950

RAMASWAMY S, KULASEKARAN D: The ganglion on the internal laryngeal nerve. Arch Otolaryngol 100:28, 1974

RAMSEY GH, WATSON JS, GRAMIAK R, WEINBERG SA: Cinefluorographic analysis of the mechanism of swallowing. Radiology 64:498, 1955

RAVEN RW: Pouches of the pharynx and oesophagus: With special reference to the embryological and morphological aspects. Br J Surg 21:235, 1933

ROBISON JM: Lymphangitis of the retropharyngeal system. Arch Otolaryngol 40:385, 1944

SCHRAMM VL, MAY M, LAORATO AS: Gelfoam paste injection for vocal cord paralysis: Temporary rehabilitation of glottic incompetence. Laryngoscope 88:1268, 1978

SCHUGT HP: The piriform sinus. Anatomic and clinical observations, with a review of the literature. Arch Otolaryngol 31:626, 1940

SEED L: The King operation for bilateral abductor paralysis of the vocal cords. Ann Otol Rhinol Laryngol 51:66, 1942

SELLARS IE, KEEN EN: The anatomy and movements of the cricoarytenoid joint. Laryngoscope 88:667, 1978

SHALLOW TA: Combined one stage closed method for the treatment of pharyngeal diverticula. Surg Gynecol Obstet 62:624, 1936

SHIRER JW: Modification of the King operation for bilateral vocal cord paralysis. Ann Surg 120:617, 1944

SNOOK T: The later development of the bursa pharyngea: Homo. Anat Rec 58:303, 1934

SOKOLOWSKY RR: Effect of the extrinsic laryngeal muscles on voice production. Arch Otolaryngol 38:355, 1943

SONESSON B: On the anatomy and vibratory pattern of the human vocal folds: With special reference to a photo-electrical method for studying vibratory movements. Acta Otolaryng (Suppl 156), 1960

STRONG LH: The mechanism of laryngeal pitch. Anat Rec 63:13, 1935

SUNDERLAND S, SWANEY WE: The intraneural topography of the recurrent laryngeal nerve in man. Anat Rec 114:411, 1952

SUZUKI M, KIRCHNER JM: Afferent nerve fibers in the external branch of the superior laryngeal nerve of the cat. Ann Otol Rhinol Laryngol 77:1059, 1968

SUZUKI M, KIRCHNER JA: Sensory fibers in the recurrent laryngeal nerve. An electrophysiological study of some laryngeal afferent fibers in the recurrent laryngeal nerve of the cat. Ann Otol Rhinol Laryngol 78:21, 1969a

SUZUKI M, KIRCHNER JA: The posterior cricoarytenoid as an inspiratory muscle. Ann Otol Rhinol Laryngol 78:849, 1969b

SUZUKI M, KIRCHNER JA, MURAKAMI Y: The cricothyroid as a respiratory muscle. Its characteristics in bilateral laryngeal nerve paralysis. Ann Otol Rhinol Laryngol 79:976, 1970

TAYLOR HM: Ventricular laryngocele. Ann Otol Rhinol Laryngol 53:536, 1944

THORNELL WC: A new intralaryngeal approach for arytenoidectomy in the treatment of bilateral abductor vocal cord paralysis. J Clin Endocrinol 10:1118, 1950

TODD TW, FOWLER RH: The muscular relations of the tonsil. Am J Anat 40:355, 1927

TUCKER G: Laryngoptosis: Ptosis of the larynx due to downward displacement of the hyoid bone resulting from fibrosis and shortening (congenital anomaly) of the left sternohyoid and sternothyroid muscles. Arch Otolaryngol 25:389, 1937

TUCKER HM: Reinnervation of the unilaterally paralyzed larynx. Ann Otol Rhinol Laryngol 86:789, 1977

TUCKER HM: Human laryngeal innervation: Long-term experience with the nerve–muscle pedicle technique. Laryngoscope 88:598, 1978

VOGEL PH: The innervation of the larynx of man and the dog. Am J Anat 90:427, 1952

WELSH LW, WELSH JJ, BEHLKE FM: Analysis of laryngeal compartments. Ann Otol Rhinol Laryngol 71:913, 1962

WHILLIS J: A note on the muscles of the palate and the superior constrictor. J Anat 65:92, 1930

WRIGHT ES: The Kelly operation for restoration of laryngeal function following bilateral paralysis of the vocal cords: Report of three cases. Ann Otol Rhinol Laryngol 52:346, 1943

YARINGTON CT JR, FRAZER JP: The surgical treatment of laryngocele. Surg Gynecol Obstet 122:308, 1966

ZUCKERBRAUN L, BAHNA MS: Cricopharyngeal myotomy as the only treatment for Zenker diverticulum. Ann Otol Rhinol Laryngol 88:798, 1979

Chapter 9
THE NECK

General Considerations

The fascia and fascial spaces of the neck, and the anatomy of the pharynx and larynx, have already been described in separate chapters on these subjects; they will be alluded to here, therefore, only insofar as is necessary in describing other anatomy of the neck. Some of the pertinent general anatomy is shown in Figure 9-1.

The skin and subcutaneous connective tissue of the neck are thin and flexible. Within the subcutaneous tissue lies the platysma muscle, extending from the upper thoracic region over the clavicle and the anterolateral surface of the neck and over the mandible toward the mouth. It is extremely thin, as its name implies, and varies considerably in development; it is generally and properly disregarded in planning incisions in the neck. Concerning the latter, transverse incisions are to be preferred when possible, since they produce more satisfactory cosmetic results than do longitudinal ones. This is based upon good anatomical and surgical principles. For one thing, it is generally recognized that incisions across flexion creases in any part of the body are apt to produce undue scarring and interference with flexion. In the second place, Langer's lines (tension lines of the skin, cleavage lines) are arranged transversely in the neck, and it is well known that incisions parallel to these lines heal without the formation of the additional scar tissue produced by incisions across them; when the tension lines are cut across, undue retraction of the lips of the wound necessarily results. Freeland and Rogers reported that the skin anterior to the trapezius is supplied by both ascending and descending arterial branches and suggested that a horizontal incision about the middle of the neck, meeting a longitudinal incision close to the anterior border of the trapezius, should yield the most satisfactory skin flap.

The cutaneous innervation of the neck is largely through the transverse cervical (transversus colli, cutaneus colli) branch of the cervical plexus, containing fibers derived from the second and third cervical nerves. This nerve passes close around the posterior border of the sternocleidomastoid muscle and then forward on its lateral surface, dividing into two main branches that run upward toward the jaw and downward toward the base of the neck. The skin of the base of the neck, like that over the upper part of the thorax, is supplied by the supraclavicular nerves (C3 and C4), which descend behind the sternocleidomastoid in the posterior triangle and fan out as they descend. Behind the ascending branch of the transverse cervical nerve, rounding the posterior border of the sternocleidomastoid muscle and running upward on it toward the front of the ear, is the great auricular nerve; it supplies skin of the neck and that over the

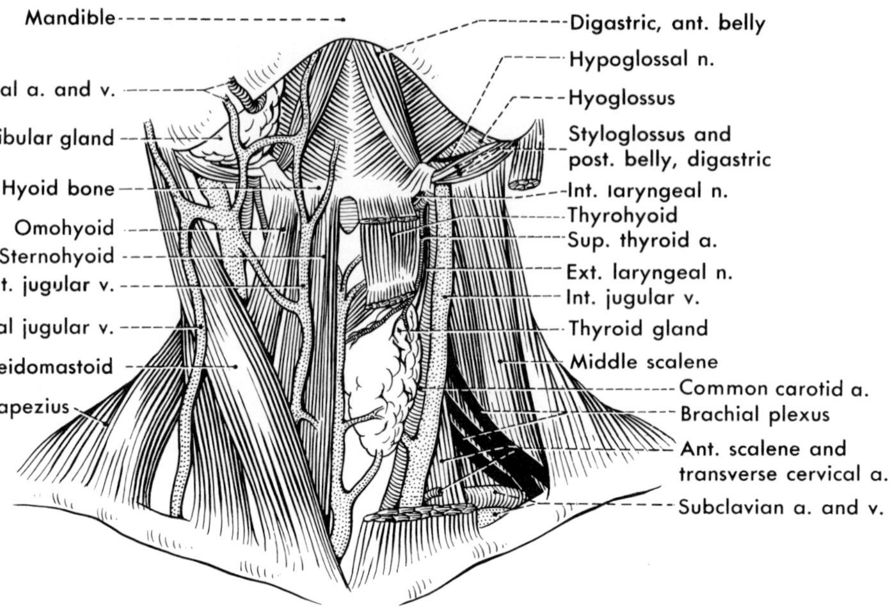

Fig. 9-1. A general view of the neck. On the left side of the neck the sternocleidomastoid and most of the infrahyoid muscles, the submandibular gland, and other superficially located structures have been removed.

angle of the mandible before it reaches the ear (see Fig. 6-19).

TRIANGLES OF THE NECK

The prominent landmarks in the neck are the hyoid bone, the thyroid cartilage and trachea, and the sternocleidomastoid muscles. The sternocleidomastoid muscles divide each side of the neck into two major triangles, anterior and posterior. The anterior triangle is delimited by the anterior border of the sternocleidomastoid muscle laterally, the midline medially, and the lower border of the mandible above; the posterior triangle is delimited by the posterior border of the sternocleidomastoid anteromedially, the anterior border of the trapezius muscle posterolaterally, and the clavicle below. These triangles are covered by the cervical fascia (deep fascia of the neck), and it is not until this fascia has been removed that the subsidiary triangles of each of these major triangles can be clearly seen. In the intact neck, the depression between the clavicular and sternal heads of the sternocleidomastoid is the lesser supraclavicular fossa, and that at the base of the posterior triangle is the omoclavicular triangle or greater supraclavicular fossa.

Two subsidiary parts of the anterior triangle are regularly described, and often four triangles are named. Those in the NA are the carotid and submandibular. The *carotid* (superior carotid) *triangle* is bounded below and anteriorly by the superior belly of the omohyoid, above by the stylohyoid and the posterior belly of the digastric, and posteriorly by the sternocleidomastoid; the *submandibular* (submaxillary) *triangle* lies between the anterior and posterior bellies of the digastric and the lower border of the mandible. The other two triangles often described are an *inferior carotid* or *muscular triangle,* below the hyoid bone and anteroinferior to the superior belly of the omohyoid, and a *submental* or *suprahyoid triangle,* bounded by the midline, the anterior belly of the digastric, and the hyoid bone (Fig. 9-2).

Subdivision of the posterior triangle into two parts is of little practical importance and is often therefore disregarded; the larger upper portion of this triangle has been called the *occipital triangle,* while a small portion above the clavicle has been termed the *subclavian (omoclavicular) triangle.* The two are described as separated by the posterior belly of the omohyoid as this crosses the posterior triangle. This muscle runs close to and almost parallel with the clavicle, so that the subclavian triangle is usually small; typically, the omohyoid muscle must be displaced upward if a surgical approach is to be through this triangle.

The brachial plexus and the subclavian artery pass through the base of the posterior triangle and therefore are usually approached through this; the transverse cervical and suprascapular (transverse scapular) arteries run across it, and the internal jugular vein, the termination of the external jugular vein, the termination of the thoracic duct, and the phrenic nerve, although actually lying under cover of the sternocleidomastoid muscle, are most easily approached through the base of the posterior triangle.

The floor of the posterior triangle is formed, from above downward, largely by the splenius capitis, the levator scapulae, and the posterior and middle scalene muscles; at the uppermost part of the triangle the occipital artery emerges from under cover of the splenius capitis to cross the semispinalis capitis and penetrate the trapezius muscle. The lesser occipital nerve, from the cervical plexus, runs upward in the triangle, emerging from under cover of the posterior border of the sternocleidomastoid, and the great auricular and transversus colli nerves turn sharply around the border of the sternocleidomastoid; the accessory nerve and the more lateral supraclavicular nerves run obliquely downward across the triangle, from the posterior border of the sternocleidomastoid toward the anterior border of the trapezius and the upper border of the clavicle.

It is important to note that the accessory nerve crosses the posterior triangle not in contact with the muscles forming its floor, but separated from the muscle on which it lies (levator scapulae) by a distinct and heavy layer of fascia. Both the superficial and prevertebral layers of the cervical fascia cross the posterior triangle (see Fig. 5-1), and the accessory nerve runs between these two layers; it therefore is embedded in dense fascia or in connective tissue and fat between dense fascial planes. The same is also true in part of the supraclavicular nerves and of the lower portion of the external jugular vein and a proximal portion of the transverse cervical artery. Deep to the prevertebral layer, and therefore immediately adjacent to the muscles, are the nerves of the brachial plexus and, more anteriorly, the phrenic nerve.

With the exception of the accessory nerve,

Fig. 9-2. Anterior (*a*) and posterior (*b*) triangles of the neck.

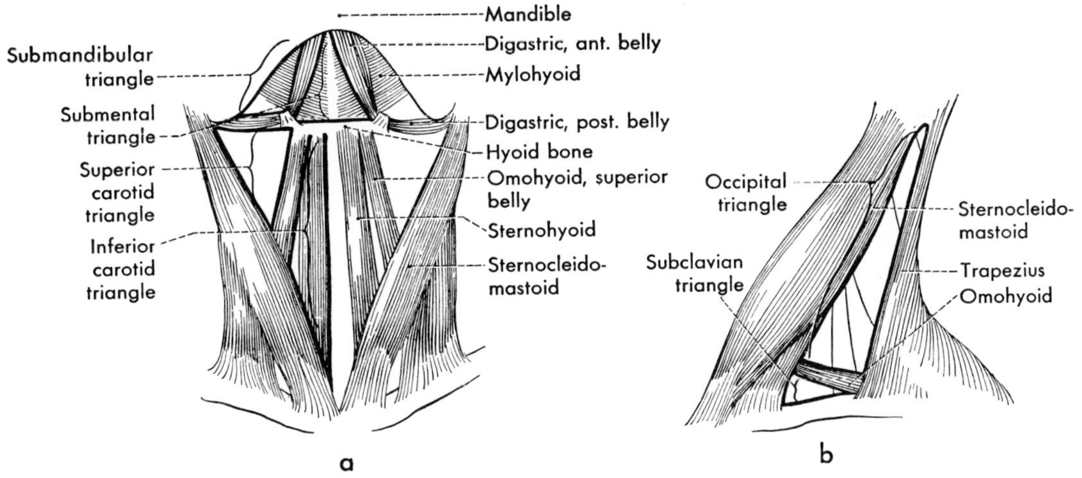

which emerges from behind or through the posterior border of the sternocleidomastoid in the middle third of the muscle and descends almost vertically to disappear deep to the anterior border of the trapezius muscle, the nerves appearing relatively superficially within the confines of the posterior triangle of the neck are all cutaneous branches of the cervical plexus.

Musculature

STERNOCLEIDOMASTOID

The sternocleidomastoid muscle (see Fig. 9-1) arises by a tendinous head from the sternum (the curved upper border of which, between the heads of the two muscles, is known as the jugular or suprasternal notch) and by a wider muscular head from the proximal portion of the clavicle. It passes upward, laterally, and posteriorly to insert upon the mastoid process and is enclosed by the superficial or anterior layer of the cervical fascia, which splits to pass around it. In addition to the platysma, which overlies it, the sternocleidomastoid is crossed superficially by twigs from the medial (anterior) supraclavicular nerve, by the transversus colli and great auricular nerves, and by the external jugular vein. In its lower part it covers the lateral borders of the infrahyoid muscles and the great vessels of the neck; farther up it crosses these and forms the posterior boundary of the carotid triangle.

In the lower part of the carotid triangle the sternocleidomastoid receives a branch from the superior thyroid artery, and at about the level of the hyoid bone it receives one from the occipital artery or directly from the external carotid. At its lower end, smaller twigs from the transverse cervical artery enter it. The muscle is supplied by the accessory nerve, which runs downward and a little posteriorly through or deep to it; this nerve typically emerges between the internal jugular and internal carotid vessels and crosses the lateral surface of the internal jugular to enter the deep surface of the sternocleidomastoid some 4 to 6 cm (1½–2½ inches) below the tip of the mastoid process. Sensory fibers to the muscle are from the cervical plexus; McKenzie has revived the older concept that the muscle is in part derived from myotomes and therefore receives some motor innervation from cervical nerves in addition to that from the accessory.

INFRAHYOID MUSCLES

The infrahyoid muscles, also commonly called "strap" muscles, lie in front of the larynx, thyroid gland, and trachea. The four muscles in this group are the sternohyoid, sternothyroid, thyrohyoid, and omohyoid (Fig. 9-3). They form the floor of the inferior carotid or muscular triangle, but in their lower portions are largely covered by the sternocleidomastoid muscle. The cervical fascia sends laminae in front of, between, and behind them; the part passing anteriorly is usually termed the superficial or anterior layer; the part behind, the pretracheal or middle layer.

The *sternohyoid* is a relatively broad, flat, and thin muscle that extends from an origin on the posterior surface of the sternum and the adjacent medial end of the clavicle to an insertion upon the hyoid bone. Posterolateral to it, and descending at first parallel with it, is the superior belly of the *omohyoid muscle;* as this runs downward from the hyoid bone it gradually diverges from the sternohyoid to become continuous with its inferior belly. The inferior belly of the omohyoid is attached to the upper border of the scapula just medial to the scapular notch; the two bellies are typically attached to the clavicle by a sling of deep fascia and normally join by a tendon. Langsam reported that in 90% of the bodies he examined, the two bellies of the omohyoid presented an intermediate tendon of some sort, although this was sometimes incomplete, sometimes constricted, and sometimes not; in 10% there was no tendinous intersection to be noted.

Deep to the sternohyoid and the superior belly of the omohyoid are the *sternothyroid* and *thyrohyoid muscles,* also thin but extending posterolaterally farther than the sternohyoid. The sternothyroid arises from the posterior surface of the sternum below the origin of the

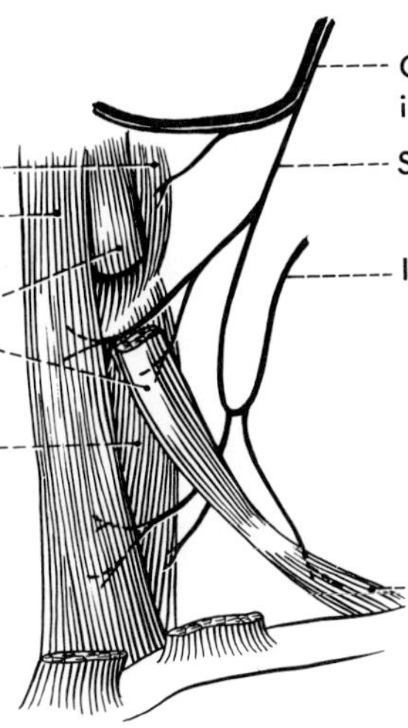

Fig. 9-3. The infrahyoid muscles and their nerve supply. A part of the superior belly of the omohyoid is shown cut away, and the hypoglossal nerve is shown diagrammatically as separated into its two components, cervical fibers and true hypoglossal fibers.

sternohyoid and inserts on the oblique line of the thyroid cartilage; the thyrohyoid arises from the oblique line and continues upward to an insertion upon the hyoid bone.

These four muscles are innervated by cervical nerve fibers that accompany the hypoglossal nerve in part. The nerve to the thyrohyoid descends from the hypoglossal as this nerve curves forward under cover of the submandibular (submaxillary) gland to pass above the hyoid bone; the nerves to the remaining muscles are derived either from the superior root of the ansa cervicalis (the descendens hypoglossi of the ansa hypoglossi) or the ansa itself. As a rule, a branch directly from the superior root supplies the upper belly of the omohyoid muscle and the upper portions of the sternohyoid and sternothyroid, while branches from the ansa cervicalis supply the inferior belly of the omohyoid and the lower portions of the sternohyoid and sternothyroid. The latter muscles are frequently cut in exposing the thyroid gland, and efforts are sometimes made to avoid injuring their nerve supply. Unfortunately, the levels at which the nerves enter the muscles are not constant. Thus Langsam found the branch to the superior belly of the omohyoid muscle passing behind the posterior edge of this muscle anywhere from 2.5 cm above the level of the superior thyroid notch to 5 cm below the level of this notch; in 91% of the subjects that he studied, however, the level of crossing lay between the level of the notch and a horizontal plane 2 cm inferior to this. Similarly, it is usual to find the upper branch to the sternohyoid and sternothyroid entering these muscles at the level of the thyroid cartilage, so that sectioning them below this level would not usually involve this nerve. The other branch or branches to the infrahyoid muscles often enter the sternohyoid and sternothyroid only a little above the level of the jugular (suprasternal) notch, so that, generally speaking, it would seem that the innervation of these muscles could best be avoided by sectioning them approximately halfway between the lower border of the thyroid cartilage and the jugular notch.

SCALENES

The muscles of the suprahyoid region have already been discussed in Chapter 7. Much of the muscle mass of the posterior and posterolateral portions of the neck is associated with the vertebral column, and is described in Volume 3. The deep musculature of the anterior and anterolateral aspects of the neck consists

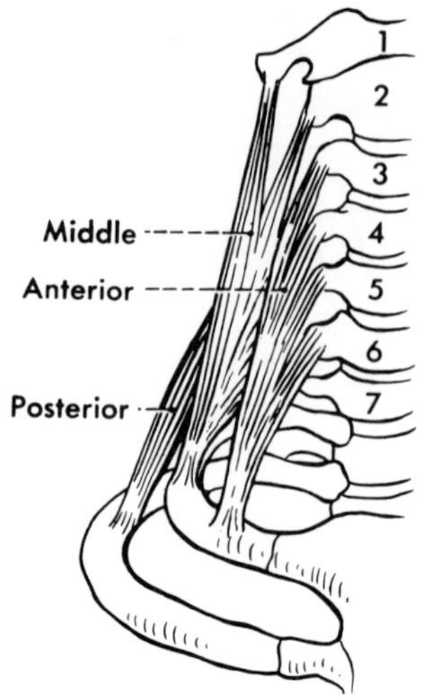

Fig. 9-4. The scalene muscles, the minimus not being shown. For some of the relations to the brachial plexus, see Figs. 9-31 and 9-35.

primarily of two muscle groups, the longus and the scalene muscles. The longus muscles are actually anterior vertebral musculature and are therefore described with the other vertebral muscles. It is only necessary to note here that the two pairs of muscles, longus colli (cervicis) and longus capitis, cover much of the anterior surface of the cervical vertebral column, that the prevertebral fascia covering them, and separating them from the pharynx and esophagus, is divisible into two layers—prevertebral fascia proper, on the surfaces of the muscles, and the "alar fascia"—between which lies very loose connective tissue forming the "danger space" of the neck (see Chap. 5), and that in front of the danger space and the alar fascia is the retrovisceral compartment, bounded anteriorly by the pharynx and esophagus.

Posterior

The scalene muscles (Fig. 9-4) extend from the transverse processes of the cervical vertebrae to the ribs and are therefore flexors and rotators of the vertebral column and elevators of the thoracic cage; their prime importance to the surgeon, however, is their relation to the brachial plexus, subclavian vessels, and phrenic nerve. Of the three scalene muscles, anterior, middle, and posterior, the posterior has no particularly important relations. It forms a part of the floor of the posterior triangle of the neck, lying between the levator scapulae (above and posteriorly) and the scalenus medius (anteriorly), and is crossed superficially by the nerve to the rhomboids and by the long thoracic nerve as this descends to the upper border of the serratus anterior. The scalenus posterior is a rather short muscle, usually arising from the transverse processes of about the fifth and sixth cervical vertebrae and attaching to the outer surface of the second rib or occasionally the third. It has been regarded as a posterior part of the scalenus medius that passes beyond the first rib; in fact, except at their insertions, these two muscles are often difficult to separate.

Middle

The middle scalene (scalenus medius), typically the largest of the scalene group, arises from the transverse processes of most of the cervical vertebrae; according to Cave, it arose from all the cervical vertebrae in about 61% of 60 subjects, from the second to the seventh vertebrae in 15%, from the third to the seventh in 10%, and from the third to the sixth in 10% (in the percentage not accounted for, the origin was not determined). The muscle inserts into the first rib behind, or posterolateral to, the groove for the subclavian artery. In its lower portion the scalenus medius is separated from the scalenus anterior by the brachial plexus and the subclavian artery; the subclavian vein, however, runs in front of the anterior scalene. Higher up, the muscle is usually pierced by the nerve to the rhomboids (dorsal scapular nerve) and by the major portion (fifth and sixth cervical portions) of the long thoracic nerve. If the insertion of the medius is farther forward than usual, it may be very close to or even blend with that of the anterior, thus lessening the room between the lower ends of the two muscles and forcing the subclavian artery and the lower part of the brachial plexus into a higher position than usual.

Anterior

The scalenus anterior (anticus) muscle, clinically the most important of the scalenes, is usually described as arising from the anterior tubercles of the transverse processes of the third or fourth to the sixth cervical vertebrae; Reed ('48) stated that many fibers arise also from the posterior tubercles of the fourth to sixth vertebrae and that the fourth to sixth cervical nerves normally emerge, therefore, through the fibers of the anterior scalene rather than between this and the middle scalene as usually described. As they extend downward, however, all the large elements of the brachial plexus come to lie between these two scalenes. From its origin the fibers of the anterior scalene run downward and slightly laterally to insert on the upper surface of the first rib between the groove for the subclavian artery and that for the vein. Although in the lower part of the neck the sternocleidomastoid is directed inferiorly and medially while the anterior scalene is directed inferiorly and slightly laterally, the lateral boundaries of the two muscles for a short distance above the clavicle correspond approximately, and the lateroposterior edge of the sternocleidomastoid may therefore be used to locate the anterior scalene muscle.

Because of its intimate relation to the brachial plexus, the anterior scalene has been thought to be responsible for certain clinical manifestations of pressure upon the plexus (scalenus anticus syndrome): spasm of the muscle has been thought to exert pressure upon the brachial plexus either directly or by elevation of the first rib against the plexus. This question is discussed later in more detail but it may be noted here that Reed deduced from his dissections that the nerves of the plexus could be compressed by spasm of only those fibers passing posterior to them. He found an especially large group of fibers posterior to the sixth nerve.

The brachial plexus and the subclavian artery emerge from behind the lateral border of the scalenus anterior muscle and are therefore the important posterior relations of this muscle; anteriorly, the most important relation is the phrenic nerve. Toward the upper end of the muscle the phrenic nerve crosses its lateral border and then descends with a slightly medial inclination, lying directly against the muscle fibers and therefore behind the prevertebral layer of cervical fascia, here usually termed "scalene fascia." Since the fibers of the scalenus anterior are inclined slightly laterally, the phrenic nerve and the muscle bundles do not run exactly parallel with each other. As the muscle reaches its insertion the phrenic nerve passes across its medial border to proceed behind the first rib and enter the thorax.

Close to its insertion, and therefore behind the clavicle, the anterior scalene is crossed anteriorly by the subclavian vein; above this, and typically behind or no more than slightly above the level of the clavicle, the suprascapular (transverse scapular) artery crosses the muscle transversely, and still higher is the transverse or superficial cervical artery, when these arteries arise medial to the scalene. The short thyrocervical trunk, which typically gives rise to these vessels, runs upward along the medial border of the muscle, and so does its main branch, the inferior thyroid artery; above these the ascending cervical branch of the inferior thyroid artery ascends along the muscle close to its origin. The cervical sympathetic trunk parallels the phrenic nerve, for which it has been mistaken, although it lies more medially, at about the level of the tips of the transverse processes. The vagus nerve also descends in front of the anterior scalene, but lies within the carotid sheath with the common carotid artery and the internal jugular vein. Finally, in operations upon either the phrenic nerve or the anterior scalene muscle of the left side, the thoracic duct is liable to injury; this attains a more superficial position in the neck by passing forward medial to the anterior scalene and then crossing the lower part of this muscle, between the muscle and the internal jugular vein, to enter approximately the angle of juncture between the internal jugular and subclavian vessels. In this terminal portion of its course the thoracic duct may be entirely hidden by the veins until it actually reaches the angle between them, or it may loop upward and outward and be particularly exposed.

Minimus

In addition to the three scalenes already described, an additional one, the scalenus minimus (scalenus pleuralis, scalenus anticus

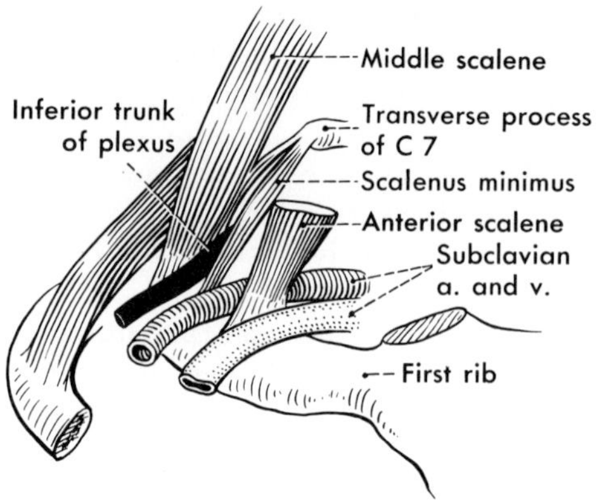

Fig. 9-5. A scalenus minimus muscle. In this case it separates the inferior trunk of the brachial plexus from the remainder of that plexus (not shown) and the subclavian artery. (After Shore LR: J Anat 60:418, 1926)

minor), is frequently present (Fig. 9-5); it is usually stated to be present, bilaterally or unilaterally, in about two thirds of bodies; Sunderland and Bedbrook found it more often, in 80% on the right side and 69% on the left in 100 specimens.

It consists typically of a small bundle of muscle fibers that arises from the anterior tubercle of the sixth or seventh cervical vertebra and usually attaches both into the first rib and into the fascia of the pleural cupula (suprapleural membrane, Sibson's fascia). It has been regarded as a detached portion of the anterior scalene. The insertion into the first rib may be close to or some distance removed from the insertion of the anterior scalene, but the two muscles are apparently always separated by the subclavian artery—that is, the scalenus minimus inserts between the subclavian artery and at least a portion of the brachial plexus (Shore; Stott); it may also separate elements of the plexus. When the insertion of the minimus is close to that of the anterior scalene, the subclavian artery may be held well above the upper surface of the first rib.

INNERVATION

The scalene muscles are innervated by short branches from the anterior rami of the nerves forming the cervical and brachial plexuses, given off close to the intervertebral foramina. The scalenus medius usually receives fibers from the third to the eighth cervical nerves; the others receive theirs through the cervical nerves that form the brachial plexus.

ANOMALIES AND PATHOLOGY

Anomalies of the infrahyoid muscles are occasionally encountered but are usually anatomical curiosities rather than of clinical importance; since these muscles come from a common mass it is easy to see how faulty cleavages of this mass may produce anomalies. Doubling of a belly of the omohyoid, attachment of the superior belly to the mandible or to the clavicle instead of to the posterior belly, or additional anomalous muscles are the ones most frequently reported (*e.g.*, Frazer, '01; Rutherford; Taylor; and Jackson). The levator glandulae thyroideae, discussed later, apparently sometimes represents a development from the infrahyoid muscles and is in any case closely associated topographically with them; it is of surgical interest merely because of its attachment to the thyroid gland. A case of congenital shortening of the sternohyoid and sternothyroid muscles has already been mentioned.

The scalenus anticus syndrome is discussed in connection with the brachial plexus.

TORTICOLLIS

Torticollis, or wryneck, is usually classified as either congenital or spasmodic. According to Hough, the incidence of the former has been variously reported as 0.3%, 0.5%, or 1:150,000 births and appears to be about one tenth as common as clubfoot. He found that involvement of the right side was only slightly more common than the left and that no sexual or racial predisposition had been shown.

Congenital Torticollis

Congenital torticollis is apparently usually limited to involvement of the sternocleidomastoid muscle, but may involve the platysma, the scalenes, or even the splenius muscles. As reviewed by Hough, this condition has been variously thought to be due to heredity, a constitutional defect, arrested development, birth injury, infective myositis, or ischemic fibrosis. Middleton stated that congenital torticollis is not hereditary (but it has been reported in twins) nor as a rule congenital and that it usually appears months or years after birth. More recent investigators (*e.g.,* Chandler and Altenberg; Larson and Rosenow) have agreed that a tumorlike ovoid swelling develops in the muscle some 10 to 14 days after birth; although it may regress and even disappear in ensuing months, the muscle fibers may be gradually replaced by fibrous tissue.

Middleton regarded a temporary acute venous obstruction, occurring during labor, as being the precipitating cause of the disease, and Eddy, noting that the several vascular supplies to the muscle have been said to lack anastomosing channels between them, expressed the belief that extreme rotation of the head might cause ischemia in the lower part of the muscle by compression of the artery supplying this portion. Chandler and Altenberg have shown, however, that this concept of the circulation to the muscle is incorrect, since they were able to demonstrate anastomoses both between the arteries and between the veins of the muscle.

The latter authors regarded congenital torticollis as resulting from any of a number of factors, acting separately or together; although many cases of torticollis are associated with difficult labor, the attractive theory that a birth injury of some type is necessarily responsible is ruled out by their report of a case of torticollis in which the diagnosis was made *in utero* and delivery was by cesarian section—despite the nontraumatic delivery, the torticollis was present at birth and the "tumor" appeared in the muscle at 14 days. As they suggest, it seems possible that the torticollis may be responsible for the difficult delivery, rather than vice versa.

Coventry and Harris reported that not only does the tumor usually disappear spontaneously at about 14 weeks after birth, but the torticollis itself may also disappear without surgical intervention. Of 32 patients in which this occurred, 26 had received physical therapy and six had received no treatment at all. They therefore advised that section of the sternocleidomastoid should be carried out only when wryneck deformity persists beyond the age of 1 year.

Copland stated that the anterior scalene may apparently be involved secondarily in torticollis and suggested that this may sometimes be responsible for failure of relief following section of the sternocleidomastoid muscle. Stewart ('35) reported a case due to abnormal attachment of the left sternocleidomastoid to the right clavicle, and Grieve warned that care should be taken to distinguish between ocular and sternocleidomastoid torticollis.

Spasmodic Torticollis

Herz and Hoefer felt that in so-called spasmodic torticollis the electrical activity in the muscles affected hardly justified the term "spasmodic," although the activity of the muscle was sustained—the defect is not in the muscle, but in the nervous system. Putnam and coauthors pointed out that spasmodic torticollis usually involves three groups of muscles: the short muscles between the atlas, axis, and skull; the larger muscles of the neck, including the sternocleidomastoid, trapezius, vertebral musculature, and scalenes; and the hyoid musculature. They concluded that favorable results can be obtained by reducing the strength of the participating muscles and reported varying degrees of success in so doing by sectioning motor branches of the cervical nerves. The spinal root of the accessory nerve has also been sectioned intracranially in attempting to relieve this condition.

Rugh expressed the belief that there are actually two types of spasmodic torticollis, one being due to a basic functional weakness of the nerve cells and the other to lesions in the basal ganglia or perhaps to disturbed labyrinthine function. He suggested that the best treatment is the use of a cast for as long as 2 years.

Arteries

SUBCLAVIAN ARTERY AND BRANCHES

At the base of the neck on the right side the right subclavian artery arises from the brachiocephalic (innominate) trunk behind the sternoclavicular articulation; it arches upward and laterally, rising usually above the level of the upper border of the clavicle and above the subclavian vein, to pass behind the anterior scalene muscle and leave the neck by crossing the first rib, thereby becoming the axillary. The left subclavian artery arises lower than the right one, since it comes typically from the arch of the aorta; it therefore ascends through the superior mediastinum to the base of the neck, where it arches laterally behind the anterior scalene and, like the right artery, crosses the first rib to become the axillary (Fig. 9-6). The anterior scalene muscle is the landmark by which the subclavian artery is divided for descriptive purposes into three portions; the first part of each vessel is medial to the scalene, the second behind it, the third lateral to it. Except for their first parts, the relations of the two subclavian arteries are essentially similar.

First Part

The first part of the right subclavian artery is fairly short; it is covered superficially by the sternal head of the sternocleidomastoid muscle and, as it arches laterally, is crossed anteriorly by the internal jugular and vertebral veins, the vagus nerve, and usually by both vagal and sympathetic cardiac branches. As the vagus passes across the subclavian it gives off the right recurrent nerve, which turns below and then behind the artery to ascend in the neck. Medial to this, sympathetic fibers, forming the ansa subclavia, loop below the artery and pass upward in front of it to form an anterior connection between the stellate ganglion and the cervical sympathetic trunk. The artery is partly overlapped by the sternohyoid and sternothyroid muscles, and in front of these, between them and the sternocleidomastoid, the anterior jugular vein runs laterally.

Posteroinferiorly the subclavian artery rests upon the pleura of the cupula of the lung. Behind it the sympathetic trunk extends from the thorax into the neck; there may be additional descending cardiac branches, and the recurrent laryngeal nerve and ansa subclavia turn upward. The stellate (cervicothoracic) or inferior cervical and first thoracic ganglia lie largely behind the subclavian artery close to the level of origin of the vertebral artery, and the vertebral or "low middle cervical" ganglion lies close to it anterosuperiorly, usually on the lower part of the vertebral artery.

The first part of the left subclavian artery as it passes into the neck is also related anteriorly to the sternocleidomastoid and the infrahyoid muscles. Behind these the vagus and phrenic nerves, and cardiac branches of the vagus and sympathetic, pass downward parallel to the vessel along its front; as on the right side, cardiac branches often also pass behind it. At the base of the neck the left internal jugular and vertebral veins lie between the nerves and the infrahyoid muscles, and the left common carotid artery lies anteromedial to the subclavian. Posteromedial to it are the esophagus, trachea, thoracic duct, and left recurrent laryngeal nerve, and lateral to it is the cupula of the left lung; as the artery reaches the neck the thoracic duct passes forward between it and the carotid and then arches laterally in a plane anterior to the subclavian artery to enter the angle between internal jugular and subclavian veins.

The most striking anomaly of the subclavian artery is the rather rare (about 1%) retroesophageal right subclavian artery; far less frequently, a retroesophageal left subclavian artery is found. These conditions are discussed in Volume 2 in connection with other anomalies of the aortic arch system.

Second Part

The second part of both subclavian arteries is typically the highest part of the arch of each and lies between the anterior and middle scalene muscles. It is here separated from the pleura by the middle scalene. The lower part

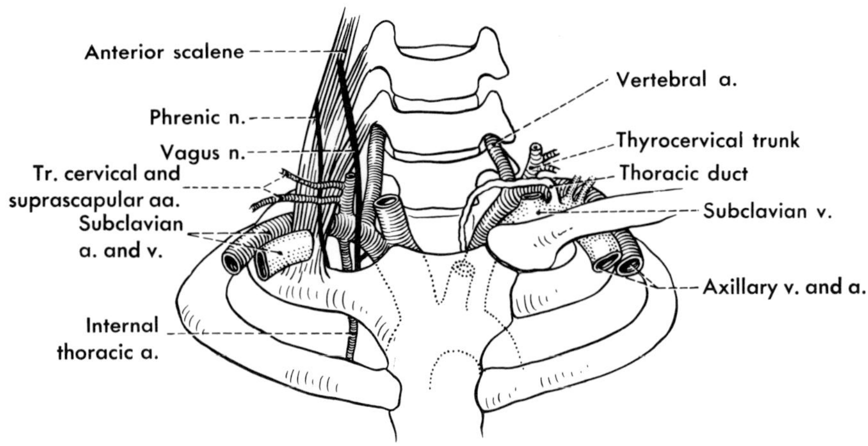

Fig. 9-6. Some relations and branches of the subclavian arteries.

of the brachial plexus usually lies immediately above this part of the artery. As already noted, the subclavian artery may be raised higher than usual by the presence of a scalenus minimus muscle or a more anterior insertion of the scalenus medius.

Third Part

The third part of the artery is related posteriorly to the scalenus medius, with the lowest trunk of the brachial plexus lying between the vessel and the muscle; above and lateral to it are the upper trunks of the brachial plexus, or their derivative cords. This part of the vessel is crossed superficially by the supraclavicular nerves and the nerve to the subclavius as these descend, and in front of it lies also the external jugular vein, which here usually receives the suprascapular, transverse cervical, and anterior jugular veins.

The third part of the artery lies within the subclavian triangle and, being free of the lateral border of the sternocleidomastoid, is the most readily accessible portion of the vessel. It has been, therefore, the favored portion for ligation of the subclavian; the chief anastomotic channels available after this is done are the anastomoses between the suprascapular (transverse scapular) and transverse cervical arteries on the one hand and the subscapular artery on the other and those of branches of the internal thoracic (internal mammary) and intercostals, with the lateral thoracic and subscapular arteries.

The important branches of the subclavian artery at the base of the neck are the vertebral artery, the thyrocervical trunk, the internal thoracic (mammary) artery, and the costocervical trunk. These branches tend to come off close together, from the first part of the subclavian artery and close to the medial border of the scalenus anterior muscle. However, origin of an artery from the second or third parts of the subclavian is very common. Usually, an artery of this origin is one that otherwise arises from the thyrocervical trunk.

VERTEBRAL ARTERY

The vertebral artery, a large and rather constant branch, is typically the first to arise; it takes its origin from the upper and posterior part of the first portion of the subclavian artery and running upward enters the foramen of a transverse process—usually that of the sixth, but sometimes as high as the fourth or as low as the seventh.

The part between the origin of the artery and its entrance into the foramen of a transverse process is called the first part of the artery and therefore varies in length (Fig. 9-7); the second portion of the artery courses through the transverse foramina, and varies in length inversely with the first part. The spi-

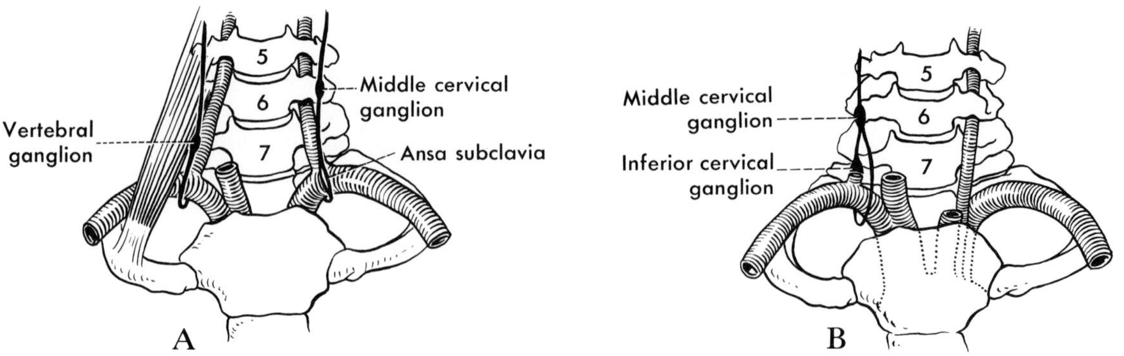

Fig. 9-7. Some relations and variations of the vertebral artery. In (*A*), the vertebral artery is shown entering the canal in the transverse process of the fifth cervical vertebra on one side, that of the sixth on the other, and the different positions of the middle cervical and vertebral sympathetic ganglia are shown. In (*B*), the right vertebral artery is shown cut away to reveal the inferior cervical ganglion (or upper part of the stellate, as the case may be), while the left vertebral is shown arising, as it does occasionally, directly from the arch of the aorta.

nal branches of the second portion of the artery are segmental twigs given off to supply the vertebrae and to accompany the roots of the spinal nerves to supply the meninges and the spinal cord. Variable tiny branches are also given off from the first two parts of the vertebral to adjacent muscles. The third and fourth parts of the artery are the part in the suboccipital region, above the atlas, and that between the point at which the artery pierces the atlantooccipital membrane and its ending to form the basilar artery. These parts are described in connection with the suboccipital region and the cranium, respectively.

At its origin the vertebral artery lies usually directly in front of or anterolateral to the inferior cervical ganglion or the upper end of the stellate, while the vertebral, or low middle cervical, ganglion typically lies on the front of the artery. Thus the two lowest ganglia of the cervical trunk are frequently at the same level, separated by no more than the diameter of the vertebral artery. The vertebral vein runs in front of the artery, and sympathetic fibers derived from the stellate ganglion run upward upon the artery; most of these are actually rami communicantes between the ganglion and some of the nerves of the brachial plexus and therefore leave the artery to join the nerves as they emerge from the intervertebral foramina. Other fibers are, strictly speaking, the vertebral sympathetic plexus, and they end on the artery. The part of the plexus that accompanies the artery into the skull is not derived from this level, but from the superior cervical sympathetic ganglion and upper cervical nerves.

The vertebral artery may arise lateral to or with the thyrocervical trunk instead of medial to it; Daseler and Anson reported origin of the left vertebral directly from the arch of the aorta in almost 2.5% of their cases and occasional instances of origin of the right vertebral at the bifurcation of the brachiocephalic (innominate) or from the right common carotid. They found a few instances in which a smaller accessory vertebral artery was present, and Neuberger reported a case in which a vertebral arose by two stems, one from the brachiocephalic and one from the subclavian.

In recent years, reversed flow in the vertebral artery as a result of stenosis or occlusion of the subclavian artery proximal to the origin of the vertebral has been shown to be a cause of some cases of cerebral ischemia. This reversed flow, which diverts to the subclavian blood intended for the brain, is usually called the "subclavian steal" syndrome; Santschi and co-workers reviewed a number of cases.

Aneurysms and arteriovenous fistulas of the vertebral vessels are apparently rare; Elkin and Harris stated that arteriovenous fistulas of the vertebral vessels are among the rarest of arterial injuries, presumably because of the location of the vessels, and that only some 100 cases had been reported. They suggested that adequate exposure of the first part of the artery is best obtained by an incision through the sternocleidomastoid muscle and of most of the second part by approaching along the anterior border of this muscle; for the upper portion of the second part of the artery, or for the third, they reflected the sternocleidomastoid from its insertion in order to reach the transverse process of the atlas.

THYROCERVICAL TRUNK

The thyrocervical trunk (thyroid axis) arises from the upper and front part of the subclavian artery, close to the medial margin of the scalenus anterior, and divides almost immediately into several branches (Fig. 9-8). These are typically the suprascapular (transverse scapular), the transverse cervical or its superficial branch, and the inferior thyroid artery. The so-called typical pattern of branching of this vessel, found to occur in only 29% of 544 bodies in a British series (Thomson), is for the suprascapular to be given off first and for the transverse cervical to be given off above this; the inferior thyroid artery is the most constant branch in its origin from the thyrocervical trunk, yet it has been reported missing in about 6% of bodies. The variations in the branches, and the exact manner of branching, of the thyrocervical trunk are numerous; they have been classified, for instance, by Thomson, and by Bean, although the figures on the occurrence of the different types do not agree. In general, either the suprascapular or transverse cervical artery, or both, may arise directly from the subclavian artery instead of from the thyrocervical trunk. Also, these two arteries may arise by a common stem, from either the thyrocervical trunk or the subclavian, or one branch of the transverse cervical may arise from the thyrocervical trunk and the other from the subclavian. Daseler and Anson found no thyrocervical trunk in 17% of sides, its place usually being taken by the inferior thyroid artery, but sometimes by another of the major branches of the trunk.

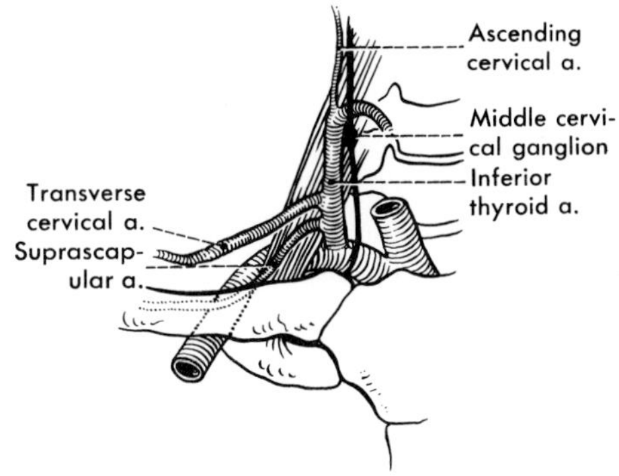

Fig. 9-8. The thyrocervical trunk.

Suprascapular Artery

The suprascapular (transverse scapular) artery, usually a branch of the thyrocervical trunk, passes transversely across the neck in front of the anterior scalene muscle, the brachial plexus, and the third part of the subclavian artery; from its origin it usually runs at first slightly downward, so as to lie behind the upper border of the clavicle. It reaches the scapula at the level of the scapular (lesser scapular) notch and is distributed to muscles on the dorsal surface of the scapula.

In a goodly number of cases the suprascapular arises from the third part of the subclavian or from the first or second parts, instead of from the thyrocervical trunk. The reported incidence varies widely; Bean apparently found an origin from the third part of the subclavian in 22% of the bodies he studied, and Read and Trotter one of 28%, while Daseler and Anson reported an origin from the subclavian together with the transverse cervical in 4.38% and independently in 13.16%; in the latter workers' cases, however, the origin was from the first part of the subclavian as frequently as it was from the third part and from the second part in a smaller number of instances. The suprascapular may also arise from the internal thoracic (internal mammary); Bean reported this origin in 10%, Read and Trotter in 5.1%, and Daseler and Anson in 3.74%. Still less frequently it arises from the axillary artery, and occasionally it shows a common stem (from the subclavian) with the costocervical trunk.

Transverse Cervical Artery

Even more variable in its origin is the transverse cervical (transversa colli) artery (Fig. 9-9). As traditionally described, this artery commonly arises from the thyrocervical trunk and runs almost transversely across the neck anterior to the anterior scalene muscle and the brachial plexus; it roughly parallels the suprascapular artery but lies somewhat higher and above the level of the upper border of the clavicle. In some accounts, when it has this origin and course, it is called the superficial cervical. It passes across the posterior triangle of the neck to reach the deep surface of the trapezius, where it divides into superficial and deep branches (once called ascending and descending, respectively). The superficial branch runs on the deep surface of the trapezius, the deep branch on the deep surface of the rhomboids, close to the medial border of the scapula.

It has long been known that separate origin of the two branches of the transverse cervical artery is not uncommon; for instance, in the British series (Thomson) the superficial branch arose from the thyrocervical trunk and the deep branch from the third part of the subclavian in 22% of 544 bodies, and this was apparently true in 30% of Bean's series. Later accounts make it appear, however, that a common origin from the thyrocervical trunk is the exception rather than the rule; in Daseler and Anson's extensive series this origin was apparently found in slightly less than one third of cases, and in Huelke's series in

Fig. 9-9. Some of the variations in the branches of the thyrocervical trunk (*T.T.*), and the approximate percentages in which they have been reported to occur. In the upper row, variations in the transverse cervical artery and its branches are shown. In (*a*) the transverse cervical artery (*T.C.*) arises from the thyrocervical trunk either with the suprascapular (*S.*) or directly (*hatched area*), and both the superficial branch or superficial cervical artery (*S.C.*) and the deep branch (dorsal scapular or descending scapular artery, *D.S.*) arise from the transverse cervical. In (*b*), the prevailing pattern, the superficial cervical artery arises from the thyrocervical trunk either with the suprascapular or directly (*hatched area*), while the dorsal scapular arises from the second or third part of the subclavian. In (*c*), the superficial cervical and the dorsal scapular arise together from the second or third part of the subclavian; (*d*) depicts origin of the suprascapular artery from the subclavian; (*e*), its origin from the internal thoracic; and (*f*), absence of the inferior thyroid as a branch of the thyrocervical trunk.

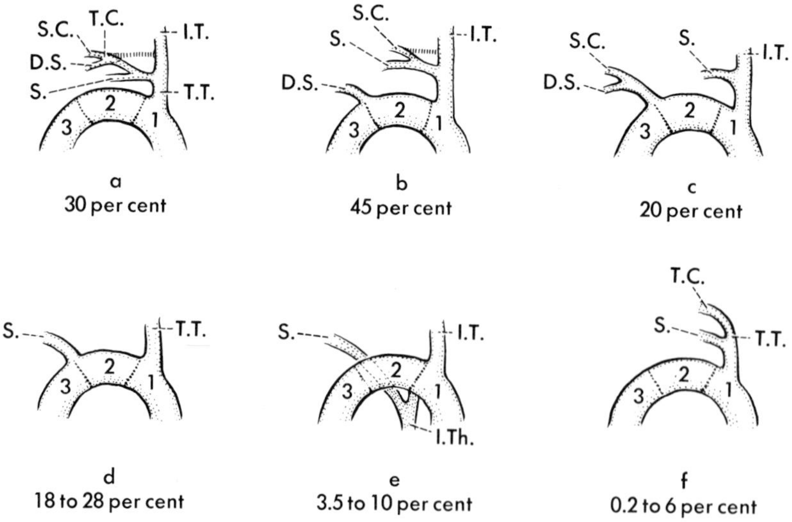

approximately the same, 30.3%. The new anatomical terminology recognizes this by listing the transverse cervical artery as having a constant superficial branch, which is also called the superficial cervical artery, and an inconstant deep branch (dorsal or descending scapular artery), the usual origin of which is from the subclavian.

In addition to the approximately 30% of sides in which the two branches of the transverse cervical artery arose together from the thyrocervical trunk, Huelke found them arising together from the subclavian in about 20%. Including this common origin, Daseler and Anson found the *dorsal* or *descending scapular artery* arising from the second or third parts of the subclavian in 65%, Huelke, in 67%; in a few instances this origin was in common with the suprascapular artery and, rarely, origin from the costocervical trunk, the thyrocervical trunk, the first part of the subclavian, and the axillary artery was observed. In contrast, the *superficial cervical artery* typically arises from the thyrocervical trunk, more commonly than not with the suprascapular artery. In Huelke's series it had this origin in 77.5% (including the 30% in which both it and the dorsal scapular shared this origin), and in Daseler and Anson's series slightly more. The next most common origin is with the dorsal scapular from the subclavian; rare origins are as an independent branch from the subclavian, the internal thoracic, or the costocervical trunk.

Inferior Thyroid Artery

The inferior thyroid artery normally runs upward in front of the vertebral artery, continuing the direction of the thyrocervical trunk. At about the level of the sixth cervical vertebra the inferior thyroid artery gives off its ascending cervical branch and then passes medially and usually somewhat downward to make an arch behind the carotid sheath, penetrate the prevertebral fascia, and reach the posterior aspect of the thyroid gland. As it makes its medial turn, the artery lies on the longus colli muscle and crosses the cervical sympathetic trunk; the relation to this varies, since the trunk may be behind or in front of the artery, may be split by the artery, or cardiac branches may cross anterior to the artery while the trunk itself passes behind. When a middle cervical ganglion is present, it usually lies a little above the artery. As the inferior thyroid artery reaches the thyroid gland it typically divides and also necessarily comes in close relation to the recurrent laryngeal nerve. The terminal portion of this artery and its relation to the recurrent nerve are discussed in detail in connection with the thyroid gland.

Occasional instances of double inferior thyroid arteries have been reported (*e.g.,* Allan); they may have similar courses, but sometimes one passes in front of the common carotid instead of behind that vessel. A more frequent variation in the course of the inferior thyroid is for it to pass behind instead of in front of the vertebral artery. Daseler and Anson quoted evidence that this occurs in approximately 50% of the cases in which the vertebral artery first enters a transverse foramen higher than that of the sixth cervical vertebra, which in their series occurred in about 7% of sides.

The inferior thyroid artery goes primarily to the thyroid gland, and its other branches are small. In its course it gives off muscular twigs to the scalenus anterior, the longus colli, the infrahyoid muscles, and the inferior constrictor of the pharynx; esophageal and tracheal branches; and the inferior laryngeal artery. The latter passes along the trachea with the recurrent laryngeal nerve to enter the larynx with it.

The largest of the branches of the inferior thyroid artery other than to the thyroid gland is the small *ascending cervical artery.* This leaves the inferior thyroid as that artery turns to pass behind the carotid sheath and runs up along the anterior tubercles of the transverse processes of the cervical vertebrae; it gives twigs to the muscles of the neck and spinal branches that, like those of the vertebral, are distributed to the bodies of the vertebrae and to the spinal cord and its membranes, and it anastomoses with the ascending pharyngeal and occipital arteries. Although most commonly from the superior thyroid, the ascending cervical may arise from other arteries at the base of the neck and is fairly frequently represented by more than one vessel.

When the inferior thyroid artery is not a branch of the thyrocervical trunk (reported in as high as 6% of sides in one series), its place may be taken by an artery that arises from the subclavian, the arch of the aorta, the common carotid, or some other source; such an aberrant vessel is usually called the *thyroid ima* artery.

INTERNAL THORACIC ARTERY

The internal thoracic (formerly internal mammary) artery arises from the lower surface of the arch of the subclavian artery close to the medial edge of the anterior scalene muscle and therefore at about the same level as the thyrocervical trunk. It passes downward with a slight inclination medialward to enter the thorax behind the first rib and continues downward on the inner surface of the thoracic wall.

Fig. 9-10. The right common carotid artery and its branches.

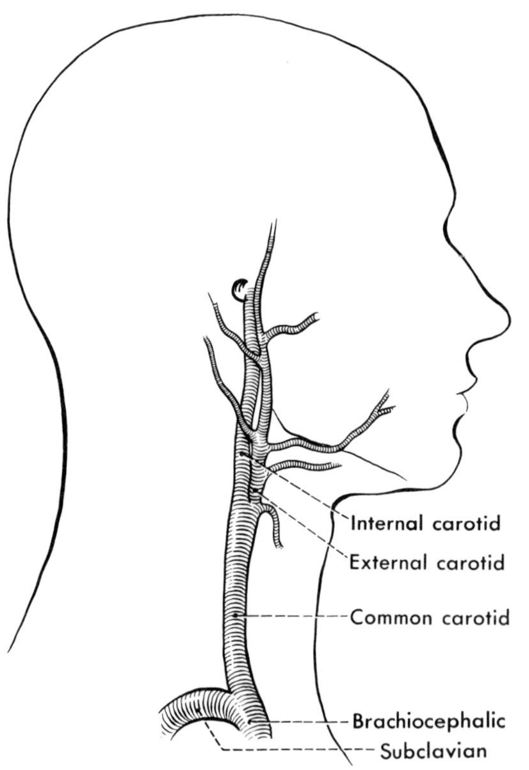

COSTOCERVICAL TRUNK

The costocervical trunk is a short vessel arising from the posterior surface of the subclavian artery, usually just medial to the edge of the anterior scalene on the left and behind the anterior scalene on the right. It runs upward and backward above the dome of the pleura and as it turns downward behind the pleura divides into two branches, the supreme intercostal and the deep cervical. There are many variations in the origin of the costocervical trunk; these include origin with other branches of the subclavian artery and separate origin of its supreme intercostal and deep cervical branches.

The supreme intercostal crosses the neck of the first rib and gives rise to the first posterior intercostal artery, with which it may terminate, but usually forms the second one or helps to form it by anastomosing with the uppermost intercostal branch from the aorta.

The deep cervical artery passes directly backward between the seventh and eighth cervical nerves, and below the transverse process of the seventh cervical vertebra, and then runs upward between the semispinalis capitis and semispinalis cervicis on the posterolateral aspect of the neck.

COMMON CAROTID ARTERY

The two common carotid arteries differ in length, because the right usually arises from the brachiocephalic (innominate) artery behind the sternoclavicular joint (Fig. 9-10), while the left arises from the arch of the aorta and has therefore a thoracic as well as a cervical course. Both arteries normally end by bifurcating into external and internal carotids slightly below the hyoid bone. At the base of the neck, as they emerge from behind the sternoclavicular articulations, the two common carotid arteries lie deeply, being covered not only by the various layers of fascia but also by the sternocleidomastoid and infrahyoid or "strap" muscles; in the carotid triangle, however, they are more readily accessible, being covered anteromedially only by skin, the platysma, and fascia, although overlapped somewhat, laterally, by the anterior margin of the sternocleidomastoid.

Each common carotid artery is enclosed in a carotid sheath, a layer of connective tissue that also contains, in separate compartments, the internal jugular vein and the vagus nerve; within this sheath the internal jugular vein lies laterally and the common carotid medially, while the vagus lies behind and somewhat between the two. Anteromedially, the common carotid artery is related to the trachea, the thyroid gland, and the larynx; posterior to the artery are the longus colli and longus capitis muscles over the transverse processes of the cervical vertebrae, and in the connective tissue between these muscles and the carotid sheath is the cervical sympathetic trunk, which therefore lies behind the artery. The inferior thyroid arteries also course behind the carotid arteries to reach the thyroid gland.

In the lower part of the neck the right recurrent nerve crosses obliquely behind the right common carotid to attain a more medial position, while the left recurrent nerve lies behind the left common carotid as both pass from the thorax into the neck. On the left side also the thoracic duct lies behind the common carotid as they enter the neck, but some 3 to 4 cm above the clavicle it turns laterally to cross the anterior surfaces of the vertebral, thyrocervical, and subclavian arteries.

Ligation of Carotid Arteries

Above the level at which it is crossed by the omohyoid muscle, the carotid is crossed by few structures of importance, and this is therefore the usual site of election for ligation of the vessel. The superior thyroid vein, the sternocleidomastoid branch of the superior thyroid artery, and sometimes a communication between the facial and the anterior jugular veins cross superficial to the artery in the carotid triangle and may be wounded in the operation for ligation; the superior root of the ansa cervicalis (formerly descendens hypoglossi to the ansa hypoglossi) can often be found here also on the front of the carotid sheath or within the sheath and may join the inferior root (formerly descendens cervicalis) on the front of the sheath to form the ansa, but these are more closely associated with the jugular vein than with the artery.

In the operation for ligation, a skin incision at the level of the hyoid bone, therefore at about the level of the carotid bifurcation, allows access to the common carotid and its branches. The internal jugular vein is best avoided in opening the carotid sheath by doing this well to the medial side; some care in passing the ligature about the artery may be necessary to avoid inclusion of the vein or of the vagus nerve. Ligation of the external carotid is most easily carried out through the same approach, identifying the bifurcation of the common carotid and placing the ligature about the first part of the external carotid, usually between the superior thyroid and lingual branches.

The collateral circulation available after ligation of the common carotid artery on one side is relatively abundant (Fig. 9-11); it includes the anastomoses between the branches of the external carotids of the two sides, especially across the face and scalp, and also the anastomoses between branches of the external carotid and subclavian arteries, especially between the superior and inferior thyroid arteries of both the same and the opposite sides and between the occipital artery and the deep and transverse cervical arteries.

In ligation of the external carotid artery above the superior thyroid, a majority of these channels are still available and may perhaps be supplemented by anastomoses beteen the ophthalmic artery and the facial. More collateral circulation would be available to the external carotid artery if the ligature were placed below the superior thyroid artery, but this is not necessary, and the superior thyroid frequently arises so low that there is not room for a ligature on the external carotid below its point of origin.

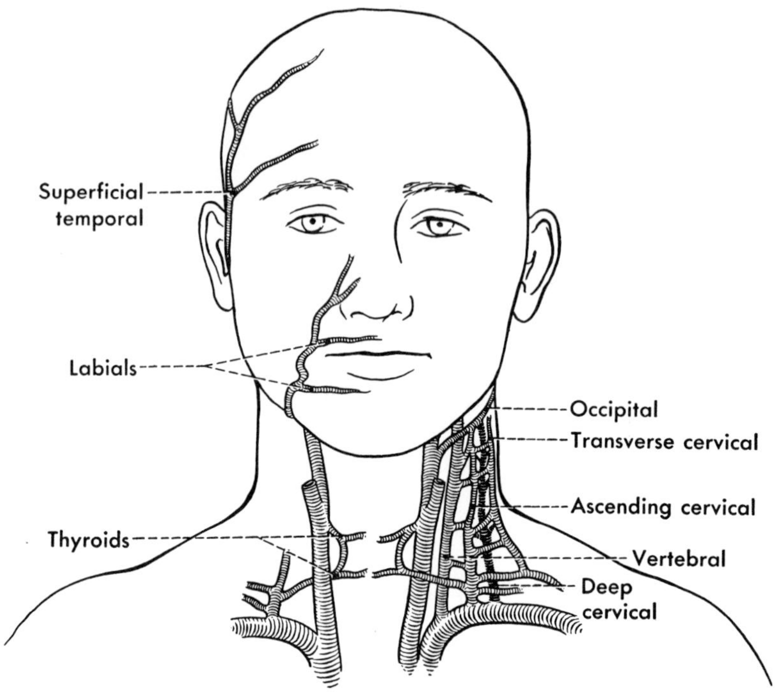

Fig. 9-11. Some of the collateral channels available after ligation of the common carotid artery. On the right side of the body are shown the chief communications between the two sides; on the left, the chief longitudinal anastomoses.

It might be noted here that although the external and internal carotid arteries lie parallel to each other at their origins, they are not situated in the relation to each other that their names would seem to imply; rather, the external carotid artery tends to lie anteromedial while the internal carotid lies posterolateral; the vessels are named not from their relations to each other but from their distributions. Any doubt as to which vessel is which can be resolved by identifying branches as arising from one of them: the internal carotid normally has no branches in its cervical course, as contrasted with the external carotid, which gives off several branches close to its origin.

Ligation of the external carotid artery has no serious sequelae unless the internal carotid is injured during the operation. Ligation of the common carotid has been said to reduce the flow in the internal carotid by about 50%, and it has been assumed that the remaining flow is sustained solely by the anastomotic channels between the external carotids of the two sides. Measurements of the blood flow in the internal carotid after occlusion of the common carotid are said to have shown, however, that in only about 50% of apparently normal persons does the external carotid deliver blood to the internal; in the remainder, the blood flow was said to be from the internal into the external, thus diverting blood from the brain (Roberts and co-workers). Ligation of the common carotid, generally regarded as a far safer procedure than ligation of the internal carotid, may, according to these findings, reduce intracranial blood flow more than the latter procedure.

Ligation of the internal carotid has always been regarded as a serious procedure. Pemberton and Livermore noted that valid statistics are not available; however, they stated that although Dandy had had a death rate of less than 4% in ligations of this artery for intracranial aneurysm, the death rate was 15.7% in a collected series of 51 cases of internal carotid ligation (for other than aneurysm) that they reviewed, and in half again as many additional patients there were cerebral complications. They found that about 30% of patients operated upon for tumors of the carotid body have in the past died as a direct result of the ligation, which was at one time a routine procedure. In cases of intracerebral aneurysm, the dangers of ligation can sometimes be mitigated by gradual occlusion through an adjustable clamp on the internal carotid.

CAROTID BIFURCATION

Beyond the rare origin of a thyroid ima artery from the common carotid at the level of the thyroid gland, and the not uncommon origin of the superior thyroid artery from the uppermost portion of the common carotid, the common carotid artery gives off no branches except for its terminal ones, the external and internal carotids. Extremely rarely, external and internal carotids arise independently from the subclavian and there is therefore no common carotid (Fig. 9-12), or the common carotid bifurcates immediately after its origin. The bifurcation of the common carotid ordinarily occurs, however, at approximately the level of the interval between the superior border of the thyroid cartilage and the inferior border of the hyoid bone; Kantor quoted Quain as having found 31 cases of low bifurcation among 295 individuals. In 26 of these 31 cases, the bifurcations lay at about the levels of the upper or middle thirds of the thyroid cartilage, while in the other five they were at the level of the cricoid. On the posteromedial surface of the common carotid artery at its bifurcation is the carotid body.

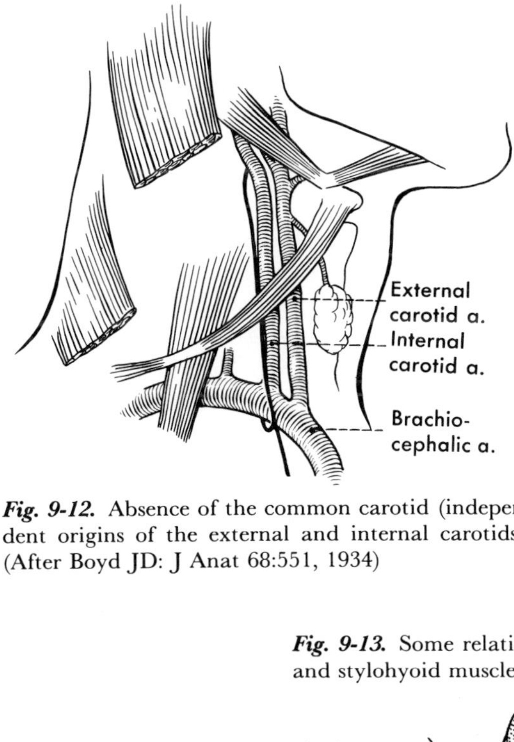

Fig. 9-12. Absence of the common carotid (independent origins of the external and internal carotids). (After Boyd JD: J Anat 68:551, 1934)

INTERNAL CAROTID ARTERY

From its origin the internal carotid artery ascends directly toward the carotid canal and is crossed laterally, in ascending order, by the hypoglossal nerve, the occipital artery, the posterior belly of the digastric and the associated stylohyoid muscle, and the posterior auricular artery (Figs. 9-13 and 9-14). Still higher, and close to the base of the skull, it has the external carotid artery situated anterolateral to it; the stylopharyngeus muscle and the associated glossopharyngeal nerve,

Fig. 9-13. Some relations of the internal carotid, the mandible and the digastric and stylohyoid muscles having been removed.

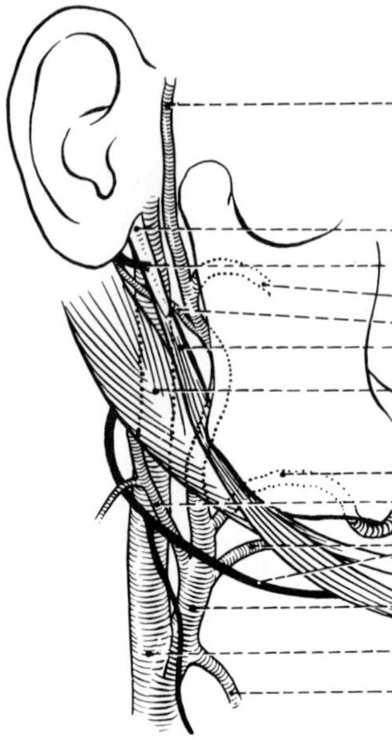

Fig. 9-14. The deep course of the external and internal carotid arteries, diagrammed as if the parotid gland had been removed.

the pharyngeal branch of the vagus, and the stylohyoid ligament all pass lateral to the internal carotid, between it and the external carotid. Eagle regarded deviation of the tip of an elongated styloid process, with impingement upon either the internal or external carotid artery, as a cause of some facial pains and headaches. In the upper part of its course the internal carotid lies deep to the parotid gland, and the internal jugular vein, which at first lay lateral to the artery, runs behind it. Medially, the internal carotid is related to the pharynx and especially therefore to the superior constrictor muscle and the tonsillar region; except when it is unusually tortuous, it does not come into direct contact with the pharyngeal wall in the tonsillar region (although it does higher), but is separated from the pharynx by the retropharyngeal space. The artery lies upon the longus capitis muscle, and between it and the muscle are the superior cervical ganglion and the vagus nerve; higher up, where the internal jugular vein lies behind the artery, the ninth through the twelfth cranial nerves all normally pass behind the artery, between it and the internal jugular vein.

EXTERNAL CAROTID ARTERY AND BRANCHES

After its origin in the carotid triangle anteromedial to the internal carotid, the external carotid artery (see Fig. 9-14) passes upward deep to the posterior belly of the digastric and the stylohyoid muscle (although a superficial course has been reported in at least two individuals), crosses the styloglossus and the stylopharyngeus muscles on their lateral side (these muscles separating it from the internal carotid), and, approximately paralleling the ramus of the mandible, passes into the deeper portion of the parotid gland. Joessel reported three instances in which the external carotid was double for a part of its course. Close to its origin the superior laryngeal nerve lies behind it at first and then crosses it medially to descend toward the larynx; at approximately the same level the hypoglossal nerve crosses the artery laterally, and within the substance of the parotid gland the facial nerve also crosses superficial to it. In addition to these nerves, the artery is crossed laterally by the superior thyroid, common facial, and lingual veins, and within the substance of the parotid gland it lies deep to the retromandibular (posterior facial) vein. In the upper part of its course the external carotid lies anterolateral to the internal carotid, being separated from this not only by a portion of the parotid gland and the stylopharyngeus and styloglossus muscles, but also by the glossopharyngeal nerve and the pharyngeal branch of the vagus.

Superior Thyroid

The superior thyroid artery is typically the first branch from the external carotid, arising usually from the anterior surface of the latter. Its origin may be so low that it is really from the common carotid rather than the external; Daseler and Anson found such an origin, from the carotid at or below the bifurcation, in 45% of 180 sides. The artery runs downward and forward to supply the adjacent muscles, the larynx, and the thyroid gland; its course is further described with the thyroid gland.

Ascending Pharyngeal

The ascending pharyngeal artery, the smallest branch of the external carotid, arises from the posterior surface of the latter, often close to its origin, and ascends vertically anteromedial to the internal carotid and upon the side of the pharynx; it supplies several branches to the pharynx, a branch to the palate and tonsil, the inferior tympanic branch to the tympanum, and several meningeal branches; of these, the larger terminal branch is usually a posterior meningeal that enters the cranium through the jugular foramen. It also supplies twigs to adjacent muscles, nerves, and lymph nodes, and anastomoses with branches of the ascending cervical artery. In approximately 14% of individuals the ascending pharyngeal arises not from the external carotid but from the occipital artery.

Lingual

The lingual artery arises from the anterior surface of the external carotid above the superior thyroid artery, usually at or above the level of the hyoid bone; depending upon the level of its origin it may run almost straight forward at the level of the upper border of the greater cornu of the hyoid, may run upward above this level and then curve downward and forward again, or may simply curve downward and forward toward the tongue. In any case, it passes deep to the digastric and the stylohyoid to lie against the lateral wall of the pharynx and runs forward to disappear medial to the hyoglossus muscle. The variations in its origin, its relationships in Lesser's triangle (the usual site of ligation of the vessel) and its further course and branches are described in connection with the tongue (see Chap. 7).

Facial

The facial (external maxillary) artery also arises from the anterior surface of the external carotid artery, at about the lower border of the digastric muscle and somewhat under cover of the angle of the mandible; it passes forward and upward deep to the digastric and stylohyoid, arching above the latter to come in contact with the submandibular (submaxillary) gland. This artery may arise in common with the lingual (Fig. 9-15) or, very rarely, with the maxillary; its further course

Fig. 9-15. Some variations in the branching of the external carotid artery. In (*a*) is shown what is usually described as the more common arrangement; in (*b*) the superior thyroid is shown arising from the common carotid and the lingual and facial arteries as having a common stem; and in (*c*) the posterior auricular artery is a branch of the occipital, while the ascending pharyngeal arises higher than usual.

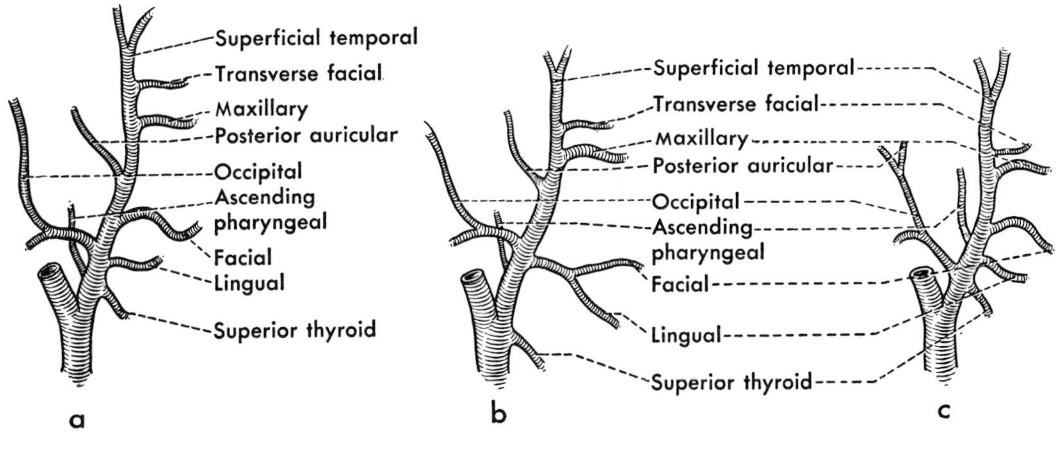

and its branches are described in connection with the face (see Chap. 6). The cervical branches of the artery are the ascending palatine, the tonsillar branch, branches to the submandibular salivary gland, the submental artery running forward upon the mylohyoid, and muscular branches to some of the masticator muscles.

Occipital

The occipital artery arises from the posterior surface of the external carotid, at about the same level as the facial artery and therefore close to the lower margin of the posterior belly of the digastric. It runs backward to cross the internal carotid artery and internal jugular vein superficially. Close to its origin the hypoglossal nerve emerges between the internal jugular and the internal carotid, therefore deep to the occipital artery, and turns forward toward the tongue lateral to the artery; thus the nerve and artery hook around each other. The occipital artery runs obliquely upward and backward among the vertebral muscles, then turns upward onto the scalp either through the cranial attachment of the trapezius or in the space between that and the sternocleidomastoid. In the latter part of its course it is joined by the greater occipital nerve.

In addition to twigs to adjacent muscles, it gives off a named muscular branch, the sternocleidomastoid, to this muscle; this branch usually passes downward and backward over the hypoglossal nerve as the nerve loops around the occipital artery. The artery's auricular branch supplies the back of the concha, and its mastoid branch, which may arise with the auricular, enters the skull through the mastoid foramen to supply diploë, mastoid cells, and dura; a meningeal branch usually ascends with the internal jugular vein to enter the skull through the jugular foramen and the condylar canal, and at the base of the skull the occipital gives off a descending branch that may anastomose with an ascending branch of the transverse cervical artery and with the vertebral and deep cervical arteries. These connections help maintain the collateral circulation after ligation of either the common carotid or the subclavian artery.

Posterior Auricular

The posterior auricular artery also arises from the posterior surface of the external carotid, but above the level at which this artery is crossed by the digastric and stylohyoid muscles; it is largely covered by the parotid gland and ascends along the styloid process to the groove between the external ear and the mastoid process. It may arise as a branch of the occipital rather than as an independent branch from the external carotid. It gives off twigs to adjacent muscles and to the parotid gland, a stylomastoid and a posterior tympanic artery to mastoid cells and the tympanic cavity, an auricular branch to the back of the auricle, and an occipital branch to the scalp above and behind the ear.

Superficial Temporal and Maxillary

The terminal branches of the external carotid artery, the superficial temporal and the (internal) maxillary, arise within the substance of the parotid gland; the superficial temporal continues the upward direction of the parent stem, while the maxillary passes forward medial to the ramus of the mandible. These vessels have been described especially in connection with the face (see Chap. 6) and with the jaws (see Chap. 7).

Carotid Body and Carotid Sinus

The region of the carotid bifurcation contains two sensory structures, different anatomically and physiologically and of different surgical importance. One of these is the carotid body; the other, the carotid sinus.

CAROTID BODY

The carotid body (glomus caroticum) is a small (from about 2.5 by 5 mm to about 4 by 7 mm) flattened structure lying at about the level of the carotid bifurcation. Although sometimes described as lying in the bifurcation, it is typically on the posteromedial side of the common carotid artery and held firmly against this by connective tissue; its upper pole may, however, extend above the carotid bifurcation (Fig. 9-16).

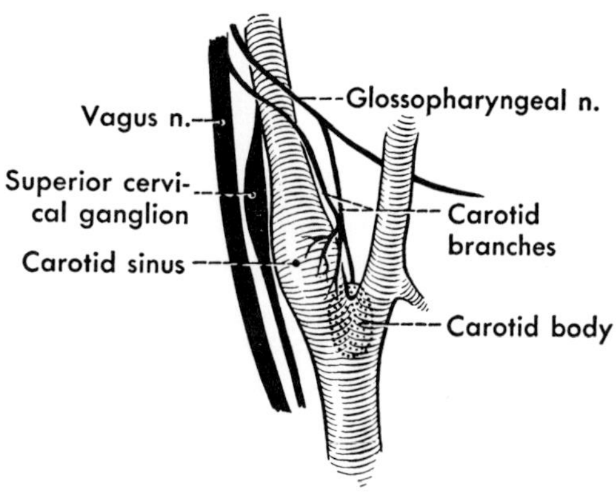

Fig. 9-16. The carotid sinus and carotid body and their innervation. What part the carotid branch of the vagus plays in this innervation is not known (see text). Note that the carotid body lies not so much in, as medial to, the carotid bifurcation.

In contrast to the carotid body of some animals, which forms a rounded and compact unit, the carotid body of the human being is divided into many lobules by the dense connective tissue that binds it together and also binds it to the adventitia of the common carotid artery. It is probably the varying development and position of these lobules that have led to the report that the carotid body tends to degenerate in adulthood; actually, there is no good evidence that it does so and no apparent reason why it should. The carotid body contains numerous nerves and nerve endings and was once regarded as a part of the chromaffin system of the body—portions of which do undergo degeneration after birth.

Following the finding that the carotid body has a sensory nerve supply, unlike the adrenal medulla, and the demonstration by Schmidt and Comroe that it is a specialized sensory organ, or vascular chemoreceptor, that responds to chemical changes in the blood (particularly hypoxia) by producing a pronounced increase in the rate and depth of respiration, other studies led to the belief that it contained no epinephrine or norepinephrine. The carotid body was then classified as a "nonchromaffin paraganglion." Later, studies of carotid body morphology with the electron microscope, such as that of Duncan and Yates, produced rather convincing evidence that the carotid body does contain catecholamines, although apparently in a different form from those found in the adrenal medulla. Clinical experience has confirmed this, for some carotid body tumors have been shown to secrete both epinephrine and norepinephrine. It is not known why the majority do not or even what percentage do (McGuirt and Harker). Lawson found that cells derived from carotid body tumors (and glomus jugulare tumors) give rise to dendritic processes in tissue culture and are probably derived, as are the cells of the adrenal medulla, from neural crest.

A tumor of the carotid body may be largely vascular or largely epithelial and is sometimes referred to as a glomus tumor. However, it is not similar to the usual glomus tumor (a tumor of an arteriovenous anastomosis); rather than causing pain when pressed upon, as does a true glomus tumor, a carotid body tumor that does not secrete catecholamines is symptomless except for those effects that might be produced by any tumor in that general location, and many have not been correctly diagnosed preoperatively. (Carotid angiography now allows more certain diagnosis.) Although only about 5% of the tumors metastasize (McGuirt and Harker), they may embarrass the circulation and at operation are frequently found to surround the common or internal carotid arteries. In such instances, complete removal by ligation and excision of the vessel concerned resulted in a rather high rate of morbidity and mortality. However, Wilson reported that most of them can be removed completely without permanently occluding the carotid circulation, by dissecting the adventitia and the tumor away from the media of the vessel, and some 10 years later McGuirt and Harker agreed that this is the preferable treatment. Garfia found, in serial sections of 120 carotid bifurcations and adjacent tissue, three instances in which small bodies of glomus tissue lay in fat adjacent to the carotid sinus.

Tissue similar to the carotid body occurs elsewhere and may also give rise to tumors; most common is the *glomus jugulare* tumor (see Chap. 3).

The carotid body has also been of interest to surgeons because of reports that its unilateral removal will often alleviate asthma, increasing markedly the low vital capacity produced by repeated asthmatic attacks. There has been no explanation of this supposed occurrence, and Segal and Dulfano found no objective changes in 15 carefully studied patients.

CAROTID SINUS

The carotid sinus is completely different in structure and function from the carotid body. It is a part of the carotid artery characterized by much elastic tissue and little muscle, and by numerous and complicated sensory nerve endings. Grossly, the carotid sinus is usually identified as a dilatation (also called carotid bulb) of the lower end of the internal carotid artery or the upper end of the common carotid, but the physiological carotid sinus does not necessarily correspond to this, since nerve endings may be situated above the gross sinus. In contrast to the carotid body, a chemoreceptor, the carotid sinus is a pressoreceptor: its nerve endings are stimulated by pressure—normally, of course, the pressure of the blood itself. Stimulation of the carotid sinus causes reflex slowing of the heart rate and reduction of general blood pressure, so that the carotid sinus represents the sensory organ of a reflex mechanism by which the blood pressure can be reduced if it begins to get too high.

Although the carotid sinuses are probably the best known of the pressoreceptor areas, they are not the only ones in the body; exactly similar areas occur on the right subclavian artery and the arch of the aorta and possibly also on blood vessels elsewhere. Denervation of both carotid sinuses and of the pressoreceptor areas at the base of the neck and in the thorax produces a prolonged and apparently permanent rise in blood pressure (*e.g.,* Heymans); however, denervation of one or both carotid sinuses has no such effect, because the fibers of pressoreceptor function in the cervical portions of the vagus nerves prevent a rise in blood pressure when the carotid sinuses alone are denervated.

The clinical, as distinguished from the physiological, importance of the carotid sinus is that it may become unduly sensitive to pressure, whereupon relatively slight pressure upon the sinus from without, as in turning the head, may produce an abnormal flow of impulses from it, with consequent slowing of the heart, reduction of blood pressure, and loss of consciousness.

The carotid sinus syndrome—that is, the syndrome of a hypersensitive carotid sinus—can be abolished by denervation of the sinus. The nerve to both the carotid body and the carotid sinus is the sinus nerve (ramus sinus carotici, intercarotid nerve, nerve of Hering or de Castro); it arises largely or entirely from the glossopharyngeal nerve after this has left the skull. The carotid sinus nerve descends in the connective tissue between the internal and external carotids or upon the internal carotid. Connections to this nerve from the vagus, from the superior cervical ganglion, and even from the hypoglossal nerve are sometimes seen in dissections; certainly neither the hypoglossal nerve nor the sympathetic system has anything in particular to do with the function of the carotid body, and it is doubtful that the tenth nerve does. Experiments upon animals have indicated that the ninth nerve is the sole nerve supply to the carotid body and the chief one to the sinus, and Ray and Stewart have apparently shown that in human beings the ninth nerve is at least the primary nerve, since they have found that the carotid sinus reflex is abolished by intracranial division of the ninth.

Although the sinus has been denervated by sectioning the carotid sinus nerve, by sectioning the ninth nerve intracranially, and by stripping the adventitia from the carotid bifurcation, the latter technique has apparently become the method of choice (Cheng and Norris; Cohen and co-workers). Pick and Wertheim have described a technique for local anesthetic block of the carotid sinus nerves.

Although carotid body tumors, because of their situation, might be thought to be a not unusual cause of the carotid sinus syndrome, this is not the case; McSwain and Spencer, in reporting two cases in which there was such an association, found only a few similar ones in the literature. Apparently the carotid sinus syndrome is due to definite but unknown physiological changes in this sinus.

Veins

The veins of the neck vary considerably in their connections with each other and in their relative sizes; those conducting blood downward from the head and face include the external jugular, the anterior jugular, the internal jugular, and the vertebral; at the base of the neck are the suprascapular (transverse scapular) and transverse cervical veins and the subclavian, which unites with the internal jugular to form the brachiocephalic or innominate vein. The subcutaneous veins, the external and anterior jugulars, are especially variable in size and course; the connections between the various veins, both superficial and deep, are also extremely variable.

SUPERFICIAL VEINS

External Jugular

The external jugular vein begins in the substance of the parotid gland, where it is most often (about 35%, according to Brown) formed by the union of the retromandibular (posterior facial) and the posterior auricular veins or, quite variably, by a single one or some combination of these and the facial, maxillary, or other veins in the parotid region; close to the parotid gland it usually receives a communication from the internal jugular vein. It runs vertically downward across the superficial surface of the sternocleidomastoid muscle (very rarely deep to it) to pierce the fascia of the posterior triangle of the neck at the posterior border of this muscle just above the clavicle (Fig. 9-17).

At about the middle of the neck the external jugular may receive a vein (posterior external jugular) from the back part of the neck, and in the posterior triangle it usually receives the transverse cervical, suprascapular and, frequently, the anterior jugular veins, although these may end directly in the subclavian. Occasionally it receives the upper end of the cephalic vein or a branch of this, which crosses the clavicle superficially to enter the posterior triangle. The external jugular vein usually terminates in the subclavian vein, but in about a third of instances ends in the internal jugular. It may be double or have a bifid termination.

Anterior Jugular

The anterior jugular vein usually begins in the suprahyoid region through the confluence of various and variable superficial veins here, or it may arise more laterally, from the retromandibular or the facial vein, or from parotid veins. It has variable communications with the internal jugular or the facial vein and descends on the infrahyoid muscles; the paired veins typically parallel each other, but when they arise in the parotid region they tend to descend along the anterior borders of the two sternocleidomastoid muscles and hence to converge. In the lower portion of the neck, above the sternum, the two veins are commonly united by a transversely disposed *jugular venous arch,* which is then one of the vessels encountered in a low tracheostomy. Below the jugular venous arch each anterior jugular curves laterally to empty into the terminal portion of the external jugular vein or into the subclavian vein between the external and internal jugulars. Sometimes, usually when one anterior jugular is missing from its more lateral location, an unpaired vein (median cervical vein) descends from the submental region in the midline of the neck. A number of the variations in the superficial veins of the

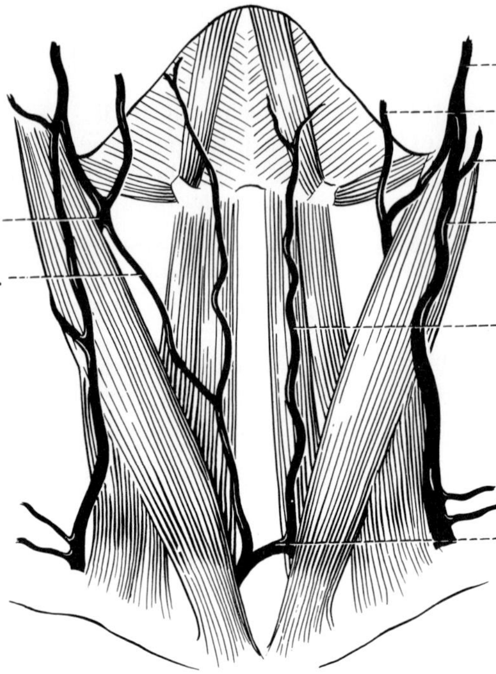

Fig. 9-17. Superficial veins of the neck.

neck have been described by Pikkieff. She found a median cervical vein in 6.2% of 96 subjects, absence of 1 anterior jugular in 13.5%, and an arcus venosus juguli in only about 58%.

DEEP VEINS

Internal Jugular

The internal jugular vein, the right usually being the larger, begins at the jugular foramen as the continuation of the sigmoid sinus; its first part is somewhat enlarged to form the superior bulb and projects into the floor of the tympanic cavity. Often close against the jugular bulb, and connected with the glossopharyngeal nerve, is the glomus jugulare, a minute bit of apparent chemoreceptor tissue from which, as already noted, tumors may arise.

In the uppermost part of its course the internal jugular lies posterior to the internal carotid artery, from which it is separated by the ninth, tenth, and eleventh cranial nerves as they emerge from the anterior portion of the jugular foramen; lower down it lies lateral to the internal carotid and also to the common carotid artery. The glossopharyngeal and hypoglossal nerves usually pass forward between the internal carotid and internal jugular, and the accessory nerve passes backward and downward across the internal jugular, usually superficial to this but sometimes deep to it. As the vein assumes a position lateral to the artery, the vagus nerve lies behind and somewhat between the two vessels, just as it does lower down in relation to this vein and the common carotid artery. During much of its course in the neck the internal jugular vein is intimately related to deep cervical lymph nodes, many of which are embedded in that portion of the carotid sheath surrounding the vein.

As its upper end the internal jugular vein receives the inferior petrosal sinus, which passes through the anterior end of the jugular foramen. The middle meningeal veins and the vein of the cochlear aqueduct also enter it here. At about the level of the hyoid bone it receives the facial vein (formerly common facial, formed by the union of "anterior facial" and a part of the retromandibular or "posterior facial"). At about this same level the internal jugular vein typically receives the superior thyroid vein (Fig. 9-18) and often receives pharyngeal and lingual veins (which may unite and empty together or may join the facial—the pattern here is so variable that the veins can be identified only by tracing them). The middle thyroid vein, often absent, typically joins the internal jugular at about the level of junction of the middle and lower thirds of the thyroid gland, passing in front of the common carotid artery to do so.

At the base of the neck the right internal jugular vein lies slightly lateral to the common carotid artery and crosses in front of the right subclavian artery to join the subclavian vein, while on the left the internal jugular usually somewhat overlaps the common carotid anteriorly. Just above its termination the internal jugular vein may show a second dilation, the inferior bulb; below the level of the subclavian artery on the right, and in front of the subclavian artery on the left, the internal jugular veins join the subclavians to form the two brachiocephalic (innominate) veins. Just above this level the vertebral vein enters the internal jugular on its posteromedial surface, or enters the first portion of the brachiocephalic vein, and the external jugular vein may also open into the lower portion of the internal jugular. The thoracic duct on the left, and the right lymphatic duct or its constituent vessels on the right, open approximately into the angle of union between the internal jugular and subclavian vessels.

Veins at the Base of the Neck

The *vertebral vein* begins in the suboccipital venous plexus, which is formed in the suboccipital triangle by numerous small vessels from the internal vertebral venous plexus, by small veins from the deeper muscles of the upper part of the neck, and by the occipital vein. The vertebral vein first descends as a plexus about the vertebral artery, in the canal

Fig. 9-18. Some of the deeper veins of the neck. These, like the superficial veins, vary in both size and connections. The inferior root of the ansa (formerly descendens cervicalis), shown here lying superficial to the internal jugular, may lie deep to it; thus the ansa cervicalis (hypoglossi) may or may not surround the vein.

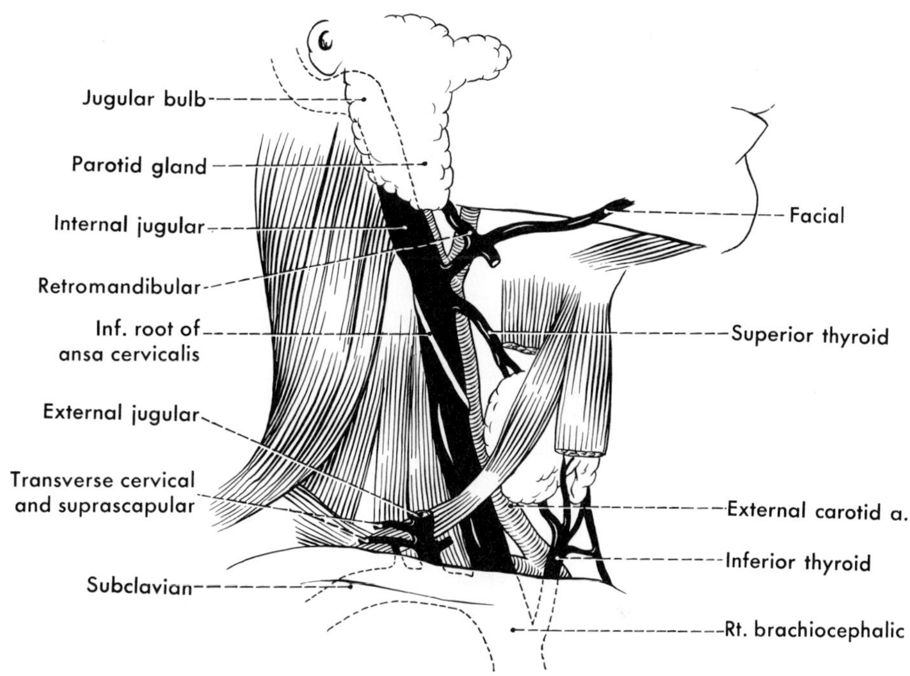

formed by the transverse foramina of the cervical vertebrae; at the lower end of the canal the plexus unites into a single vein that lies in front of the vertebral artery and opens into the posterior surface of the first part of the brachiocephalic (innominate) vein or the lower end of the internal jugular. Just before its termination it receives the *anterior vertebral vein,* which runs across the anterior surfaces of the transverse processes of the vertebrae with the ascending cervical artery.

The *deep cervical vein,* like the vertebral, begins in the suboccipital plexus and drains the posterior external vertebral venous plexus and the deep muscles of the neck. Running with the artery of the same name, it passes beneath the transverse process of the seventh cervical vertebra; it opens into the brachiocephalic close to the vertebral, or into the vertebral.

The thyroid veins are discussed later in connection with the thyroid gland.

The *subclavian vein,* the continuation of the axillary, begins at the outer border of the first rib. Close to its origin it may receive one or more tributaries corresponding to branches of the thoracoacromial artery. The subclavian usually receives the external jugular, may receive the suprascapular, transverse cervical, and other deep veins, and terminates by uniting with the internal jugular to form the brachiocephalic close to the sternoclavicular joint. Each subclavian vein lies anterior to the corresponding artery and at a slightly lower level. Anteriorly it is related to the clavicle; at the level of the anterior scalene muscle it passes in front of this muscle, by which it is then separated from the artery. Normally the phrenic nerve passes behind it, but sometimes the nerve, or an accessory phrenic nerve, crosses the vein anteriorly. The right lymphatic duct and the thoracic duct, or tributaries to these, may empty into the terminal portions of the subclavian veins.

The *suprascapular* (transverse scapular) and *transverse cervical* veins, and sometimes also a superficial cervical vein, accompany the arteries of the same name; like these, they tend to pass in front of or through the brachial plexus. They end variably, in the subclavian vein, the external jugular, or both.

Lymph Nodes and Lymphatics

The lymphatic system of the neck consists of numerous lymph nodes intimately connected with each other by lymphatic channels, and of the terminations of the thoracic and right lymphatic ducts.

The lymph nodes of the neck are usually divided for convenience of description into some four or five groups, although these groups are continuous with each other (Fig. 9-19). A convenient classification is that which describes the lymph nodes as being di-

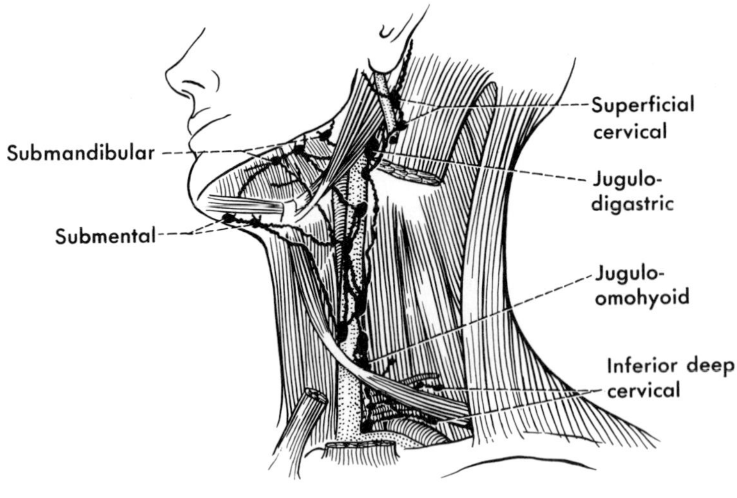

Fig. 9-19. The principal lymph nodes of the neck. Tracheal nodes are not shown here, but pretracheal ones of this group appear in Fig. 9-53.

vided into submandibular (submaxillary), submental, superficial cervical (or external deep cervical), and deep cervical (or internal deep cervical) groups; in addition, a tracheal (paratracheal and pretracheal) group is usually also described.

The submandibular and the submental (suprahyoid) nodes have already been described in Chapter 7, and it need only be pointed out here that they are intimately connected with the fascia over the anterior belly of the digastric and the mylohyoid muscles and with the submandibular gland itself. The *submental nodes* receive their afferent lymphatics from the skin of the chin, the middle portion of the lower lip, and the tip of the tongue, and from mucous membrane of the mouth; their efferents pass in part to the submandibular nodes and in part into the deep cervical chain directly. The *submandibular nodes* receive afferents not only from the submental nodes but from the lower portion of the nasal cavity, the upper lip, the gums, much of the anterior part of the tongue, and the lateral parts of the lower lip; their efferent vessels pass into the upper deep cervical nodes.

The *superficial cervical nodes* lie along the external jugular vein as it emerges from under cover of the parotid gland and therefore upon the external surface of the sternocleidomastoid muscle. Their afferents are said variously to consist of cutaneous lymphatics from the face, especially the auricular and parotid regions; of the efferents of retroauricular (mastoid) and parotid nodes; and sometimes also of occipital nodes. Their efferents drain around the anterior border of the sternocleidomastoid muscle into the deep cervical nodes at about the level of the hyoid bone.

The *tracheal* or anterior cervical nodes are an irregular and inconstant group placed in front of and alongside the larynx and trachea; some of them are along the anterior jugular vein, while others lie deeper on the cricothyroid ligament and the front of the trachea (see Fig. 9-53) or along the recurrent laryngeal nerves. They drain the lower part of the larynx and the upper part of the trachea and a medial part of the thyroid gland. Their efferents pass to lower deep cervical nodes.

DEEP CERVICAL NODES

The deep cervical lymph nodes are numerous and prominent, and many of them are large. They form a chain embedded in the connective tissue of the carotid sheath; the majority are in that portion of the sheath about the internal jugular vein. They extend from the base of the skull to the base of the neck.

The nodes of the deep cervical chain are usually described as constituting two groups, although actually there is no subdivision; the upper or superior deep cervical nodes are those lying above the level at which the omohyoid muscle crosses the carotid artery and jugular vein, and the inferior or supraclavicular deep cervical nodes are those lying below this level.

The *superior deep cervical nodes* are especially closely related to the internal jugular vein; they receive their afferents from both superficial and deep structures of the head and upper portion of the neck, some of these afferents reaching these nodes primarily, others being derived from buccal and mandibular (facial), parotid, retroauricular, occipital, submental, submandibular, and retropharyngeal nodes. The superior deep cervical nodes are largely under cover of the sternocleidomastoid muscle; the especially large nodes of this group are the jugulodigastric (subdigastric or tonsillar) node, situated on the internal jugular and internal carotid vessels where the posterior belly of the digastric crosses these, and the juguloomohyoid node, forming the lower end of the superior cervical group, at the point at which the omohyoid crosses the vessels.

The *inferior deep cervical lymph nodes* lie largely posterior and lateral to the internal jugular vein and extend into the subclavian triangle. The more anterior ones are adjacent to the internal jugular vein and therefore under cover of the posterior margin of the sternocleidomastoid, but the more posterolateral ones are more closely associated with the subclavian vessels and the brachial plexus and with the accessory nerve as it courses between the sternocleidomastoid and trapezius muscles. (The nerve has been severed in biopsy of nodes in the posterior triangle.)

The inferior deep cervical nodes receive afferents from the back of the scalp and of the neck, from many of the superior deep cervical nodes, from some of the axillary nodes, from some lower tracheal nodes that drain the lung, and even from some vessels directly from the skin of the arm and from the pectoral region. Altogether, therefore, the inferior cervical nodes receive a great deal of the lymphatic drainage from the entire head and neck, plus some drainage from the arm and the thorax. The efferent vessels of the inferior cervical nodes unite to form the jugular lymphatic trunk, which may be reinforced by efferents from the superior deep cervical nodes and on the left side typically joins the thoracic duct, while on the right it joins the right lymphatic duct or empties independently at the junction of internal jugular and subclavian veins. The lymphatic drainage of the head and neck has been summarized by Trotter and by Looney, among others.

Smith reported failure in attempting to visualize cervical lymph nodes by injection of radiopaque material into lymphatics of the hand, but said injection into the scalp had proved useful in locating the nodes before radical neck dissection.

Because of their wide connections, the deep cervical nodes may be involved in carcinoma originating anywhere in the head or neck; in addition, the supraclavicular or lower deep cervical nodes (also known as the nodes of Virchow or of Trosier or as the signal nodes—terms sometimes applied more specifically to a prescalene node) may be involved by carcinoma originating within the abdominal or thoracic cavities. In a study by Viacava and Pack, the lower deep cervical nodes were found to be involved (commonly by retrograde extension) in 2.8% of patients with tumors in the abdomen and thorax. The nodes on the left were more frequently involved, probably because of their relationship to the thoracic duct, while the nodes on the right were usually involved only when there were tumors in the thorax. In 41 patients, however, the supraclavicular metastatic growths were the first clinical signs of a malignant tumor and led to a search for the primary site of this tumor. Connar, in a later study, pointed out that some of the tracheal nodes in the thorax drain directly into lower deep cervical nodes, and he emphasized the value of biopsy of the latter nodes in diagnosing intrathoracic lesions.

BLOCK RESECTIONS

Since it is impossible to dissect out individually all the cervical lymph nodes, the technique of block resection of lymph nodes, or radical dissection of the neck, has been developed. This operation apparently offers the best hope of "5-year cures" of metastatic carcinoma, and the operation itself is associated with a low mortality rate.

Essentially, the technique of block dissection includes removal of the sternocleidomastoid muscle and the internal jugular vein, with all the connective tissue about them and all that overlying the muscles of the floor of both the posterior and the anterior triangles of the neck. Brown and McDowell and Beahrs and co-workers, among others, have described the operation. It may be started from above or below, as desired, but the fundamentals are that the sternocleidomastoid muscle and the internal jugular vein be sectioned both high in the neck and at the base of the neck and removed as far as possible in a block that includes the connective tissue harboring the lymph nodes (Figs. 9-20 to 9-23). As the block of tissue is lifted, the omohyoid muscle is included, the brachial plexus is cleaned of connective tissue, and so also are the common carotid artery and the vagus and phrenic nerves; these structures are of course left intact. The scalene and levator scapulae muscles are carefully cleaned, but the cutaneous branches of the cervical plexus are sacrificed and lifted with the block, leaving only the stumps of this plexus and its muscular branches. The portion of the accessory nerve passing through the posterior triangle to the trapezius is sometimes saved and sometimes sacrificed because of the lymph nodes along its course in the posterior triangle. Becker and

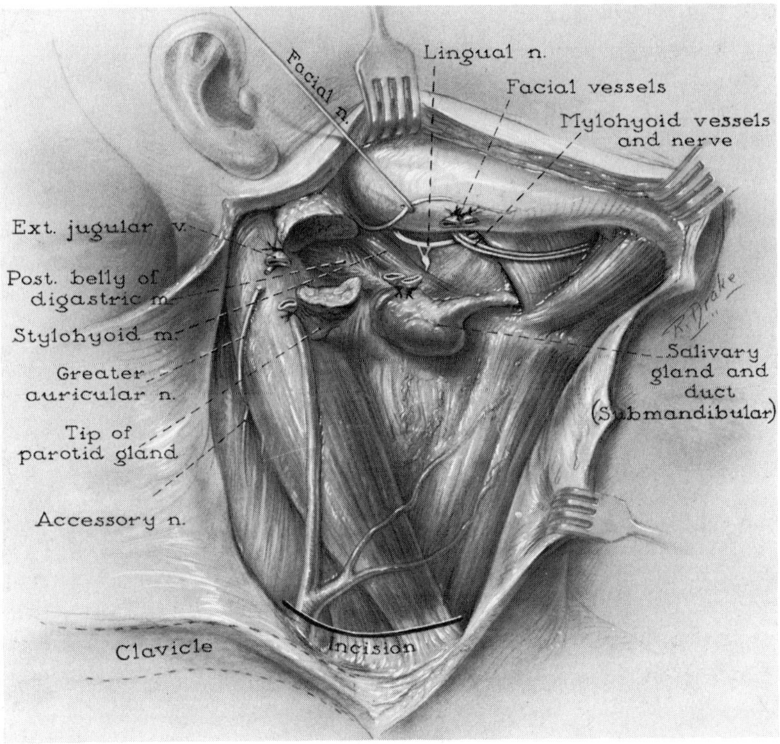

Fig. 9-20. An early stage in a radical neck dissection. The marginal mandibular division of the facial nerve is retracted upward; the external jugular vein and the facial vessels have been divided; the submandibular gland, mobilized; and the lower end of the parotid gland, amputated. (Beahrs OH, Gossel JD, Hollinshead WH: Am J Surg 90:490, 1955)

Parell preferred not to sacrifice the nerve even when it passed through the sternocleidomastoid muscle and have described their technique for preserving it.

The posterior belly of the digastric is detached from the hyoid bone and reflected, and so is the stylohyoid (this is omitted by some surgeons), in order to afford access to the upper nodes of the superior cervical group; in the submental region the deep cervical fascia over the mylohyoid muscle is removed with the anterior belly of the digastric, leaving the bare muscle of the mylohyoid, thus ensuring removal of the nodes here; the submandibular (submaxillary) gland and its associated nodes are removed after ligation of the facial artery and vein. Two nerves are especially watched for here, the hypoglossal nerve passing deeply across the submandibular triangle, and the marginal mandibular branch of the facial running approximately along the lower margin of the mandible. The lingual nerve also appears, for although normally well under cover of the mandible, it is dragged downward by traction on the submandibular gland.

The dissection is completed by removing the lower tip of the parotid gland with its associated lymph nodes, ligating the upper end of the internal jugular, and sectioning the posterior belly of the digastric, the stylohyoid, and the sternocleidomastoid muscles close to their origins. Obviously, this type of careful anatomical dissection down to the surfaces of the deeper muscles, removing all overlying muscles and connective tissue and sparing only the important vessels and motor nerves, offers the nearest obtainable approach to a complete removal of the lymph nodes of the neck.

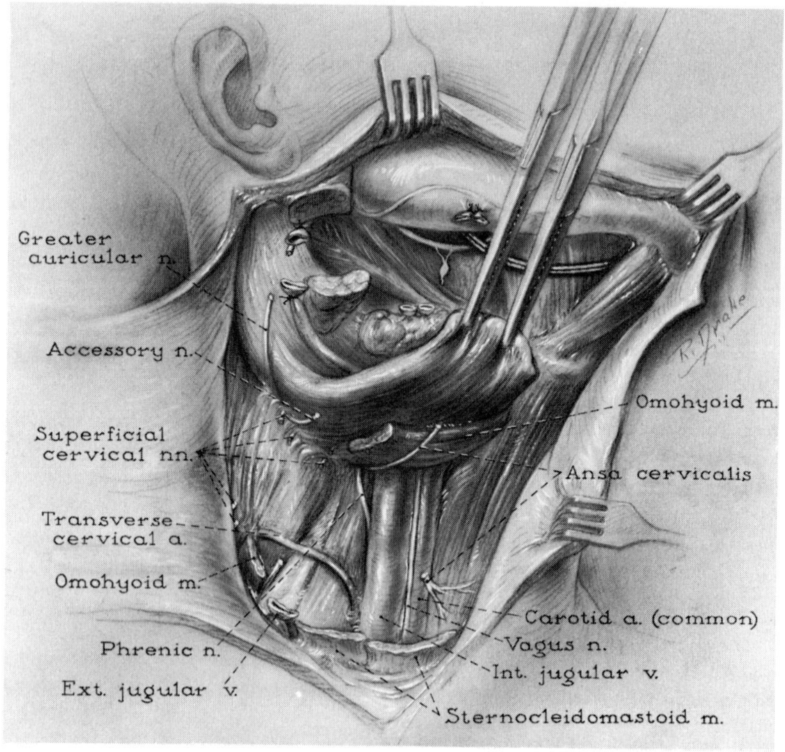

Fig. 9-21. Radical neck dissection, continued. The sternocleidomastoid and omohyoid muscles have been divided and retracted medially and upward, exposing the brachial plexus, scalene muscles, and large vessels of the neck. Usually the transverse cervical artery is also included in the specimen. (Beahrs OH, Gossel JD, Hollinshead WH: Am J Surg 90:490, 1955)

When nodes on both sides of the neck have been involved, the former practice was to do an incomplete neck dissection on the second side, sparing the internal jugular, or to do the second dissection after what was considered a fairly safe interval of time, often about 3 months, to allow the development of a collateral circulation.

Since the emissary veins, making their exits through foramina in the skull, cannot be expected to enlarge very much, and since in any case they empty mostly into tributaries of the external and internal jugular veins and both are removed in a radical neck dissection, the effective collateral circulation from the cranial venous sinuses after bilateral radical neck dissection is largely through the internal vertebral venous plexuses. Morfit was among the earlier workers to indicate that this pathway may be adequate, or nearly so, when both sides of the neck are dissected at one operation, although he advised that rises in the pressure of the cerebrospinal fluid be counteracted by withdrawal of fluid. Schweizer and Leak were less impressed with the changes in cerebrospinal fluid pressure. Since bilateral dissection does not allow development of the collateral circulation from the soft tissues of the nose and throat (through deep veins, such as the deep cervical, the pterygoid plexus, and the pharyngeal plexus), edema of these tissues must be expected. It apparently varies much among patients; Morfit regarded it as usually threatening death through suffocation, while Schweizer and Leak usually found it not to be a problem. Nevertheless, the latter workers agreed that a tracheostomy tube should be inserted before the second side is dissected. Moore and Frazell, reporting more recently on a larger series, described complications following the operation as being frequent, but agreed that it should be used when carcinoma is so advanced as to make staging impractical. Weingarten suggested that sparing the external jugular veins helps to relieve the condition.

Feldman and Applebaum expressed the opinion that a radical neck dissection should not include the submandibular triangle if the primary tumor would not normally drain into lymph nodes located here and if no palpable nodes are present; if either of these conditions is not met, they would remove the platysma and the marginal mandibular nerve in the dissection of the triangle. Skolnik and coworkers found no involvement of lymph nodes of the posterior triangle in 51 patients in whom they serially sectioned the contents of the triangle removed in a radical neck operation; they therefore suggested that the accessory nerve might be spared by not dissecting the posterior triangle, but since the nerve passes through the sternocleidomastoid in about 80% of persons, it could usually be preserved only by dissecting it out of the muscle. Saunders and Johnson suggested exercises to improve shoulder movements when the nerve has been sacrificed.

THORACIC AND RIGHT LYMPHATIC DUCTS

The large lymphatic channels that terminate at the base of the neck are the thoracic duct on the left (Fig. 9-24) and the right lymphatic

Fig. 9-22. Radical neck dissection, continued. The internal jugular vein has been divided and drawn upward, the accessory nerve sacrificed, the submandibular duct sectioned, and the digastric muscle and contiguous structures separated from the mandible and retracted laterally. (Beahrs OH, Gossel JD, Hollinshead WH: Am J Surg 90:490, 1955)

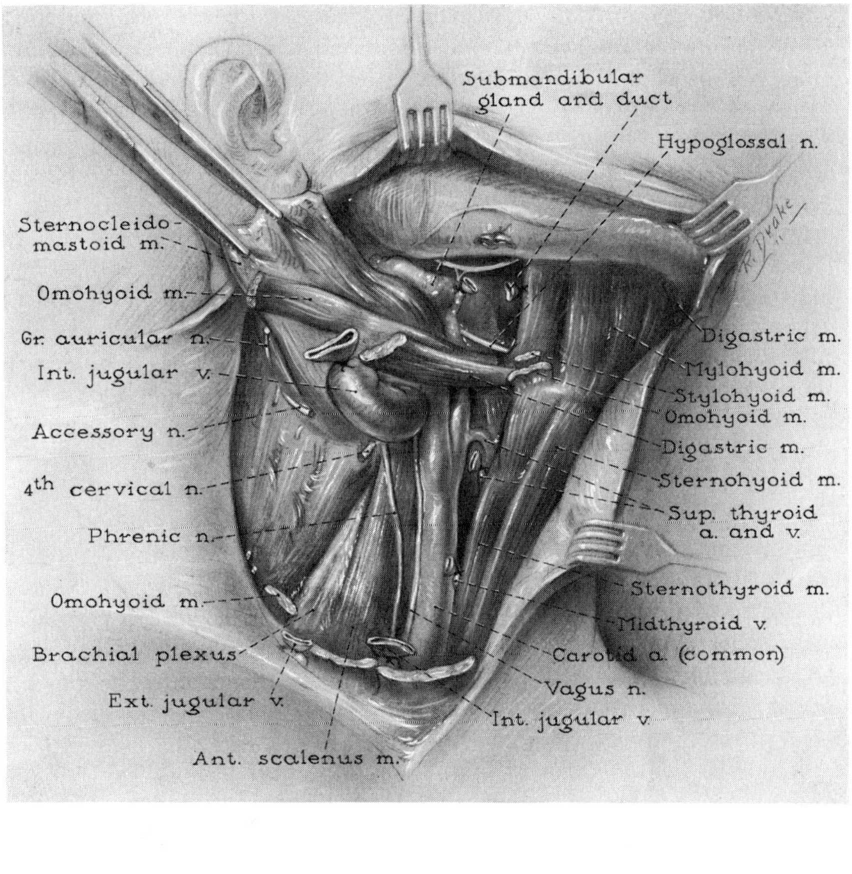

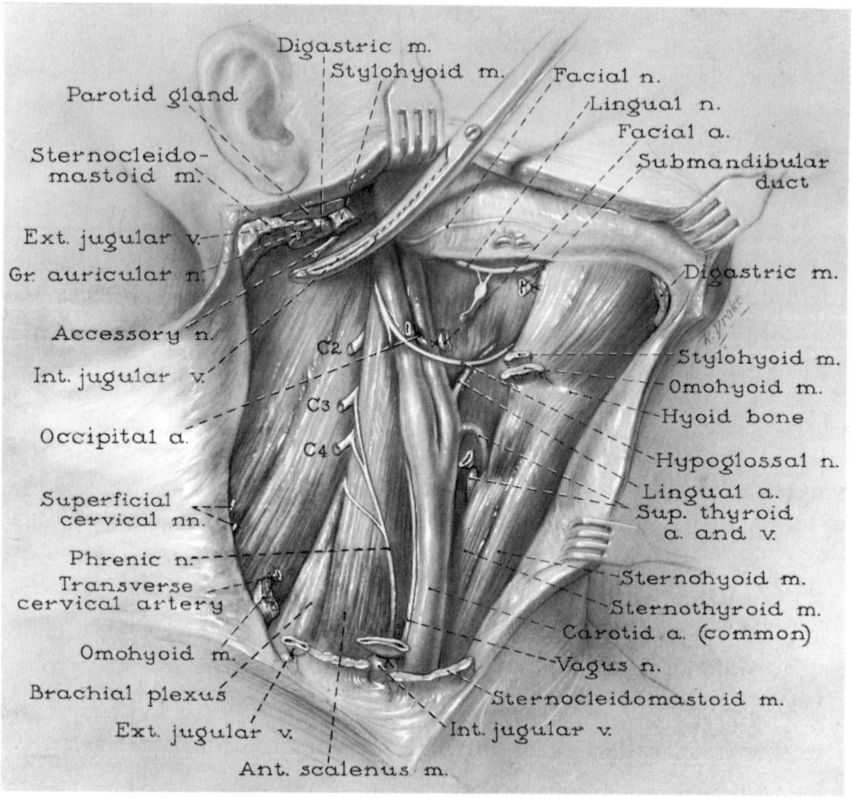

Fig. 9-23. Radical neck dissection, concluded. The internal jugular vein is divided as high as possible, and the sternocleidomastoid and other muscles are sectioned to allow removal of the tissue to be sacrificed. (Beahrs OH, Gossel JD, Hollinshead WH: Am J Surg 90:490, 1955)

duct on the right. In the upper part of the thorax the thoracic duct lies at first behind the first part of the left subclavian artery, but as it enters the base of the neck it lies somewhat to the right of this and behind the left common carotid artery. From this deep position, between the esophagus and the left pleura and behind the left common carotid and left vagus, it arches upward, forward and laterally; in so doing it emerges between the left common carotid and left subclavian arteries; as it turns laterally it runs in front of the vertebral artery and the thyrocervical trunk, and arching above the level of the subclavian artery passes between the internal jugular vein and the anterior scalene muscle to the (lateral) angle of junction between the left internal jugular and subclavian veins, crossing the subclavian artery superficially (Fig. 9-25). The duct may loop high into the neck and then downward or may not rise above its level of termination; Lissitzyn stated that the high-looping duct is typical of individuals with narrow thoracic apertures; the lower duct, of individuals with broad thoracic apertures.

At its termination it is typically joined by the left jugular and left subclavian trunks and may also receive the left bronchomediastinal (mediastinal) trunk; the bronchomediastinal trunk and, less often, a jugular or subclavian trunk may empty into a vein independently of the thoracic duct. Typically, however, the thoracic duct drains both lower limbs, most of the abdomen and its contents, and the left side of the head and neck and may drain the left side of the thorax.

The exact point of ending of the thoracic

duct varies, since it may open into the internal jugular, the subclavian, or the angle of junction between these; at its mouth it is provided with a pair of valves that apparently prevent the passage of venous blood into the duct during life, but in the cadaver the terminal portion of the duct may be filled with blood. Davis described the duct as opening into the subclavian vein in 59% of his cases, into the jugulosubclavian angle (angulus venosus) in 22.7%, into the left brachiocephalic, left internal jugular, and right internal jugular in 4.5% each, and, bifid, into the left internal jugular and left angle or into the left internal jugular and left vertebral in 4.5% each. The upper end of the duct may be doubled, tripled, or quadrupled; Davis found a single termination in only 77% of the thoracic ducts studied.

At least sometimes the thoracic duct communicates with the right lymphatic duct; Simpson and Graham apparently believed that a potential anastomosis regularly exists, although often so small that it functions in only one of five patients. They stated, however, that an anastomosis here becomes functional if the thoracic duct itself is closed in the neck, and more recently Larson and co-workers have demonstrated such drainage in two patients by lymphangiography.

Fig. 9-24. Lymphangiogram of the upper end of the thoracic duct as it deviates to the left and enters the neck. (Courtesy Dr. W. Eugene Miller)

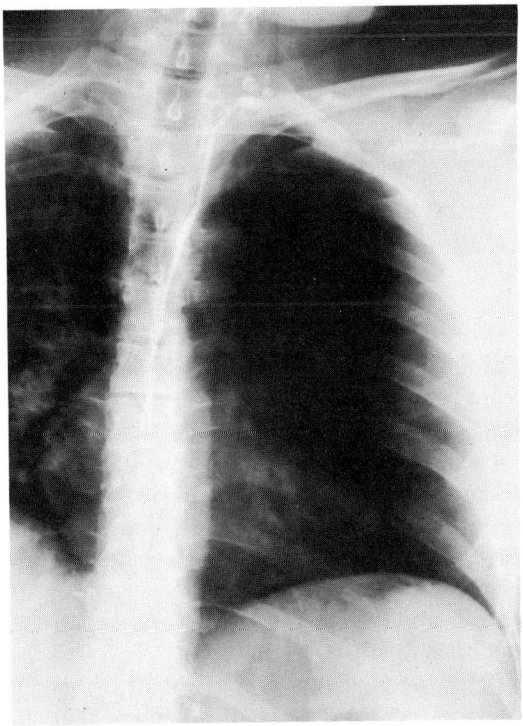

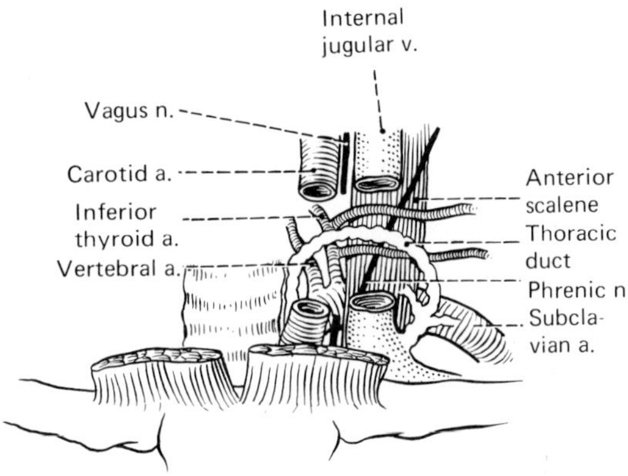

Fig. 9-25. Relations of the upper end of the thoracic duct. A segment of the contents of the carotid sheath has been cut away.

The terminal portion of the thoracic duct may be injured in accident or in operations at the base of the neck. Injuries to it in the neck can, if recognized at the time of operation, frequently be successfully treated by temporary pressure. Ligation can be carried out safely, however, and should be done if a postoperatively discovered leakage does not disappear spontaneously. Smith used lymphangiography, injecting the radiopaque material into the foot, to locate a leaking thoracic duct. Presumably, the lymph is fairly quickly rerouted after ligation, either around the ligature or to the other side.

RIGHT LYMPHATIC DUCT

The right lymphatic duct is usually defined as being formed by the union of the right jugular and right subclavian trunks, which may be joined by the right bronchomediastinal trunk; therefore, in its fullest development (when it is sometimes then called a right thoracic duct) it receives lymph from the right side of the head and neck, from the right

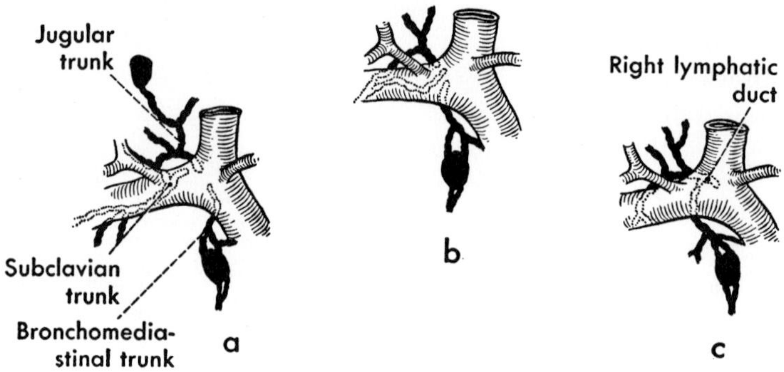

Fig. 9-26. Various manners of ending of the lymphatic ducts on the right side of the neck. In (*a*) all three trunks are shown ending separately; in (*b*) a right lymphatic duct is formed by the union of jugular and subclavian trunks, but the latter trunk branches and ends also independently; in (*c*) all three major ducts unite to form a right lymphatic duct. (After Poirier P, Charpy A: Traité d'Anatomie humaine, Tome 2, Fasc 4, Paris, Masson, 1909)

upper extremity, and from the right side of the thorax including the right lung (and part of the left lung), the right side of the heart, and a portion of the convex surface of the liver. The three channels that go to make up the right lymphatic duct usually empty separately or in various combinations into the angle of union between the subclavian and internal jugular veins or into one of these veins (Fig. 9-26); a single vessel draining the right side is apparently rare. Like the thoracic duct, the right lymphatic duct is provided with two semilunar valves at its opening into the venous system.

RIGHT THORACIC DUCT

It has already been mentioned that there may be a communicating channel between the thoracic duct and the right lymphatic duct; in the thorax, where the thoracic duct was once bilateral, parts of both ducts may remain and parallel each other. In cases of retroesophageal right subclavian artery or of a persistent right aortic arch with a left descending aorta, the development of the upper end of the (left) thoracic duct is apparently frequently interfered with, since in these individuals the thoracic duct commonly deviates to the right instead of the left in the posterior mediastinum and empties into the right side of the venous system instead of into the left as it does normally. This occurred in the one instance of retroesophageal right subclavian seen by the author and was thought by Szawlowski to be always true. It is apparently not, for Hammer and Meis reported that in two of their three cases of anomalous right subclavian artery, the thoracic duct emptied on the left side.

Nerves

CERVICAL PLEXUS

The cervical plexus is formed by the union of the ventral (anterior) rami of the second, third, and fourth cervical nerves, with sometimes also a contribution from the first. (Of the dorsal or posterior rami of these nerves, to skin and musculature of the dorsum, the one from the second cervical has an especially large cutaneous component, which chiefly forms the greater occipital nerve.) Close to their origins the ventral rami of these nerves receive rami communicantes from the cervical sympathetic trunk and give off muscular branches to the adjacent muscles—the longus capitis and colli, the levator scapulae, and the scalenus medius.

The first, or more frequently the second, cervical nerves also gives off fibers that join the hypoglossal and run for some distance with this, part of them leaving it as the superior root (descendens hypoglossi) of the ansa cervicalis (ansa hypoglossi) and the remainder running farther forward to supply the thyrohyoid and geniohyoid muscles (Fig. 9-27). Similarly, fibers from the second or second and third cervical nerves form the inferior root of the ansa (formerly descendens cervicalis), which in turn unites with the superior root to supply the infrahyoid muscles. These branches, with the contribution to the phrenic (from the third and fourth cervical) and fibers sent either directly or by way of the accessory nerve to the sternocleidomastoid and trapezius muscles (second, third, and fourth cervical), constitute the so-called deep branches (except for the latter, largely motor) of the plexus.

The superficial or cutaneous branches of the plexus are derived from a series of simple loops between the second, third, and fourth cervicals; they are the lesser occipital, great auricular, transverse cervical, and supraclavicular nerves. The *lesser occipital nerve* is derived from the second or the second and third cervicals, passes upward around the posterior border of the sternocleidomastoid muscle, and follows approximately this border of the muscle upward to the skin over the side of the head behind the ear.

The *great auricular nerve* is typically derived with the transverse cervical nerve from the loop between the second and third cervical nerves; it also rounds the posterior border of the sternocleidomastoid and passes almost directly upward toward the ear, where it supplies skin over the parotid gland, almost all the auricle, and skin over thc mastoid process.

The *transverse colli* or *transverse cervical nerve,* also called cutaneous colli, anterior cutaneous nerve of the neck, and superficial cervical nerve, typically arises by a common stem with the great auricular, from C2 and C3, and turns sharply around the posterior border of the sternocleidomastoid near its middle, running at first almost transversely but subsequently dividing into branches that ascend and descend to supply most of the skin of the anterior portion of the neck. Upper branches

Fig. 9-27. Schema of the cervical plexus. The short, unlabeled branches arising from the nerves close to their origins are muscular branches to the longus muscles, the scalenes, and the levator scapulae.

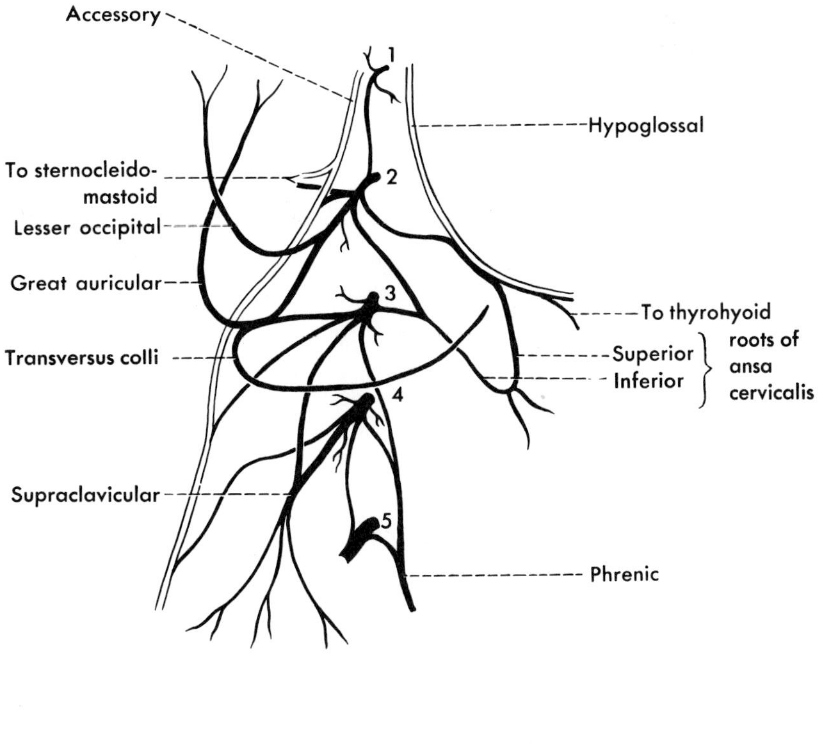

of this nerve anastomose with the cervical branch of the facial.

The *supraclavicular nerves* arise from a loop between the third and fourth cervical nerves and divide typically into three sets, medial (anterior), intermediate (middle), and lateral or posterior; the medial supraclavicular nerve supplies skin over the anterior portion of the clavicle and the adjacent portions of the neck and thorax, and also filaments to the sternoclavicular joint; the intermediate supraclavicular nerve supplies especially skin over the anterior portion of the deltoid and the clavicular head of the pectoralis major, while the lateral supraclavicular crosses the outer surface of the trapezius and the acromion and supplies skin on the upper and posterior parts of the shoulder. It is the superficial branches of the plexus that are routinely sacrificed in block dissections of the neck.

The branches that the cervical plexus sends to the sternocleidomastoid and trapezius muscles have been a source of much confusion, both as to their component fibers and their distribution in the muscles as compared with the accessory. It used to be maintained that the innervation of the sternocleidomastoid and trapezius muscles by both accessory and cervical nerve fibers indicated an origin of these muscles from both branchial arches and myotomes. Corbin and Harrison have shown that in the monkey these connections from the upper cervical nerves to the accessory nerve or directly to the sternocleidomastoid and trapezius muscles contain sensory fibers only, and one would expect this to be true also of human beings. However, McKenzie stated that the developing sternocleidomastoid muscle of human beings, and probably the trapezius, receive contributions from myotomes, while the sternocleidomastoid of the monkey does not; he therefore maintained that there should be motor fibers in the cervical nerve supply to these muscles in man but not in the monkey.

Fig. 9-28. The right phrenic nerve and some of its relations. On the left side these include the thoracic duct (Fig. 9-25).

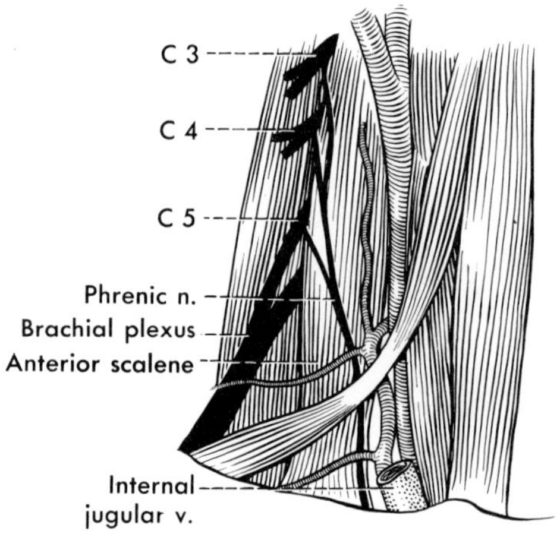

PHRENIC NERVE

The phrenic nerve usually arises from the third, fourth, and fifth cervical nerves, so that it is derived from both cervical and brachial plexuses (Fig. 9-28); it passes downward with a slight medial inclination on the front of the anterior scalene muscle, between this muscle and its covering (prevertebral) fascia, and is commonly approached surgically in this position. The vagus nerve lies within the carotid sheath anterior to the phrenic nerve, and the cervical sympathetic trunk lies somewhat more medially, posterior to the carotid sheath but not upon the anterior scalene muscle; both these nerves have at operation been mistaken for the phrenic nerve. The important point in the identification of the phrenic nerve is that it lies behind the fascia and directly on the anterior surface of the muscle fibers of the anterior scalene muscle.

The phrenic nerve, or a part of it, may pass in front of the subclavian vein as it leaves the neck. Hughes reported that this occurred in 10 of 138 sides, and that in 8 of those the entire nerve passed anterior to the vein.

The roots of the phrenic nerve vary much in length, and one of them, or a part of it, may join the rest of the nerve low in the neck or in the thorax. At one time, crushing or section of the phrenic nerve as it lies on the anterior scalene, in order to paralyze one side of the diaphragm, was a fairly common procedure, and in such an operation a long root of the phrenic that does not closely parallel the

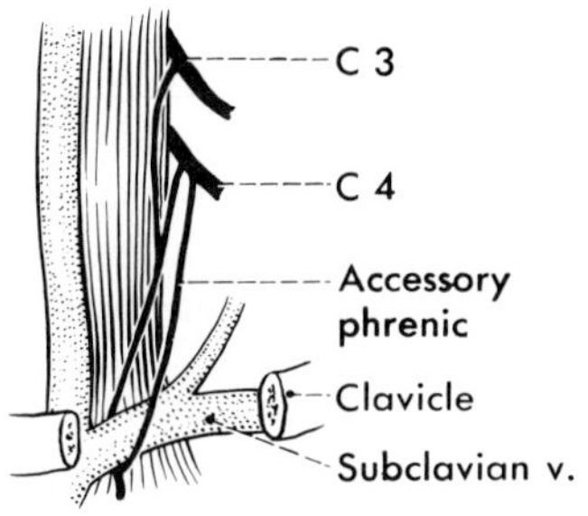

a

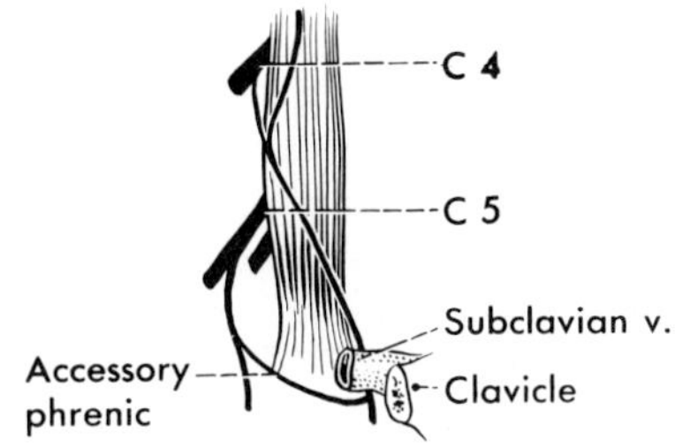

b

Fig. 9-29. Two types of accessory phrenic nerve. In (*a*) a long accessory from C4 loops in front of the subclavian vein to join the phrenic proper; in (*b*) the accessory phrenic arises from the fifth cervical nerve in common with the nerve to the subclavius and, after leaving this, runs almost horizontally to join the phrenic. (After Decker HR: J Thoracic Surg 2:538, 1933)

rest of the nerve can be overlooked, so that only a part of the nerve is interrupted. In consequence, it became customary to refer to such roots as *accessory phrenic nerves.*

A phrenic nerve root that arises with the nerve to the subclavius rather than directly from C5, one that passes in front of the subclavian vein while the main phrenic passes behind (Fig. 9-29), or one that arises from the ansa cervicalis (Turner), thus indirectly from C3, is easily categorized as an accessory phrenic nerve, but otherwise there is no unanimity on what constitutes this entity, and therefore the reported incidence varies much. Thus Decker found accessory phrenic nerves in 30% of 72 dissected sides, while Kelley, reviewing 309 operations on the phrenic nerve, reported that there was a second phrenic at the level of operation in 99.2% of patients, and if one excluded the instances in which the two nerves paralleled each other on the anterior scalene, an accessory phrenic was present in 75.7%. Most nerves that are called accessory phrenics lie lateral to the main phrenic, and although some arise from the third or fourth cervical nerve, the majority arise from the fifth cervical, either directly or with the nerve to the subclavius.

BRACHIAL PLEXUS

The brachial plexus is formed primarily by the union of ventral (anterior) rami of the lower four cervical and the first thoracic nerves. At its origin the plexus is related anteriorly to the anterior scalene muscle and posteriorly to the middle scalene; Reed ('48) stated that the fourth to sixth cervical nerves typically first pass through the anterior scalene muscle, but the plexus as a whole makes its appearance between the anterior and middle scalenes. The ascending cervical branch of the inferior thyroid artery ascends across the anterior surface of the plexus, and the suprascapular and transverse cervical arteries cross it, either anteriorly or among its branches, as they run laterally and posteriorly. The distal portion of the plexus is in close relation to the subclavian and axillary arteries.

The ventral rami of the nerves forming the brachial plexus lie almost completely behind the anterior scalene muscle, and the union of these rami to form trunks occurs close to the lateral edge of this muscle. Passing in front of the brachial plexus and onto the anterior surface of the anterior scalene muscle is the phrenic nerve, typically receiving a branch from the fifth cervical. The variations in this have already been discussed.

TRUNKS, DIVISIONS, AND CORDS

In the formation of the brachial plexus, the ventral rami of the fifth and sixth cervical nerves unite at about the lateral edge of the anterior scalene to form an upper trunk, the seventh nerve emerges by itself as a middle trunk, and the eighth cervical and first thoracic nerves, usually uniting behind the anterior scalene muscle, form a lower trunk (Fig. 9-30). Each of these trunks in turn forms two divisions, an anterior and a posterior one; the splitting of the superior and middle trunks into anterior and posterior divisions usually occurs only a little lateral to the anterior scalene, while that of the lower trunk occurs farther distally, usually either behind the clavicle or in the axilla. Also, whereas the superior and middle trunks split into divisions of approximately equal size, the divisions of the lower trunk are markedly unequal; the anterior division of the inferior trunk is always much larger, while the posterior division of the trunk, contributed to the radial nerve, is usually very small. It therefore follows that the radial nerve contains relatively few fibers of eighth cervical and first thoracic origin, and apparently few of these are from the first thoracic.

The divisions of the upper, middle, and lower trunks typically unite in a standard pattern to form the cords of the brachial plexus. The anterior divisions of upper and middle trunks unite to form the lateral cord, formed at about the level of the clavicle; as this passes between the clavicle and first rib it becomes associated with the anterolateral aspect of the axillary artery (Figs. 9-31 and 9-32). The anterior division of the lower trunk continues alone to form the medial cord of the plexus, lying at first somewhat behind the subclavian artery as they cross the first rib together, but subsequently passing slightly forward to lie medial to the axillary artery. The posterior divisions of all three trunks unite to form the posterior cord of the brachial plexus, which then lies behind the axillary artery.

The level at which the cords are formed varies somewhat; the divisions of the upper and middle trunks, and the unions of their component divisions to form the lateral cord and the greater part of the posterior cord, typically take place in the lower part of the neck, or the union of their posterior divisions occur in the upper part of the axilla. The inferior trunk usually divides relatively late, as or after it crosses the first rib (lying here typically behind the subclavian artery, between it and the first rib); at this point of division, of course, the medial cord is formed, while the

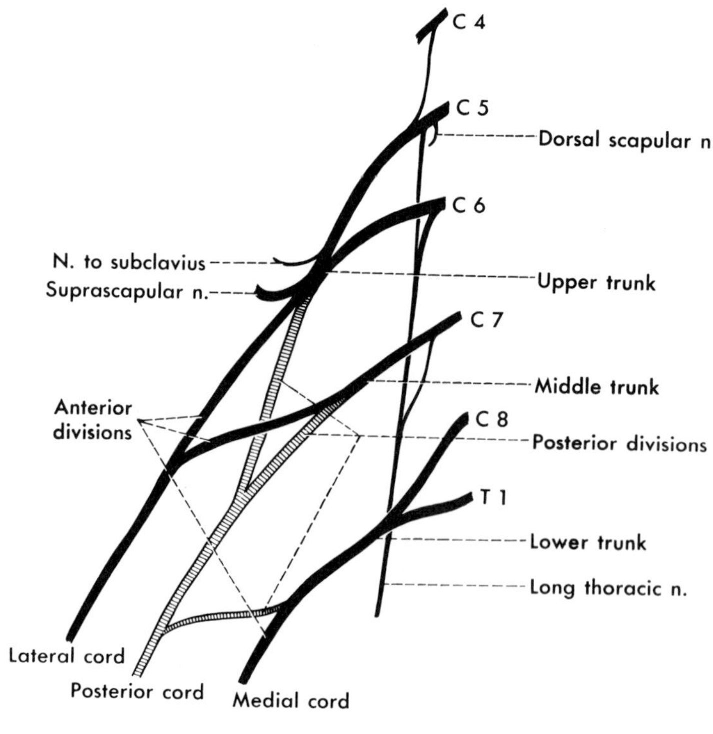

Fig. 9-30. Schema of the formation of the brachial plexus, and its branches in the neck. The twigs to the longus and scalene muscles are not shown.

Fig. 9-31. The relation of the brachial plexus to the subclavian and axillary arteries. The suprascapular and transverse cervical arteries are also closely related to the plexus at the base of the neck, passing anterior to or through the plexus. The posterior cord is exaggeratedly prolonged in order to show its position.

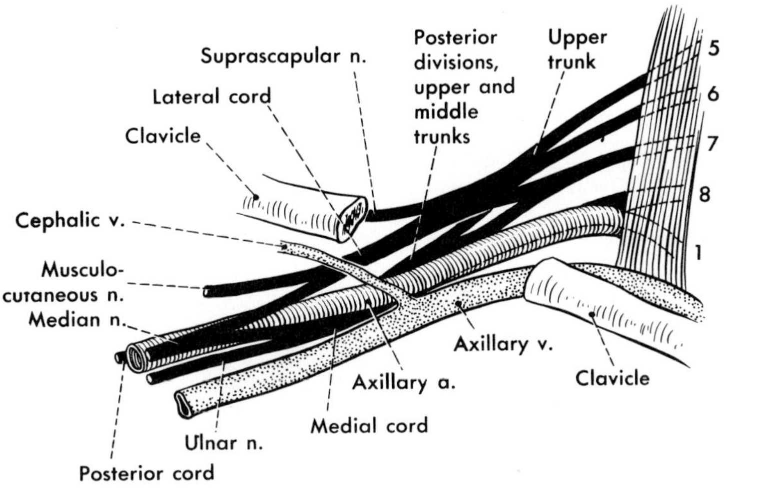

definitive posterior cord is not completely formed until the posterior division of the lower trunk joins the already united posterior divisions of the middle and upper trunks. The posterior cord is usually formed therefore in the axilla.

The nerves given off from the three cords typically arise in the axilla and are described in connection with the upper limb (see Volume 3). Not infrequently, however, the lateral pectoral (lateral anterior thoracic) nerve to the pectoralis major, usually the first branch from the lateral cord, arises in the lower part of the neck, sometimes even from the anterior divisions of the upper and middle trunks before these have united.

BRANCHES IN THE NECK

Nerve to the Subclavius

Although most of the branches of the brachial plexus arise, as noted, in the axilla, several have higher origins and leave the brachial plexus in the neck. Thus, in addition to its contribution to the phrenic nerve, the fifth nerve or the upper trunk gives rise to the small nerve to the subclavius, which passes downward in front of the lower portion of the brachial plexus and in front of the subclavian artery and vein to reach this muscle. It is important only because it may contribute to the origin of the phrenic nerve.

Suprascapular Nerve

In addition, the upper trunk before its division gives off the larger and important suprascapular nerve (to the supraspinatus and infraspinatus muscles) that at first parallels the upper trunk of the brachial plexus, running lateral to it, but subsequently swings laterally at about the level of the clavicle to parallel the suprascapular (transverse scapular) vessels and pass backward toward the scapular notch.

Dorsal Scapular Nerve

Of the other supraclavicular branches of the brachial plexus, several are twigs to the longus colli and to the scalenes and arise from the lower four cervical nerves as these leave the intervertebral foramina. They are thus completely hidden by the scalene muscles. The dorsal scapular nerve (to the rhomboidci) also has an origin close to an intervertebral foramen, arising from the fifth cervical nerve under cover of the anterior scalene, turning backward through the scalenus medius, and running downward and backward in the neck on or through the levator scapulae muscle.

Long Thoracic Nerve

The long thoracic nerve (to the serratus anterior muscle) is the only one of the deeper branches of the brachial plexus to be obvious in dissections in the anterior region of the

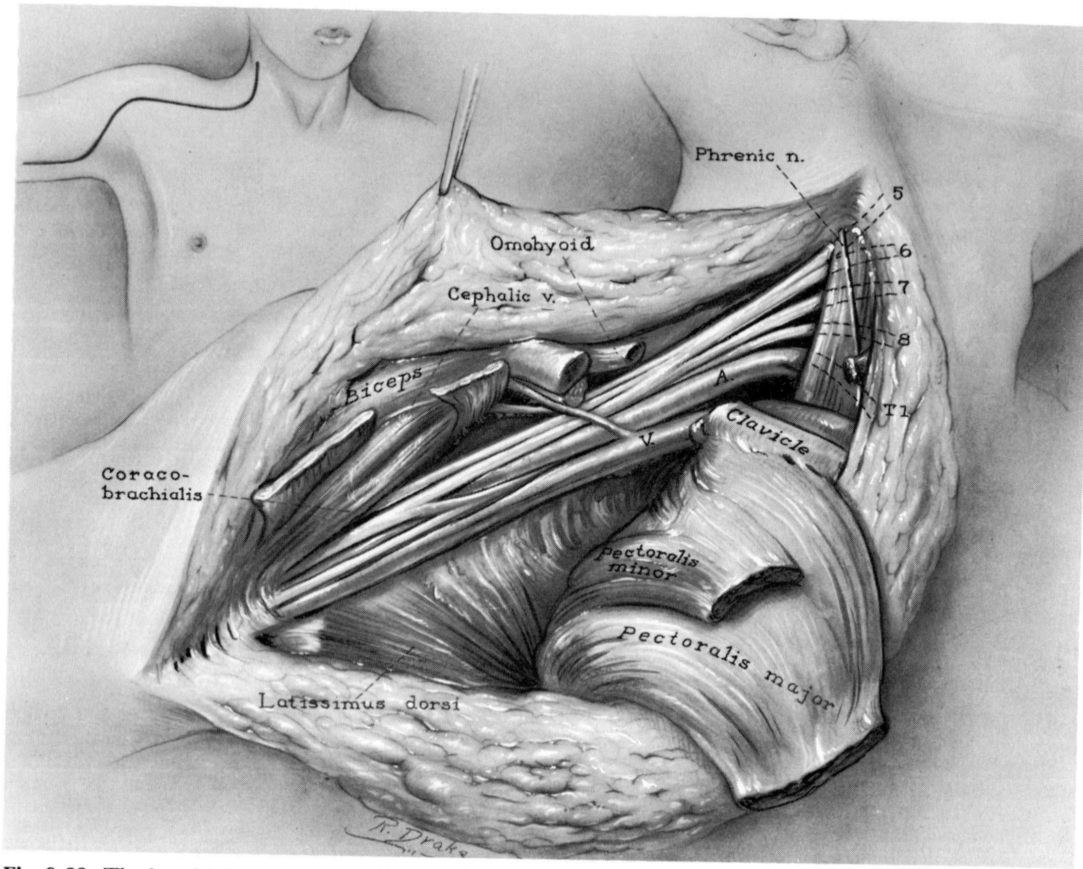

Fig. 9-32. The brachial plexus exposed by excision of a part of the clavicle. (Craig W McK, MacCarty CS: In Walters W: Lewis' Practice of Surgery, Vol 3, Chap. 7. Hagerstown, Prior, 1955)

neck. It is usually said to arise from the fifth, sixth, and seventh cervical nerves, close to the exits of these nerves from the intervertebral foramina. The long thoracic's origin from the sixth cervical nerve seems to be constant and relatively large; those from the fifth and the seventh are more variable, the root from the seventh being least frequently found. The roots from the fifth and sixth cervicals pass through, that from the seventh nerve usually behind, the middle scalene muscle, to emerge upon the anterolateral surface of the muscle (where they unite if they have not done so previously) and descend on this surface behind the brachial plexus and onto the superficial surface of the serratus anterior muscle.

It has been stated that the long thoracic nerve has sometimes been crushed mistakenly for the phrenic, since its relation to the middle scalene is somewhat like that of the phrenic to the anterior scalene; however, the phrenic nerve lies far medial to the long thoracic and, moreover, the long thoracic nerve passes downward behind the brachial plexus, while the phrenic is of course in front of the plexus, separated from this by the anterior scalene muscle. There would seem to be relatively little excuse for confusing these two nerves.

VARIATIONS

The brachial plexus is rather constant in the gross anatomy of its formation, although varying somewhat in the size of the contribution that it receives from the fourth cervical and the first thoracic nerves, and also in the fact that it may sometimes receive fibers from the second thoracic nerve. Occasionally, anomalies such as the formation of a single anterior cord (instead of a lateral and medial cord) or the union of all elements of the plexus into a single cord such as described by Singer may occur.

The more common variations in the brachial plexus have been described by Harris and Kerr. Kerr, in a study of 175 plexuses, divided them into 3 groups: (1) those receiving fibers from C4, (2) those receiving no fibers from C4 but all the ventral ramus of C5, and (3) those receiving only a part of the ventral ramus of C5. In his series, more than 62% of the brachial plexuses belonged to group 1, 30% to group 2, and 7% to group 3.

Normally, therefore, there is little shift in origin of the brachial plexus, although plexuses are sometimes classified, according to the segments that contribute to them, as prefixed, normal, or postfixed. Although these variations affect somewhat the segmental innervation of the muscles to which the nerves of the brachial plexus are distributed, the range is, as will be noted, no more than a segment. There may be a small communication from the second thoracic nerve to the brachial plexus; Todd described this communication as consisting largely of small myelinated and unmyelinated fibers and expressed the belief that these were sympathetic in nature, although some cutaneous sensory or voluntary motor fibers probably run through this connection also, on occasion.

OTHER RELATIONS

Kolodny pointed out that although birth injuries of the brachial plexus are usually either of the upper (Erb-Duchenne) or lower (Déjerine or Klumpke) type, since the nerves entering into the plexus are angulated sharply over the upwardly directed transverse processes, this is not usually true in the adult; here the transverse processes run downward, and traction usually traumatizes the tissue and blood vessels about the nerve trunks, so that the plexus as a whole becomes embedded in scar tissue.

Since the brachial plexus passes in the relatively narrow space between the clavicle and the first rib and in the upper part of the axilla is related to the coracoid process, the tendon of insertion of the pectoralis minor on this process, and the head of the humerus, it is apparent that extreme postures of the upper limb might bring the brachial plexus into contact with some of these structures, pressing it against them or stretching it across them. Clausen pointed out that the danger of injury through such positions is increased when the patient is under general anesthesia, where the muscular protection is largely absent, and discussed some of the positions to be avoided if such damage is not to occur.

Brachial plexus anesthesia has apparently been most commonly carried out by injection through the greater supraclavicular fossa (omoclavicular trigone), the first rib serving as a landmark and the plexus being injected as it crosses this rib. At the level at which the plexus crosses the first rib, most of it is lateral to and above the subclavian artery, and the site of the injection can be therefore somewhat guided by placing a finger upon the artery or starting laterally on the first rib and proceeding medially. The lateral and posterior cords or the divisions entering into them tend to lie somewhat anterior and posterior to the artery, respectively, but on its lateral aspect; the medial cord (or more commonly the lower trunk, which consists largely of fibers that form the medial cord) lies more directly behind it, between the artery and the middle scalene muscle, or between the artery and the rib.

The subclavian artery, the subclavian vein, and the dome of the pleura are all liable to injury in a brachial plexus block; according to some authors, neither puncture of the pleura nor of the subclavian artery carries any particular danger, but others (*e.g.*, de Pablo and Diez-Mallo) warned that injection into the vein may be fatal and thus must be avoided. Moore and co-workers listed pneumothorax, stellate ganglion block, and phrenic nerve block as complications of supraclavicular block. They regarded the axillary approach as being easier to learn, but said that it often cannot substitute for the supraclavicular one and that neither approach is free from danger.

One of the difficulties in attaining successful brachial plexus blocks is undoubtedly the connective tissue surrounding the brachial plexus, carried downward around it from the prevertebral fascia, which presumably somewhat protects the nerves from the action of the anesthetic unless the latter is delivered within this sheath. The position of the lower trunk behind the subclavian artery also makes it difficult to obtain good anesthesia in the area of distribution of the ulnar nerve.

NEUROVASCULAR COMPRESSION

Syndromes of pain, paresthesia, or numbness and evidence of disturbed circulation to the limb, apparently resulting from compression of the brachial plexus and the subclavian–axillary vessels, have been reported under a large number of names. Perhaps the most inclusive terms are "thoracic outlet syndrome" (also called cervicothoracic outlet syndrome and thoracic inlet syndrome) and "shoulder girdle syndrome." Other terms, implying that the anatomy of the neurovascular compression is known, are cervical rib syndrome, scalenus anticus syndrome, costoclavicular syndrome, and hyperabduction or subcoracoid–pectoralis minor syndrome; these syndromes are all essentially similar, and the terms are not mutually exclusive, since the compression may be caused by more than the structure named—for instance, by compression between the anterior scalene and a cervical rib.

Fig. 9-33. Two types of cervical ribs—a short one ending freely and a longer one articulating with the first thoracic rib. Observe in the latter case the very limited space that may exist between the rib and the anterior scalene muscle for the passage of the subclavian artery and brachial plexus into the axilla.

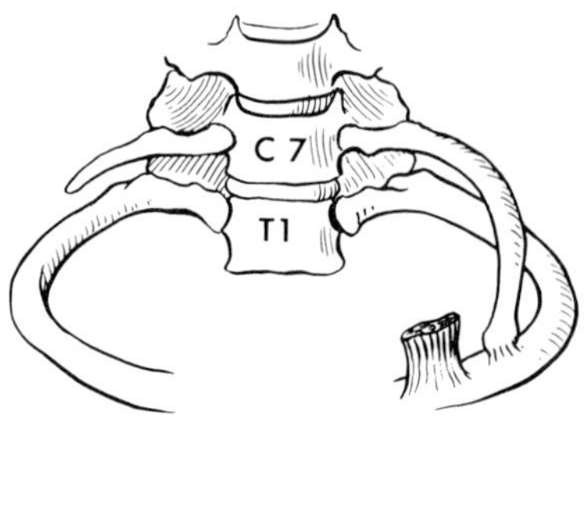

Cervical Rib

The incidence of cervical rib is said to be between 0.5% and 1% of individuals (Hill), or 6 patients per 1000 (Adson, '47), but symptoms are said to occur in only about 10% of such individuals (Hill). Cervical ribs have been said to be bilateral in from about 50% to 80% of cases and, when unilateral, to be more common on the left (Hill); Adson ('51) quoted their occurrence as being twice as frequent in females as males, bilateral in 50%, and, if unilateral, as common on one side as the other.

Cervical ribs vary considerably in length (Fig. 9-33). They may end freely or articulate with the first rib through a cartilage, but apparently the attachment to the first thoracic rib is more frequently by a fibrous band. This band may in turn act in the same manner as a rib, by affording a firm structure against which the brachial plexus can be compressed.

Cervical ribs and their fibrous bands may interfere with the brachial plexus in one of two ways: they may narrow the space available behind the insertion of the anterior scalene muscle, through which the brachial plexus and subclavian artery must pass over the first rib, or they may so project that a portion of the brachial plexus must pass across them. If the rib is short, the seventh nerve alone may lie in contact with the rib; if it is long, the entire brachial plexus may pass over it. In the latter case the lower trunk of the plexus rests against the cervical rib and, in order to reach its point of exit between the normal first thoracic rib and the clavicle, has to run upward to pass over this extra obstruction. In such a case, obviously, the angulation of the plexus produced by a cervical rib would be increased by any lowering of the shoulder.

Rudimentary First Thoracic Rib

Rudimentary first thoracic ribs, although perhaps not as well known as cervical ribs, occur almost as frequently (Etter) and may be associated with exactly the same symptoms as those commonly attributed to a cervical rib (Dow; White and associates). A rudimentary first thoracic rib is one that fails to reach the sternum, its anterior end being usually fixed to the second rib by ligaments. Typically associated with a rudimentary first thoracic rib is a downward shift in the segmental origin of the brachial plexus, so that the second thoracic nerve contributes largely to this plexus (Dow; Hertslet and Keith). Since the vertebral relations between the second thoracic nerve and the first rib are the same as those between the first thoracic nerve and a cervical rib—that is, in each case the lowest nerve must ascend in order to cross above the rib—the angulation produced by a first thoracic rib with a postfixed plexus is somewhat similar to that produced by a cervical rib with a normal plexus (Fig. 9-34).

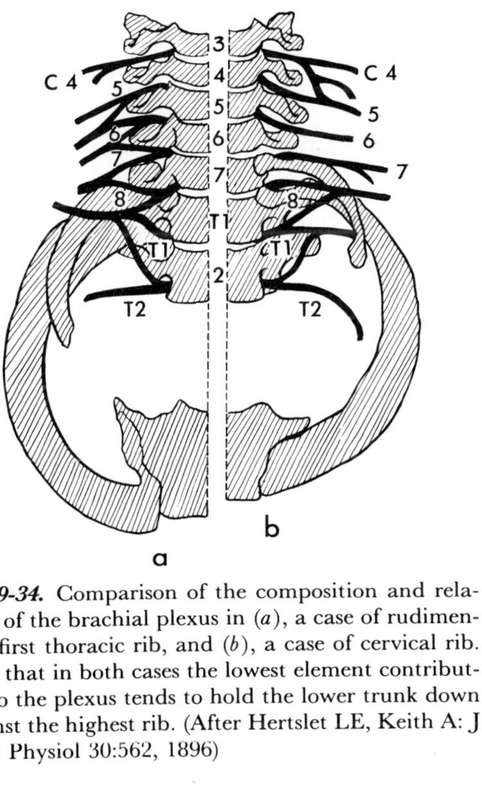

Fig. 9-34. Comparison of the composition and relations of the brachial plexus in (*a*), a case of rudimentary first thoracic rib, and (*b*), a case of cervical rib. Note that in both cases the lowest element contributing to the plexus tends to hold the lower trunk down against the highest rib. (After Hertslet LE, Keith A: J Anat Physiol 30:562, 1896)

Scalenus Anticus Syndrome

Although most writers have abandoned the concept that neurovascular compression is caused by spasm of the anterior scalene muscle (unless this in turn results from a protruded cervical disk), there is abundant evidence that the anterior scalene may be a factor in the compression. When there is a cervical rib present, the anterior scalene may compress the neurovascular bundle against this rib (Fig. 9-35), or a particularly broad and tendinous insertion of the muscle, narrowing the gap that the artery and nerves must traverse between anterior and middle scalenes, may present a hard edge against which the softer structures are forced during certain movements. Symptoms are identical, whether or not a cervical rib is present, and since section of the anterior scalene is a much simpler operation than removal of a cervical rib, scaleniotomy has been performed whether or not a cervical rib was present. Adson ('51) described the subclavian artery and the lower part of the plexus as dropping downward and forward after section of the anterior scalene and found that such section frequently so increases the space available that removal of a cervical rib, when present, is unnecessary.

Fig. 9-35. How contraction of the anterior scalene muscle in the presence of a cervical rib can produce compression of the subclavian artery and brachial plexus, and how scaleniotomy alone may relieve this compression by allowing the vessel and nerves to drop forward (Adson AW: J Int Coll Surgeons 16:546, 1951)

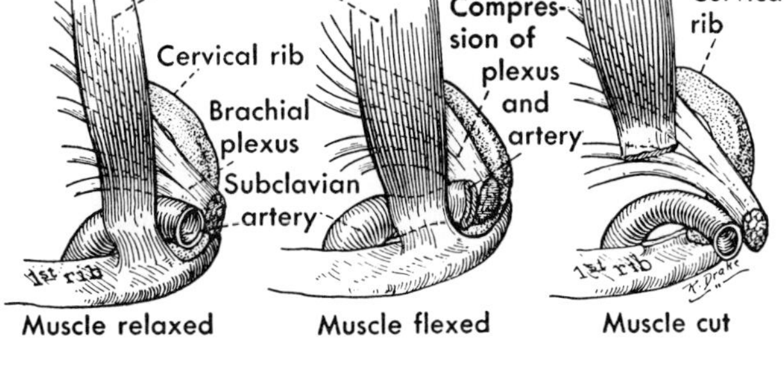

Adson's test or maneuver for compression by the anterior scalene, whether or not a cervical rib is present, is obliteration of the pulse in the affected limb when the contracted muscle is forced backward. This test must be carried out while the patient is in a sitting position with the hands resting on the knees and consists of having the patient take a deep breath, extend his neck as much as possible, and turn his face toward the affected side. Adson warned that scaleniotomy for cervical rib or scalenus anticus syndrome would accomplish little unless the result of this test is positive. A positive test, however, cannot be regarded as pathognomonic for anterior scalene–cervical rib syndrome, since it may occur in persons with no clinical signs of such compression. Telford and Mottershead advanced evidence that sometimes obliteration of the radial pulse in normal persons during the Adson maneuver results from constriction of the axillary artery between the two roots of the median nerve, and they, along with Adson ('51), Jones and Randall, Walshe and colleagues, and others, have emphasized that there may be many different causes of compression—Adson, for example, reported apparent instances of compression by the omohyoid muscle.

The use of scaleniotomy to relieve apparent scalenus anticus–cervical rib syndrome has markedly declined for at least two reasons. One reason is that the neurological symptoms once thought to be caused by neurovascular compression at the lower cervical level have often been found to be caused by compression in the carpal canal ("carpal tunnel syndrome") or by compression of a nerve by a protruded cervical disk, neither of which was recognized in the early days of scaleniotomy. Thus Spurling and Scoville reported that reflex spasm of the anterior scalene as a result of compression of a nerve root may produce neurocirculatory signs that overshadow the typical clinical features of the disk, and Kristoff and Odom warned that scaleniotomy should never be done until the presence of a protruded cervical disk has been ruled out.

The second reason for the decline of scaleniotomy is that nonsurgical treatment has proved effective in relieving a majority of cases of scalenus anticus–cervical rib syndrome. It has been recognized by a number of workers that the symptoms more commonly appear on the right side and more commonly in women. Spurling and Grantham attributed this to a gradual dropping of the shoulder girdle, more pronounced on the dominant side and also more pronounced in women. This would obviously tend to increase traction of the neurovascular bundle on a cervical rib, the anterior scalene, or any other structure that it crosses in its downward course. Thus exercise and physical therapy that improve the tone of the elevators of the shoulder might be expected to be helpful in alleviating the symptoms; Naffziger and Grant reported as early as 1938 that this was so, and Spurling and Grantham reported only a few years later that such conservative treatment gave relief in 300 of 400 patients. The percentage of patients requiring surgical treatment has declined much more since these earlier reports. Part of the success of scaleniotomy has been attributed to the fact that it allows the first rib and the remainder of the thoracic cage to drop (Naffziger and Grant); in addition, it enlarges the space behind the muscle and therefore decompresses the neurovascular bundle if it is compressed here.

Origin of Vascular Symptoms

Vascular symptoms associated with neurovascular compression may include coldness, numbness or paresthesia, and cyanosis of the affected limb and may somewhat simulate Raynaud's disease. The question of their origin has long been a source of argument and still is not settled. Mechanical constriction of the subclavian artery has been felt by many workers to be an unsatisfactory explanation, the primary argument being that since sudden occlusion of the subclavian artery does not usually produce marked vascular changes in the limb, the collateral circulation should be even more adequate when there is only partial and often intermittent compression (*e.g.,* Telford and Stopford). Actually, vascular symptoms may be minimal. However, the

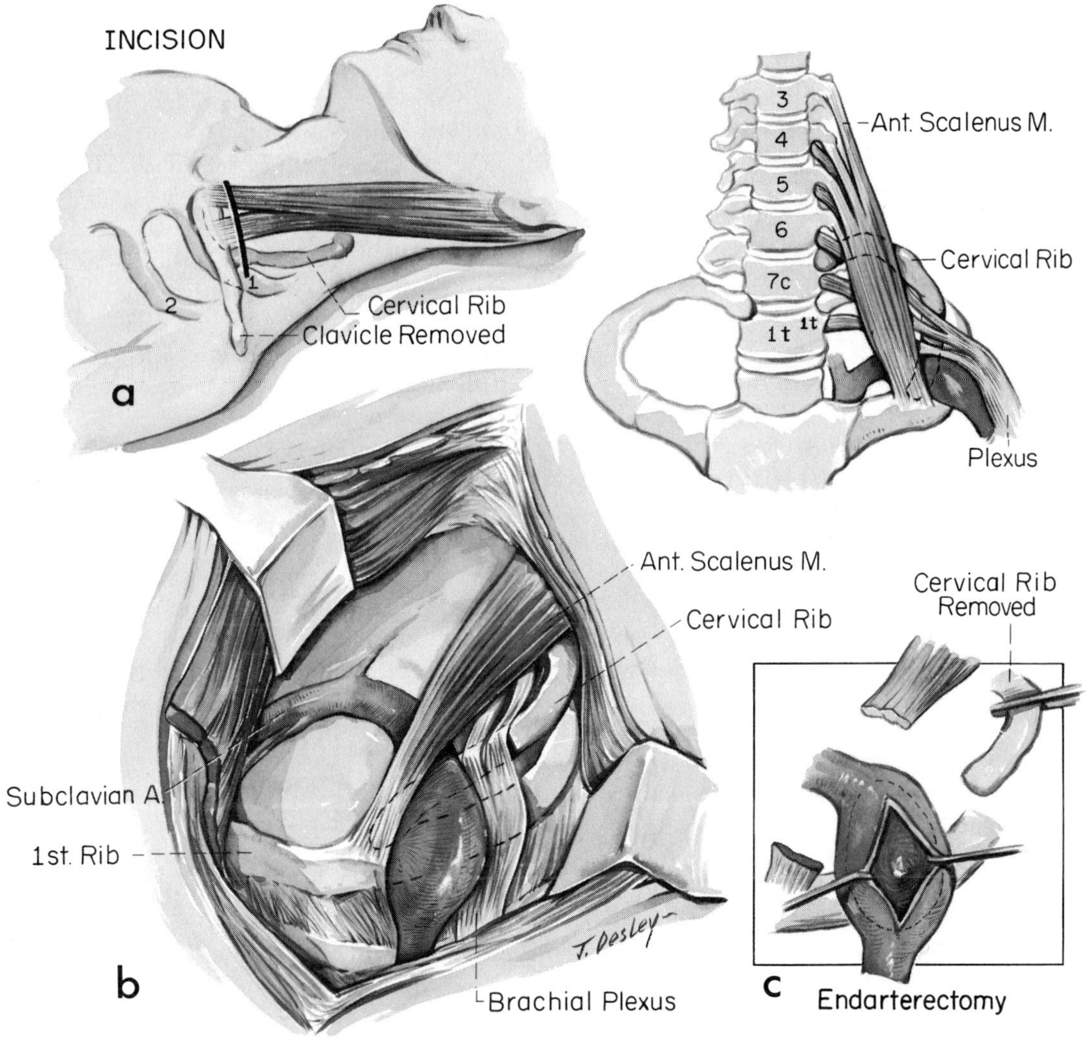

Fig. 9-36. Removal of a cervical rib and repair of the artery. (Courtesy Dr. P. E. Bernatz)

compressed subclavian artery is not necessarily otherwise normal, for Adson ('51) reported arteriosclerotic plaques in it as it coursed behind the anterior scalene, and such involvement of the artery at the medial border of the muscle could easily affect the collateral circulation to the limb. Further, poststenotic aneurysmal dilations (Fig. 9-36) may occur on the artery just distal to the scalene (in 100 operations associated with cervical rib that they reviewed, Lindskog and Howes found 27 in which there were aneurysms); Eden regarded the peripheral vascular changes as being caused by intimal damage, such as might occur in an aneurysm, with the formation of multiple emboli.

Telford and Stopford described a case in which the sympathetic fibers in the first thoracic nerve were concentrated especially in the lower border of the nerve, where they would be easily subject to pressure from a cervical rib, and suggested that stimulation of the sympathetic fibers to the limb through such pressure might cause the peripheral vascular symptoms. Blair and co-workers and Sunderland were unable to confirm such a concentration of sympathetic fibers, since they described them as distributed throughout the nerve, but the former workers regarded chronic inflammation of the nerve, extending throughout it, as stimulating the sympathetic fibers. Although Eden objected that chronic irritation of a nerve is usually succeeded by paralysis (with therefore, in the case of cervical rib, disappearance of the peripheral vascular symptoms if they were caused by this), many workers have argued that sympathetic stimulation may be a contributing or the sole cause of the vascular symptoms. Adson ('51) accepted both mechanical constriction and sympathetic stimulation as contributors. Rather than direct stimulation of the sympathetic fibers in the nerves of the brachial plexus, it also seems possible that the pressure on the subclavian artery might stimulate sensory fibers that reflexly influence the sympathetic outflow to the limb.

Possibly, therefore, the peripheral vascular symptoms may result from varying combinations of mechanical obstruction, pathology, and sympathetic overactivity. In cases in which the last-mentioned is apparently prominent, cervicothoracic ganglionectomy, or removal of the second thoracic ganglion, has been combined with scaleniotomy.

Other Sites of Obstruction

Although Telford and Mottershead argued that lowering the shoulder should not pinch the neurovascular bundle between the clavicle and the first rib because the clavicle normally moves forward in such movement, Lord and Rosati regarded compression here as a major source of symptoms, and Gilroy and Meyer agreed that the ischemia from such compression is the predominant and usually only finding in most patients with thoracic outlet syndrome. The costoclavicular maneuver, having the patient draw his shoulders backward and downward, will diminish the radial pulse not only in persons with apparent costoclavicular syndrome, but in many persons with no symptoms. Deformation of the first rib or the clavicle may, of course, be a cause of compression here. Roos and Owens regarded conservative measures, such as exercise, as being ineffective in the more severe cases of compression and advocated removal off the first rib through the axillary approach, with resection of the clavicle also if vascular surgery is necessary.

When the arms are raised above the head (hyperabducted), the neurovascular bundle may be forced against the coracoid process and the insertion of the pectoralis minor muscle. Symptoms are usually slight and temporary and may require no treatment other than, for instance, advice to discontinue sleeping in this position. Sometimes however, according to Stone and Lord, it may be advisable to section the pectoralis minor tendon.

Finally, venous obstruction at the thoracic outlet, with no other findings, has been attributed to, among other causes, compression of the subclavian vein between the tendon of the subclavius muscle and the first rib or between the anterior scalene and the clavicle (McCleery and co-workers) and to compression by a phrenic nerve that passes in front of the vein (Hughes).

CERVICAL SYMPATHETIC TRUNK

The cervical sympathetic trunk (chain) consists of two to four ganglia (superior, middle, vertebral, and inferior or stellate) and the trunk connecting these (Fig. 9-37). While a few postganglionic fibers may descend in the cervical trunk (Foley), its fibers consist largely of ascending preganglionic ones that emerge through the ventral roots of upper thoracic nerves and pass upward to make synapse in higher ganglia of the trunk, although some may pass beyond the superior cervical ganglion to end in ganglia of the carotid plexus. It is generally believed that few or no sensory fibers of spinal nerve origin run in the upper part of the trunk (cardiac pain fibers probably do descend in the lower part), but it is not uncommon, at least in the cat, for vagal fibers (probably sensory ones) to be incorporated in relatively large numbers in the sympathetic trunk (*e.g.*, Hinsey and coauthors; Foley).

THE GANGLIA

Superior Cervical Ganglion

The large superior cervical ganglion lies on the longus capitis muscle at about the levels of the bodies of the second and third cervical vertebrae; it is therefore above the carotid bifurcation and lies somewhat posteromedial to the first portion of the internal carotid artery, being related also anteriorly to the inferior (nodose) ganglion of the vagus. The gray rami communicantes (postganglionic fibers) given off from the superior cervical ganglion are variable, but regularly reach at least the first three cervical nerves and often the fourth. They are typically multiple for each nerve and run through the longus capitis muscle, where they may join the nerves' muscular branches instead of their main trunks (Kimmel, '59).

The superior cervical ganglion has communications with the vagus, the hypoglossal, and other nerves in this neighborhood; and it gives off external carotid nerves to form the external carotid plexus that accompanies the branches of this vessel, and the large and important internal carotid nerve that leaves the upper end of the ganglion as a broad sheet of fibers that subsequently breaks up to form a plexus about the internal carotid artery. The very slender superior cervical cardiac nerve likewise typically arises from the superior cervical ganglion and descends behind the great vessels.

The internal carotid plexus, the chief continuation from the superior cervical ganglion, consists primarily of postganglionic fibers derived from the cells of this ganglion. Within the cavernous sinus many of the fibers of the plexus leave the artery to be distributed to the orbit (in the cat, fibers for the dilator pupillae leave earlier, joining the tympanic plexus), and others leave it, as the deep petrosal nerve, to join the greater (superficial) petrosal, course to the pterygopalatine (sphenopalatine) ganglion, and be distributed through the branches of this ganglion. Perhaps some fibers also reach the skin of the upper part of the face, which is otherwise, however, supplied

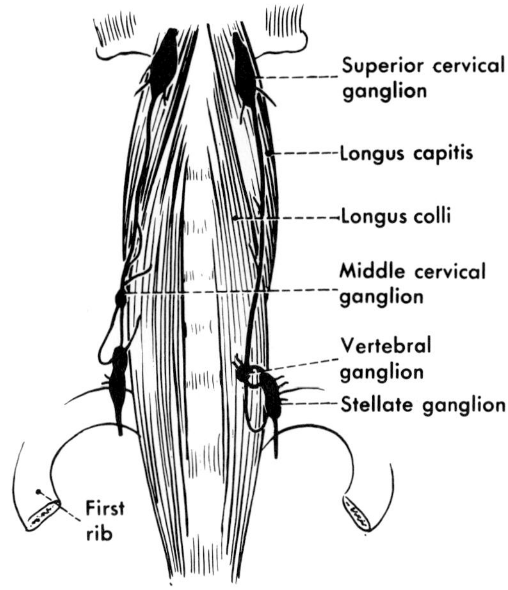

Fig. 9-37. Cervical sympathetic trunk. Both a middle and a vertebral ganglion are illustrated.

through the external carotid plexus. Still others are distributed along the intracranial branches of the internal carotid. It has already been mentioned that the internal carotid plexus may contain ganglia, although whether they are all simply misplaced portions of the superior cervical ganglion, as usually supposed, or some are parasympathetic ganglia is not known. It has likewise been pointed out that there is evidence that the internal carotid plexus may contain some afferent fibers from upper thoracic spinal nerves.

The preganglionic fibers reaching the superior cervical ganglion are derived largely from the upper three thoracic nerves by means of their white rami communicantes and enter the upper thoracic trunk to run upward through this and the cervical trunk to end in the superior cervical ganglion. Interruption of the cervical trunk anywhere along its course will therefore interrupt all impulses to the superior cervical ganglion and be followed by the temporary flushing, the absence of sweating, the miosis, and the ptosis of the upper lid that characterize Horner's syndrome. However, the majority of preganglionic fibers destined for the innervation of the orbit are derived from the first thoracic nerve, so that section of the sympathetic trunk below the first thoracic ganglion, sparing the fibers from the first thoracic nerve, leaves the eye apparently unaffected.

Middle and Vertebral Ganglia

Two ganglia, the middle cervical and the vertebral, ordinarily occur on the sympathetic trunk between the superior cervical ganglion and the inferior cervical or stellate ganglion, although both are not necessarily present at once; occasional smaller or intermediate ganglia may be associated with either. The *middle cervical ganglion* typically lies at the level of the transverse process of the sixth cervical vertebra (the carotid tubercle), where it is usually associated with the inferior thyroid artery as this vessel loops medially toward the thyroid gland. Becker and Grunt found a middle cervical ganglion in 62.3% of 114 cervical trunks, usually just above the inferior thyroid artery, but sometimes higher and sometimes in the loop of the artery. Wrete described the middle cervical ganglion as giving rise to a gray ramus communicans that bifurcates and goes to nerves C5 and C6, with sometimes another to C4 and C5, and on the basis of those connections he reported that the middle cervical ganglion is sometimes indistinguishably fused with the lower end of the superior cervical ganglion and, less frequently, with the vertebral. Kimmel ('59) described both the middle cervical and the stellate ganglion as contributing fibers to a hitherto undescribed plexus on the acending cervical artery and reaching the second to the fourth cervical nerves. The middle cervical cardiac nerve arises from the middle cervical ganglion or from the sympathetic trunk directly. In its course downward in the neck it may be joined by other cardiac nerves.

The *vertebral ganglion,* also known as the intermediate ganglion or as a low middle cervical ganglion, is about the same size as the middle cervical ganglion but typically lies lower than that, on the front of the vertebral artery close to the latter's origin from the subclavian. It is then connected to the upper end of the inferior cervical or stellate ganglion, which lies at about the same level but behind or medial to the vertebral artery, by fibers that pass on both sides of the artery (and vein) and usually also by the ansa subclavia (see below). Becker and Grunt found a vertebral ganglion in 87.7% of their dissections and both vertebral and middle cervical ganglia in 55.3%. In contrast, Jamieson and co-workers found neither ganglion in 47% of 100 sides, and only two instances in which both were present. Apparently, the vertebral ganglion may give rise to no rami communicantes or may send a ramus to nerve C8.

Stellate or Inferior Cervical

The stellate or cervicothoracic ganglion is formed by the fusion of the lowest cervical ganglion (inferior cervical) and the first thoracic, with sometimes the second or even the third thoracic ganglion added at the lower end. Statistics on the occurrence of a separate inferior cervical ganglion vary; Perlow and Vehe reported it in only 8 of 48 sides, and Jamieson and co-workers found a similar proportion, 18 in 100 sides; on the basis that the ganglia should not be considered fused if there was between them a constriction no greater in diameter than the trunk elsewhere, Becker and Grunt described a separate ganglion in 62.3% of sides. It is not particularly important whether one is dealing with an inferior cervical ganglion or the upper end of a stellate ganglion—connections and relations are similar, and the first thoracic part has often been resected whether or not they were fused. The inferior cervical cardiac nerve arises from the stellate ganglion.

The inferior cervical ganglion or the upper end of the stellate ganglion typically lies posterior or posteromedial to the vertebral artery (Fig. 9-38), while the first thoracic ganglion or the lower part of the stellate extends downward and backward behind the subclavian artery on the first thoracic vertebra, usually reaching the level of the emerging first thoracic nerve. Perlow's and Vehe's report that the ganglion lay anterior or anteromedial to the vertebral artery in 18 cases, medial in 23, and posteromedial or posterior in only seven may have been based upon a failure to distinguish between the vertebral ganglion, not generally recognized as a separte ganglion at that time, and the true upper end of the inferior cervical or stellate.

The fibers leaving the stellate ganglion to ascend in the cervical sympathetic trunk do so by several routes. If there is a vertebral ganglion, some fibers usually pass on each side of the vertebral artery to reach this ganglion, before continuing in the trunk above it; if there is no vertebral ganglion, the trunk simply ascends from thc upper end of the stellate to the middle cervical or, if that is lacking, to the superior cervical ganglion. In addition, some of the preganglionic fibers that traverse the stellate run below the subclavian artery and loop upward in front of it, forming the *ansa subclavia* (see Fig. 9-38). Sometimes the ansa (also called ansa Vieussens) joins the upper end of the stellate ganglion or the cervical sympathetic trunk directly; usually, however, it joins the vertebral ganglion or the middle cervical ganglion (70.9% and 23.3%, respectively, in the series of Becker and Grunt).

The gray rami communicantes leaving the inferior cervical–stellate complex are directed toward the first thoracic and the lower cervical nerves as these emerge from the intervertebral foramina. The rami to the lowest three cervical nerves typically divide, one part going directly to a nerve, the other running upward to reach the next nerve above (Kimmel, '59; and others), or sometimes also the second nerve above. The ascending branches from the rami to C6 and C7, and sometimes some fibers of the ramus to C8, regularly run upward along the vertebral artery and form the *vertebral nerve.* In the ten fetuses he examined, Kimmel found divided rami at the C7 and C6 levels contributing to the vertebral nerve in all cases and at C5 in seven cases. Fibers from the vertebral nerve occasionally ascended higher to join nerves C4 and C3.

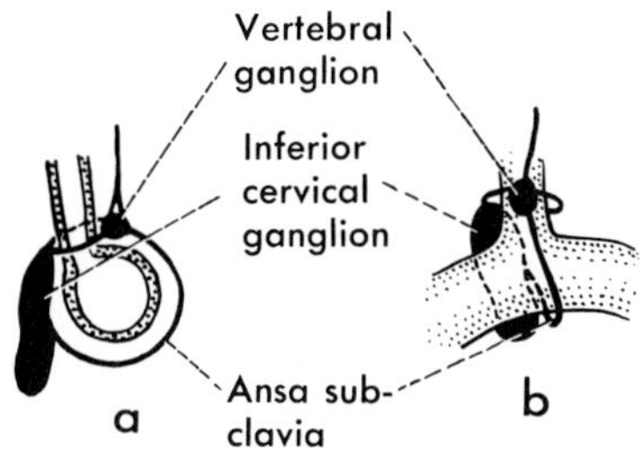

Fig. 9-38. Relations of the vertebral and inferior cervical (or stellate) ganglia to each other and to the subclavian and vertebral arteries: (*a*) in a sagittal section and (*b*) from the front.

FUNCTIONAL CONNECTIONS

As evidenced by the accounts above and by the careful dissections of Axford and of Siwe, the ganglia of the cervical sympathetic trunk show considerable variation in the number of spinal nerves with which they are connected and in the courses that their rami communicantes take. The rami may have small ganglia upon them (Skoog) and therefore contain a few preganglionic fibers, but otherwise they are composed of postganglionic fibers from the ganglia of the trunk.

The vertebral nerve, although consisting of rami communicantes to spinal nerves, also contributes fibers to a plexus on the vertebral artery. The vertebral plexus thus formed, however, does not follow the vertebral artery into the skull, but is replaced in the upper part of the neck by another plexus whose sensory fibers are derived from the third or second and third spinal nerves and whose sympathetic fibers arise from the superior cervical ganglion (Kimmel, '59).

The general distribution of postganglionic fibers from the cervical trunk, as demonstrated by the anhydrosis produced by injuries to it, has been reported for human beings by List and Peet and by Guttmann; it should be emphasized, however, that the variations in the trunk and its connections are so numerous that any blanket statements about it are apt to be misleading.

Since the postganglionic fibers to the upper limb are derived from the stellate or inferior cervical ganglion and upper ganglia of the thoracic sympathetic trunk, sympathetic impulses to the limb may be largely blocked by injection of the stellate ganglion or surgical interference with it. In ganglion blocks, the position of the ganglion in relation to the subclavian and vertebral arteries must be clearly borne in mind; the bony landmark for this injection is the transverse process of the seventh cervical vertebra. Murphrey described a technique that he found successful in 95.7% of the patients in whom it was tried. In the anterior approach for removal of the stellate ganglion, now seldom used, the anterior scalene muscle must be cut and the subclavian artery retracted downward in order to allow an approach to the ganglion.

As discussed in more detail in connection with the thoracic sympathetic trunk (see Volume 2), removal of the stellate ganglion was for many years the operation of choice in attempts to alleviate the symptoms of Raynaud's disease in the upper limb. Stellectomy and removal of the upper several segments of the thoracic trunk result in a postganglionic or anatomical denervation of the blood vessels of the upper limb and, in view of the ensuing sensitivity to epinephrine that has been said to occur after postganglionic denervation, was for a time abandoned by many surgeons. In recent years, however, a number of workers have reverted to the postganglionic denervation as the one giving the most favorable and lasting results.

Since the preganglionic fibers to the upper limb are derived largely from below the first thoracic nerve (*e.g.*, Hyndman and Wolkin), an approximate preganglionic denervation such as that described by Smithwick can be carried out through an upper thoracic approach. This usually includes sectioning the thoracic trunk below at least the third thoracic ganglion, then reflecting the trunk cranially and severing the rami communicantes uniting the more cranial ganglia with the spinal nerves, up to but not including the connection to the first thoracic nerve. Apparently the first thoracic nerve distributes most of its preganglionic fibers to the orbit, and few or none to the limb; thus if its rami communicantes are left intact, the limb is at least almost completely denervated and yet the disagreeable sequelae of Horner's syndrome are avoided.

Similarly, removal of the second and third thoracic ganglia instead of the stellate spares the eye and, although probably producing a mixed preganglionic–postganglionic denervation, has an advantage over stellectomy in that it should interrupt fibers of the gray rami communicantes that ascend posterior to the necks of the ribs to reach the nerves of the brachial plexus (Kimmel, '59; and others). Finally, Palumbo has differed from most workers in maintaining that the first thoracic ganglion, or the lower part of the stellate, can usually be removed without producing Horner's syndrome, although that does occur if the inferior cervical ganglion or the equivalent upper end of the stellate is removed.

CRANIAL NERVES IN THE NECK

In addition to the cervical branch of the facial nerve, which descends into the neck to supply the platysma muscle, the ninth through the twelfth cranial nerves are all related to the upper portions of the great vessels of the neck, and the tenth and eleventh run through the neck for a considerable distance.

NINTH NERVE

Just outside the jugular foramen, thus in the jugular fossa, the glossopharyngeal nerve bears its two ganglia (superior, which may lie in the foramen, and inferior or petrous) and gives off its tympanic branch to the middle ear cavity and its vagal communicating (auricular) branch to the external ear. Since section of the nerve below the fossa is useless in abolishing the pain of glossopharyngeal neuralgia when it involves the ear, and since the approach to the nerve as it leaves the foramen is difficult, this nerve is typically sectioned intracranially for glossopharyngeal neuralgia. The ramus communicans with the auricular branch of the vagus (also called the auricular branch of the ninth nerve) is distributed with the vagus to the ear. Also outside the jugular foramen, the ninth nerve sometimes has a branch of communication with the facial nerve; the functional importance of this branch is not known, but it has been thought to contain preganglionic fibers from the ninth nerve that instead of traversing the lesser petrosal, reach the otic ganglion through the chorda tympani branch of the facial and the communicating branch from the chorda to the ganglion.

As it emerges from the jugular foramen, the glossopharyngeal nerve lies anterolateral to the vagus and accessory nerves, medial to the internal carotid artery, and anteromedial to the jugular bulb. Almost immediately upon its exit it passes laterally between the jugular vein and the internal carotid artery (Fig. 9-39) and then forward between the internal and external carotids, to lie deep to the styloid process and the muscles attached to it. It is most easily located here by looking along the medial surface of the lower border of the stylopharyngeus muscle. It supplies this muscle and curves around its lateral border to lie on the lateral surface of this and subsequently on the middle constrictor of the pharynx as it runs forward toward the tongue; deep to the hyoglossus muscle it divides into tonsillar and lingual branches.

While the glossopharyngeal is related to the internal carotid artery it gives off its carotid branch (carotid sinus branch, nerve of Hering, intercarotid nerve), which descends between internal and external carotids to unite with the carotid branch of the vagus and supply the carotid body and carotid sinus. Also in this location it gives off its pharyngeal branch or branches which pass forward between internal and external carotid vessels, below the main trunk of the ninth nerve, to unite with the pharyngeal branches of the vagus on the middle constrictor muscle. The pharyngeal branches of the ninth are sensory to the pharynx, and the motor distribution of the nerve is to the stylopharyngeus muscle only.

TENTH NERVE

The vagus nerve also makes its exit through the jugular foramen, lying between the glossopharyngeal and accessory nerves. Within the foramen the vagus bears its superior (jugular) ganglion, and at or just distal to this ganglion it receives the internal ramus (accessory branch to vagus) of the accessory nerve, a branch consisting of the so-called cranial or bulbar rootlets of the accessory, the fibers of which are thereafter distributed entirely with the vagus. The cranial rootlets and internal ramus of the accessory contribute to the vagus the majority of its voluntary motor fibers. Just outside the foramen the vagus expands again to form the large inferior or nodose ganglion (Fig. 9-40). The vagus at first lies medial to both the internal jugular vein and the internal carotid artery, but as it descends, and as the vessels shift slightly so that the vein comes to lie lateral to the artery, the vagus lies behind and somewhat between the two. Below the bifurcation of the common carotid it is related in a similar fashion to this vessel and to the internal jugular vein. Since the vagus is enclosed within the carotid sheath, and therefore anterior to the prevertebral fascia, there is no necessity of exposing it in seaching for the phrenic nerve; however, it has been mistaken for the phrenic when the carotid sheath was unintentionally opened.

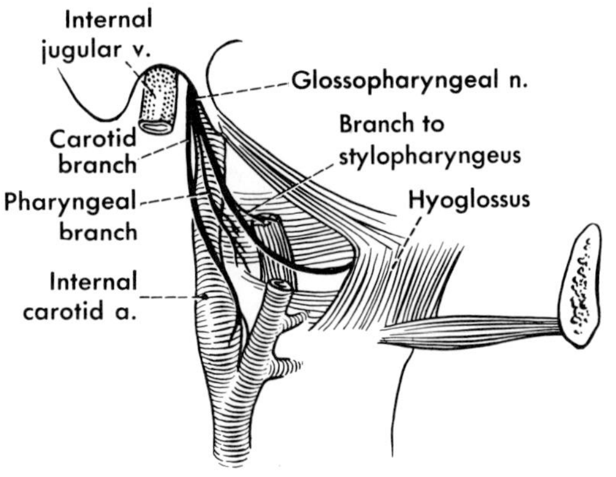

Fig. 9-39. Course of the glossopharyngeal nerve.

Anomalous positions of the cervical vagus are very rare; Gibson described one of the few cases in which the nerve has been found in front of the common carotid instead of behind it. Partial doubling of the trunk in the neck has been reported, according to Taguchi (1888), who himself described a case in which an abnormal branch of the vagus helped to form an ansa "hypoglossi." Von Lippmann described a similar case, in which the origin of

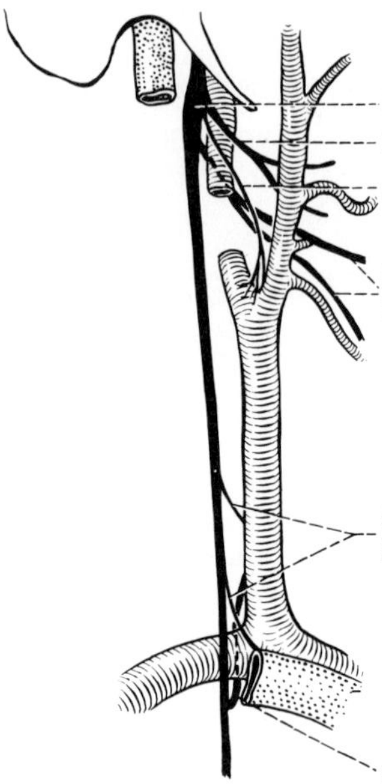

Fig. 9-40. Course of the right vagus nerve in the neck.

the abnormal branch was seen to be from cervical fibers that first joined the vagus and then separated from it, and the present author has also seen one.

Meningeal, Auricular, and Pharyngeal Branches

The meningeal and auricular branches of the vagus arise from this nerve within the jugular fossa; the pharyngeal branch arises from the upper part of the inferior (nodose) ganglion and passes across the lateral side of the internal carotid artery, between this and the external carotid, to join the pharyngeal branch of the glossopharyngeal in forming the pharyngeal plexus. The pharyngeal branch of the tenth nerve is the chief motor supply to the pharynx. The vagus also can be shown to contribute fibers to the carotid sinus nerve, although this is primarily a glossopharyngeal branch. The inferior vagal ganglion usually has communications with the accessory and hypoglossal nerves, the superior cervical ganglion, and the upper two spinal nerves, but the nature of these is not understood.

Superior Laryngeal Nerve

The superior laryngeal nerve arises from the lower end of the inferior ganglion of the vagus (nodose ganglion), usually receives a branch from the superior cervical ganglion of the sympathetic, descends behind the internal carotid artery, and divides into its two terminal branches, external and internal. The smaller external branch (external laryngeal nerve) descends and runs forward on the larynx, being closely associated with the superior thyroid artery; it usually gives a branch to the inferior pharyngeal constrictor and ends by supplying the cricothyroid muscle. The larger internal branch (internal laryngeal nerve) curves forward on the thyrohyoid membrane and enters the larynx with the superior laryngeal artery, to supply the mucous membrane of the larynx. The superior laryngeal nerve is discussed more fully in connection with the larynx.

Cervical Cardiac Branches

The vagus also contains fibers destined for chemoreceptor tissue and vasodepressor areas in the lower part of the neck and the upper part of the thorax; in some animals these run separately and constitute the depressor nerves. After their connections and branches high in the neck, the vagus nerves have no other branches until they approach the middle of the neck; between this and the base of the neck each nerve usually gives off two, or two groups of, small cardiac branches, which run downward and medially, often joining sympathetic cardiac branches, to the cardiac plexus. They vary much, but are called the superior and inferior cervical cardiac branches.

Recurrent Laryngeal Nerves

The recurrent laryngeal nerves (recurrent nerves) are especially important branches of the vagi in the neck, although only the right recurrent arises in the neck; this leaves the right vagus in front of the subclavian artery and loops below and behind the artery to ascend obliquely to the posterolateral aspect of the trachea and pass upward toward the larynx. The left recurrent nerve arises from the left vagus as this passes across the arch of the aorta and therefore has a mediastinal origin; it likewise runs upward alongside the trachea to reach the larynx. The somewhat variable

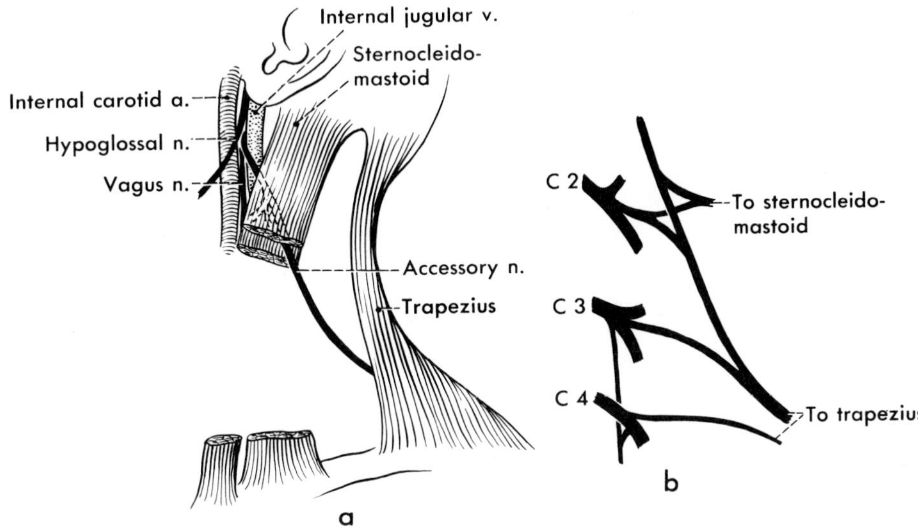

Fig. 9-41. Usual course and relations of the accessory nerve (*a*) and its connections with the cervical plexus (*b*).

relations of the recurrent nerves are described in connection with the thyroid gland; their distribution within the larynx is described in connection with this structure.

ELEVENTH NERVE

The accessory (spinal accessory) nerve divides at the jugular foramen into its internal and external rami, and the internal ramus, also called the accessory branch to the vagus, joins the vagus nerve just outside the foramen. The accessory nerve of gross anatomy is therefore purely the spinal portion, derived from the external ramus and taking origin from upper segments of the spinal cord; it consists entirely of voluntary motor fibers that are distributed to the sternocleidomastoid and trapezius muscles. It is the most posterior of the three nerves leaving the jugular foramen; it at first lies medial to and somewhat between the internal jugular and the internal carotid vessels and typically passes between the vein and the artery to cross the lateral surface of the vein (Fig. 9-41). This is not a fixed relation, however, for the nerve sometimes passes behind rather than in front of the vein. Tandler found the accessory nerve passing anterior to the vein in two-thirds of 150 bodies, and passing behind it in the other third; similarly, Parsons and Keith summarized it as anterior in 70%, posterior in 26.8%, and passing through the vein in 3.2% of 415 sides (Fig. 9-42). This apparently depends upon the fact that in the embryo a venous ring surrounds the nerve, and the relation of the latter to the vein varies according to which segment of the ring persists.

After crossing the internal jugular vein, the accessory nerve descends obliquely downward and backward to the upper part of the sternocleidomastoid; it gives off its branch into the deep surface of this muscle some 4 cm or more below the tip of the mastoid process and may pass downward and backward either deep to the sternocleidomastoid or through it (about 18% and 82%, respectively, according to Parsons and Keith) to course across the posterior triangle and enter the deep surface of the trapezius. In the posterior triangle it lies between the superficial and prevertebral layers of the cervical fascia. The accessory nerve typically joined before it enters the sternocleidomastoid by a branch from the second cervical nerve and is also joined as it enters the trapezius by branches from the third and fourth cervical nerves. These are usually believed to be entirely sensory.

Coleman and Walker indicated their belief that the accessory nerve is highly variable and

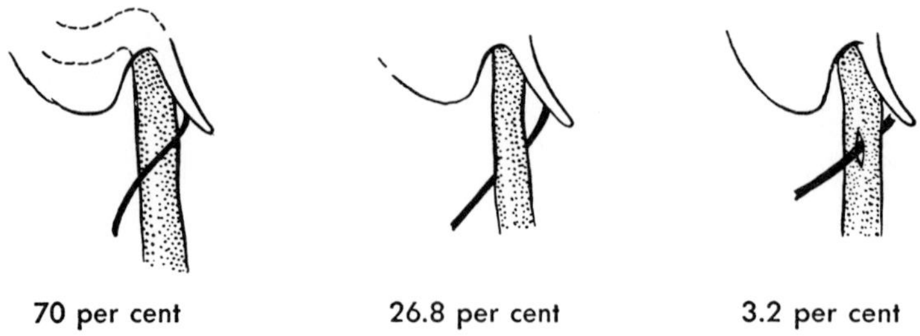

Fig. 9-42. Variations in the relations of the accessory nerve to the internal jugular vein. The percentages given are those of Parsons FG, Keith A: J Anat Physiol 32:177, 1897.

described it as dividing into sternocleidomastoid and trapezius branches so high that one branch passed behind, and the other in front of, the internal jugular vein. They also stated that the nerve may fail to reach the trapezius, which is then supplied by cervical nerves only. As they have given no detail of their anatomical researches—for instance, the number of cadavers investigated and whether the observations were on students' dissections or on dissections carried out personally—it is impossible to assess these statements. However, sensory fibers only have been demonstrated in the cervical nerve supply to the trapezius in carefully controlled experiments in animals, and claims of a motor supply from this source in human beings rest upon indirect embryological and debatable clinical evidence. The variation in the obvious crippling of the shoulder following section of the accessory nerve, which these observations purport to explain, can undoubtedly usually be explained upon the basis of variation in the degree to which other muscles compensate for paralysis of the trapezius. The accessory nerve is sometimes purposefully sectioned so that it can be used in a spinofacial anastomosis for alleviation of facial paralysis.

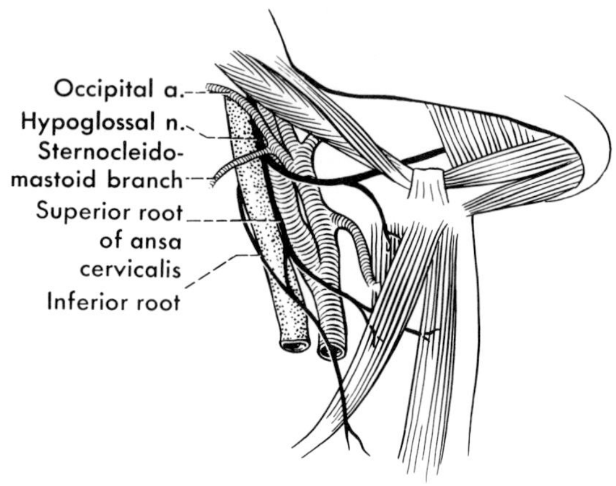

Fig. 9-43. Hypoglossal nerve in the neck. The branches to the infrahyoid muscles are not actually composed of hypoglossal fibers (see Fig. 9-3).

TWELFTH NERVE

The hypoglossal nerve emerges through the hypoglossal canal (anterior condylar canal) and therefore medial to the internal jugular vein and internal carotid artery. It passes laterally and downward to lie against the posterior aspect of the vagus nerve, where it exchanges branches of communication with this nerve and with the sympathetic, and then runs downward and outward between the jugular vein and the internal carotid artery (Fig. 9-43). Close to its exit from the skull it receives branches of communication from upper cervical nerves, usually the second; it is these fibers that leave it as the superior root of the ansa cervicalis (once descendens hypoglossi), the nerve to the thyrohyoid, and the nerve to the geniohyoid. The hypoglossal nerve itself probably supplies only muscles of the tongue.

As the nerve emerges between the jugular and internal carotid vessels or, rarely, posterior to the internal jugular—Löwy found this in only 8 of 100 bodies—it lies under cover of the posterior belly of the digastric muscle. It passes forward across the lateral surfaces of both the internal and the external carotid arteries, giving off the superior root of the ansa as it does so, and passing lateral to the origin of the occipital artery from the external carotid, so that it loops forward about this vessel and its sternocleidomastoid branch. Below the angle of the jaw it is accessible as it lies deep to the posterior end of the submandibular gland, before it passes forward to the tongue. Its further course has been described in connection with the tongue. Like the accessory nerve, the hypoglossal is sometimes sectioned and united to the distal stem of the facial nerve in paralysis of the facial muscles.

The two apparent branches of the hypoglossal nerve in the neck are the superior root of the ansa cervicalis (descendens hypoglossi to the ansa hypoglossi) and the nerve to the thyrohyoid muscle. The superior root leaves the hypoglossal stem as this turns forward around the origin of the occipital artery and descends on the internal carotid and then the common carotid arteries, either in front of or within the carotid sheath, gives off a branch to the superior belly of the omohyoid and the upper parts of the sternohyoid and sternothyroid, and then joins the inferior root (descendens cervicalis) to form the ansa cervicalis. Since both roots of the ansa are composed of cervical nerve fibers, the infrahyoid muscles are innervated entirely through cervical nerves and not by the hypoglossal nerve itself. The ansa cervicalis may be either a short or a long loop—that is, the union between the two roots may be situated high or low in the neck; also, the inferior root may pass medial to the internal jugular vein to join the superior root between the internal jugular and the common carotid, or posterior and then lateral to the vein to form a loop (ansa) across the front of the internal jugular vein. Both these relationships occur frequently; Grant reported that the inferior root passed medial to the internal jugular vein in 43% of 188 bodies and lateral to it in 57%.

Visceral Structures of the Neck

The visceral structures of the neck include the thyroid and parathyroid glands, a portion of the pharynx, the larynx, trachea, and esophagus, and sometimes portions of the thymus. The pharynx and larynx have already been described in Chapter 8. Only the other structures in the neck will therefore be described here.

THYROID GLAND

The thyroid gland lies below and on the sides of the thyroid cartilage, covered anteriorly by the infrahyoid or strap muscles with the exception of the thyrohyoid muscle. In thyroidectomy (Figs. 9-44 to 9-49), the sternohyoid and sternothyroid muscles must be separated from the gland or cut and reflected; if they are sectioned transversely at about the level of the isthmus of the gland, their innervation is probably least apt to be injured, although the levels at which the nerves enter these muscles vary considerably. If the gland is much enlarged, the muscles may be so thinned as to be barely identifiable.

The thyroid gland consists of right and left lobes situated anterolateral to the larynx and trachea and united across the front, just below the cricoid cartilage, by an isthmus. It therefore partially surrounds the trachea, and enlargements of it may early produce dyspnea. Sometimes the gland expands also somewhat behind the trachea; this is important clinically, since it brings the gland into close contact with the recurrent laryngeal nerve even when this is situated deeply in the tracheoesophageal groove. Else regarded this relationship as often responsible for the recurrence of goiter, owing to incomplete removal of this part of the gland. From the isthmus, but not always in the midline, a process of thyroid tissue may project upward; this is the pyramidal process and may be a short stump or rise high enough to be attached to the hyoid bone (Fig. 9-50).

Thyroid Fascia and Capsule

The thyroid gland lies in the visceral compartment of the neck and is of course invested by connective tissue. The thin layer of connective tissue between the posterior surfaces of the strap muscles and the thyroid gland is usually referred to as the pretracheal fascia; connective tissue in closer relation to the thy-

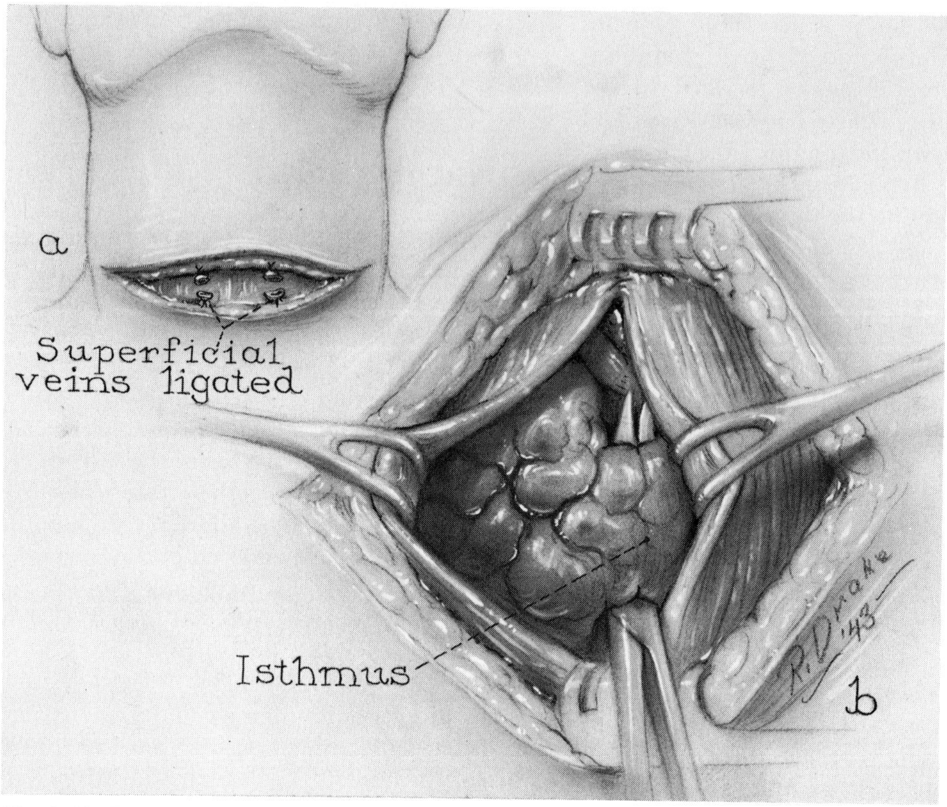

Fig. 9-44. Thyroidectomy. Exposure of the thyroid gland by retraction of infrahyoid muscles, and freeing of the isthmus from the trachea. (Dixon CF: Surg Clin North Am 23:1083, 1943)

roid has been called the perithyroid sheath, thyroid fascia, and false thyroid capsule. The terminology here as used in the clinical literature is confusing, however, because this delicate connective tissue layer, united loosely to the thyroid gland, has been variably described and even called *the* capsule of the thyroid gland. Thus, for instance, Fowler and Hanson identified the capsule of the thyroid gland as the tissue usually described as pretracheal fascia and failed to find it as definitely developed as it is often described. They also found that it did not envelop the thyroid gland posteriorly but, rather, ran behind the esophagus; they stated that the recurrent nerves lie inside this so-called capsule when it is well developed. Ziegelman described the false capsule as failing to envelop the gland posteriorly, while others have described or diagrammed it as completely surrounding the gland. There thus appears to be no clear agreement on what constitutes the thyroid fascia.

It is presumably also the varying development of the connective tissue adjacent to the thyroid gland that accounts for the various statements concerning the relations of the parathyroid glands to this layer. For instance, Walton stated that the normal superior parathyroid glands always lie between the thyroid fascia and the gland itself, while the inferior parathyroids may lie either within or outside this fascia or be embedded in it. Most anatomy texts describe the parathyroid glands as lying between the fascia and the thyroid gland itself, except of course for aberrant ones that are not so closely connected with the thyroid gland.

Even though it seems reasonable, therefore, to regard the perithyroid sheath or thyroid

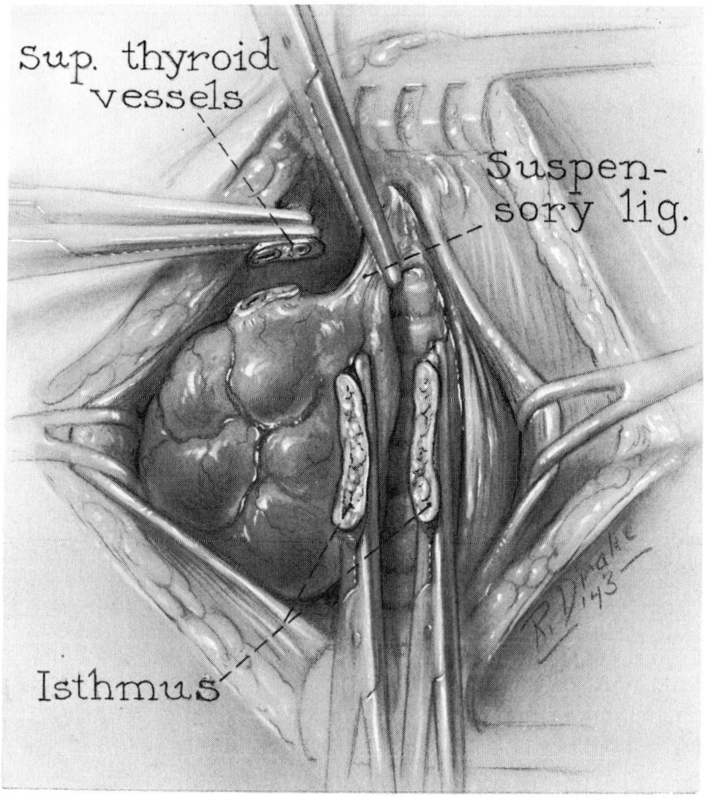

Fig. 9-45. Division of the superior thyroid vessels, the suspensory ligament, and the isthmus. (Dixon CF: Surg Clin North Am 23:1083, 1943)

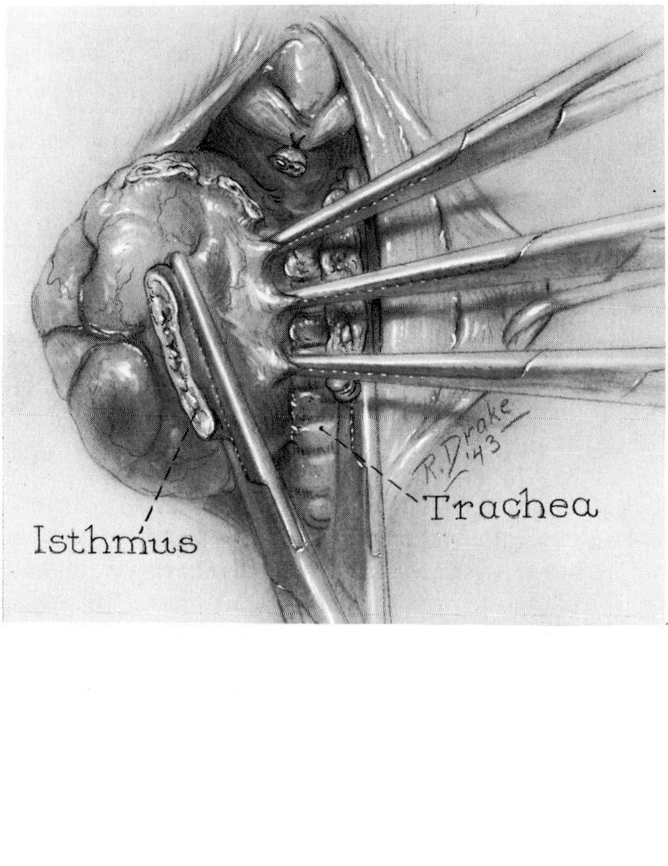

Fig. 9-46. Isthmus retracted laterally to allow clamps to be placed on the right lobe behind the planned line of incision. (Dixon CF: Surg Clin North Am 23:1083, 1943)

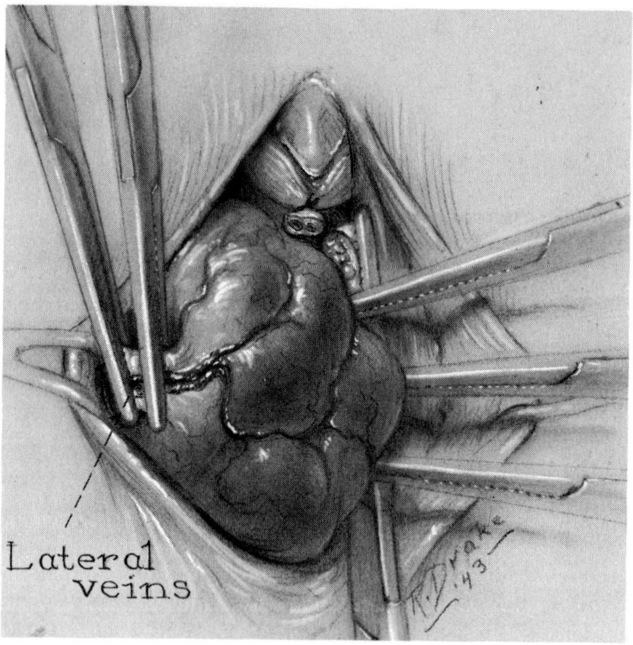

Fig. 9-47. Right lobe rolled medially, and lateral veins, contributing to the middle thyroid vein, clamped for ligation. (Dixon CF: Surg Clin North Am 23:1083, 1943)

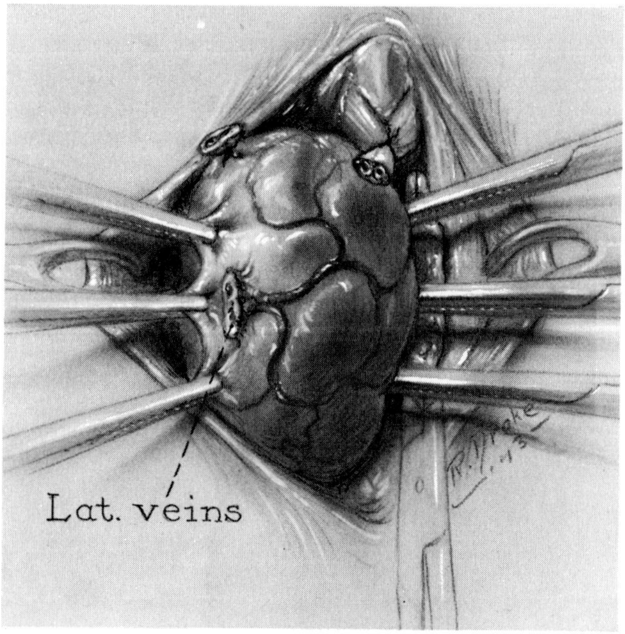

Fig. 9-48. Right lobe rolled medially, and clamps placed posteriorly to include capsular and glandular vessels. (Dixon CF: Surg Clin North Am 23:1083, 1943)

fascia as being the more or less well-developed condensation of connective tissue about the thyroid gland after the visceral compartment has been entered, this "fascia" should not be confused with the true capsule of the thyroid gland. The latter term should be reserved for the connective tissue forming the surface of the gland, binding it together, and sending septa between the lobules; the true capsule is thus an integral part of the gland, not separable from the parenchyma except by sharp dissection; in it run the larger branches of the

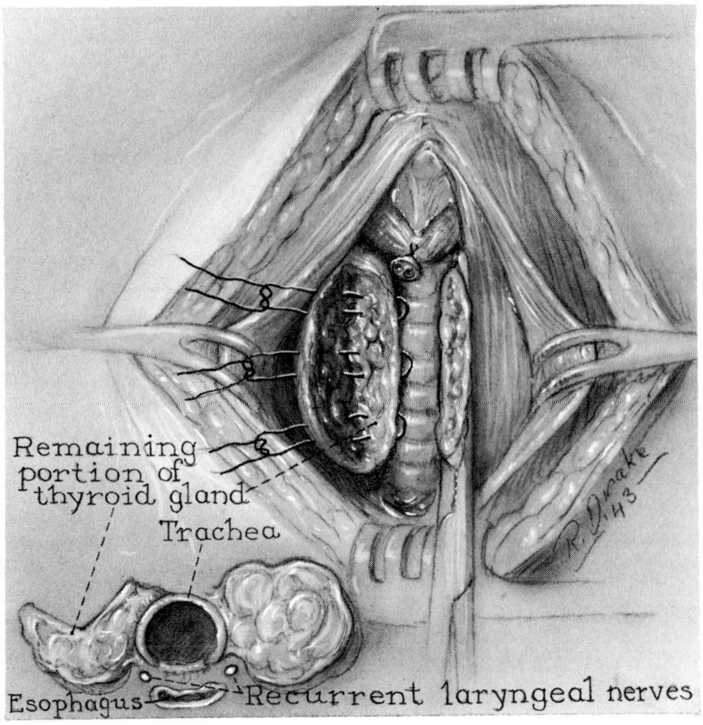

Fig. 9-49. The resection of the right lobe completed, and sutures in place for closure. Inset shows the V-shaped resection. (Dixon CF: Surg Clin North Am 23:1083, 1943)

vessels as they enter the gland, and subcapsular dissections are made by incising it.

Variations in Shape

Although the two lobes of the thyroid are usually approximately symmetrical, Marshall recorded a number of variations in the form of the gland in children; he reported upon 60 cases, a great majority of which he apparently regarded as in some degree abnormal, but these were apparently selected cases. The percentages he quoted are certainly not encountered routinely in the dissecting laboratory. In 7% of his cases one lobe was markedly larger than the other, and in one instance the right lobe was absent entirely, while in six (10%) the isthmus was entirely absent (Fig. 9-50). In these children, also, a pyramidal process was present in 43%, but in not all of these was it symmetrically placed. The pyramidal process was sometimes found connected to the isthmus and sometimes to one of the lobes; similarly, the isthmus was sometimes found to be symmetrically placed, and sometimes quite broadly confluent with one of the two lobes.

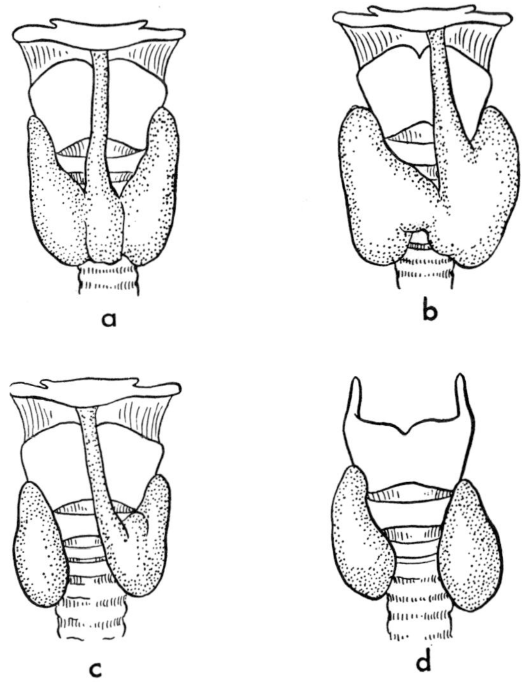

Fig. 9-50. Some unusual forms of the thyroid gland. (*a*), (*b*), and (*c*) all have large pyramidal processes extending to the hyoid bone; in (*b*) and (*c*) the process is asymmetrically attached to the thyroid gland; in (*c*) and (*d*) the isthmus is lacking. (After Marshall CF: J Anat Physiol 29:234, 1895)

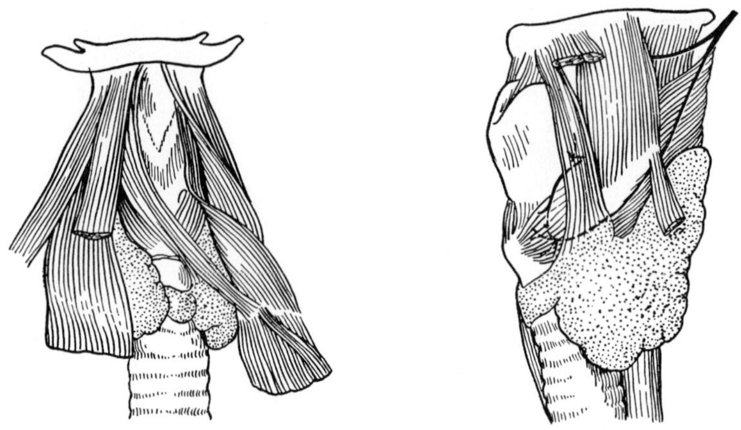

Fig. 9-51. Two forms of the levator glandulae thyroideae. In (*a*) it is pretracheal and runs obliquely from one side to the other; its nerve supply was not identified. In (*b*) it is confined to one side and associated with the infrahyoid muscles, but is innervated by the external branch of the superior laryngeal nerve. (After Eisler P: Anat Anz 17:183, 1900)

Levator Glandulae Thyroideae Muscle

An occasional anomaly encountered in connection with the thyroid gland is the occurrence of a muscular slip, or paired slips, which attach to it, descending from the region of the pharynx or the hyoid bone. Such a slip, known as the levator glandulae thyroideae muscle, is more frequent on the left side, but may be more or less bilaterally symmetrical or occur only on the right (Fig. 9-51). It is usually assumed to be a derivative of the infrahyoid muscles, but its innervation is not always in accord with this. In the case recorded by Keyes, for instance, a left levator of the thyroid gland was supplied by a branch from the external laryngeal ramus of the right superior laryngeal nerve, while no branch was found to the right levator.

Eisler, in an extensive study of these muscles and their innervations, pointed out that they probably represent any one of several discrete muscle bundles. He stated that the levator of the thyroid gland may be innervated either through the ansa cervicalis (hypoglossi) or through the vagus, usually in the latter case by the superior laryngeal branch. He regarded levators of the thyroid as divisible into three groups—anterior levators, derived from the cricothyroid and innervated by the external branch of the superior laryngeal nerve; lateral levators, derived from the thyrohyoid muscle or from the infrahyoid muscle group as a whole, attaching to the hyoid bone and innervated from the ansa cervicalis; and posterior levators, innervated like the anterior ones by the superior laryngeal branch of the vagus and apparently derived from the inferior constrictor of the pharynx. In a case that he described, the muscle, although attached to the hyoid bone, was innervated through the superior laryngeal branch of the vagus, and he expressed the belief that it represented a derivative of the cricothyroid muscle.

BLOOD SUPPLY

The thyroid gland and the suprarenal gland have each been stated to have the most abundant blood supply of any organ in the body for their size. Whichever really deserves this honor, the thyroid is a very vascular structure, being supplied by four main arteries—two superior thyroid and two inferior thyroid (Fig. 9-52 and see Fig. 9-61).

Superior Thyroid Artery

The superior thyroid artery, although smaller than the inferior, is more constant. It arises usually as the first anterior branch of the external carotid, but may originate from the uppermost part of the common carotid. Smith and Benton quoted the incidence of this origin as 5% in the Swiss and 45% in Americans; they described what they believed to be only the fourth case in which the origin was more than 2 cm below the bifurcation.

In its descent toward the apex of a lobe of the thyroid gland, the superior thyroid artery lies close against (usually just superficial to but sometimes slightly below) the external branch of the superior laryngeal nerve; the nerve can therefore be damaged when a clamp is placed upon the artery, and, if the upper pole of the gland is particularly high, the internal branch of the superior laryngeal may also be liable to damage in this manner. These nerves are discussed in connection with the larynx.

In its course toward the gland, each vessel gives off a small infrahyoid branch that runs along the lower border of the hyoid bone, a branch to the sternocleidomastoid muscle, a superior laryngeal artery that accompanies the internal branch of the superior laryngeal nerve across the thyrohyoid membrane to enter the larynx, and a cricothyroid artery that runs upon the cricothyroid ligament (membrane) close to the lower border of the thyroid cartilage. As the superior thyroid artery nears the gland it divides into an anterior and a posterior branch that then distribute numerous branches to the gland. A part of the anterior branch typically runs close to the medial side of the thyroid lobe to anastomose above the isthmus with its fellow from the opposite side.

Delitzen described one of the rare cases of an abnormal superior thyroid artery—in this case, it continued inferiorly after supplying the thyroid gland and ended in the upper part of the pectoralis major.

Inferior Thyroid Artery

The usual origin of the inferior thyroid artery from the thyrocervical trunk has already been described. Although the superior thyroid artery is variable primarily only in its exact origin, the inferior thyroid is much more variable in both its origin and relations. Typically, it ascends along the medial side of the anterior scalene muscle behind the prevertebral fascia and loops downward and medially on the anterior surface of the longus colli, here being variously related to the cervical sympathetic trunk and the middle cervical ganglion; it penetrates the prevertebral fascia and, at about the point of its branching,

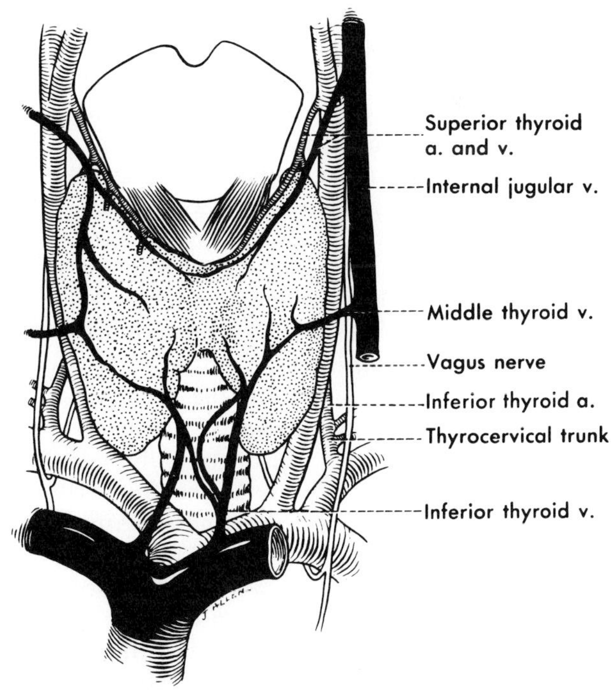

Fig. 9-52. Arteries and veins of the thyroid gland, from the front. (Hollinshead WH: Surg Clin North Am 32:1115, 1952)

crosses the more vertically directed inferior laryngeal nerve. The artery typically divides into two major branches, an upper and a lower, but may divide into a larger number; the lower branch runs toward the lower pole, the upper lies on the posterior surface of the gland at about the junction of its upper two thirds and lower third and is, according to Bachhuber, frequently the major branch of the vessel. He found it absent in only three of 200 sides.

Smaller branches are the ascending cervical artery, typically arising at about the highest part in the course of the inferior thyroid and running upward in the groove between the scalene and longus muscles; tracheal, pharyngeal, and esophageal branches; and the inferior laryngeal artery, which accompanies the nerve of the same name into the larynx.

One inferior thyroid artery is sometimes absent; Faller and Schärer found this to be true in 6% of 100 sides that they investigated and reported that the incidence of absence has been given in the literature as from 0.2% to 5.9%. When an inferior thyroid artery is absent, its place is usually taken by a branch from the superior thyroid artery of the same side or from the inferior thyroid artery of the other side; less frequently, its place is taken by a thyroid ima artery (see the following section). Rarely, doubling of an inferior thyroid artery may occur, as in the case reported by Allan. Allan found that double thyroid arteries may take the usual course of a single artery, but that sometimes, as in the case he reported, one of the vessels passes in front of instead of behind the common carotid.

Normally, anastomoses of considerable size exist between the superior and inferior thyroid arteries, of the same and the opposite sides. Stewart ('32) stated that it has been shown that the superior and inferior thyroid arteries on the same side always communicate, that the two superior, or one superior and a contralateral inferior, anastomose in 80% of glands, and that direct communications between all 4 vessels occur in 20% of patients. Presumably he was describing large superficially placed anastomoses; whether large anastomoses exist or not, the smaller branches of the various vessels anastomose abundantly with each other; moreover, there are also numerous fine anastomoses between the thyroid vessels and those of the trachea and esophagus, and any or all of the thyroid arteries may be ligated without fear of producing necrosis of the gland.

The relationship of the inferior thyroid artery to the recurrent laryngeal nerve is of considerable importance; this is recounted in more detail in a following section, but it may be stated here that the nerve more commonly passes behind the inferior thyroid artery or its main branches on the left than on the right, but may pass in front of these, or between the major branches, on either side.

Thyroid Ima Artery

Occasionally there is present, as either accessory to or replacing the inferior thyroid artery, a vessel known as the thyroid ima artery. This has been variously defined, so much so that Pratt concluded that it would never have received a special name were it not for its importance in tracheostomy, but would have been regarded simply as an aberrant inferior thyroid. As usually described, the thyroid ima artery arises from the brachiocephalic (innominate; probably the most common origin), from the right common carotid (reported as occurring in as much as 2% of bodies), from the aortic arch directly (reported to have an incidence of 0.36% or more), or, even more rarely, from the internal thoracic (mammary), an anterior mediastinal vessel (Kimmel, '49), or either subclavian.

An artery to the thyroid gland that arises from the subclavian is more commonly regarded as an inferior thyroid of aberrant origin, but any vessel to the thyroid from the other origins listed is usually designated as a thyroid ima, regardless of whether an inferior thyroid proper is present. The vessel more commonly arises on the right and tends to ascend in front of the trachea—hence its importance in tracheostomy. The reported incidence of occurrence of the thyroid ima artery varies widely, presumably in part because of the ways the vessel has been defined. Faller and Schärer found a reported incidence in the literature of from 1.5% to 12.2% and 2 cases (4%) in the 50 bodies they themselves investigated. The thyroid ima may be a large trunk fully capable of replacing an inferior thyroid or merely an accessory twig.

Venous Drainage of Thyroid

The venous drainage of the thyroid gland is by two or three pairs of veins, which anastomose freely on the gland. The superior thyroid vein emerges from the upper part of the thyroid gland and accompanies the superior thyroid artery, emptying into the internal jugular vein or the lower end of the facial (common facial) at about the level of the carotid bifurcation. The middle thyroid vein, variable both in its presence and its size, arises from the lateral surface of the thyroid at about the level of junction of the upper two thirds and lower third and drains laterally into the internal jugular vein, crossing in front of the common carotid artery; it may be torn in separation of the gland from the carotid sheath. Bachhuber found a middle thyroid vein in only slightly more than half of the 100 bodies he investigated. When it is not present, its place is taken by a vein that drains into either the superior or inferior thyroid vein.

The inferior thyroid veins typically form two trunks that descend from the lower part of the thyroid gland; the right one usually passes anterior to the brachiocephalic (innominate) artery to empty into the right brachiocephalic vein or in front of the trachea to empty into the left brachiocephalic vein, while the left passes obliquely over the trachea to join the left brachiocephalic vein. When, as sometimes happens, the right vein joins the left so that both veins drain by a common stem into the left brachiocephalic, or when branches of both unite to form a common stem between right and left veins, the common stem is often referred to as the thyroid ima vein. Even when this junction does not occur, the two inferior thyroids may anastomose freely with each other, forming a plexus (plexus thyroideus impar) in front of the trachea. They are always, therefore, a definite potential source of bleeding in a tracheostomy. The thyroid veins also receive branches from the strap muscles and from the larynx and trachea.

NERVES

Nerve fibers have been described as reaching the thyroid gland from both the superior cervical ganglion and the superior laryngeal branch of the vagus (Nonidez, '31), but degeneration experiments have apparently shown that the great majority of the fibers are sympathetic ones (Holmgren and Naumann). The vagal fibers are said to form sensory arborizations upon the blood vessels (Nonidez, '35).

Within the gland the nerves form perivascular and interfollicular (or perifollicular) plexuses; a few observers (*e.g.*, Rhinehart) have reported nerve endings about the bases of the follicular cells, but the majority of workers have failed to find such endings. Most observations have indicated that the interfollicular plexus is really part of the perivascular one and is not in contact with the follicular cells. Similarly, a variety of experiments (Smith and Moloy; Holmgren and Naumann) have indicated that the nerves are entirely related functionally to the blood vessels and that, as is usual in the case of endocrine glands, their stimulation or ablation does not materially affect the function of the glandular tissue.

LYMPHATICS

The lymphatic drainage of the thyroid gland is chiefly by vessels that accompany the arterial blood supply (Fig. 9-53). The superior lymphatic channels drain the cranial border of the isthmus, much of the medial surface of the lobes, and the ventral and dorsal surfaces of the upper part of the lobes. The inferior channels drain the major portion of the isthmus and the lower portions of the lobes. The upper channels empty into the upper deep cervical nodes; the lower ones, into the lower deep cervical nodes including the supraclaviculars and into pretracheal and prelaryngeal nodes. Mahorner and co-workers

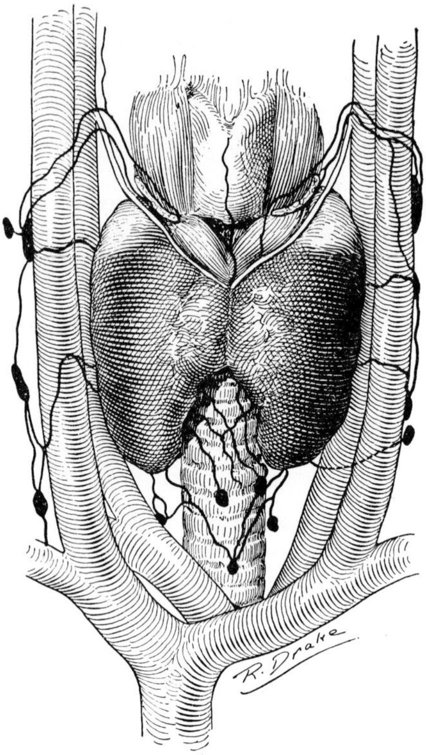

Fig. 9-53. Usual lymphatic drainage of the thyroid gland. Deep cervical and tracheal nodes are shown. (Mahorner HR, Caylor HD, Schlotthauer CF, Pemberton, J de J: Anat Rec 36:341, 1927)

described an additional drainage, from the middle of each lobe, which passes directly laterally to enter the deep cervical nodes. Although they found the majority of the effer ents from the thyroid to end in regional nodes as expected, in 3 of 23 individuals they traced a lymph channel from the right lobe directly into the right subclavian vein, without passing through a node.

RELATIONS OF THE RECURRENT LARYNGEAL NERVE

Among the most important relations of the thyroid gland are those to the recurrent laryngeal nerves and the parathyroid glands. The right recurrent laryngeal nerve arises from the vagus as this trunk passes in front of the subclavian artery, curves below and behind the subclavian, and then passes upward, angling somewhat medially, to run in the visceral compartment of the neck. The left recurrent laryngeal arises from the left vagus as this passes in front of the arch of the aorta, the nerve passing below and upward behind the aorta at the point of attachment of the ligamentum arteriosum. Its course in the neck is essentially similar to that on the right side, since it ascends more or less in the angle between the esophagus and trachea, in the visceral compartment.

Close to their origins the two recurrent laryngeal nerves usually give rise to cardiac branches, and as they ascend on either side of the trachea they give off branches to the trachea and the esophagus. The terminal portion of each nerve, entering the larynx, is the inferior laryngeal nerve; its distribution has already been discussed in connection with the larynx.

As the recurrent laryngeal nerves ascend in the neck they necessarily cross the inferior thyroid arteries (which arise lateral to them and turn medially to reach the thyroid gland); the nerves also come in more or less direct contact with the posterior or posteromedial surface of the thyroid gland. The relationship between the nerve on the one hand and the artery and gland on the other is of considerable importance to surgeons—especially if they plan to attempt a total thyroidectomy or removal of tumors from the posterior surface of the gland. Unfortunately, these relations are not constant.

Relationship to Trachea

The two recurrent nerves are not necessarily tucked into the groove between the trachea and esophagus; they may run considerably farther laterally than this, and the right is especially likely to do so (see Fig. 9-61). Berlin, for instance, noted that the right recurrent nerve may be as much as a centimeter lateral to the trachea; in dissections upon 70 cadavers, he found the right recurrent nerve to be in the tracheoesophageal sulcus in only 42 of these, the left to be similarly situated in 49; in a similar series of 70 operations, in only some of which the nerve was exposed, he concluded that the nerve was in the sulcus in 41 cases on the right and in 45 on the left.

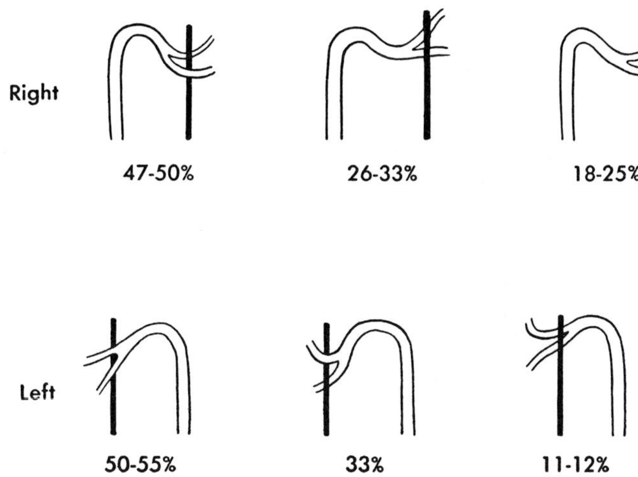

Fig. 9-54. Three chief relations between the recurrent laryngeal nerve (*black*) and the inferior thyroid artery, and the approximate percentages in which these have been described as occurring on each side. (Hollinshead WH: Surg Clin North Am 32:1115, 1952)

Relationship to Inferior Thyroid Artery

The relationship of the recurrent nerve to the inferior thyroid artery is also variable (Fig. 9-54). Depending upon the distance from the tracheoesophageal groove at which the recurrent nerve lies, and upon the point at which the artery divides, the nerve may cross either the main artery or its branches. Further, it may pass behind the artery, in front of it, between its two major branches, or among minor branches of these; or the nerve may divide before reaching the level of the artery and a part pass in front of and another part behind the artery or some of its branches. These variations have been the subject of a number of investigations that have sought to establish the relative frequency of the different patterns so constantly observed in the dissecting room, but the tabulations of various workers do not agree very closely.

From the combined series of Taguchi (1889) and Fowler and Hanson, totaling 522 sides, it would appear that the nerve is more commonly posterior to the artery than anterior to it (in 299 instances it was posterior; in 114, anterior); Taguchi described it as passing between the branches of the artery in 75 of his 122 cases, however, while Fowler and Hanson, disregarding the smaller branches of the artery, reported the nerve as between branches of the artery in only 34 of their series of 400.

Neither of these papers indicated a possible difference between the right and left sides, since sides were not recorded. Nordland, in a short series of 62 sides, found the nerve anterior to the artery more often than posterior, but no marked difference in sides, while Berlin and Lahey, and Berlin, the latter in a series of 70 dissections, found that the left nerve was more commonly posterior to the artery than was the right. Bachhuber's series of 200 sides and Reed's ('43) series of 506 sides support this conclusion and are also in fair agreement with each other. On the right side, Bachhuber found the nerve posterior to the artery in 33%, anterior to it in 18%, and passing between branches of the vessel in 48%, while Reed found it posterior in 26.9%, anterior in 25.7%, and between branches in 47.6%; on the left side, Bachhuber found the nerve posterior in 55%, anterior in 11%, and between in 33%, while Reed found it posterior in 51.4%, anterior in 11.5%, and between in 33.1%. Bachhuber found the nerve dividing into branches while related to the artery only twice (1%), while Reed found it thus dividing in 5.4%.

From the various figures quoted above, it is obviously impossible to establish any statistically valid conclusions; the situation appears even more complicated by the fact that in 253 bodies, Reed found a total of 28 different types, and that in only 17% of bodies were the relations of the nerve to the artery alike on both sides. However, it appears obvious that

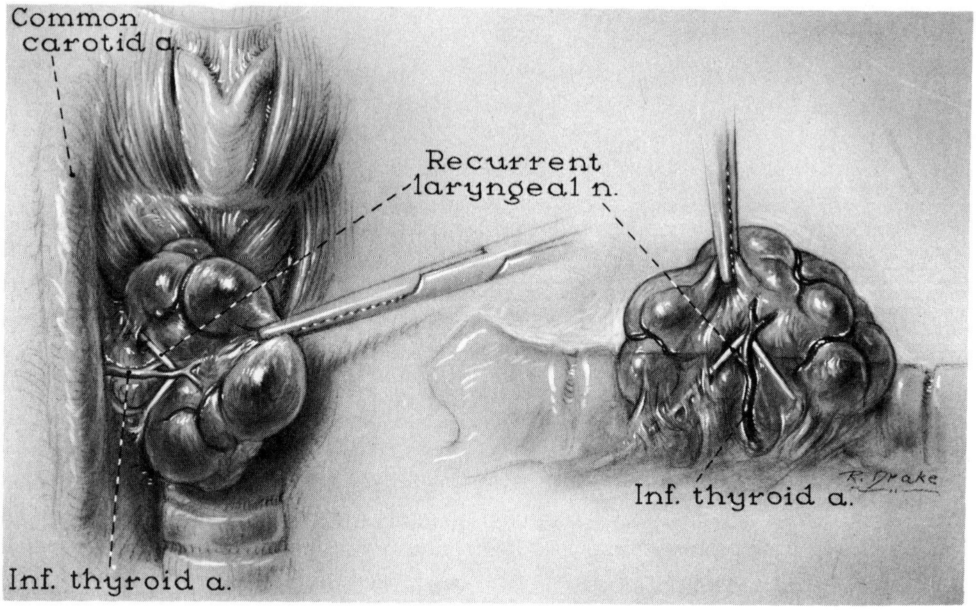

Fig. 9-55. The recurrent laryngeal nerve pulled forward by the inferior thyroid artery during thyroidectomy. (Black BM: The Cyclopedia of Medicine, Surgery, Specialties, 3rd ed, Vol 14, p 173. Philadelphia, FA Davis, 1962)

the nerve rather frequently passes between either the main or minor branches of the artery and that this is somewhat more common (about 50%) on the right; on the left the most common position of the nerve (approximately 50%) is behind the artery, and only relatively rarely (about 10%–12%) does it pass in front, while on the right side it is in front and behind in more nearly equal proportion (roughly 25%).

It is the varying but close relation between the inferior thyroid artery and the recurrent nerve that makes it dangerous to attempt to control bleeding at this level by the use of hemostats, and that has led to the practice of ligating the artery well laterally. Also, the fact that the nerve may pass in front of the artery or some of its branches means that it is held relatively close to the gland and is necessarily dragged forward as the lobe is rotated forward at operation (Fig. 9-55).

Mountain and co-workers reported that their routine attempts to expose the nerve during thyroid operations resulted in a very low incidence of paralysis. Flisberg and Lindholm located the nerve by recording from the vocalis muscle while they moved stimulating electrodes in its vicinity and varied the intensity of the stimulus.

Relationship to Thyroid Gland

The closeness of the relationship between the recurrent laryngeal nerve and the thyroid gland during operation varies in part according to whether the nerve lies deeply in the tracheoesophageal groove or more laterally, where it is necessarily more closely applied to the posterior surface of the gland, and in part according to whether the nerve passes in front of the inferior thyroid artery or some of its branches. There is also an additional factor affecting this relationship: At the level of the upper two or three tracheal rings, the thyroid gland is attached to the trachea by a considerable quantity of connective tissue (variably called the adherent zone or the lateral or suspensory ligament); the recurrent nerve at this level—that is, at about the middle third of the thyroid gland—is here in closest contact with the gland and may lie against its posterior surface, pass through the adherent zone, or even penetrate the gland itself (Fig. 9-56).

Of 140 nerves dissected in cadavers, Berlin found 10% actually penetrating the gland for a short distance and 25% passing through the adherent zone; in a similar number of operations he identified the recurrent nerves in 52% and found the recurrent penetrating the thyroid gland in 7% and passing through the adherent zone in 31%. Thus in a total thyroidectomy or removal of a tumor from the posterior surface of the thyroid it is entirely possible to elevate the recurrent nerve with the lobe of the gland and therefore to section it inadvertently in removing thyroid tissue.

Extralaryngeal Division of Nerve

The recurrent or inferior laryngeal nerve is usually described as dividing into anterior and posterior branches as it enters the larynx, but may do so outside the larynx. It has already been noted that Bachhuber found this to be true in at least 1%, and Reed found it in at least 5.4% of cases (they apparently listed only the extralaryngeal divisions that occurred while the nerve was in contact with the inferior thyroid artery), but Weeks and Hinton found extralaryngeal division to be the rule in the small number of cases they investigated—in only 1 of 20 sides did the nerve fail to divide before entering the larynx, and the level of division varied from 1 to 4 cm below this entrance.

Such a high percentage has not been found in more extensive series, but extralaryngeal division of the nerve is certainly common rather than rare. King and Gregg described the nerve as dividing while in relation to the thyroid gland in 8 of 32 sides (25%); they stated that in some of these cases the two branches of the nerve were a full 0.6 cm apart and called attention to the significance of this fact, which may allow differential injury to branches of the nerve, in the interpretation of the behavior of paralyzed vocal cords. Armstrong and Hinton reported extralaryngeal division of the nerve in 73% of 100 sides; Morrison found the nerve divided 0.5 cm or more below the level of the cricothyroid joint on at least one side in 43% of 100 cadavers and stated that in the majority of these the divided nerve lay in the region involved in operations upon the thyroid gland. There is apparently no significant difference in the incidence of extralaryngeal division of left and right nerves. The consequences of injury to the recurrent or inferior laryngeal nerve are discussed in connection with the larynx (see Chap. 8).

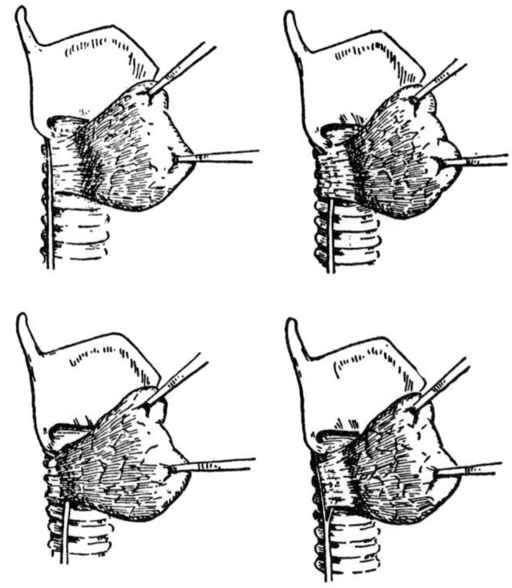

Fig. 9-56. Varying relations of the recurrent nerve to the adherent zone (suspensory or lateral ligament) of the thyroid gland. (Berlin DD: Surg Gynecol Obstet 60:19, 1935, by permission of Surgery, Gynecology & Obstetrics)

Anomalous Recurrent Nerve

Finally, it should be mentioned that the recurrent laryngeal nerve may be anomalous, in that it may arise from the cervical portion of the vagus at about the level of the larynx or thyroid gland, and run rather directly to the larynx without looping about the subclavian artery and thus without ascending through the major part of its usual course. An anomalous inferior laryngeal (which is not, strictly speaking, a recurrent nerve, although it may loop around the inferior thyroid artery) is found far more frequently on the right side, since it is always associated with an anomalous (retroesophageal) subclavian artery (Fig. 9-57). Only when there is a right instead of a left arch of the aorta *and* ligamentum arteriosum, and at the same time an anomalous or retroesophageal left subclavian artery, can a similar disposition of the inferior laryngeal

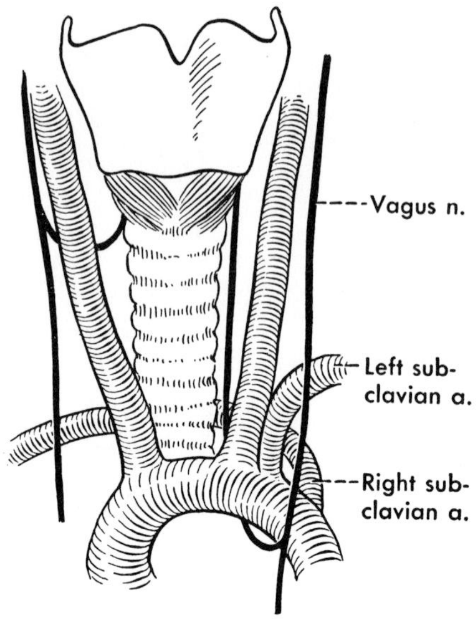

Fig. 9-57. Anomalous right "recurrent" nerve. Note the anomalous (retroesophageal) right subclavian artery, the occurrence of which is reponsible for the anomaly of the nerve.

nerve occur on the left side. Retroesophageal right subclavian arteries are believed to occur in about 1% of individuals. They are discussed in connection with the thorax, and the embryology that is responsible for the associated anomaly of the inferior laryngeal nerve is indicated in Figure 9-58.

THYROGLOSSAL CYSTS AND FISTULAS

Since the thyroid gland, or at least the major portion of it, develops as a diverticulum from the floor of the pharynx at the level of the first pharyngeal pouches, ectopic thyroid glands and remains of the one-time connection of the gland to the pharynx (thyroglossal duct) may be located anywhere along the original pathway of this duct between the foramen cecum on the tongue and the thyroid gland itself. A common occurrence of thyroid tissue outside the major body of the gland is the pyramidal lobe, usually in the midline and sometimes extending as high as the hyoid bone; although this is generally not regarded as ectopic thyroid tissue, because of its common occurrence, failure to remove it may sometimes be the cause of a recurrence of goiter. The most striking type of ectopic thyroid, which may represent the entire thyroid anlage or may be coexistent with thyroid tissue in the neck, is the lingual thyroid (see Fig. 7-51). A very rare type is midline cervical ectopia, apparently often mistaken for thyroglossal duct cyst. Meyerowitz and Buchholz added two cases in which this tissue represented the total thyroid mass to the 18 cases previously reported.

Persistence of portions of the thyroglossal duct itself, which normally begins to atrophy about the fifth to sixth week of development, may give rise to fistulous tracts connected to the tongue or to cysts along the original course of the duct. External fistulous openings on the neck, if developed from the thyroglossal duct, are necessarily secondary openings, since the duct does not, of course, open to the outside.

Thyroglossal cysts and fistulas have been discussed in connection with the tongue, for the cysts may have a suprahyoid position; the large majority, however, are infrahyoid (*e.g.,* Hubert; Marshall and Becker). They occur most commonly immediately below the hyoid bone, and according to Marshall and Becker cysts at this level are often displaced to the left side. Some retrosternal cysts have also been thought to be thyroglossal ones, but if so must represent an abnormal migration of the remains of the duct. Finally, it should be emphasized again that thyroglossal cysts and fistulas require complete surgical removal; if the tract extends both above and below the hyoid bone a midsection of the bone itself should be removed and the tissue about the fistulous tract should be dissected with care. The duct may pass behind the hyoid bone, in front of it, or through it.

LATERAL ABERRANT THYROID GLANDS

Small lateral paired diverticula from the bases of the fourth pharyngeal (branchial) pouch, known as the ultimobranchial or postbranchial bodies (Fig. 9-59), typically become associated with the thyroid gland in an early stage of development and are incorporated into it. Whether they contribute true thyroid follicles has been debated for years, but their

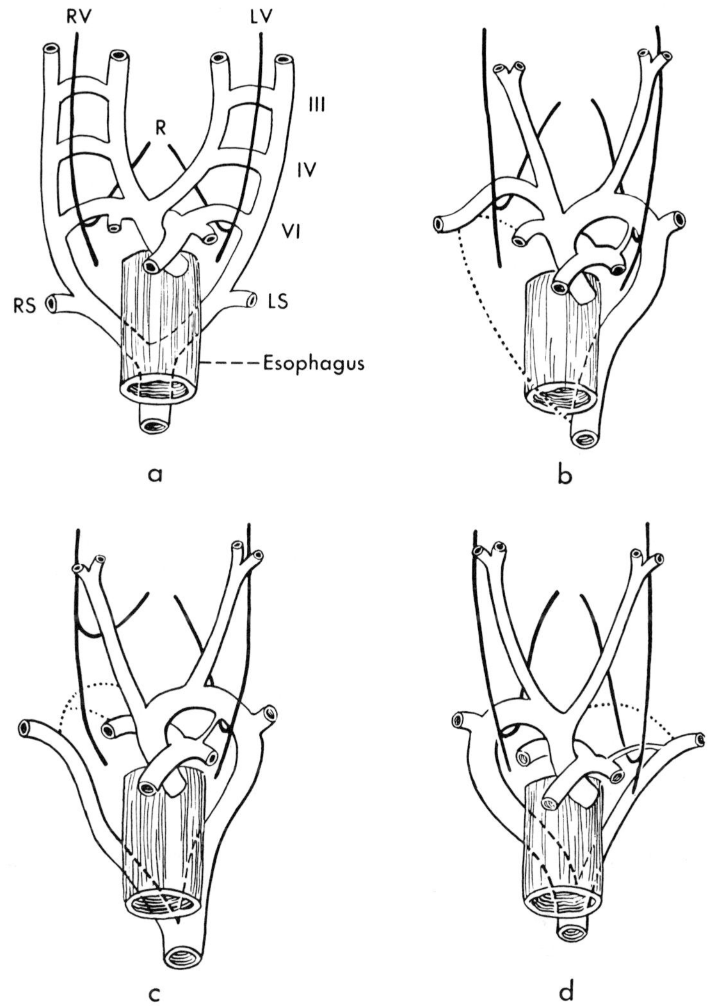

Fig. 9-58. Development of retroesophageal subclavian arteries and the relations of the recurrent branches of the vagus nerves. (*a*), The caudal part of the aortic arch system in an early stage of development; (*b*), the normal adult condition. *RV* and *LV* denote the right and left vagi, respectively; *R,* the recurrent nerves; *RS* and *LS,* the aortic segmental arteries that form the major parts of the two subclavians; the *Roman numerals* identify the aortic arches. *Broken lines* in (*b*), (*c*), and (*d*) represent parts of the aortic arch system that have disappeared. In (*a*) both recurrent nerves pass caudal to the sixth aortic arches. In (*b*) the disappearance of the distal part of the right sixth aortic arch has left the right recurrent nerve in contact with the fourth arch, or base of the subclavian artery, while the left nerve turns around the junction of the ligamentum arteriosum (sixth arch) and the arch of the aorta. In (*c*) the disappearance of the fourth arch and a proximal part of the dorsal aortic root has resulted in the right subclavian artery arising retroesophageally from the aorta and in the right recurrent nerve taking a rather direct course to the larynx instead of looping around the subclavian. In (*d*) the right fourth arch and the dorsal aortic root form the arch of the aorta, and the disappearance of the left fourth arch has resulted in a retroesophageal origin of the left subclavian; however, the persistence of the distal portion of the left sixth arch as the ligamentum arteriosum retains the recurrent nerve in the thorax.

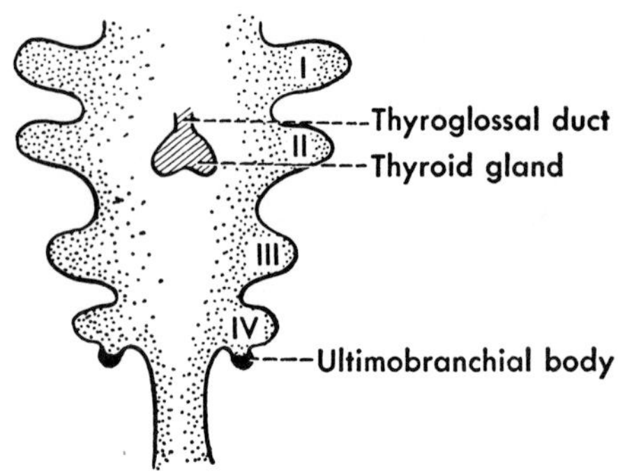

Fig. 9-59. Ventral view of the pharynx showing the origin of the thyroid gland and the position of the ultimobranchial body (which has been regarded as contributing to the thyroid—see text). The *Roman numerals* identify the pharyngeal pouches.

interest to the surgeon lay in Weller's statement that when they do not join the thyroid gland they give rise to aberrant, laterally lying, thyroid tissue in which tumors may develop. The evidence against this seems to be conclusive. Although Godwin thought the ultimobranchial tissue that becomes incorporated in or closely associated with the thyroid gland formed both true thyroid follicles and parafollicular or interstitial cells (and apparently thought either could be transformed into the other), he found no follicles in ultimobranchial derivatives not associated with the thyroid gland; Ramsay transplanted from mouse embryos the branchial pouch bearing the ultimobranchial body and found thymus and parathyroid but no thyroid tissue developing therefrom. It now seems well established that so-called lateral aberrant thyroid tumors are actually metastatically involved lymph nodes (*e.g.,* King and Pemberton; Warren and Feldman). The matter becomes even more academic with the finding of Polak and co-workers that the parafollicular cells, the secretors of calcitonin whose origin has been definitely traced to the ultimobranchial bodies, are actually derived from neural crest cells that migrate into the ultimobranchial body.

Fraser and Duckworth found cysts of the ultimobranchial body in fetuses to be associated mostly with the upper half of the thyroid gland, the part where calcitonin is usually found. In one instance a cyst was closely associated with a parathyroid gland; apparently it is known that parathyroid glands sometimes contain calcitonin.

INTRATHORACIC GOITER

In addition to the undoubted occasional occurrence of thyroid tissue along the path of the thyroglossal duct, the thyroid gland may also be displaced downward so as to become "substernal" (really retrosternal) or actually intrathoracic. This is apparently not due, however, to excess caudal migration of the thyroid gland during its development, but simply represents a goitrous expansion of the gland behind the sternum. Lahey and Swinton found that the sex incidence of intrathoracic goiter corresponds to the occurrence of goiter of any type—that is, that most of the individuals showing this are females. An intrathoracic goiter normally exerts pressure upon the trachea, thus interfering with breathing; Lahey and Swinton stressed the fact that any patient with mechanical interference with breathing or with dilation of superficial thoracic veins (from pressure upon the brachiocephalic veins or superior vena cava) should be regarded as possibly goitrous.

Sharer reviewed 212 patients in whom the lower border of the thyroid gland upon swallowing remained definitely below the jugular (suprasternal) notch; in 4% of these, some part of the gland extended below the arch of the aorta. He stated that most aberrant retrosternal goiters probably have some connection with the parent gland, although this may be merely a fibrous cord. As Lahey and Swinton pointed out, a goitrous expansion of the thyroid in a retrosternal position must extend farther into the thorax if its diameter is increased, and therefore enlargement tends to displace it downward still more. Sweet suggested that the occasional intrathoracic goiter in the posterior mediastinum arises from lateral or posterior parts of a lobe, while the more common one in the anterior mediastinum arises from the isthmus or the lower pole of a lobe.

OTHER ABERRANT THYROID TISSUE

Thyroid tissue has been reported in the larynx, trachea, pons, pleura, pericardium, and ovaries; in some of these cases it is obviously impossible to conceive of the thyroid gland of pharyngeal origin as having participated in the formation of these misplaced bits of thyroid tissue; for instance, the thyroid tissue occasionally reported as present in the ovaries (*e.g.,* Masson and Mueller) must be due to a teratoma or metaplasia of the tissues there.

Krafka expressed the belief that intratracheal thyroid is due to an abnormal early fixation of the tip of a lobe to the trachea, so that the cartilages grow around and enclose it—he did not regard the trachea as actually being invaded by the thyroid. At the time of his writing, he was able to find 39 cases reported in the literature. Meyers and Pantango stated that thyroid tissue accounts for approximately 7% of all primary intratracheal tumors.

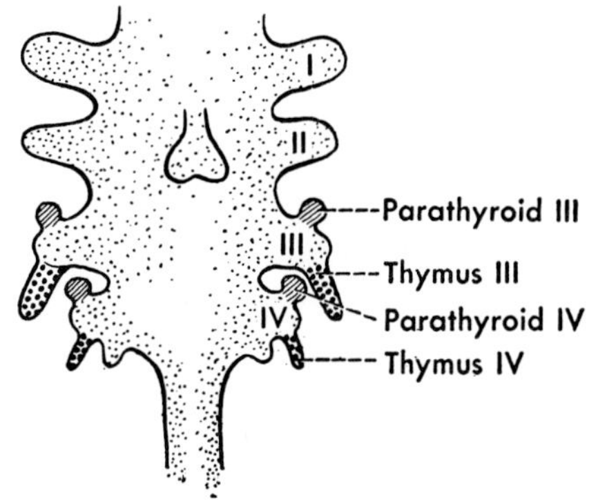

Fig. 9-60. Ventral view of the pharynx showing the origin of the parathyroid glands from the pharyngeal pouches (identified by the *Roman numerals*). The thymus, arising chiefly from the third pouch, and the thyroid gland (unlabeled) are also shown.

PARATHYROID GLANDS

Development

The parathyroid glands, typically four in number, develop from the dorsal extremities of the third and fourth pharyngeal (branchial) pouches (Fig. 9-60) and are sometimes therefore designated as parathyroid glands III and parathyroid glands IV. In the course of their development the parathyroids derived from the third pouches soon become associated with the thymus, which develops primarily from these pouches, while those from the fourth pouches become associated with the developing thyroid gland. It is the association between the parathyroids III and the thymus that induced Weller to refer to these glands as the "parathymus"—justifiable embryologically perhaps, but a needless addition to anatomical synonyms.

As the thyroid and thymus, with their associated parathyroid glands, move caudally from the region in which they originate, the thymus normally descends beyond the level at which the thyroid halts. The parathyroids derived from the third branchial pouches are therefore typically carried farther caudally than are those derived from the fourth branchial pouches. Thus the parathyroids from the fourth pouches are typically located more cranially on the thyroid gland and are called the superior parathyroid glands, while those derived from the third pouches are typically freed from the thymus and become associated with the thyroid gland toward its lower pole, becoming the inferior parathyroid glands.

Number

From this method of development, one would expect exactly four parathyroids in each individual, but both more and fewer have been reported. Thus Heinbach, for instance, found an average of only 3.2 parathyroids in 25 individuals and expressed the belief that four parathyroids occur in no more than 50% of persons. On the other hand Norris ('37), working on embryos and fetuses, found no instances in which there were fewer than four parathyroids and suggested that reports of fewer are due to the difficulty of finding them in gross dissection. It might be remarked that the parathyroid glands are usually difficult to recognize in the cadaver with any certainty and may easily be overlooked, or even improperly identified, unless they are checked histologically; some of the data upon these glands are probably inaccurate because of

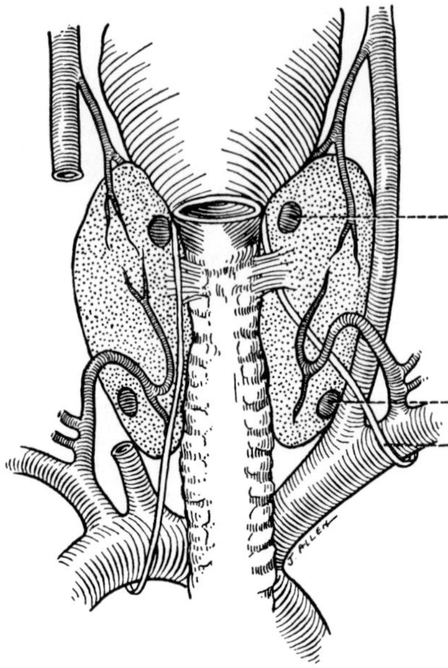

Fig. 9-61. Common positions of the parathyroid glands on the posterior aspect of the thyroid gland. (Hollinshead WH: Surg Clin North Am 32:1115, 1952)

failure to do this. They are somewhat more easily recognized in the living individual, since their pinkish or yellowish color contrasts somewhat with the darker red of the thyroid gland itself.

Gilmour and Martin, in a detailed anatomical study, found four parathyroids in about 80% of 527 persons, more than four in 6%, and fewer than four in only 14%; they attributed the finding of fewer than four to failure to find all that were present and stated that the combined weight of the glands in these cases suggested that one had been overlooked. Supernumerary parathyroids apparently arise by division of a single parathyroid; Godwin described this as a regular occurrence in the dog, and Norris ('37) found evidence of its occasional occurrence in the human fetus.

Location

Both sets of parathyroid glands are typically located upon the posterior aspect of the lateral lobes of the thyroid gland (Fig. 9-61) but this is only a general rule with many exceptions; because of the manner in which they arise and migrate into the neck, the glands are often displaced and may be located in other positions on the thyroid gland or lie above or below it (Figs. 9-62 and 9-63). MacCallum, for instance, in one of the early experimental

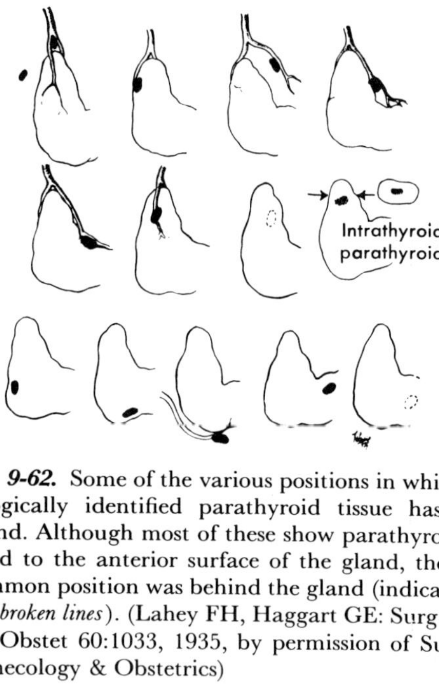

Fig. 9-62. Some of the various positions in which histologically identified parathyroid tissue has been found. Although most of these show parathyroids related to the anterior surface of the gland, the most common position was behind the gland (indicated by the *broken lines*). (Lahey FH, Haggart GE: Surg Gynecol Obstet 60:1033, 1935, by permission of Surgery, Gynecology & Obstetrics)

and anatomical studies of these glands, found the "typical" arrangement in only 20 of 64 carefully studied necropsy specimens.

Although parathyroid tissue usually lies in the thyroid region, it may occur anywhere between the level of the bifurcation of the common carotid and the mediastinum; in the one case, the parathyroid tissue has remained essentially in the location in which it first developed, while in the second case it has probably failed to free itself from the thymus and has therefore traveled with this tissue into the mediastinum. A search for a parathyroid tumor may therefore be a task of considerable difficulty; however, parathyroid glands are not found scattered indiscriminately through the tissues of the neck, but are confined to the visceral compartment or the tissue connected with the thymus.

According to Lahey and Haggart the most constant position for the superior parathyroid gland (parathyroid IV) is on the posterolateral aspect of the upper pole of the thyroid gland, where it rests against the "laryngeal cartilage"; that for the inferior is close to the entrance of the inferior thyroid artery and below the lateral exits of the middle and superior thyroid veins. Black, more specifically, described the upper parathyroid gland as normally lying more posteriorly, upon the pharynx or esophagus rather than directly on the thyroid gland; it then lies dorsal to the recurrent nerve and the inferior thyroid artery and is usually not as near the upper pole as about the level of junction of middle and upper thirds of the lobe. It is, in his experience, more easily found than is the inferior parathyroid gland. He stated that the inferior parathyroid gland usually lies near the inferior pole of the thyroid, in a more ventral plane than the superior, and usually anterior to the recurrent laryngeal nerve and the inferior thyroid artery. He found it most commonly associated with the posterolateral surface and inferior pole of the thyroid gland, usually near, but anterior to, branches of the inferior thyroid artery and somewhat medial to branches of the inferior thyroid vein. In contrast, Heinbach, among the glands that he identified, found 7% opposite the cranial third of the thyroid, 57% opposite the middle third, 25.6% opposite the caudal third, and 10.4% entirely caudal to the thyroid gland (Fig. 9-63); he found only one gland lying partly on the anterior surface and none inside the thyroid gland. Because of the varying positions, he questioned the advisability of trying to designate the glands as superior and inferior.

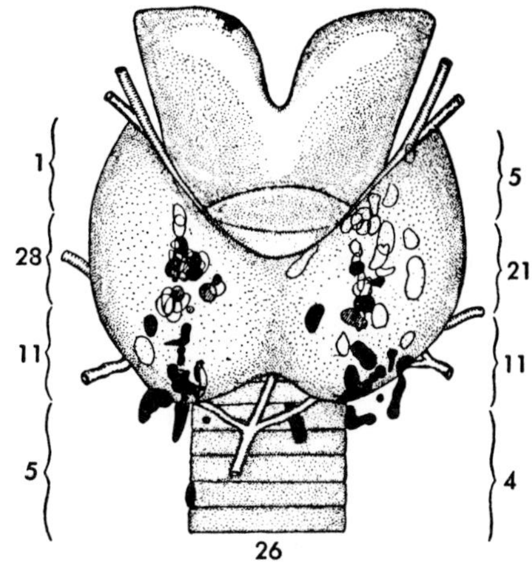

Fig. 9-63. Composite diagram indicating the positions of all the parathyroid glands found in 25 bodies, represented as if the thyroid gland were transparent. The *brackets* along the sides indicate, from above downward, the upper, middle, and lower thirds of the thyroid gland and a region below the gland; the *figures* indicate the numbers of parathyroid glands found in each of these regions. Those glands that appeared to be superior ones are shown in *outline,* those that appeared to be inferior ones are shown *solid,* and those that could not be assigned to either group are *shaded.* (Heinbach WF Jr: Anat Rec 57:251, 1933)

In consequence of the usual positions posterior to the thyroid gland, it has long been an axiom among surgeons particularly concerned with the thyroid gland that the posterior part of the capsule of the gland should be left intact if possible, in order to preserve sufficient parathyroid tissue to avoid tetany. As indicated in the following paragraph, the parathyroid glands have such variable positions that some parathyroid tissue may be removed in subtotal thyroidectomy and some may remain after total thyroidectomy. However, Chamberlin and co-workers regarded postoperative tetany after total thyroidectomy as a fairly common occurrence; they advocated that a part of the procedure of total thyroidectomy should be careful isolation of the parathyroid glands with their blood supply and an adjacent part of the thyroid capsule.

Millzner found parathyroid glands on the anterior or lateral part of the capsule of the thyroid in approximately 30% of all bodies and actually on the anterior capsule in approximately 10%; in these observations, made upon 740 surgical specimens and 56 cadavers, he found that the inferior parathyroids assume such positions approximately twice as often as do the superior. Norris ('37) found parathyroid III (the lower parathyroid) lying in the tissue lateral to the trachea and back of the posterior edge of the thyroid lobe in 75% of fetuses, embedded in the posterior edge of the lobe in 15%, embedded in the lateral surface of the gland in 5%, and anterior to or below the lower thyroid pole in 5%. He found parathyroid IV between the posterior edge of the lobe and the lateral aspect of the esophagus (in the loose areolar tissue here) in 60%, in much the same position except against the posterior edge of the thyroid in 25%, in close relationship to the anterolateral wall of the esophagus in 10%, and lateral to the gland in 5%. Lahey and Haggart have illustrated the many positions (posterior, lateral, anterior) in which they have identified parathyroid tissue, including the rare intrathyroid position of the parathyroid.

In view of these varying positions of the parathyroid gland, it is obviously useless to describe them as having a fixed relation to fascial layers associated with the thyroid gland—that is, as being either routinely embedded in the true capsule of the gland or between the true and the false capsule. Walton stated that in his experience the superior parathyroids always lie between the "thyroid fascia" and the gland, but superior parathyroids may in fact be located some distance upward, along the superior artery, for instance. He described the inferior parathyroids as having two normal positions: upon the thyroid gland below the inferior thyroid artery as it lies on the posterior surface of the gland, in which case, he stated, they lie anterior to the thyroid fascia; and above the inferior thyroid artery, in which case they lie outside of the thyroid fascia. Rarely, he noted, they are actually embedded in the gland. He regarded the relationship of the parathyroid to the fascia as important, from the standpoint of the directions in which tumors of it may be displaced, but to what degree this is so is not clear.

Black stated that an abnormally placed parathyroid tumor in the mediastinum (Figs. 9-64 and 9-65) may be in either the anterior or posterior portions of the superior mediastinum if derived from the inferior gland, while mediastinal tumors of the superior gland usually lie in the posterior portion of the mediastinum. He suggested that this displacement may be due either to abnormal migration depending upon their embryology or to a consequence of their enlargement, in which the same forces that cause intrathoracic thyroid tumors play a part. Reinhoff stated that approximately 20% of parathyroid tissue is displaced as a result of these two factors; Norris ('47), in an extensive review of parathyroid adenomas, found that 10.7% of 322 of them were reported to be definitely aberrant in position—the majority being located in the mediastinum.

Blood Supply

However close their relation to the thyroid gland and the thyroid arteries, there need be no hesitation about ligating the blood supply to the thyroid gland for fear this will interfere with the function of the parathyroids. These glands, when located upon the thyroid gland, share with it the abundant anastomoses that exist between the thyroid arteries and the arteries of the larynx, pharynx, trachea, and esophagus; Curtis stated that each gland normally receives a single artery that is usually derived from the inferior thyroid (for both the inferior and superior parathyroids), but may be derived from anastomotic vessels between the two. Chamberlin and co-workers agreed that the superior parathyroid artery may arise either from a major branch of the inferior thyroid artery or from an anastomosis between superior and inferior thyroid arteries and described the inferior parathyroid artery as regularly arising from the inferior thyroid. Curtis was able to inject the parathyroid glands after ligation of the inferior or of all four thyroid arteries. However, such injection apparently depends upon the anastomoses among the thyroid arteries and the anastomoses of these

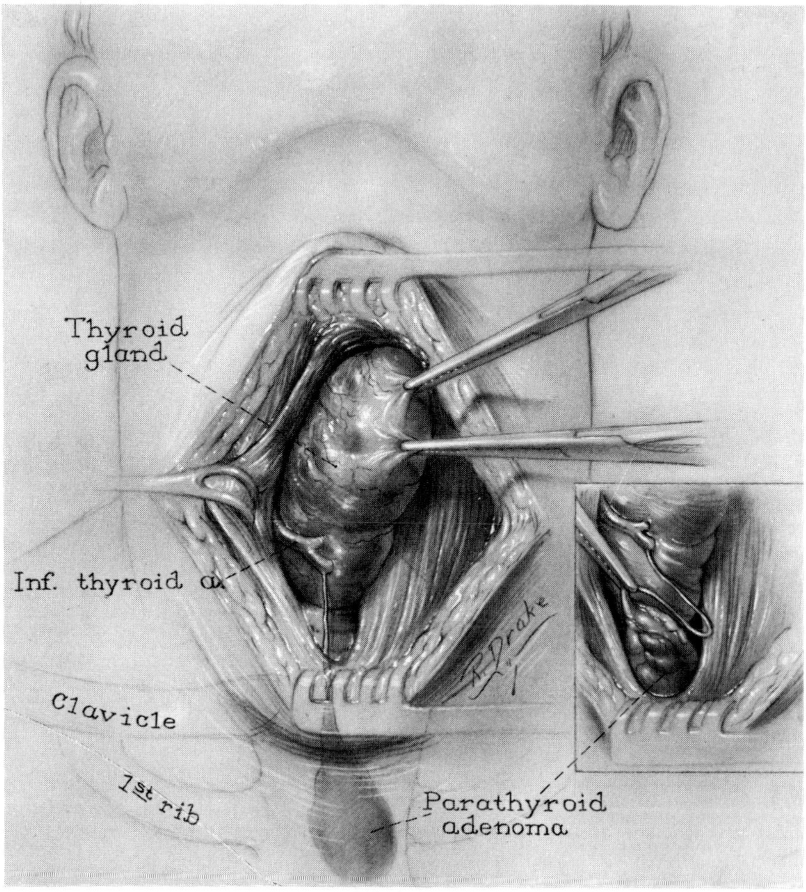

Fig. 9-64. A mediastinal parathyroid adenoma with its blood supply from the inferior thyroid artery. (Black BM: Surg Gynecol Obstet 87:172, 1948, by permission of Surgery, Gynecology & Obstetrics)

with adjacent vessels such as those of the trachea and esophagus, because the parathyroid arteries are end arteries, according to Chamberlin and co-workers. Consequently, they advocated that in total thyroidectomy the smaller branches of the inferior thyroid artery be ligated as the gland is mobilized, and that as soon as the parathyroid glands are recognized they and their blood supply should be moved away from the thyroid gland.

CERVICAL THYMUS

Because of the origin of the thymus from the branchial region (the third pharyngeal pouches), it sometimes occurs that during its migration into the mediastinum accessory bits of thymic tissue may be disconnected from the body of the thymus and remain in the neck, or processes from the lobes of the thymus, marking the path of migration, may extend upward through the superior thoracic aperture to the thyroid cartilage or higher. Harman, for instance, described a case in which two processes from the thymus passed up into the neck, the longer one continuing above the level of the thyroid cartilage to lie under cover of the angle of the mandible; there was also an accessory thymus, detached from the main body, at about the lower border of the thyroid gland. Similarly, Bien described two cases in which in newborn infants a portion of the thymus extended upward

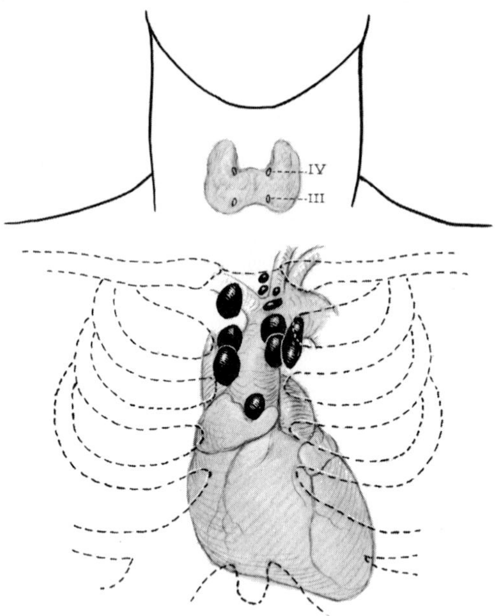

Fig. 9-65. Locations of 11 mediastinal parathyroid adenomas. (Courtesy Dr. B. M. Black)

into the neck to come into close relation to the hyoid bone.

TRACHEA AND ESOPHAGUS

Trachea

The pharynx and larynx have been discussed in the preceding chapter. The trachea begins at the lower border of the cricoid cartilage, extending downward therefore from about the level of the sixth cervical vertebra. Its walls are kept from collapse, and largely formed, by a series of cartilages having a somewhat C-shaped form, with the open ends directed posteriorly; these rings are united to each other by elastic membranes, the annular (anular) ligaments, and the posteriorly lying space between the ends of the cartilages is occupied by a thin muscular coat, the trachealis muscle.

The important anterior relationships of the trachea in the neck include the thyroid gland, the isthmus of which crosses the anterior surface of the trachea at about the level of the second, third, and fourth cartilages; below this are the inferior thyroid veins, tracheal lymph nodes, and sometimes a thyroid ima artery, and anterior to all these are the infrahyoid or strap muscles (see Figs. 9-52 and 9-44). Laterally are the lobes of the thyroid gland, which may extend somewhat behind, and the great vessels of the neck; posterolaterally are the recurrent nerves lying more or less in the tracheoesophageal groove. Posteriorly, the trachea is related throughout its length to the esophagus. The brachiocephalic (innominate) artery usually crosses the trachea obliquely in a retrosternal position, but sometimes it appears anterior to the trachea in the lower portion of the neck. Close to the jugular (suprasternal) notch, the jugular venous arch (between the two anterior jugular veins) frequently passes transversely superficial to the strap muscles.

The relations of the structures in front of the trachea must be borne in mind in doing a tracheostomy; because of the attendant dangers of subsequent stenosis of the larynx high tracheostomies, that is, above the level of the thyroid gland, are not considered safe; tracheostomies are done below the level of the thyroid isthmus, at about the level of the fourth, fifth, and sixth tracheal rings, or through or behind the isthmus, at about the levels of the second, third, and fourth rings. Many surgeons much prefer this higher approach, finding that the lower ones are more difficult in their subsequent management, and in general less satisfactory. In incisions at the lower level, moreover, the inferior thyroid veins are necessarily encountered, and even if an uncosmetic incision in the anterior midline is made, large midline vessels may be observed. Further, longitudinal incisions carried down too close to the sternum may easily damage the anterior jugular venous arch or even the left brachiocephalic vein. The short transverse incision at the proper level, with due regard for the positions of the external and internal jugular vessels, would seem to be a better anatomical approach, with ligation of vessels as necessary on the front of the trachea.

The most important anomaly connected with the trachea is tracheoesophageal fistula; this is discussed in connection with the thorax (see Volume 2).

Esophagus

The esophagus begins at the lower border of the cricoid cartilage, where it is continuous with the pharynx. This is said to be its narrowest point, and spasm of the transverse fibers of the cricopharyngeal part of the inferior pharyngeal constrictor, further narrowing the esophageal orifice, has been held responsible for the occurrence of hypopharyngeal diverticula. Unlike the trachea, which runs straight downward in the midline, the cervical portion of the esophagus is somewhat curved, with its convex side to the left; it therefore projects somewhat to the left of the trachea, and incisions for approach to the esophagus are commonly made on this side. On its anterior surface the esophagus is largely related to the trachea; anterolaterally are the lateral lobes of the thyroid gland, especially the left lobe, the inferior thyroid arteries, and the carotid sheath with its contents. Because of the lateral curvature of the esophagus, the left recurrent nerve tends to lie on the anterior surface of the esophagus, the right recurrent along the right border of the esophagus, when these nerves are in the tracheoesophageal sulcus. Posteriorly, the esophagus lies upon the vertebral column, being separated from it however by the two layers of fascia usually termed prevertebral fascia, and by the longus colli muscles.

The muscular coat of the cervical portion of the esophagus is largely voluntary muscle, arranged in an inner circular and an outer longitudinal layer; the fibers of the inner layer are continuous with the inferior constrictor of the pharynx. The longitudinal fibers of the esophagus, which over most of this organ form an even layer, diverge at the upper end of the esophagus to form two bands between which, posteriorly, are exposed the inner circular fibers (see Fig. 8-9); these bands swing laterally and then anteriorly around the esophagus to attach to a common tendon (cricoesophageal tendon) inserted on the midline of the back of the cricoid cartilage. On the posterior aspect of the esophagus the circular fibers blend with those of the inferior constrictor, but anteriorly some of the highest fibers attach to the edge of the cricoesophageal tendon and thus into the cricoid cartilage, and above this point cease (Fig. 9-66).

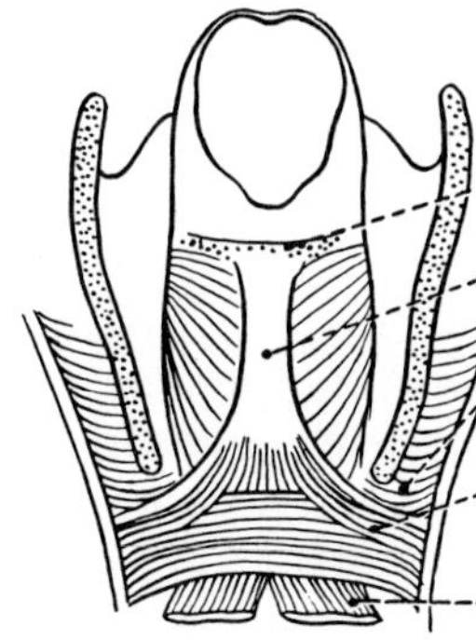

Fig. 9-66. Attachment of the esophagus to the cricoid cartilage, the pharynx and esophagus having been opened from behind.

In addition to its esophageal glands proper, the cervical portion of the esophagus fairly frequently (in perhaps 70% of individuals) contains also superficially located glands similar to those found in the cardia or even the fundus of the stomach and therefore referred to as cardiac, fundic, or peptic glands; the secretion of these glands may give rise to erosion of the wall of the esophagus, just as that of similar glands in the stomach apparently influences the formation of gastric ulcers.

NERVE AND BLOOD SUPPLY

The cervical portions of both the trachea and esophagus have similar nerve and blood supplies; the nerve supply is derived from the recurrent laryngeal nerves, which give off branches to both structures as they ascend to the larynx, while the blood supply is primarily through the inferior thyroid arteries. The arteries are multiple and small; Swigart and co-workers, for instance, found esophageal arteries derived from the terminal branches of the inferior thyroids and from the ascending or the descending segments of these vessels; they also found, more rarely (12.8%), branches to the esophagus from the transverse cervical, subclavian, ascending pharyngeal, vertebral, and common carotid arteries, and the costocervical trunk. In addition, of course, the thoracic portions of the trachea and esophagus, and the pharynx and larynx, have their own blood supply that through anastomoses may aid in supplying the cervical portions.

Shapiro and Robillard found a somewhat similar origin of the arteries to the esophagus in the 50 cases they dissected; they described major arteries to the cervical portion of the esophagus as originating from the inferior thyroid in 34, from the subclavian in 12, from a thyroid ima in three, and from the common carotid once, with accessory arteries derived variously from the subclavian, aorta, carotid, and vertebral. They also described the gross anastomoses that they observed in the vessels of this region and commented upon the fact that separation of the esophagus from the trachea here may jeopardize the blood supply, since the main vessels tend to run in the tracheoesophageal groove and supply both structures. According to Demel the left wall of the cervical segment of the esophagus receives fewer blood vessels than does the right side, and Shapiro and Robillard found the main branches on the right to be larger or more numerous in about two thirds of their series.

ANOMALIES AND PATHOLOGY

Congenital anomalies of the esophagus are fortunately relatively rare, but their incidence is not really known; they may account for a number of deaths in infants and never be reported. Phelps, in reviewing the various types of congenital anomalies, listed nine varieties: (1) absence, (2) a solid cord throughout, (3) doubling of the esophagus, complete or partial, (4) spasmodic contractions, (5) diverticula, either traction or pulsion, (6) tracheoesophageal fistula, (7) cysts of the esophagus, (8) atresia, varying from a solid cord between segments to a fold, valve, web or diaphragm, and (9) stenosis, either from a development narrowing of the lumen or from pressure from without. None of these anomalies is limited to the cervical portion of the esophagus, and the majority are associated especially with the longer thoracic segment and are therefore discussed in connection with the thorax (see Volume 2).

Since both the trachea and the esophagus are located in the visceral compartment of the neck, perforations of either of these lead to swelling within the cervical fascial layers. Lichtenstein stated that tracheostomy may be necessary in such cases, but that usually incision into the fascial spaces of the neck is sufficient to relieve the signs of obstruction. It may be noted again here that anterior perforations of the esophagus tend to lead to spread into the anterior portion of the superior mediastinum, while posterior perforations tend to spread into the posterior mediastinum.

BRANCHIAL CYSTS AND FISTULAS

Cysts and fistulas in the neck may be similar in origin to cysts and fistulas found elsewhere in the body—for instance, dermoid cysts—or may be connected especially with the development of the pharyngeal region and its derivatives. Thyroglossal cysts and fistulas have already been discussed; so also have laryngoceles and pharyngoesophageal diverticula, which, although not strictly congenital, may be associated with congenital weaknesses. The other type of cysts and fistulas found more frequently in the neck are those referred to as branchial or branchiogenic cysts and fistulas.

Origin

Branchial cysts and fistulas have been attributed to persistence of parts of the branchial pouches, the lateral outpouchings of the embryonic pharynx; to persistence of parts of the branchial clefts, the ectodermal grooves that come in contact with the branchial pouches; to persistence of a part of the cervical (precervical) sinus, an ectodermally-lined cavity formed by growth of an operculum from the second branchial arch and its fusion with the fourth arch, which thus comes to receive the external openings of the last three branchial clefts (Fig. 9-67); and to persistence of a thymic or thymopharyngeal duct, the original connections between the thymus and the third branchial pouches from which it originates. The latter concept, that "branchial" cysts and fistulas are often of thymic duct origin, was based on Wenglowski's belief that no true branchial remains can exist below the level of the hyoid bone. That is obviously a false assumption (the muscles and cartilages of the larynx, for instance, are of branchial origin, and both the third and fourth branchial pouches are situated below the arches that contribute to the hyoid bone). It is difficult to account for this theory's persistence in view of the total lack of evidence to support it.

Although it is hard to follow embryologically the shifting branchial pouch-branchial cleft relationships, reference to Figure 9-67 will indicate that pouch, cleft, cervical sinus or a combination of them could contribute to a cyst or fistula of the neck, and that a complete fistula should represent all three. Anatomical relationships may also be a clue as to which particular pouch or cleft has contributed; thus, although cysts may be displaced from their original locus, a cyst or sinus that lies between the external and internal carotid arteries should be derived from the second branchial pouch–cleft, and internal fistulas from the various branchial pouches should be distinguishable because of the difference in the level of their openings into the pharynx. Note, however, that in contrast to the opinion of some earlier writers, the position of the external opening of a fistula would ordinarily give no clue to the particular branchial cleft involved, since the opening is presumably that of the cervical sinus into which the second, third, and fourth clefts all open. The cervical sinus is normally closed from within outward, according to Frazer ('26) and Garrett, and failure of it to close would therefore typically produce an external fistula rather than a cyst.

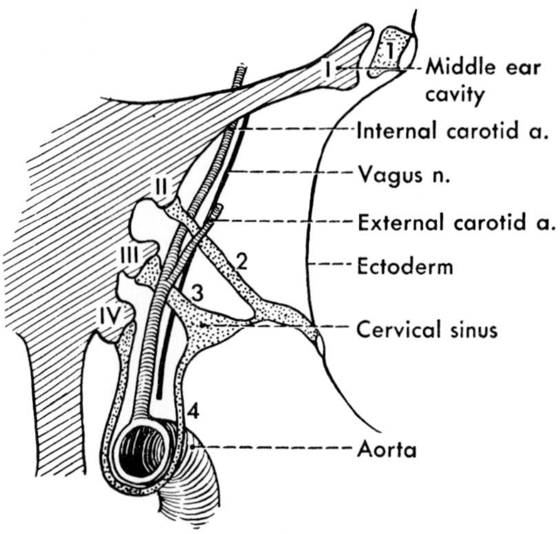

Fig. 9-67. Schema of the branchial pouches and clefts and their important vascular relations, in a ventral view. The branchial pouches are identified by *Roman numerals;* the clefts, by *Arabic ones.* The second to fourth clefts are here shown drawn out into tubes communicating with the cervical sinus, which in turn opens to the exterior. (After Frazer JE: Br J Surg 11:131, 1923)

Types of Fistulas

According to Bailey's excellent review of branchial fistulas, the most common variety is that which is blind internally, having no connection with the pharynx; the next most common, although rare, is the complete fistula (Fig. 9-68), while the least common is an incomplete internal fistula. Branchial fistulas occur about three times as often in females as in males; in 80% of persons, according to Bailey, the external opening is in the lower third of the neck, at about the anterior border of the sternocleidomastoid muscle—although cysts are more typically present in the upper third of the neck. (Bailey also pointed out that the rare "branchial cartilages" and cervical auricles are typically likewise found in the lower third of the neck.) The internal opening, when present, is usually into the pharynx close to the tonsil, thus indicating the origin of these fistulas from the second branchial pouch.

Relations

Apparently, all complete branchial fistulas, as reported, for instance, by Bolman, Kleinert, Berry, and others, open into the tonsillar region. This is understandable because, as pointed out by Frazer ('26) and emphasized by Raven, complete connections from the third and fourth branchial pouches to the skin would necessarily have to pass across the hypoglossal nerve and would most likely be severed by the upward movement of that nerve during development; further, a connection from the fourth pouch to the cervical

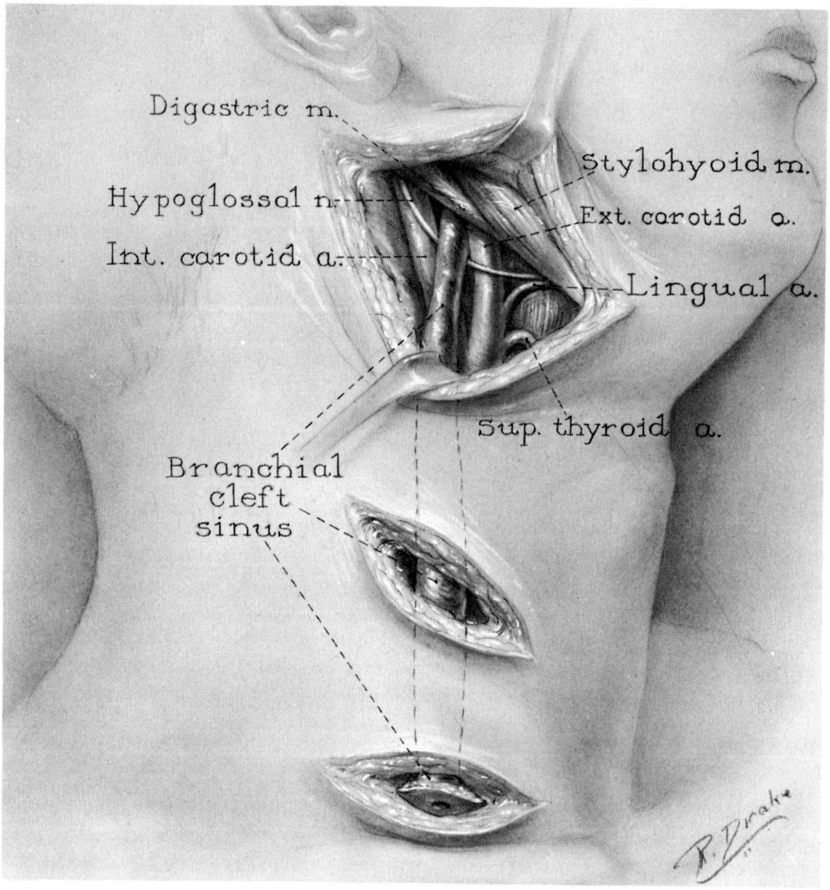

Fig. 9-68. A complete branchial cleft sinus. Note its passage between the external and internal carotids. (ReMine WH: Surg Clin North Am 43:1033, 1963)

sinus would have to pass below the subclavian artery on the right or the arch of the aorta on the left. Fistulas and sinuses arising from the second pouch necessarily pass between the external and internal carotid vessels if they are long enough to reach this level, and the majority of fistulas do so. Raven has discussed branchiogenic pouches and fistulas in considerable detail; in addition to tracts apparently derived from the second pharyngeal pouch, he also identified pharyngeal diverticula or internal fistulas that he considered to be derivatives of the third and fourth pharyngeal pouches (Fig. 9-69). Apparently, a derivative from the third pouch should open into the piriform recess, while one of the fourth should open into the pharynx below this level. Branchiogenic fistulas derived from the third pouch necessarily run posterior and lateral to the internal and external or common carotids.

Stingle and Priebe reported a rare case of an ectopic salivary gland opening low in the neck and resembling somewhat the opening of a branchial fistula. They stated that the ectopic salivary gland is believed to develop from the second branchial pouch.

Cysts

Branchiogenic cysts (Fig. 9-70) are said to be lined by stratified squamous epithelium or by columnar ciliated or nonciliated epithelium resembling that of the respiratory tract; some authors have described them as being apparently primarily endodermal, although how one identifies "endodermal epithelium" is not clear to this writer. They may contain lymphoid tissue, and this has been interpreted to mean that they are not truly branchiogenic cysts but are, rather, derived from thymic tissue; Ladd and Gross pointed out that the presence of lymphoid tissue may result from chronic infection.

According to Ladd and Gross, branchiogenic cysts tend to lie higher in the neck than the cutaneous opening of a branchial fistula and are apt to be located opposite the middle third of the sternocleidomastoid muscle. Bailey pointed out the various relationships of branchiogenic cysts—they may lie close against the sternocleidomastoid, close against the internal jugular, within the bifurcation of the carotid, or be small and situated against the pharyngeal wall. It is important to realize that branchiogenic cysts and fistulas are very often closely associated with the great vessels of the neck and also with the lower cranial nerves, especially the vagus, accessory, and hypoglossal.

The rare cysts or fistulas in or closely related to the parotid gland have been generally said to be remains of the ventral portion of the first branchial cleft, thus essentially a doubling of the external auditory meatus; although most of these fistulas end blindly internally close to the meatus, some open into it, according to Work, or even into the middle ear cavity (Olsen and co-workers).

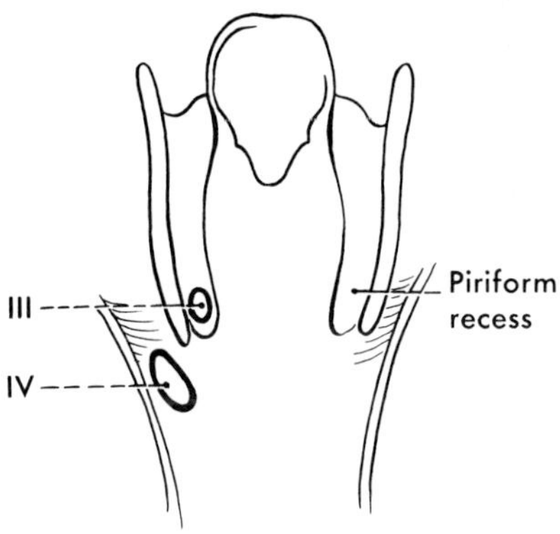

Fig. 9-69. Positions of the openings of two pouches of the pharynx, one presumably derived from the third pouch, one from the fourth, according to Raven RW: Br J Surg 21:235, 1933.

Fig. 9-70. A branchial cyst. (Neel HB, Pemberton J de J: Surgery 18:267, 1945, by permission of the CV Mosby Company)

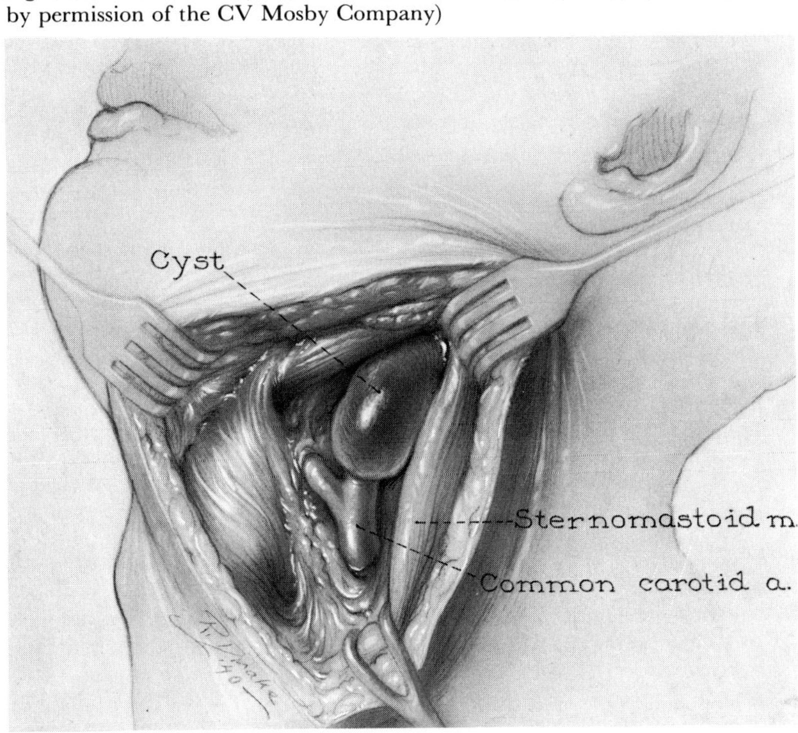

Work, and Aronson and co-workers, agreed that the fistulas may have an ectodermal (epithelial) component only or may be associated with mesodermal tissue—cartilage. Their relation to the facial nerve while it is in the parotid gland varies but is always close, and removal of the formation therefore requires special care; Miglets reported an instance of spontaneous facial paralysis associated with a parotid branchial cyst. It has been estimated that cysts and fistulas derived from the first branchial cleft represent less than 1% of all branchial cleft anomalies. Schwartz and Calcaterra reported what they believed to be only the second instance of bilateral first branchial cleft remains.

References

ADSON AW: Surgical treatment for symptoms produced by cervical ribs and the scalenus anticus muscle. Surg Gynecol Obstet 85:687, 1947

ADSON AW: Cervical ribs: Symptoms, differential diagnosis and indications for section of the insertion of the scalenus anticus muscle. J Int Coll Surgeons 16:546, 1951

ALLAN FD: An accessory or superficial inferior thyroid artery in a full-term infant. Anat Rec 112:539, 1952

ARMSTRONG WG, HINTON JW: Multiple divisions of the recurrent laryngeal nerve: An anatomic study. Arch Surg 62:532, 1951

ARONSON RS, BATSAKIS JG, RICE DH, WORK WP: Anomalies of the first branchial cleft. Arch Otolaryngol 102:737, 1976

AXFORD M: Some observations on the cervical sympathetic in man. J Anat 62:301, 1928

BACHHUBER CA: Complications of thyroid surgery: Anatomy of the recurrent laryngeal nerve, middle thyroid vein and inferior thyroid artery. Am J Surg ns60:96, 1943

BAILEY H: The clinical aspects of branchial fistulae. Br J Surg 21:173, 1933

BEAN RB: A composite study of the subclavian artery in man. Am J Anat 4:303, 1905

BEAHRS OH, GOSSEL JD, HOLLINSHEAD WH: Technic and surgical anatomy of radical neck dissection. Am J Surg 90:490, 1955

BECKER GD, PARELL GJ: Technique of preserving the spinal accessory nerve during radical neck dissection. Laryngoscope 89:827, 1979

BECKER RF, GRUNT JA: The cervical sympathetic ganglia. Anat Rec 127:1, 1957

BERLIN DD: The recurrent laryngeal nerves in total ablation of the normal thyroid gland: An anatomical and surgical study. Surg Gynecol Obstet 60:19, 1935

BERLIN DD, LAHEY FH: Dissections of the recurrent and superior laryngeal nerves: The relation of the recurrent to the inferior thyroid artery and the relation of the superior to abductor paralysis. Surg Gynecol Obstet 49:102, 1929

BERRY G: Branchial cyst: Two instructive cases. Ann Otol Rhinol Laryngol 43:287, 1934

BIEN G: Ueber accessorische Thymuslappen im Trigonum caroticum. Anat Anz 29:325, 1906

BLACK BM: Surgical aspects of hyperparathyroidism: Review of sixty-three cases. Surg Gynecol Obstet 87:172, 1948

BLAIR DM, DAVIES F, MCKISSOCK W: The etiology of the vascular symptoms of cervical rib. Br J Surg 22:406, 1935

BOLMAN RM: Bilateral branchial cleft cyst. Am J Surg ns71:96, 1946

BROWN JB, MCDOWELL F: Neck dissections for metastatic carcinoma. Surg Gynecol Obstet 79:115, 1944

BROWN S: The external jugular vein in American whites and Negroes. Am J Phys Anthropol 28:213, 1941

CATTELL RB, WELCH ML: The carotid sinus syndrome: Its surgical treatment. Surgery 22:59, 1947

CAVE AJE: A note on the origin of the M. scalenus medius. J Anat 67:480, 1933

CHAMBERLIN JA, FRIES JG, ALLEN HC JR: Thyroid carcinoma and the problem of postoperative tetany. Surgery 55:787, 1964

CHANDLER FA, ALTENBERG A: "Congenital" muscular torticollis. JAMA 125:476, 1944

CHENG LH, NORRIS CW: Surgical management of the carotid sinus syndrome. Arch Otolaryngol 97:395, 1973

CLAUSEN EG: Postoperative ("anesthetic") paralysis of the brachial plexus: A review of the literature and report of nine cases. Surgery 12:933, 1942

COHEN FL, FRUCHAN CT, KING RB: Carotid sinus syndrome. Report of five cases and review of the literature. J Neurosurg 45:78, 1976

COLEMAN CC, WALKER JC: Technic of anastomosis of the branches of the facial nerve with the spinal accessory for facial paralysis. Ann Surg 131:960, 1950

CONNAR RG: Prescalene and deep cervical lymph node biopsy. Surg Gynecol Obstet 101:732, 1955

COPLAND SM: The scalenus anticus factor in congenital torticollis. Surgery 11:624, 1942

CORBIN KB, HARRISON F: The sensory innervation of the spinal accessory and tongue musculature in the Rhesus monkey. Brain 62:191, 1939

COVENTRY MB, HARRIS LE: Congenital muscular torticollis in infancy. Some observations regarding treatment. J Bone Joint Surg 41-A:815, 1959

CURTIS GM: The blood supply of the human parathyroids. Surg Gynecol Obstet 51:805, 1930

DASELER EH, ANSON BJ: Surgical anatomy of the subclavian artery and its branches. Surg Gynecol Obstet 108:149, 1959

DAVIS HK: A statistical study of the thoracic duct in man. Am J Anat 17:211, 1915

DECKER HR: The results of phrenic nerve operations in 222 cases with a discussion of the technic of the operations. J Thorac Surg 2:538, 1933

DELITZIN S: Ueber eine Varietät des Verlaufes der Arteria thyroidea superior. Arch Anat Physiol p 105, 1892

DEMEL R: Die Gefässversorgung der Speiseröhre: Ein Beitrag zur Oesophaguschirurgie. Arch Klin Chir 128:453, 1924

DOW DR: The anatomy of rudimentary first thoracic ribs, with special reference to the arrangement of the brachial plexus. J Anat 59:166, 1925

DUNCAN D, YATES R: Ultrastructure of the carotid body of the cat as revealed by various fixatives and the use of reserpine. Anat Rec 157:667, 1967

EAGLE WW: Symptomatic elongated styloid process: Report of two cases of styloid process-carotid artery syndrome with operation. Arch Otolaryngol 49:490, 1949

EDDY HC: Vascular origin of congenital torticollis: Case report. Am J Surg ns17:100, 1932

EDEN KC: The vascular complications of cervical ribs and first thoracic rib abnormalities. Br J Surg 27:111, 1939

EISLER P: Der M. levator glandulae thyreoideae und verwandte praelaryngeale Muskelbildungen. Anat Anz 17:183, 1900

ELKIN DC, HARRIS MH: Arteriovenous aneurysm of the vertebral vessels: Report of ten cases. Ann Surg 124:934, 1946

ELSE JE: Anatomical variations in the thyroid as a cause of recurrent goiter. Am J Surg ns8:92, 1930

ETTER LE: Osseous abnormalities of the thoracic cage seen in forty thousand consecutive chest photoroentgenograms. Am J Roentgenol 51:359, 1944

FALLER A, SCHÄRER O: Über die Variabilität der Arteriae thyreoideae. Acta Anat 4:119, 1947

FELDMAN DE, APPLEBAUM EL: The submandibular triangle in radical neck dissection. Arch Otolaryngol 103:705, 1977

FLISBERG K, LINDHOLM T: Electrical stimulation of the human recurrent laryngeal nerve during thyroid operations. Acta Otolaryngol (Suppl 263):63, 1970

FRASER BA, DUCKWORTH JWA: Position of ultimobranchial body cysts in the human fetal thyroid gland. Acta Anat 105:269, 1979

FOLEY JO: The components of the cervical sympathetic trunk with special reference to its accessory cells and ganglia. J Comp Neurol 82:77, 1945

FOWLER CH, HANSON WA: Surgical anatomy of the thyroid gland with special reference to the relations of the recurrent laryngeal nerve. Surg Gynecol Obstet 49:59, 1929

FRAZER JE: Anomaly of omo-hyoid. J Anat Physiol 35:494, 1901

FRAZER JE: The nomenclature of diseased states caused by certain vestigial structures in the neck. Br J Surg 11:131, 1923

FRAZER JE: The disappearance of the precervical sinus. J Anat 61:132, 1926

FREELAND AP, ROGERS JH: The vascular supply of the cervical skin with reference to incision planning. Laryngoscope 85:714, 1975

GARFIA A: Glomus tissue in the vicinity of the human carotid sinus. J Anat 130:1, 1980

GARRETT FD: Development of the cervical vesicles in man. Anat Rec 100:101, 1948

GIBSON A: Bilateral abnormal relationship of the vagus nerve in its cervical portion. J Anat Physiol 49:389, 1915

GILMOUR JR, MARTIN WJ: The weight of the parathyroid glands. J Pathol Bacteriol 44:431, 1937

GILROY J, MEYER JS: Compression of the subclavian artery as a cause of ischaemic branchial neuropathy. Brain 86:733, 1963

GODWIN MC: The development of the parathyroids in the dog with emphasis upon the origin of accessory glands. Anat Rec 68:305, 1937

GRANT JCB: An Atlas of Anatomy 2nd ed. Baltimore, Williams & Wilkins, 1947

GRIEVE J: The relative incidence of sternomastoid and ocular torticollis in aircrew recruits. Br J Surg 33:285, 1946

GUTTMANN L: The distribution of disturbances of sweat secretion after extirpation of certain sympathetic cervical ganglia in man. J Anat 74:537, 1940

HAMMER DL, MEIS AM: Thyroid arteries and anomalous subclavian in the white and the Negro. Am J Phys Anthropol 28:227, 1941

HARMAN NB: "Socia thymi cervicalis," and thymus accessorius. J Anat Physiol 36:47, 1901

HARRIS W: The true form of the brachial plexus, and its motor distribution. J Anat Physiol 38:399, 1904

HEINBACH WF JR: A study of the number and location of the parathyroid glands in man. Anat Rec 57:251, 1933

HERTSLET LE, KEITH A: A comparison of the anomalous parts of two subjects, the one with a cervical rib, the other with a rudimentary first rib. J Anat Physiol 30:562, 1896

HERZ E, HOEFER PFA: Spasmodic torticollis. I. Physiologic analysis of involuntary motor activity. Arch Neurol Psychiatr 61:129, 1949

HEYMANS C: Some aspects of blood pressure regulation and experimental arterial hypertension. Surgery 4:487, 1938

HILL RM: Vascular anomalies of the upper limbs associated with cervical ribs: Report of a case and review of the literature. Br J Surg 27:100, 1939

HINSEY JC, HARE K, WOLF GA JR: Structure of the cervical sympathetic chain in the cat. Anat Rec 82:175, 1942

HOLMGREN H, NAUMANN B: A study of the nerves of the thyroid gland and their relationship to glandular function. Acta Endocrinol 3:215, 1949

HOUGH G DE N JR: Congenital torticollis: A review and result study. Surg Gynecol Obstet 58:972, 1934

HUBERT L: Thyroglossal cysts and sinuses: Analysis of forty-three cases. Arch Otolaryngol 45:105, 1947

HUELKE DF: A study of the transverse cervical and dorsal scapular arteries. Anat Rec 132:233, 1958

HUGHES ESR: Venous obstruction in the upper extremity. Br J Surg 36:155, 1948

HYNDMAN OR, WOLKIN J: Sympathectomy of the upper extremity: Evidence that only the second dorsal ganglion need be removed for complete sympathectomy. Arch Surg 45:145, 1942

JACKSON JL: An infra-hyoid muscular anomaly in man. Anat Rec 59:41, 1934

JAMIESON RW, SMITH DB, ANSON BJ: The cervical sympathetic ganglia: An anatomical study of 100 cervicothoracic dissections. Q Bull Northwestern Univ Med School 26:219, 1952

JOESSEL: Neue Anomalien der Carotis externa und der Maxillaris interna. Arch Anat Physiol p 433, 1878

JONES C, RANDALL DH: Thoracic inlet neurovascular interference. Br J Surg 40:128, 1952

KANTOR H: Tiefe Teilung der Arteria carotis communis. Anat Anz 26:492, 1905

KELLEY WO: Phrenic nerve paralysis: Special consideration of the accessory phrenic nerve. J Thorac Surg 19:923, 1950

KERR AT: The brachial plexus of nerves in man, the variations in its formation and branches. Am J Anat 23:285, 1918

KEYES EL: Demonstration of the nerve to the levator glandulae thyreoideae muscle. Anat Rec 77:293, 1940

KIMMEL DL: Anterior mediastinal origin of the thyroidea ima artery (abstr). Anat Rec 103:544, 1949

KIMMEL DL: The cervical sympathetic rami and the vertebral plexus in the human fetus. J Comp Neurol 112:141, 1959

KING BT, GREGG RL: An anatomical reason for the various behaviors of paralyzed vocal cords. Ann Otol Rhinol Laryngol 57:925, 1948

KING WLM, PEMBERTON J DEJ: So-called lateral aberrant thyroid tumors. Surg Gynecol Obstet 74:991, 1942

KLEINERT MN: Branchial fistula. Arch Otolaryngol 18:510, 1933

KOLODNY A: Traction paralysis of the brachial plexus. Am J Surg ns51:620, 1941

KRAFKA J JR: Intratracheal thyroid occurring in a seven months human fetus. Ann Surg 106:457, 1937

KRISTOFF FV, ODOM GL: Ruptured intervertebral disk in the cervical region: A report of twenty cases. Arch Surg 54:287, 1947

LADD WE, GROSS RE: Congenital branchiogenic anomalies: A report of 82 cases. Am J Surg ns39:234, 1938

LAHEY FH, HAGGART GE: Hyperparathyroidism: Clinical diagnosis and the operative technique of parathyroidectomy. Surg Gynecol Obstet 60:1033, 1935

LAHEY FH, SWINTON NW: Intrathoracic goiter. Surg Gynecol Obstet 59:627, 1934

LANGSAM CL: M. omohyoideus in American whites and Negroes. Am J Phys Anthropol 28:249, 1941

LARSON DL, BOND TP, RODIN AE, COERS CR, LEWIS SR: Clinical and experimental obstruction of the thoracic duct. Surgery 60:35, 1966

LARSON LM, ROSENOW JH: Fibrous sternomastoid "tumors" of infancy: Their role in the etiology of muscular torticollis. Minn Med 31:1243, 1948

LAWSON W: The neuroendocrine nature of the glomus cells: An experimental, ultrastructural, and histochemical tissue culture study. Laryngoscope 90:120, 1980

LICHTENSTEIN ME: Acute injuries to the neck involving the food and air passages. Surg Gynecol Obstet 85:734, 1947

LINDSKOG GE, HOWES EL: Cervical rib associated with aneurysm of the subclavian artery: Report of a case and review of the recent literature. Arch Surg 34:310, 1937

VON LIPPMANN R: Abnormer Ursprung des Ramus descendens n. hypoglossi aus dem N. vagus. Anat Anz 37:1, 1910

LISSITZYN MS: Ductus thoracicus. Arch Klin Chir 128:215, 1924

LIST CF, PEET MM: Sweat secretion in man. II. Anatomic distribution of disturbances in sweating associated with lesions of the sympathetic nervous system. Arch Neurol Psychiatr 40:27, 1938

LOONEY WW: Lymphatic drainage of head and neck—emphasizing special structures. Ann Otol Rhinol Laryngol 44:33, 1935

LORD JW JR, ROSATI LM: Neurovascular compression syndromes of the upper extremity. Ciba Clin Symposia 10:35, 1958

LÖWY R: Ueber das topographische Verhalten des Nervus hypoglossus zur Vena jugularis interna. Anat Anz 37:10, 1910

MACCALLUM WG: The surgical relations of the parathyroid glands. Br Med J 2:1282, 1906

MAHORNER HR, CAYLOR HD, SCHLOTTHAUER CF, PEMBERTON J DEJ: Observations on the lymphatic connections of the thyroid gland in man. Anat Rec 36:341, 1927

MARSHALL CF: Variations in the form of the thyroid gland in man. J Anat Physiol 29:234, 1895

MARSHALL SF, BECKER WF: Thyroglossal cysts and sinuses. Ann Surg 129:642, 1949

MASSON JC, MUELLER SC: Ovarian tumors of thyroid tissue. Surg Gynecol Obstet 56:931, 1933

MCCLEERY RS, KESTERON JE, KIRTLEY JA, LOVE RB: Subclavius and anterior scalene muscle compression as a cause of intermittent obstruction of the subclavian vein. Ann Surg 133:588, 1951

MCGUIRT WF, HARKER LA: Carotid body tumor. Arch Otolaryngol 101:58, 1975

MCKENZIE J: The morphology of the sternomastoid and trapezius muscles. J Anat 89:526, 1955

MCSWAIN B, SPENCER FC: Carotid body tumor in association with carotid sinus syndrome: Report of two cases. Surgery 22:222, 1947

MEYEROWITZ BR, BUCHHOLZ RB: Midline cervical ectopic thyroid tissue. Surgery 65:358, 1969

MEYERS EN, PANTANGO P JR: Intratracheal thyroid. Laryngoscope 85:1833, 1975

MIDDLETON DS: The pathology of congenital torticollis. Br J Surg 18:188, 1930

MIGLETS AW: Parotid branchial cleft cyst with facial paralysis. Report of a case. Arch Otolaryngol 101:637, 1975

MILLZNER RJ: The normal variations in the position of the human parathyroid glands. Anat Rec 48:399, 1931

MOORE DC, BRIDENBAUGH LD, EATHER KF: Block of the upper extremity: Supraclavicular approach versus axillary approach. Arch Surg 90:68, 1964

MOORE OS, FRAZELL ES: Simultaneous bilateral neck dissection. Am J Surg 107:565, 1964

MORFIT HM: Simultaneous bilateral radical neck dissection: Total ablation of both internal and external jugular venous systems at one sitting. Surgery 31:216, 1952

MORRISON LF: Recurrent laryngeal nerve paralysis: A revised conception based on the dissection of one hundred cadavers. Ann Otol Rhinol Laryngol 61:567, 1952

MOUNTAIN JC, STEWART GR, COLCOCK BF: The recurrent laryngeal nerve in thyroid operations. Surg Gynecol Obstet 133:978, 1971

MURPHREY DR JR: Stellate ganglion block: A new anterior approach. Ann Surg 120:759, 1944

NAFFZIGER HC, GRANT WT: Neuritis of the brachial plexus mechanical in origin: The scalenus syndrome. Surg Gynecol Obstet 67:722, 1938

NEUBERGER H: Über einige Arterienvariatäten am Hals. Anat Anz 41:618, 1912

NONIDEZ JF: Innervation of the thyroid gland. II. Origin and course of the thyroid nerves in the dog. Am J Anat 48:299, 1931

NONIDEZ JF: Innervation of the thyroid gland. III. Distribution and termination of the nerve fibers in the dog. Am J Anat 57:135, 1935

NORDLAND M: The larynx as related to surgery of the thyroid based on an anatomical study. Surg Gynecol Obstet 51:449, 1930

NORRIS EH: The parathyroid glands and the lateral thyroid in man: Their morphogenesis, histogenesis, topographic anatomy and prenatal growth. Contrib Embryol 26:247, 1937

NORRIS EH: The parathyroid adenoma: A study of 322 cases. Int Abstr Surg 84:1, 1947

OLSEN KD, MARAGOS NF, WEILAND LH: First branchial cleft anomalies. Laryngoscope 90:423, 1980

DE PABLO JS, DIEZ-MALLO J: Experience with three thousand cases of brachial plexus block; its dangers: Report of a fatal case. Ann Surg 128:956, 1948

PALUMBO LT: A new concept of the sympathetic pathways to the eye. A new technique to avoid a Horner's syndrome. Surgery 42:740, 1957

PARSONS FG, KEITH A: Seventh report of the Committee of Collective Investigation of the Anatomical Society of Great Britain and Ireland, for the year 1896–97. Question III. The position of the spinal accessory nerve. Whether it passes outward between the jugular vein and internal carotid artery, or between the jugular vein and the atlas? Whether it perforates the sterno-mastoid or not: If so, does the whole nerve perforate, or only a part? Which division of the sterno-mastoid does it perforate? J Anat Physiol 32:177, 1897

PATTERSON RH: Cervical ribs and the scalenus muscle syndrome. Ann Surg 111:531, 1940

PEMBERTON J DEJ, LIVERMORE GR JR: Surgical treatment of carotid body tumors: Value of anticoagulants in carotid ligation. Ann Surg 133:837, 1951

PERLOW S, VEHE KL: Variations in the gross anatomy of the stellate and lumbar sympathetic ganglia. Am J Surg ns30:454, 1935

PHELPS KA: Congenital anomalies of the esophagus with a report of nine cases. Ann Otol Rhinol Laryngol 39:364, 1930

PICK J, WERTHEIM H: A technic for blocking the carotid sinus nerves. Ann Surg 127:144, 1948

PIKKIEFF E: On subcutaneous veins of the neck. J Anat 72:119, 1937

POLAK JM, PEARSE AGE, LELIÈVRE L, FONTAINE J, LEDOUARIN MM: Immunocytochemical confirmation of the neural crest origin of avian calcitonin-producing cells. Histochemistry 40:209, 1974

PRATT GW: The thyroidea ima artery. J Anat 50:239, 1916

PUTNAM TJ, HERZ E, GLASER GH: Spasmodic torticollis. III. Surgical treatment. Arch Neurol Psychiatr 61:240, 1949

RAMSAY AJ: Experimental studies on the developmental potentialities of the third pharyngeal pouch in the mammalian embryo (mouse) (abstr). Anat Rec 106:234, 1950

RAVEN RW: Pouches of the pharynx and oesophagus with special reference to the embryological and morphological aspects. Br J Surg 21:235, 1933

RAY BS, STEWART HJ: Role of the glossopharyngeal nerve in the carotid sinus reflex in man: Relief of carotid sinus syndrome by intracranial section of the glossopharyngeal nerve. Surgery 23:411, 1948

READ WT, TROTTER M: The origins of transverse cervical and of transverse scapular arteries in American whites and Negroes. Am J Phys Anthropol 28:239, 1941

REED AF: The relations of the inferior laryngeal nerve to the inferior thyroid artery. Anat Rec 85:17, 1943

REED AF: Anatomic relations of the anterior scalene muscle. (abstr). Anat Rec 100:706, 1948

RHINEHART DA: The nerves of the thyroid and parathyroid bodies. Am J Anat 13:91, 1912

RIENHOFF WF JR: The surgical treatment of hyperparathyroidism: With a report of 27 cases. Ann Surg 131:917, 1950

ROBERTS B, HARDESTY WH, HOLLING HE, REIVICH M, TOOLE JF: Studies on extracranial cerebral blood flow. Surgery 56:826, 1964

ROOS DB, OWENS JC: Thoracic outlet syndrome. Arch Surg 93:71, 1966

RUGH JT: Spasmodic torticollis: Its cause and treatment. Am J Surg ns49:490, 1940

RUTHERFORD NC: A curious arrangement of the retro-clavicular musculature. Anat Anz 37:148, 1910

SANTSCHI DR, FRAHM CJ, PASCALE LR, DUMANIAN AV: The subclavian steal syndrome: Clinical and angiographic considerations in 74 cases in adults. J Thorac Cardiovasc Surg 51:103, 1966

SAUNDERS WH, JOHNSON EW: Rehabilitation of the shoulder after radical neck dissection. Ann Otol Rhinol Laryngol 84:812, 1975

SCHMIDT CF, COMROE JH JR: Functions of the carotid and aortic bodies. Physiol Rev 20:115, 1940

SCHWARTZ HE, CALCATERRA TC: Bilateral first branchial cleft. Ann Otol Rhinol Laryngol 86:849, 1977

SCHWEIZER O, LEAK GH: A study of spinal fluid pressures in operations requiring removal of both internal jugular veins. Ann Surg 136:948, 1952

SEGAL MS, DULFANO MJ: Glomectomy in the treatment of chronic bronchial asthma: A report of fifteen unsuccessful cases. N Engl J Med 272:57, 1965

SHAPIRO AL, ROBILLARD GL: The esophageal arteries: Their configurational anatomy and variations in relation to surgery. Ann Surg 131:171, 1950

SHARER RF: Substernal thyroid. Am J Surg ns32:56, 1936

SHORE LR: An example of the muscle scalenus minimus. J Anat 60:418, 1926

SIMPSON WL, GRAHAM DG, Injuries of the cervical thoracic duct. Ann Otol Rhinol Laryngol 52:834, 1943

SINGER E: Human brachial plexus united into a single cord: Description and interpretation. Anat Rec 55:411, 1933

SIWE SA: The cervical part of the gangliated cord, with special reference to its connections with the spinal nerves and certain cerebral nerves. Am J Anat 48:479, 1931

SKOLNIK EM, YEE KF, FRIEDMAN M, GOLDEN TA: The posterior triangle in radical neck surgery. Arch Otolaryngol 102:1, 1976

SKOOG T: Ganglia in the communicating rami of the cervical sympathetic trunk. Lancet 2:457, 1947

SMITH IH, MOLOY HC: The effect of nerve stimulation and nerve degeneration on the mitochondria and histology of the thyroid gland. Anat Rec 45:393, 1930

SMITH JR: Lymphatic cannulation of the head and neck. Plast Reconstr Surg 32:607, 1963

SMITH SD, BENTON RS: A rare origin of the superior thyroid artery. Acta Anat 101:91, 1978

SMITHWICK RH: The problem of producing complete and lasting sympathetic denervation of the upper extremity by preganglionic section. Ann Surg 112:1085, 1940

SPURLING RG, GRANTHAM EG: The painful arm and shoulder with especial reference to the problem of scalenus neurocirculatory compression. J Missouri Med Assoc 38:340, 1941

SPURLING RG, SCOVILLE WB: Lateral rupture of the cervical intervertebral discs: A common cause of shoulder and arm pain. Surg Gynecol Obstet 78:350, 1944

STEWART JD: Circulation of the human thyroid. Arch Surg 25:1157, 1932

STEWART WJ: Torticollis due to the aberrant sternal portion of the sternocleidomastoid muscle. J Bone Joint Surg ns17:493, 1935

STINGLE WH, PRIEBE CJ: Ectopic salivary gland and sinus in the lower neck. Ann Otol Rhinol Laryngol 83:379, 1974

STONE PW, LORD JW JR: The clavicle and its relation to trauma to the subclavian artery and vein. Am J Surg 89:834, 1955

STOTT CF: A note on the scalenus minimus muscle. J Anat 62:359, 1928

SUNDERLAND S: The distribution of sympathetic fibres in the brachial plexus in man. Brain 71:88, 1948

SUNDERLAND S, BEDBROOK GM: Narrowing of the second part of the subclavian artery. Anat Rec 104:299, 1949

SWEET RH: Intrathoracic goiter located in the posterior mediastinum. Surg Gynecol Obstet 89:57, 1949

SWIGART LL, SIEKERT RG, HAMBLEY WC, ANSON BJ: The esophageal arteries: An anatomic study of 150 specimens. Surg Gynecol Obstet 90:234, 1950

SZAWLOWSKI J: Über das Verhalten des Ductus thoracicus bei Persistenz der rechten absteigenden Aortenwurzel. Anat Anz 3:839, 1888

TAGUCHI K: Ueber eine seltene Anomalie des Verlaufes des Vagusstammes und eines seiner Aeste. Arch Anat Physiol, p 365, 1888

TAGUCHI K: Die Lage des Nervus recurrens nervi vagi zur Arteria thyreoidea inferior. Arch Anat Physiol, p 309, 1889

TANDLER J: Die Entwickelung der Lagebeziehung

zwischen N. accessorius and V. jugularis interna beim Menschen. Anat Anz 31:473, 1907

TAYLOR J: An unusual variation of the omohyoid muscle. J Anat 59:331, 1925

TELFORD ED, MOTTERSHEAD S: Pressure at cervicobrachial junction: An operative and anatomical study. J Bone Joint Surg sB30:249, 1948

TELFORD ED, STOPFORD JSB: The vascular complications of cervical rib. Br J Surg 18:557, 1931

THOMSON A: Second annual report of the Committee of Collective Investigation of the Anatomical Society of Great Britain and Ireland for the year 1890–91. J Anat Physiol 26:76, 1891

TODD TW: The hinder end of the brachial plexus in man and mammals. Anat Anz 42:129, 1912

TROTTER HA: The surgical anatomy of the lymphatics of the head and neck. Ann Otol Rhinol Laryngol 39:384, 1930

TURNER W: A phrenic nerve receiving a root of origin from the descendens hypoglossi. J Anat Physiol 27:427, 1893

VIACAVA EP, PACK GT: Significance of supraclavicular signal node in patients with abdominal and thoracic cancer: A study of one hundred and twenty-two cases. Arch Surg 48:109, 1944

WALSHE FMR, JACKSON H, WYBURN–MASON R: On some pressure effects associated with cervical and with rudimentary and "normal" first ribs, and the factors entering into their causation. Brain 67:141, 1944

WALTON AJ: The surgical treatment of parathyroid tumours. Br J Surg 19:285, 1931

WARREN S, FELDMAN JD: The nature of lateral "aberrant" thyroid tumors. Surg Gynecol Obstet 88:31, 1949

WEEKS C, HINTON JW: Extralaryngeal division of the recurrent laryngeal nerve: Its significance in vocal cord paralysis. Ann Surg 116:251, 1942

WEINGARTEN CZ: Simultaneous bilateral radical neck dissection: Preservation of the external jugular vein. Arch Otolaryngol 97:309, 1973

WELLER GL JR: Development of the thyroid, parathyroid and thymus glands in man. Contrib Embryol 24:93, 1933

WENGLOWSKI R: Ueber die Halsfisteln und Cysten: Zweiter Theil. Die seitlichen Halsfisteln und Cysten. Arch Klin Chir 100:789, 1913

WHITE JC, POPPEL MH, ADAMS R: Congenital malformations of the first thoracic rib: A cause of brachial neuralgia which simulates the cervical rib syndrome. Surg Gynecol Obstet 81:643, 1945

WILSON H: Carotid body tumors: Surgical management. Ann Surg 159:959, 1964

WORK WP: Newer concepts of first branchial cleft defects. Laryngoscope 82:1581, 1972

WRETE M: The anatomy of the sympathetic trunks in man. J Anat 93:448, 1959

ZIEGELMAN EF: Laryngeal nerves: Surgical importance in relation to the thyroid arteries, thyroid gland and larynx. Arch Otolaryngol 18:793, 1933

INDEX

With few exceptions, the individual structures and parts listed in this index are arranged only according to the noun which identifies them, rather than according to descriptive adjectives. Thus, for instance, the various muscles are listed alphabetically under the heading *Muscle(s)*, but, throughout the index, are not listed in accordance with the qualifying adjective by which they are often known—the sternocleidomastoid muscle, for example, occurs in the index under *M,* but not under *S.*

The index is also a regional one, listing under the specific part of the body concerned the page numbers on which discussions of the nerves, blood vessels, muscles, and so forth, of that part will be found. The page references in *italics* identify the more complete of two or more descriptions or discussions; those in **boldface** indicate illustrations.

ISBN 0-06-141264-3

90000

9 780061 412646